Polar Mechanics 2018

Fifth All-Russian Conference with International Participation

IOP Conference Series: Earth and Environmental Science
Volume 193

Novosibirsk, Russia
9 – 11 October 2018

ISBN: 978-1-5108-7378-0
ISSN: 1755-1307

Printed from e-media with permission by:

Curran Associates, Inc.
57 Morehouse Lane
Red Hook, NY 12571

CURRAN ASSOCIATES INC. proceedings.com

Some format issues inherent in the e-media version may also appear in this print version.

Printed by Curran Associates, Inc. (2018)

Institute of Physics
Dirac House, Temple Back
Bristol BS1 6BE UK

Phone: 44 1 17 929 7481
Fax: 44 1 17 920 0979

techtracking@iop.org

Additional copies of this publication are available from:

Curran Associates, Inc.
57 Morehouse Lane
Red Hook, NY 12571 USA
Phone: 845-758-0400
Fax: 845-758-2633
Email: curran@proceedings.com
Web: www.proceedings.com

Table of contents

Volume 193

The Fifth All-Russian Conference with International Participation POLAR MECHANICS

9–11 October 2018, Novosibirsk, Russian Federation

Accepted papers received: 3 October 2018
Published online: 30 October 2018

Preface

Papers

Preface

The present volume represents contributions from participants of the Fifth All-Russian Conference with International Participation "POLAR MECHANICS" held at the Technopark of Novosibirsk Akademgorodok, Russia, October, 9-11 2018, organized by the Lavrentyev Institute of Hydrodynamics, Trofimuk Institute of Petroleum Geology and Geophysics, and Novosibirsk State University.

The conference is traditional and attracts the attention of many scientists. The objective of the conference is to coordinate work and to share information on the present state of studies in the field of polar mechanics and related sciences performed by scientific, scientific-and-technological and other groups and to define the priority areas of fundamental and oriented scientific research aimed to solve the problems of development and efficient use of Arctic and Antarctic resources.

The actuality of topics, which is traditional for the Conference "Polar Mechanics", is defined by the problems of development of northern regions. The development of equipment and technologies, mining, the reconstruction of the Northern Sea Route under modern conditions, the construction and development of infrastructure of Arctic and Antarctic regions require to improve the methods of numerical and full-scale simulation of behavior of materials and structures, to develop the methods of predicting the natural and technological processes in high latitudes and under low temperatures. It is therefore necessary to perform large-scale fundamental, applied and engineering research to develop techniques used to predict and control phenomena, which are specific for the polar mechanics at the stage of developing a new technology and at the stage of designing new technical devices to develop northern regions.

The main topics of the conference are

- Mechanical, tribological and rheological properties of ice;
- Mathematical simulation in ice mechanics;
- Simulated ice and ice model basins;
- Geological and geophysical research in the Arctic;
- Permafrost study and gas hydrates;
- Singularities of construction and use of engineering structures under Arctic conditions: under permafrost conditions, in coastal and shelf zones, underwater construction work;
- Mining and transporting of mineral resources under Arctic conditions;
- Navigation under ice conditions;
- Arctic material science;
- Hydrodynamics and computational fluid dynamics in Arctic technologies;
- Environment: Polar/Arctic and Ocean.

The Organizer Committee
Novosibirsk, October 2018

IOP Conf. Series: Earth and Environmental Science **193** (2018) 011002 doi:10.1088/1755-1315/193/1/011002

Peer review statement

All papers published in this volume of *IOP Conference Series: Earth and Environmental Science* have been peer reviewed through processes administered by the proceedings Editors. Reviews were conducted by expert referees to the professional and scientific standards expected of a proceedings journal published by IOP Publishing.

Phase analysis of hydrated Portland cement

Yu A Abzaev, A I Gnyrya, S V Korobkov and K S Gauss

Tomsk State University of Architecture and Building, Tomsk, Russian Federation

E-mail: korobkov_1973@mail.ru

Abstract. The paper studies the quantitative phase composition in cement paste using the Rietveld refinement method. Investigations are carried out within 0–67 h time period at 40, 50 and 70 ^{0}C. It is found that tobermorite, deliate and mCanH$_2$O$_{10}$Si$_2$ are major phases in cement phase, where m = 5 or 3, and n = 3 or 1. The phase composition is determined by the contribution from estimated peak intensities to the integrated peak intensity which, in turn, is compared to the experimental diffraction patterns. The volume fractions of the studied phases are measured. It is found that the time and temperature of isothermal curing significantly affect the structure of the cement paste.

1. Introduction

In different types of concrete, the basic binding material is Portland cement. The main components of Portland cement are calcium silicates C_3S and C_2S, aluminate C_3A and ferrite C_4AF. Such components as calcite, calcium oxide, magnesium oxide, sodium and potassium sulphates are also present [1]. During the hydration process, various products form in Portland cement, namely: calcium–silicate–hydrate (C–S–H), Portlandite, ettringite, and others [2–6]. Information about the structure of solid materials in initial phases and hydration products attracts the interest of many researchers in practical applications of Portland cement and estimation of its performance characteristics and strength properties, in particular. This information includes the crystal lattice parameters, space group, grade, relative coordinates of atoms, *etc*. The analysis of the rapid achievement of high performance characteristics cannot be carried out without structural information. Using GEMS software it is possible to describe the chemical interaction between the solid and liquid phases in Portland cements as a thermodynamic system having the certain content of products at different stages of hydration process [2, 4, 7]. GEMS allows considering the influence of such external and internal factors as temperature and its gradients, curing time, quantification of clinker phases, water-cement ratio, mineral additives, *etc*. GEMS also allows predicting the quantitative content of hydration products in various conditions within longtime periods. Along with the obtained structural information on the hydration product parameters, it is possible to predict diverse performance characteristics of binding materials in concretes. It seems to be rather relevant to study the quantitative phase composition in Portland cements at different curing stages using the Rietveld refinement method and GEMS software.

The aim of this work is to identify the structure and phase composition of Portland cement using the Rietveld refinement method and curing mechanisms in terms of the hydration model proposed by Parrot and Killoh [3] detected by GEMS modelling within 0–67 h time period at 40, 50 and 70 ^{0}C.

2. Material and model approximation

The type CEM I 42.5B Portland cement was used in this experiment to determine its weight content. This type of Portland cement is manufactured at OOO 'Topkinskii cement', Novosibirsk, Russia [10]. Its chemical composition and the weight content are summarized in Table 1. The chemical composition allows us to obtain the mineralogical composition of Portland cement using the Bogue equations [1]. The obtained composition of major phases includes 61.06 g alite, 13.62 g belite, 12.79 g ferrite and 6.50 g aluminate per 100 g. The water-cement ratio necessary for the formation of a binding material is selected to be 0.41.

Table 1. Chemical composition of the type CEM I 42.5B Portland cement.

Chemical composition	Weight content (wt.%)	Weight (g)	
SiO_2	20.49	C_3S	61.06
Al_2O_3	5.06	C_2S	13.62
Fe_2O_3	4.14	C_4AF	12.79
CaO	62.83	C_3A	6.50
MgO	1.75	Aqueous solution	41
Na_2O	0.34	O_2	0.34
K_2O	0.65	MgO	1.78
SO_3	3.10	K_2O	0.66
–	–	Na_2O	0.34
–	–	SO_3	3.15

The X-ray diffraction (XRD) analysis was carried out by DRON-4-07 diffractometer (Russia) which was modified for digital signal processing. Measurements were conducted using copper radiation (K_α) and Bragg-Brentano X-ray optical scheme. Specifications for the DRON-4-07 included 0.02^0 scanning step; $14–71^0$ range for angles to be scanned; 30 kV voltage; and 25 mA current. The phase composition in Portland cement was determined by the Rietveld refinement [11, 12]. A tobermorite-like model of the cement paste was created in accordance with the GEMS modeling results and literature information [5]. Based on the C–S–H detailed analysis, Richardson [5] presented structural information of a number of tobermorite crystal lattices which include water molecules. The primary analysis of diffraction patterns shows both overlapping reflections and non-monotonic background behavior on the dependences between the X-ray reflection and Bragg reflection angle, which indicate to the contribution from the amorphous phase to the X-ray reflection intensity (Figure 1). The Rietveld refinement utilizes alite, belite, aluminate and alumoferrite for the identification of clinker phases. These minerals are selected from the Crystallography Open Database [13]. It is interesting that the amount of aluminate and alumoferrite in Portland cement is insignificant (table 1). Also, it is important to note that we estimate the contributions of crystalline phases to the integrated intensity (Table 2). The approximation of the background pattern is addressed separately over the estimated peak intensity of the hydrated amorphous phase.

The qphase analysis (QPA) based on the Rietveld method (Table 2), shows a satisfactory agreement between the theoretical calculations and experimental findings of integrated intensities. The convergence criterion is $R_{wp} \approx 14\%$ (see Figure 1, curve *3*). The identified quantitative composition of clinker phases and hydration products predominates because the total phase intensity is close to unity (~98%) as can be seen from table 2.

In order to identify the amorphous hydration product within the molecular dynamics, the C–S–H crystal lattice is amorphized in the Amorphous software code, the operation of which was described in [11]. The results obtained are then used for the background approximation of the experimental XRD patterns with the Rietveld method (see Figure 2). The simulation modeling is performed in a force field, in which we consider only non-valence pair contributions, *i.e.* Coulomb interactions and van der Waals forces. For the latter case, we apply the Lennard-Jones potential. The truncation radius

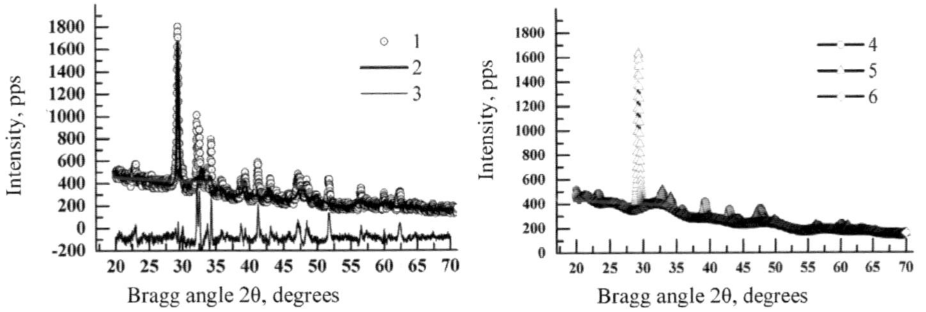

Figure 1. Peak intensities of hydration products: *1* – experimental; *2* – integrated; *3* – difference between experimental and integrated intensities; *4* – alite phase; *5* – belite phase; *6* – Portlandite phase.

of electrostatic forces is 1.85 nm. The measurement accuracy of the van der Waals forces is 10^{-5} kcal/mole. The molecular modeling of the atom paths is performed at a room temperature, at 1 fsec time step. The initial atom velocity is selected using the Boltzmann distribution and periodic boundary conditions for elementary cells. Calculations of the atom paths are carried out in terms of the microcanonical ensemble in which the specified number of atoms, the volume and the temperature range of 10 K remain. 1000 iterations are studied. The initial conditions for calculations include the atomic density and temperature. The energy density for the amorphous mineral is calculated depending on the atomic density range of 1.5–3.5 g/cm^2 with a view to reach the optimum state, when the total energy of domain is the lowest, *i.e.* the convergence criterion is –142.158 kcal/mole at 3.2 g/cm^2 atomic density. Such a density determines the size of the cubic domain which concentrates atoms of the mineral of interest. As a result of this procedure, C–S–H achieves the amorphous state with the established spatial distribution of atoms with the achieved force equilibrium on them. Unlike the short-range order which is retained for amorphous oxides, the long-range order is disturbed. If the primary space group P1 is assigned to the domains, they are identified as an elementary C–S–H cell, the total structural information for which is known. Hence, for the background identification of the experimental XRD pattern, the structural factors and the integrated intensity can be calculated for C–S–H. The results of the background approximation are presented in Figure 1, the convergence criterion is $R_{wp} \approx 1.73\%$.

Table 2. Structural phase parameters.

Phases	*a*, (Å)[a]	*b*, (Å)[a]	*c*, (Å)[a]	α, (degrees)[b]	β, (degrees)[b]	γ, (degrees)[b]	Symmetry groups	Phase composition, %
Tobermorite	11.350	7.300	20.840	90.00	97.30	90.00	C1, Triclinic	47.72
Amorphous	11.6152	11.6152	11.6152	90.00	90.00	90.00	P1, Triclinic	–
Ettringite	11.230	11.230	10.720	90.00	90.00	120.00	P63/mcm, Hexagonal	37.40
Portlandite	3.5862	3.5862	4.8801	90.00	90.00	120.00	P-3m1, Trigonal	0.30
Alite	12.235	7.073	9.298	90.00	116.31	90.00	C1m1, Monoclinic	7.96
Belite	5.5041	6.7622	9.3281	90.00	94.172	90.00	P21/n, Monoclinic	5.12

[a] lattice parameters (1Å = 10^{-10} m).
[b] lattice angles.

3. Results and discussion

Based on the results of GEMS modeling, major phases which predominate in Portland cement are identified. The experimental XRD patterns are complicated by the distribution of reflections, for which the estimated intensity peaks overlap. Therefore, the structure identification using the Rietveld refinement is divided into two stages. The first stage includes the analysis of contributions only from such hydration products as crystalline cement paste (C–S–H), ettringite, C–S–H amorphous component and Portlandite. The results of the analysis of the lattice structural parameters and phase composition are given in Table 2. It is shown that the cement paste and ettringite predominate and are respectively 47.72 and 37.4 % (see Table 2, Figure 1). The difference between the theoretical calculations and experimental data on the peak intensities (Figure 1) shows a good approximation of XRD patterns. The second stage includes a study of peak intensities without hydration products as presented in Figure 2. Using the Rietveld refinement, the contributions from clinker minerals are studied. According to Table 2, the contributions from alite and belite are respectively 7.96 and 5.12 %. The difference between the peak intensities also indicates to a good approximation of XRD patterns.

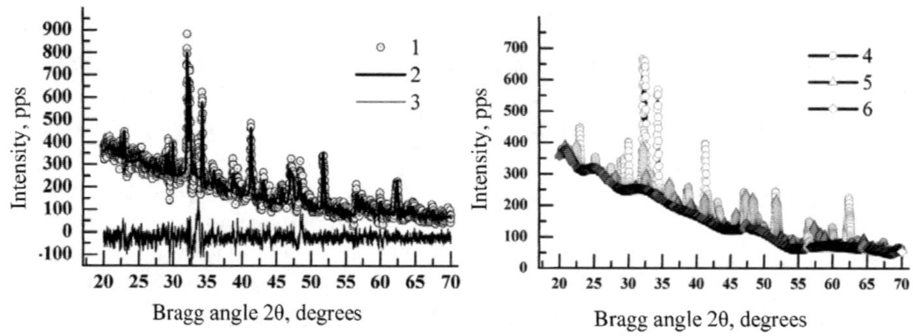

Figure 2. Peak intensities of hydration products without tobermorite and ettringite contributions: *1* – experimental; *2* – integrated; *3* – difference between experimental and integrated intensities; *4* – alite phase; *5* – belite phase; *6* – Portlandite phase.

With a view to determine the stability Δ*E* of major phase lattices at 0 K, the *ab initio* analysis of their energy *E* is performed [6]. The results are summarized in Table 3.

Table 3. Energy and stability of phase lattices.

Phases	Tobermorite	Amorphous	Ettringite	Alite	Belite	Portlandite
E, eV	–61216.54	–65463.99	–72504.21	–31868.08	–934.85	–1911.72
ΔE, eV	–2335.69	–2281.27	–934.85	–934.85	–391.20	–65.02

Energies of the studied phases, including the amorphous, are negative. Moreover, they are stable relative to the delamination into pure elements. The binding energy Δ*E* reaches several hundred electron volts, that indicates their high stability. The phases stabilize during the hydration process.

Thus, the QPA performed with the Rietveld refinement allows us to determine the amount of contributions from both cement paste and ettringite hydration products and clinker minerals. Over 85% of the Portland cement composition belong to cement paste (C–S–H) and ettringite. Complete structural information is obtained for all the minerals, and the phase stability is achieved. In a reliable manner, our investigations show the dominance of the short-range order C–S–H phase, the amorphous component of Portland cement.

4. Conclusions

Based on the research results it can be concluded that the behavior of the stress-strain curves includes three stages, namely: transition, elastic and plastic strain ranges. The temperature and time of isothermal curing substantially affected the intensity of strength generation in the plastic strain range. At the end of the isothermal curing, the values of the limit strength and the yield stress almost coincided. Stress-strain curves in the plastic strain range are non-monotonic, with a high density of breaks, and a catastrophic failure of specimens occurred at the end of the investigated temperature range. Within 0–67 h curing, the yield stress increased almost by an order to magnitude, and the elasticity modulus increased also. It was shown that with the growth in the curing temperature, these parameters also increase, and a more intensive strength generation was observed. The yield stress was found to be the main estimation parameter of the strength properties at different curing temperatures. After 67 h curing the strength properties of Portland cement achieved not less than 75% of the design strength values corresponding to 28-day curing.

Acknowledgement
Authors wish to acknowledge the financial support from the Russian Foundation for Basic Research (Grant N 18-08-01025).

References

[1] Pashenko A A 1991 *Cement Theory* (Kiev: Budivel'nik) p 166
[2] Lothenbach B and Winnerfeld F 2006 Thermodynamic modelling of the hydration of Portland cement *Cem. Concr. Res.* **36** 2 209–26
[3] Parrot L J and Killoh D C 1984 Prediction of cement hydration. *Proc. Br. Ceram. Soc.* **35** 41–54
[4] Jansen D, Neunhoeffer F G, Lothenbach B and Neubauer J 2012 *Cem. Concr. Res.* 42 134–8
[5] Richardson J G 2014 *Acta Crystallogr.* **B70** 6 903–923
[6] Thomas J J, Biernacki J J, Bullard J W, Bishnoi S, Dolado J S, Scherer G W and Luttge A 2011 *Cem. Concr. Res.* 41 1257–1278
[7] GEMS: Terms and Conditions of Use, available: http://gems.web.psi.ch/termsofuse.html
[8] Gnyrya A, Abzaev Y, Korobkov S, Mokshin D, Gauss K and Boyarintsev 2017 *AIP Conf. Proc.* **1899** pp 040005-1–8
[9] Gnyrya A I, Abzaev Yu A, Korobkov S V, Boyarintsev A A, Mokshin D I and Gauss K S 2017 *Izv. Vyssh. Uchebn. Zaved., Stroit.* **8** pp 34–45
[10] Standard cements, available: www.sibcem.ru/index.php/production/cement/obshchestroitelnye-tsementy
[11] Abzaev Yu A, Kopanitsa N O, Klimenov V A, Sarkisov Yu S, Gorlenko N P Dem'yanenko O V and Zav'yalov A P 2015. *Vestnik TSUAB* 3 pp. 121–33(in Russian)
[12] Klopotov A A, Abzaev Yu A, Potekaev A I and Volokitin O G 2012 *Fundamentals of X-ray Diffraction in Materials Science* (Tomsk: TSUAB) p 275
[13] Crystallography Open Database, available: www.crystallography.net (accessed: January 17, 2017).

Optimization methods in the design of heat-saving functional devices

G V Alekseev[1,2] and D A Tereshko[1]

[1] Institute of Applied Mathematics FEB RAS, 7, Radio St., Vladivostok, Russia
[2] School of Natural Sciences, Far Eastern Federal University, Vladivostok, Russia

E-mail: alekseev@iam.dvo.ru

Abstract. Inverse problems associated with designing cylindrical heat manipulation devices are studied. Using the optimization method these inverse problems are reduced to corresponding control problems. Thermal conductivities of device materials play the role of passive controls. Numerical algorithm based on the particle swarm optimization is proposed. Results of numerical experiments for shielding and heat concentration problems are presented and discussed.

1. Introduction

In the polar latitudes the heat-saving and heat-shielding problems for industrial and residential objects play an important role [1, 2]. An innovative solution to these problems leads to the problems of heat flux manipulation. One of the ways to solve these problems is based on the extensive use of special materials, so-called thermal metamaterials. The need to use metamaterials for physical fields manipulation became clear after the publication of pioneer works [3, 4, 5] devoted to design of invisibility devices. The idea of cloaking objects by placing them inside a three-dimensional cloaking shell filled with an inhomogeneous anisotropic medium with specially selected parameters was proposed in these works.

Based on this idea, a method to design invisibility devices was developed first for electromagnetic fields, and then for acoustic, magnetic, electrical, thermal and other static fields [6, 7, 8]. Further, this method was also used to design other functional devices such as concentrators, rotators, inverters, etc. (see [9]). It should be noted that technical implementation of these devices is associated with great difficulties due to the lack of materials that possess the above properties. Therefore another design strategy began develop recently. It obtained the name of inverse design because it is related with solving inverse problems (see [10, 11]). The optimization method forms the core of the inverse design methodology. A growing number of papers is devoted to applying the inverse design for physical fields manipulation. Among them we mention [12, 13] where numerical optimization algorithms are applied for finding the unknown material parameters of cloak and papers [14, 15, 16] devoted to theoretical analysis of cloaking problems using the optimization approach.

In recent years numerical design of heat manipulation devices is associated with the use of topology optimization and discrete material optimization. In [17] topology optimization approach was applied to design anisotropic thermal-composite materials for heat flux shielding, focusing, and reversal. In [18] topology optimization was used to design thermal cloaking shell composed of iron and aluminium. A series of papers is devoted to the application of discrete material optimization approach for numerical designing heat flux manipulation devices. Among them we mention works where a laminate of predefined materials with controlled orientation was used to facilitate the fabrication of designed thermal cloaks and concentrators. In [19] the material at a given point of the heat flux manipulation device is chosen between two prescribed materials with highly different conductivity.

In this paper optimization method is applied to solve inverse problems for the 2D stationary model of heat transfer. These problems arise in designing heat flux manipulation devices that have the form of the cylindrical shell. It is assumed that the desired shell consists of finite number of subdomains filled with homogeneous isotropic medium. Thermal conductivities of the medium filling subdomains play the role of controls. As a result the heat flux manipilation problems are reduced to solving the corresponding finite-dimensional extremum problems. We propose a numerical algorithm based on the particle swarm optimization (see [20]) according to the scheme of [21, 22] and discuss some results of numerical experiments.

2. Statement of direct problem

We begin with statement of the direct heat conductivity problem, considered in a rectangle $D = \{\mathbf{x} \equiv (x, y) : |x| < x_0, |y| < y_0\}$ with specified numbers $x_0 > 0$ and $y_0 > 0$ (see Figure 1a). We will assume that an external temperature field T^e is created by two vertical plates $x = \mp x_0$ which are kept at different values T_1 and T_2, while the upper and lower plates $y = \pm y_0$ are thermally insulated. We assume that there is a material shell (Ω, κ) inside D. Here, Ω is a circular layer $a < |\mathbf{x}| < b$ where $b < \min\{x_0, y_0\}$ and κ is the thermal conductivity of the inhomogeneous medium filling the domain Ω. We assume also that the interior Ω_i: $|\mathbf{x}| < a$ and exterior Ω_e: $|\mathbf{x}| > b$ of Ω are filled with the same homogeneous medium having constant thermal conductivity $k_0 > 0$ (see Figure 1b).

In this case, the direct conduction problem consists of determination of a triplet of functions, namely T_i in Ω_i, T in Ω, and T_e in Ω_e, which satisfy the equations

$$k_0 \, \Delta T_i = 0 \ \text{ in } \ \Omega_i, \tag{1}$$

$$\operatorname{div}(\kappa \operatorname{grad} T) = 0 \ \text{ in } \ \Omega, \tag{2}$$

$$k_0 \, \Delta T_e = 0 \ \text{ in } \ \Omega_e, \tag{3}$$

and obey the following boundary conditions on the boundary ∂D of D

$$T_e|_{x=-x_0} = T_1, \ \ T_e|_{x=x_0} = T_2, \ \ \frac{\partial T_e}{\partial y}|_{y=\pm y_0} = 0 \tag{4}$$

and the matching conditions on internal Γ_i and external Γ_e components of the boundary Γ of the shell Ω having the form

$$T_i = T, \ \ k_0 \frac{\partial T_i}{\partial n} = (\kappa \nabla T) \cdot \mathbf{n} \text{ on } \Gamma_i, \ \ T_e = T, \ \ k_0 \frac{\partial T_e}{\partial n} = (\kappa \nabla T) \cdot \mathbf{n} \text{ on } \Gamma_e. \tag{5}$$

In the particular case when $\kappa = k_0$ (so that the homogeneous and isotropic medium filling the domain D as shown in Figure 1a) the solution of the problem (1)–(5) is described by the temperature field T^e and the constant heat flux $\mathbf{q}^e$ by the formulas

$$T^e(x) = \frac{T_2 - T_1}{2} \frac{x}{x_0} + \frac{T_1 + T_2}{2}, \ \ \mathbf{q}^e \equiv -k_0 \nabla T^e = \left(\frac{T_1 - T_2}{2} \frac{k_0}{x_0}, 0 \right). \tag{6}$$

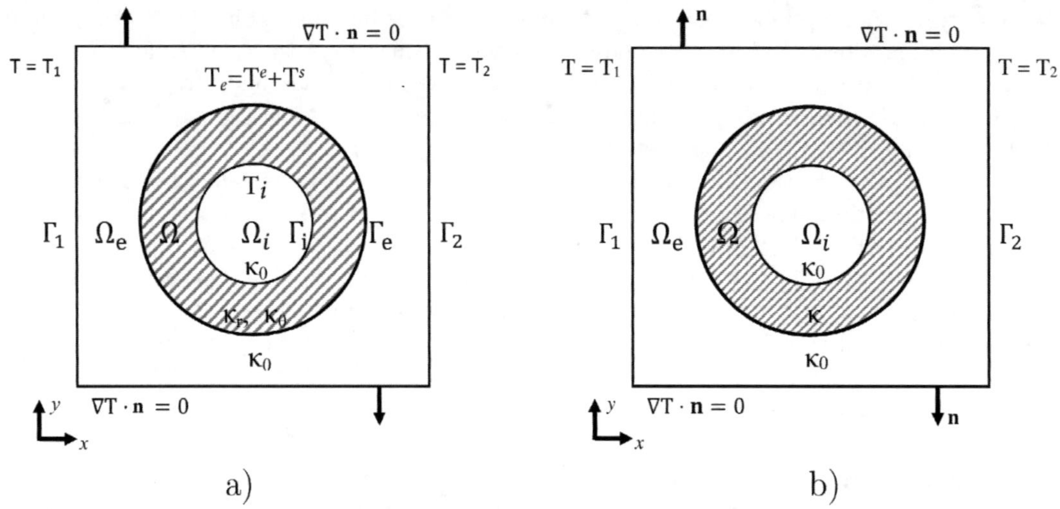

Figure 1. Two geometries of the problem: a) without a shell; b) with a shell.

3. Control problems for complex shell

Now we assume that (Ω, κ) is a shell consisting of M subdomains Ω_j, $j = 1, 2, \ldots, M$. Each of these subdomains is filled with a homogeneous and isotropic medium described by constant thermal conductivity $k_j > 0$, $j = 1, 2, \ldots, M$. The parameter κ of this shell is given by

$$\kappa(\mathbf{x}) = \sum_{j=1}^{M} k_j \chi_j(\mathbf{x}). \tag{7}$$

Here $\chi_j(\mathbf{x})$ is a characteristic function of subdomain Ω_j which is equal to 1 inside of Ω_j and to 0 outside of Ω_j while k_j are some constant coefficients.

We remind that our purpose is the numerical analysis of inverse problems for model (1)–(5) arising in designing heat flux manipulation devices. In the case of a composed shell these problems consist of finding unknown coefficients k_j, $j = 1, 2, ..., M$ in (7) forming a M-dimensional vector $\mathbf{k} = (k_1, k_2, ..., k_M)$. We will refer to $\mathbf{k}$ as the conductivity vector.

By optimization method our inverse problems are reduced to extremum problems of minimization of certain cost functionals which adequately correspond to the inverse problems under consideration. In order to formulate these control problems we denote by $T[\mathbf{k}] \equiv T[k_1, k_2, \ldots, k_M]$ the solution to the direct problem (1)–(5) corresponding to parameters k_j in Ω_j, $j = 1, 2, \ldots, M$, and to conductivity k_0 in Ω_i and Ω_e. By $\mathbf{q}[\mathbf{k}]$ we define corresponding heat flux $\mathbf{q}[\mathbf{k}] = -k_0 \nabla T[\mathbf{k}]$ in Ω_i. We will assume below that the vector $\mathbf{k}$ belongs to the bounded control set

$$S = \{\mathbf{k} \equiv (k_1, k_2, \ldots, k_M) : k_{min} \leq k_j \leq k_{max}, j = 1, 2, \ldots, M\}. \tag{8}$$

Here given positive constants k_{min} and k_{max} are lower and upper bounds of the control set S. Let us define the cost functional

$$J(\mathbf{k}) = \frac{\|\mathbf{q}[\mathbf{k}] - \mathbf{q}^d\|_{L^2(\Omega_i)}}{\|\mathbf{q}^e\|_{L^2(\Omega_i)}} \tag{9}$$

where

$$\|\mathbf{q}[\mathbf{k}] - \mathbf{q}^d\|^2_{L^2(\Omega_i)} = \int_{\Omega_i} |\mathbf{q}[\mathbf{k}] - \mathbf{q}^d|^2 d\mathbf{x}, \quad \|\mathbf{q}^e\|^2_{L^2(\Omega_i)} = \int_{\Omega_i} |\mathbf{q}^e|^2 d\mathbf{x},$$

and formulate following control problem:

$$J(\mathbf{k}) \to \inf, \quad \mathbf{k} \in S. \tag{10}$$

Here $\mathbf{q}^d$ is the desired heat flux in domain Ω_i. In particular, we assume that $\mathbf{q}^d = \alpha \mathbf{q}^e$ where heat flux $\mathbf{q}^e$ is defined in (6) while the constant $\alpha = 0$ in shielding problem, $\alpha = -1$ in the heat flux inversion problem and $\alpha > 1$ in heat concentration problem.

To solve problem (10) we use an algorithm based on the particle swarm optimization [20]. Within this method, the design parameters are presented as coordinates of the radius-vector $\mathbf{k} \equiv (k_1, k_2, \ldots, k_M)$ of some abstract particle. A particle swarm is considered to be any finite set of particles $\mathbf{k}_1, \ldots, \mathbf{k}_N$. Within the particle swarm optimization, one sets the initial swarm position $\mathbf{k}_j^0$, $j = 1, 2, \ldots, N$ and the iterative displacement procedure $\mathbf{k}_j^{i+1} = \mathbf{k}_j^i + \mathbf{v}_j^{i+1}$ for all particles $\mathbf{k}_j$, which is described by the formulas

$$\mathbf{v}_j^{i+1} = w\mathbf{v}_j^i + c_1 d_1 (\mathbf{p}_j^i - \mathbf{k}_j^i) + c_2 d_2 (\mathbf{p}_g - \mathbf{k}_j^i). \tag{11}$$

After each displacement we calculate the value $J(\mathbf{k}_j^{i+1})$ of the functional J for the new position $\mathbf{k}_j^{i+1}$, compare it to the current minimum value and, if necessary, update the personal and global best positions $\mathbf{p}_j$ and $\mathbf{p}_g$. In the end of this iteration process all particles have to come at the global minimum point.

Here $\mathbf{v}_j$ is the displacement vector; w, c_1 and c_2 are constant parameters of algorithm; d_1 and d_2 are random variables which are uniformly distributed over the interval $(0, 1)$. The subscript $j \in \{1, 2, \ldots, N\}$ in (11) denotes the particle number, and the superscript $i \in \{0, 1, \ldots, L\}$ indicates the iteration number. The choice of main parameters was considered in more detail in [20, 21, 22].

4. Numerical results

In this section we discuss numerical results obtained by the particle swarm optimization algorithm for some model heat flux manipulation problems. The most time-consuming part of the above-described algorithm is the calculation of the values $J(\mathbf{k}_j^i)$ of the functional J for the particle position $\mathbf{k}_j^i$ at different i and j values. This procedure includes two stages. At the first stage, the solution $T[\mathbf{k}_j^i]$ to direct problem (1)–(5) is calculated using the FreeFEM++ software package (www.freefem.org). After determining $T[\mathbf{k}_j^i]$, at the second stage we calculate the mean squared integral norms entering the definition of the functional J in (9) using the formulas of numerical integration.

Numerical experiments were performed for values $N = 25$, $L = 50$, $w = 0.4$, $c_1 = 1$, $c_2 = 1.5$ of the particle swarm optimization parameters and at the following given data:

$$x_0 = y_0 = 4.5 \text{ cm}, \; a = 1 \text{ cm}, \; b = 3.25 \text{ cm}, \; T_1 = 321.85 \text{ K}, \; T_2 = 283.15 \text{ K}, \; k_0 = 1 \text{ W/(m} \cdot \text{K)}.$$

The role of the external field was played by the field T^e determined in (6).

Below we assume that the background material in Ω_i and Ω_e is glass ($k_0 = 1$ W/(m · K)) while lower and upper bounds in (8) are determined by $k_{min} = 3 \times 10^{-2} k_0$ (polystyrene) and $k_{max} = 403\, k_0$ (copper).

Different types of inverse problems require different partitioning of the shell into subdomains Ω_j. In our first group of tests we solve the shielding problem for multilayer shell with isotropic layers. Denote by M the even number of layers Ω_i and by k_j the thermal conductivity of the j-th layer, $j = 1, \ldots, M$. Optimal values k_1^{opt}, k_2^{opt} and corresponding value J^{opt} of the cost functional J obtained by solving problem (10) with $\mathbf{q}^d = \mathbf{0}$ using particle swarm optimization are presented in Table 1 for five different scenarios corresponding to $M = 2, 4, 6, 8$ and 10.

IOP Conf. Series: Earth and Environmental Science **193** (2018) 012002 doi:10.1088/1755-1315/193/1/012002

Table 1. Numerical results for shielding problem.

M	k_1^{opt}/k_0	k_2^{opt}/k_0	J^{opt}
2	3×10^{-2}	403	2.44×10^{-3}
4	3×10^{-2}	403	8.55×10^{-6}
6	3×10^{-2}	403	9.81×10^{-8}
8	3×10^{-2}	403	2.31×10^{-9}
10	3×10^{-2}	403	8.23×10^{-11}

Isolines of corresponding field T^{opt} for a four-layer shell ($M = 4$) are plotted in Figure 2a. It is very interesting that all found parameters k_j^{opt} with odd indices $3, 5, \ldots, M-1$ coincide with k_1^{opt} which is equal to k_{min} while optimal parameters $k_4^{opt}, \ldots, k_M^{opt}$ with even indices coincide with k_2^{opt} which is equal to k_{max}. Thus we have

$$k_1^{opt} = k_3^{opt} = \ldots = k_{M-1}^{opt} = 3 \times 10^{-2}\, k_0 \text{ and } k_2^{opt} = k_4^{opt} = \ldots = k_M^{opt} = 403\, k_0.$$

Analysis of Table 1 shows that the optimal values J^{opt} of the cost functional J decrease with increasing the number of layers M. In particular, $J^{opt} = 8.23 \times 10^{-11}$ at $M = 10$ that clearly corresponds to very high shielding efficiency. In this case the heat flux inside the domain Ω_i is close to zero. Such a shell allows us to protect this area from very high or very low temperatures using only two materials.

Another important problem is related to the heat concentration. First, we try to solve the problem (10) with $\mathbf{q}^d = \alpha \mathbf{q}^e$, $\alpha = b/a = 3.25$ using a four-layer shell ($M = 4$). In this case we obtain following optimal values of controls k_j and the functional J:

$$k_1^{opt} = 0.775, \quad k_2^{opt} = 1.108, \quad k_3^{opt} = 1.098, \quad k_4^{opt} = 1.588, \quad J^{opt} = 2.208.$$

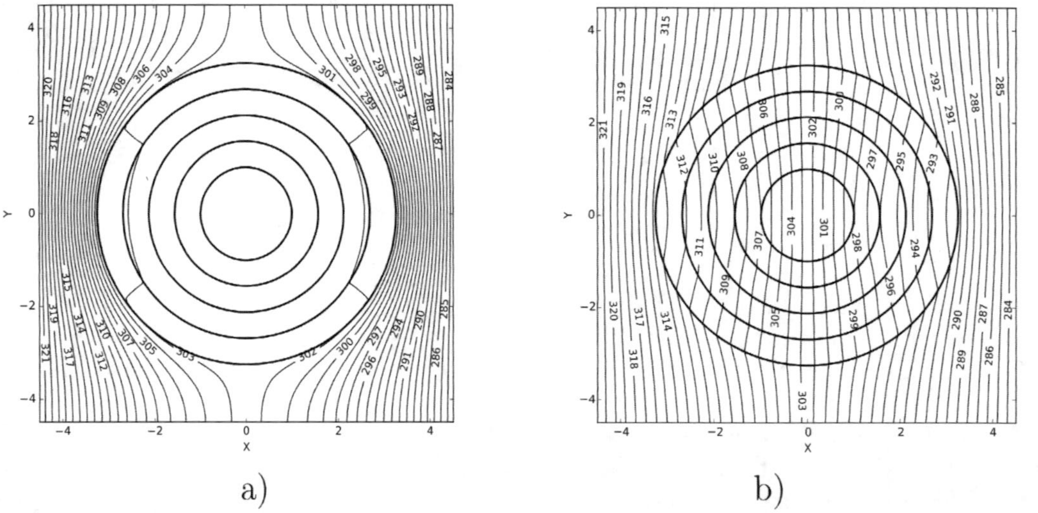

Figure 2. Isolines of temperature field for a four-layer shell with isotropic layers: a) shielding problem; b) concentration problem.

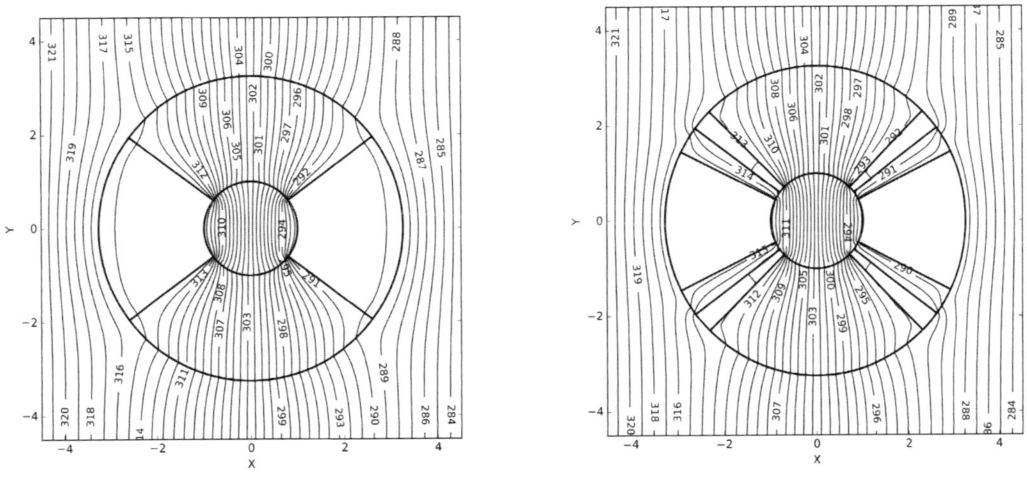

a) b)

Figure 3. Isolines of temperature field for the heat concentration problem: a) two materials; b) four materials.

Isolines of corresponding temperature field T^{opt} are presented in Figure 2b. It is clearly seen that the layered shell has a low shielding efficiency due to improper partition on subdomains. Therefore in our second group of tests we will use sectors as subdomains Ω_j, $j = 1, \ldots, M$. We assume that the distribution of materials in the shell is symmetric.

First we will try to design a concentration shell using only two materials. For this purpose we split the shell into four sectors (left, right, upper and lower) using lines with angles φ_1 and φ_2 so that $\tan\varphi_1 = 0.75$ and $\tan\varphi_2 = -0.75$. Taking into account the symmetry, we assume that the thermal conductivity in the left and right sectors is equal to k_1 while the thermal conductivity in the upper and lower sectors is equal to k_2. Numerical solution of concentration problem by particle swarm optimization give us the following optimal values of controls k_j and the functional J:

$$k_1^{opt} = 41.66, \quad k_2^{opt} = 0.03, \quad J^{opt} = 0.99.$$

Isolines of corresponding temperature field T^{opt} are presented in Figure 3a. Let us note that this version of the shell has a higher concentration efficiency compared to the layered shell. We can see high thermal conductivity in the left and right sectors and low value of the thermal conductivity in the upper and lower sectors. This distribution of materials in the shell allow us to concentrate the heat in domain Ω_i.

Now we split the shell into twelve sectors using lines with angles $\varphi_1, \ldots, \varphi_6$ so that

$$\tan\varphi_1 = 0.5, \quad \tan\varphi_2 = 0.75, \quad \tan\varphi_3 = 1, \quad \tan\varphi_4 = -0.5, \quad \tan\varphi_5 = -0.75, \quad \tan\varphi_6 = -1.$$

In this case we have four controls $k_1, \ldots, k_4$. Solving problem (10) with $\mathbf{q}^d = 3.25\,\mathbf{q}^e$ we obtain following optimal values:

$$k_1^{opt} = 225, \quad k_2^{opt} = 0.03, \quad k_3^{opt} = 188, \quad k_4^{opt} = 0.03, \quad J^{opt} = 0.81.$$

Isolines of corresponding temperature field T^{opt} are plotted in Figure 3b. As in the previous case, left and right sectors have high thermal conductivity to ensure the heat concentration in domain Ω_i.

Using a larger number of controls has increased the efficiency of the concentration shell. Unfortunately, it is not always possible to find materials corresponding to the optimal values of thermal conductivity. Therefore, a more promising approach is related to the situation when we control the geometric characteristics of the shell for a given set of available materials. This new approach will be considered in the future work of the authors.

Acknowledgments
The first author was supported by the Basic Research Program "Far East" (project no. 18-5-064), the second author was supported by the Federal Agency for Scientific Organizations in the framework of the state task (subject no. 0263-2018-0001).

References

[1] Buznik V M and Kablov E N 2017 Arctic materials science: current state and prospects *Herald Russ. Acad. Sci.* **87** 397–408
[2] Yan J-B, Liu X-M, Liew J Y R, Qian X and Zhang M-H 2016 Steel-concrete-steel sandwich system in Arctic offshore structure: Materials, experiments, and design *Mater. Des.* **91** 111–21
[3] Dolin L S 1961 On a possibility of comparison of three-dimensional electromagnetic systems with nonuniform anisotropic filling *Izv. Vuz. Radiofiz.* **4** 964–7
[4] Leonhardt U 2006 Optical conformal mapping *Science* **312** 1777–80
[5] Pendry J B, Schurig D and Smith D R 2006 Controlling electromagnetic fields *Science* **312** 1780–2
[6] Gomory F et al. 2012 Experimental realization of a magnetic cloak *Science* **335** 1466–8
[7] Guenneau S, Amra C and Veynante D 2012 Transformation thermodynamics: Cloaking and concentrating heat flux *Opt. Express* **20** 8207
[8] Narayana S and Sato V 2012 Heat flux manipulation with engineered thermal materials *Phys. Rev. Lett.* **108** 214303
[9] Han T et al. 2014 Full control and manipulation of heat signatures: cloaking, camouflage and thermal metamaterials *Adv. Mater.* **26** 1731–4
[10] Xu S, Wang Y, Zhang B and Chen H 2013 Invisibility cloaks from forward design to inverse design *Sci. China Inf. Sci.* **56** 120408
[11] Alekseev G V 2016 *Invisibility Problems in Acoustics, Optics and Heat Transfer* (Vladivostok: Dalnauka)
[12] Xi S, Chen H, Zhang B, Wu B-I and Kong J A 2009 Route to low-scattering cylindrical cloaks with finite permittivity and permeability *Phys. Rev. B* **79** 155122
[13] Popa B I and Cummer S A 2009 Cloaking with optimized homogeneous anisotropic layers *Phys. Rev. A*, **79** 023806
[14] Alekseev G V and Levin V A 2014 Optimization method of searching parameters of an inhomogeneous liquid medium in the acoustic cloaking problem *Doklady Phys.* **59** 89–93
[15] Alekseev G V and Levin V A 2016 An optimization method for problems of thermal cloaking of material bodies *Doklady Phys.* **61** 546–50
[16] Alekseev G V 2018 Analysis of a two-dimensional thermal cloaking problem on the basis of optimization *Comput. Math. Math. Phys.* **58** 478–92
[17] Dede E M, Nomura T and Lee J 2014 Thermal-composite design optimization for heat flux shielding, focusing, and reversal *Struct. Multidisc. Optim.* **49** 59–68
[18] Fujii G, Akimoto Y and Takahashi M 2018 Exploring optimal topology of thermal cloaks by CMA-ES *Appl. Phys. Lett.* **112** 061108
[19] Fachinotti V D, Ciarbonetti A A, Peralta I and Rintoul I 2018 Optimization-based design of easy-to-make devices for heat flux manipulation *Int. J. Therm. Sci.* **128** 38–48
[20] Poli R, Kennedy J and Blackwel T 2007 Particle swarm optimization: an overview *Swarm Intel.* **1** 33–57
[21] Alekseev G V, Levin V A and Tereshko D A 2017 Optimization analysis of the thermal cloaking problem for a cylindrical body *Doklady Phys.* **62** 71–5
[22] Alekseev G V, Levin V A and Tereshko D A 2017 The optimization method in design problems of spherical layered thermal shells *Doklady Phys.* **62** 465–9

Monitoring and modelling issues of the thermoabrasive coastal dynamics

D M Aleksyutina, N N Shabanova, O V Kokin, A P Vergun, A V Novikova and S A Ogorodov

Lomonosov Moscow State University, Moscow, Russia

E-mail: aleksyutina@gmail.com

Abstract. In view of the active development of the northern territories the study of the coastal dynamics of the Arctic seas remains very relevant. Despite the fact that scientists have been studying the coastal dynamics since the last century, there are many issues of monitoring of the dynamics of the coasts, complicating the creation of reliable models of this dynamics. Monitoring of the coast dynamics is carried out with the help of field and remote methods having their nuances, errors and conditions of applicability. There are some difficulties associated with the severity of the climatic conditions of the region and its inaccessibility. The main problems related to methodological approaches are considered in the article and some ways of solving these problems are offered, from the positions that seem to the authors to be feasible.

1. Introduction

The study of the northern seas and oceans coasts began in the 14th-16th centuries during the major geographical discoveries in the Arctic. Later in the 20th century, the stage of active economic development of the Arctic coast was connected to the exploration of raw material, the development of the Northern Sea Route, which was accompanied by research due to the construction of port facilities and other infrastructure on the coast.

The coastal zone of the Arctic seas is characterized by high dynamism due to the presence of permafrost, where the state and stability of soils depend on the temperature regime. Many environmental, climatic, geologic, biologic and anthropogenic factors influence the coastal erosion rates (soils erodibility, fractures, adjacent bathymetry, cliff lithology, waves, wind, vegetation, etc.). In Arctic region, additional factors relate directly to the presence of permafrost and cold weather conditions which influence the coastal destruction processes. According to Grigoriev [1], some of these additional factors include: the presence of sea ice, the permafrost (temperature) regime and cryostratigraphy of the area. The combination of hydrometeorological factors determines the conditions for soil thawing and removal of thawed material in the sea. Year by year these conditions change. Hydrometeorological factors of the Arctic coastal retreat include first of all the thermal regime of the territory (air temperature, sea water temperature, surface temperature) which determines the thermodenudation intensity, and the energy (wind-wave) regime that determines the thermal abrasion intensity and removal of thawed material from the beach. The wind-wave regime depends on the frequency and intensity of storms directing to the coast from the sea side and on the duration of the ice-free period [2, 3]. Arctic seas are covered by sea-ice during about 9-10 month per year [4, 5].

During this period, the sea ice cover protects the coastal slopes from destruction. Frozen and thawed soils have different physical and mechanical properties, including erosion resistance. The coastal erosion rate varies in time and space. The presence of permafrost causes a lot of degradation processes, linked to thermal degradation of the soils, e.g. thermal abrasion, thermal denudation, or thermokarst subsidence [6-9]. Every year the Arctic coastline composed by frozen unconsolidated deposits retreats into the continent by a value of several centimeters to several meters.

Despite the fact that scientists have been studying the Arctic coastal dynamics since the last century, still there are many problems [10] of coastal dynamics monitoring complicating the creation of reliable models of this dynamics. The purpose of the paper is to identify contemporary issues of Arctic coastal dynamics in order to attract attention of the scientific community to this and to solve these problems in the future.

2. Study area and background

Coastal dynamics monitoring on the key areas at the Western part of Russian Arctic (Barents and Kara Seas regions) has been carried out by Laboratory of Geoecology of the North at the Geography Faculty (Lomonosov Moscow State University) together with Zubov State Oceanographic Institute (Russian Federal Service for Hydrometeorology and Environmental Monitoring) for more than 35 years (figure 1). Kara Sea key-sites are characterized by continuous permafrost, with annual ground temperature of -4…-6 °C and permafrost thickness up to 50-100 m. Varandey key-site is located in the zone of sporadic permafrost. Frequent layers of unfrozen highly saturated soil interbed the permafrost.

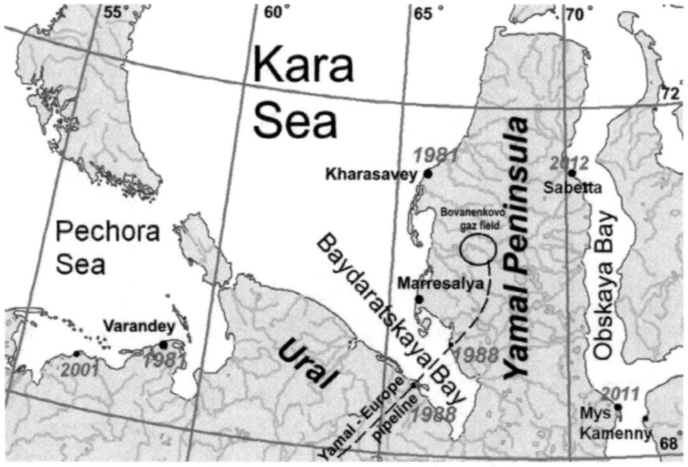

Figure 1. Key sites of studies; years of the start of observations [11].

3. Monitoring of the coastal dynamic

Monitoring of the coast dynamics is carried out with the help of field work (geodetic survey) and remote sensing methods.

3.1. A subsection Geodetic survey

For better understanding the coastal retreat, approximately 120 profiles perpendicular to the coastline were installed in the 1980-s, the 2000-s, the 2010-s. From the start of the investigation period the monitoring technology has been developed. At the start of study the coastal retreat was investigated by tachometric measurements along profiles perpendicular to the coastline (figure 2). Measurements were done on benchmark and on the distinctive relief points (coastal cliff position, position of coastline, beach berm, etc.), where repeated measurements were made in 10-15 cm increments for better accuracy.

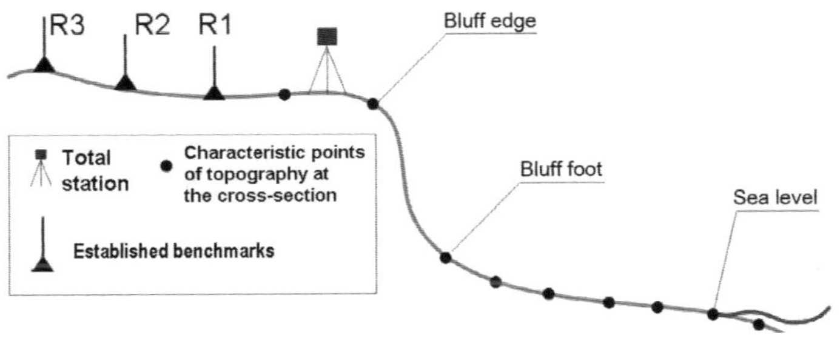

Figure 2. Coastal dynamics monitoring by trigonometric levelling [12].

Currently, a reliable method of studying coastal dynamics for relatively short time periods by using the differential GPS (DGPS) with a Trimble GeoXH® (2008 series) rover connected to an external Zephyr® antennae [5] is used. For our research we utilized two-frequency satellite receivers Trimble® (Trimble R8 GPS Receiver, Trimble tsc2 Controller, Trimble HPB450). Reference (base) station was installed, navigational data was collected and corrections using the known coordinates were performed. Together with the reference station, the Trimble HPB450 modem transmission equipment was installed, which broadcasted corrections in the CMR + format to a mobile satellite receiver, which internal modem received the correction data. The using of the DGPS positioning in our conditions allowed estimating the accuracy of about 0.2 m both horizontally and vertically. The operator moved close to the coastal edge and moved the receiver strictly above the cliff, then the track was recorded. When leveling profiles, the operator installed a mobile receiver on the profile benchmark selected on the terrain. The survey was performed in real time kinematic (RTK) mode and was carried out in the WGS 84 UTM42N coordinate system.

The typical for tundra landscape peat moss vegetation covered the cliff making the identification of the exact position on the cliff edge complicated. The use of high-precision equipment in fields (levels, theodolites, DGPS, etc.) did not solve this problem. So, for example, on high coast during their retreat there is a gradual dangling and slipping of blocks of soil-vegetation cover up to several meters in diameter, which is close to the average annual rate of displacement of the edge of the coastal cliff (figure 3). At the same time, on the low shores, the mapping of the brow was complicated by the accumulation of sand resulting from wind impact and storm surges.

Figure 3. Typical coast of the Kara Sea (hard to identify the position of the cliff near the borehole because of sliding vegetation cover).

In addition, during field works there were some difficulties associated with the severity of the climatic conditions of the region and its inaccessibility. Episodic short-term periods of field work hampered the identification of a quantitative contribution of individual main factors that change during one season of the coastal retreat (in the beginning of summer thermodenudation caused by the thermal exposure of positive air temperatures in the sea-blocked sea prevails, and in autumn period, thermal abrasion is caused by storm activity on the sea).

3.2. Remote setting data
Archive remote sensing data is particularly important for investigation of coastal dynamics, since it allows analysing data over a long-time period. A reliable method of studying coastal dynamics for relatively long time periods is the analysis of multitemporal aerial and space images [13, 14]. For our study area, the best and earliest available image was Corona from 1964 or 1968 from the United States Geological Survey with a resolution 2-7 m. More recent images used in the coastal dynamics study were the aerial photographs obtained by the Geography Faculty in the late 1980s, which had resolution of 0.7 m. For short-term observation, ultra-high resolution images are needed. QuickBird-2, WorldView-2 and WorldView-3 with a resolution of 0.7-0.3 m were used.

However, work with remote sensing data also has its own nuances and features. Space images were provided with the reference files containing the satellite's orbit parameters, but the accuracy of the georeference was not enough to analyse the annual variability of the coastal dynamics. The images should be georeferenced with ground control points, collected during the field work. However, in territories outside of the technogenic development there is the problem of absolute geo-referencing of images, due to the lack of objects with spatially stable outlines, as a result, it is necessary to use conditionally stable objects (for example, the contours of lakes which water level, and, accordingly, the planned outlines vary from season to the season). Part of this problem is solved by referencing a lot of control points, thereby reducing the error.

Another issue with remote sensing data is recognising the position of the coast line. There are not many of images of Russian Arctic coast ready for analysis. For example, this may be due to the snow covering the ledge (figure 4). Also, the visual border of the edge cannot always be identified on the low terraces, which are covered with sand under surges. The difficulties described above in the identification of the cliff position (when the block are sliding) occurs also during field works.

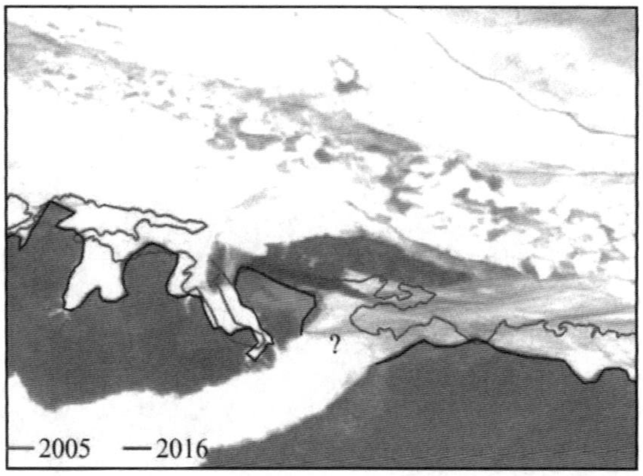

Figure 4. Snow cover on top of the marine terrace at relief settling the Ural coast, complicating coastline digitization on image (2016).

Also space image reflects a one-time moment with specific characteristics that can change during one season (for example, the width of the beach changes during tides and autumn storm surges). In addition, remote sensing does not allow assessing the changes in the coastal profile, but only the planned outlines of the edge of the cliff. This problems can be solved with help of new approaches and methods, like the stereomutlipairs from drone mapping or survey by 3D laser scanner. Among other things, the human factor is another problem: even with experienced researchers who work with remote sensing data and know preliminary methodological agreements, researchers decode the images in different ways.

3.3. Erosion rate estimations

Field work observation data and satellite imagery analysis were used for estimation of coastal retreat rates at the key-sites. Results processing are usually performed using ArcGIS 10.2 software [Kricuc, 2014]. Comparison between the cliff position from DGPS mapping and the cliff contour obtained from space images allowed us to determine the general pattern of coastal destruction within the studied key-sites and also to quantitatively estimate the coastal retreat rates at the different time periods. To estimate the cliff position changes Thieler et al. [15] produced Digital Shoreline Analysis System (DSAS), available as an extension to ArcGIS. The program automatically creates transections perpendicular to the general direction of the shorelines (which is chosen) with an optional spacing along the baseline. However, another problems associated with remote sensing data do not allow analyze two neighboring years of space images, because coastal retreat is not very large and determinations coincide and sum up errors.

4. Modeling of Arctic coastal dynamic

Currently, there are a lot of models trying to predict the arctic coastal dynamics. However, these models, as a rule, are regional and are focused on modeling the shoreline for particular part of the selected coast. Early studies with the prediction of the coastal dynamics were conducted for the regions outside the permafrost, and did not take into account the permafrost characteristics. F.A. Are [16] proposed a model for processing the frozen coast of water-collecting area based on thermophysical calculations of the subsidence of frozen sediment during the thawing. Vasiliev et al. [17] considered the modeling of the dynamics of thermal abrasion from the position of a stochastic model when the natural process of retreat of the coast was partly random. The morphodynamic model of development of the Arctic coasts of Russia developed by Leontiev [18, 19] made it possible to quantify the transport of material influencing the rate of destruction of the coast while the model described in detail the characteristics of the storm impact and only the sediment during ice thawing in frozen soil. For the shores of the East Siberian Sea, mathematical model of the sandy accumulative bar dynamics under the conditions of long-term changes in open water and sea level but under unchanged permafrost, geological conditions and atmospheric circulation have been developed and tested [20]. S. O. Razumov [21, 22] developed models for the development of thermal abrasion of ice banks with a stationary indicator of abrasion activity and a stable sea level, as well as under a changing climate conditions.

All the models considered were calibrated for specific parts of the coast, for example, for Alaska and the Canadian part of the Beaufort Sea with specific features of coastal destruction [23]. Coastal cliffs of these areas are exposed to active erosion which entails virtually no beaches on the shore, and the retreat of the edge of the coastal ledges is associated with a "block" collapse of frozen rocks. In connection with such conditions the Kobayashi model [24] taking into account the thermal characteristics of water, sea level, time of impact, estimates the rate of penetration of the thermoabrasive niche along the base of the frozen cliff, which is always in contact with a constant sea water level. In addition, the thawing of frozen soil blocks falling into the sea water was evaluated by various researchers [25, 26, 27] based on empirical equations for the Alaska coast, which took into account the height of the cliff, the period wave, water temperature and freezing point of water. Further development of the Kobayashi model showed the need to take into account the convective heat flux which depends on the rate of sea water movement and its temperature which strongly affect the frozen soil [28, 29]. This approach was embodied in the COSMOS program [30], which is still used for modeling [31].

Thus, the above mentioned features of monitoring the coastal dynamics complicate the possibilities of comparing and analyzing the data obtained. In that context there are many problems in simulating and modeling the process of coastal retreat which associated with a lack of conditioned factual material, which compels modeling on empirical dependencies. The ambiguity of the leading factors complicates the creation of an universal model that takes into account all parameters of the retreating shore, so at this stage only a description of the qualitative characteristics of the coastal dynamics is possible. In addition, the coasts of the Western sector of the Russian Arctic are composed of unconsolidated sediments with varying cryolithological composition, which additionally complicates the modeling.

5. Conclusion
Thus, the paper considers the main problems which have scientists involved in the Arctic coastal dynamics. Some problems could be solved now with a thoughtful approach and with the use of modern technologies.

Acknowledgments
The present study was prepared with the financial support of the Russian Science Foundation project No.16-17-00034.

References
[1] Grigoriev M N, Razumov S O, Kunitzkiy V V and Spektor V B 2006 *Kriosf. Zemli* **10** 74-95
[2] Ogorodov S A et al. 2016 *Geography, Environment, Sustainability* **3** 53-73
[3] Shabanova N N, Ogorodov S A and Romanenko F A 2017 *Proc. Coastal Dynamics 2017* (Helsingør: Coastal Dynamics) pp 108-16
[4] GEOS 1997 *Baydaratskaya Bay Environmental Conditions. The Basic Results of Studies for the Pipeline "Yamal-Center" Underwater Crossing Design* (Moscow: GEOS Publishing House) p 432
[5] Jones B M, Arp C D, Jorgenson M T, Hinkel K M, Schmutz J A and Flint P L 2009 *Geophys. Res. Lett.* **36** p 5
[6] Are F E 2012 *Coastal Erosion of the Arctic Lowlands* (Novosibirsk: Academic Publishing House "Geo") p 291
[7] Zhigarev L A 1997 *Submarine Permafrost* (Moscow: Moscow State University Press) p 318
[8] Reimnitz E and Are F E 2000 *Polarforschung* **68** 231-35
[9] Günther F, Overduin P P, Sandakov A V, Grosse G and Grigoriev M N 2013 *Biogeosciences* **10** 4297-18
[10] Vasiliev A A 2001 *Arctic Coastal Dynamics, Report of an International Workshop (Potsdam, Germany) Ber. Polarforsch. Meeresforsch.* ed V Rachold, J Brown and S Solomon pp 77
[11] Ogorodov S A et al. *Geography, Environment, Sustainability* 2016 **3** 53-73
[12] Baranskaya A, Belova N, Kokin O, Shabanova N, Aleksyutina D and Ogorodov S 2017 *Proc. 13 Int. MEDCOAST Congress on Coastal and Marine Science, Engineering, Management and* Conservation *(Mellieha, Malta)* vol. 1 (Mugla: Publisher Mediterranien Coastal Foundation Dalyan) pp 281-92
[13] Ogorodov S A, Belova N G, Kuznetsov D E and Noskov A I 2011 *Earth from Space – the Most Important Solutions* **10** 66-70
[14] Kritsuk L N, Dubrovin V A and Yastreba N V 2014 *Kriosf. Zemli* **4** 59-69
[15] Thieler E R, Himmelstoss E A, Zichichi J L and Ergul A 2009 *Digital Shoreline Analysis System (DSAS) Version 4.0 Open-File Report 2008-1278* (U.S. Geological Survey)
[16] Are F E 1985 *Principles of Share Thermoabrasion Forecast* (Novosibirsk: Nauka) p 172
[17] Vasiliev A A, Ostroumov V E, Gubin S V and Sorokovikov V A 2007 *Kriosf. Zemli* **2** 60-7
[18] Leont'ev I O 2002 *Geomorphologiay* **1** 53-64
[19] Leont'ev I O 2004 *Oceanologiya* **44** 457-68
[20] Pavlidis Y A and Leont'ev I O 2002 *Bull RFBR* **4** 53–7
[21] Razumov S O 2000 *Geomorphologiya* **3** 88-94
[22] Razumov S O 2001 *Kriosf. Zemli* **1** 53-60

[23] Frederick J M, Thomas M A, Bull D L, Jones C and Roberts J 2016 *AGU Fall Meeting Abstract (San Francisco)* EP13C-1043
[24] Kobayashi N 1985 *Journal of Geophysical Research* **90** 11983-8
[25] White F M, Spaulding M L and Gominho L 1980 *U. S. Coast Guard Report no CG-D-62-80* p 126
[26] Russell-Head D D 1980 *Annals of Glaciology* **1** 119-22
[27] Holland P R, Jenkins A and Holland D M 2008 *Journal of Climate* **21** 2558-72
[28] Kobayashi N and Aktan D 1986 *Journal of Waterway, Port, Coastal, and Ocean Engineering* **112** 140-58
[29] Kobayashi N and Vidrine J 1995 *Research Report no. CACR-95-12* (Newark Delaware: Center for Applied Coastal Research)
[30] Baird and Associates 1995 *Final Report Prepared for Geological Survey of Canada* (Oakville, Ontario, Canada)
[31] Pearson S, Lubbad R, Le T M H and Nairn R 2016 *Proc. Int. Conf. on Scour and Erosion* (United Kingdom: CRC Press) pp 1-10

IOP Conf. Series: Earth and Environmental Science **193** (2018) 012004 doi:10.1088/1755-1315/193/1/012004

Seismogeological analysis and petroleum potential of the Achimov Formation of the Gydan Peninsula

V N Bardachevskiy, S V Ershov and N I Shestakova

Trofimuk Institute of Petroleum Geology and Geophysics, Novosibirsk, Russia

E-mail: BardachevskyVN@ipgg.sbras.ru, ErshovSV@ipgg,sbras.ru, ShestakovaNI@ipgg.sbras.ru

Abstract. Considered questions of structure and formation conditions of the Berrias–Lower Aptian deposits of the Gydan Peninsula of Western Siberia. Their seismogeological characteristics are given. Recognized eight seismic sequences, most of which have a wedge shape and correspond to third-order sequences. The main prospects for the search for hydrocarbon traps are associated with the Achimov petroleum complex, lenticular sandy-aleuritic beds of which have deep-water genesis. Two types of promising zones are identified for the search for lithological traps in the deposits of the Achimov Formation. One of them is associated with depocenters accumulation of clinoforms, the second – with the regional wedging distal Achimov strata in the direction of the Taimyr paleolandmass.

1. Introduction

The Gydan Peninsula, located in the Arctic regions of Western Siberia, is a promising area in search of hydrocarbon deposits. A number of large oil and gas fields such as Salmanovskoe, Geophysicheskoe, Tota-Yakhinskoe, Gydanskoe, and Soletskoe, with deposits in the Jurassic and Cretaceous sediments, have already been discovered on this territory. Nevertheless, the study drilling remains extremely low. A little more than 180 wells were drilled, which are mainly concentrated on a small number of areas. Most of them are located near the coast of the Ob and Taz Bays. A significant part of the wells uncover the Cretaceous deposits on the bottom hole and only about twenty wells reached the Jurassic complex.

In recent years, seismic exploration works become more intense in the Arctic regions. On the territory of Gydan, the total length of seismic profiles is currently about 60 thousand linear kilometers, and their average density is about 0.60 km/km^2.

In the Jurassic and Cretaceous of Western Siberia is allocated a number of petroleum complexes. Each of them has its own features of geological structure, litho-facies composition of rocks, regularities in the allocation of hydrocarbon deposits, etc. One of such promising petroleum complexes in the territory of Gydan is Achimov petroleum complex. The huge reserves of oil and gas of the Priobskaya and East-Urengoy oil and gas bearing zones of Western Siberia are associated with the Achimov Formation. On the territory of the Gydan Peninsula hydrocarbon deposits in the Achimov strata were found at the Deryabinskoe field. However, most of the resources of this complex remain undetected.

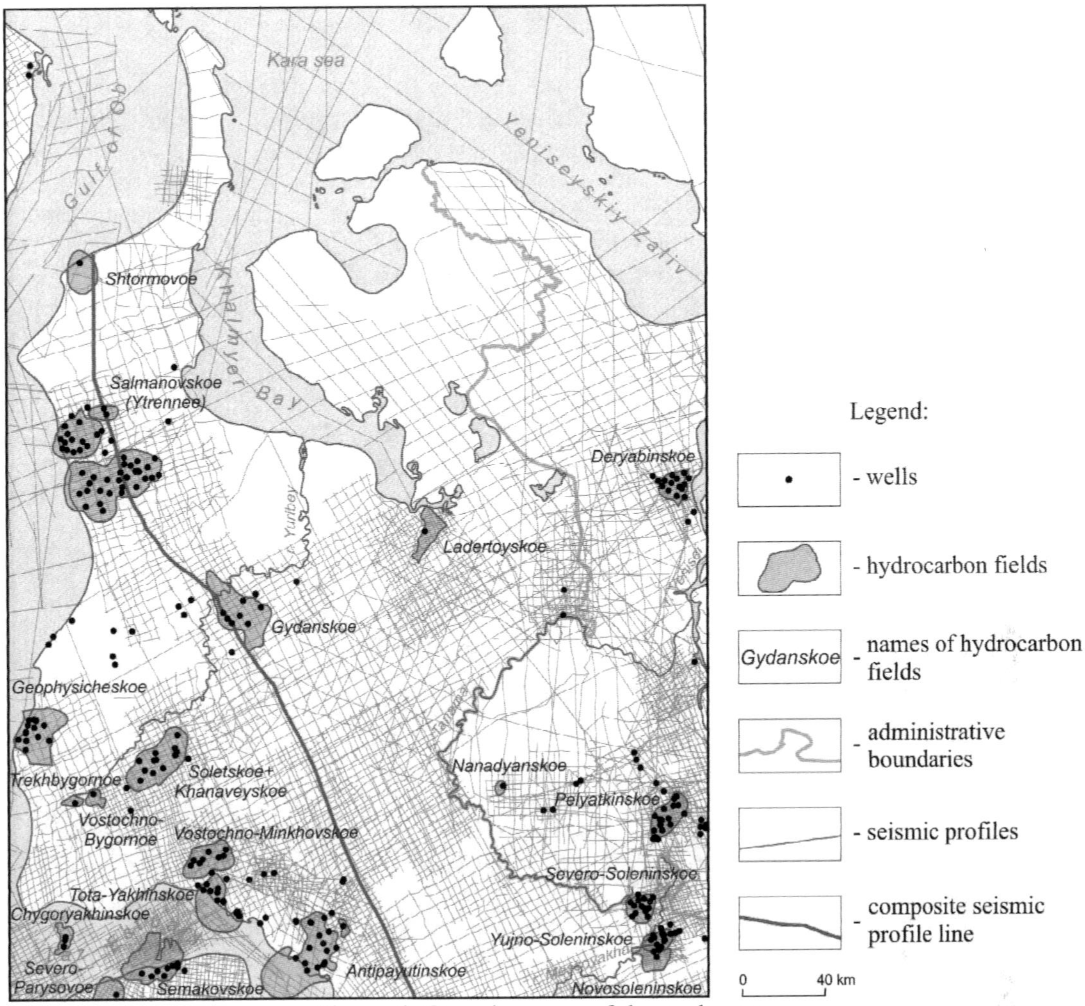

Figure 1. Overview map of the study area.

2. Seismogeological characteristics

The Gydan peninsula is located in the north-east of the West Siberian plate. In the section of the Mesozoic sedimentary cover, five seismogeological megacomplexes can be recognized: Triassic, Jurassic, Berrias–Lower Aptian, Aptian–Cenomanian and Turonian-Maastrichtian. Of special interest is the Berrias–Lower Aptian megacomplex. Its distinctive feature is the clinoform structure of the section, which was caused by the lateral filling of a relatively deep-water basin as a result of avalanche sedimentation. At the time seismic sections, the Berrias–Lower Aptian megacomplex of Gydan is defined by reflecting horizons B (top of Bazhenov and Gol'chikha Formations) и M (bottom of Neite shale member).

To construct a seismogeological model of the Berrias–Lower Aptian deposits, including the Achimov Formation of the Gydan Peninsula, the authors carried out a complexation of geophysical and geological methods. First of all, they include seismic and sequence-stratigraphic methods, paleogeomorphological reconstructions, interpretation of the results of well logging (WL).

The principle of selection seismic sequences (SS) was based on comparison of wavefield features of seismic sections, WL data, spatial location of plug shale member, boundaries of second- and third-order sequences, maximum flooding surfaces. Segregation and correlation of SS controlled by paleontological data. Stratigraphic finding of reflecting horizons assign to bottom of regional shale members was carried out by comparing WL data embedded in temporary seismic sections through vertical seismic profiling and seismic logging.

Within the Gydan Peninsula, eight seismic sequences of the Valanginian–Lower Aptian age have been identified: Savui (sv), Cheuska (ch), Sarman (srm), Urengoi (urn), Pim (pm), Seakha (sh), Arkticheskiy (ark) and Lower Alym (nal) (figure 2). The names are given by the name of regional shale members lying at the base of seismic sequences. All SS except for the Lower Alym are wedge-shaped bodies (clinoforms), which correspond to third-order sequences. The time of their formation varies from 0.4 to 2.5 million years. The formation of the Lower Cretaceous clinoforms of Western Siberia should be considered in terms of the Depositional Sequence III model [1].

In the transgressive phases of SS formation, mainly shale sediments accumulated, regressive – sandy-aleuritic horizons. The boundaries of SS are controlled by reflecting horizons at the base of the sequences of the third-order and are marked by the extremum of the negative phase.

The clinoform part of the Berrias–Lower Aptian megacomplex is displayed on the stacked data by a series an oblique reflective horizons, which are flattening as the horizon B approaches (figure 2). In the regressive part of the wedge-shaped seismic sequence (WSS), an oblique boundaries stop tracing up the uprising according to the pattern of toplap, downwards by the type of downlap. In the zone of greatest thickness, a wave pattern with a chaotic, low-amplitude form of recording is singled out. The transgressive parts of WSS represented mainly dynamically expressive negative extremes of seismic recording.

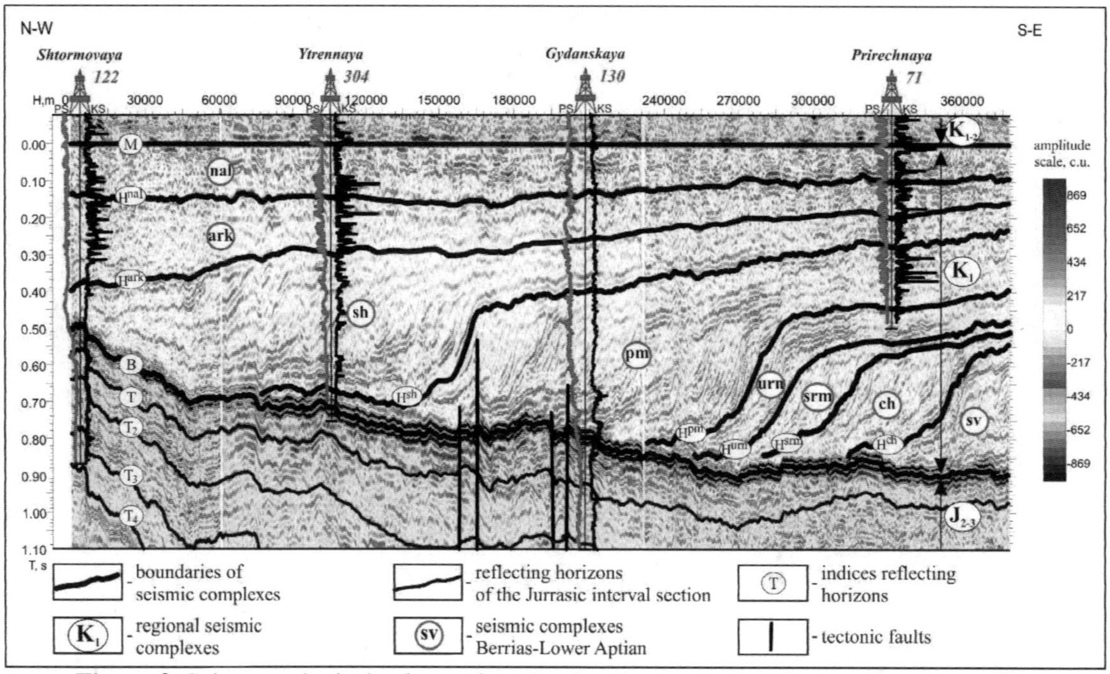

Figure 2. Seismogeological paleosection Berrias–Lower Aptian along composite profile.

The upper part of the section of the Berrias–Lower Aptian megacomplex is a cover horizontally layered formation. In the regressive parts of seismic sequences, the wave pattern has a low-amplitude, chaotic form of recording, with an increase in thickness, a weak oblique can be appear. The transgressive, mainly shale part of seismic sequence is characterized by a dynamically expressed wave pattern.

3. Formation conditions of the clinoform complex

The accumulation of the Berrias–Lower Aptian deposits of Western Siberia occurred under the conditions of a relatively deep-water paleobasin regressing. According to paleobathimetric reconstructions conducted by one of the authors of the article [2], the depth of the Early Berrias basin to the beginning of the formation of the Lower Cretaceous clinoform complex in the territory of Gydan did not exceed 160 m. During the Berriasian, Valanginian and Hauterivian as a result of tectonic subsidence of the bottom of the basin in areas of uncompensated sedimentation, the depths of the paleobasin continued to increase.

In the Cretaceous period, there were two source of terrigenous material, which formed two systems of clinoforms on Gydan, respectively. For the first system of clinoforms, the Siberian Platform was the source area. The second system of clinoforms was formed due to terrigenous material carried down from the Taimyr landmass. However, the volumes of sediment drifted from the Siberian platform were many times higher than similar amounts of material carried from the northern source area. This was the reason for the asymmetry of the structure of the clinoform complex of the Yenisei-Khatanga regional depression and adjacent areas of the West Siberian plate.

The relatively deep-water environments in the area under investigation existed almost until the late Hauterivian [3]. The West Siberian clinoform complex was progradated cyclically, interrupted by short-duration transgressions, during which regional shale members were formed. On the territory of the Gydan peninsula, the compensation of the paleobasin by sediments occurred during the late Valanginian and the Hauterivian. In the extreme north of the peninsula, this process also affected the beginning of the Barremian. The rate of progradation of the sedimentary complex was irregular in time. The maximum value it reached in the early Hauterivian. The authors associate this with the time of compensation of precipitation in the basin of the adjacent Yenisei-Khatanga regional depression from the east. As a result of which sedimentary material coming from the Siberian platform to the western regions of the depression began to move by rivers in transit to the territory of Gydan.

4. Seismogeological model and oil and gas prospects

Within the Gydan Peninsula, the depths of the Berrias-Lower Aptian megacomplex top reach 2400-2500 m. The maximum value are recorded in depressions along the southern and western coasts of Gydan. In general, for the relief there is a tendency of uplift in the eastern and northeast directions towards Taimyr. In the relief, we can conditionally distinguish two zones, differing in the gradients of the submergence of the top of the megacomplex. The outer Pritaymyr zone is characterized by elevated gradients, averaging 10–20 m/km, and the presence of small-scale local structures that can be mapped only with a dense network of seismic profiles. In the inner zone, the gradients of immersion of the top of the Berrias–Lower Aptian megacomplex decrease several times and a number of positive structures of the second-, third- and lower-orders appear in the relief.

The maximum thickness of the Berrias–Lower Aptian deposits is noted in the southeast of Gydan and reaches 1800–1900 m. If to consider each WSS separately, it is possible to allocate three zones. The first two are zones of reduced thickness, associated respectively with its undatheme and fondotheme. The third zone, within which WSS depocenter are allocated, is represented primarily by the clinotheme and, to a lesser extent, by unda- and fondotheme.

In the formation of the Gydan clinoform complex, the depocenters of the clinoforms along the Valanginian, the Hauterivian and the Barremian were successively displaced in the north-west direction. In figure 3, as an example, a map of the thickness of the Pim WSS of the Lower Hauterivian age is presented. During its formation, source area participated, located both on the Siberian platform and on the Taimyr. As a result, on the seismic profiles it is possible to observe counter clinoforms [4]. At the time of accumulation of the Pim WSS, the precipitation of the Yenisei-Khatanga sedimentary basin was compensated and the clinoform breakpoint acquired a submeridional orientation.

In the structure of the clinoforms, two levels of sandy-aleuritic bodies are distinguished. The first level is associated with the uppermost bed of shallow-marine and continental genesis. The second is

IOP Conf. Series: Earth and Environmental Science **193** (2018) 012004
doi:10.1088/1755-1315/193/1/012004

confined to the distal relatively deep-water deposits of the Achimov Formation. Sandy-aleuritic bodies of shallow and deep-water sediments have different morphology and reservoir properties, differ in structural plans. These and a number of other factors have caused the dominance of various types of trap in different parts of the clinoforms, which requires a differentiated approach to their study.

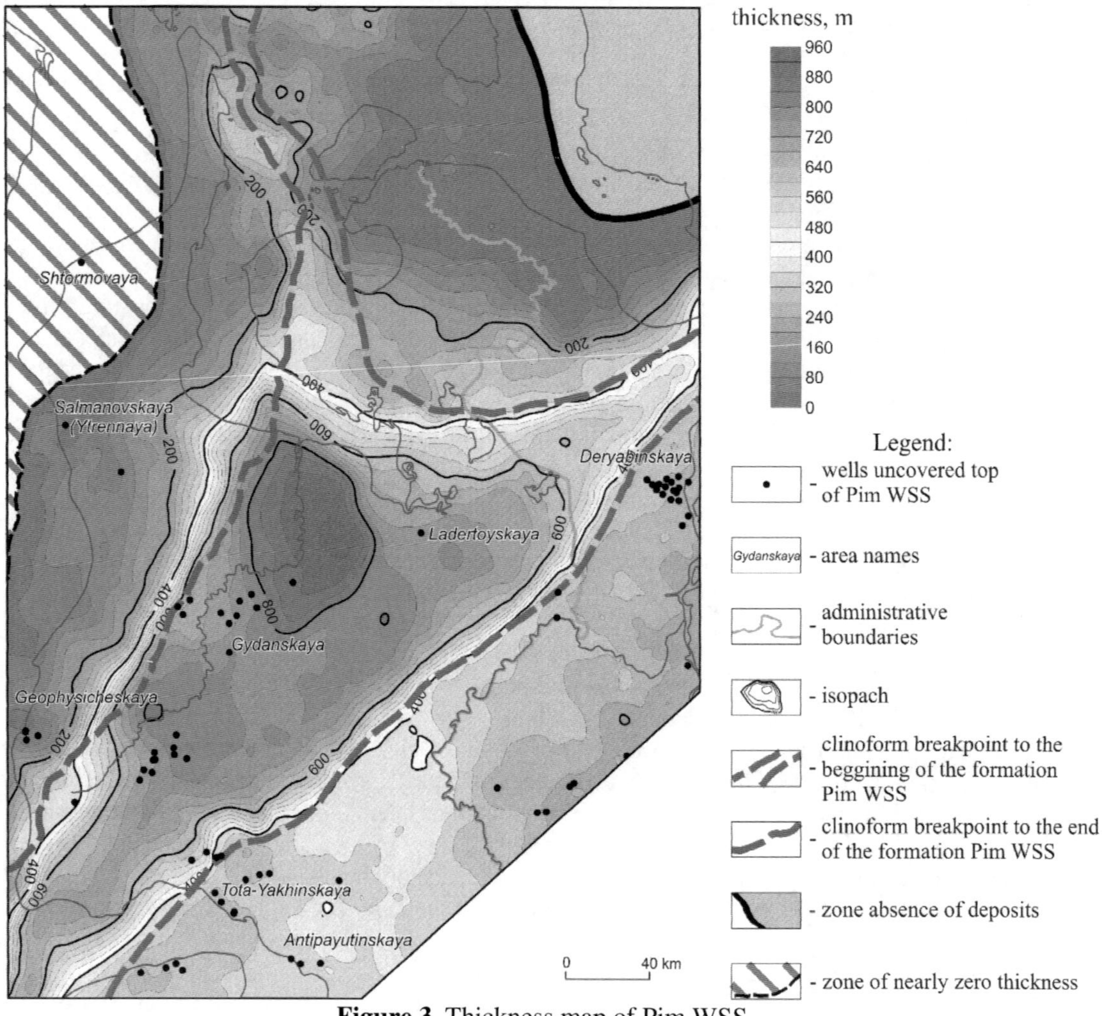

Figure 3. Thickness map of Pim WSS.

The structural plan of the Berrias–Lower Aptian megacomplex of Gydan is not very favorable for the search for anticline oil and gas traps. Large structures have now been drilled. Further search for anticline deposits of hydrocarbons can be attributed only to small and small-amplitude traps. Therefore, the search for lithological traps is of particular relevance. From this point of view, one of the prospective intervals is the Achimov Formation lying at the base of the Cretaceous. It represents a series of lenses relatively deep-sea genesis, rejuvenating in the north-west direction. The lenticular form of the sandy bodies of the Achimov Formation allows to consider both positive and negative structures as promising for discovering new deposits of oil and gas.

The degree of exploration of the resources of the Achimov Formation remains extremely low. Its deposits on Gydan have uncovered about 40 wells at depths from 2380 to 3440 m. All wells except Gydanskaya well 130 drilled along the perimeter of the peninsula.

There are two types of promising zones for the search for lithological traps in the deposits of the Achimov Formation. The first type of zones is confined to the depocenter of clinoforms. As the practice of oil and gas prospecting in Western Siberia has shown, the increased effective thicknesses of Achimov sandstones and zones with improved reservoir properties are usually confined to the depocenter of clinoforms [5]. On the territory of the Gydan Peninsula in the Lower Cretaceous clinoform complex, two major depocenters of accumulation of sedimentary material are allocated, confined to the sediments of the Pim and Seakha WSS (figure 4). Within the depocenters, the most promising are the areas raised in the structural sense.

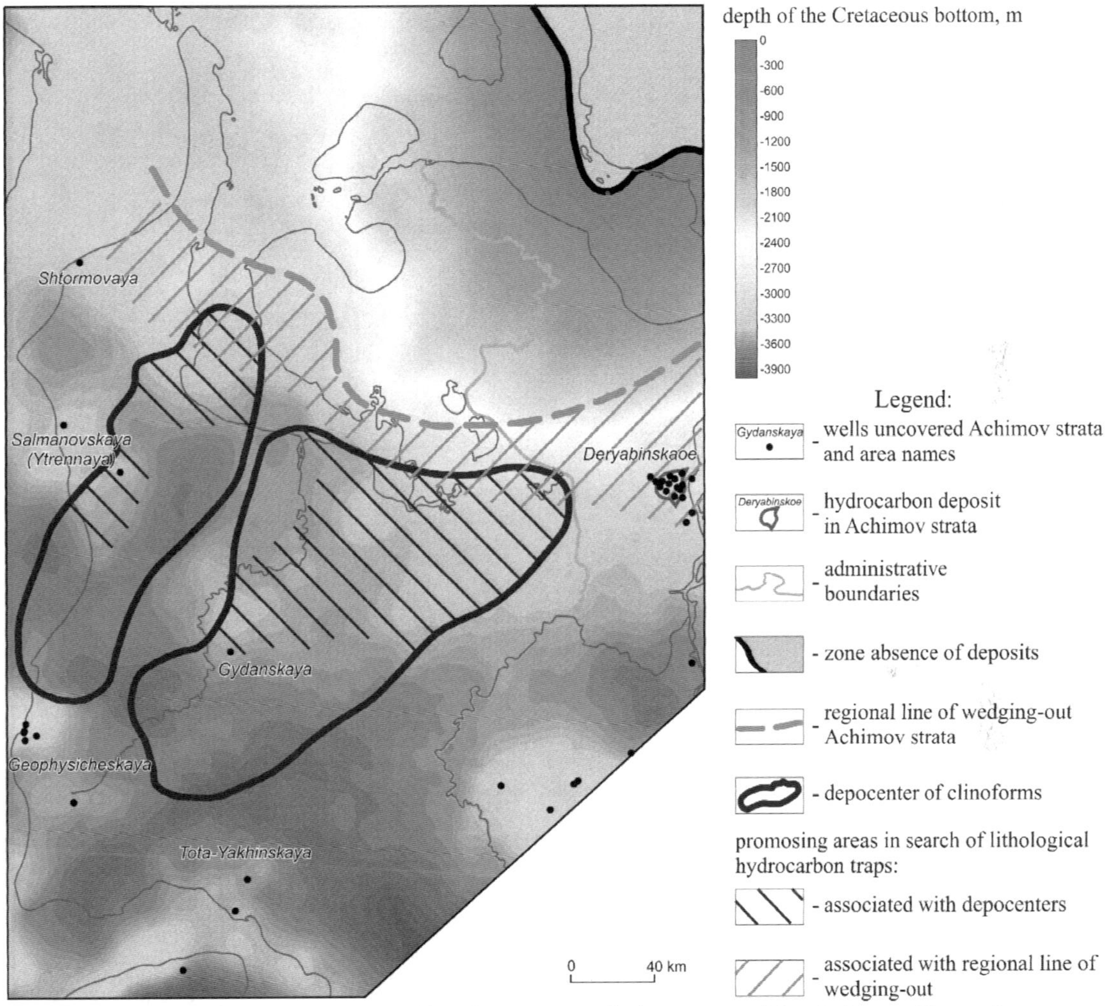

Figure 4. Map of prospective oil and gas accumulation zones of the Achimov Formation.

The second type of zones is associated with the regional attenuation of the distal strata of the Achimov Formation toward the Taimyr paleolandmass. This zone includes the Deryabinskoe gas condensate field, discovered in the east of the Gydan Peninsula. East within the area identified gas deposits in Khabeyskoe field.

The selected prospective oil and gas accumulation zones require further detailed study. These areas should be given priority in setting up prospecting for hydrocarbon deposits on the territory of the Gydan Peninsula.

Acknowledgments
The reported study was funded by RFBR according to the research project № 18-05-00210.

References

[1]. Ershov S 2017 *Russian Geology and Geophysics* **58** 963–72
[2]. Ershov S 2016 *Russian Geology and Geophysics* **57** 1221-38
[3]. Kontorovich A et al. 2014 *Russian Geology and Geophysics* **55** 582-609
[4]. Kontorovich V 2011 *Russian Geology and Geophysics* **52** 804-24
[5]. Nezhdanov A, Ponomarev V, Turenkov N and Gorbunov S 2000 *Geology and petroleum potential of the Achimov Formation of Western Siberia* (Moscow: Academy of Mining Sciences) p 247

One-dimensional vertical model of permafrost dynamics

V M Belolipetskii and S N Genova

Institute of Computational Modeling SB RAS, Krasnoyarsk, Russia

E-mail: belolip@icm.krasn.ru

Abstract. Simple description of the temperature field in soils during their freezing or thawing is considered with the help of solutions of Stefan problems. A mathematical model is based on the equations of thermal conductivity for the frozen and thawed layers. In the vertical structure of the permafrost zone, three layers are distinguished: thawed soil, frozen soil, snow. A simplified numerical algorithm for solving one-dimensional (in the vertical direction) thermal conductivity problems with moving phase transition boundaries with the formation of new and cancellation of existing layers is proposed.

1. Introduction

In connection with the change in global temperature the response of permafrost to climate change is interested. Both the thickness of a layer of thawed soil and the duration of its existence are increased by the climate warming. There is extensive literature on the mathematical modeling of permafrost (see [1-7]). In this paper the small scale numerical model is proposed for description vertical temperature distribution in thawed and frozen layers taking into account the formation of new and revocation of the existing layers (proposed in [8] for swamp-lake landscapes). Since the vertical temperature gradients are usually larger than the horizontal one all physical process are assumed one-dimensional in the vertical direction in the description of the heat transfer. We distinguish the following layers in the vertical direction: the snow, thawed soil, frozen soil. The theoretical description of the temperature field in soils during their freezing and thawing is carried out using solutions of Stefan problem. A mathematical model based on the equations of thermal conductivity for frozen and thawed areas. At the borders of phase transition (freezing-thawing) the conditions of equality of temperatures to the phase transition temperature and the Stefan condition are used. The formulated mathematical model of vertical temperature distributions in thawed and frozen layers takes into account the formation of new and cancellation of existing layers.

There are various options for the location of frozen and thawed layers. When switching from one variant to another the layers are added or deleted. Five options are considered (table 1). The conditions of switching from one option to another are determined. Figure 1 illustrates a vertical structure for variants 3 and 4.

Table 1. Variants of the location for frozen and thawed soil.

Variant	Snow	Frozen soil	Thawed	Frozen	Thawed	Frozen	Thawed

nubmer			soil	soil	soil	soil	soil
1	+					+	+
2					+	+	+
3	+			+	+	+	+
4			+	+	+	+	+
5	+	+	+	+	+	+	+

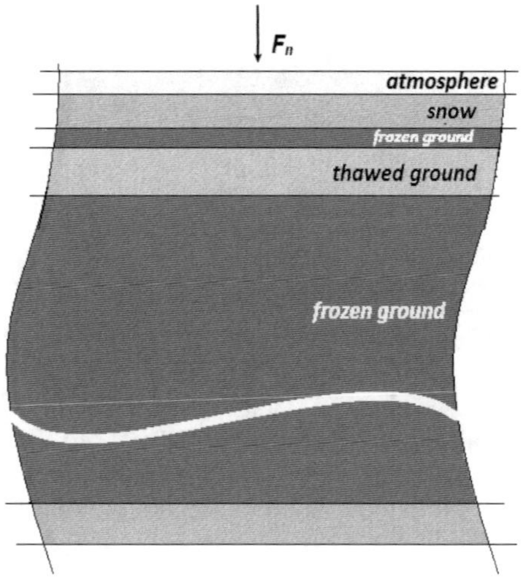

a)

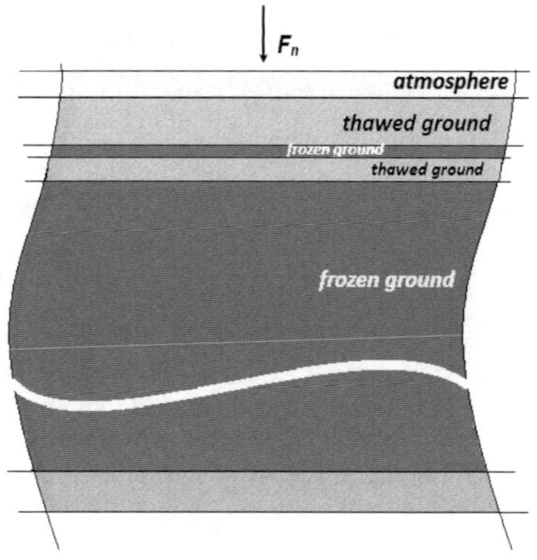

b)

Figure 1. Variants 3 (a) and 4 (b).

2. Mathematical model of dynamics of freezing and thawing of permafrost

The vertical temperature distribution in every layer are determined by solving the heat equation, satisfying the appropriate boundary conditions:

$$\frac{\partial T_i}{\partial t} = K_i \frac{\partial^2 T_i}{\partial z^2}. \tag{1}$$

Here $T_i(t, z)$ is a temperature of i-th layer $h_{i-1} \le z \le h_i$), K_i is a thermal diffusivity, t is time, z is a vertical coordinate (downward).

Boundary conditions.

The condition on the boundary of the atmosphere–soil in summer (z=0):

$$\left(K \frac{\partial T}{\partial z} \right) = -\frac{F_n}{c_w \rho_w}, \tag{2}$$

the condition on the boundary of the atmosphere-snow (z=0): the snow temperature is set

$$T = T_{sn}. \tag{3}$$

The condition on the boundary of a snow and frozen ground (z=0):

$$T_i = T_{i+1}, \qquad \left(\lambda \frac{\partial T}{\partial z} \right)_i = \left(\lambda \frac{\partial T}{\partial z} \right)_{i+1}. \tag{4}$$

On moving boundaries of phase transition ($z = h_i$)

$$T_i = T_{i+1} = T_{ph},$$

$$\rho_i L_i \frac{dh_i}{dt} = \left(\lambda \frac{\partial T}{\partial z}\right)_i - \left(\lambda \frac{\partial T}{\partial z}\right)_{i+1}$$, if the *i*-th layer is frost and the (*i*+1)-th is thawed one, (5)

$$\rho_{i+1} L_{i+1} \frac{dh_i}{dt} = \left(\lambda \frac{\partial T}{\partial z}\right)_{i+1} - \left(\lambda \frac{\partial T}{\partial z}\right)_i$$, if the *i*-th layer is thawed and the (*i*+1)-th is frost one. (5^1)

Here ρ_w is water density, c_w is the specific heat capacity of water, F_n is the total heat flow on the boundary of the atmosphere-ground, T_s is temperature on soil surface, $z = h_i$ is coordinate of the boundary between *i*-th and (*i*+1)-th layers, λ is coefficient of heat conductivity, T_{ph} is phase transition temperature, ρ_i is density of *i*-th layer, $L_i = L_w \cdot W_{Vi}$ is volumetric latent heat of melting environment in i – th layer, $W_{Vi} = \Omega_{wi} \cdot (\Omega_i)^{-1}$ is soil humidity in *i*-th layer, Ω_{wi} is the volume of water in the soil, Ω_i is the volume of the soi, $W_{Vi} = 10^{-3} W_i$, W_i (mm/m) – is soil humidity.

The initial conditions:

$$T_i(0,z) = T_i^0, \qquad \delta_i = \delta_i^0.$$

Parameterization of heat flow through the underlying surface.

Heat flows are important parameters that affect soil temperature. The total heat flow F_n is defining by known formulas [9,10]:

$$F_n = (1-\alpha)F_I - (F_{ef} + F_t + F_{ev}),$$

where F_{ef} is effective long-wave radiation, F_{ev} is heat transfer by evaporation, F_t is convective heat exchange, F_I is the incoming short wave radiation.

Short-wave radiation is calculated by the formula [11]

$$F_I = 0.94 \cdot Q(h_c)(1 - 0.65n^2),$$

where

$$Q(h_c) = \left(0.66 + 0.34\frac{\gamma - 0.9 + 0.4\sin h_c}{0.1 + 0.4\sin h_c}\right)\frac{\kappa_n \sin^2 h_c}{\rho^2(\sin h_c + 0.107)},$$

$$h_c = \arcsin\left(\sin\varphi_k \sin\gamma_1 + \cos\varphi_k \cos\gamma_1 \cos\left((t - t_n)\frac{\pi}{12}\right)\right),$$

$$\gamma_1 = 0.4 + 23.4 \cdot \cos\left(\frac{2\pi}{365}(d+192)\right) - 0.4 \cdot \cos\left(\frac{2\pi}{365}(d-192)\right),$$

κ_n=1.11÷1.23 depending on the humidity content of the atmosphere, n is total cloud score in fractions of a unit, h_c is height of the sun in degrees, ρ is air density, γ = 0.94, φ_k is latitude in degrees, t = 0,1,…,23 is the local astronomical time; t_n=12 is the noon local time; γ_1 is the sun declination, d is the ordinal day number since the beginning of the year.

The semi-empirical formula was used to calculate the effective radiation of the F_{ef} [9,10]:

$$F_{ef} = F_{ef}^0(1 - 0.79n) + 3.8\sigma(T_a + 273.15)^3(T_s - T_a),$$

$$F_{ef}^0 = 0.95\sigma(T_s + 273.15)^4(0.39 - 0.058\sqrt{e}).$$

Here T_a is the surface air temperature (°*C*), T_s is the temperature of the underlying surface (°*C*), n is total cloud score (fractions of a unit), $\sigma = 5.67 \cdot 10^{-8}$ is the Stefan-Boltzmann constant (W/(m^2 degree^4)), e is the water vapor pressure in the atmosphere (in milibar).

Turbulent exchange between the underlying surface and the atmosphere is determined by the formula [9,10]

$$F_t = \rho_a c_a D(T_s - T_a),$$

where ρ_a is the air density, c_a is the specific heat of air, D =0.0063 is the external diffusion coefficient (m/s).

Evaporation E (mm/m) is calculated by the formula [9,10]

$$E = \begin{cases} E_0 & for\ W > W_0, \\ E_0 \dfrac{W}{W_0} & for\ W \le W_0, \end{cases}$$

here W (mm/m) is the ground (soil) moisture, the critical soil moisture content W_0 is assumed equal to 200 mm/m. The evaporability of E_0 is determined by the [9,10]

$$E_0 = \rho_a D(e_0 - e),$$

e_0 - the saturating water-vapour pressure near the earth's surface is found by the Magnus formula

$$e_0 = 6.11 \cdot \exp\left(\frac{17.57 \cdot T_s}{241.9 + T_s}\right).$$

To determine the moisture content of the top layer of soil, a method based on the solution of the moisture balance equation is used [9,10]:

$$\delta_T \frac{dW}{dt} = r - f - E \quad , \qquad (6)$$

$$f = \begin{cases} r \dfrac{W}{W_0} \sqrt{m_0^2 \left[1 - \left(1 - \dfrac{E_0}{r}\right)^2\right] + \left(1 - \dfrac{E_0}{r}\right)^2} & \text{for} \quad r > E_0, \\ m_0 r \dfrac{W}{W_0} & \text{for} \quad r \le E_0, \end{cases}$$

where r is is rainfall (mm/h), f is the water runoff (mm/h), δ_T is the thickness of the upper melt layer of soil (m), $m_0 = 0.2$.

Heat expenditure on evaporation F_u is calculated by the ratio

$$F_{ev} = L_{ev} E,$$

where $L_{ev} = 600$ (kcal / kg) is latent heat of evaporation.

The snow cover is an important factor of interaction in a system of the atmosphere and underlying surface. One of the main parameters of the snow cover, which determines the heat-shielding properties, is the coefficient of the snow thermal conductivity (λ_{sn}).The dependence of the coefficient of snow thermal conductivity on its density (ρ_{sn}) can be estimated by the Abels formula $\lambda_{sn} = 0.0068 \rho_{sn}^2$ (cal (cm s degree)$^{-1}$).

The influence of snow on soil freezing was estimated by using the quasi-stationary approximation for the two layers of snow with different densities and thermal conductivity. On the border atmosphere-snow set snow temperature equals to air temperature; at the boundary of snow layers the conditions of continuity of temperature and heat flow are fulfilled. It is assumed that the temperature in the frozen soil layer (T_2^{fr}) is known ($z=z^*$). To calculate the temperature of the soil on the border of snow – soil, the formula is obtained:

$$T_1^{fr} = \frac{T_a + \mu \cdot T_2^{fr}}{1+\mu}, \qquad \mu = \frac{\delta_{sn1}}{\lambda_{sn1}} + \frac{\delta_{sn2}}{\lambda_{sn2}} \cdot \frac{\lambda_{fr}}{z^*}.$$

where T_a is the air temperature, $\lambda_{sn1}, \lambda_{sn2}$ - snow thermal conductivity coefficients in layers 1 and 2, respectively; $\delta_{sn1}, \delta_{sn2}$ are the the thickness of the layers 1 and 2, λ_{fr} is the thermal conductivity coefficient of frozen soil.

Snow melting begins when the temperature of the snow rises to zero in the spring. The expenses of heat for melting snow with thickness δ_{sn} is $F_{sn} = 80\rho_{sn}\delta_{sn}$ (cal/cm^2). Heat exchange at the snow – atmosphere boundary is $F_{a-sn} = \alpha(T_{sn} - T_a)$, heat transfer on the border of a snow – frozen ground is $F_{sn-fr} = 0.5\lambda_{fr} \cdot (T_2^{fr} - T_{sn})(\delta_{fr})^{-1}$, $\alpha = 6.44 \cdot (W_2 + 0.3)^{0.5}$ in cal/(cm^{2}s), W_2 is the wind speed. The time spent on the snow melting is determined by the ratio of $\tau_{sn} = F_{sn} \cdot (F_{a-sn} - F_{sn-fr})^{-1}$.

3. Numerical algorithm

It is assumed that the transition of moisture in the soil from the frozen state to the melt and vice versa does not change the overall moisture content. The numerical solution of the equation (6) has the form

$$\hat{W} = W^n + \Delta t \frac{(r - f - E)^n}{\delta_T},$$

$$W^{n+1} = \frac{\hat{W}\delta_T^n + W_{fr}(\delta_T^{n+1} - \delta_T^n)}{\delta_T^{n+1}} \quad \text{for} \quad \delta_T^{n+1} > \delta_T^n,$$

$$W_{fr}^{n+1} = \frac{W_{fr}^n \delta_{fr}^n + W_T^n(\delta_{fr}^n - \delta_{fr}^{n+1})}{\delta_{fr}^{n+1}} \quad \text{for} \quad \delta_T^{n+1} \le \delta_T^n, \qquad W^{n+1} = \hat{W}.$$

Where δ_{fr} is the thickness of adjacent layer of frozen soil, W_{fr} is moisture reserves of frozen soil. Consider an arbitrary i-th layer: $h_{i-1} \le z_i \le h_i$, $\delta_i = h_i - h_{i-1}$, where δ_i is the thickness of i-th layer. We introduce new independent variables (t, ξ):

$$t = t, \qquad \xi_i = \frac{z_i - h_{i-1}}{\delta_i}, \qquad 0 \le \xi_i \le 1.$$

Since

$$\frac{\partial}{\partial z_i} = \frac{1}{\delta_i}\frac{\partial}{\partial \xi_i}, \qquad \frac{\partial^2}{\partial z_i^2} = \frac{1}{\delta_i^2}\frac{\partial^2}{\partial \xi_i^2}, \qquad \frac{\partial}{\partial t} = \frac{\partial}{\partial t} + w_i \frac{\partial}{\partial \xi_i},$$

$$w_i = \frac{\partial \xi_i}{\partial t} = \frac{1}{\delta_i}\left[(\xi_i - 1)\frac{dh_{i-1}}{dt} - \xi_i \frac{dh_i}{dt}\right],$$

then equation (1) and boundary conditions (2) – (5) written as

$$\frac{\partial T_i}{\partial t} + w_i \frac{\partial T_i}{\partial \xi_i} = \frac{K_i}{\delta_i^2}\frac{\partial^2 T_i}{\partial \xi_i^2}, \tag{7}$$

$$\left(\frac{K_i}{\delta_i}\frac{\partial T_i}{\partial \xi_i}\right)_{\xi_i=0} = -\frac{F_n}{c_w \rho_w}, \qquad T_i\big|_{\xi_i=0} = T_{ice}, \qquad T_i\big|_{\xi_i=1} = T_{i+1}\big|_{\xi_{i+1}=0}, \tag{8}$$

$$\left.\left(\frac{\lambda_i}{\delta_i}\frac{\partial T_i}{\partial \xi_i}\right)\right|_{\xi_i=1} = \left.\left(\frac{\lambda_{i+1}}{\delta_{i+1}}\frac{\partial T_{i+1}}{\partial \xi_{i+1}}\right)\right|_{\xi_{i+1}=0}, \qquad T_i\big|_{\xi_i=1} = T_{i+1}\big|_{\xi_{i+1}=0} = T_{ph}, \tag{9}$$

$$\rho_i\, L_i \frac{d\,h_i}{d\,t} = \left.\left(\frac{\lambda_i}{\delta_i}\frac{\partial T_i}{\partial \xi_i}\right)\right|_{\xi_i=1} - \left.\left(\frac{\lambda_{i+1}}{\delta_{i+1}}\frac{\partial T_{i+1}}{\partial \xi_{i+1}}\right)\right|_{\xi_{i+1}=0} . \tag{10}$$

Let us consider the solution of the formulated problems on coarse (three-point) grid for each of layers: $\xi_{i1}=0,\quad \xi_{i2}=0.5,\quad \xi_{i3}=1.$

We approximate the equation (7) using the implicit difference scheme and and directional differences for the convective terms. Grid equations corresponding to the differential equation (7) for i–th layer are of the form:

$$\frac{T_{i,2}-T^n_{i,2}}{\Delta t} + w^- T_{i,3} + 2w^0 T_{i,2} - w^+ T_{i,1} = \frac{K_i(T_{i,1}-2T_{i,2}+T_{i,3})}{(0.5\delta^n_i)^2}, \tag{11}$$

$$w_{i,2} = -\left(\frac{dh_i}{dt}+\frac{dh_{i-1}}{dt}\right)/(2\delta^n_i), \qquad w^0 = |w_{i,2}|, \qquad w^- = w_{i,2} - |w_{i,2}|, \qquad w^+ = w_{i,2} + |w_{i,2}|,$$

here Δt is time step, $t_{n+1} = t_n + \Delta t$, $T^{n+1}_{i,k} = T_i(t_{n+1}, \xi_{ik})$.

Boundary conditions (8) – (10) for the difference grid are represented as:
on the border of the atmosphere – soil

$$\left[1+\frac{8K_i\Delta t}{(\delta^n_i)^2}\right]T_{i,1} - \frac{8K_i\Delta t}{(\delta^n_i)^2}T_{i,2} = T^n_{i,1} + \frac{4\Delta t F_n}{c_w\rho_w} , \tag{12}$$

on the border of the atmosphere - snow+soil

$$T_{i,1} = T_{sn}, \tag{13}$$

at the borders of phase transition $z = h_i$

$$T_{i,3} = T_{i+1,1} = T_{ph} \tag{14}$$

$$\rho_i L_i \frac{h^{n+1}_i - h^n_i}{\Delta t} = \lambda_i \frac{T_{i,3}-T_{i,2}}{0.5\delta^n_i} - \lambda_{i+1}\frac{T_{i+1,2}-T_{i+1,1}}{0.5\delta^n_{i+1}} . \tag{15}$$

If δ_i becomes less than a predetermined small value ε ($\delta_i < \varepsilon$), the corresponding layer will be cancelled. If due to melting or freezing a new k–th layer is formed, the initial thickness $\delta^0_k = \varepsilon$ and temperatures are set for this layer.

The following algorithm is used for the obtained tasks. If all parameters (temperature distributions in layers under study and the positions of phase transition) are known on n-th time step, then a finding the unknown parameters at time t_{n+1} is performed in two stages:
The first step is to define the temperature distribution in the selected layers (taking into account the relations (11) – (14) by solution of systems of linear algebraic equations of small dimension.
In the second phase it is clarified the position of the phase boundary by the numerical solution of equation (15).

4. Calculations examples

Calculations were performed for a permafrost soil using weather data of 2010-2012 years for the weather station Dudinka (69^{0}24' North latitude). 2010 year could be considered as "cold" (about 240 days with mean daily negative temperature, the average temperature for the period from 1 October to 30 April was -20.72°C); 2011 – "warm" (190 days with mean daily negative temperature and the average temperature for October to April was -15.22°C). The depth of seasonal thawing of permafrost soils, the temperature of the melt and frozen layers, the time intervals of the existence for layer of thawed soil were determined. Figure 2 shows the calculated boundaries of thawing-freezing of the soil.

Table 2. The thickness of the layer of thawing frozen soil (cm) and the time of its existence

year	thickness	time existence
2010 г.	99.87	4.06.10 – 7.12.10
2011 г.	138.4	22.04.11 – 15.02.11
2012 г.	136.5	21.04.12 –

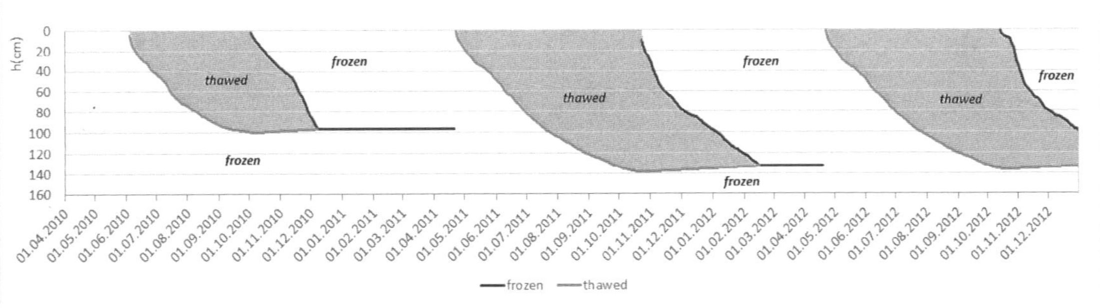

Figure 2. Dynamics of soil thawing-freezing.

Thus, a relatively simple one-dimensional model can be used to assess the current state and forecast of permafrost in the land regions.

The work is executed with financial support of RFBR grant No. 17-45-240884 p_a.

References

[1] Anisimov O A and Nelson F E 1998 Forecast of changes in permafrost conditions in the Northern hemisphere: application of results of balance and transitive calculations on models of General atmosphere circulation *Earth Cryosphere* **2** 53–7

[2] Malevsky-Malevich S P, Molkentin E K and Nadezhina E D 2000 Model estimates of the evolution of permafrost and the distribution of the layer of seasonal thawing, depending on the climatic conditions in the Northern regions of Western Siberia *Earth Cryosphere* **4** 49–57

[3] Arzhanov M M, Eliseev A V, Demchenko P F and Mokhov I I 2007 Modeling of changes in temperature and hydrological regimes of near-surface permafrost using climate data (reanalysis) *Earth Cryosphere* **4** 65–9

[4] Sazonova T S and Romanovsky V E 2003 A model for regional scale estimation of temporal and spatial variability of active layer thickness and mean annual ground temperatures *Permafrost and Periglacial Processes* **2** 125–40

[5] Goodrich L E 1982 The influence of snow cover on the ground thermal regime *Canadian Geotechnical Journal* **19**(4) 421–32

[6] Nicolsky D J, Romanovsky V E, Alekseev V A and Lawrence D M 2007 Impreved modeling of permafrost dynamics in a GCM land-surface scheme *Geophysical Research Letters* **34** L08501

[7] Zhang T J, Frauenfeld O W, Serreze M C, Etringer A, Oelke C, McCreight J, Barry R G, Gilichinsky D, Yang D Q, Ye H C, Ling F and Chudinova S 2005 Spatial and temporal variability in active layer thickness over the Russian Arctic drainage basin *Journal of Geophysical Research-Atmospheres* **110** D16101

[8] Belolipetskii V M and Genova S N 2016 A numerical model of the seasonal thawing of permafrost in the bog-lake landscapes *Journal of Siberian Federal University. Mathematics & Physics* **9**(2) 158–165

[9] Budyko M I 1974 *Climate Change* (Leningrad: Gidrometeoizdat) p 280

IOP Conf. Series: Earth and Environmental Science **193** (2018) 012005 doi:10.1088/1755-1315/193/1/012005

[10] Anisimov O A and Belolutskaya M A 2003 The Impact of climate change on permafrost: predicting and evaluating the uncertainty *Problems of ecological monitoring and modeling of ecosystems* XIX (SPb: Gidrometeoizdat) pp 21-38
[11] Timofeev N A 1983 *Radiative regime of the oceans* (Kiev: Naukova Dumka)

Mathematical modelling of level ice with continental shelf structures interaction

A S Bolshev and S A Frolov
Peter the Great Saint-Petersburg Polytechnic University, St.Petersburg, Russia

Abstract. Development of hydrocarbons deposits on the continental shelf is connected to the construction of floating and fixed objects functioning in high seas conditions continuously. For the safety analysis of such structures not only the knowledge of ice load values is needed, but also such of their frequency and time-dependent characteristics. The purpose of the present work is further development of methodology for interaction dynamics analysis between ice bodies and Arctic shelf structures presented previously in [1-4]. The paper illustrates the realization of dynamic method of calculation of ice floe deformation and movement upon contact with inclined planes of floating or fixed structures. The presented methodology can be used in the research or project work connected to finding time-dependent characteristics of ice loads, their statistical consistencies and dependency on structures' reaction. The ice field behaviour description algorithm is implemented in the *Anchored Structures* program. The paper presents some results of the comparison study which prove working efficiency of the presented methodologies.

1. Problem statement and restrictions

When designing methodological procedures, we allow the following assumptions:

1. The ice field is presented as an even flat plate of the length L_0, width B_0 and thickness h. Next, the field in the plan is presented as a discreet spatial model where the elementary rectangular fragments interact with each other along the lines connecting the characteristic points (nodes) placed in the intersecting of the rectangular grid (figure 1).
2. While analyzing the strain-stress state the ice field is regarded as a flexible plate of medium thickness upon the elastic foundation loaded along the contact area. During the interaction ice can be deformed longitudinally and cross-sectionally, as well as bend.
3. Ice is also considered an elastic/fragile feature allowing plastic hinges formation at nodal points when being bent. At the same time, ice crack formation can be caused by its compression, bending, torsion or by a combination of deformations. After cracking the elementary ice floe possesses some residual strength causing the load capacity of the crack to reduce gradually.

2. The mathematical ice field deformation model

To analyze ice field deflection and deformation under vertical and horizontal loads of the inclined surface planes of the structure we use the von Karman equation for deflections with higher values than the plate thickness with the addition of membrane forces [5].

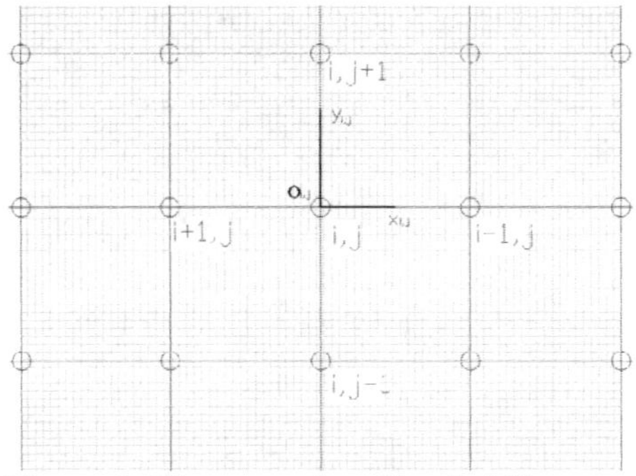

Figure 1. The example of ice field lay-out.

$$\begin{cases} q(x,y) = q_m + D\left(\dfrac{\partial w^4}{\partial x^4} + \dfrac{\partial w^4}{\partial y^4} + 2\dfrac{\partial w^4}{\partial x^2 \partial y^2} \right) \\ q_m = h\left(T_x \dfrac{\partial w^2}{\partial x^2} + T_y \dfrac{\partial w^2}{\partial y^2} + 2S\dfrac{\partial w^2}{\partial x \partial y} \right) \\ T_x = K(\varepsilon_x + \mu\varepsilon_y) \\ T_y = K(\varepsilon_y + \mu\varepsilon_x) \\ S = K_1\omega \end{cases} \tag{1}$$

where $q(x,y)$ is an external distributed vertical load per area unit on the plate (from the ice weight and the buoyancy force); D – bending rigidity of a plate which equals to $D = \dfrac{E_I h^3}{12(1-\mu^2)}$; q_m – addition to the vertical load connected to the vertical projection of the membrane forces at the final deflection; T_x, T_y и S – longitudinal, transversal and shear membrane stresses; E_I – Young's modulus of ice; μ – Poisson's ratio; h – ice thickness; K and K_1 – plate rigidity to spraining/compression and shearing; $K = \dfrac{E_I}{1-\mu^2}$, $K_1 = \dfrac{K(1-\mu)}{2}$.

For discretization purposes the ice filed is divided into rectangular elements (ice floes) of the length *L*, width *B* and thickness *h* (the element's height equals to the ice field thickness *h*). The number of elements to be divided in is chosen so that the *L/h* proportion is within the range of 3-6 and so that the element is the plate of average thickness.

Inside each element we place a point node where the inertia characteristics of this element and its external loads are concentrated. Each node has three degrees of freedom. Node numbering about X-axis is opposite the ice movement direction, about Y-axis – about the axis (the corresponding indexes are *i* and *j*). The adjacent nodes about X-axis and Y-axis are connected by links (line segments) which pass the internal forces in the ice field, described in the equation (1).

Let us assume that at each moment in time we know the deflection form and, correspondingly, the coordinates of each node $\mathbf{X_{ij}} = \left| x_{ij}\, y_{ij}\, z_{ij} \right|^T$. Then the distributed external vertical load from the buoyancy load and ice weight for the node *ij* is:

$$F^e_{zij} = \int_{y_{ij}-B/2}^{y_{ij}+B/2} \int_{x_{ij}-L/2}^{x_{ij}+L/2} q(x,y)\mathrm{d}x\mathrm{d}y, \tag{2}$$

here the index *e* means *external*. At the same time, $q(x,y)$ equals to

$$q(x,y)=\begin{cases} g\left(\rho_h(\frac{h}{2}-z)-\rho_l h\right), -\frac{h}{2}<z<\frac{h}{2}, \\ gh(\rho_h-\rho_l), z<-\frac{h}{2}, \\ -gh\rho_l, z>\frac{h}{2}, \\ z=f(x,y), \end{cases} \quad (3)$$

ρ_l и ρ_h – densities of ice and water correspondingly. Three occurrences are 1) ice floe intersecting the water surface, 2) full submersion, 3) full emersion.

The integral equation (2) is found numerically, for instance, using Gauss integration method (4x4 points). The z coordinates of the required points inside a specific ice floe are calculated as linear combinations of the coordinates of the considered node and four adjacent ones. This level of detail is required as some part of an ice floe can emerge from water and some other part can be fully submerged.

Let us examine the internal forces in the nodes. From equation (1) we get:

$$\hat{F}_{ij}=D\int_{y_{ij}-B/2}^{y_{ij}+B/2}\int_{x_{ij}-L/2}^{x_{ij}+L/2}\left(\frac{\partial w^4}{\partial x^4}+\frac{\partial w^4}{\partial y^4}+2\frac{\partial w^4}{\partial x^2\partial y^2}\right)\mathrm{d}x\mathrm{d}y. \quad (4)$$

The force $\hat{F}_{ij}$ is presented as a sum of the three components: $\hat{F}_{ij}=\hat{F}_{1ij}+\hat{F}_{2ij}+\hat{F}_{3ij}$, according with the number of the additive component. The first two additive components are written as follows:

$$\begin{cases} \hat{F}_{1ij}=\hat{F}_{Bi-1j}-\hat{F}_{Bij}, \\ \hat{F}_{2ij}=\hat{F}_{Rij+1}-\hat{F}_{Rij}, \end{cases} \quad (5)$$

where $\hat{F}_{Bij}$, $\hat{F}_{Rij}$ - normal forces, influencing the back and the right faces of the ice floe ij:

$$\hat{F}_{Bij}=\int_{y_{ij}-B/2}^{y_{ij}+B/2} p_B(y)\mathrm{d}y\,,\ \hat{F}_{Rij}=\int_{x_{ij}-L/2}^{x_{ij}+L/2} p_R(x)\mathrm{d}y\,. \quad (6)$$

Here $p_B(y)$ и $p_R(x)$ are the loads per unit of length on the back and the right face of element, at the same time:

$$p_{Bij}=\frac{M_{Yij}-M_{Yi+1j}}{L},\ p_{Rij}=\frac{M_{Xij}-M_{Xij-1}}{B}. \quad (7)$$

Here $M_{Yij}=D_{ij}\mathbf{N_{ij}}\frac{\partial^2}{\partial x^2}\mathbf{X_{ij}}, M_{Xij}=D_{ij}\mathbf{N_{ij}}\frac{\partial^2}{\partial y^2}\mathbf{X_{ij}}$ are bending moments per unit of length in a node about Y and X axes; $\mathbf{N}_{ij}$ is a normal vector to the mean plane of the section. Using the normal vector allows calculating moments for any final deflections (inclinations) of the section. The second derivative is calculated numerically using three adjacent nodes $\mathbf{X_{ij}}$ in the corresponding direction. The same numerical method (using three adjacent nodes) is used to calculate integral equations (6).

Then, the third component equals to:

$$\hat{F}_{3ij}=2D\int_{y_{ij}-B/2}^{y_{ij}+B/2}\int_{x_{ij}-L/2}^{x_{ij}+L/2}\frac{\partial w^4}{\partial x^2\partial y^2}\mathrm{d}x\mathrm{d}y=\tilde{F}_{3i-1j}-\tilde{F}_{3i-1j-1}-\tilde{F}_{3ij}+\tilde{F}_{3ij-1}, \quad (8)$$

$$\text{where } \tilde{F}_{3ij}\approx\frac{4D_{ij}}{LB}\mathbf{N_{ij}}\left(\mathbf{X_{ij+1}}-\mathbf{X_{ij}}-\mathbf{X_{i+1j+1}}+\mathbf{X_{i+1j}}\right). \quad (9)$$

Let us examine the membrane forces and longitudinal, transversal and shear deformations ε_{xij}, ε_{yij} и ω_{ij} connected to them. Assuming the first two deformations to be constant between the adjacent nodes, for the element ij we get:

$$\begin{cases} \varepsilon_{xij} = \left|\boldsymbol{\tau}_{\mathbf{xij}}\right| - 1, \\ \varepsilon_{yij} = \left|\boldsymbol{\tau}_{\mathbf{yij}}\right| - 1, \\ \omega_{ij} = \boldsymbol{\tau}_{\mathbf{xij}} \cdot \boldsymbol{\tau}_{\mathbf{yij}}, \end{cases} \quad (10)$$

where $\boldsymbol{\tau}_{\mathbf{xij}} = \frac{1}{L}(\mathbf{X}_{\mathbf{ij}} - \mathbf{X}_{\mathbf{i\text{-}1j}})$ и $\boldsymbol{\tau}_{\mathbf{yij}} = \frac{1}{B}(\mathbf{X}_{\mathbf{ij}} - \mathbf{X}_{\mathbf{ij\text{-}1}})$ – direction vectors between the adjacent nodes.

Membrane forces per unit of length at the back (index B) and right (index R) node face:

$$\begin{cases} p_{Bij} = -hK(\varepsilon_{xij} + \mu\varepsilon_{yij}), \\ p_{Rij} = -hK(\varepsilon_{yij} + \mu\varepsilon_{xij}), \\ p_{SBij} = \frac{1}{2}hK_1(\omega_{ij} + \omega_{i+1j}), \\ p_{SRij} = \frac{1}{2}hK_1(\omega_{ij} + \omega_{ij-1}). \end{cases} \quad (11)$$

Then we use the interpolation of the integral expression from the membrane force per unit of length using the three points, similarly to the bending forces, to calculate the total membrane forces $F_{Bij}, F_{Rij}, S_{Bij}, S_{Rij}$.

The additional normal force from q_m in equation (1) equals to:

$$F_{aij} = \int_{y_{ij}-B/2}^{y_{ij}+B/2} \int_{x_{ij}-L/2}^{x_{ij}+L/2} q_m(x,y)\mathrm{d}x\mathrm{d}y = h\int_{y_{ij}-B/2}^{y_{ij}+B/2} \int_{x_{ij}-L/2}^{x_{ij}+L/2} (T_x\frac{\partial w^2}{\partial x^2} + T_y\frac{\partial w^2}{\partial y^2} + 2S\frac{\partial w^2}{\partial x\partial y})\mathrm{d}x\mathrm{d}y = F_4 + F_5 + F_6, \quad (12)$$

$$F_4 = h\int_{y_{ij}-B/2}^{y_{ij}+B/2} \int_{x_{ij}-L/2}^{x_{ij}+L/2} T_x\frac{\partial w^2}{\partial x^2}dxdy = F_{Bij}\tau_{xij} - F_{Bi-1j}\tau_{xi-1j}. \quad (13)$$

Evidently, the additional q_m only requires using membrane forces as vectors, considering having real tangent lines in the membrane force calculation point. Then the sum vectors of the membrane and the intersecting forces influencing the sector in fixed coordinate system equal to:

$$\begin{cases} \mathbf{F}_{\mathbf{ij}} = \mathbf{F}_{\mathbf{Bi\text{-}1j}} - \mathbf{F}_{\mathbf{Bij}} + \mathbf{F}_{\mathbf{Rij+1}} - \mathbf{F}_{\mathbf{Rij}} + \mathbf{S}_{\mathbf{Bij}} - \mathbf{S}_{\mathbf{Bi\text{-}1j}} + \mathbf{S}_{\mathbf{Rij}} - \mathbf{S}_{\mathbf{Rij+1}}, \\ \hat{\mathbf{F}}_{\mathbf{ij}} = \hat{\mathbf{F}}_{\mathbf{Bi\text{-}1j}} - \hat{\mathbf{F}}_{\mathbf{Bij}} + \hat{\mathbf{F}}_{\mathbf{Rij+1}} - \hat{\mathbf{F}}_{\mathbf{Rij}} + \hat{\mathbf{F}}_{\mathbf{3ij}}. \end{cases} \quad (14)$$

Each vector component in equation (14) is based on scalar values calculated above and on the direction vectors:

$$\begin{cases} \mathbf{F}_{\mathbf{Bij}} = F_{Bij}\boldsymbol{\tau}_{\mathbf{xij}}, \\ \mathbf{F}_{\mathbf{Rij}} = F_{Rij}\boldsymbol{\tau}_{\mathbf{yij}}, \\ \mathbf{S}_{\mathbf{Bij}} = \frac{1}{4}S_{Bij}(\boldsymbol{\tau}_{\mathbf{yij}} + \boldsymbol{\tau}_{\mathbf{yi+1j}} + \boldsymbol{\tau}_{\mathbf{yij+1}} + \boldsymbol{\tau}_{\mathbf{yi+1j+1}}), \\ \mathbf{S}_{\mathbf{Rij}} = \frac{1}{4}S_{Rij}(\boldsymbol{\tau}_{\mathbf{xij}} + \boldsymbol{\tau}_{\mathbf{xij\text{-}1}} + \boldsymbol{\tau}_{\mathbf{xi\text{-}1j}} + \boldsymbol{\tau}_{\mathbf{xi\text{-}1j\text{-}1}}), \end{cases} \quad (15)$$

for the membrane forces and

$$\begin{cases} \hat{\mathbf{F}}_{\mathbf{Bij}} = \frac{1}{2}\hat{F}_{Bij}(\mathbf{N}_{\mathbf{ij}} + \mathbf{N}_{\mathbf{i+1j}}), \\ \hat{\mathbf{F}}_{\mathbf{Rij}} = \frac{1}{2}\hat{F}_{Rij}(\mathbf{N}_{\mathbf{ij}} + \mathbf{N}_{\mathbf{ij-1}}), \\ \hat{\mathbf{F}}_{\mathbf{3ij}} = \hat{F}_{3ij}\mathbf{N}_{\mathbf{ij}} \end{cases} \tag{16}$$

for the intersecting forces.

In case of a very tight bending, a closer approximation of the shear forces direction vector is needed than averaging along the four direction vectors in equation (15). And vice versa for the ice floes which are more remote from the ice/structure contact area the bending stresses and vertical movements are insignificant so all the normal forces can be dismissed, leaving only the membrane ones. At the same time, the equations for deformations (15) also significantly simplified (considering the small level of the deformation):

$$\begin{cases} \varepsilon_{xij} = \frac{x_{ij} - x_{i+1j}}{L} - 1, \\ \varepsilon_{yij} = \frac{y_{ij} - y_{ij-1}}{B} - 1, \\ \omega_{ij} = \frac{x_{ij+1} - x_{ij-1}}{2B} + \frac{y_{ij} - y_{i+1j}}{L}. \end{cases}$$

This allows to substantially simplify the calculations for the majority of elements and significantly increase the modelling efficiency.

Dissipative forces required to exclude unrestrained resonance oscillation of a specific element equal to:

$$\mathbf{F}_{\mathbf{ij}}^{\mathbf{d}} = \frac{K_d}{4}(\mathbf{V}_{\mathbf{i-1j}} + \mathbf{V}_{\mathbf{i+1j}} + \mathbf{V}_{\mathbf{ij-1}} + \mathbf{V}_{\mathbf{ij+1}} - 4\mathbf{V}_{\mathbf{ij}}). \tag{17}$$

That suggests that the element decelerates if it moves differently from the adjacent ice floes.

Here $\mathbf{V}_{\mathbf{ij}}$ – velocity vector of the element in the coordinate system fixed with it, $\mathbf{V}_{\mathbf{ij}} = |v_{1ij} v_{2ij} v_{3ij}|^T$. The dissipation factor is introduced, for instance, from the 1% damping condition of the element at the resonance frequency:

$$K_d = 0.01\sqrt{cm} = 0.01Bh\sqrt{E_I \rho_I}. \tag{18}$$

As the deformations and the elastic forces of the plate's element act in directions of natural axis of this element, the set of simultaneous equations of its movement is written in the finite displacement terms:

$$\begin{cases} \rho_I BLh(\mathbf{E} + \boldsymbol{\lambda})\dot{\mathbf{V}}_{\mathbf{ij}} + \frac{\rho_h}{2} BL\mathbf{C}_v \mathbf{V}_{\mathbf{ij}} |\mathbf{V}_{\mathbf{ij}}| = \mathbf{P}_{\mathbf{ij}}^{\mathbf{T}}(\hat{\mathbf{F}}_{\mathbf{ij}} + \mathbf{F}_{\mathbf{ij}} + \mathbf{F}_{\mathbf{ij}}^{\mathbf{c}} + \mathbf{F}_{\mathbf{ij}}^{\mathbf{e}}) + \mathbf{F}_{\mathbf{ij}}^{\mathbf{d}}, \\ \dot{\mathbf{X}}_{\mathbf{ij}} = \mathbf{P}_{\mathbf{ij}}\mathbf{V}_{\mathbf{ij}}, \end{cases} \tag{19}$$

where $\mathbf{E}$ – unit diagonal matrix; $\boldsymbol{\lambda}$ – added masses matrix; $\mathbf{C}_v$ – diagonal matrix of viscous coefficients of ice resistance in liquid; $\mathbf{F}_{\mathbf{ij}}^{\mathbf{c}}$ – structure contact force vector (if this element is in contact), is considered in detail in [1]; $\mathbf{F}_{\mathbf{ij}}^{\mathbf{e}}$ – external distributed forces vector (buoyancy force and weight force), $\mathbf{F}_{\mathbf{ij}}^{\mathbf{e}} = \mathbf{k}F_{zij}^{e}$; $\mathbf{k}$ – vertical ort; $\mathbf{P}_{\mathbf{ij}}$ – element's rotation tensor relative to the fixed coordinate system, $\mathbf{P}_{\mathbf{ij}} = |\boldsymbol{\tau}_{\mathbf{xij}} \boldsymbol{\tau}_{\mathbf{yij}} \mathbf{N}_{\mathbf{ij}}|$.

The additive components with the product of coefficients of speeds intersection in the equation (19) are disposed of, as the ice movement is slow. The equation (19) will also describe viscous resistance and the added mass more accurately, which are tied to the ice floe dimension.

3. Ice failure

Let us write down the assumptions caused by this model, the ones which we will hereafter consider in the ice failure model construction.

1. Within any specific ice floe the bending rigidity D_{ij} is constant and initially equals D.
2. The crack formation in each ice floe resulting from bending begins after bending stress exceeds the stress limit R_f in the node.
3. The decrease in ice floe load capacity after cracking and further bending results from decrease in bending rigidity D_{ij} irrespective of the direction of bending or its prominence. Otherwise the using of equation system (1) is impossible.
4. For the system cohesiveness let us assume that ice floes, remaining interconnected by hinges, cannot separate even as a result from the total loss of the load capacity due to bending. This is applicable as the ice floe with no strength does not have a significant impact on the general strain-stress state of the field.
5. The separation of the ice floe from the ones in adjacent rows is possible after exceeding the separation stress limit R_t
6. For the sake of the same cohesiveness the separation is only possible if the separation of the preceding ice floes in the row took place.
7. The separation and the failure of the first floe in a row take place in the following cases:

a. The back of the ice floe touched the surface of the structure, which it glides along upwards the structure (or downwards in case of the inverse cone).

b. The ice floe inclination about the X and Y axes exceeded the accepted value.

c. The ice floe failed completely due to shattering (local failure).

4. Example tasks

To check ice floe deformation modelling, an example of loading of an ice field row with vertical load along the front plane of this field is provided below.

Ice parameters: thickness – 1 m, Young's modulus – 1 GPa, bending stress limit - R_f=0.1 MPa.

Two loading cases are considered: load per unit of length 10 kPa /m (the ice floats on the water surface, there is an analytical solution) and 30 kPa /m (the field is partly submerged, failure is prohibited), figure 2 (a) and (b) correspondingly.

The ice field bending moment (figure 3a) and deflection (figure 3b) are analyzed. The provided graphs show the ice longitudinal length and deflection are normalized by the characteristic length of the fragment (corresponds at the peak point of the bending moment), bending moment per unit of length – by the set load per unit of length and the fragment length, which is [6]:

$$L_1 = \frac{\pi}{4}\left(\frac{4D}{\rho_h g}\right)^{0.25} \tag{20}$$

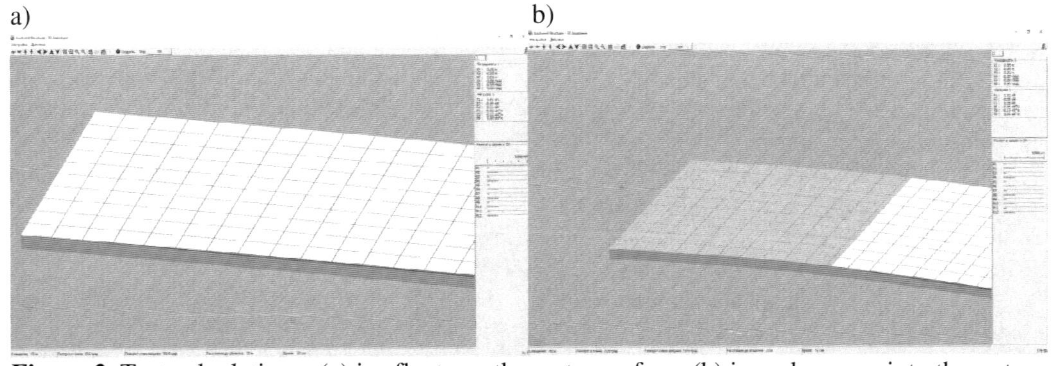

Figure 2. Test calculation – (a) ice floats on the water surface, (b) ice submerges into the water.

For the first loading case the received values of deflection and bending moment are 0.5 - 1% accurate comparing to the theoretical solution. If ice is submerged into the water the situation changes radically: peak point of the moment moves twice further away from the ice floe edge, and the increase in deflection is much more substantial than the increase in the load. Figure 4 shows further ice failure.

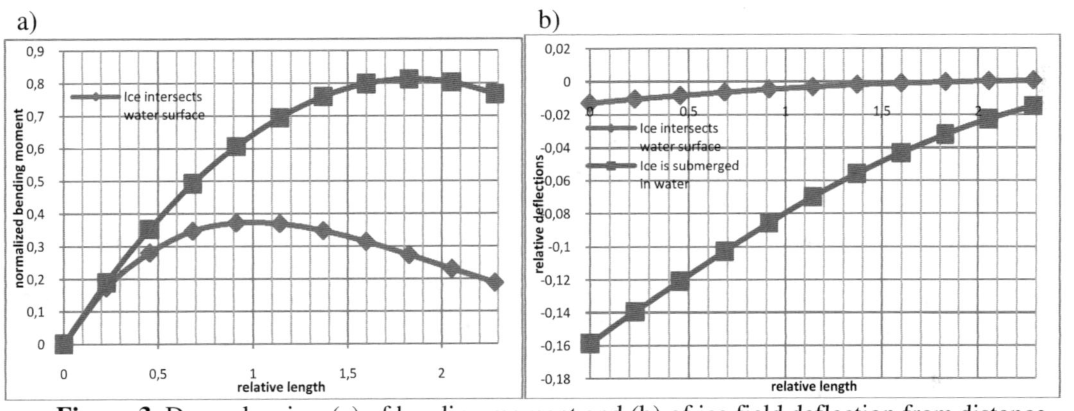

Figure 3. Dependencies: (a) of bending moment and (b) of ice field deflection from distance to the edge of the field (load application point).

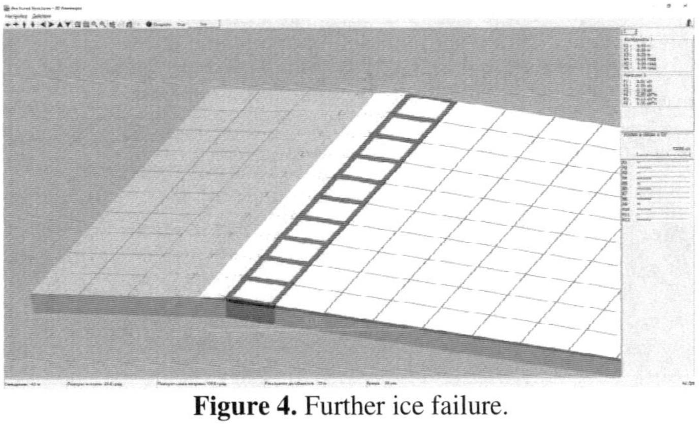

Figure 4. Further ice failure.

IOP Conf. Series: Earth and Environmental Science **193** (2018) 012006 doi:10.1088/1755-1315/193/1/012006

5. Conclusions

The main advantages of this mathematical model of an ice field are the following:

1) High computational efficiency – dynamic analysis of ice loading of a structure in the time domain with statistically-valid results can be performed on a personal computer, taking no more than about ten minutes.

2) The method works even with significant bending and inclinations of an ice field, including the cases of full submersion and full emersion of ice.

3) High cohesiveness of the model – even the fields that are considerably damaged by the contact with a structure behave quite stably.

4) Versatility – there are possibilities for floating as well as fixed structures, having generatrixes in forms of right cone as well as an inverse cone and wall-sided ones.

5) Unlike the FEM methods, having a specific set of elements' characteristics, this method allows flexible adjustment of ice failure models to refine the design model.

References

[1] Bolshev A S, Andryushin A V, Kuteynikov M A and Frolov S A 2016 Computational simulation of floating shelf structures behavior under the influence of level ice *Proc. of the 3rd international Polar Mechanics symposium* pp 795–806

[2] Bolshev A S, Kuteynikov M A and Frolov S A 2013 Mathematical modeling of offshore floating objects in Anchored Structures program complex *Research and technology collection RRS* **36** 68–90

[3] Bolshev A S, Frolov S A and Shkhinek K N 2012 Mathematical Modelling of Floating Anchored Objects Behavior under Ice Influence *Proc. of the 10th ISOPE Pacific / Offshore Mechanics Symposium* (Vladivostok, Russia) pp 68–72

[4] Bolshev A S, Frolov S A and Shkhinek K N 2011 Mathematical modeling of floating anchored objects behaviour under the level ice of ridges influence *Society of Petroleum Engineers - Arctic and Extreme Environments, Conference and Exhibition* **2** 934–45

[5] Sachenkov A A 2012 A lecture course on plate bending theory *Lobachevskii Institute of Mathematics and Mechanics, Kazan Federal University* (Kazan) p 53

[6] Timoshenko S P 1972 *Course of elastic theory* (Kiev: Naukova Dumka) p 507

Alternative to mathematical models of gas production: numerical experiment

E A Bondarev[1], **I I Rozhin**[1] **and K K Argunova**[1]

[1]Laboratory of Technological Gas Hydrates, Institute of Oil & Gas Problems, of Siberian Branch, RAS, 1 Oktyabrskaya str., Yakutsk, 677980, Russia

E-mail: bondarev@ipng.ysn.ru

Abstract. Computational experiment for comparison of quasi-one dimensional and two dimensional mathematical statements of the conjugate problem of heat exchange between gas producing well and the rocks has been carried out. The problem consists of solution of differential equations describing non-isothermal gas flow in the well and the heat conduction equation for surrounding rocks. The dependence of the coefficient of heat exchange of the gas with the hydrate layer on the time-varying flow area is taken into account in a quasi-stationary mathematical model of hydrates formation (dissociation) in gas well. The results of computational experiment correspond to two cases of production of wet and dry gas at the Otradninsky gas-condensate field.

1. Introduction

In the extraction and transportation of natural gas in the Northern regions, such natural factors as low climatic temperature and the presence of a thick layers of permafrost largely determine the technological modes of gas production. This is caused by the fact that under certain thermodynamic conditions, natural gas, when combined with water, forms solid crystalline compounds, namely gas hydrates, which can be formed in both bottom-hole zone and the wells. The formation of hydrates in the bottom-hole zone leads to a reduction in well productivity, whereas their formation in the wells can lead to a complete cessation of gas extraction. Such emergency situations can have dire consequences. Currently the only means of combating this unwanted phenomenon is the injection of methanol or other hydrate inhibitors into wells. This measure is ineffective, since the methanol is removed from the wells along with the produced gas, and, in addition, it significantly increases the cost of gas production.

Modern mathematical models of gas production consist of gas dynamic equations (tube hydraulics) as well as heat conduction equation for surrounding rocks. In Arctic regions their upper part is permafrost. So for appropriate description of gas production and control of possible accidents (well head damage because of ground thawing, hydrate plugs) one needs to solve the conjugated Stefan problem. The set of differential equations should be added with corresponding initial and boundary conditions where the most important ones are conditions of 1) heat exchange between rocks close to wellhead and atmospheric air and 2) heat exchange between flowing gas and surrounding rocks. The model should be more complicated in the case of hydrate formation in a well.

Up till now the second of heat exchange conditions was stated using approximation based on the fact that convective heat transfer in gas well has much shorter characteristic time than conductive one in rocks. Then only radial components will remain in the equation of heat conduction for each of well sections while their connection is implemented through boundary condition at outside surface of a well [1, 2]. The paper is devoted to numerical experiment to test validity of this approximation for real cases of gas production.

2. Problem statement

The authors use quasi-stationary mathematical model of real gas flow in tubes [1, 2] and mathematical model of hydrate formation [1 – 6] based on the generalized Stefan problem where phase transition temperature depends on gas flow pressure. The model consists of two ordinary differential equations.

$$\frac{dp}{dx} = -\rho_{\mathrm{g}} g \sin\varphi - \frac{\sqrt{\pi}\psi M^2}{4\rho_{\mathrm{g}} S^{2.5} S_0^{2.5}}, \tag{1}$$

$$\frac{dT}{dx} - \varepsilon\frac{dp}{dx} = \frac{\pi D\alpha}{c_p M}(T_{\mathrm{e}} - T) - \frac{g}{c_p}\sin\varphi, \tag{2}$$

where c_p – specific heat capacity of gas, D – inner tube diameter, g – gravity acceleration, $M = \rho_{\mathrm{g}} \upsilon S S_0$ – constant mass flow rate, p – pressure, S – dimensionless cross-section, S_0 – initial cross-section, T – gas temperature, T_{e} – temperature of rocks around well, υ – gas flow velocity, x – distance along tube axe, α – total heat transfer coefficient, ρ_{g} – gas density, φ – angle of tube inclination to horizon, ψ – hydraulic resistance, ε – throttle coefficient.

Gas equation of state is written in the form

$$\rho_{\mathrm{g}} = \frac{p}{ZRT}, \quad \mathrm{Z} = \mathrm{Z}(p, T), \quad \varepsilon = \frac{RT^2}{c_p p}\left(\frac{\partial \mathrm{Z}}{\partial T}\right)_p, \tag{3}$$

where Z – real gas compressibility, empirical function of reduced pressure and temperature; R – gas constant; for real gas compressibility the Latonov-Gurevich formulae $\mathrm{Z} = (0.17376 \ln T + 0.73)^p + 0.1p$ is used [7].

Dynamics of well cross-section S can be found from the dimensionless equation [6]:

$$\frac{dS}{d\tau} = b_2 \frac{T_{\mathrm{e}} - T_{\mathrm{h}}(p)}{1 - b_2 \ln S} - b_1\sqrt{S}(T_{\mathrm{h}}(p) - T), \tag{4}$$

where $b_1 = \alpha_1 D_0 / 4\lambda_{\mathrm{h}}$, $b_2 = \alpha_2 D_0 / 4\lambda_{\mathrm{h}}$, α_1 – coefficient of heat transfer between gas and hydrate layer; α_2 – coefficient of heat transfer between hydrate and rocks, D_0 – initial tube diameter, $\tau = \frac{\lambda_{\mathrm{h}} T_{\mathrm{c}}}{\rho_{\mathrm{h}} q_{\mathrm{h}} D_0^2} t$ – dimensionless time, λ_{h}, ρ_{h} – heat conduction and density of hydrate, q_{h} – specific latent heat of hydrate formation, t – time, $T_{\mathrm{h}}(p) = a \ln p + b$ – equilibrium temperature of hydrate formation. Coefficients a and b can be found from a curve built according to the D. Sloan method [8], properties of mixture components being known. Critical temperature and pressure are determined by the W.B. Kay relations [9]: $p_{\mathrm{c}} = \sum_{i=1}^{n} y_i p_{\mathrm{c}i}$, $T_{\mathrm{c}} = \sum_{i=1}^{n} y_i T_{\mathrm{c}i}$; where y_i – volume content and $p_{\mathrm{c}i}$, $T_{\mathrm{c}i}$ – critical pressure and temperature of i-th component of natural gas. Coefficient α_1 depends on tube cross-section, which in turn is a function of time. In the sections of well tube where hydrate formation (dissociation) takes place, it should be calculated by the following relation:

$$\frac{\alpha_1 D_0}{\lambda_g} = 0.023\mathrm{Pr}^{0.43}\left(\frac{M}{D_0 \eta_g}\right)^{0.8}\left(\frac{4}{\pi}\right)^{0.8}\frac{1}{S^{0.9}}, \tag{5}$$

where λ_g, η_g – coefficients of heat conduction and viscosity of gas, Pr – the Prandtl number. In so doing rock temperature T_e should be substituted with equilibrium temperature of hydrate formation T_h.

Initial conditions for the equations (1), (2) and (4) are:

$$p(0) = p_0,\ T(0) = T_0,\ S(0) = \mathrm{const}. \tag{6}$$

Equations (2) and (4) include rock temperature T_e, which are found from the solution of heat conduction equation

$$\tilde{C}(T_e)\frac{\partial T_e}{\partial t} = \nabla(\lambda(T_e)\nabla T_e),\ 0 < x < L,\ r_b < r < r_k,\ t > 0, \tag{7}$$

where $\lambda(T_e)$ – heat conductivity and $\tilde{C}(T_e)$ – volume heat capacity of rocks which are piece-wise temperature functions.

In quasi-one dimensional problem, that is when heat flow takes place only along radial coordinate, for connection between layers along well one should use the equation (2) and boundary condition on outer well surface

$$\lambda(T_e)\frac{\partial T_e}{\partial r} = \alpha(T_e - T),\ r = r_b. \tag{8}$$

At conditional radius the heat flow is zero:

$$\frac{\partial T_e}{\partial r} = 0,\ r = r_k. \tag{9}$$

At well bottom hole rock temperature is constant and equals reservoir temperature T_{e0}:

$$T_e = T_{e0},\ x = 0. \tag{10}$$

At daylight surface condition of convective heat transfer with atmosphere includes radiation heat inflow [10, 11]:

$$\lambda(T_e)\frac{\partial T_e}{\partial x} = \alpha_L\left(T_e - T_a + \frac{Q_s(1-A)}{\alpha_a}\right),\ x = L, \tag{11}$$

where T_a – air temperature, Q_s – total solar radiation, A – albedo of the Earth surface; $\alpha_a = 15.12\sqrt{\upsilon_w}$ – coefficient of heat transfer between soil surface and atmosphere, dependent on wind velocity υ_w [12]; $\alpha_L = (1/\alpha_a + \delta_{sn}/\lambda_{sn})^{-1}$ – total heat transfer coefficient taking into account thickness of snow layer δ_{sn} with heat conductivity $\lambda_{sn} = 1.163\cdot(0.03 + 0.303\rho_{sn} - 0.177\rho_{sn}^2 + 2.25\rho_{sn}^3)$, where ρ_{sn} – snow density (g/cm^3) [10]. To describe cyclic variations of Q_s, A, υ_w, δ_{sn}, и ρ_{sn}, their month averaged values [12] were approximated by linear functions.

Initial rocks temperature:

$$T_e = \begin{cases} T_{e0} - \Gamma_1 x, & 0 < x < L - H \\ T_{ph} - \Gamma_2 x, & L - H < x < L \end{cases}, \tag{12}$$

where T_{e0} – temperature at bottom hole, $T_{ph} = 273.15$ K – temperature at lower permafrost boundary, Γ_1 and Γ_2 – geothermal gradient of thawed and frozen rocks, L – well depth, H – permafrost thickness.

Algorithm of numerical solution of the problem (1) – (12) looks in the following way:

I. Choose geometrical and physical characteristics and state conditions (6) and (12).

II. Initial value of cross-section is given, solve equations (1) – (3) to find pressure $p(x)$ and temperature of gas $T(x)$.

III. Making a time-step and looking at x as a parameter, from equations (4) and (6) find a new value of cross-section.

IV. Solve the problem (7) – (12) to find rocks temperature. Since smoothed coefficients in the equation (7) are temperature dependent, the finite-difference problem will be non-linear, simple iterations method in combination with a sweeping algorithm should be used.

Tasks II – IV are repeated at each time step.

3. Calculation results

Given data correspond to Otradninsky gas-condensate field in Sakha (Yakutia) Republic [13]: $a = 6.635$ K, $b = 182951$ K, $c_p = 2300$ J/(kg·K), $D_0 = 0.146$ m, $H = 680$ m, $L = 2480$ m, $p_0 = 188.35 \cdot 10^5$ Pa, $p_c = 44.71 \cdot 10^5$ Pa, $q_h = 510000$ J/kg, $q_{ph} = 334400$ J/kg, $R = 438.3$ J/kg·K), $T_0 = 286.35$ K, $T_c = 195.376$ K, $T_{e0} = 286.48$ K, $T_{ph} = 273.15$ K; $\alpha = 5.82$ W/(m^2·K), $\varphi = 90°$, $\Gamma_1 = 0.0074$ K/m, $\Gamma_2 = 0.0073$ K/m, $\eta_g = 1.3 \cdot 10^{-5}$ Pa·s, $\lambda_g = 0.0307$ W/(m·K), $\lambda_h = 1.88$ W/(m·K), $\rho_h = 920$ kg/m^3, $\psi = 0.02$.

The formulae for atmospheric temperature:

$$T_a = \frac{T_{aw} - T_{as}}{2} \cdot \sin\left(\frac{\pi(30(m-1) + t + 75)}{180}\right) + \frac{T_{aw} + T_{as}}{2},$$

where T_{aw}, T_{as} – minimal and maximal air temperature according to [14], t – time, days, m –month number, starting at well start-up.

At first, consider a gas producing well without adding inhibitors which make hydrate formation and even well plugging possible. Results of calculations are shown at Figure 1 – Figure 3.

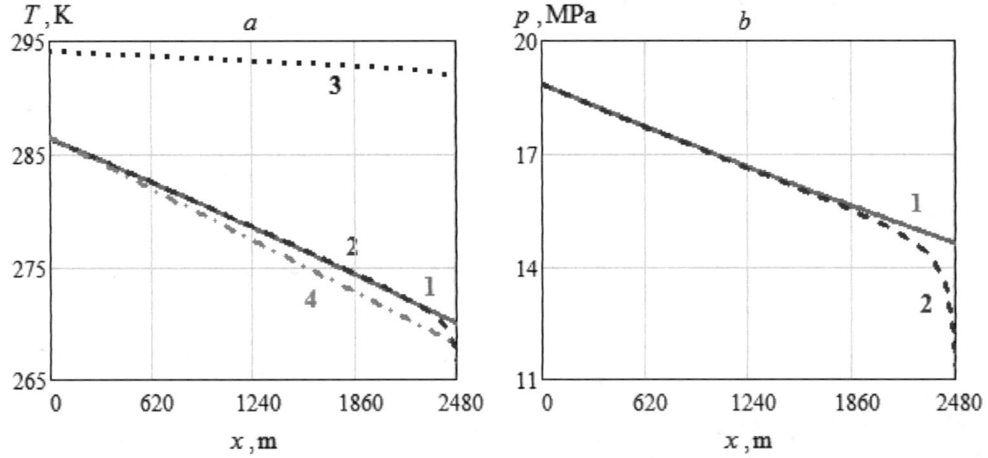

Figure 1. Temperature (a) and pressure (b) distribution along a well: 1 – 5 min., 2 – 5 hours, 3 – equilibrium temperature, 4 – temperature at outer well surface.

At Figure 1 curves 1 and 2 for both models are indistinguishable, so they are marked with the same figures. It is worth to notice that hydrate plugs only the upper part of a well, which is entirely in permafrost rocks (compare curves 1 and 2). Another important feature of the process is very short time of complete plugging (about 5 hours). The calculations lead to the following conclusion: if hydrates plug producing well then both models give identical results for all technological parameters (temperature (Figure 1a) and pressure (Figure 1b) distribution along a well, dynamics of free cross-section (Figure 2) and phase transition boundary in rocks (Figure 3). It is due to a very short period of thermal interaction between gas flow and surrounding rocks (see, for example, a very small thawing zone at Figure 3)

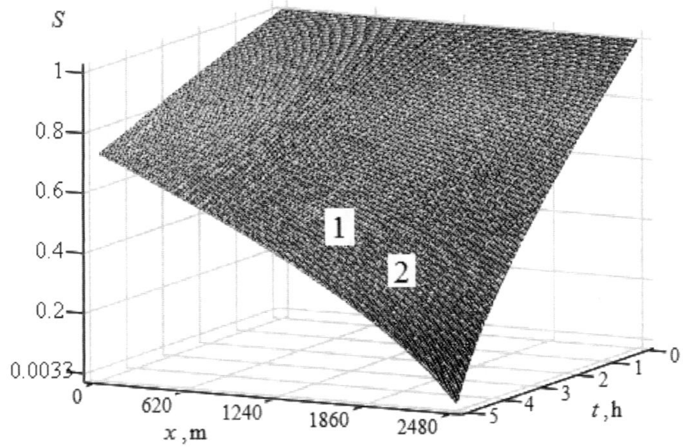

Figure 2. Dynamics of free cross-section distribution in well: 1 – quasi-one dimensional model, 2 – two dimensional model.

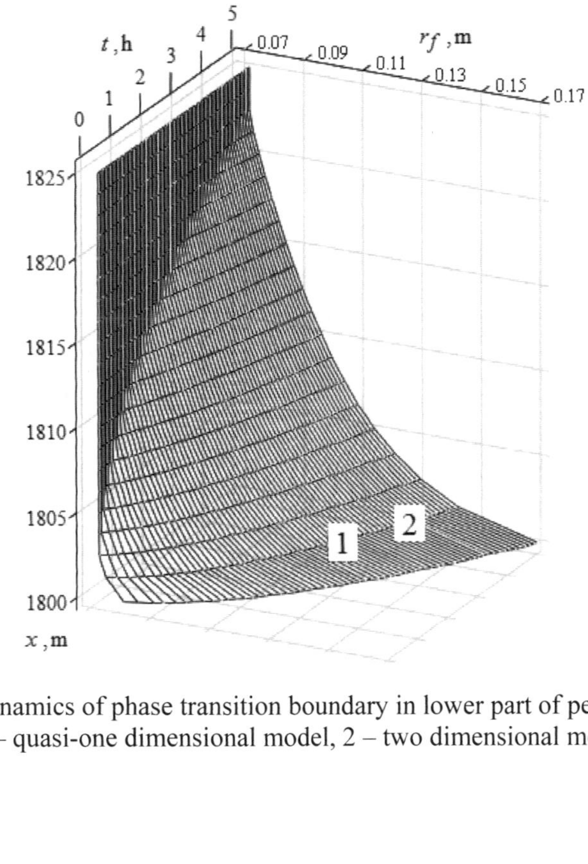

Figure 3. Dynamics of phase transition boundary in lower part of permafrost rocks: 1 – quasi-one dimensional model, 2 – two dimensional model.

Second case corresponds to injection of inhibitor into well. The results of calculation can be seen at Figure 4 – Figure 5. Figure 4 shows that both models give equal temperature and pressure distributions along gas well (see curves 1 and 2 at Figure 4a, Figure 4b). However, phase transition boundary value for quasi-one dimensional model exceeds that for two-dimensional model (compare curves 1 and 2 at Figure 5). However, the difference can be seen only near the lower part of permafrost rocks, about 60 m which is shown at Fig. 5, Still, it can be important while one estimates the danger of well head damage due to permafrost thawing.

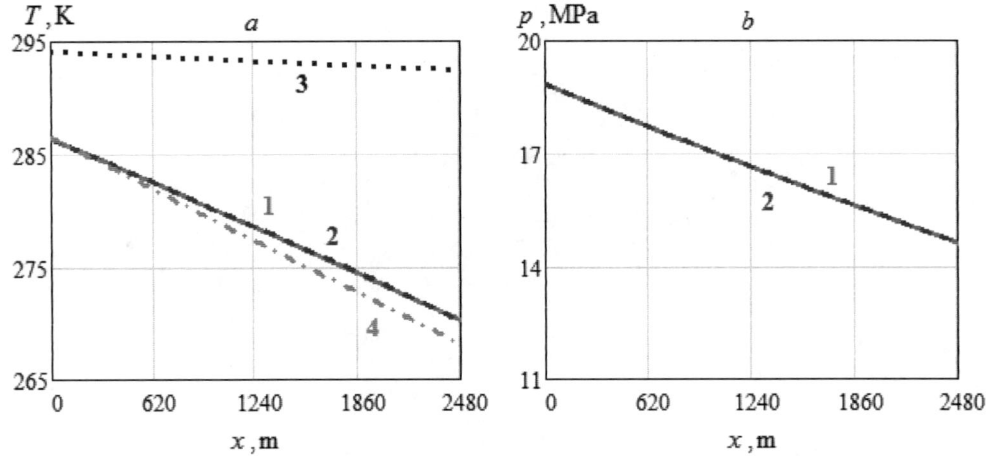

Figure 4. Temperature (a) and pressure (b) distribution along a well after a year of gas producing:
1 – quasi-one dimensional model, 2 –two dimensional model,
3 – equilibrium temperature, 4 – temperature at outer well surface.

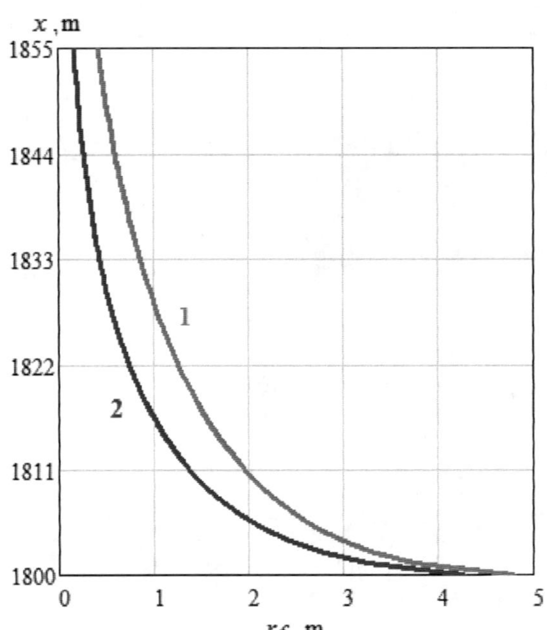

Figure 5. Phase boundary position along permafrost zone:
1 – quasi-one dimensional model, 2 – two dimensional model.

Acknowledgments
The work is carried out within the framework of the State Order № 0377-2016-003. The reported research was partly funded by Russian Foundation for Basic Research and the government of the Republic of Sakha (Yakutia) of the Russian Federation, grant № 18-45-140035.

References

[1] Bondarev E A and Voevodin A F 2017 *Hydraulic Problem Solutions for Natural Gas Production and Transport Systems* (Novosibirsk: RAS Siberian Div. Publishing house) p 209

[2] Bondarev E A, Rozhin I I and Argunova K K 2017 Characteristics of mathematical models of natural gas production and transport in Russian Arctic regions *Notes of Mining Institute* **228** 705–16

[3] Bondarev E A, Vasiliev V I, Voevodin A F, Pavlov N N and Shadrina A P 1988 *Thermohydrodynamics of Gas Production and Transport Systems* (Novosibirsk: Science publishing house Siberian Div. RAS) p 272

[4] Argunova K K, Bondarev E A, Nikolaev V E and Rozhin I I 2008 Determination of hydrate formation intervals for well in permafrost *Scientific Journal Oil and Gas* p 11 http://www.ogbus.ru/authors/Argunova/Argunova_2.pdf

[5] Argunova K K, Bondarev E A and Rozhin I I 2011 Mathematical models of hydrate formation in gas wells *Earth's Cryosphere* **XV(2)** 65–9

[6] Bondarev E A, Rozhin I I and Argunova K K 2014 Modeling the formation of hydrates in gas wells in their thermal interaction with rocks *J. Eng. Phys and Thermophys.* **87(4)** 900–7

[7] Latonov V V and Gurevich G R 1969 Calculation of natural gas compressibility coefficient *Gas Industry* **2** 7–9

[8] Sloan E D and Koh C A 2008 *Clathrate Hydrates of Natural Gases* (Boca Raton USA: Taylor&Francis Group/CRC Press) p 720

[9] Kay W B 1936 Density of hydrocarbon gases and vapors at high temperature and pressures *Industrial & Engineering Chemistry Research* **28** 1014–9

[10] Pavlov A V 1975 *Heat Exchange Between Soil and Atmosphere in the Northern and Temperate USSR Regions* (Yakutsk: Yakutsk publishing house) p 304

[11] Samokhin A V, Sleptsov V I, Vychuzhin T A, Shkulev S P and Mestnikov A E 1994 Protection of permafrost rock outcrops during open cast of diamonds in Yakutia *Preprint* (Yakutsk: YaSC Siberion Div RAS Publishers) p 44

[12] Porkhaev G V and Schelokov V K 1980 *Prediction of Permafrost Ground Temperature at Build-on Areas* (Leningrad: Stroiizdat Leningrad Div.) p 112

[13] Bondarev E A, Rozhin I I and Argunova K K 2014 Hydrate formation at development of the Otradninsky gas condensate field *SOCAR Proceedings* **4** 46–53

[14] 1989 *Applied and Scientific Climate of the USSR Directory Series 3 Long-standing data Parts 1–6 Release 24 YaSSR Book 1* (Leningrad: Hydrometheo publishers) p 608

Significance evaluation of effect of operational changes in material on impact strength of various zones in leaf spring at low temperature

I I Buslaeva[1] and S P Yakovleva[1,2]

[1] Yakut Scientific Center, Siberian Branch of the Russian Academy of Sciences, 2, Petrovskogo street, Yakutsk, 677980, Russia
[2] Institute of Physical-Technical Problems of the North, Siberian Branch of the Russian Academy of Sciences, 1, Oktyabr'skaya street, Yakutsk, 677980, Russia

E-mail: buslajeva@mail.ru

Abstract. Significance of the effect of structural changes in spring steel arising during operation in various zones of the leaf springs on zonal resistance of the steel to a brittle fracture at temperatures of a climatic range of a permafrost region is studied. Structural damage was determined at the temperatures of 20, –20 and –60 °C and impact strength of the different zones of the leaf spring failed during the operation was estimated. Features of a fine structure of fracture surfaces of Charpy specimens and micromechanisms of their failure are investigated. To determine variability of the impact strength in the zones of the spring, analysis of variance (ANOVA) of test data was carried out. It was found that at temperatures of 20 and –20 °C the effect of the zone damage of a material on the impact strength is insignificant, and at –60 °C the effect of the zonal damage on the resistance to the brittle fracture becomes significant. The fractographic features of the fine structure of the fracture surfaces at the various test temperatures are discussed. Results can be used in a prediction of a life time of elastic elements in vehicle suspension: it is necessary to take into account zoning of the operational damage of the steel under conditions of the low climatic temperatures.

1. Introduction

One of actual problems while attempting to ensure operability of machinery in the North is a low of reliability of elastic elements of vehicle suspension at low climatic temperatures (–50 °C and below). The leaf springs of motor transport under low temperatures exhibits life time reduced by 1.5-5 times [1]. The operating life of the elastic elements of the suspension at the low temperatures is decreased because of shock cyclic loads transferred to the transport vehicles when driving on roads with uneven surfaces (winter, ice and dirt roads). These loads lead to formation of the micro- and mesodamages in a material structure, integration of which leads to emergence and development of micro- and macrocracks [2, 3, 4]. Therefore, research of accumulation of the structural damage, the formation and propagation of the micro- and macrocracks in the material of the elastic elements at the low temperatures is of great importance for diagnostic tests of their current state and forecast of the residual life time. A necessary stage in such studies is a determination of the significance of zonal heterogeneity of the structural changes arising from operation in details and the effect of the structural damage on physico-mechanical characteristics of the material, which are associated with the operability of the elastic elements at the low climatic temperatures. A structural steel ability to resist the brittle fracture is evaluated by the results of the tests of notched specimens for impact bending at the climatic range temperatures. The aim of this article is to study the significance of the effect of the material changes occurring in different zones of the spring during the operation under the climatic conditions on the zonal resistance of the spring steel to the brittle fracture when the temperature decreases to –60 °C.

2. Material and investigation methods

The object of the research is the main spring leaf of the front suspension of a KAMAZ truck. The main leaf made of steel 60S2 underwent fatigue failure during the operation in the road-climatic conditions of Yakutia with 62,000 miles to failure [5, 6]. The fatigue crack passed near a front bracket attaching the spring to a frame of the truck. The crack divided the spring leaf into two fragments. To determine the impact strength, the Charpy specimens were made from the long fragment. The longitudinal specimens were cut from three zones of the spring, ranked according to intensity of the operating loads. Zone I is located near the fracture, i.e. near an attachment of the main leaf to the truck frame through the bracket, zone II is in the middle between the fracture and center of the leaf, where the spring is attached to a front axle, zone III is near the spring center. The transverse loads from the vehicle frame and longitudinal pushing and braking forces, as well as the lateral forces during the turn, are transferred to the main leaf when the vehicle is moving. Theoretically, zone III near a place of attachment of the spring to the front axle is the most loaded one, where the greatest internal bending moment arises from the transverse load. Near the attachment of the front spring to the frame (zone I) there is a multiaxial stress state associated with the transfer of the transverse, longitudinal and lateral forces to the spring. Zone II is considered less loaded.

Three groups of the specimens were tested for the impact bending in accordance with the standard method on a pendulum impact tester Amsler RKP 450 Zwick within a temperature chamber providing the specimen temperatures in the range of –80 °C to the room temperature. The tests were carried out at 20, –20 and –60 °C. A width of the spring allowed to cut out 6 specimens within each zone with dimensions of 10 × 10 × 55 mm (two specimens for each test temperature).

An electron fractographic investigation of the specimen fractures was also performed to reveal the fine structure of the fracture surfaces in detail (scanning electron microscopes TM 3030 Hitachi and JEOL JSM-6480LV were used).

3. Results and discussion

The test results and dispersion parameters of the impact strength KV in each zone are shown in Table 1. From Table 1 it can be seen that the decrease of the impact strength occurs more intensely in the range of the temperatures from 20 to –20 °C than in the one from –20 to –60 °C. The largest value range of the impact strength KV is observed in locations where the main leaf is attached to the frame and front axle of the truck (the most loaded zones I and III). The considerable spread of the impact strength in each zone indicates the significant differences of the metal structural and mechanical states not only along a length of the spring, but also across its width. The greatest spread of the KV values is observed at 20 °C. As the temperature decreases, the variation reduces; it will be presented below that this should be associated with a reduction in the ability of the material to deform.

To assess the effect of operational changes in the material on the impact strength of different zones of the main leaf, the analysis of variance was carried out. A level of the damage in the spring metal was taken as a dependent variable (a factor) affecting the response variable (the impact strength at the different temperatures). Considering that each zone of the spring leaf exhibits its level of the structure damage to the metal, we range the test results into three groups. A procedure of the analysis of variance of the data consists in the determination of a ratio of the systematic variance between the groups to the random variance within the groups.

Table 1. Impact strength of specimens of material from various spring zones at different temperatures.

Test temperature and parameters	Impact strength in zones, kJ/m^2		
	Zone I	Zone II	Zone III
20 °C	207.4	211.7	225.6
20 °C	303.9	237.7	295.8
Absolute value of difference at 20 °C	96.4	26	70.3
Average at 20 °C	255.6	224.6	260.7
–20 °C	164.2	151.5	205.9
–20 °C	138.4	160.3	170.9
Absolute value of difference at –20 °C	25.9	8.9	35
Average at –20 °C	151.3	155.9	188.4
–60 °C	112.1	112.7	146.9
–60 °C	126.8	113.2	138.5
Absolute value of difference at –60 °C	14.7	0.5	8.4
Average at –60 °C	119.4	112.9	142.7

When applying the single-factor ANOVA, there should be at least three levels of the independent variable and at least two observations at each level. The results of the impact strength tests at the different temperatures satisfy these requirements: a number of the levels $m = 3$ (according to the three zones) and the number of observations at each level $n = 2$. For the results at each temperature for the total number of observations $N = 6$, we obtain the following degrees of freedom: between the groups $f_{BG} = m - 1 = 2$ and within the groups $f_{WG} = N - m = 3$.

We will perform the analysis of variance at 20 °C. A subject to verification is a null hypothesis: the mean values of the impact strength within the three zones differ insignificantly, i.e. all the groups of the test results are random samples from the same population. The data for a calculation are given in Table 1. The total average for all the impact strength values is equal to 247 kJ/m^2. We will find a sum of the squared differences of the impact strength values from the mean within the groups $SS_{WG} = 7458.2$ and the "between-group" sum of the squared differences of the values from the total average $SS_{BG} = 1520.6$. The "within-group" mean square value is $MS_{WG} = \frac{SS_{WG}}{f_{WG}} = \frac{7458.2}{3} = 2486.1$ and the between-group mean square value is $MS_{BG} = \frac{SS_{BG}}{f_{BG}} = \frac{1520.6}{2} = 760.3$. The ratio of the between-group variance to the within-group one has an F-distribution. To verify the null hypothesis, we use F-test. An F-test statistic is calculated using the following formula: $F = \frac{MS_{BG}}{MS_{WG}} = \frac{760.3}{2486.1} = 0.3$. With a desired false-rejection probability $\alpha = 0.05$ and the degrees of freedom $f_{BG} = 2$ and $f_{BG} = 3$, we determine the critical value $F_{cr} = 9.6$. Since the test statistic F is less than the critical value F_{cr}, there is no reason to reject the null hypothesis: the group averages differ insignificantly (the factor of the zone affects insignificantly at the test temperature of 20 °C) [7].

Similarly, we will perform the calculations for the test temperatures of –20 and –60 °C. The results of the calculations are given in Table 2. It can be seen that the empirical values of the F-test increase as the temperature decreases, and at the temperature of -60 °C this parameter exceeds the critical value $F = 10.2 > F_{cr} = 9.6$, which allows rejecting the null hypothesis and accepting that the group averages differ significantly. Thus, at the temperature of –60 °C, the variability of the impact strength is due to the studied independent variable – the factor of the spring zone with the different material damage. A physical nature of this effect is discussed below in the analysis of the parameters of the microdamage (the substructure changes are estimated by the microhardness measurements) and mesodamage (porosity) in conjunction with the microstructure features of the fractures. A damage evaluation was carried out in [5], in which it was revealed that the metal structure of the three zones differs in the mesodamage the most. Appearance and role of the porosity in the fatigue are considered, for example, in [8, 9]. Zones I and II are characterized by the almost identical volume ratio of pores, but the small pores prevail in zone I. Zone III has the largest volume ratio of the mesodamages due to prevalence of the coarse pores.

Table 2. Analysis of variance of test results.

Temperature	x_m	SS_{wg}	SS_{bg}	MS_{WG}	MS_{BG}	F
20 °C	247	7458.2	1520.6	2486.1	760.3	0.3
–20 °C	165.2	985.3	1636.9	328.4	818.5	2.5
–60 °C	125	144.3	981.2	48.1	490.6	10.2

The secondary cracks were found in the fractures of all the Charpy specimens (Figures 1 and 2), the microstructure of which reflects the mechanisms and energy intensity of the development of the specimen fracture as a whole.

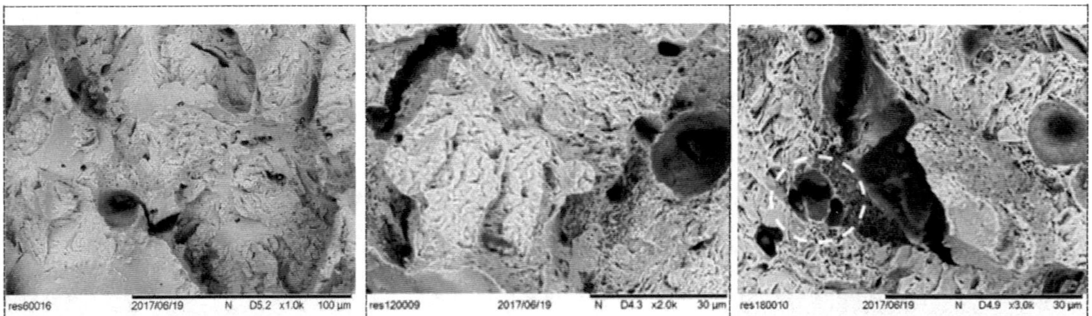

(a) (b) (c)

Figure 1. Microstructure of fracture surfaces of metal specimens from zone I (a), zone II (b) and zone III (c), tested at –20 °C

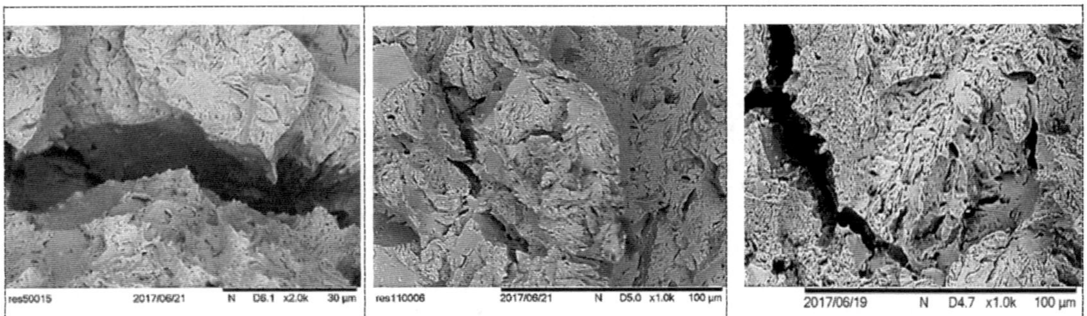

(a) (b) (c)

Figure 2. Microstructure of fracture surfaces of metal specimens from zone I (a), zone II (b) and zone III (c), tested at –60 °C

As can be seen in figures 1 and 2, the cracks are formed by coalescence of the pores. The pore coalescence is particularly well traced within a dashed circle in figure 1(c). For zone III with the coarse pores, this process was accompanied by a noticeable plastic deformation (stretching the pores till their coalescence, deforming cavity walls) even at –60 °C. The deformation of the material by stretching the pores and their fusion is also traced in the fractures of the specimens of zones I and II tested at –20 °C, but the smaller length of the secondary cracks is noticeable. At –60 °C, their trajectory becomes more broken (Figures 2(a) and 2(b)) that indicates the decrease in the ability of the material to deform. Smoothness of the walls of the resulting secondary cracks confirms the brittle mechanism of their opening by splitting crystallites. Thus, the changes in the values of the impact strength and their dispersion, depending on the spring zone and test temperature, are due to the effect of the existing structural damage on a possibility of microdeformations as the mechanism of material adaptation to the loads and, accordingly, the resistance to failure.

4. Conclusion

The analysis of variance of the data of the test on the impact strength for the Charpy specimens from the different zones of the main leaf of the KAMAZ truck spring was carried out. Comparison of the test statistic F with the calculated critical value showed that with the desired false-rejection probability $\alpha = 0.05$, the factor of the zone with the different damage at the test temperatures 20 and –20 °C affects the impact strength insignificantly. At –60 °C, the differences are reliable in a statistical sense that indicates the significant effect of the structural damage of the spring zone on the zonal resistance to the brittle fracture at this temperature. The mechanism of the effect is determined by the possibility of the microdeformations as a way of the material adaptation to the loads. The obtained results point out the necessity of the zoning of the spring leaf along the length and width during the evaluation of the metal structural damage and can be used in a prediction of life time of the elastic elements in the vehicle suspension under conditions of the low climatic temperatures.

References

[1] Buslaeva I I, Ishkov A M and Levin A I 2014 *Proc. Eurasian Symp. on Reliability of Materials and Machines for Cold Climate Regions (St.Petersburg)* (St.Petersburg: Polytechnic University Publishing) 347–51

[2] Murakami Yu 2002 *Metals Fatigue: Effects of Small Defects and Nonmetallic inclusions* (London: Elsevier) 298

[3] Sakai T, Li W, Lian B and Oguma N 2011 *Proc. 5th Int. Conf. on Very High Cycle Fatigue: VHCF-5 (Berlin)* (Berlin: DVM) 19–26

[4] Romanov A N 2014 Role of the structural state in the formation of deformation, strength, tribological, and technology properties of structural materials *J. of Machinery Manufacture and Reliability* **43** 233–41

[5] Yakovleva S P, Buslaeva I I, Makharova S N and Levin A I 2017 Operational damage to the structure and failure of the KAMAZ truck spring in the temperature-load conditions of the North *J. of Machinery Manufacture and Reliability* **46** 488–93

[6] Chetan J, Khushbu P and Nauman M 2012 The fatigue analysis of a vehicle suspension system-A review article *Int. J. of Advanced Computer Research* Vol 2 **4**

[7] Scheffe H 1999 *The Analysis of Variance* (New York: John Wiley & Sons)

[8] Bhat S and Patibandla R 2011 *Metal Fatigue and Basic Theoretical Models: A Review, Alloy Steel - Properties and Use* ed E V Morales (InTech) 203–36

[9] Saedi N, Ashrafizadeh F, Niroumand B, Forouzan M R and Barlat F 2014 Damage mechanism and modeling of void nucleation process in a ferrite-martensite dual phase steel *Engineering Fracture Mechanics* **127** 97–103

IOP Conf. Series: Earth and Environmental Science **193** (2018) 012009 doi:10.1088/1755-1315/193/1/012009

Features of bitumoids composition of bottom sediments of the coastal zone of the East Siberian Sea

Yu S Glyaznetsova, I N Zueva, O N Chalaya and S Kh Lifshits

Institute of Oil and Gas Problems, Yakutsk, Russia

E-mail: gchlab@ipng.ysn.ru

Abstract. An anthropogenic influence, in particular, oil contamination on the Arctic ecosystems is studied. For this composition of the chloroform bitumoids extracted from the samples of bottom sediments is explored. The samples were taken near seaport Kolyma River and a coastal zone of the East Siberian Sea. In bottom sediments of the seaport the yeild of bitumoid reached 25500 mkg g^{-1}. As appeared, their samples were oil contaminated. Hydrocarbons prevailed significantly over the sum of resinous components in their composition as aliphatic compounds dominated over oxygen groups in chemical structure. N-alkanes maked up 28% of the sum of the identified hydrocarbons with a distribution maximum in the area $nC_{15,16}$. CPI is typical for oil and oil products. The values of ratio of isoprenoids to n-alkanes and the distribution of isoprenoids point to processes of biodegradation of oil hydrocarbons. In the sea samples the yield of bitumoid much less 193-1577 mkg g^{-1} and oil contamination traces aren't found. The future of chemical structure and distribution of individual saturated hydrocarbons reveal a strong relationship with terrigenous organic matter of recent deposits. The prevalence of odd hydrocarbons nC_{23}-nC_{35} and high concentrations nC_{27}, nC_{29}, nC_{31} is characteristic for lipids of highest land plants.

1. Introduction

The East Siberian Sea (ESS) is one of the least studied aquatorium of the Russian Federation. Kolyma River is the largest flowing into the ESS, the river flow of which is one of the determining factors of sedimentation in the coastal zone of the ESS [1-3]. Oil and oil products are the most dangerous sources of contamination and are an environmental danger to the Arctic nature environment [3-7]. Along the Kolyma River only in the territory of the Yakutia there are three large oil depots and the Green Cape sea port. As a result of spills during transportation and operation of oil depots there is a contamination of waters and coastal part of land. Oil contamination becomes a source of oil hydrocarbons flow through the river into the sea and its coastal part.

Data on the study of the feature of the geochemical background of hydrocarbons in sediments are given in a number of works performed by various researchers [3, 6, 8-13]. Mainly, the investigations connect with determining of the origin and mechanism of distribution of hydrocarbons in bottom sediments and estimation of level of contamination. In the analysis of the hydrocarbon content in bottom sediments there is a problem of differentiated assessment of the anthropogenic constituent and natural which is due to a natural geochemical background, because often geochemical anomalies associate only with oil hydrocarbons. However, it is known the spread of anthropogenic hydrocarbons occurs on a natural hydrocarbon background, that doesn't ignore.

The purpose of researches was a determination of content and composition natural and technogenic hydrocarbons in the top layer of bottom sediments of the Green Cape seaport and in coastal zone ESS, on border of the Kolyma River – ESS.

2. Data and Methods

Material of a research were the samples of river superficial (0-5 cm) and bottom sediments of the Green Cape seaport 68°47.32N 161°22.13E and also sea sediments selected in the area 69°40.120N 162°20.38E.

The complex of analytical researches included extraction by chloroform with receiving chloroform bitumoid. For study chemical structure the selected bitumoids we applied IR Fourier spectroscopy (Protege 460 FTIR spectrometer, Nicolet) in the range of 500-4000 cm^{-1}, liquid chromatography and mass spectrometry (GC/MS). The liquid adsorption chromatography method was used to determine the group component composition of the extracts. The method consists in the initial deposition of asphaltenes with petroleum ether followed by the washing out of hydrocarbon fractions and resins with various organic solvents on a sorption column with silica gel [14]. GC/MS studies on the individual composition of saturated hydrocarbon fractions were carried out using a system that includes the Agilent 6890 gas chromatograph which has an interface with the highly efficient mass-selective detector Agilent 5973N. The chromatograph is equipped with a quartz capillary column (30 m long, 0.25 mm in diameter, impregnated with HP-5MS phase). The carrier gas is helium. The flow rate is 1 ml/min. The temperature of the evaporator is 320 °C. The temperature rise was programmed to range from 100 °C to 300 °C at a rate of 6 °C min^{-1}. The ionizing voltage of the source is 70 eV. In more detail techniques of researches are given in work [13, 15-19]. In comparison with the control, the obtained results are statistically reliable.

3. Results and Discussion

The conducted researches have shown that in sample R-1 of the sediments selected in the Kolyma River seaport, content of bitumoid reached 25500 mkg g^{-1}. As appeared, the sample with the maximum bitumoid content is oil contaminated. Hydrocarbons prevailed significantly over the sum of resinous components in their composition (table 1) as well as hydrocarbon groups (CH_2, CH_3) dominate over content of oxygen groups and bonds. The type IR spectrum is characteristic for oil contaminated samples that confirms the appearance of absorption bands of aromatic structures – 750, 810, 880 and 1600 cm^{-1} (figure 1).

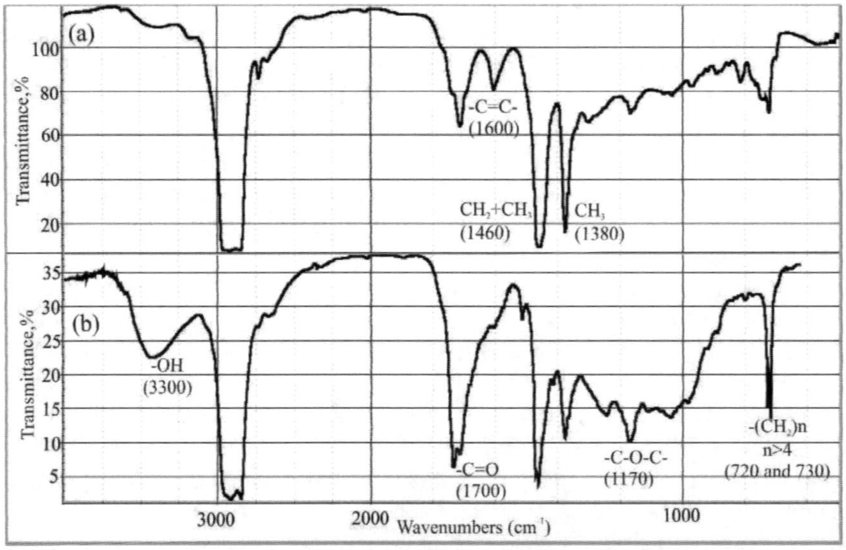

Figure 1. IR-spectra of extracts of sediment samples: (a) R-1; (b) S-1.

Table 1. Geochemical characteristic of bottom sediments.

Parameters	Parameter values				
№ station	R-1	R-2	S-1	S-2	S-3
Sample set	River sediments		Sea sediments		
Yield bitumoid, mkg g^{-1}	25500	710	1577	318	193
Group component composition bitumoid, %:					
Hydrocarbons	75.60	11.50	14.7	22.2	21.9
Benzol resins	6.40	9.30	40.3	18.3	17.5
Alcohol-benzol resins	16.10	31.70	28.5	45.4	46.4
∑resins	22.50	71.1	68.8	63.7	63.9
Asphaltenes	1.90	17.4	16.5	14.1	14.2
Individual composition of saturated hydrocarbons:					
$\sum_{b.b.t.}$-nC_{20}/$\sum nC_{21}$-$_{e.b.t}$ [a]	2.58	0.21	0.04	0.06	0.08
Maximum n-alkanes	$nC_{15,16}$	$nC_{27,29}$	$nC_{27,29}$	$nC_{27,29}$	$nC_{27,29}$
Isoprenoids/n-alkanes	1.67	0.11	0.01	0.03	0.04
CPI	0.70	3.44	4.77	4.59	4.47
Pr/nC_{17}	61.00	0.74	0.67	1.12	0.83
Ph/nC_{18}	3.41	0.89	1.25	1.37	1.18
Pr+Ph/нC_{17}+нC_{18}	8.25	0.83	0.86	1.25	0.92

[a] b.b.t. temperature of beginning boiling; e.b.t. temperature of end boiling.

Presence of oil hydrocarbons shows features of composition and distribution of saturated hydrocarbons (figure 2). N-alkanes make up 28% of the sum of the identified hydrocarbons and are characterized by prevalence of relatively low-molecular n-alkanes ($\sum_{b.b.t.}$-nC_{20}/$\sum nC_{21}$-$_{e.b.t.}$) with a distribution maximum in the area $nC_{15,16}$. The relation of the sum of even alkanes to odd (CPI) is typical for oil and oil products [20, 21] High content of the isoprenoids prevailing over n-alkanes is detected (table 1). Pristane (Pr) and phitane (Ph) dominate in isoprenoids composition and prevail over a near eluated n-heptadecane (nC_{17}) and n-octadecane (nC_{18}). The latest is characteristic in distribution of the saturated hydrocarbons which underwent influence of processes of biodegradation. According to the received results, as well literate data, the biodegradation processes caused transformation oil hydrocarbons proceed in the Arctic conditions at low temperatures [22, 23].

Studying of the sea sediments samples S-1, S-2 and S-3 from a coastal zone ESS has shown that yield of bitumoid changes in abroad limits from traces to values 193 - 1577 mkg g^{-1} (table 1). The composition of their samples is close to sample R-2, which is characteristic for natural background of the river [24]. In group composition of bitumoids resines components dominate over hydrocarbons and asphaltenes. According literate data and own results a character of IR-spectra of bitumoids is typical for terrigenous organic matter of recent deposits. Compounds with long methylene chains dominate as bitumoids as their fraction (hydrocarbons, resins and asphaltenes) in chemical structure (figure 1). As rule absorption of doublet band in the field of 720-730 cm^{-1} prevails in spectra as absorption bands of aromatic rings practically are absent. Presence in spectra absorption band of high intensity in the field of 1700-1740 cm^{-1} is equal absorption of methylenes chains 1465 cm^{-1}. In this case it indicates a prevalence resinous components over hydrocarbons.

Saturated hydrocarbons are presented generally by alkanes of a normal structure, high-molecular homologs prevail in their composition (table 1, figure 1). A distribution maximum is on $nC_{27,29}$. N-alkanes are characterized by prevalence of homologs with odd number of atoms of carbon in a molecule, CPI values are rather high. The content of isoprenoids is very low it shows the ratio isoprenoids/n-alkanes. Pr and Ph are prevailing in isoprenoids composition.

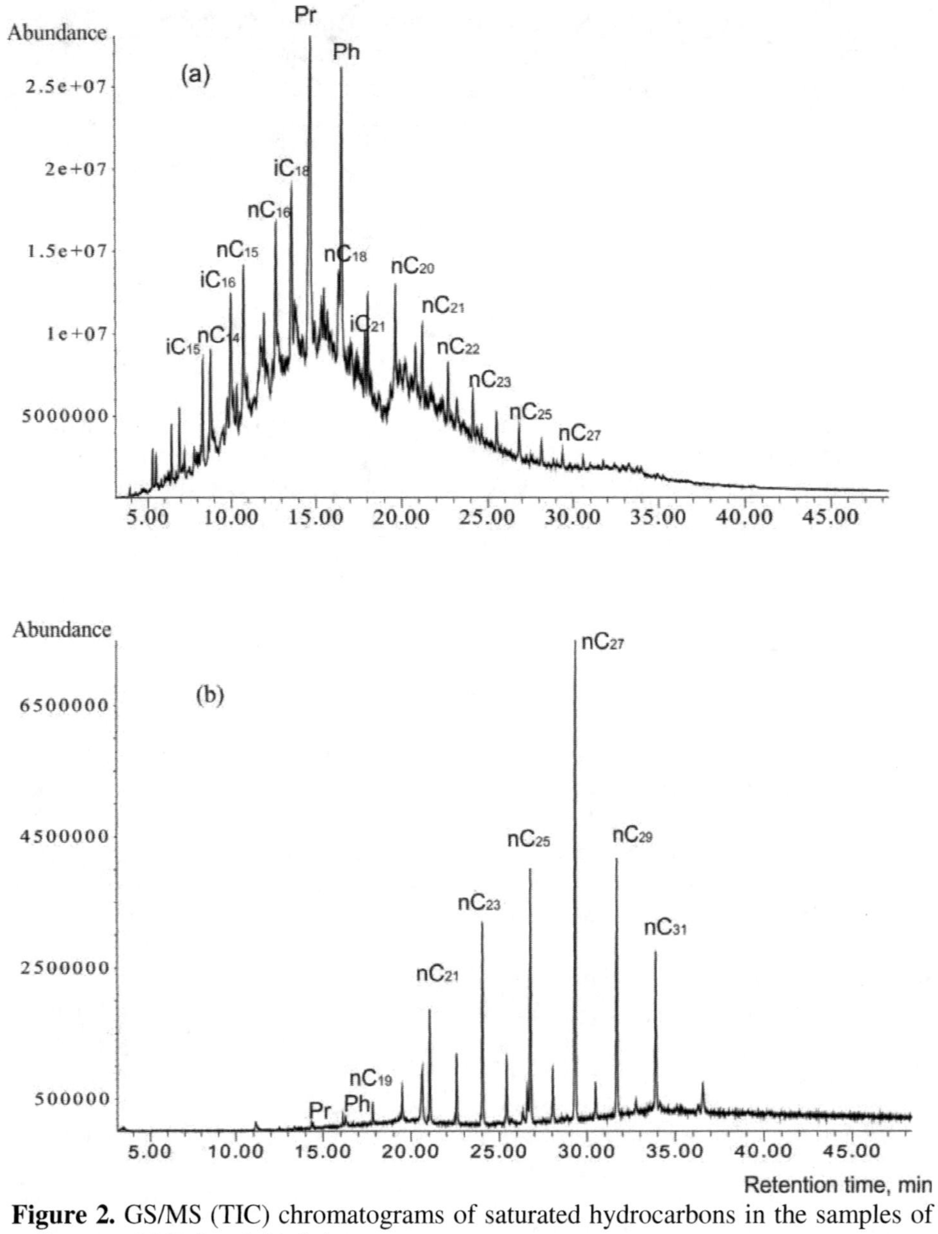

Figure 2. GS/MS (TIC) chromatograms of saturated hydrocarbons in the samples of sediments: (a) R-1 and (b) S-1.

The same structure and features of distribution of individual saturated hydrocarbons for all studied samples of sea bottom sediments find strong relationship with the terrigenous organic matter derived from high land plants [20, 21]. The prevalence of odd hydrocarbons in the range nC_{23}-nC_{35} and high concentrations nC_{27}, nC_{29}, nC_{31} is characteristic for lipids of organic matter of this type.

4. Conclusion

Thus, the conducted researches of sea and river bottom sediments have shown the differences in their genesis established on a complex of geochemical indicators, including on composition of molecules biomarkers.

It is interesting to note that composition and features of chemical structure of bitumoids of bottom sediments of a coastal sea zone, didn't reveal close connection with aquagene organic matter.

Moreover, a genetic relationship with terrigenous organic matter is detected as on a composition of relic hydrocarbons and the character of their distribution, as features of chemical structure of bitumoids. Thus, sea bottom sediments from a coastal zone were formed, mainly, as a result of a drift of organic matter of the highest land plants by river flow. In the studied samples oil contamination traces aren't found.

On the contrary in the river bottom sediments of seaport a presence of oil hydrocarbons is established. Typical for oil contaminated samples very high values of the yield of bitumoids, their chemical composition and other geochemical indicators indicate oil contamination of river bottom sediments.

Acknowledgements
The reported research was funded by Russian Foundation for Basic Research and the government of the region of the Russian Federation, grant No. 18-45-140009 p_a.

References

[1] McClelland J W, Dery S J, Peterson B J, Holmes R M and Wood E F 2006 *Geophysical Research Letters* **33** L06715
[2] Dachs J, Bayona J M, Fillaux J and Albaiges J 1999 *Marine Chemistry* **65** 195–210
[3] Nemirovskaya I A 2004 *Hydrocarbons in the Ocean* (Moscow: Scientific World) p 328
[4] Safronov A F 2015 *Economy of the East of Russia* **2(004)** 21-5
[5] Patin S A 2017 *Oil and Continental Shelf Ecology* Vol 1 (Moscow: Vniro publishing) p 326
[6] AMAP (Arctic Monitoring and Assessment Programme) 2004 (Oslo: AMAP) p 310
[7] Filler D M, Snape I and Barnes D L 2008 *Bioremediation of Petroleum Hydrocarbons in Cold Regions* (Cambridge University Press) p 273
[8] Petrova V I, Batova G I, Kursheva A V, Litvinenko I V, Savinov V M and Savinova T N 2008 *Oceanology* **48** 196-203
[9] Vetrov A A, Semiletov I P, Dudarev O V, Peresypkin V I and Charkin A N 2008 *Geochemistry International* **46** 156-67
[10] Pugach S P and Pipko I I 2012 *Proc. of the Russian Academy of Sciences* **47** 671-74
[11] Karlsson E S, Bruchert V, Tesi T, Charkin A, Dudarev O, Semiletov I and Gustafsson O 2015 *Marine Chemistry* **170** 11–22
[12] Boitsov S, Petrova V, Jensen H K B, Kursheva A, Litvinenko I, Chen Y and Klungsoyr J 2011 *Marine Environmental Research* **71** 357-68
[13] Yunker M B, Macdonald R W, Snowdon L R and Fowler B R 2011*Organic Geochemestry* **42** 1109-46
[14] Uspensky V A, Radchenko O A, Gorskaya A I and Shishkova A P 1975 *Methods of Bituminological Studies* (Leningrad: Nedra) p 123
[15] Kashirtsev V A 2003 *Organic Geochemistry of the Naphtides of the East of Siberian Platform* (Yakutsk: Publisher of the SB RAS) p 160
[16] Jovančićević B, Pavlović I, Vrvić M, Beškoski V, Antić M, Kronimus A and Schwarzbauer J 2008 *Water Air Soil Pollution* **190** 299-307
[17] Brakstad O G, Nonstad I, Faksness L G and Brandvik J 2008 *Microbial Ecology* **55** 540-52
[18] Dahle S, Savinov V, Petrova V, Klungsøyr J, Savinova T, Batova G and Kursheva A 2006 *Norwegian Journal of Geology* **86** 41–50
[19] Wang Z, Fingas M and Page D S 1999 *Journal of Chromatography A* **843** 369-411
[20] Kashirtsev V A, Gayduk V V, Chalaya O N and Zueva I N 2012 *Geology and Geophysics* **53** 1027-39
[21] Zueva I N, Chalaya O N, Kashirtsev V A, Gayduk V V and Safronov A F 2014 *Proc. Cretaceous System of Russia and Neighboring Countries. Problems of a Stratigraphy and Paleogeography* (Vladivostok: Dalnauka) 127-30

[22] Filler D M, Snape I and David L 2008 Barnes *Bioremediation of Petroleum Hydrocarbons in Cold Regions* (Cambridge University Press) p 273
[23] Glyaznetsova Yu S, Zueva I N, Chalaya O N and Lifshits S Kh 2012 *Arktika and North* **5** 97-108
[24] Glyaznetsova Yu S, Zueva I N, Lifshits S Kh and Chalaya O N 2018 *Proc. Conf. Geology and Mineral Resources of the Northeast of Russia* vol 1 (Yakutsk: Publisher Permafrost Institute) pp 279-82

Mechanical properties of cement paste curing at different isothermal conditions

A I Gnyrya[1], Yu A Abzaev[2], S V Korobkov[1] and K S Gauss[1]

[1] Construction Engineering Technology Department, Tomsk State University of Architecture and Building, 2, Solyanaya Sq., 634003, Tomsk, Russian Federation
[2] Higher Mathematics Department, Tomsk State University of Architecture and Building, 2, Solyanaya Sq., 634003, Tomsk, Russian Federation

E-mail: korobkov_1973@mail.ru

Abstract. The paper studies the intensity of strength generation in the concrete paste during the time period of 0–67 hours, at 40, 50 and 70 °C isothermal curing. It is shown that at these temperatures the yield stress increases by one order of magnitude. The elastic properties of the cement paste also significantly increase. The plastic strain range reduces with curing time, and the catastrophic failure of specimens occurs at the end of the investigated temperature range. Stress-strained curves in the plastic range are non-monotonous, with a high density of breaks. The yield stress is the main parameter for evaluating the strength properties of the cement paste.

1. Introduction

In different types of concrete, the main binding material is cement paste (Portland cement). Local stresses under the external load distribute in concretes non-homogeneously. Aggregates in heavy concretes usually exhibit high strength properties. Therefore, strength properties of concrete are determined particularly by heavy fractions which concentrate local stresses in their interfacial regions which substantially exceed the average stress values for concrete. On the contrary, mechanical properties of concretes with a high porosity are determined by elastic properties of the cement paste [1–10].

Concrete hardening is a multifactorial process influenced by the water/cement ratio, temperature, moisture conditions, phase composition, *etc.* As a consequence, the achievement of performance properties is individual for concretes. Although the literature in this field evaluates the strength properties of concretes, there are merely normalized yield values for deformed concretes which correspond to 28 days of curing [4]. According to the authors, this parameter is insufficient for detailed research into the strength properties of cement paste. Also of interest is a study of mechanical properties of hydrated cement paste depending on time and temperature of isothermal curing with a view of the quantitative estimation of changes in the Portland cement yield stress and the intensity of strength generation which develop during curing up to in-service values.

The aim of this work is to investigate mechanical properties of hydrated cement paste depending on time and temperature of isothermal curing. As a result, the quantitative estimation is provided in relation to changes in the yield stress and the strength generation intensity in Portland cement during its curing up to in-service values.

In this paper, we continue research presented in [11-12].

2. Material and model approximation

Portland cement manufactured at Topkinskii cement plant, Kemerovo region, Topki, Russia, was studied in this experiment [13]. The test machine INSTRON-3382 was used to measure the ultimate compressive strength of the cement paste. Tests were carried out on 20×20×20 mm specimens at a room temperature and 1 mm/min deformation rate. The load determination accuracy was 5 N, the maximum load equaled 100 kN. Isothermal tests were conducted in a curing room. The values of the yield stress and the module of elasticity were averaged over those obtained for five dependences.

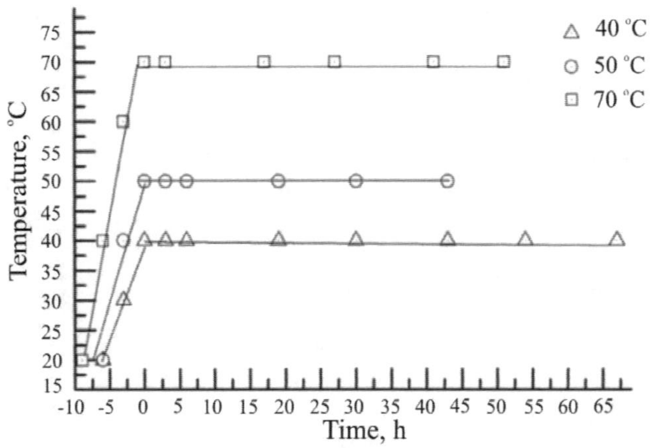

Figure 1. Isothermal curves of cement paste curing at 40, 50 and 70 °C.

Figure 1 plots isothermal curves of cement paste curing. The cement paste was warmed up to 40, 50 and 70 °C for isothermal curing during 3, 5 and 8 hours, respectively. At the next stage, the cement paste was cured in a moist curing room at a constant temperature. As can be seen from Fig. 1, the following time values were selected respectively for 40, 50 and 70 °C isothermal curing: 0, 3, 6, 19, 30, 43, 54, 67; 0, 3, 6, 19, 30, 43 and 0, 3, 17, 27, 41, 51 hours.

A chromel-copel thermocouple with 0.5 mm diameter was used for more accurate temperature measurements (±0.5 °C). Measurements were performed on-line, using a temperature recorder 'Teplograf' equipped with a general-purpose multichannel adapter 'Terem'.

3. Results and discussion

Figure 2 contains plots of the typical stress-strain (σ–ε) dependences which correspond to 40 °C initial curing at 0 h (Fig. 2a) and 67 h (Fig. 2b).

According to Fig. 2b, the dependence is characterized by the initial transition area, elastic strain (long linear) range and plastic strain range.

The elastic strain range is restricted by the linear stress-strain dependence. In the plastic strain range, this dependence is complex. The length of this range depends on the curing time. At the initial stages, the plastic strain range turns to be long and comparable with that of the elastic strain range. A gradual increase in the yield stress up to the strength limit values is normally observed, followed by the long area of the stress decrease. This order of σ–ε dependence behavior is characteristic for the initial stages of curing at all the investigated temperatures. The temperature has an effect only on the quantitative parameters, such as the yield stress, strength and curing intensity on σ–ε dependences. With increasing curing time and temperature, the plastic strain range significantly reduces, and a high density of breaks is normally observed beyond the yield stress. In some cases, the strength limit values coincide with that of the yield stress, *i.e.* the difference between σ–ε curves composes shares of percent of strain. It does not appear feasible to determine how curing temperature and time affect the yield stress in the plastic strain range at advanced stages. The yielding curves show both the jump-like growth and decrease in stresses. Portland cement specimens tend to a catastrophic failure

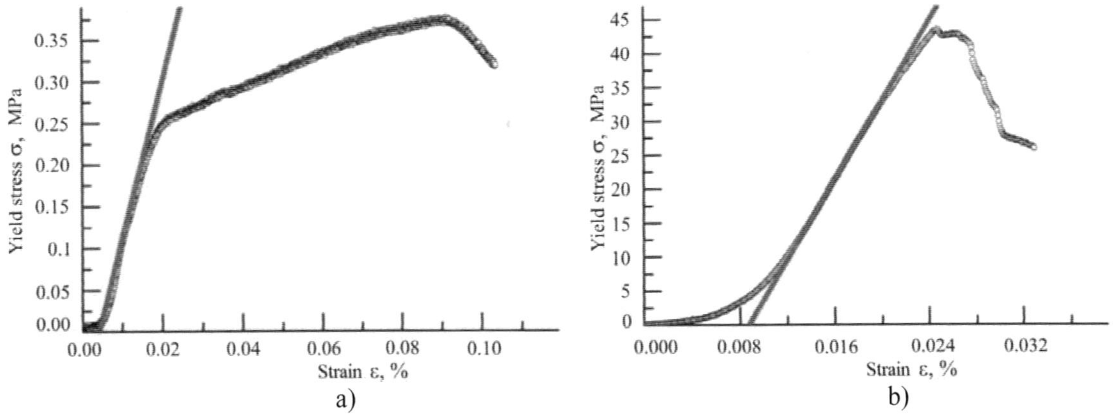

Figure 2. Stress-strain (σ–ε) dependences at 40 °C curing and different curing time: a) – 0 h; b) – 67 h. Red lines indicate approximations of linear curve sections.

with curing time. In the plastic range, they deform individually. During the isothermal warming, the transition area varies a little at the initial curing stages. The increase in isothermal warming and heating time leads to a significant reduction in the transition area on the yielding curves.

The analysis of the elasticity modulus and the yield stress on σ–ε curves shows that the former characterizes the intensity of strength generation within the elastic strain range. The yield stress σ_o shown on σ–ε curves corresponds to the transient stress between the linear and plastic ranges. The red line shown in Fig. 2 indicates the approximation of the linear curve section as

$$\sigma = A + G \cdot \varepsilon, \tag{1}$$

where σ is the stress applied; ε is the strain; *A* is the constant; *G* is the elasticity modulus. The linear correlation coefficient is close to unity (not less than 0.998) for all specimens of interest.

Figures 3–5 present dependences between the yield stress and the elasticity modulus of the cement paste depending on the curing time or the time of strength generation. These dependences are obtained at different curing temperatures.

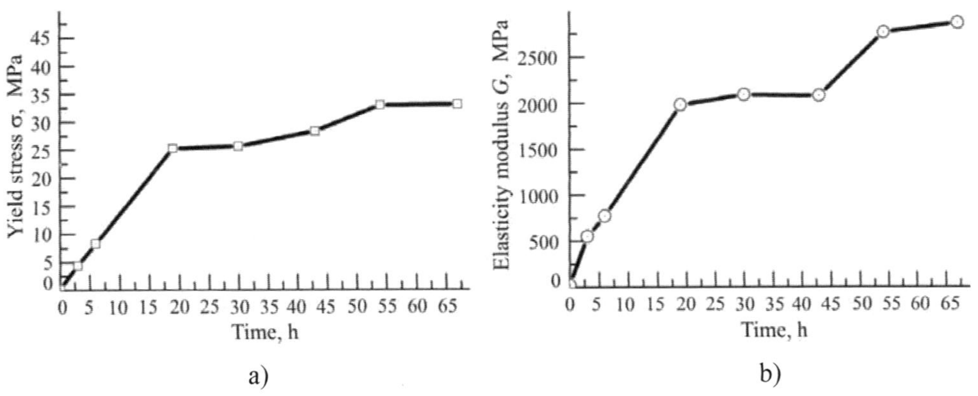

Figure 3. Time dependences for 40 °C curing: a) – yield stress; b) – elasticity modulus.

IOP Conf. Series: Earth and Environmental Science **193** (2018) 012010 doi:10.1088/1755-1315/193/1/012010

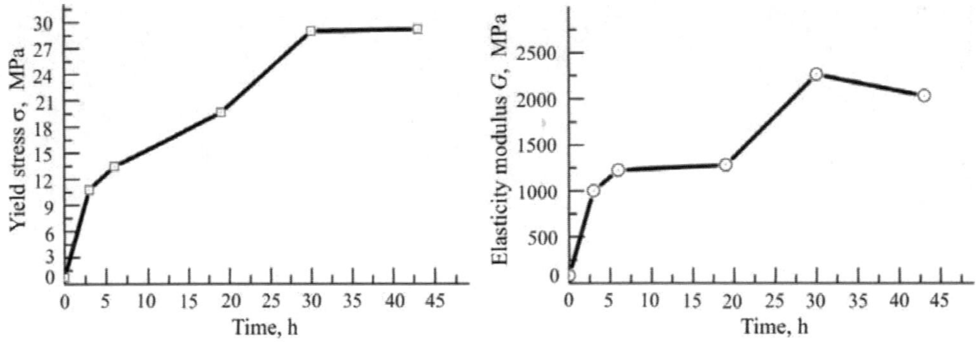

a) b)

Figure 4. Time dependences for 50 °C curing: a) – yield stress; b) – elasticity modulus.

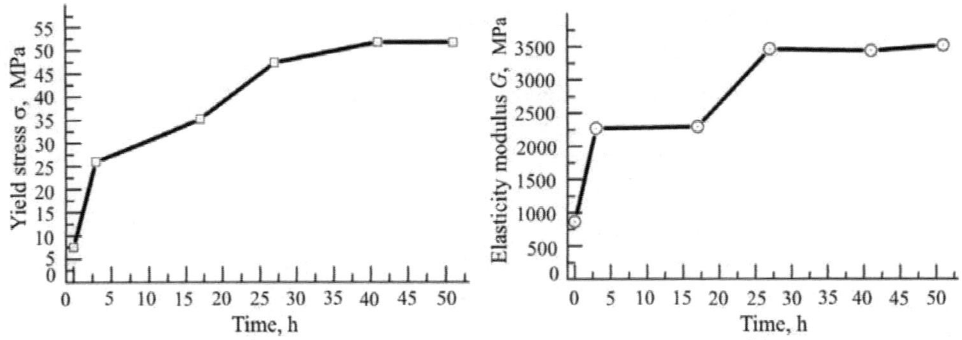

a) b)

Figure 5. Time dependences for 70 °C curing: a) – yield stress; b) – elasticity modulus.

As shown in Figs 3–5, mechanical properties substantially increase with the curing time. Thus, starting from the isothermal warming time point, the yield stress increases by more than one order of magnitude at 40 °C and 50 °C curing (Figs 3–4), respectively during 0–67 h and 0–43 h time periods. At 70 °C (Fig. 5) curing, the yield stress increases by almost one order of magnitude during 0–51 h time period. The elastic modulus also considerably grows. As can be seen from Fig. 6, the warming temperature has a great effect on the intensity of strength generation and the yield stress. This figure contains the temperature dependences for the yield stress and the elasticity modulus during 0, 3, 30 and 43 h curing.

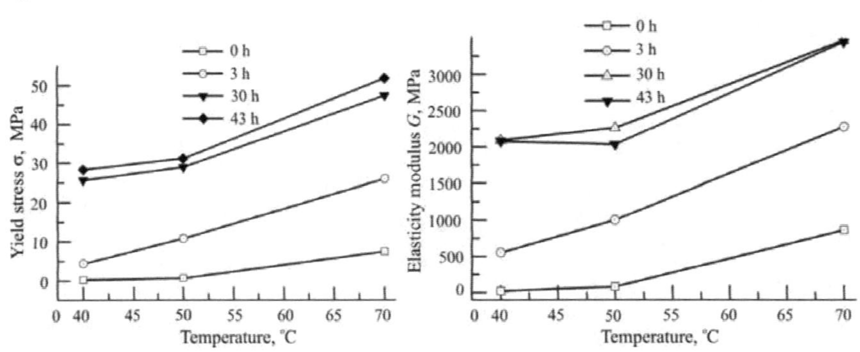

a) b)

Figure 6. Temperature dependences: a) – yield stress; b) – elasticity modulus.

According to Fig. 6, the values of the yield stress and elasticity modulus increase with the temperature growth. During a rather short period of time (not over 67 h), the cement paste generates a greater deal of strength as compared to 28-day curing at a room temperature. The values obtained for the yield stress and the elasticity modulus after 28 days match the design parameters. It is found that the elasticity modulus is ~3630.6 MPa, while the yield stress is 51.75 MPa. Apparently, after 67 h curing strength properties come to not less than 75% of the design strength values. With the temperature growth, this percentage is achieved during considerably shorter time of curing.

4. Conclusions

In conclusion, it is evident that the behaviour of the stress-strain curves includes three stages, namely: transition, elastic and plastic strain ranges. The temperature and time of isothermal curing substantially affected the intensity of strength generation in the plastic strain range. At the end of the isothermal curing, the values of the limit strength and the yield stress almost coincided. Stress-strain curves in the plastic range are non-monotonous, with a high density of breaks, and a catastrophic failure of specimens occurred at the end of the investigated temperature range. Within 0–67 h curing, the yield stress increased almost by an order to magnitude, and the elasticity modulus increased also. It was shown that with the growth in the curing temperature, these parameters also increase, and a more intensive strength generation was observed. The yield stress was found to be the main estimation parameter of the strength properties at different curing temperatures. After 67 h curing the strength properties of Portland cement achieved not less than 75% of the design strength values corresponding to 28-day curing.

Acknowledgement

Authors wish to acknowledge the financial support from the Russian Foundation for Basic Research (Grant N 18-08-01025).

References

[1] Berg O Ya 1962 *Physical Basics of Concrete and Reinforced Concrete Strength Theory* (Moscow: Gosstroyizdat) p 96 (in Russian)
[2] Berg O Ya, Shcherbakov E N and Pisanko G N 1971 *High-Strength Concrete* (Moscow: Stroyizdat) p 208 (in Russian)
[3] Akhverdov I N 1981 *Basics of Concrete Physics* (Moscow: Stroyizdat) p 644 (in Russian)
[4] Akimova T N and Vasil'ev Yu E 2007 *Cemented Concrete* (Moscow: MADI). p 46 (in Russian)
[5] Mironov S A 1948 *Temperature Factor of Concrete Hardening* (Moscow: Stroyizdat) p 46 (in Russian)
[6] Bazhenov Yu M 2007 *Concrete Technology* (Moscow: ASV) p 528 (in Russian)
[7] Nevil' A M 1972 *Concrete Properties* (Moscow: Stroyizdat) p 343 (in Russian)
[8] Pashenko A.A. 1991 *Cement Theory* (Kiev: Budivel'nik) p 166 (in Russian)
[9] Adamtsevich A O, Pashkevich S A and Pustovgar A P 2013 Calorimetry for fore-casting strength growth in cement systems of accelerated hardening *Inzhenerno-stroitel'nyi zhurnal* **3** 36–42 (in Russian)
[10] Stepanova V F 2014 *Durability of Concrete* (Moscow: ASV) p 126 (in Russian)
[11] Gnyrya A, Abzaev Y, Korobkov S, Mokshin D, Gauss K and Boyarintsev A 2017 High-temperature structure formation in cement depending on curing time *AIP Conf. Proc.* **1899** 040005-1–8.
[12] Gnyrya A I, Abzaev Yu A, Korobkov S V, Boyarintsev A A, Mokshin D I and Gauss K S 2017 Influence of time and increased temperature on structural creation of cement stone *Izv. Vyssh. Uchebn. Zaved., Stroit.* **8** 34–45
[13] Standard cements, available: www.sibcem.ru/index.php/production/cement/obshchestroitelnye-tsementy (in Russian)

Design optimization of hydraulic fracturing

B N Starovoitova[1], S V Golovin[1,3], G V Paderin[2], E V Shel[2], E A Kavunnikova[3] and A M Krivtsov[4]

[1] Lavrentyev Institute of Hydrodynamics, Lavrentyev pr. 15, Novosibirsk, Russia
[2] Gazpromneft Science & Technology Centre, Moika River emb. 75–79, St.Petersburg, Russia
[3] Novosibirsk State University, Pirogova str. 1, Novosibirsk, Russia
[4] Peter the Great St.Petersburg Polytechnic University, Polytechnicheskaya str. 29, St.Petersburg, Russia

E-mail: botagoz@hydro.nsc.ru, golovin@hydro.nsc.ru

Abstract. The approach to the problem of hydraulic fracturing design optimization is proposed which combines three modules: a fracture geometry module, an oil production module and an economic module. The approach is based on the genetic algorithm — the universal multivariate and multiobjective optimization method. The optimization process involves subsequent determination of the fracture geometry, calculation of the oil production depending on the number of fractures, and description of the investment–return cash flows. Iterated by the genetic algorithm for different set of input parameters, this approach allows determination of the optimal design of a multiply fractured well for a given set of geological and economical information. The results show that the proposed method is plausible and can be useful for achieving projected objectives.

1. Introduction

One of the most efficient methods for enhancement of oil recovery is the hydraulic fracturing of reservoirs. Modern technologies imply creation of multiply fractured horizontal wells (MFHW) in low-permeable reservoirs. In light of the high cost of this method, the problem of optimization of the MFHW design is of great interest.

It is necessary to distinguish between physical optimization, when the criterion of success is to achieve the maximum production rate, and the economic one — when the maximum production should be reached under certain economic constraints, for example, the limitations to the total costs of the hydraulic fracturing jobs.

The concept of the Unified Fracture Design (UFD) is proposed in [1], where the size of the fracture is determined by the proppant mass pumped into the reservoir. A linear relationship is assumed between the well flow rate and the pressure drawdown, whereas the proportionality coefficient is called the Productivity Index (PI). It was found that for any proppant mass pumped into the reservoir, there is a unique fracture geometry corresponding to the maximum PI. All other combinations of lengths and widths of the fracture would result in lower PI values. This is called the physical optimization [1]. In [2], an advanced model describing the fracture width profile was added to the physical optimization. This addition has improved the accuracy of the net pressure and the fracture geometry calculation.

The procedure linking physical and economic optimization for a gas reservoir is offered in [3]. This optimization algorithm can be summarized as follows: for the fixed proppant mass, authors are looking for the fracture geometry (length and propped width) providing the maximum of PI. Then, by variation of the proppant mass, authors select the fracture giving the maximum of the Net Present Value (NPV). However, it remains unclear, how to determine the fracturing job schedule that would lead to the desired fracturing geometry.

In this article, a different approach is considered: for a given set of fracturing parameters, we use a two-dimensional fracture propagation model for predicting fracture geometry. Then, we couple the geometry with a productivity module to predict the MFHW flow rate. And finally, we use the economic module for computation of NPV. The optimization is performed by the Genetic Algorithm (GA) — Non-dominated Sorting Genetic Algorithm-II (NSGA-II) proposed in [4].

The problem close to our is considered in [5, 6], where authors reduce the multi-objectives optimization problem to single-objective one.

2. The optimization algorithm

Formally, any optimization problem can be stated as: find values of parameters $x_1, \dots, x_n$ to maximize or minimize objective functions

$$(f_1(\mathbf{x}), ..., f_k(\mathbf{x})), \quad \mathbf{x} = (x_1, \dots, x_n), \tag{1}$$

subject to bounds

$$\mathbf{x} : x_{L,i} \leqslant x_i \leqslant x_{U,i}.$$

Here $\mathbf{x}$ represents the vector of free optimization parameters; $x_{L,i}$, $x_{U,i}$ are lower and upper bounds; n is the number of free optimization parameters. We will consider only minimization problem. Maximization of a function $f(\mathbf{x})$ can be achieved by minimizing $-f(\mathbf{x})$.

In the article, the real–coded GA consisting in the direct representation of the actual values of variables is used. The optimization procedure starts with generating the initial population in a random way by using the uniform distribution in each variable. Each set of optimization parameters is named as the individual. The size of the population is N.

It is said that individual $\mathbf{x1}$ is better than $\mathbf{x2}$ for the i–th objective function if $f_i(\mathbf{x1}) < f_i(\mathbf{x2})$. It is said that the individual $\mathbf{x1}$ dominates individual $\mathbf{x2}$ by Pareto, if $\mathbf{x1}$ is not worse than $\mathbf{x2}$ by all objective functions and at least by one criterion is better than $\mathbf{x2}$. On the other hand, if $\mathbf{x1}$ is better than $\mathbf{x2}$ for one criterion and $\mathbf{x2}$ is better than $\mathbf{x1}$ for another one, then it is said that individuals $\mathbf{x1}$ and $\mathbf{x2}$ are non–dominant. The set of non–dominant individuals is called Pareto optimal set or Pareto front.

Since in general case, it is impossible to find $\mathbf{x}$ minimizing all objective functions simultaneously, the solution of multiobjective optimization problem is the Pareto front.

Once the initial population is created the population is sorted on the non-domination criterion. The non–dominant set in the current population is called the first front, whereas individuals in it become the unit rank value. Individuals in the second front being dominated by the individuals of the first front only, are assigned the rank value of 2 and so on. In addition to the rank, a crowding distance [7] is calculated for each individual. The crowding distance is a measure of how close an individual is to its neighbours. Large average crowding distance results in better diversity in the population.

"Parents" are selected from the population by using binary tournament selection on the base of the rank and the crowding distance. An individual is selected if its rank is smaller than the other or if the crowding distance is greater than the other. The selected "parents" generate "children" by the crossingover and mutation operators. In this article, we use the arithmetical crossingover and the mutation of a polynomial type [8].

After the operations of crossingover and mutation, an intermediate population of size $2N$ is created by gathering the "parent" and "child" populations. The joint population is sorted again based on the rank value and the crowding distance, and the new "parent" population is obtained by keeping N best individuals.

3. Problem definition

The problem of hydraulic fracture optimization can be formulated as follows. For a given reservoir one must select a set of fracturing parameters, such as the viscosity of fracturing fluids, the injection rate of the fluid, the injection time and the proppant concentration, such that a hydraulic fracture geometry (propped width, length and permeability) is obtained to provide the design objectives: maximum of cumulative well production (Q_{tot}) and Net Present Value (NPV) at minimum treatment costs (C_{HF}):

$$Q_{tot} \to \max, \quad NPV \to \max, \quad C_{HF} \to \min . \tag{2}$$

In practice, the most valuable customer information is the proppant mass required for the hydraulic fracturing [1, 3], therefore, we choose

(i) M_p (proppant mass for single fracturing) and
(ii) N_f (number of fractures)

as free optimization parameters under the following bound constraints:

(i) $4000\ [\text{kg}] \leq M_p \leq 20000\ [\text{kg}]$,
(ii) $3 \leq N_f \leq 10$.

The injection rate q_i, the proppant concentration C_p (C_p is the volume concentration) and the effective viscosity of the fracturing fluid μ_0 are considered to be given. By the proppant concentration we mean an average value for the whole injection process.

To calculate the objective functions, it is necessary to include the hydraulic fracture model for obtaining fracture geometry, the post fracturing production model and the economical model.

3.1. Fracture geometry

The fracture geometry is defined by length, width and permeability of a fracture. The height of fracture is assumed to be equal the height of the reservoir. Concerning the properties of the reservoir, standard assumptions of the linear theory are made and the proppant slurry is considered as Newtonian fluid with an effective viscosity defined by the rheological parameters. In this paper, the Khristianovich – Girtsma – de Klerk (KGD) model with Carter's equation for fluid leak–off, is used to predict the fracture geometry. In [9] a fast algorithm for calculating the fracture parameters is proposed. In order to apply this algorithm it is necessary to define the injection time t_i and the viscosity of proppant slurry μ_f.

If the injection rate is constant, the time required to pump a given proppant mass with a given concentration of the slurry is calculated as

$$t_i = \frac{M_p}{\rho_p(1 - m_p)q_i C_p}, \tag{3}$$

where ρ_p is the proppant density, and m_p is the proppant pack porosity.

Once the injection time is determined, the total volume of the injected slurry is found: $V_i = q_i \cdot t_i$.

Using the proppant mass and the proppant volume concentration, we obtain the volume of "pure" fluid as

$$V_F = \frac{M_p}{\rho_p(1 - m_p)} \cdot \frac{(1 - C_p)}{C_p} = V_i \cdot (1 - C_p).$$

IOP Conf. Series: Earth and Environmental Science **193** (2018) 012011 doi:10.1088/1755-1315/193/1/012011

The viscosity of the proppant slurry is determined by the viscosity of the carrier fluid and the proppant concentration [10]:

$$\mu_f(C_p) = \mu_0 \Big(1 - \frac{C_p}{C_{p*}}\Big)^{-2.5}, \tag{4}$$

where C_{p*} is the maximum of the proppant concentration. Note, that formula (4) holds at concentrations $C_p \leq 0.5$.

Following [9], we find the fracture parameters: the fracture half-length; the fracture width and the fluid efficiency η. The latter is determined as the ratio of the fracture volume V_{Frac} to the volume of the total injected slurry V_i ($\eta = V_{Frac}/V_i$).

The proppant concentration in the propped fracture is considered to be equal to the maximum value, hence, the propped width of fracture is $w_F = (wC_p)/(\eta C_{p*})$.

3.2. Calculation of the post-fracture production rate

We consider a single-phase fluid flow in a rectangular reservoir Ω with a fractured horizontal well located in the centre of the reservoir. The fractures are assumed to be identical and placed uniformly and symmetrically along the well.

The express assessment of the production rate of the horizontal well with multiple fractures is proposed in [11]. According to [11], the flow rate Q is a sum of inflows of internal Q_{in} and external Q_d fractures:

$$Q_{in} = \frac{2kHL}{b\mu_n(R-l)}\Big(p_p - \frac{p_0}{2}\Big(\frac{1+2a}{1+a}\Big) - \frac{p_z}{2}\Big(\frac{1}{1+a}\Big)\Big), \tag{5}$$

$$Q_d = \frac{4kHl(p_p - p_z)}{b\mu_n(R-l)}, \tag{6}$$

$$Q = Q_{in} + Q_d. \tag{7}$$

Here k is the reservoir permeability, H is the reservoir height, L is the well length, b is the oil volumetric factor, μ_n is the viscosity of oil, R is the drainage radius, l is the fracture half-length, p_p is the reservoir pressure, p_z is the bottomhole pressure and p_0 is some intermediate pressure.

The parameter a is determined as:

$$a = \frac{2(k/k_f)l^2}{wL}(N_f - 1) = \frac{2l(N_f-1)}{LF_{cd}}. \tag{8}$$

Here w is the fracture width, k_f is the permeability of the propped fracture, N_f is the number of fractures, $F_{cd} = (k_f w)/(kl)$ is the dimensionless fracture conductivity.

The intermediate pressure p_0 is defined as:

$$p_0 = \frac{p_p(1+a) - (0.5 - \bar{b})p_z}{0.5 + \bar{b} + a}, \quad \bar{b} = \frac{4(N_f-1)^2 l(R-l)}{L^2}. \tag{9}$$

Considering Q as the initial production rate, we assume that the post–fracture production rate declines exponentially [12]:

$$Q_t = Qe^{-\alpha t}, \tag{10}$$

where t is time, Q_t is the production rate in time t, α is an empirical decline rate.

The cumulative production Q_{tot} to the time t is given by: $Q_{tot} = \int Q_t dt = \int Q_t \exp(-\alpha t)dt = (Q - Q_t)/\alpha$. The cumulative yearly production $N_{p,t}$ for the t–year is $N_{p,t} = (Q_{t-1} - Q_t)/\alpha$.

3.3. Economic criteria

The economic criterion NPV, commonly used for estimating the hydraulic fracturing efficiency, is calculated as follows [3, 13]:

$$NPV = \sum_{t=1}^{T_{max}} \frac{\Pi_t - A_t}{(1+D)^t} - C_{HF}. \tag{11}$$

Here T_{max} is the number of years for which a revenue is calculated, t is the current year, Π_t is the cash inflow at t-th year, A_t is current expenses, D is the discount rate, C_{HF} is the cost of hydraulic fracturing with N_f fractures.

The cash inflow Π_t in the t–th year is obtained as the product of the cumulative oil production $N_{p,t}$ and an average oil price. We propose to estimate the fracturing cost C_{HF} as follows:

$$C_{HF} = N_f(FC + Pr_p \cdot M_p + V_F \cdot Pr_F) + AC, \tag{12}$$

where M_p is the proppant mass [kg], Pr_p is the proppant price [\$/kg], Pr_F is the fluid price [\$/m^3], V_F is the volume of the fluid [m^3], N_f is the number of fractures, FC is the cost of equipment, e.g. pumping [\$], AC is the fixed and miscellaneous costs [\$].

Current expenses A_t include operational costs, taxes, costs of oil treatment and transportation. The treatment and transportation costs are proportional to the volume of the oil produced and can be included into the oil price, the taxes are proportional to the revenue of the t-th year, operational costs do not depend on the volume of the oil produced, so we consider them to be constant.

4. Examples and results

A horizontal well located in the centre of the oil reservoir is used to illustrate an application of the proposed approach. Multi–objective optimization problem for three objective functions

$$-Q_{tot} \to \min, -NPV \to \min, C_{HF} \to \min$$

is solved. We use a population of size 100 and run NSGA II for the 100 generations. The parameters characterizing the reservoir and wellbore properties are presented in Table 1, which also contains fracture mechanics data for obtaining the fracture geometry. The fracture permeability is assumed to be determined only by the proppant pack. The economic data are taken as follows : the oil price is 80 [\$/m^3]; the carrier fluid price is 80 [\$/m^3]; the proppant price is 2.2 [\$/kg]; the discount rate D is 0.2; taxes are 0.24; fixed costs are AC=10000 [\$]; equipment costs are FC = 1000 [\$]. The oil price includes the oil treatment and transportation costs. We calculate the NPV and the cumulative production for 3 and 5 years and for various reservoir permeability k = 1 mD and k = 10 mD. It is clear that the NPV depends on the cumulative production but the dependence is non-linear.

Figure 1 shows Pareto front of 100-th generation for reservoir permeability k = 1 mD. It represents the dependencies of NPV and cumulative production on fracturing costs.

As one can see, the maximum of NPV does not coincide with the maximum of cumulative production. Figure 2 shows parameters of fracturing design providing the NPV values close to the maximum and the total inflow values close to the maximum. In order to receive high NPV, the number of fractures from 6 to 9 and lengths up to 100m should be taken. The larger the number of fractures one takes, the smaller should be their length. On the contrary, to obtain the maximum of the cumulative production, one should take the fractures with lengths from 120 to 140 m independently on the number of fractures. Pareto fronts for various T_{max} and reservoir permeability k = 1mD are presented in Figure 3. It can be seen that the NPV values do not differ much, but the total inflows do.

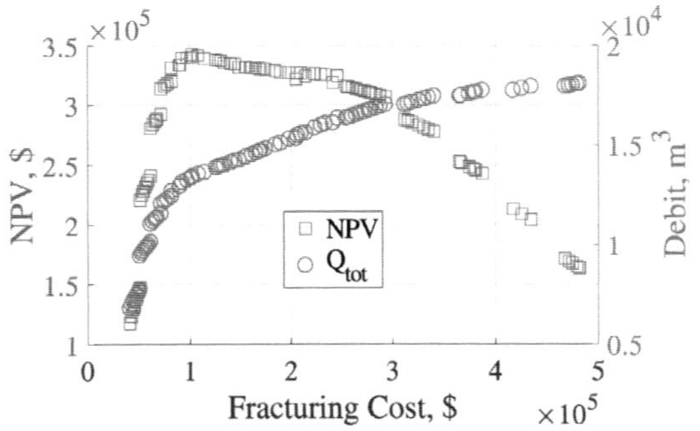

Figure 1. Pareto front for reservoir permeability $k = 1$ mD.

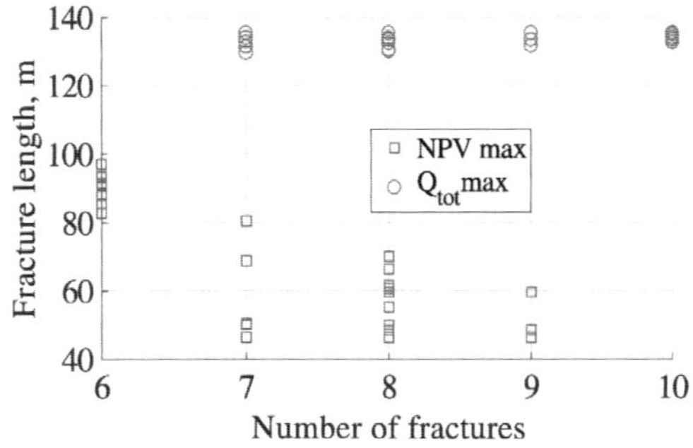

Figure 2. Parameters plane.

Table 1. Reservoir and well parameters.

Parameter	Value	Parameter	Value
reservoir height	$H = 20$ [m]	oil volumetric factor	b =1.189
oil viscosity	$\mu_n = 2 \cdot 10^{-3}$ [Pa s]	drainage radius	R =400[m]
fracture permeability	$k_f = 30$ [D]	bottomhole pressure	$p_z = 5$ [MPa]
well length	L =600 [m]	reservoir pressure	$p_p = 15$[MPa]
annual decline rate	$\alpha_a = 0.587$[year^{-1}]	Poisson ratio	ν =0.25
leak-off coefficient	$C_l = 10^{-5}$ [m/s$^{0.5}$]	Young's modulus	$E = 3.5 \cdot 10^{10}$[Pa]
fracture toughness	$K_I = 3 \cdot 10^6$ [Pa m$^{0.5}$]	proppant density	$\rho_p = 2645$ [kg/m^3]
proppant concentration	C_p=0.3	injection rate	$q_i = 1$ [m^3/s]
maximum proppant concentration	C_{p*}=0.65	fracturing fluid viscosity	$\mu_0 = 0.1$ [Pa s]

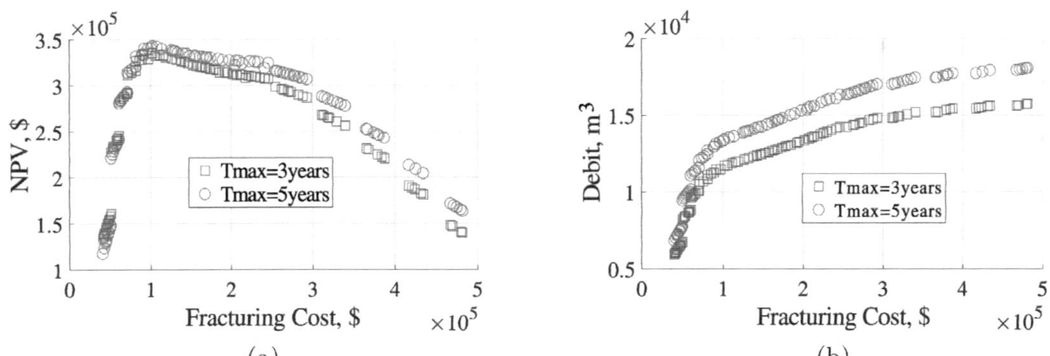

(a) (b)

Figure 3. Pareto fronts NPV (a) and total inflow (b) for various values of T_{max} and $k = 1$mD.

Pareto front for the reservoir with higher permeability $k = 10$mD is shown in Figure 4. For this case, the dependence of NPV on cumulative production is almost linear. It means that the fracture geometry providing the maximum of NPV at the same time provides the maximum of cumulative production. Pareto fronts for various T_{max} are presented in Figure 5.

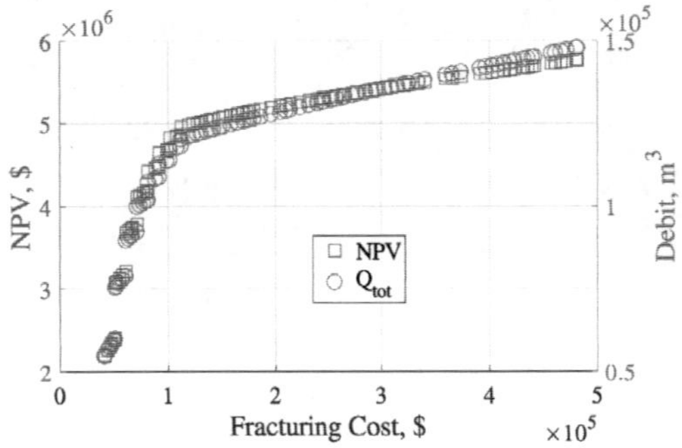

Figure 4. Pareto front for the reservoir permeability $k = 10$mD.

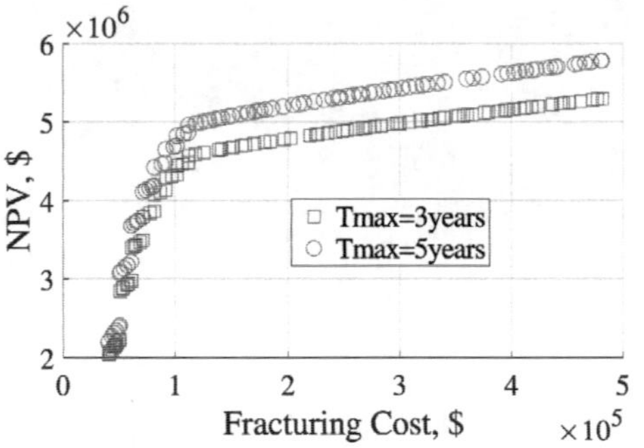

Figure 5. Pareto fronts for various values T_{max} and $k = 10$mD.

5. Conclusion

The approach to determine an optimization design of a multiply-fractured horizontal well is proposed. The numerical tests demonstrate reasonable results on a synthetic set of variables close to real. Although the results are qualitative, they are useful for hydraulic fracturing design and for choosing wells best suited for the fracturing.

Acknowledgments

This work was supported by Ministry of Education and Science of the Russian Federation within the framework of the Federal Program "Research and development in priority areas for the development of the scientific and technological complex of Russia for 2014 – 2020" (activity 1.2), grant No. 14.575.21.0146 of September 26, 2017, unique identifier: RFMEFI57517X0146. Authors are grateful to the Industrial partner of the project: Gazpromneft Science & Technology Centre for providing the data used in numerical experiments.

References

[1] Economides M J, Oligney R E and Valko P 2002 *Unified Fracture Design: Bridging the Gap Between Theory and Practice* (Texas: Orsa Press Alvin)
[2] Paderin G V 2016 *Proc. SPE Russian Petroleum Technology Conf. Exhib., 24-26 October, Moscow, Russia* (SPE)
[3] Marongiu-Porcu M, Economides M J and Holditch S A 2013 *J. Nat. Gas Sci. Eng.* **14** 91–107
[4] Deb K, Agrawal S, Pratap A and Meyarivan T 2002 *IEEE Trans. Evol. Comput.* **6** 182–197
[5] Rahman M M, Rahman M K and Rahman S S 2001 *J. Petrol. Sci. Eng.* **31** 41–62
[6] Rahman M M, Rahman M K and Rahman S S 2003 *Pet. Sci. Technol.* **21** 1721–58
[7] Goldberg D E 1989 *Genetic Algorithms in Search, Optimization and Machine Learning* (Boston: Addison–Wesley)
[8] Beyer H G and Deb K 2001 *IEEE Trans. Evol. Comput.* **5** 250–270
[9] Dontsov E V 2017 *Int. J. Fracture* **205** 231–237
[10] Mueller S, Llewellin E W and Mader H M 2010 *Proc. R. Soc. Lond.* A **466** 1201–28
[11] Elkin S V, Aleroev A A, Veremko N A and Chertenkov M V 2016 *Neft. Khoz. – Oil Ind. J.(in Russ.)* **12** 110–113
[12] Belyadi H, Fathi E and Belyadi F 2016 *Hydraulic Fracturing in Unconventional Reservoirs: Theories, Operations, and Economic Analysis* (Elsevier Science) pp 305–23
[13] Vinogradova I A 2004 *Neft. Khoz. – Oil Ind. J.(in Russ.)* **4** 56–58

IOP Conf. Series: Earth and Environmental Science **193** (2018) 012012 doi:10.1088/1755-1315/193/1/012012

Model of ultrasonic cavitation depolymerization of oil for evaluation of optimum modes and conditions providing reducing viscosity at low temperatures

R N Golykh[1]

[1]Biysk technological institute (branch) of Altai State Technical University named after I.I. Polzunov

Abstract. The paper presents the model of ultrasonic cavitation depolymerization which is destruction of oil macromolecules under cavitation. The model allows to evaluate optimum modes and conditions of ultrasonic influence, providing reducing oil viscosity at low temperatures. The model is based on probability approach which allows to simplify the task of cavitation depolymerization calculation in order to solve the system of self-similar equations for macromolecules with different weights concentration. The equations include probability of macromolecules destruction. The probability is determined by cross-section of collision of macromolecules pair and possibility of macromolecule destruction at contact with another molecule. The calculations have shown that the cavitation reduces the modal molecular weight up to 2 times. It causes reducing the oil viscosity up to 4…10 times. This has great actuality for oil transferring at low temperatures.

1. Introduction

The transferring of oil at North, Antarctic (feasibly in future) and tundra conditions represents big problem due to very high viscosity (up to 1 Pa·s). The high viscosity makes oil extraction from bowels of Earth and transferring through pipes impossible. The problem requires preliminary reducing the oil viscosity. One of promising ways for reducing oil viscosity is ultrasonic cavitation depolymerization. The ultrasonic action in the cavitation mode allows us to concentrate the energy of the oscillations propagating in the medium into the energy of cavitation bubbles in which local temperature rises up to 5000 K, and when they explode, shock waves with pressure amplitude up to 1000 MPa occur. This leads to the destruction of molecular bonds, ensures the homogeneity of the structure of a oil, the lower weight of macromolecules, and can reduce the viscosity [1, 2]. Purposeful obtaining of an oil with the necessary viscosity and productivity of ultrasonic treatment of oil for a particular application requires selection of ultrasonic cavitation influence modes and conditions. While the effect of cavitation on the structure of oil macromolecules has not been studied in detail until today. In connection with this, a model of cavitation effect on the molecular structure of a polymer should be developed. The model will make it possible to reveal the dependence of the change in the structure of molecules and the oil viscosity on the modes, conditions and exposition time of cavitation influence.

The developed model is described in following section.

2. The postulates of oil cavitation depolymerization model

Oil is the liquid medium consisting of molecular chains of monomers connected by hydrogen or carbon bonds. Cavitation leads to the destruction of molecular chains (cavitation depolymerization). At the same time, the fractional composition of molecular chains by the concentration of molecules with different amounts of monomeric units determines the rheological properties of the oil. Thus, the proposed model is aimed to determine fractional composition of molecular chains and rheological properties during cavitation depolymerization in dependency on starting viscosity, influence intensity and time. The proposed model of cavitation depolymerization based on following physical mechanism of macromolecules destruction under cavitation. The shock waves generated by cavitation cause increasing the dispersion of velocity of macromolecules, because around cavitation bubble the pressure rises sharply. Due to this the amount of molecules pairs with relative velocity more than the critical one (at the critical velocity one of the molecules is destroyed at collision) increases.

The assumptions used for the model development are as follows:

1. The macromolecule has a linear structure consisting of serially connected monomer units through equal distances r_e (the model of a Gaussian sub chain or "spring beads").
2. The distribution of macromolecules over the velocities of the centers of mass obeys a Gaussian law (the Maxwell's distribution).
3. The pairing of collisions of macromolecules (the probability of simultaneous collision of 3 or more macromolecules is negligible).
4. For each elementary act of collision of macromolecules, it is possible to destroy an individual molecule (with the number of monomer units i) by only two particles with the numbers of monomer units M and i–M.
5. The disruption of the macromolecule occurs only at the point of its contact with another molecule during the collision.
6. The probability of rupture of a macromolecule depends weakly on the position of its point of contact with another molecule, since the number of monomeric units $i >> 1$.
7. The cavitation bubble size is much more than single macromolecule size.

The model is based on the presented assumptions. The model has four stages presented as follows:

1. The determination of characteristic of cavitation area depending on modes and conditions of ultrasonic influence. Results of the stage are n_{bub} (concentration of cavitation bubbles, m^{-3}), $R(t)$ (bubble radius, m), $P(t)$ (shock wave pressure near the bubble, Pa).
2. The determination of probability of macromolecules breakup β_{ij} at collision of macromolecules pair with number of monomeric units i and j respectively.
3. Calculation of evolution of fraction composition of molecules c_i (c_i is the numeric concentration of molecules with number of monomeric units i).
4. The calculation of oil viscosity dependent on fraction composition of molecules.

Each stage allow to obtain own results which is input data for next stage. The first stage is for determination of characteristic of cavitation area dependent on US oscillations intensity and polymer viscosity at the current time moment. The last stage allows to determine the polymer viscosity of fraction composition of molecules and finally to determine the polymer viscosity dependency on time and intensity influence.

For the 1st stage (determining of characteristics of cavitation area) the sub model of cavitation bubbles formation was presented in [1, 3].

The sub model is based on Gilmor's equation of cavitation bubble dynamics [3] and the probability models of cavitation bubbles coalescence and breakup which allow determining cavitation bubbles concentration according to the expression

$$n_\infty = \frac{2(j-1)}{25\pi R_{MAX}{}^2 i\left|\mathbf{d}_{12}(T)-\mathbf{d}_{12}(0)\right|}, \tag{1}$$

j is the number of cavitation nuclei formed after single bubble collapse [3]; R_{MAX} is the maximal cavitation bubble radius, m; i is number of periods of oscillations required for single expansion-collapse of bubble cycle; T is period of US oscillations, s; $\mathbf{d}_{12}$ is the vector of distance between centers of bubble pair, m.

The forms and locations of advanced cavitation area (green zones in Fig. 1)are determined by modified wave equation taking into account influence of cavitation bubbles on acoustic properties of liquid [1]:

$$\Delta\left(\sqrt{2\rho c I(\mathbf{x})}e^{i\varphi(\mathbf{x})}\right)+\frac{\omega^2}{c_0^{\ 2}}\left(1-\frac{\rho_0 c_0^{\ 2}\,\overline{\delta_1}\left(\sqrt{2\rho c I(\mathbf{x})}e^{i\varphi(\mathbf{x})}\right)}{\sqrt{2\rho c I(\mathbf{x})}e^{i\varphi(\mathbf{x})}}\right)\sqrt{2\rho c I(\mathbf{x})}e^{i\varphi(\mathbf{x})}=0\,, \qquad (2)$$

I is the intensity of ultrasonic vibrations, W/m^2; φ is the phase shift of vibrations of acoustic pressure in liquid; ω is the circular frequency of the initial ultrasonic field, s^{-1}; c_0 is the acoustic speed in a continuous liquid, m/s; ρ_0 is the steady-state density of continuous liquid, kg/m^3; ρ is the density of cavitating liquid, kg/m^3; c is the acoustic speed in cavitating liquid, m/s; $\overline{\delta_1}\left(\sqrt{2\rho c I(\mathbf{x})}e^{i\varphi(\mathbf{r})}\right)$ is the complex amplitude of the change of volume content of the cavitation bubbles relative to mean value, $\mathbf{x}$ is the radius-vector of liquid, m.

The color distributions of cavitation zones at different conditions are presented in Fig. 1.

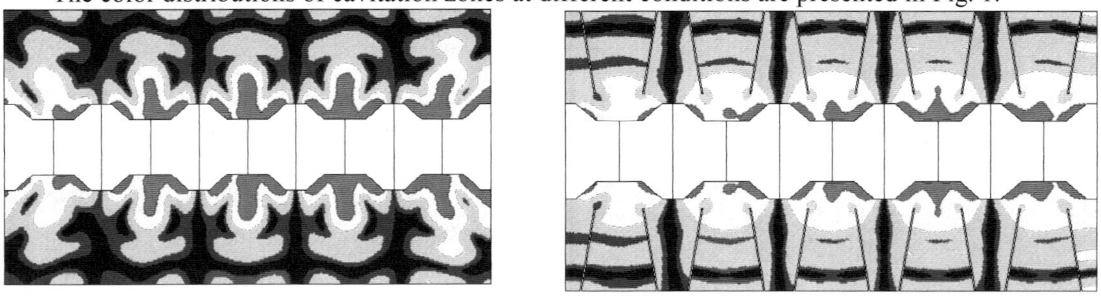

a) distributions of cavitation zones in cylindrical volume without reflectors

b) distributions of cavitation zones in cylindrical volume with reflectors (the oil flow is perpendicular to cylinder axis)

Figure 1. Color distributions of cavitation zones.

As follows from the Fig. 1, the cavitation zones distribution at circular plate reflectors is more uniform than without reflectors.

Thus, the optimum conditions of ultrasonic influence are presence of circular plate reflectors because these provide uniformly of oil viscosity reducing.

The sub model allowing to determine radius of bubble, concentration of bubbles and shock wave pressure. Founded values are used for determination of macromolecules breakup probability and fraction composition evolution. The sub model for determining fraction composition evolution is described in following section.

3. The postulates of the oil cavitation depolymerization model

The sub model for determining oil macromolecules fraction composition evolution is based on probability approach. According to the approach the intensity of molecules breakup is characterized by breakup probability $\beta_{ij,k,M}$ at unity concentration of molecules with weights (numbers of monomer units) i and j per unit of time. $\beta_{ij,k,M}$ is the probability of breakup of molecule with weight i (i-weight molecule) on two molecules having weights k and i–k. The breakup occurs at collision between i-weight and j-weight molecules on points separating i-weight molecule on k-weight and (i–k)-weight molecule, j-weight molecule on M and (j-M) molecule.

The obtained concentrations evolution equation (3) including the probability is

$$\frac{\partial c_i}{\partial t}=\frac{1}{2}\sum_{j>i,\,k>i}\left(\sum_{M_2=1}^{k-1}\beta_{jk,i,M_2}+\sum_{M_2=1}^{k-1}\beta_{jk,j-i,M_2}+\sum_{M_1=1}^{j-1}\beta_{kj,i,M_1}+\sum_{M_1=1}^{j-1}\beta_{kj,k-i,M_1}\right)c_jc_k+$$
$$+\sum_{j>i,\,k\le i,}\left(\sum_{M_2=1}^{k-1}\beta_{jk,i,M_2}+\sum_{M_2=1}^{k-1}\beta_{jk,j-i,M_2}\right)c_jc_k-\sum_{l,M_1=1\ldots i-1,M_2=1\ldots l-1}\beta_{il,M_1,M_2}c_ic_l \quad , \qquad (3)$$

where c_i is number concentration of i-weight molecule, mol/m^3.

At the large average molecular weight (much more than 1) the $\beta_{ij,k,M}$ is independent on k and M (assumption 6 in section 2) and equals to β_{ij}. According to it the equation (3) is reduces to (4) which likes a Smoluchovski equation [4]:

$$\frac{\partial c_i}{\partial t}=\sum_{j>i,\,k>i}(k+j-2)\beta_{jk}c_jc_k+2\sum_{j>i,\,k\le i}(k-1)\beta_{jk}c_jc_k-\sum_{l}(i-1)(l-1)\beta_{il}c_ic_l\,. \qquad (4)$$

The probability included in equation (4) is determined by the analysis of pair molecules collision, presented in following section.

4. The determining of probability of macromolecules destruction

The molecules breakup occurs after the molecules collision when the relative molecules speed exceed critical value v_{min}. The process is schematically presented in Fig. 2.

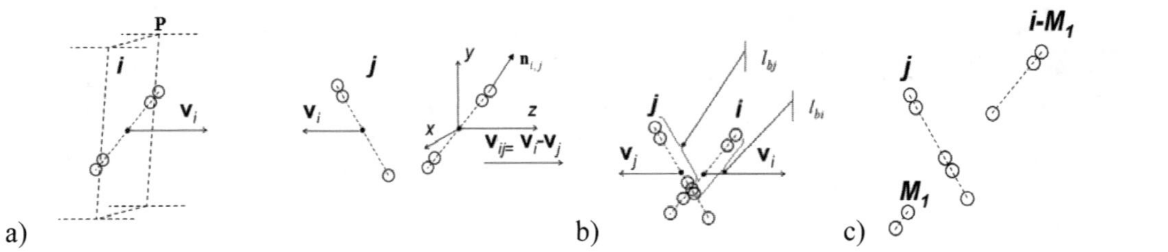

Figure 2. Schematic diagram of macromolecules breakup after collision.

Each macromolecule participating in collision characterized by velocity $\mathbf{v}_{i,j}$, number of monomeric units i, j and orientation vector $\mathbf{n}_{i,j}$.

The determining of probability of macromolecules destruction is proportional to intersection during relative motion of macromolecules probability and to i-weight macromolecule breakup at collision. According to the accepted linear model of macromolecule ("spring-beads"), the macromolecules collision cross-section is like a parallelogram (Fig. 2a).

The square of collision cross-section (5) is

$$S=\frac{1}{2}N_iN_jr_e^2\left|\left(\mathbf{n}_i-\frac{(\mathbf{n}_i,\mathbf{v}_{ij})}{|\mathbf{v}_{ij}|^2}\mathbf{v}_{ij}\right)\times\left(\mathbf{n}_j-\frac{(\mathbf{n}_j,\mathbf{v}_{ij})}{|\mathbf{v}_{ij}|^2}\mathbf{v}_{ij}\right)\right|, \qquad (5)$$

where $\mathbf{n}_i$ is orientation vector of i-th molecule; $\mathbf{n}_j$ is orientation vector of j-th molecule; N_i is number of monomeric units in i-th molecule; N_j is number of monomeric units in j-th molecule; r_e is equilibrium distance between monomeric units, m.

According to the Sommerfeld model [5-7] and due to the expression for collision cross-section square the expression for macromolecules breakup probability is as follows:

$$P\left(C_{N_i,\mathbf{v}_i,\mathbf{n}_i},C_{N_j,\mathbf{v}_j,\mathbf{n}_j},\left|\mathbf{v}_i-\mathbf{v}_j\right|,d\mathbf{v}_i,d\mathbf{v}_j,d\Omega\left(\mathbf{n}_i\right)d\Omega\left(\mathbf{n}_j\right),\Delta V\Delta t\right)=$$
$$=H\left(\left|\mathbf{v}_i-\mathbf{v}_j\right|-v_{\min}\left(\mathbf{n}_i,N_i,\mathbf{n}_j,N_j\right)\right)C_{N_i,\mathbf{v}_i,\mathbf{n}_i}C_{N_j,\mathbf{v}_j,\mathbf{n}_j}\times \tag{6}$$
$$\times\Delta V\Delta t\left|\mathbf{v}_i-\mathbf{v}_j\right|Sd\mathbf{v}_i d\mathbf{v}_j d\Omega\left(\mathbf{n}_i\right)d\Omega\left(\mathbf{n}_j\right),$$

P is number of molecules pairs destructed in volume ΔV during time Δt; $C_{N,\mathbf{v},\mathbf{n}}\Delta V d\mathbf{v} d\Omega(\mathbf{n})$ is number of macromolecules in volume ΔV having velocities from range $[v_x;v_x+dv_x]\times[v_y;v_y+dv_y]\times[v_z;v_z+dv_z]$, orientations with the solid angle $d\Omega(\mathbf{n})$ and number of monomeric units N; $H(v)$ is Heaviside's function which equals 1, if $v \geq 0$, and 0 if another case; $\mathbf{n}_i$ is i-th macromolecule orientation vector; $\mathbf{n}_j$ is j-th macromolecule orientation vector; N_i is number of monomeric units in *i*-th macromolecule; N_j is number of monomeric units in j-th macromolecule; $v_{\min}(\mathbf{n}_i,N_i,\mathbf{n}_j,N_j)$ is minimal (critical) velocity necessary for breakup of i-th macromolecule when it collides with j-th macromolecule, m/s.

According to the made assumptions, the distribution by velocities of centers of molecules masses obeys Maxwell's law and the distribution by macromolecules orientation is uniform (it is assumed, that oil is isotropic liquid). Therefore the probability of molecules collision and breakup was integrated by each velocities components and orientations.

$$\beta_{ijmicro}c_i c_j \Delta V\Delta t=\frac{N_A}{64\pi^3}\Delta V\Delta t\cdot\int\limits_{\|\mathbf{n}_1\|=1}\int\limits_{\|\mathbf{n}_2\|=1}\int\limits_{R^3}\int\limits_{R^3}Pd\mathbf{v}_1 d\mathbf{v}_2 d\Omega(\mathbf{n}_1)d\Omega(\mathbf{n}_2), \tag{7}$$

where i, j are weights of considered macromolecules ensembles; $\mathbf{v}_1$ is macromolecules velocity with weight i, m/s; $\mathbf{v}_2$ is macromolecules velocity with weight j, m/s; $\mathbf{n}_1$ is orientation vector of macromolecule with weight i, m/s; $\mathbf{n}_2$ – is orientation vector of macromolecule with weight j, m/s; N_A is Avogadro number, $6{,}02\cdot 10^{23}$ mol^{-1}.

Because the during cavitation ensemble pressure field $p(t,\mathbf{r})$ in oil is non-uniform and non-constant, the probability is averaged by ultrasonic oscillations period T and local volume ΔV with size much more than cavitation bubble diameter but much less than ultrasonic wave length:

$$\beta_{ij}=\left\langle\beta_{ijmicro}\right\rangle=\frac{1}{T\Delta V}\int\limits_0^T\int\limits_{\Delta V}\beta_{ijmicro}(p(t+t_1,\mathbf{r}+\mathbf{r}_1))dV(\mathbf{r}_1)dt_1\,. \tag{8}$$

The expression (8) gives the specific probability of macromolecules collision and breakup included in equation (4).

The final expression of the probability is following:

$$\beta_{ij}=N_A\frac{n_{bub}}{T}\int\limits_0^T\int\limits_{R(t)}^{\infty}\int\limits_{-1}^{1}\int\limits_0^{\pi}\int\limits_0^{\pi}8r^2r_e^2\sin^2\theta_i\sin^2\theta_j\times$$
$$\times\frac{\sqrt{2\delta_i^2+2\delta_j^2}\,e^{-\frac{v_{\min}^2(i,j,\theta_i,\theta_j,\alpha)}{2\delta_i^2+2\delta_j^2}}\left[\frac{v_{\min}^2\left(i,j,\theta_i,\theta_j,\alpha\right)}{2\delta_i^2+2\delta_j^2}+1\right]}{\sqrt{\pi}}\partial\theta_i\partial\theta_j\partial\alpha\partial r\partial t, \tag{9}$$

$$\alpha=\cos\left(\phi_i-\phi_j\right),\ \delta_i=C_i p\left(r\right),$$

where C_i is proportionality coefficient determined by thermodynamically conditions in the polymer liquid, m/s; r_e is equilibrium distance between monomer units, m; N_A is number of Avogadro, mol^{-1};

n_{bub} is cavitation bubbles concentration, m^{-3}; v_{min} is minimal relative macromolecules pair velocity necessary for macromolecules breakup, m/s; $P(t)$ is the shock wave pressure, Pa; $R(t)$ is the radius, m.

The minimal relative macromolecules pair velocity v_{min} obeys energy conservation law (sum of molecules kinetic energy and hydrogen bonds potential energy is constant) and can be determined empirically by indirect methods (approximation of viscosity dependence of time at fixed cavitation energy). In the energy conservation laws the Morse potential function of monomeric unit interaction $V(r) = D_\sigma\left(1-e^{-a(r-r_e)}\right)^2$ is used [8].

The substitution of probability (9) in equation (8) allows to calculate macromolecules fraction composition evolution. Results of calculation are presented in the next section.

5. The calculating of macromolecules fraction composition evolution

The calculated mass fractions of oil molecules with different monomer units are presented in Fig. 3, 4. In each Figure the model oscillations frequency is 22 kHz.

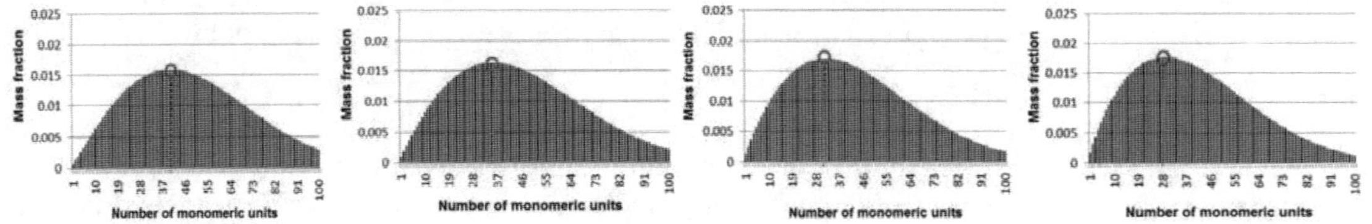

a) 0 s b) 100 s c) 200 s d) 300 s

Figure 3. Macromolecules dispersed composition at different times (intensity of ultrasonic influence 25 W/cm^2).

According to the presented histograms, at the intensity 10 W/cm^2 the modal number of monomer units is reduces up to 40%. It is enough to reduce viscosity up to more than 3.5 times.

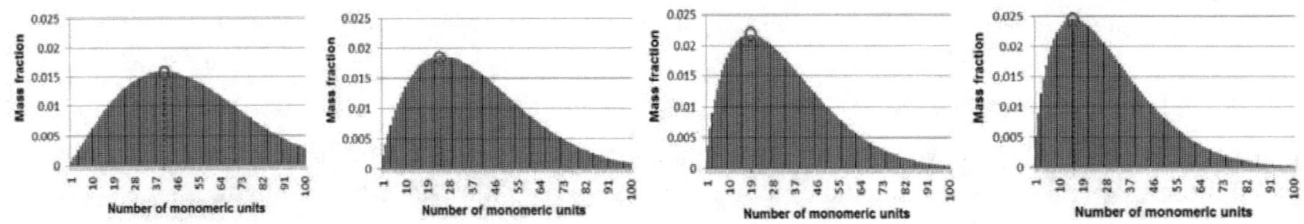

a) 0 s b) 100 s c) 200 s d) 300 s

Figure 4. Macromolecules dispersed composition at different times (intensity of ultrasonic influence 25 W/cm^2).

According to the distributions presented in Fig. 4, higher intensity causes the faster maximum shift. For example, through 300 s, the modal number of monomer units reduces up to more than 2 times. It causes viscosity reducing up to more than 10 times. In the following section the results of calculation of oil viscosity changing were presented.

6. The calculating of oil viscosity evolution

The oil like a polymeric liquid, because viscosity is calculated by empirical law, obtained by Rouse P.E., Khokhlov A.R., Pyshnograi G.V. [9–10]:

$$\eta(t) = \eta(0)\frac{\sum_{i=1}^{\infty} c_i^{2\delta+1}(t)\, i^{3\delta+2}}{\sum_{i=1}^{\infty} c_i^{2\delta+1}(0)\, i^{3\delta+2}}, \tag{10}$$

t is time, s; δ is empirical parameters which equals to 2.4 for most part of polymers.

According to expression (9) and equation (6) for concentrations c_i, the dependences for viscosity on time were obtained (Fig. 5).

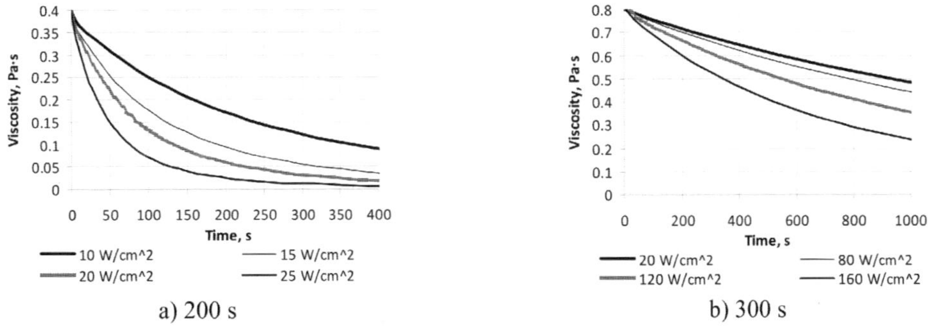

Figure 5. Macromolecules dispersed composition at different times (intensity of ultrasonic influence 10 W/cm^2).

According to the dependences, the ultrasonic with intensity 25 W/cm^2 reduces the viscosity up to 10 times and more (for polymer with starting viscosity 0.4 Pa·s). As well as, the possibility of ultrasonic cavitation depolymerization of high viscous oils at low temperatures (0.8 Pa·s) was shown. Ultrasonic cavitation influence can reduce viscosity up to 3.5 times during 1000 s at starting viscosity 0.8 Pa·s (US oscillations intensity is 160 W/cm^2).

For ultrasonic depolymerization of high viscous oil at low temperatures the larger intensities is required due to viscosity prevents cavitation.

Obtained results can be used for choose the operating modes and conditions of industrial ultrasonic equipment for cavitation depolymerization of oil for its transferring at low temperatures.

Acknowledgements
The reported study was supported by Grant of President of Russian Federation No. MK-2813.2018.8.

References

[1] Khmelev V N, Golykh R N, Shalunova A V, Nesterov V A and Ilchenko E V 2015 *Proc. 16th International Conference of Young Specialist on Micro / Nanotechnologies and Electron Devices EDM 2015 (Novosibirsk)* (Novosibirsk: NSTU) p 249
[2] Golykh R N 2016 *Journal of Physics: Conference Series* **722** 012010
[3] Khmelev V N, Golykh R N, Shalunov A V, Khmelev S S and Karzakova K A 2014 *Proc. 15th International Conference of Young Specialist on Micro / Nanotechnologies and Electron Devices EDM 2014 (Novosibirsk)* (Novosibirsk: NSTU) p 203
[4] Khmelev V N, Shalunov A V, Golykh R N and Shalunova K V 2010 *Proceedings of 11th Annual International Conference and Seminar on Micro/Nanotechnologies and Electron Devices (Novosibirsk)* (Novosibirsk: NSTU) p 328
[5] Sommerfeld M A and Blei S 2002 *ILASS-Europe: Conference Proceedings (Zaragoza)*
[6] Sommerfeld M 2001 *International Journal of Multiphase Flow* **27** 1829–58
[7] Hoffman T L 2006 *Journal of Aerosol Science* **28** 919–36
[8] Legon A C 2014 *Physical Chemistry Chemical Physics* **16(24)** 12415–21
[9] Altukhov Y A, Gusev A S and Pyshnograi G V 2012 *Introduction in Mesoscopic Theory of Flowable Polymeric Systems* (Barnaul)
[10] Rouse P E 1953 *The Journal of Chemical Physics* **21(7)** 1272–80

Application of cohesive zone model to the fracture process of freshwater polycrystalline ice under flexural loading

I Gribanov, R Taylor and R Sarracino
Memorial University of Newfoundland, St.John's, NL, Canada
E-mail: ig1453@mun.ca

Abstract. Cohesive zone modeling is a promising technique for modeling fracture processes in polycrystalline solids. The Park-Paulino-Roesler (PPR) formulation for cohesive zones is a flexible method that allows one to reproduce behaviours of a wide range of materials. In this work, an implicit finite element simulation is performed to capture the fracture of freshwater, polycrystalline ice beams in 4-point bending tests. The distribution of the damaged cohesive zones and the indentation force are recorded as a function of time. Several simulations are performed to establish the relationship between the grain size and the flexural strength of the beams.

1. Introduction

While ice is a beautiful natural material, it can pose a risk to ships and structures in the marine environment, as well as to people who may use winter ice cover as a transportation route. During the formation process, prior to establishing a level ice cover, sea ice may partially form, break up and refreeze multiple times. Its microstructural composition depends largely on its age and growth conditions. In general, sea ice contains inclusions of brine and air, often having a granular microstructure in the uppermost layer, which transitions to a columnar microstructure within a few centimeters below the top surface [1]. In dynamic ocean environments, deformation processes in the sea ice can result in more complex microstructures as the sea ice is repeatedly broken, reoriented and refrozen. Icebergs, on the other hand, calve off glacial ice masses formed from compacted snow and generally are nearly free of solid particles and soluble substances [2]. Iceberg ice typically has a polycrystalline structure that can be characterized by grain size. Over large volumes, significant variation in grain size may be expected, as well as the occurence of pre-exisiting cracks which may remain in the ice or may heal over time, resulting in large veins of ice with a different microstructure in the healed crack regions [3].

In the vicinity of a large iceberg (Figure 1), one can occasionally hear cracking sounds – a sign of gradual fracture processes that can culminate in sudden catastrophic splitting. For interactions between ice and structures, fracture is highly important both in limiting loads and in localizing contact into zones of intense local pressures through which the majority of loads are transmitted to a structure [4]. Consequently, understanding and modelling fracture mechanisms in such ice features is highly important. While full-scale ice structure interactions are highly complex, insights into certain aspects of ice fracture behaviour can be studied through the measurement of ice strength under controlled conditions. Such tests are performed and experimental data is made available for compressive indentation conditions [5, 6, 7, 8, 9, 10, 11], as well as uniaxial compression tests on cylinders [12, 13] and flexural tests on beams [3, 14]. In the present paper, emphasis is placed on modelling freshwater polycrystalline (e.g. glacial) ice subjected to flexural loading conditions.

1.1. Related work on modeling

Finite element approaches to modeling damage can be roughly classified into two groups. In the first group, the damage is treated as continuous and expressed as a function of material coordinates. The crack surfaces are not explicitly tracked, and the material is not fragmented [15]. In the second group, the grains, cracks and other individual constituents are explicitly resolved. Concepts that are specific to the material, such as wing cracks, can be introduced into latter models [16]. Such approaches usually require more computational resources, because the shapes of the individual constituents are recorded and processed.

Cohesive zone models (CZM), where individual grains in the material are held together by cohesive tractions (Figure 2), fall into the second category of methods. Several applications of this method to the mechanics of ice have been proposed [17, 18, 19].

Figure 1. An iceberg sighting on May 28, 2017, near Topsail Beach, Conception Bay South, NL. Photo courtesy of http://mynewfoundlandkayakexperience.blogspot.ca.

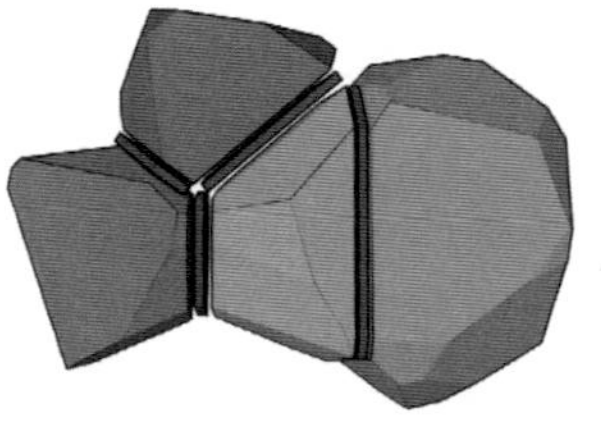

Figure 2. Polyhedral grains connected with cohesive zones (CZ). The rest volume of CZs can be zero or non-zero. Grains may be deformable or rigid.

2. Description of the method

To capture the dynamic behaviour of the modeled sample we utilize the implicit finite element method (FEM) with Newmark-beta integration. The discretized non-linear equation of motion is solved via Newton-Raphson iterations. The modeled geometry consists of separate grains which are connected by cohesive zones at the beginning of the simulation. The grains have linear elastic response, with the rotational component eliminated from the deformation gradient tensor to allow arbitrary rotations of the elements [20].

The traction-separation relations for the cohesive zones are as formulated by Park-Paulino-Roesler (PPR) [21]. All cohesive zones are triangular, with zero rest volumes. The normal separation is distinguished from the tangential separation; both normal and tangential forces are found as the gradient of the common potential Ψ:

$$\begin{aligned}\Psi(\Delta_n, \Delta_t) = \min(\phi_n, \phi_t) + \\ \left[\Gamma_n \left(1 - \frac{\Delta_n}{\delta_n}\right)^{\alpha} \left(\frac{m}{\alpha} + \frac{\Delta_n}{\delta_n}\right)^{m} + \langle \phi_n - \phi_t \rangle\right] \times \\ \left[\Gamma_t \left(1 - \frac{|\Delta_t|}{\delta_t}\right)^{\beta} \left(\frac{n}{\beta} + \frac{|\Delta_t|}{\delta_t}\right)^{n} + \langle \phi_t - \phi_n \rangle\right],\end{aligned} \tag{1}$$

where $\langle \cdot \rangle$ is the Macaulay bracket, i.e.

$$\langle x \rangle = \begin{cases} 0 & x < 0, \\ x & x \geq 0. \end{cases}$$

Separation parameters in the normal direction Δ_n and tangential direction Δ_t are independent, allowing material properties in these directions to be set separately. The parameters $(\phi_n, \phi_t; \delta_n, \delta_t; \alpha, \beta; \lambda_n, \lambda_t)$ determine the properties of the material, such as fracture energy, cohesive strength and brittleness. Energy constants Γ_n and Γ_t are defined by the expressions

$$\Gamma_n = (-\phi_n)^{\langle \phi_n - \phi_t \rangle / (\phi_n - \phi_t)} \left(\frac{\alpha}{m}\right)^{m}, \; \Gamma_t = (-\phi_t)^{\langle \phi_t - \phi_n \rangle / (\phi_t - \phi_n)} \left(\frac{\beta}{n}\right)^{n}$$

when the normal and tangential fracture energies are not equal ($\phi_n \neq \phi_t$), and

$$\Gamma_n = -\phi_n \left(\frac{\alpha}{m}\right)^{m}, \; \Gamma_t = \left(\frac{\beta}{n}\right)^{n}$$

when they are equal ($\phi_n = \phi_t$). Non-dimensional exponents m and n are related to initial slope indicators λ_n and λ_t by the following expressions

$$m = \frac{\alpha(\alpha - 1)\lambda_n^2}{(1 - \alpha\lambda_n^2)}, \; n = \frac{\beta(\beta - 1)\lambda_t^2}{(1 - \beta\lambda_t^2)}.$$

A typical traction-separation law is illustrated on Figure 3.

Because of the exponential terms, the gradient of $\Psi(\Delta_n, \Delta_t)$ takes a substantial amount of time to compute, especially for a large number of CZs. For best performance, these functions are evaluated in parallel on a GPU. The matrix in the resulting linear system is indefinite symmetric and cannot be resolved by GPU-optimized algorithms, therefore a CPU-based linear solver is used. Additionally, unloading/reloading relations are implemented as described by Spring [22], as in the implicit method they perform better than those described by Park et al. [21].

To account for the interaction of fragmented material, the contact detection and response is implemented, which is based on the penetration distance. The implementation is distributed as open-source[1].

3. Results and discussion

The flexibility of the PPR traction-separation laws allows to tune the model to a wide range of materials, including ice. A set of parameters (Table 1) was selected to match the experimental data for uniaxial tensile and compressive tests [19]. Some of the parameters have physical meaning, i.e. fracture energy and cohesive strength. Other parameters, such as α, β, λ_n and λ_t, determine whether the fracture will be brittle or ductile.

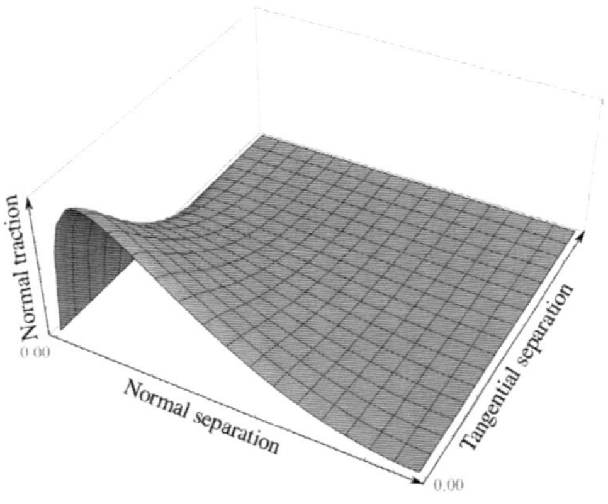

Figure 3. Normal component of the traction-separation relations as a function of normal and tangential separations Δ_n, Δ_t. Tangential component may be similar or different, depending on the selected parameters.

Table 1. Parameters of the material.

Parameter	Notation	Value
Young's modulus	E	10 GPa
Poisson's ratio	ν	0.3
Density	ρ	$916.2\,\mathrm{kg/m^3}$
Mode I and II fracture energies	ϕ_n, ϕ_t	$30\,\mathrm{J\,m^{-2}}$
Normal cohesive strength	σ_{max}	1.0 MPa
Tangential cohesive strength	τ_{max}	1.5 MPa
Shape of traction-separation curve	α, β	3
Non-dimensional slope indicators in PPR model	λ_n, λ_t	0.02
Magnitude coefficient for potential-based collision response	k	5000

The four-point bending simulations are performed on 28 different beam models. The geometries with random arrangements of grains were generated in Neper [23], and the dimensions are summarized on Figure 4. The beams measure $10 \times 10 \times 50$ cm and consist of 500, 750, 1000, 1250, 1500 and 2000 grains. Each grain is subdivided into tetrahedral elements, having 100 elements per grain on average. The indenters measure 5 cm in diameter, are rigid, and move with the speed of 0.23 cm/s. The initial time step is 0.025 s. All initial setups are available along with the open-source implementation of this project.

Nodal displacements and the forces acting on the indenters are recorded at each step. Simulations perform 300 steps, passing the point of beam failure. One of the tests is illustrated in Figures. 5-7, where the initiation and propagation of the crack is shown in a beam with 500 grains. The corresponding force-time curve is nonlinear (Figure 8), with the slope gradually increasing. Displacements that occur during the loading stage are small and not visible on the images. However, the strain energy is high and the accumulation of damage in the cohesive zones is visible. In this test, the force on the indenters reaches 5349 N at the moment of fracture.

After the fracture occurs, the newly formed fragments acquire substantial velocity due to

[1] https://github.com/igorg520b/icFlow

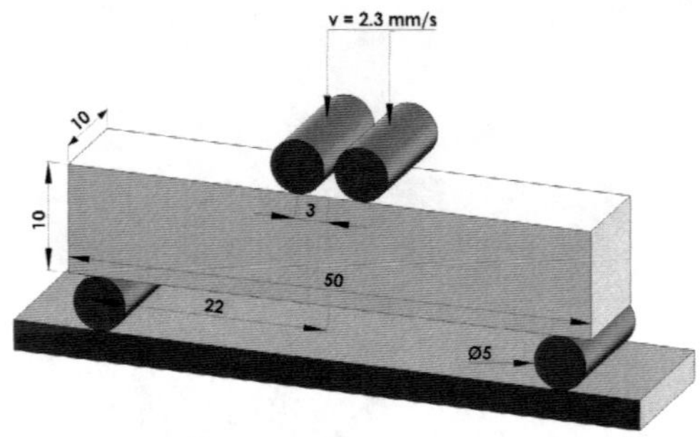

Figure 4. Setup of the four-point beam bending. Dimensions are shown in centimeters.

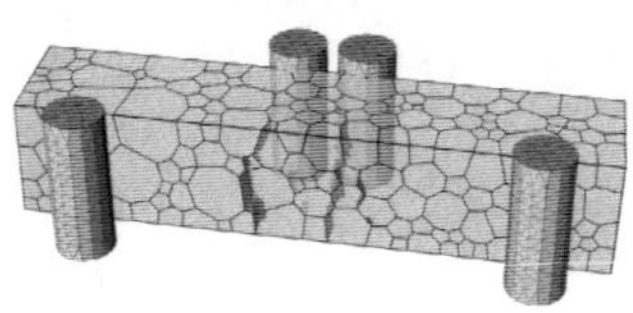

Figure 5. Simulation step 193; $t = 4.2797$ s. Damaged CZs shown in green, fracture surface shown in red.

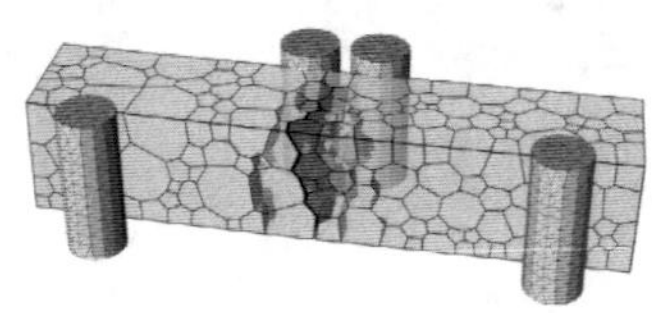

Figure 6. Simulation step 221; $t = 4.5106$ s.

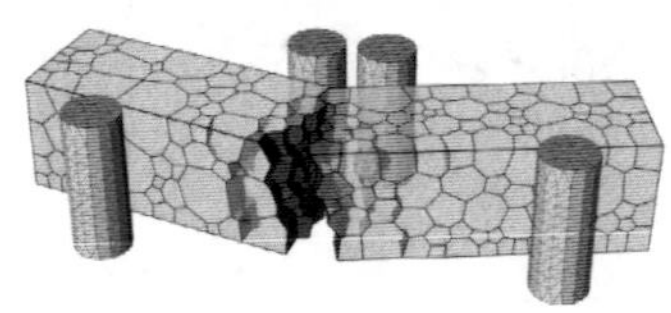

Figure 7. Simulation step 256; $t = 4.5216$ s.

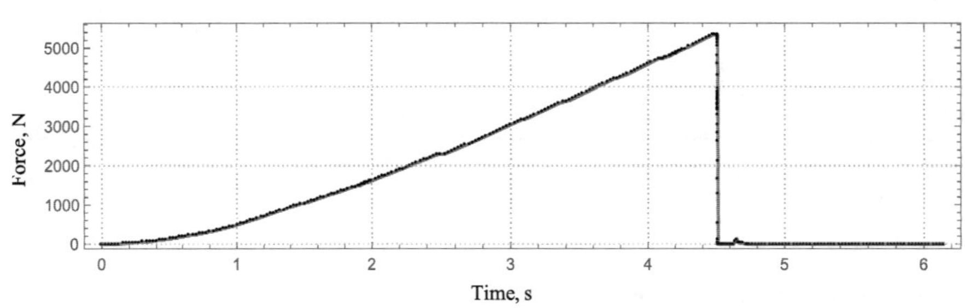

Figure 8. Indentation force vs time.

the release of elastic strain energy. Before they start moving in opposite directions, an impact occurs along the formed boundary near the indenters. In the current model, only intergranular fracture and elastic deformation are considered. In nature, however, ice would be expected to also experience local crushing at the contact points. The interplay between fracture and crushing during full-scale compressive ice-structure interactions is highly important for modelling loads of interest for engineering design, but is beyond the scope of the present paper. Accounting for such mechanisms to better model interactions involving progressive failure, which occurs during processes such as indentation, is the direction of the future work.

In each performed test the maximum force is determined before the beam fails. The corresponding flexural strength is computed as

$$\sigma = \frac{3F(L - L_i)}{2bd^2},$$

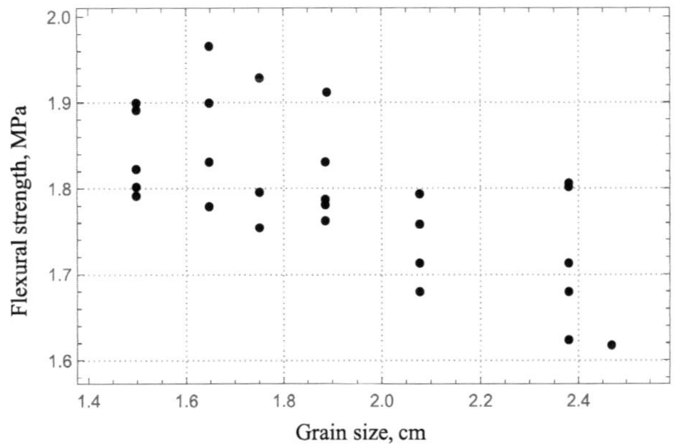

Figure 9. Relationship between the grain size and the flexural strength of the beam obtained from 28 simulations on beams with differing grain sizes.

where F is the load force at the fracture point, L is the length of the support (outer) span between the indenters, L_i is the length of the loading (inner) span, b is the width of the beam and d is the thickness of the beam. The results are summarized in Figure 9 as a plot of flexural strength vs. grain size. As may be observed, the strength of the blocks increases with decreasing grain size, as expected from Hall-Petch relationship. The flexural strengths obtained from these simulations are in accord with several experimental results [24, 25, 26], however experimental measurements of the flexural strength of ice vary significantly [27] and the influence of scale effects is highly important [28]. Such effects will be explored further in future work.

4. Conclusion

The complexities of ice-structure interactions and the inherent variability in ice strength, coupled with the need for risk-based approaches in engineering design, result in a continued strong emphasis on empirical approaches in many branches of ice mechanics. Nevertheless, numerical modelling is an important tool for studying certain aspects of ice behavior, and the importance of numerical modelling will undoubtedly increase in the future as newer methods which capture the unique processes that take place in ice continue to be developed. In particular, detailed accurate modelling of ice fracture would be highly valuable in assessing load limiting processes and potential strategies for improving design.

In nature, deformation, fracture and motion of ice are affected by numerous factors, making the simulation of full-scale processes extremely challenging. The work presented here, although employed for the specialized case of polycrystalline ice under flexural loading, provides an improved technique which can be extended to model the behavior of less homogeneous ice under more general conditions. Limitations of the approach at this stage of development include the absence of processes associated with crushing; specifically microcracking, dynamic recrystallization and local pressure melting. Neither does the model account for viscous deformation, intragranular fracture and variabilities in grain-grain cohesion. However, it does capture grain size dependence, accumulation of crack-related damage and well represents experimental force-time curves. It also captures discrete fracture events, fragmentation and allows for the integration of more complex models of ice continuum behaviour in future work. Being a new and flexible model, the PPR technique provides an improved approach for modeling ice using cohesive zones.

Acknowledgments

The authors gratefully acknowledge core funding for the Centre for Arctic Resource Development (CARD) program at C-CORE by Hibernia Management and Development Company, Ltd. (HMDC) and Terra Nova Development (Suncor Energy Inc. - Operator). Financial support from the Research and Development Corporation (RDC) of Newfoundland and the donation of Quadro M5000 GPU by NVIDIA are gratefully acknowledged. The authors also thank C-CORE for providing office space and creating a productive environment.

References

[1] Weeks W 2010 *On Sea Ice* (University of Alaska Press) ISBN 9781602231016
[2] Mellor M 1980 *Physics and Mechanics of Ice* ed Tryde P (Berlin, Heidelberg: Springer Berlin Heidelberg) pp 217–245 ISBN 978-3-642-81436-5 978-3-642-81434-1 dOI: 10.1007/978-3-642-81434-1_17
[3] Barrette P D and Jordaan I J 2002 *Journal of Glaciology* **48** 587–591
[4] Taylor R and Jordaan I 2015 *Engineering Fracture Mechanics* **134** 242–266 ISSN 00137944
[5] Browne T, Taylor R, Jordaan I and Gurtner A 2013 *Cold Regions Science and Technology* **88** 2–9 ISSN 0165232X
[6] Habib K B, Taylor R S, Bruneau S and Jordaan I J 2015 (ASME) p V008T07A014 ISBN 978-0-7918-5656-7
[7] O'Rourke B J, Jordaan I J, Taylor R S and Gurtner A 2016 *Cold Regions Science and Technology* **124** 25–39 ISSN 0165232X
[8] O'Rourke B J, Jordaan I J, Taylor R S and Gurtner A 2016 *Cold Regions Science and Technology* **124** 11–24 ISSN 0165232X
[9] Birajdar P, Taylor R, Habib K and Hossain R 2016 (Offshore Technology Conference)
[10] Birajdar P, Taylor R and Hossain R 2017 (San Francisco, California, USA)
[11] Taylor R, Browne T, Jordaan I and Gurtner A 2013 (ASME) p V006T07A024 ISBN 978-0-7918-5540-9
[12] Lee R W and Schulson E M 1988 *Journal of Offshore Mechanics and Arctic Engineering* **110** 187 ISSN 08927219
[13] Schulson E 1990 *Acta Metallurgica et Materialia* **38** 1963–1976 ISSN 09567151
[14] Gagnon R and Gammon P 1995 *Journal of Glaciology* **41** 103–111
[15] Moore P, Jordaan I and Taylor R 2013
[16] Kolari K 2017 *International Journal of Solids and Structures* **115** 27–42
[17] Lu W, Lubbad R and Løset S 2014 *Journal of Offshore Mechanics and Arctic Engineering* **136** 031501
[18] Wang F, Zou Z J, Zhou L, Ren Y Z and Wang S Q 2018 *Cold Regions Science and Technology* **149** 1–15
[19] Gribanov I, Taylor R and Sarracino R 2018 *Engineering Fracture Mechanics* **196** 142–156
[20] Gribanov I, Taylor R and Sarracino R 2018 *International Journal for Numerical Methods in Engineering*
[21] Park K, Paulino G H and Roesler J R 2009 *Journal of the Mechanics and Physics of Solids* **57** 891–908 ISSN 0022-5096
[22] Spring D W, Giraldo-Londoo O and Paulino G H 2016 *Mechanics Research Communications* **78** 100–109 ISSN 00936413
[23] Quey R, Dawson P R and Barbe F 2011 *Computer Methods in Applied Mechanics and Engineering* **200** 1729–1745 ISSN 0045-7825
[24] Timco G W and Frederking R 1983 *Cold Regions Science and Technology* **8** 35–41
[25] Tatinclaux J C and Wu C Y 1978 *Applied Techniques for Cold Environments* (ASCE) pp 295–306
[26] Parsons B, Lal M, Williams F, Dempsey J, Snellen J, Everard J, Slade T and Wllliams J 1992 *Philosophical Magazine A* **66** 1017–1036
[27] Lainey L and Tinawi R 1984 *Canadian Journal of Civil Engineering* **11** 884–923
[28] Aly M, Taylor R S, Bailey E and Turbull I 2018 *Proceedings of OMAE 2018*

Compression waves in porous media containing gas hydrate

A A Gubaidullin[1,2], O Yu Boldyreva[1] and D N Dudko[1]

[1]Tyumen Branch of Khristianovich Institute of Theoretical and Applied Mechanics SB RAS, Tyumen, Russia
[2]Tyumen State University, Tyumen, Russia

E-mail: gubai@tmn.ru

Abstract. Wave processes in a porous medium containing water, gas and gas hydrate are numerically investigated. The two-velocity model of porous medium is used. The nonlinearity of the oscillations of the bubbles is taken into account in the equation of gas state and the equation for bubble radius. The transmission of compression of step waves from the water into hydrate-containing porous medium is investigated. Beyond the line of phase equilibrium the association of gas hydrate is possible. The influence of the medium parameters and the intensity of the incident wave on the evolution of waves in this porous medium is studied. When entering the porous medium, the initial pulse is divided into fast (deformation) and slow (filtration) waves. In a slow wave damped oscillations associated with pulsations of bubbles are observed. The increase in the initial equilibrium pressure in the medium leads to an increase in the velocities of the deformation and filtration waves. With the initial volume content of the gas in the pore space unchanged, an increase in the initial hydrate content leads to a small increase in the velocities and amplitudes of fast and slow waves. At constant initial volume content of hydrate, the increase of gas content leads to a noticeable decrease of slow wave speed and decrease of amplitudes of both fast and slow waves. It was found that the accounting of the processes of formation of gas hydrate in a porous medium at the considered time intervals does not have a noticeable effect on the character of wave propagation.

1. Introduction

Currently, the scientific literature presents the results of experimental studies on the association and dissociation of hydrates in porous media [1] and the association of hydrate in water with bubbles of hydrate-forming gas during the passage of shock waves [2]. The theoretical study of the process of hydrate association in the propagation of a shock wave in water with bubbles of a hydrate-forming gas was carried out in [3]. The pore space of hydrate-containing porous media usually includes water and gas bubbles, so the presence of gas bubbles should be taken into account in the study of wave processes in such media. Wave propagation in porous media saturated with liquid and gas bubbles was studied experimentally in [4-7]. The numerical study of wave processes in partially saturated porous media was carried out in [8].

In this paper, wave processes in a porous medium, whose pore space contains water, gas hydrate and hydrate-forming gas in the form of bubbles, are numerically investigated. The nonlinearity of bubble oscillations is taken into account in the gas state equation and the Rayleigh-Plesset equation for a gas bubble in a porous medium.

2. Basic Equations

To study wave propagation in a porous medium containing gas hydrate and a bubbly liquid, two velocity, two stress tensors model of saturated porous medium is used [9, 10]. The pore space is filled with three-phase mixture of hydrate and liquid with gas bubbles, bubble size is calculated using the equation of Rayleigh-Plesset for a gas bubble in a porous medium [4].

The equations of mass and momentum balance are in the form:

$$\frac{\partial \rho_l}{\partial t} + \nabla^i\left(\rho_l v_l^i\right) = 4\pi a_b^2 n_b j_l, \quad \frac{\partial \rho_g}{\partial t} + \nabla^i\left(\rho_g v_l^i\right) = 4\pi a_b^2 n_b j_g,$$

$$\frac{\partial \rho_h}{\partial t} + \nabla^i\left(\rho_h v_l^i\right) = 4\pi a_b^2 n_b j_h, \quad \frac{\partial \rho_s}{\partial t} + \nabla^i\left(\rho_s v_s^i\right) = 0, \quad \frac{\partial n_b}{\partial t} + \nabla^i\left(n_b v_l^i\right) = 0, \tag{1}$$

$$\rho_{l+g+h}\frac{d_l v_l^i}{dt} = -\alpha_{l+g+h}\nabla^i p_l - F^i, \quad \rho_s \frac{d_s v_s^i}{dt} = -\alpha_s \nabla^i p_l + \nabla^j \sigma_{s*}^{ji} + F^i, \tag{2}$$

where ρ_j, v_j, α_j are the density, velocity, volumetric content of the j - th phase, $v_h = v_g = v_l$, the subscripts $j = s, l, g, l{+}g{+}h$ relate to the skeleton of the porous medium, liquid, gas or mixture of liquid and bubbles and hydrate; σ_{s*}, p_l are effective stress in the skeleton and the pressure in the liquid, respectively, a_b is bubble radius, n_b is the number of bubbles per unit volume.

Interphase force interaction is a sum of viscous friction force F_μ and added masses force F_m:

$$F = F_m + F_\mu, \quad F_m = \frac{1}{2}\eta_m \alpha_s \rho_{l+g+h}\left(\frac{d_l v_l}{dt} - \frac{d_s v_s}{dt}\right), \quad F_\mu = \eta_\mu \alpha_s \alpha_{l+g+h}\mu_l a_{s*}^{-2}\left(v_l - v_s\right), \tag{3}$$

where a_{s*} is the characteristic size of the grains of the skeleton, μ_l is viscosity of the fluid, η_m, η_μ are the coefficients of interaction between phases dependent on the structure of the pore space.

The temperatures of the skeleton, water and hydrate are assumed constant and equal to the initial temperature T_0. The heat equation of inflow for gas phase is taken in the form [10]:

$$\rho_g c_{gV}\frac{d_l T_g}{dt} = \alpha_g p_g \frac{1}{\overset{\circ}{\rho}_g}\frac{d_l \overset{\circ}{\rho}_g}{dt} - 4\pi a_b^2 n_b q_{gl} - 4\pi a_b^2 n_b j_g c_{gV}\left(T_g - T_0\right), \tag{4}$$

$$q_{gl} = \mathrm{Nu}\,\lambda_g \frac{T_g - T_0}{2a_b}. \quad \mathrm{Nu} = \begin{cases} 10, \ \mathrm{Pe} < 10^2 \\ \sqrt{\mathrm{Pe}}, \ \mathrm{Pe} \ge 10^2 \end{cases}, \quad \mathrm{Pe} = 12\left(\gamma_g - 1\right)\frac{T_0}{\left|T_g - T_0\right|}\frac{a_b\left|\dot{a}_b\right|}{\nu_g}, \tag{5}$$

where λ_g, ν_g are thermal conductivity and thermal diffusivity.

If the pressure and (or) temperature deviate from the phase equilibrium line "gas + water ↔ gas hydrate" given by the equation

$$\frac{p_s}{p_{s0}} = \exp\left(\frac{T_s - T_{s0}}{T_*}\right), \tag{6}$$

then phase transitions are possible. In the compression wave the association of gas hydrate may occur at the gas bubble surface. The intensity of the phase association is given by the following equation [9, 10]:

$$j_g = -\frac{\beta\left(p_g - p_s\left(T_0\right)\right)}{\sqrt{2\pi R_g T_0}}, \quad j_h = -j_g/G, \quad j_l = -j_h - j_g. \tag{7}$$

Here G is mass fraction of gas in gas hydrate, β is empirical coefficient characterizing the intensity of hydrate association.

The skeleton of the porous medium is assumed to be elastic with effective moduli of elasticity λ_{s*}, μ_{s*}:

$$\sigma_{s*}^{ij} = \alpha_s \left(\lambda_{s*} \delta^{ij} \varepsilon_s^{mm} + 2\mu_{s*} \varepsilon_s^{ij} + \gamma_{s*} \delta^{ij} p_l \right), \quad \gamma_{s*} = \frac{\lambda_{s*} + 2/3\,\mu_{s*}}{K_s}, \tag{8}$$

$$\frac{d_s \varepsilon_s^{ij}}{dt} = \frac{1}{2}\left(\nabla^i v_s^j + \nabla^j v_s^i \right),$$

where ε_s is strain of the skeleton, K_s is the bulk elastic modulus of the skeleton material, γ_{s*} is the ratio of bulk elastic moduli of the skeleton and its material.

The state equations for solid and liquid phases are given in the acoustic approximation:

$$p_s - p_{s0} = C_s^2 \left(\rho_s^\circ - \rho_{s0}^\circ \right), \quad p_l - p_{l0} = C_l^2 \left(\rho_l^\circ - \rho_{l0}^\circ \right), \quad p_h - p_{h0} = C_h^2 \left(\rho_h^\circ - \rho_{h0}^\circ \right) \tag{9}$$

here subscript 0 indicates the unperturbed value, ρ°_j, C_j, p_j are the true density, the sound velocity and the true pressure in the material of the j - th phase.

The behavior of gas in the bubbles follows the Mendeleev – Clapeyron equation:

$$p_g = \rho_g^\circ R_g T_g , \tag{10}$$

bubble radius a_b changes according to the equation of Rayleigh-Plesset for the bubble in a porous medium [4]:

$$\frac{da_b}{dt} = w_R + w_A + \frac{j_g}{\rho_{l0}^\circ}, \qquad w_A = \frac{p_g - p_l - \dfrac{2\Sigma}{a_b}}{\rho_{l0}^\circ C_l x_g^{1/3}}, \tag{11}$$

$$\rho_{l0}^\circ \left(\dot{w}_R a_b + \frac{3}{2} w_R^2 \right) = p_g - p_l - \frac{2\Sigma}{a_b} - 4\mu_l \frac{w_R}{a_b} \left(1 + \frac{1}{4} \alpha_s \eta_\mu \left(\frac{a_b}{a_{s*}} \right)^2 \right)$$

here x_g is the gas volume fraction in the bubbly liquid, Σ is the coefficient of surface tension at the liquid and gas boundary, w_A takes into account the compressibility of the liquid [11].

To close the system of equations, the relations between true and effective densities and volume fractions, and the relations between true pressures p_l, p_s, p_h in the phases and the effective pressure p_{s*} in the skeleton are used

$$\rho_j = \alpha_j \rho_j^\circ, \quad \rho_{l+g+h} = \alpha_l \rho_l^\circ + \alpha_g \rho_g^\circ + \alpha_h \rho_h^\circ, \quad \alpha_s + \alpha_l + \alpha_g + \alpha_h = 1, \tag{12}$$

$$\alpha_g = \frac{4}{3}\pi a_b^3 n_b, \qquad p_{s*} = \alpha_s \left(p_s - p_l \right), \qquad p_{s*} = -\frac{1}{3}\sigma_{s*}^{mm}.$$

3. Propagation of stepwise wave

Methodology for calculating the motion of a porous medium containing liquid with gas bubbles based on the method of Lax–Wendroff was developed, and the numerical study of pressure wave propagation in such medium was carried out. Perturbation in the porous medium (x>0) is generated by a wave incident from the pure liquid (x<0).

We have investigated the influence of medium parameters and of the initial pulse on the wave evolution in a porous medium containing the water, gas and gas hydrate.

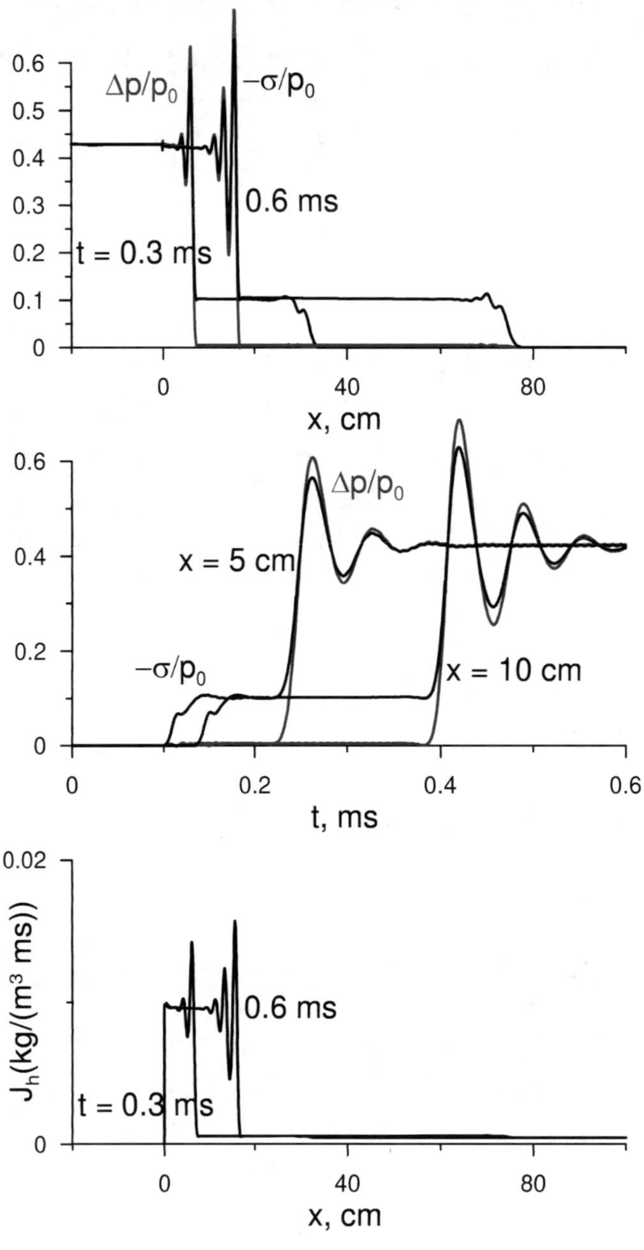

Figure 1. The pressure in the liquid, the total stress and the intensity of hydrate association in the step wave of the amplitude equal to 0.5· p_0 from a liquid into a porous medium at the distances of 5 cm and 10 cm from the boundary and at the moments 0.3 and 0.6 ms.

Figure 1 shows the variation of the pressure in the liquid p, the total stress $\sigma = \sigma_{s*} - p$ and the intensity of hydrate association $J_h = 4\pi a_b^2 n_b j_h$ in the step wave of the amplitude equal to 0.5·p_0 passing from the water into the porous medium containing water with freon bubbles and freon hydrate inclusions. The skeleton material is quartz. The initial pressure and temperature are p_0 =20 bar, T_0 = 294 K; these pressure and temperature are on the line of phase equilibrium (6). The initial volume fractions are $\alpha_s = 0.6$, $\alpha_l = 0.352$, $\alpha_g = 0.008$, $\alpha_h = 0.04$; grain size $a_* = 1$ mm, the initial bubble radius $a_b = 0.5$ mm. The parameters of Equations (6, 7) for freon and freon hydrate $p_{s0} = 0.42$ bar, $T_{s0} = 274$ K, $T_* = 5.2$ K, $\beta=3\cdot10^{-5}$ $G = 0.3$ are taken from [3]. The other parameters of the porous medium are $\lambda_{s*} = \mu_{s*} = 2\cdot10^9$ Pa, $\eta_m = 1$, $\eta_\mu = 100$.

When entering a porous medium initial pulse splits into fast (deformational) and slow (filtrational) compression waves. In the fast wave, the pressure amplitude in the liquid is close to zero, i.e. the compression is observed only by the skeleton of the porous medium, there is a monotonic increase in the total stress in the medium. In the slow wave, damped oscillations of the pore pressure associated with the pulsation of gas bubbles are observed. If you increase the amplitude of the original incident wave, pressure oscillations in the slow wave become more noticeable. In a medium with a higher equilibrium pressure, there is a higher propagation velocity of both fast (deformational) and slow (filtrational) waves.

The effect of phase transitions on the propagation of compression waves in a porous medium containing gas hydrate was also studied. With increasing pressure above the phase equilibrium line, hydrate formation occurs at the surface of the bubbles. The calculations show that the process of hydrate formation during the compression wave passage is rather slow and does not have a noticeable effect on the wave propagation in the medium at the considered time intervals.

References

[1] Tohidi B, Anderson R, Masoudi A, Arjmandi J, Burgass R and Yang J 2003 *Russian Chemical Journal* **XLVTI(3)** 49–58
[2] Dontsov V E, Nakoryakov V E and Chernov A A 2007 *Journal of Applied Mechanics and Technical Physics* **48(3)** 346–60
[3] Shagapov V Sh, Lepihin S A and Chiglintsev I A 2010 *Thermophys. Aeromech.* **17(2)** 229–41
[4] Dontsov V E, Kuznetsov V V and Nakoryakov V E 1987 *Izv. USSR Academy of Sciences, MZhG* **4** 85–92
[5] Grinten J G M van der 1987 *An Experimental Study of Shock-Induced Wave Propagation in Dry, Water-Saturated, and Partially Saturated Porous Media,* Ph.D. Thesis (Proefschrift Eindhoven)
[6] Smeulders D M J 1992 *On Wave Propagation in Saturated and Partially Saturated Porous Media,* Ph.D. Thesis (Proefschrift Eindhoven)
[7] Smeulders D M J and van Dongen M E H 1997 *J. Fluid Mech.* **343** 351–73
[8] Dunin S Z, Mikhailov D N and Nikolaevskii V N 2006 *J. Appl. Math. Mech.* **70(2)** 251–63
[9] Nigmatulin R I 1978 *Fundamentals of Mechanics of Heterogeneous Media* (Moscow: Nauka)
[10] Nigmatulin R I 1990 *Dynamics of Multiphase Media. Part 1* (New York: Hemisphere Publ)
[11] Nigmatulin R I, Shagapov V S and Vakhitova N K 1989 *Dokl. Akad. Nauk SSSR* **304** 1077–81

Diagnostics of critical states of constructions operated under low temperature conditions

V P Gulyaev[1], P P Petrov[2] and K V Stepanova[2]

[1]Ammosov North-Eastern Federal University, Yakutsk, Russia
[2]Larionov Institute of the Physical-Technical Problems of the North, Yakutsk, Russia

E-mail: uekztddg@mail.ru

Abstract. Characteristic changes in the stress-strain state of the material in the region in front of the propagating brittle crack in samples of steels and cast irons of machines and constructions by X-ray diffractometry was established. The angular displacements of the peak and the change in other characteristics of the profile of the diffraction lines reflected from the elastically deformed regions of the material are stably fixed. The characteristics of the diffraction lines profiles reflected from the samples surface of steels and cast irons of different chemical composition differ qualitatively and indicate the instrumentally determined influence of the substructure of steels and cast irons on the processes of their resistance to nucleation and propagation of cracks. The efficiency of using the X-ray strain measurement method for nondestructive testing of the stress concentration region to determine the bearing capacity and the residual life of constructions was experimentally established.

1. Introduction

The operation of machines and equipment, main gas and oil pipelines and other technical systems in the climatic conditions of the arctic and subarctic regions of the Russian Federation is associated with the fulfillment of a number of special technical requirements for machines and constructions, and for maintaining their reliable operability during the guaranteed period [1-4]. Theories and practical applications of the fracture mechanics of high-strength structural materials used in critical constructions are the scientific basis of special technical requirements. However, the extension of the provisions and methods of fracture mechanics to high-duty constructions and products is hampered by the poor knowledge of the processes of material deformation at the microstructural level, the absence of acceptable tools for the full utilization of the service properties of materials.

2. Problem Statement

According to experts [4-10], the required reliability of machines and constructions during their operation can be achieved by operative diagnostics of the loss of bearing capacity of machine parts and structural elements using nondestructive testing methods. An expansion of the economic activities of enterprises and organizations in the arctic and subarctic regions of the country with a long period of low climatic temperatures increases the importance of the methods of nondestructive testing and assessment of the state of pre-destruction of metal structures and machinery.

However, a number of methodological issues related to the use of X-ray diffractometry methods for diagnosing the bearing capacity of the exploited constructions of various technical systems, determining the state of pre-destruction of elements of metal structures remain poorly studied [5-8]. In particular, problems of instrumental determination of deformed regions by substructural changes in the metal of the exploited constructions require a careful consideration.

The aim of this work is to establish a qualitative connection between the stress-strain state of samples of structural materials and the characteristics of the profile of diffraction lines that represent microplastic deformation at the substructure level.

3. Research Methods

To achieve this purpose, we measured the characteristics of the X-ray diffraction line profiles reflected from the blocks of the polycrystalline structure of laboratory samples of structural materials (30KHGSA, 60S2A, 65G steels, 08 ps steel; SCH20, CHVG40 cast irons). The nature, levels and types of loading of the test samples corresponded to the action of operational loads. Measurements of the characteristics of X-ray diffraction line profiles were made on two groups of samples in which a stress-strain state was created with special types and loading levels similar to the actual operational state.

The first group consists of plane rectangular samples of 30KHGSA, 60S2A, 65G steels and SCH20, CHVG40 cast irons with an edge fatigue crack loaded with a three-point bending before failure. The fatigue crack on these samples was grown by a cantilever shock-cyclic transverse bending simulating the operating mode of loading hammers of crushing machines [11]. This method of loading laboratory samples was adopted to obtain the characteristics of profiles of X-ray diffraction lines, indicating the propagation of plastic deformations from the fracture surface. The profile characteristics were measured at certain points on the lateral face of the samples (figures 1, 2).

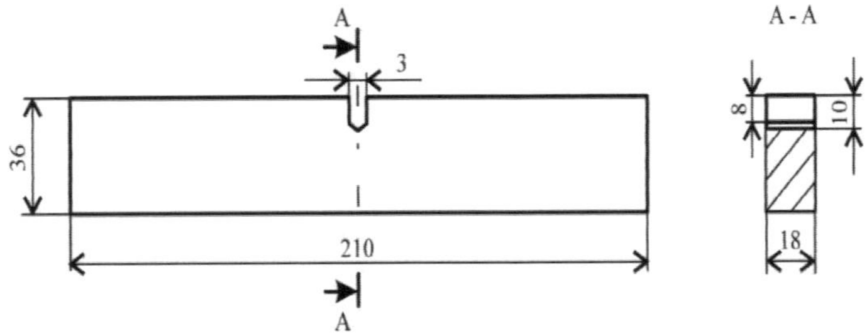

Figure 1. The type and geometric dimensions of the samples for investigating the propagation of microplastic deformations with the growth of a fatigue crack and subsequent failure.

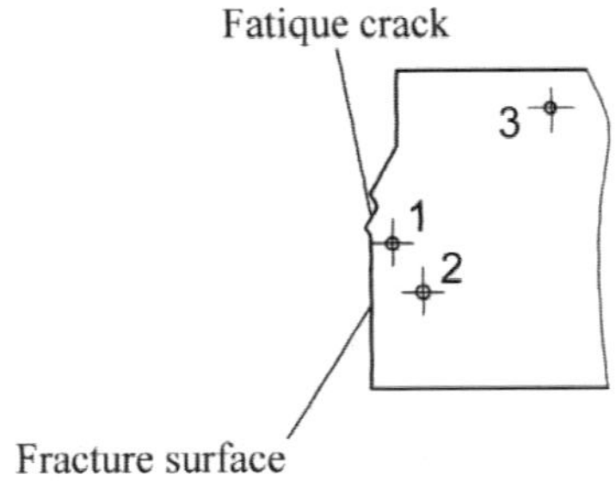

Figure 2. Measuring points of the characteristics of the profiles of X-ray diffraction lines on samples to study the propagation of microplastic deformations with the growth of the fatigue crack and subsequent destruction.

The static loading of the second group of samples is carried out in an open atmospheric platform and indoors to establish the effect of climatic factors on the change in the characteristics of the X-ray diffraction line profiles of sample material in the range of elastic stresses that constanly act for a long time. As the structural material studied, the widely used structural steel 08ps is used, the crystal structure of which has a body-centered cubic (bcc) lattice. Primary acts of microplastic deformation in such a structure are observed over close-packed slip planes of the second order, in particular, *hkl* = (220).

Static loads developed by loads of different weights are applied to assemblies consisting of 3 identical samples connected in series (figure 3). The assembly of 3 identical samples into a single line loaded with one load is caused by the need for an X-ray analysis of the central region of the gage area of each of the 3 samples. Weights masses are chosen so that in the cross-section of each sample of one assembly tensile stresses equal to $\sigma = 0.3\ \sigma_{0.2}$ (yield strength). For the second assembly, the load mass provides stresses in the cross-section of the samples $\sigma = 0.5\ \sigma_{0.2}$, and, respectively, for the third assembly $\sigma = 0.7\ \sigma_{0.2}$, for the fourth assembly $\sigma = 0.9\ \sigma_{0.2}$. Thus, from January 2013 to the present, under the action of tensile static loads on an open atmospheric platform, there are 4 assemblies of steel samples, indoors in a similar way to a set of 4 sample assemblies.

Figure 3. The scheme of loading the assembly from 3 samples tensile by loads of various masses (rotated by 90°).

For the X-ray analysis the sample assembly is unloaded and disassembled for the period of time necessary for X-ray analysis. Previously samples are visually inspected, the measurement of geometric dimensions is carried out. X-ray analysis of statically loaded samples is performed periodically every 6 months since June 2013.

Characteristics of X-ray diffraction line profiles are recorded on a Rigaku Ultima IV X-ray diffractometer manufactured by Rigaku International Corporation (Japan) with a high-precision horizontal goniometer. Analyzed sample is fixed horizontally in a goniometer, scanned by an X-ray generator and a scintillation counter that rotates in a vertical plane along the goniometer guide.

The diffraction spectrum is analyzed according to the scanning scheme θ-2θ with Bragg-Brentano focusing in Co radiation ($\lambda = 0.179020$ nm wavelength). A secondary beam uses a graphite monochromator. The width of the slit bounding the incident beam: horizontally 10 mm, vertically 0.5 mm. Soller's slit on the incident and diffracted beams are 5°; the width of the slit that bounds the diffracted beam vertically in front of the receiving slit is 0.5°.

Operating mode of the X-ray source: U = 40 kV; I = 40mA; scanning speed: 0.1 deg/min; step: 0.01 degree; scanning axis: 2θ/θ. The diffraction line profile in the coordinates "intensity-angle 2θ", the angle 2θ (in degrees) of the profile peak, the interplanar spaces d (in angstroms), the maximum intensity Imax of the reflected radiation (in imp/s), the half-width of the FWHM profile (in degrees) and other characteristics are output automatically by the PDXL-2 program.

Instrumental broadening was recorded on reference samples of 30KHGSA, 60S2A, 65G, 08 ps steels and SCH20, CHVG40 cast irons, the reference microstructure of which was obtained by recrystallization annealing in a vacuum furnace.

Based on the stated goal of the study, the first group included laboratory samples of steels and cast irons widely used for the manufacture of parts of machines and other technical systems experiencing sufficiently high operational loads that can cause the fatigue cracks initiation and propagation. Fatigue cracks initiation, as a rule, is observed in the stress concentration zones of those present in any construction and being observation objects of safekeeping of the constructions operability by methods of nondestructive testing.

4. Findings

Characteristics of profiles of X-ray diffraction lines indicating the propagation of plastic deformations developing upon the failure of laboratory samples with pre-grown fatigue cracks were obtained by X-ray analysis of individual points on the lateral face of plane samples of 30KHGSA, 60S2A, 65G steels and SCH20, CHVG40 cast irons (figure 2). The point of analysis 1 is adjacent to the surface of the fatigue crack propagation, so the crystal structure deformations revealed by the analysis will characterize the stress-strain state at which the fatigue crack continued its propagation. At point 2, the deformations of the crystal structure are likely to be due to the stress-strain state caused by the application of the load at the fracture of sample by a three-point bending. For comparison, an X-ray analysis was carried out at point 3, in which residual plastic deformations achieved at the maximum level of the stress-strain state of the sample with known loading conditions and geometry should be observed.

Numerical values of the characteristics of the diffraction line profiles obtained as a result of X-ray diffractometry of these points on the lateral face of samples 30KHGSA, 60S2A, 65G steels and SCH20, CHVG40 cast irons, automatically processed by the program PDXL-2, are given in table 1. As can be seen, the numerical values of the characteristics of the diffraction line profiles: profile maximum 2θ°, interplanar space *d*, nm and *FWHM* broadening vary depending on the position of the X-ray analysis point relative to the fatigue tip crack and the distance from the fracture surface (fracture). Such a pattern of microstructural changes determined by the stress-strain state of the material was detected in [8-10].

The features of the substructure of the deformed material are expressed most clearly by the actual graphical profiles of the diffraction lines. A joint study of the diffraction line profiles obtained during survey at different points of the lateral face of the studied sample makes it possible not only to reveal the inhomogeneity of the stressed-deformed state, but also to determine the effect of microstructural features on the course of elementary acts of plastic deformation.

Table 1. Characteristics of the profiles of the diffraction lines from the planes (220) at points 1, 2 and 3 of the lateral face of the investigated samples 30KHGSA, 60S2A, 65G, SCH20 and CHVG40.

Grade	Point number	2θ°	*d*, nm	*FWHM*
30KHGSA	1	123,839	0,101382	0,492
	2	123,842	0,101380	0,482
	3	123,876	0,101364	0,471
	standard	123,928	0,101339	0,275
60S2A	1	124,012	0,101300	0,494
	2	124,049	0,101283	0,477
	3	124,017	0,101298	0,471
	standard	124,067	0,101276	0,256
65G	1	123,815	0,101393	0,607
	2	123,796	0,101402	0,494
	3	123,803	0,101399	0,477
	standard	123,949	0,101333	0,229
SCH20	1	123,949	0,101333	0,229
	2	124,079	0,101269	0,558
	3	124,121	0,101249	0,431
	standard	124,234	0,101197	0,181
CHVG40	1	124,066	0,101275	0,491
	2	124,097	0,101261	0,559
	3	124,119	0,101250	0,423
	standard	124,127	0,101247	0,159

Figure 4 shows the profiles of the diffraction lines recorded at points with different stress-strain states and superimposed on each other, obtained on laboratory samples of cast irons SCH20 and

CHVG40. Comparison of the general form of diffraction profiles of cast iron samples SCH20 and CHVG40 reveals a significant difference in the intensity of the peaks. Vermicular iron has a higher intensity of reflection from the crystallographic plane (220) both in the annealed state (standard) and at point 3 without obvious traces of plastic deformation. At the same time, the intensity of the cast iron profile of CVG 40 at a given point is much higher than in the diffraction profile of the same region of the sample of grey cast iron SCH20.

It should be noted that the diffraction profiles taken from the planes (220) at point 2 have a coincidence in their type and characteristics for cast iron with flake and vermicular graphite (curves 2 in figure 4, table 1). Probably, this is due to the common ferritic-pearlitic metal base of the investigated cast irons, with an equal ratio of the areas of ferrite and perlite. On the contrary, the profiles of the crystallographic planes (220) at point 1, far from the surface of the fatigue crack at a distance of 1 ... 2 mm, are significantly broadened in comparison with the *FWHM* standards (samples in the annealed state) and point 3. This difference revealed by diffractometry is explained by high gradients of stresses and strains, under which the influence of the form of graphite inclusions in the common ferritic-pearlitic base of cast irons is intensified.

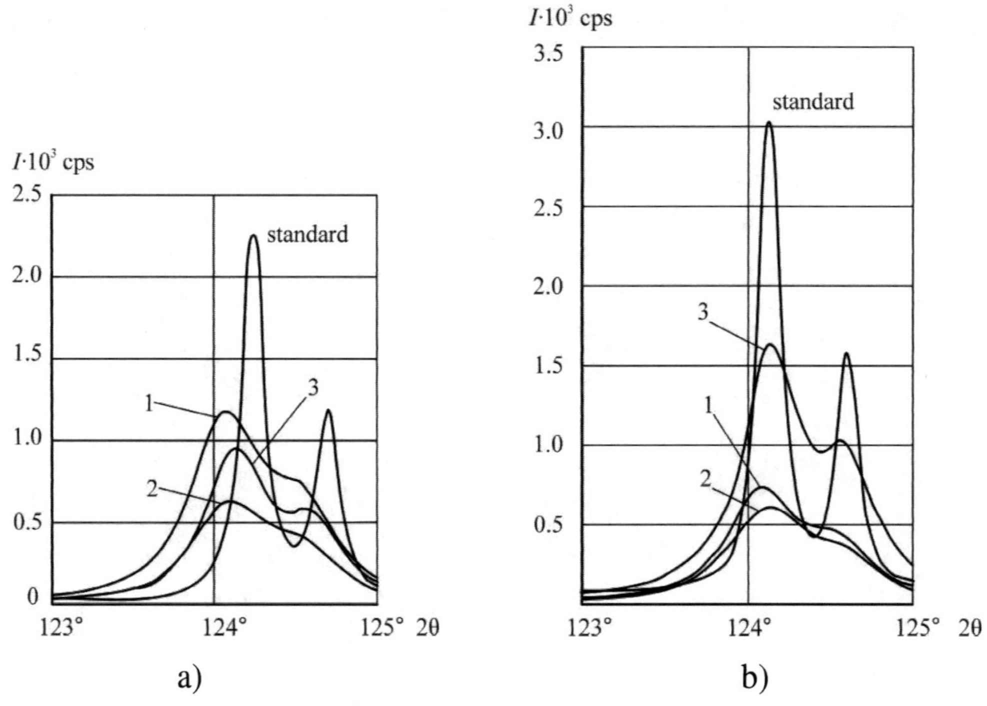

Figure 4. The profiles of the diffraction lines of deformed regions of cast iron samples SCH20 (a), CHVG40 (b): 1 - the area adjacent to the propagation plane of the fatigue crack front; 2 – the area adjacent to the fracture surface; 3 – the area with residual plastic deformations.

It is noticeable that the intensity of the diffraction profile of the crystallographic c planes (220) in the ensemble with the vermicular shape of the graphite inclusions is much lower than for the ensemble of a ferritic-pearlitic base with a flake form of graphite inclusions. This significant difference is probably due to a lower energy expenditure on the crack propagation at a single level of stressed-deformed state. The small shift in the peak of the profile of the diffraction line reflected from the planes (220) in the region 3 of the frontal plane of the cast iron sample of CHVG40 is probably due to microstructural features, namely, the vermicular shape of the graphite inclusions because the quantitative ratio of the flake and vermicular shapes in the investigated cast iron grades is the same.

There is a significant change in the interplanar spaces d depending on the location of the analysis area on the frontal plane of the samples and depending on the cast iron microstructure (table 1). The interplanar spaces d increase in accordance with the stress-strain state of the investigated regions. The increase in interplanar spaces due to a combination of distortions of crystals from microstresses and their dispersion is 30 percent or more.

The obtained results not only correspond to the assumptions concerning the role of flake graphite in cast irons, but also give grounds for a possible quantitative assessment of expenditure of energy of crack propagation in such structures [9-11].

X-ray diffractometry methods also provide insights into of plastic deformation processes by changing the dislocation density, which may reflect a change in the stress relaxation mechanisms arising in the elements of metal constructions under operational loads [8, 12, 13].

In this paper, in accordance with the goal, change in the dislocation density calculated from the broadening of the profile of diffraction lines reflecting the inhomogeneity of the elastic fields in the samples structure of structural steel 08ps is presented [14, 15]. Characteristics of the profiles of the diffraction lines were obtained periodically every 6 months by X-ray analysis of 08 pc samples, which were subjected to a static (for 5 years) loading on an open atmospheric platform and indoors (see the description above).

Figure 5 shows the curves of the change in the dislocation density ρ depending on the level of elastic stresses σ acting in the cross section of the central part of the 08 pc steel samples from January 2013.

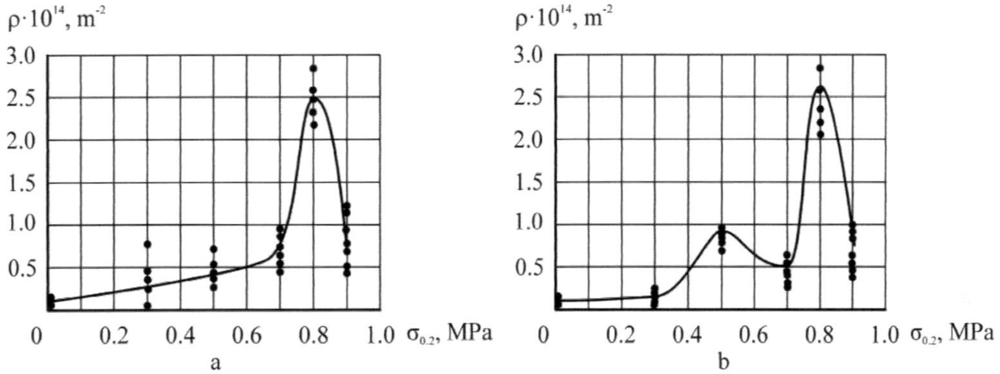

Figure 5. Change in the dislocation density on samples of 08pc steel in the range of elastic stresses: a-indoors, b-on an open atmospheric platform; •-numerical values of ρ calculated by X-ray analysis in the period 2013-2017.

The change in the dislocation density ρ in steel 08 ps samples subjected to static stretching for 5 years is evidence of a nonmonotonic character of resistance to microplastic deformations as the stress level increases from $\sigma = 0$ to $\sigma = 0.9\ \sigma_{0.2}$ (figure 5, a and b). For samples statically loaded on an open atmospheric platform, an increase in the dislocation density is observed at elastic stresses $\sigma = (0.3 \ldots 0.5)\ \sigma_{0.2}$ and $\sigma = (0.7 \ldots 0.8)\ \sigma_{0.2}$. In samples of steel 08 ps experiencing the effect of static elastic stresses under stable temperature conditions (18°C ... 19°C) for 5 years at low stresses $\sigma = (0.3 \ldots 0.5)\ \sigma_{0.2}$ the growth of the dislocation density is clearly not expressed.

5. Conclusion

The new laws of the occurrence of elementary acts of microplastic deformation in structural steels in the range of elastic stresses that correspond to real operational loads in metal constructions discovered by the ongoing research program require further analysis.

The characteristics of the diffraction line profiles of deformed structural steels and cast irons reliably reflect not only the deformed regions, but also the relative level of the distortion of crystal.

Concerning the example of cast iron with a single metal base and different forms of graphite inclusions, it is shown that the features of the microstructure of heterophase alloys have a significant effect on the main characteristics of the profiles of the diffraction lines.

X-ray diffractometry of samples of a number of structural steels and cast irons in different stress-strain states caused by a three-point bending to failure, a prolonged static stretching for 5 years in an open atmospheric platform and indoors, reveals the influence of the substructure features of polycrystalline materials on their service properties.

References

[1] Arabey A 2010 *Chernaya Metallurgiya* **7** 3-10
[2] Vorob'yov Y U, Akimov V and Sokolov Y U 2012 *System Accidents and Disasters in the Technosphere of Russia* (Moscow: FGBU VNII GOCHS (FTS)) 473
[3] Fortov V and Makhutov N 2010 *Mechanical Engineering in Russia: Condition and Development* (Moscow: RAS) 72
[4] Kushnarenko V, Chirkov Y U, Materinko K, Lukashov A and Shchepinov D 2016 *Intellekt. Innovatsii. Investitsii* **7** 117-23
[5] Aleshin N 2007 *Svarka i Diagnostika* **2** 4-10
[6] Klyuyev V, Artem'yev B and Matveyev V 2015 *Industrial Laboratory. Diagnostics of Materials* **81** 73-8
[7] Aleshin N 2011 *Svarka i Diagnostika* **6** 44-7
[8] *Normative Technical Document R 50-54-52-88* 1988 (Moscow: VNIINMASh) 24
[9] Kile Ye, Korchevskiy V and Syuy A 2011 *Fundamental'nyye Problemy Sovremennogo Materialovedeniya* **4** 7-10
[10] Korchevskiy V, Zhukov Ye, Kile Ye and Syuy A 2012 *Vestnik TOGU* **1** 13-8
[11] Cogut M, Caraim V, Lebid N and Panko O 2003 *Problemy Prochnosti* **2** 147-55
[12] Schmidt G 1983 *Mehanicheskaya Ustalost Metallov: Materialy VI Megdunarodnodo* Kollocviuma (Kiev: Naukova Dumka) p 110
[13] Balagurov A, Blagov A, Sanaveskina I, Kovalchuk M, Pisarevski U and Targonski A 2015 *Vestnik RFFI* **2** 85-9
[14] Crivoglaz M 1967 *Theory of X-ray and Thermal Neutron Scattering by Real Crystals* (Moscow: Nauka) 336
[15] Ivanov A, Megennyi Y U, Ostrov A and Fomicheva E 1987 *Zavodskaya Laboratoria* **2** 43-8

Reconstruction of hydrocarbon generation and accumulation on the Gydan Peninsula and adjacent offshore areas using basin modeling approach

A A Deshin[1], P I Safronov[2] and L M Burshtein[2]

[1]Novosibirsk State University, Novosibirsk, Russia
[2]Trofimuk Institute of Petroleum Geology and Geophysics, Novosibirsk, Russia

E-mail: DeshinAA@ipgg.sbras.ru, BurshteinLM@ipgg.sbras.ru

Abstract. Based on structural and lithological grid models, a three-dimensional numerical dynamic model of the Mesozoic-Cenozoic sedimentary cover of the Gydan source kitchen and adjacent territories was built. The thermal history of sediments was reconstructed. A map of effective deep heat flow was compiled. A good agreement was found between the measured and calculated degrees of catagenetic transformation for Upper Jurassic rocks. The possible types of kerogens from the main Jurassic and Cretaceous source rock sequences were justified. The study also provides assessments of the volumes of hydrocarbon generation from the main source rock sequences, timing of formation of regional seals and possible extent of hydrocarbon accumulation in structural traps.

1. Introduction

Because of significant exploration efforts now being devoted to the hydrocarbon resource potential of the Russian Arctic, there is considerable interest in the magnitude and structure of its resource base, which requires further quantification. One of the modern methods for quantifying petroleum resources is basin modeling, which has been earlier applied to the Gydan source kitchen and adjacent areas [1-10]. However, the most recent geological and geochemical data may provide the basis for the development of more detailed models of petroleum generation processes within the Gydan source kitchen and more precise estimates of its hydrocarbon potential.

The study area is located within the Gydan source kitchen bounded by the Antipayutin-Tadebeyakha megasyneclise and the Kara megasyneclise [11]. The Kara megasyneclise covers the southern part of the Kara Sea and the northern tip of the Yamal Peninsula. The Antipayuntin-Tadebeyakha megasyneclise is a negative closed super-order structure that covers the eastern part of the Yamal Peninsula, the Ob Bay and a significant part of the Gydan and Taz Peninsulas.

From a standpoint of petroleum zonation, the study area comprises the Gydan petroleum area, the eastern part of the Yamal petroleum area, and the western part of the Yenisei-Khatanga petroleum area. The largest hydrocarbon accumulations are concentrated at the Utrennee (Salmanovskoe), Geofizicheskoe, and Gydanskoe fields.

The thickness of the sedimentary cover in the study area reaches 7-8 km and the age appears to range from the Triassic through the Quaternary. The proved net-pay zone includes the Upper Jurassic and Cretaceous.

The numerical model of the study area is based on extensive geological, geochemical and geophysical data, including well logs, as well as determinations of petrophysical properties in reservoir and seal formations.

2. Parameters of model

The surface area of the model block is 375,000 km2. The basic structural grid models had a size of 500x700 nodes (with 1 km increment). Note that the simulation is a computationally intensive and time-consuming iterative process. Therefore, for preliminary calculations, the dimensions of the model block were reduced to 250x375 nodes (with a 2 km horizontal step).

The section was subdivided into 34 horizons with isochronous boundaries (35 mutually consistent structural surfaces), which are correlated with the formations or regional clinoforms. For each horizon, a specific lithology was prescribed and determined as a mixture of input lithological types. Petrophysical parameters such as thermal conductivity, heat capacity, etc. were set using the TemisFlow (BeicipFranlab) software database.

An essential step in modeling the evolution of the sedimentary cover is the reconstruction of the amplitude of erosion and paleo-depths. Paleobathymetry and erosion estimations are crucial for reconstructing a thermal history, and, consequently, the timing of hydrocarbon generation within the main source rock sequence (SRS) of the study area. The paleo-depths were determined using the available paleogeographic reconstructions [12, 13].

3. Results

The thermal history of sedimentation was restored based on the data on present-day temperatures, position of the lower permafrost boundary, and vitrinite reflectance [14], which reflects the maximum paleotemperatures.

The calibration of the thermal history took into account a changing density of the effective heat flow at the base of the sedimentary cover. A series of computational experiments using individual model wells and the entire sedimentary cover in the study area allowed a good fit between the calculated and actual values of vitrinite reflectance (figure 1) and present-day temperature profiles [15-17].

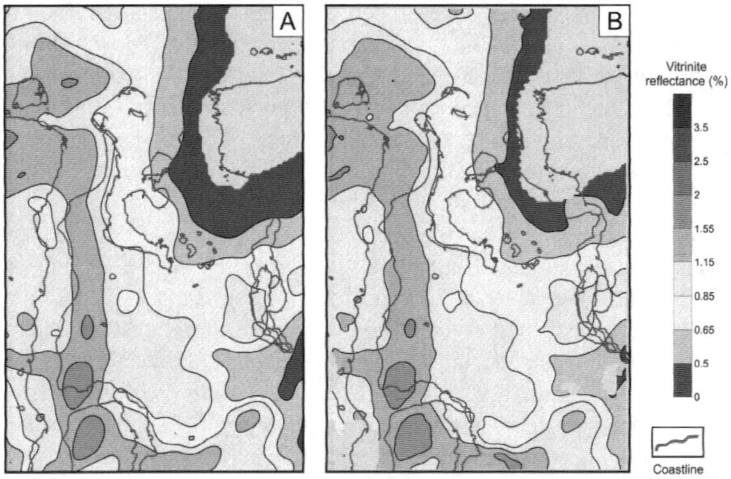

Figure 1. Katagenesis scheme comparison in the roof of Upper Jurassic based on real data (A) [18] and modeling scheme (B).

The data on organic geochemistry of sediments on the Gydan Peninsula and in the western part of the Yenisei-Khatanga trough have been presented in many previous studies [19-21]. The most recent data on organic geochemistry of sediments from the western part of the Yenisei-Khatanga trough and adjacent areas are discussed in detail in [16, 22].

In the model, for each source rock, the input parameters used were kerogen type (which specifies its kinetic characteristics), initial organic carbon content, and initial hydrogen index.

The characteristics of dispersed organic matter (DOM content in sedimentary rocks, hydrocarbon potential, kinetic characteristics of kerogen, etc.) were determined based on the results of geochemical studies conducted at the Institute of Petroleum Geology and Geophysics, Siberian Branch, Russian Academy of Sciences [16, 22]. It is commonly accepted that type I kerogens originate from aquatic lacustrine organic matter, type II kerogens from aquatic marine organic matter, and type III kerogens from terrigenous organic matter. However, the standard characteristics of these types of kerogens needed to be refined based on pyrolysis data [10].

Shale stringers of the Golchikha, Malyshev, Laida, Zimnyaya and Kiterbyut Formations were regarded as potential source rocks. The model predicts that organic matter in the Bazhenov and Malyshev source rocks can be correlated with mixed type II/III kerogens, organic matter in the Laida and Zimnyaya source rocks can be correlated with type III/IV kerogens, and organic matter in the Kiterbyut source rocks can be correlated with type III kerogen.

Using a 3D numerical geological model of the sedimentary cover of the Gydan source kitchen allowed us to reconstruct the extent and dynamics of hydrocarbon (both liquid and gaseous) generation from each source rock sequence (SRS).

The average density of hydrocarbon generation from Zimnyaya source rocks was 11 thousand t/km^2 for liquid hydrocarbons and 90 million m^3/km^2 for gaseous hydrocarbons. The highest generation densities are calculated for the Vnutrennyaya mesodepression and the Southern Messoyakha mesodepression in the south of the study area. The total hydrocarbon generation was 4.4 billion tons for liquid hydrocarbons and 33.6 trillion m^3 for gaseous hydrocarbons.

The average density of hydrocarbon generation from Kiterbyut source rocks was 1260 thousand t/km^2 for liquid hydrocarbons and 1100 million m^3/km^2 for gaseous hydrocarbons. The highest densities of hydrocarbon generation are calculated for the central and southeastern part of the study area. Source kitchens are confined to a series of first-order negative structures: Tadebeyakha megatrough, Yaptik-Sale, East Antipayutin, Yenisei, and North-Taz megadepressions (figure 2). The total hydrocarbon generation from the Kiterbyut source rocks was 472 billion t for liquid hydrocarbons and 415 trillion m^3 for gaseous hydrocarbons.

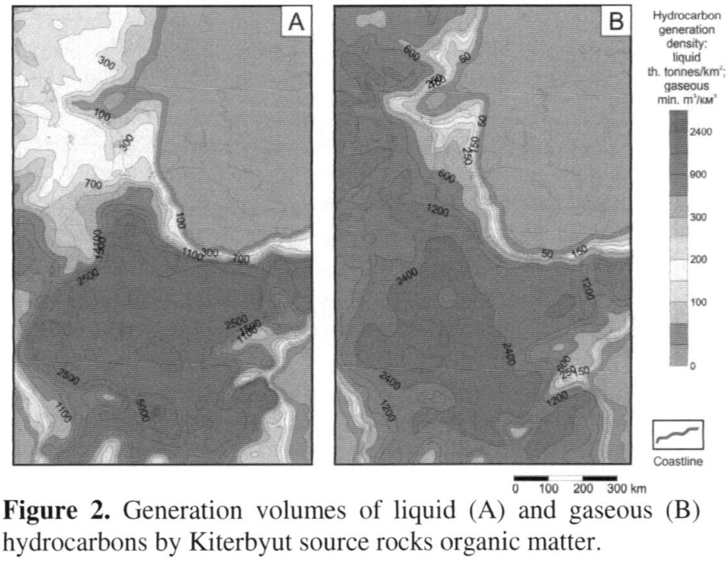

Figure 2. Generation volumes of liquid (A) and gaseous (B) hydrocarbons by Kiterbyut source rocks organic matter.

The average density of hydrocarbon generation from Laida source rocks was 190 million m^3/km^2 for gaseous hydrocarbons and 156,000 t/km^2 for liquid hydrocarbons. The highest densities of hydrocarbon generation are calculated for the central and eastern parts of the study area. Source kitchens are confined to the Yaptik-Sale, Tadebeyakha megadepressions, and East Antipayutin depression. The total hydrocarbon generation from Laida source rocks was 58 billion t for liquid hydrocarbons and 71 trillion m^3 for gaseous hydrocarbons.

The average density of hydrocarbon generation from Malyshev source rocks was 617 thousand t/km^2 for liquid hydrocarbons and 300 million m^3/km^2 for gaseous hydrocarbons. Source kitchens are confined to the central, eastern, and southern parts of the area, which include the Tadebeyakha megatrough, Anitipayutin-Tadebeyakha and Yenisei mega depressions (figure 3). The total hydrocarbon generation from Malyshev source rocks was 230 billion t for liquid hydrocarbons and 112 trillion m^3 for gaseous hydrocarbons.

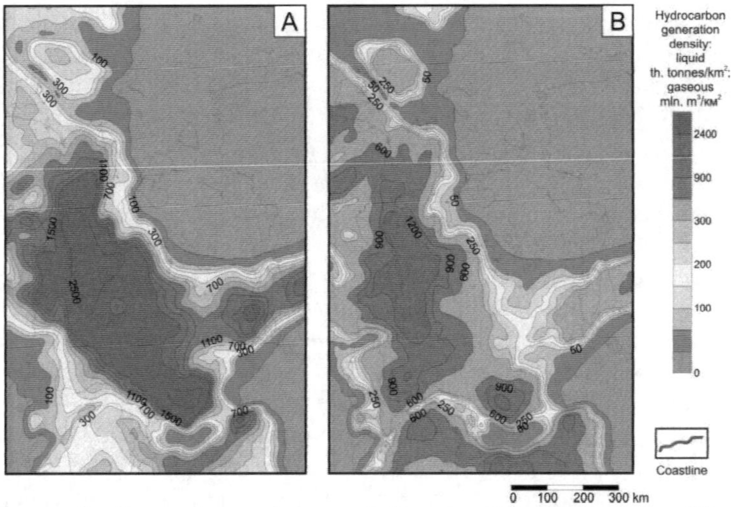

Figure 3. Generation volumes of liquid (A) and gaseous (B) hydrocarbons by Makyshev source rocks organic matter.

The average density of hydrocarbon generation from Bazhenov source rocks was 70 million m^3/km^2 for gaseous hydrocarbons and 680,000 t/km^2 for liquid hydrocarbons. Source kitchens were relocated from the center to basin's periphery, being confined to the Central Kara megadepression in the north, Yenisei megadepression in the east, North Taz megadepression in the south (figure 4). The total hydrocarbon generation from Bazhenov source rocks was 18.6 billion t for liquid hydrocarbons and 18 trillion m3 for gaseous hydrocarbons.

The cumulative hydrocarbon generation from all source rock sequences in the study area was estimated to be 950 billion t of liquid hydrocarbons and 650 trillion m^3 of gaseous hydrocarbons. It should be noted that at this stage secondary cracking processes were not taken into account by our model. But taking these into account, the actual ratio of generated volumes should be shifted toward gaseous hydrocarbons. In view of the above, the amounts generated from Kiterbyut, Malyshev, and Bazhenov source rock sequences account for >50% ~24%, and 20% of total liquid hydrocarbons, respectively. The generation from Laida and Zimnyaya source rocks accounts for only 6% and <1% of total liquid hydrocarbons, respectively.

The highest generation of gaseous hydrocarbons was from Kiterbyut source rocks, which accounts for 64% of the total amount. The amounts generated from Malyshev, Laida, Zimnyaya, and Bazhenov source rock sequences account for ~17%, 11%, 5%, and 3% of total gaseous hydrocarbons, respectively.

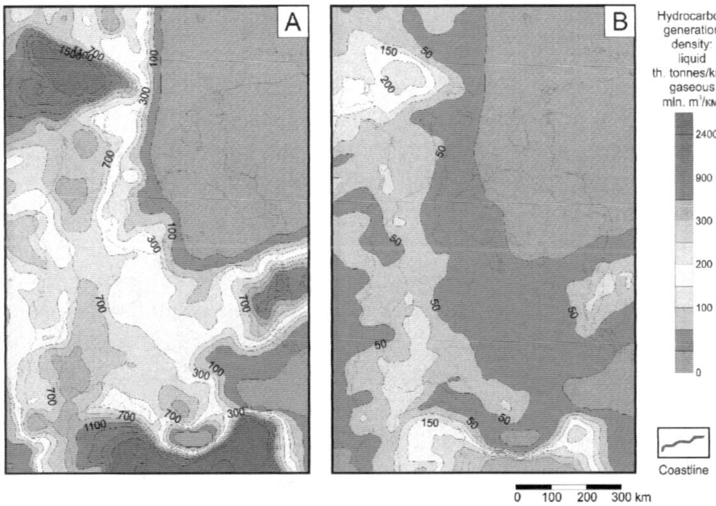

Figure 4. Generation volumes of liquid (A) and gaseous (B) hydrocarbons by Bazhenov source rocks organic matter.

The formation of hydrocarbon fields requires a combination of a number of factors. For example, the preservation of hydrocarbon accumulations, the volume and phase composition of hydrocarbons in traps depends on the timing of generation and lithification of seals. In view of this, we estimated the timing of lithification of the main seals within the Gydan source kitchen area. The boundary value of input porosity at which the rock is capable of retaining hydrocarbons was set to be 30% [7]. The Kimmeridgian-Volgian seal composed of shales of the Bazhenov horizon was lithified at the end of the Early Cretaceous- and beginning of the Late Cretaceous, at ~99 Ma (figure 5). The Lower Hauterivian seal represented by shales of the Ur'yev clinoform was lithified by the beginning of the Late Cretaceous, i.e., at ~100 Ma. The Lower Aptian seal represented by shales of the Alym horizon was lithified at ~80 Ma, in Campanian times. The Albian seal was lithified throughout the netire area at 70 Ma during the Maastrichtian. The Turonian seal represented by shales of the Kuznetsov horizon was lithified only locally in the mid-Paleogene, at 45 Ma.

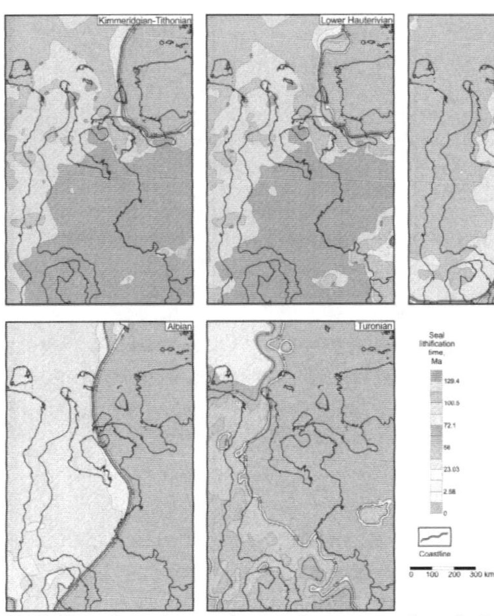

Figure 5. Lithification time scheme for general seals in the study area.

To estimate the possible extent of accumulation of hydrocarbons, which is planned to be implemented at the next stage of our study, we evaluated the rock volumes of traps within the main petroleum-bearing complexes of the study area (figure 6). The results were exemplified by the Valanginian-Aptian petroleum-bearing complex. The location of traps throughout the section shows that the actual trap geometries agree satisfactorily with the model ones. In addition to geometries, the calculated volumes of model traps are well consistent with the volumes of actual traps in the corresponding complex (figure 6).

To summarize, the study provides a quantitative assessment of the generation of liquid and gaseous hydrocarbons and possible extents of hydrocarbon accumulation. To evaluate the amounts of hydrocarbons accumulated in the study area, it is necessary to account for their losses during expulsion and migration. Based on earlier assessments [1, 7], it is assumed, in a first approximation, that the accumulation coefficients in these regions may be equal to ~1% of the total generation volumes. Then the initial geological resources of hydrocarbons associated with the Gydan source kitchen would be not less than 16 billion t of conventional hydrocarbons.

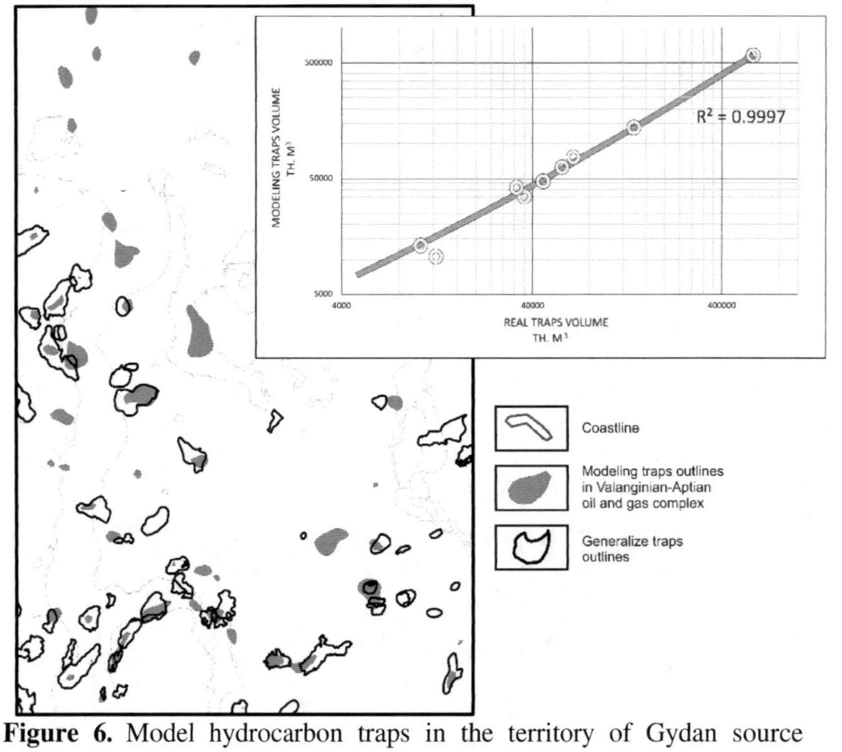

Figure 6. Model hydrocarbon traps in the territory of Gydan source kitchen.

The simulation results were used to estimate the extent of hydrocarbon generation from the main source rock sequences within the sedimentary cover of the Gydan source kitchen area, the timing of seal formation and possible volumes of hydrocarbon accumulation. The next step will involve evaluation of the phase composition of hydrocarbons within the trap and description of individual traps in the main petroleum-bearing complexes.

Acknowledgments

The reported study was funded by RFBR according to the research project № 18-302-00001 mol_a, as part of the basic project of the Russian Academy of Sciences IX.131.4.1. and with the technical support of the Russian Representative Office of BeicipFranlab.

References

[1] Fjellanger E et al. 2010 *Petroleum Geology Conference Proceedings* **7** 659-68
[2] Littke R et al. 1999 *AAPG Bulletin* **83** 1642-65
[3] Schaefer R 1999 *Chemical Geology* **156** 41-65
[4] Bostrikov O, Larichev A and Fomichev A 2011 *Neftegazovaya Geologiya. Teoriya i Praktika* **6** p 19
[5] Malysheva S 2012 *Oil Industry* **12** 14-47
[6] Malysheva S 2011 *Oil Industry* **3** 13-7
[7] Kontorovich A 2013 *Geology and Geophysics* **54** 1179-226
[8] Stupakova A et al. 2014 *Prospect and Protection of Mineral Resources* 47-51
[9] Safronov P, Ershov S, Kim N and Fomin A 2011 *Geology of Oil and Gas* 48-55
[10] Deshin A, Safronov P and Burshteyn L 2017 *Geology of Oil and Gas* 33-44
[11] Kontorovich V 2001 *Geology and Geophysics* **42** 1832-1845
[12] Kontorovich A 2014 *Geology and Geophysics* **55** 745-76
[13] Kontorovich A 2013 *Geology and Geophysics* **54** 972-1012
[14] Fomin A 2011 *Catagenesis of Organic Matter and Oil and Gas Content of the Mesozoic (Jurassic, Triassic) and Paleozoic Deposits of the West Siberian Megabasin* (Novosibirsk: Trofimuk Institute of Petroleum Geology and Geophysics Press) p 331
[15] Duchkov A 2015 (http://maps.nrcgit.ru/geoterm/)
[16] Kim N and Rodchenko A 2013 *Geology and Geophysics* **54** 1236-52
[17] Kurchikov A and Stavitsky B 1987 *Geothermy of Oil and Gas Bearing Areas of Western Siberia* (Moscow: Nedra) p 134
[18] Kontorovich A 2009 *Geology and Geophysics* **50** 1191-200
[19] Filitsov Yu 2006 *Geology, Geophysics and Development of Oil and Gas Fields* 52-7
[20] Larichev A et al. 2003 *Geology, Geophysics and Development of Oil and Gas Fields* 4-13
[21] Boldushevskaya L 2001 *Geochemical Criteria for Forecasting the Oil and Gas Content of the Mesozoic Deposits of the Yenisei-Khatanga Regional Trough and the Northeast of the West Siberian Plate* (Author's abstract of dissertation of the candidate of geologo-mineralogical sciences, Krasnoyarsk) p 22
[22] Kim N 2013 *Materials of the 1st All-Russian Youth Scientific and Practical Conference on Sciences about the Earth* (Novosibirsk: Novosibirsk State University Press) pp 96-7

Large ship motion mechanics in "narrow" ice channel

A A Dobrodeev, N Yu Klementyeva and K E Sazonov

Krylov State Research Centre, St.Petersburg, Russia

E-mail: A_Dobrodeev@ksrc.ru

Abstract. The paper considers the motion mechanics of a large-size ship sailing in a "narrow" channel behind an icebreaker whose breadth is not as wide as that of the ship. Conditions when the ship is heading asymmetrically with respect to the channel axis are analyzed; ship ice resistance is estimated for the symmetric as well as asymmetric in-channel motion scenarios.

1. Introduction

The motion mechanics of large-size ships in ice conditions have practical importance in development of marine transportation systems to export hydrocarbons from the Russian Arctic regions. Recently, heavy-tonnage vessels have been increasingly used for cargo carrying purposes in the Arctic seas. The latest example is newly commissioned LNG carriers of "Christophe de Margerie" type. The advent of such vessels poses a range of challenges in the field of marine ice engineering, with the need to update the ice sailing tactics for the vessels of that size considered as one of the most important tasks [1].

The problem why traditional methods of ice sailing tactics are no longer fit for large-size vessels is that their hulls are much wider than the breadth of any icebreaker in operation or under construction. At present two methods of icebreaker-assisted operation are practiced in Arctic shipping, namely, a traditional approach when vessels are led by an icebreaker and an alternative option when two icebreakers are used [2].

A large-size carrier led by one icebreaker has to shoulder through a "narrow-width" channel grinding ice edges with her own hull. This scenario has been repeatedly examined in theoretical studies on Arctic shipping [3, 4]. Usually it is assumed that the ship's centerline is aligned with the channel axis, while interactions of ship sides with the ice cover is symmetrical. However, sea trials as well as self-propelled model tests suggest that in some cases there is no such symmetric pattern for a large size vessel running in ice channel. In figure 1 there is a photo (borrowed from [5]) showing the *Propontis* oil tanker (beam 44 meters) led by the *Taimyr* nuclear icebreaker (beam 28 meters). It is well seen that the tanker is positioned in the channel to break ice cover with one side, her other side rubbing the ice edge. The purpose of our study is to consider the motion mechanics of a large ship sailing in a "narrow" ice channel.

Figure 1. *Propontis* oil tanker led by the *Taimyr* nuclear icebreaker

2. Object and methods of investigation

Calculations of the ship motion were performed to find out the factors that influence the position of a large-size ship in a narrow-width channel as well as to determine ship's ice resistance.

Two hypothetical large-size vessels of ice class were taken for the study. They are assumed to have the same main particulars, except the flare angle of parallel middle body. The first ship has no flare angle, while the second one has a flare of 10^0. The principal dimensions are given in the table 1 below:

Table1. Main data of large-size vessels

Length on waterline, m	240
Beam on waterline, m	50
Draft, m	12
Middle body length, m	140
Waterline slope at the bow, deg.	45
Angle of 0 frame, deg.	70

Forces induced by interaction with ice channel edges on ship hull were calculated in function of hull's lateral offset from the channel axis. Also, net ice resistance of ship and total lateral ice force were estimated. The ice resistance was found using a semi-empirical formula of B.P. Ionov [6] providing quite a straightforward procedure to take into account the narrow ice channel as well as asymmetric pattern of ice breaking by ship sides. Under symmetric ship motion with respect to the ice channel axis a new hull is "constructed" with its breadth reduced by the channel width. The stem angle with respect to the running waterline is defined by the actual waterline angle at the point of contact with the channel edge, while the functions depending on the angles of ship's frames are calculated for the hull length from the contact point of hull side/ice edge to the point of maximum hull beam. Appropriate corrections are introduced into the forebody length. At the asymmetric ship orientation with respect to the channel axis the above-described procedure is used to estimate the resistance of each hull side separately and the result should be reduced by half.

The method suggested by B.P. Ionov can be used only for estimating the ship ice resistance. Lateral forces on the opposite ship sides were found by applying one of the Shimansky coefficients η_2 expressed as follows [7]:

$$\eta_2 = \frac{\sum_I}{\sum_{III}}; \ \sum_I = \int_0^{L/2} \frac{\cos x' \cos y'}{\cos\alpha'}; \ \sum_{III} = k\int_0^{L/2} \frac{\cos^2 x'}{\cos\alpha'} dx; \qquad (1)$$

where $\sum_I, \sum_{III}$ are total lateral and longitudinal forces on one ship side $\cos x', \cos y'$ are cosines of angles between the normal to the waterline point under consideration and the O_x –axis and O_y -axis (x -axis is positive from midship to bow, y -axis is positive to starboard); $\cos\alpha'$ is the waterline slope angle with respect to the point under consideration; L' - entrance length; k - proportionality factor.

This procedure was used to estimate ice-induced forces on the above-described hypothetical hulls of large-capacity vessels sailing straight with examination of variable offsets between ship's centerline and the ice channel axis, starting from the symmetric case. The results are presented as plots of the total lateral force in figures 2 and 3. In these graphs the ice resistance of asymmetric cases is referred to the ice resistance found for the same vessel in the symmetric position.

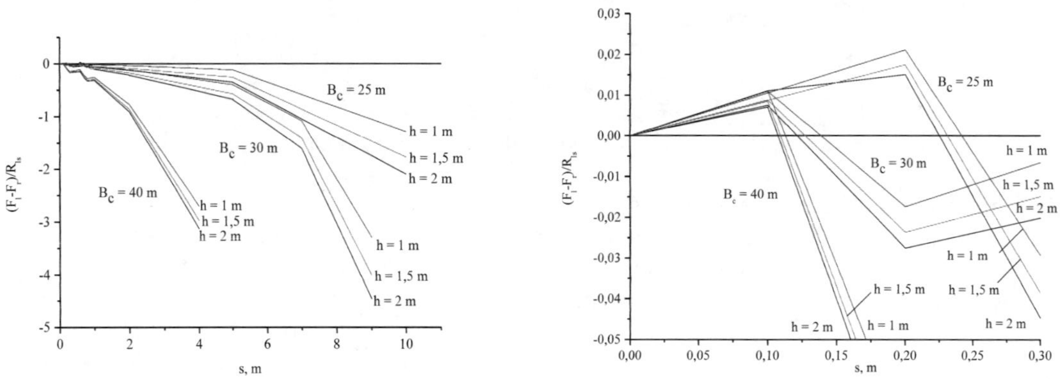

Figure 2. Relative total lateral force versus ship CL offset from channel axis, channel width and ice thickness. Flare angle in way of parallel middle body is 0^0. a – full curve; b – initial curve section at small offsets.

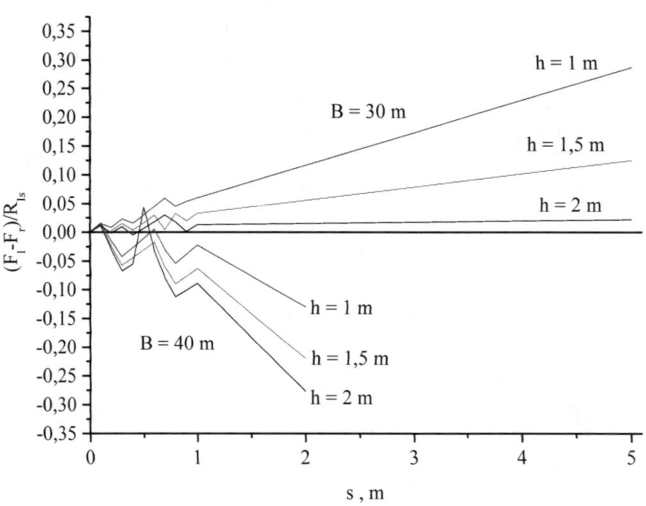

Figure 3. Relative total lateral force versus ship CL offset from channel axis, channel width and ice thickness. Flare angle in way of parallel middle body is 10^0

Ship ice resistance for asymmetric case was calculated as:

$$R_I = \frac{R_{mI}}{2}\left(1 + f_{Id}\eta_{m2}\right), \tag{2}$$

where R_{mI} - ice resistance of the ship side breaking the channel ice edge; f_{Id} - hull/ice friction coefficient; η_{m2} - Shimansky coefficient determined for the hull side part breaking the ice edge. Figure 4 presents the results as plots of the relative ice resistance versus the width of "narrow" ice channel.

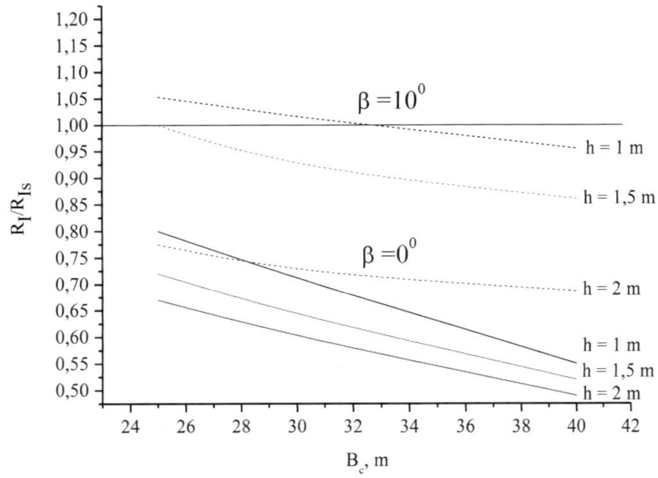

Figure 4. Relative ice resistance of ship versus channel width for asymmetric case.

3. Discussion

The data presented in figure 2 suggest the following mechanism causing a large-size vessel to deviate from symmetric motion in narrow ice channels. Firstly, as it is seen from figure 2b, the symmetric motion is stable at small excursions of the ship centerline from the channel axis. At small offsets a lateral restoring force is generated, returning the ship to her initial position with respect to the channel axis. However, stronger disturbances due to local ice thickness and/or strength variations may disrupt this equilibrium. Then the ship is exposed to a disturbing force (figure 2a) growing with the offset, and the ship position becomes unstable. Transition to another kind of stable state occurs, i.e. to an asymmetric ship position with respect to the channel axis, where the total lateral (restoring) force is zero. The asymmetric ship position is stable and it takes rather large forces to alter this ship orientation.

The above-described mechanism of loss in position stability may arise only if the ice resistance of starboard side, which is cutting into ice, happens to be less than that of the port side, withdrawing from ice. This situation is quite likely in case of large-size vessels, featuring vertical sides in way of the parallel middle body. At given ship dimensions her ice resistance depends on the bow shape [8] defined by angles of the waterline slope, angles of frames and bow lines. Most of the large-size vessels feature very small angles of waterline slope and frames in way of the entrance/middle body transition. If these parts of the hull come into contact with ice, the ice resistance is increased. If the ice contact area is shifted forward to the stem, the icebreaking performance is noticeably improved. For this reason it can be expected that there is a hullform solution for the large-size vessels enabling them to avoid or mitigate the above-discussed effect.

This assumption is confirmed by estimations for a large-size vessel whose frames in way of the middle body are inclined to 10^0. The results illustrated in figure 3 show that at small excursions, when the angle of frames in way of the middle body have no practical effect on ice resistance, the ship position proves to be stable. At larger excursions the position stability depends of the "narrow" channel width. If the channel width is relatively small (30 m in our case study), then the symmetrical position of ship with respect to the channel axis is always unstable. It can be explained by the fact that in this case ice edges interact with those parts of the ship sides that tend to break ice by bending due to their geometrical characteristics. For this reason the ship side cutting into ice has larger resistance as compared to the ship side withdrawing from ice, which gives rise to a restoring lateral force.

In a wider ice channel its edges interact with those parts of ship sides whose geometric characteristics are not as good for this job, therefore the symmetry of position can be disrupted. It should be noted that in this case the disturbing force is much weaker than in the case when there is zero angle of side inclination in way of the parallel middle body.

Greater disturbing forces in case of vertical sides with zero flare in way of the parallel middle body are determined by a change of ice breaking pattern from bending to splitting and crushing. At 10^0 angle of hull side inclination there may be a situation when the ship would break ice in a relatively wide channel by bending combined with splitting and crushing patterns. It is for this reason that the level of ice forces is found to be significantly lower.

Fig.4 gives an idea of how the ice resistance of ship changes when the symmetry of motion with respect to the channel axis is disrupted. It is seen that a large vessel oriented asymmetrically in an ice channel has a noticeably lower ice resistance as compared to the symmetric scenario. In thick ice this reduction may reach 50%. However, when ship sides are inclined this resistance reduction is much less reaching up to 25%. This effect is explained by particulars of the ship hullforms. Anyway, the identified changes in ice resistance can be used for increasing the speed of vessels sailing behind icebreakers.

4. Conclusions

The loss of motion symmetry is described for large-size vessels sailing in ice channels behind an icebreaker whose hull breadth is not as wide as that of the vessels. Simple calculation formulas was used to qualitatively explain the mechanism behind this loss of symmetry related to specific hullform features of large-size vessels. It is found that when a large-size vessel is sailing through a "narrow" ice channel there can be one or two stable positions for the vessel. The symmetric position of vessel with respect to the channel axis is stable, but for vessels with vertical sides in way of parallel middle body this stability can be easily disrupted. In this case the vessel would take a stable asymmetric position, with one of her sides breaking ice, while the other side rubbing the channel edge. For vessels with inclined sides in way of parallel middle body the symmetric position could be always stable or the above-described asymmetric scenario could happen.

According to the calculations performed in this study the ice resistance of a large vessel sailing in ice thicker than 1 m is reduced under asymmetric orientation. Depending on the hullform this reduction may reach up to 50% of the ice resistance under asymmetric orientation. This fact should be taken into account in assigning the safe distances for large-size vessels led by icebreakers.

References

[1] Pustoshnyi A V, Sazonov K E 2015 Problems of shipbuilding science at the current stage of Arctic exploration *Herald of the Russian Academy of Science* **85**(4) 316-20

[2] Sazonov K E, Dobrodeev A A 2014 Different technologies for making a wider channel in ice for large-size ships *Proceedings of the International Offshore and Polar Engineering Conference ISOPE* (*Busan*) pp 1171-6

[3] Sazonov K E 2011 Navigation challenges for large-size ships in ice conditions *Ship and Offshore Structures* **6** 231-8
[4] Dobrodeev A A, Sazonov K E 2018 Fast sailing in ice – the new goal of model studies *The Naval Architect* pp 22-4
[5] Ruksha V V, Belkin M S, Smirnov A A and Arutyunyan V G 2015 Structure and dynamics of cargo shipments via Northern Sea Route: history, present day and future prospects *Arktika: ekologia i ekonomica* **4**(20) 104-10
[6] Ionov B P, Gramuzov E M 2001 *Ship propulsion performance in ice* (St.Petersburg : Sudostroenie) p 512
[7] Kashtelyan V I, Ryvlin A Ya, Fadeev O V, Yagodkin B Ya 1972 *Icebreakers* (Leningrad: Sudostroenie) p 286
[8] Sazonov K E 2010 *Theoretical principles of ship sailing in ice* (St.Petersburg: Krylov Shipbuilding Research Institute) p 274

Analysis of unfrozen water in suspensions of permafrost bacteria with NMR technique at the temperature of permafrost occurrence

L S Podenko, N S Molokitina, A O Drachuk, O A Simonov and E O Simonova

Institute of Earth's Cryosphere, Tyumen, Russia

Abstract. The data of investigation of microorganisms influence on water phase state in frozen suspensions of from permafrost and frozen solutions of their metabolites were obtained using nuclear magnetic resonance (NMR) relaxation spectroscopy. The measurements were taken at the temperature of permafrost occurrence. For the first time, NMR relaxation parameters were determined for suspension of bacteria extracted from permafrost and solution of their metabolites. Quantity of unfrozen water was compared and contrasted in frozen suspension of bacteria extracted from permafrost, frozen solution of their metabolites and frozen water solutions of high-molecular cryoprotectors. According to the given results, it was found out that the weight of unfrozen water might be much more than the weight of biomaterial contained in the suspensions of bacteria extracted from permafrost and solutions of their metabolites at the temperature of permafrost occurrence. It was shown that microorganism metabolites extracted from permafrost exerted higher influence on unfrozen water content than higher-molecular cryoprotectors.

1. Introduction

At the present time there is overwhelming evidence that some of microorganisms preserved in ice or permafrost can not only live but keep their functioning [1-4].

Mechanisms of this survival have been staying unknown up to the present day. One of the most important factors affected on germinative faculty of biological objects is water phase state (liquid or solid) of their habitat where nutritious matter and metabolism products are transported. At the temperatures of permafrost occurrence some quantity of intracellular and extracellular water might stay unfrozen for an unlimited period of time. However, the influence of the presence of biomaterial extracted from permafrost on water phase state has not been yet estimated.

Microorganism survival at negative temperatures is associated with an outflow of peculiar substance named a cryoprotector [5]. Presence of cryoprotectors let microorganisms keep functioning for a long time over a wide temperature range. However, the influence of metabolites of microorganisms extracted from permafrost on the quantity of unfrozen water in frozen bacterial suspensions has not studied yet.

This paper presents the first results of our study on the influence of microorganisms extracted from permafrost on the quantity of unfrozen water in frozen bacterial suspensions and frozen solutions of their metabolites. To determine water phase state the method of NMR relaxation spectroscopy was used.

Content of unfrozen water in frozen bacterial suspension, frozen solution of their metabolites and frozen water solution of high-molecular cryoprotector measured to compare and contrast the influence of microorganisms extracted from permafrost and their metabolisms on the quantity of unfrozen water.

2. Experimental section

Bacteria *Bacillus sp (*in accordance with the 16s RNA sequence analysis) strain MG8, soilborn sporogenous, gram-positive organisms were used. This bacteria strain was extracted from permafrost of late Neogene (Central Yakutia, Mammoth Mountain). A suspension of microorganism cells and solutions of metabolites were prepared by the staff of the Tyumen Scientific Center, SB, RAS using previously developed methods [6,7].

The proportion of solid residue (the ratio of the dry residue mass to the initial sample mass) suspensions of *Bacillus sp.* strain MG8 and the solutions of their metabolites (MG8) (further bacterial suspensions and metabolites solutions correspondingly) was 0.1 wt % and 0.015 wt % correspondingly.

Polyvinyl alcohol (PVA) was used as a higher-molecular cryoprotector [5]. Polyvinyl alcohol [-CH2-CH(OH)-] n – artificial, water-soluble, nontoxical, biocompatible polymer. In this work, it was used PVA 16/10 GOST 10779-78. The PVA concentration for PVA solution was 0.1 wt % in all experiments.

The samples of bacterial suspensions and PVA solution were loaded into the refrigerating chamber. Then, the samples were held there at the temperature of –20 °C no less than 24 hours.

Nuclear magnetic relaxation measurements were carried out with the Bruker Minispec mq NMR relaxometer operating at 20 MHz. The Carr-Purcell-Meiboom-Gill (CPMG) pulse sequences were used to determine the spin-spin relaxation time T_2. The interval between pulses was set at 3.5 ms. During the CPMG pulse sequence realizing the first radio frequency deflected nuclear magnetization at an angle of 90°. Further 180° pulses led to the formation of spin echo [8,9]. Quantity of unfrozen water was determined by comparison of NMR signal of a calibration sample and NMR signal of the measured sample. Polyethylsiloxane liquid PES-5 was used as the calibration sample. PES-5 is polydiethylsiloxane polymer of linear structure.

As shown exponential approximation nuclear magnetic decay, one-exponential decay occurred for all measured samples:

$$A = A_0 exp\ (-t/T_2), \quad (1)$$

where A_0 – the maximum amplitude of the nuclear magnetization.

3. Results and discussion

Sedimentation processes of biomaterial in suspensions may decrease their spatial homogeneity in time. Samples of initial just-prepared suspensions had approximately the volume of 100 ml. However, a sample of suspension that extracted for NMR relaxation measurements had the volume no more than 2 ml. Within this framework, spatial homogeneity might lead to great spread of NMR measurement results. Thus, to exclude the presence of spatial homogeneity in bacterial suspensions and the influence of the sedimentation of biomaterial on the NMR measurements, time T_2 was determined for samples taken step by step at a height (top to bottom) with the 1 ms step in just-prepared suspensions. Obtained data of the relaxation time T_2 was shown in table 1.

According to the received results, T_2 for various samples collected from one initial just-prepared suspensions reproduced well (measurement errors are close to the instrument measurement error T_2 of 5%). That indicated a lack of influence of the processes of biological material sedimentation to the results of NMR measurements for NMR measurements just-prepared samples of bacterial suspensions.

NMR measurements of T_2 were started over with the same samples at the same conditions after two days. During these two days the sample were being stored at the constant temperature of 4 °C. The average relaxation time of T_2 for just-prepared microbial suspensions did not differ (within the error) from the average relaxation time of T_2 measured after two days of storage. Samples of suspensions intended for NMR measurements were not stored for more than two days.

Table 1. Spin-spin relaxation time T_2 at the temperature of 40 °C for suspensions of *Bacillus sp.* strain MG8.

Sample number	T_2 (ms)	$\lvert (T_2 - T_{2av}{}^a)100/T_{2av} \rvert$ (%)
1	1986	1
2	2064	4
3	1864	6
4	1987	1
5	2003	1
6	1985	0
7	1977	0
8	1936	2
Average relaxation time	1975	2

[a] T_{2av} – arithmetic average of a relaxation time T_2.

The relaxation time T_2 of the samples of bacterial suspensions and metabolite solutions stored at the temperature of –20 °C was measured at the temperature of –5 °C. Likewise, the suspension samples that were measured at the temperature of 40 °C, all suspension samples that were measured at the temperature of –5 °C, the nuclear magnetization decay was mono-exponential. Experimental data of measured relaxation time T_2 is given in the table 2.

Table 2. Spin-spin relaxation time T_2 at the temperature of –5 °C for frozen bacterial suspensions and frozen metabolite solutions.

Sample number	T_2 (ms)	$\lvert (T_2 - T_{2av})100/T_{2av} \rvert$ (%)
	Suspensions *Bacillus sp.* strain MG8	
1	194	33
2	188	29
3	172	18
4	109	25
5	109	25
6	102	30
	146	27
	Metabolites MG8	
1	90	8
2	101	22
3	74	11
4	74	11
5	66	20
6	93	12
	83	14

According to the data presented in the table 1 and table 2 it might be conclusion that the freezing process dramatically decreased relaxation time T_2.

The proton-containing components in the frozen bacterial suspensions were ice, bacteria cells, unfrozen water. It is known that the relaxation time T_2 of ice protons is no more than several microseconds. Hence, amplitude of the nuclear spin echo was measured after the period of time of several milliseconds. For this time, the signal related to ice almost declined up to the zero. Thus, the ice proton contribution might be neglected for the samples of bacterial suspensions.

Intracellular water mass that remained unfrozen at the temperature lower than –20 °C was more than biomaterial mass. Coincidentally, the specific content of protons in water was several times higher than in the biomaterial. Consequently, the contribution of biomaterial into the NMR signal at the temperate of –20 °C might be neglected as well.

Whence, NMR signal of the frozen bacterial suspensions at the temperature lower than –20 °C related only with the presence of unfrozen water in the samples.

For the relaxation time of the nuclear magnetization of liquid water protons in biological systems usually take place the following correlation [8]:

$$T_2^{-1} = pT_{2s}^{-1} - (1 - p)\, T_{2v}^{-1}, \tag{2}$$

where T_{2s} – spin-spin relaxation time of water connected with biomaterial, T_{2v} – spin-spin relaxation time of bulk water, p – portion of water connected with biomaterial. Relaxation time T_{2s} was much lower than relaxation time T_{2v}.

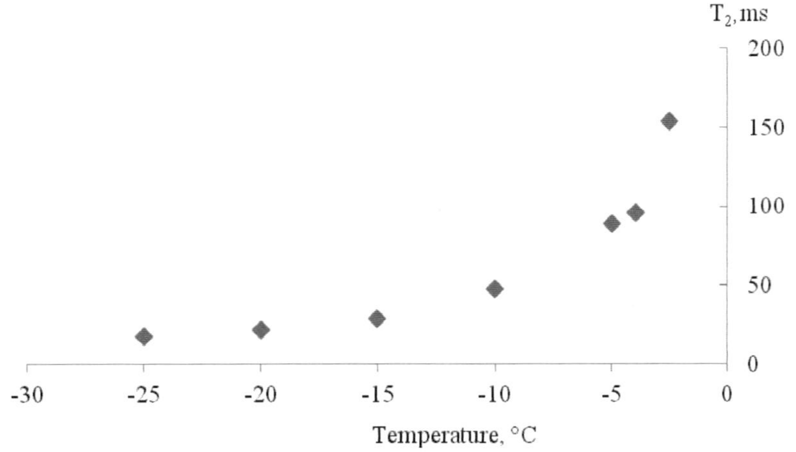

Figure 1. Changing relaxation time T_2 for the frozen metabolite suspensions of MG8.

In figure 1, it was shown the data that demonstrated changes in relaxation time T_2 of metabolite suspensions with the temperature. With the temperature decrease from 0 to –15 °C the relaxation time T_2 decidedly decreased (figure 1). According to the expression (2), the portion of water connected with biomaterial enhanced. With the further temperature decrease the relaxation time T_2 did not significantly change. Ordinarily, it is a reflection of the fact that water is in a well-connected with biomaterial [8].

The data of unfrozen water quantity measurements for the frozen MG8 metabolite solution and frozen water solutions of high-molecular cryoprotector (0.1 wt % PVA solutions) was shown in figure 2. According to the presented results there might be seen that with temperature decrease up to –15 °C the quantity of unfrozen water significantly decreased for metabolites.

Further temperature decrease up to –25 °C showed far less changing of unfrozen water quantity. There was evidence that unfrozen water was inseparable from biomaterial at the temperature no lower than –15 °C. It is well coincided with the results of relaxation time measurements. As to frozen solutions of PVA, unfrozen water in it was detected only at the temperature no less than –5 °C.

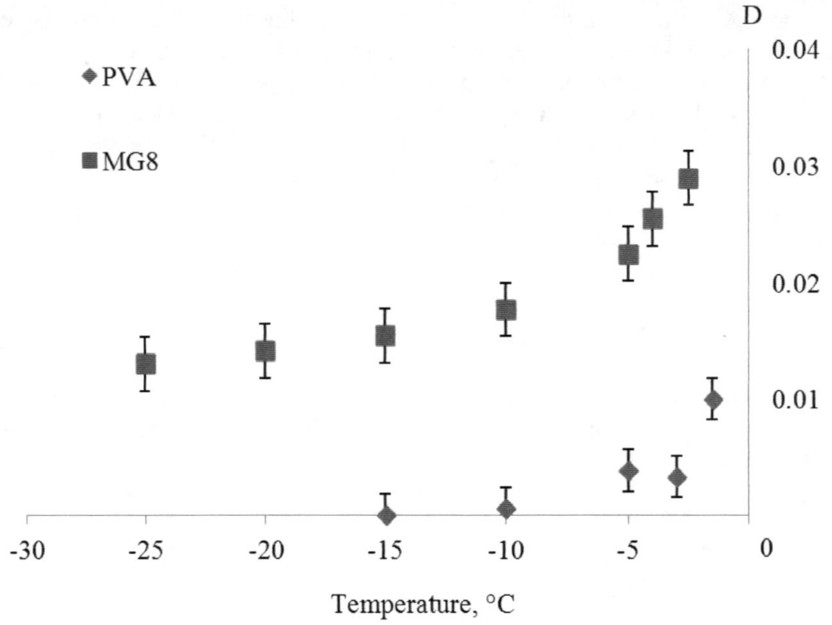

Figure 2. Changing of unfrozen water quantity in frozen suspension of metabolites MG8 and water solution of PVA (0.1 wt %).

To estimate the influence of substances loaded into water on the presence of unfrozen water in it at the negative temperatures, parameter D_m was used. D_m meant the relation of unfrozen water mass to substances mass contained in samples. The parameter D_m Increasing was evidence the fact that biomaterial had an influence on the quantity of unfrozen water and vice versa. Dry solid mass of the samples was designated as biomaterial mass of the samples. The calculated values of the D_m parameter for the studied systems are given in table 3.

Table 3. Parameter D_m of frozen samples.

System	Parameter D_m	
	–5 °C	–10 °C
Bacterial suspension *Bacillus sp.* strain MG8	32	23
Metabolite suspension MG8	220	190
PVA water solution (0.1 wt %)	4	0

According to the data presented in table 3, the unfrozen water mass in frozen suspensions of microorganisms extracted from permafrost and frozen suspensions of their metabolites might be much more than the mass of biomaterial contained in the samples.

4. Conclusion

1. It was established that the mass of unfrozen water connected with the microorganisms extracted from permafrost might be ten times higher than the mass of biomaterial containing in the samples at the temperatures of permafrost occurrence. Thereby, for frozen bacterial suspensions of *Bacillus sp.* strain MG8 extracted from permafrost the unfrozen water mass was 32 times higher than biomaterial mass at the temperature of –5 °C.

2. It was shown that metabolites of microorganisms extracted from permafrost are capable of having great influence on water phase freezing. Moreover, this influence turned out to be much higher than the influence of high-molecular cryoprotector. For example, D_m relation of unfrozen water mass to metabolite mass (MG8) was 55 higher than D_m relation of unfrozen water mass to dry solid PVA mass at the temperature of –5 °C.

Acknowledgements
The present work was supported by the Basic Research Program of RAS No. IX.135.2.3. 502 One of us, N S Molokitina, expresses her grateful thanks to the Council of Russian President grant for President's scholarship SP-3113.2018.1.

References

[1] Clein J S and Schimel J P 1995 *Soil Biology and Biochemistry* **27** 1231-4
[2] Gilichinsky D, Rivkina E, Shcherbakova V, Laurinavichuis K and Tiedje J 2003 *Astrobiology* **3** 331-41
[3] El-Registan G I, Mulyukin A L, Nikolaev YU A, Gal'chenko V F, Suzina N E and Duda V I 2006 *Microbiology* **75** 446-56
[4] Brushkov A V 2008 *Vestnik VOGiS* **12** 534-44
[5] Gibson M I 2010 *Polymer Chemistry* **1** 1141-52
[6] Subbotin A M, Narushko M V, Bome N A, Penrov S A, Malchevsky V A and Gabdullin M A 2016 *Vavilovskii Zhurnal Genetiki i Selectsii* **20** 666-72
[7] Narushko M V, Simonova E O and Subbotin A M 2016 *Proc. Conf. on Nauchnaya i Proizvodstvennaya Deyatel'nost' - Sredstvo Formirovaniya Sredy Obitaniya Chelovechestva* (Tyumen: Tyumen Industrial University Press) pp 231
[8] Mank V V and Lebovka N I 1988 *Spectroscopy of a Nuclear Magnetic Resonance of Water in Heterogeneous Systems* (Kiev: Naukova Dumka) p 204
[9] Podenko L S, Drachuk A O, Molokitina N S and Nesterov A N 2017 *Kriosphera Zemli* **21** 43-51

Methane hydrate formation in dispersed polyvinyl alcohol cryogel

L S Podenko, A O Drachuk and N S Molokitina

Institute of Earth's Cryosphere, department of Tyumen Scientific Center, SB, RAS, 86 Malygin Street, Tyumen, 625026, Russian Federation

Abstract. Kinetics of methane hydrate formation in frozen water solutions (FWS) and polyvinyl alcohol (PVA) cryogel stabilized with hydrophobic silica nanoparticles was first studied. PVA cryogel is a macroporous heterogeneous gel formed at the special conditions of the freezing and subsequent melting processes of the PVA water solution. Such parameters as a rate and a conversion of water to hydrate were investigated for dispersed FWS, PVA cryogel, and dispersed ice at the same thermodynamics conditions using the same equipment. It was found out that the transformation the main part of water contained in FWT and PVA cryogel into hydrate was more than ten times faster than that for dispersed ice. The resistance to hydrate formation/dissociation processes was studied for dispersions of PVA cryogel. Gas hydrate formation and dissociation were carried out at isochoric conditions using temperature cycling. It was shown that cryogels of PVA water solution stabilized with hydrophobic silica nanoparticles were resistant to gas hydrate formation/dissociation processes prompting suggestions that cryogels of PVA water solution might be reused for gas hydrate formation/dissociation multiple times without significantly decrease in the conversion water to hydrate.

1. Introduction

Several approaches of natural gas transportation have been considered worldwide. One of the most perspective alternatives to the well-known approaches of natural gas transport is storing and transporting natural gas in clathrate hydrate (gas hydrate technologies) [1,2]. According to the fact that the natural gas production and consumption in the Far North is growing, there is an ever-increasing need to develop technologies that might be effective in a cold climate. In works [3-5] it was presented that "dry water" could show a great promise as a great material for gas hydrate formation with the purpose of its using in the technologies of natural gas transportation, storage and utilization. "Dry water" is a free-flowing powder containing up to 98 wt % of water in droplet form and produced by rapidly mixing of water and hydrophobic silica nanoparticles in air [6]. However, insufficient resistance of water containing in "dry water" to phase transformations (freezing/thawing and gas hydrate formation/dissociation) limits its using for gas hydrate formation/dissociation and applying at conditions of seasonal temperature fluctuations near the ice melting point.

Dispersions of the gelling gum aqueous solution of polysaccharide stabilized with hydrophobic silica nanoparticles are more resistant to gas hydrate formation/dissociation [4] and remain stable even after gas hydrate formation/dissociation in it. Whereas, to produce the polysaccharide dispersions it is required a rather great amount of the gellan gum (more than 20 wt %). Moreover, in case of replacing water to the gellan gum solution it might be observed the significant decrease in storage capacity (30 %) [4]. A macroporous heterogeneous gel formed at the special conditions of the freezing and subsequent melting processes such as PVA cryogel (further cryogel) might be formed at the less concentration than that for polysaccharide (for example, 5 wt %) [7]. In the previous work we showed that dispersed cryogel stabilized with hydrophobic silica nanoparticles could remain stable after freezing/thawing cycles [8]. Therefore, it is reasonable to suggest that the dispersed cryogels will be stable to gas hydrate formation/dissociation cycles as well.

However, gas hydrate formation in cryogels stabilized with hydrophobic silica nanoparticles and the influence the gas hydrate formation/dissociation cycles on stability of cryogel dispersions have not studied yet. Moreover, there is no information about gas hydrate formation in FWS of PVA at the temperature near the ice melting point (at the conditions at which cryogel forms).

In this paper the results of the kinetics of methane hydrate formation in the dispersions of PVA FWS stabilized and data on cryogel stability to the processes of methane hydrate formation/dissociation.

2. Experimental section

Dispersed PVA frozen water solutions were produced by grinding the frozen and storing at the temperature of 258 K no less than ten hours PVA solutions or distilled water. Mechanical grinding of frozen PVA solutions and ice were carried out in the thermostatic camera Teledor for 30 s using the household blender Braun VX2050 with the speed of 18700 rotations per minute. To prepare the dispersed PVA frozen water solution the hydrophobic silica nanoparticles (HDK® H17 (Wacker Chemical AG, German), further H17) were added to the solutions in amount of from 5 till 8 wt % before the process of mechanical grinding. To form cryogel, gas hydrates, and studying the gas hydrate growth kinetics the fraction of 80–140 micrometers or less than 140 micrometers of the ground ice particles that were separated with sifting the ground PVA frozen water solutions with nanoparticles through a sieve. The sifting process was carried out with the electrodynamic vibration stand PE-6700 with a frequency of the desktop oscillation of 20 Hz at an ambient temperature of 253 K. For forming dispersed cryogel the separated fractions were stored at the room temperature up to full melting. Then, the free-flowing powder contained cryogel particles and about 95 wt % of water was observed.

"Dry water" was prepared according to the [9]. For preparing "dry water" H17 was used in amount of 5 wt %.

As a hydrate-forming gas methane was used (99.9 mol %). Gas hydrates were carried out in an isothermal reactor with a volume of 60 sm^3 made of stainless steel. The experimental apparatus for gas hydrate formation and studying of kinetics of the gas hydrate growth and dissociation presented in [10]. To prepare gas hydrate the studied sample in amount of 7 g was loaded into the reactor. The sample was placed into the reactor in the thermostatically controlled chamber Teledor at the temperature of 253 K. "Dry water" and dispersed cryogel were loaded into the reactor at the room temperature. Then, the reactor with the sample was placed into the cryostat at a given temperature (further, the temperature of gas hydrate formation). After that the reactor with the sample was stored at the constant temperature up to the moment when the sample temperature was equal to the desired temperature. Then the reactor was evacuated and charged with methane up to the pressure of 4.9 MPa. The gas charging led to a slight heating of the reactor contents (not more than 0.7 degrees). In the absence of hydrate formation, the temperature of the gas after the reactor charging restored to the initially set temperature. The pressure in the reactor changed slightly in the absence of hydrate formation.

In all cases gas hydrate formation was carried out at isochoric conditions without any mixing of the contents of the reactor. Gas hydrate formation was prolonged for 1000 or 2000 min depending on the hydrate formation rate.

The quantity of mole gas transformed to hydrates (n) was determined from the pressure changing in the reactor at the temperature using the equation of Peng-Robinson for a gas phase [11].

Assuming that the composition of methane hydrates is determined by the stoichiometric ratio $CH_4 6.0H_2O$ [12], we calculated the degree of transition of water to hydrate using the following equation:

$$\alpha = (6.0 \cdot M_w \cdot n)/m_w,$$

where M_w – a water mole weight; m_w – an initial water mass in the investigation sample.

Several cycles of gas hydrate formation/dissociation were conducted under the cycling temperature changing at isochoric conditions for dispersed cryogels (5 wt % of PVA and 8 wt % of H17; 7 wt % of PVA and 5 wt % of H17) to analyze their resistance to gas hydrate formation/dissociation processes. For comparing the results there were carried out several gas hydrate formation/dissociation cycles in "dry water" under the cycling temperature changing at isochoric conditions.

In figure 1 there was shown a typical view of temperature changing curves and pressure in the reactor after charging the reactor. According to the presented data the significant pressure decrease was obtained for all studied samples indicated that gas hydrates were being formed in the reactor. The delay of gas hydrate formation named the induction time for "dry water" samples changed stochastically was no more than 15 min. For frozen dispersion systems and cryogels were obtained the significant pressure decrease and insufficient temperature increase in the reactor after charging the reactor (figure 1), indicating that there was no any gas hydrate formation induction time. For dispersions of PVA frozen water solutions the pressure quickly decreased for about 300 min and then it was stabilized. For ground ice the pressure decreased fluently during the period of observation (figure 1).

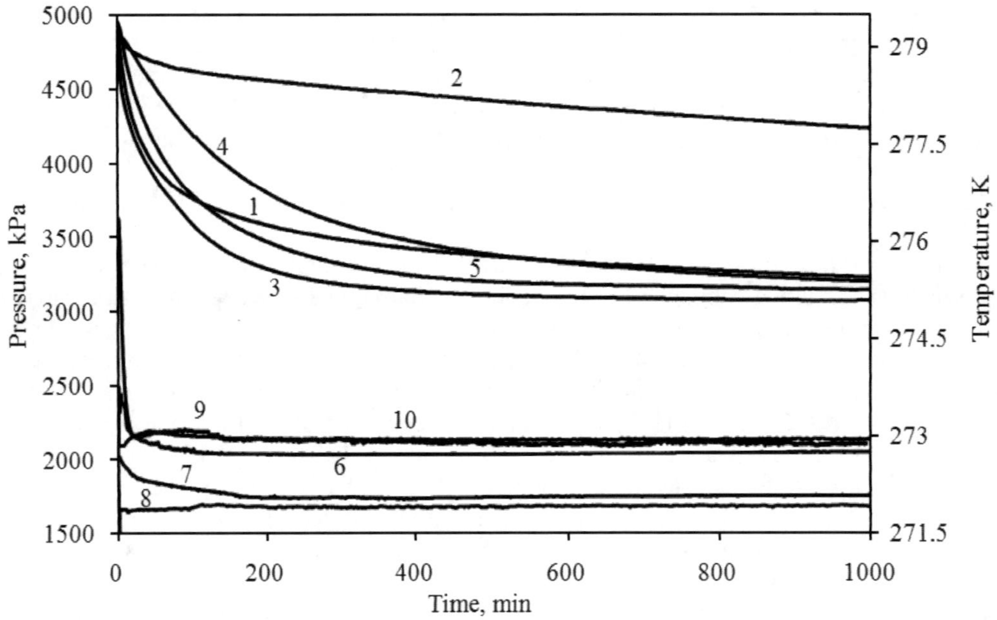

Figure 1. Pressure changing (1-5) and temperature of methane (6-10) in the reactor at isochoric conditions immediately after the reactor charging. "Dry water" (5 wt % of H17) (1 and 6); ground ice (2 and 7); frozen water solution (5 wt % of PVA and 5 wt % of H17) (3 and 8); cryogel (5 wt % of PVA and 8 wt % of H17 (4 and 9); cryogel (7 wt % of PVA) (5 and 10).

3. Results and discussion

In figure 2 it was demonstrated the typical character of changing of the water to hydrate conversion in the investigated frozen water solution dispersions of PVA, cryogel and ground ice systems immediately after the moment the reactor had been charged. From the presented data it might be assumed that the replacing ground ice by frozen water solution dispersions of PVA or cryogel dispersions let significantly decrease the conversion time of the main part water to hydrate. Based on the experimental kinetic curves of the water to hydrate transition the value of $t_{0.8}$ and α_f were calculated for the frozen water solutions of PVA, cryogel. To compare the results the parameters $t_{0.8}$ and α_f were also calculated for ground ice. The obtained results were presented in table 1. The time $t_{0.8}$ for the ground ice correlated with a transition 0.8 of mass fraction of initial water to hydrate was more than 1000 min, and the water to hydrate conversion α_f was only 0.35 after 1000 min of the gas hydrate formation. At the same conditions the time $t_{0.8}$ for frozen water solutions of PVA and cryogel (with the content of 7 wt % of PVA and 5 wt % of H17) was no more than 200 min, and the final conversion α_f was almost 0.95.

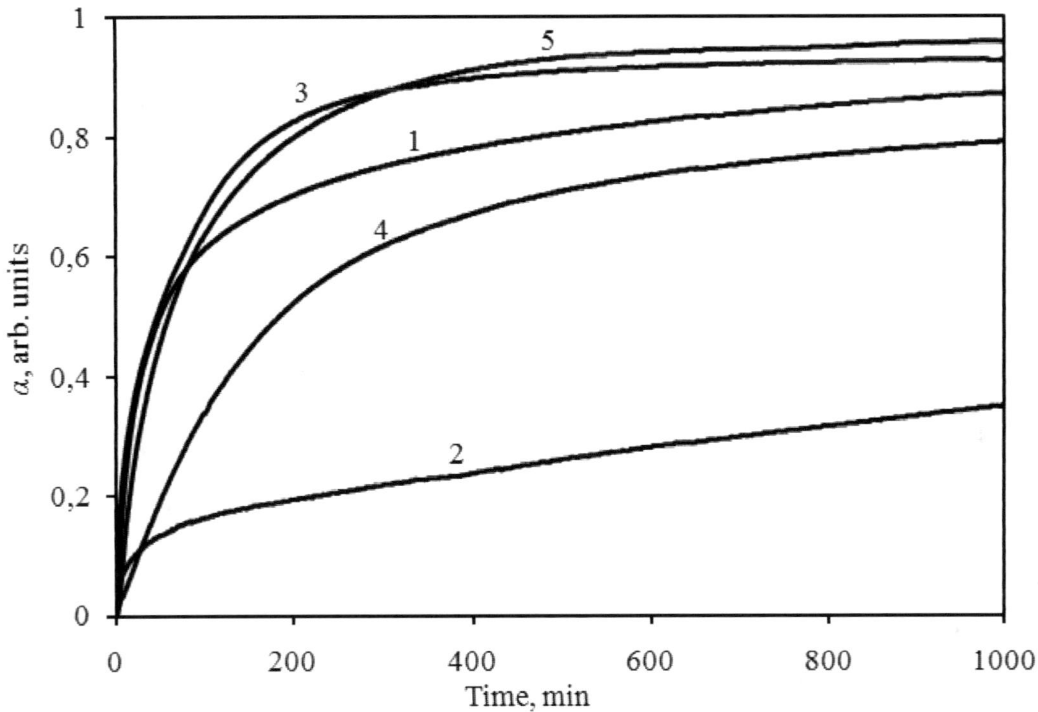

Figure 2. Changing of water to hydrate conversion during the process of gas hydrate formation for samples of "dry water" (1), ground ice (2), dispersions of PVA frozen water solution (5 wt % of PVA and 5 wt % of H17) (3), cryogel (5 wt % of PVA and 8 wt % of H17) (4), cryogel (7 wt % of PVA and 5 wt % H17) (5) at isochoric conditions. The desired temperature for gas hydrate formation in the samples of ground ice, dispersions of PVA frozen water solution was 272.2 K, for the samples of cryogel and "dry water" the temperature of gas hydrate formation was 273.2 K. The initial pressure was 4.9 MPa.

Table 1. Values of a parameter $t_{0.8}$ and the final water to hydrate conversion α_f in methane hydrates at isochoric conditions. The initial pressure was 4.9 MPa.

Sample	PVA concentration (wt %)	H17 concentration (wt %)	Temperature of gas hydrate formation (K)	$t_{0.8}$ (min)	α_f (in 1000 min)
PVA FWS	5	5	272.2	170	0.93
Ground ice	0	0	-//-	> 2000	0.35
Cryogel	5	8	273.2	1100	0.79
-//-	7	5	-//-	200	0.96
“Dry water”	0	5	-//-	460	0.87

From the given results the transition water to hydrate rate $d\alpha/dt$ was calculated for studied samples of the PVA frozen water solutions and ground ice. In figure 3 the results on changing the transition water to hydrate rate and the temperature gauge readings in the samples of the PVA frozen water solutions and ground ice (at the given temperature of gas hydrate formation of 272.2 K).

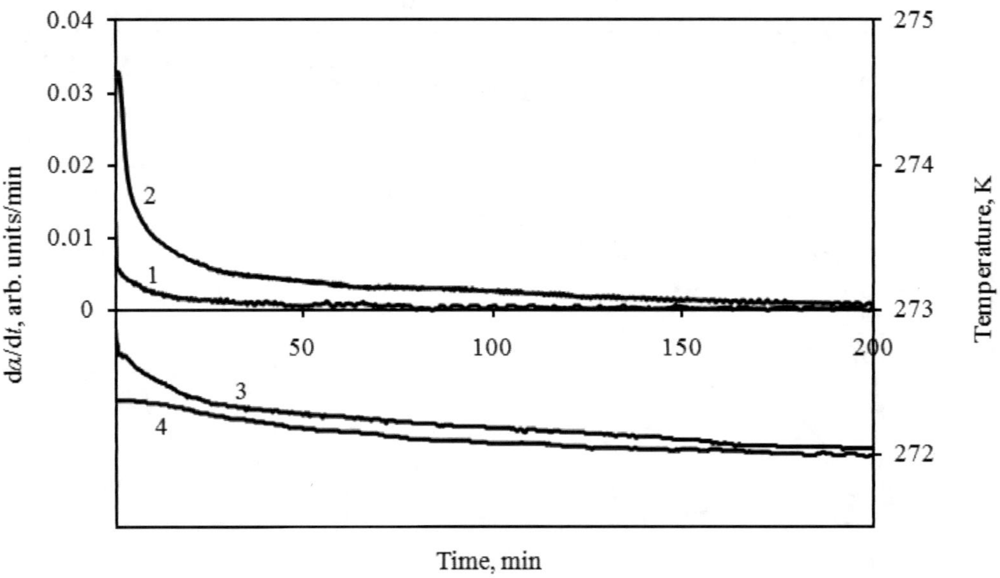

Figure 3. Changing of the water to methane hydrate conversion rate (1-2) and temperature (3-4) during the process of gas hydrate formation in the samples of ground ice (1 and 3), frozen water solutions (5 wt % of PVA and 5 wt % of H17) (2 and 4) at isochoric conditions. The initial pressure was 4.9 MPa.

According to the data, the gas hydrate formation rate of PVA frozen water solutions after no less than 200 min period of time was several times higher than that for ground ice. Moreover, the temperatures of the samples were not higher than the temperature of ice melting (272.7 K taking into account the pressure in the reactor) for all period of gas hydrate formation. It seems reasonable to say that the presence of micro-inclusions of a concentrated solution of polymer in the PVA frozen water solutions inhibited the aggregation of hydrate crystals and this made the formation of the hydrate crust impervious to gas.

In figure 4 the typical curves of pressure and temperature changing in the reactor during the gas hydrate formation/dissociation cycles were presented for cryogel dispersions. According to the data, the cooling of the reactor from 293.2 to 273.2 K was followed with the pressure decrease in it. Near the temperature of 291.2 K there was observed the significant increase of the falling pressure rate that was due to the gas hydrate formation. For the period of time approximately 1000 min the temperature in the reactor was maintained up to the 273.2 K.

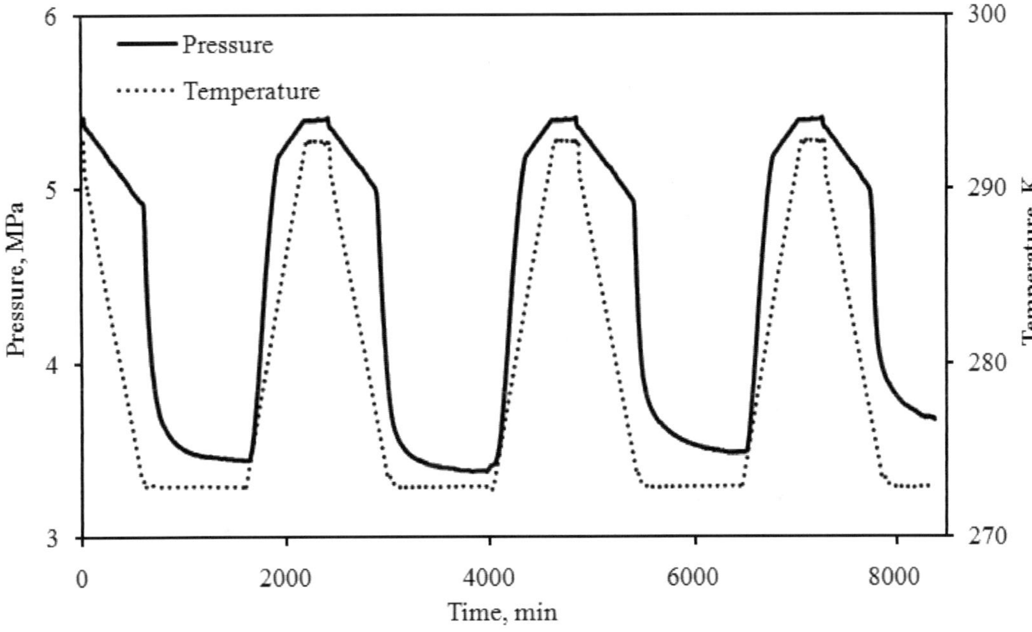

Figure 4. Methane pressure changing in the reactor using the cycling temperatutr changing at isochoric conditions for cryogel dispersions (7 wt % of PVA and 5 wt % of H17).

The pressure declining in the reactor during that time attested that the process of gas hydrate formation was continued. Heating of the reactor from 273.2 K to 293.2 K led to the pressure increase. When the temperature of the reactor was close to 275.2 K there was observed a marked rate increase of the pressure growth that was evidence of gas hydrate dissociation. For approximately 300 min the temperature in the reactor was hold equal to 293.2 K. Then, the cycle of pressure changing in the reactor was repeated. The same cycle of gas hydrate formation/dissociation was carried out for "dry water". In table 2 there were presented the data on the reduction (in percent) of the final water to hydrate conversion after gas hydrate formation/dissociation cycles. It is known that the processes of gas hydrate formation/dissociation in "dry water" lead to the loss in stability (splitting into layers of a phase of bulk water and dispersed water stabilized with hydrophobic nanoparticles of silica dioxide) [4]. That was the reason for a considerable decline (about 20 %) of water to hydrate conversion that was obtained even after one gas hydrate formation/dissociation cycle. According to the data presented in table 2, the cryogel dispersions might be more stable to the processes of gas hydrate formation/dissociation as compared with "dry water" in the event that both systems had the comparable initial water to hydrate conversion. For "dry water" the water to hydrate conversion was down by 50 % after three cycles of gas hydrate formation/dissociation. Although, the water to hydrate conversion was down by 17 % in case of using cryogel.

Table 2. The value of reduction (in percent) of the final water to hydrate conversion after cycles of gas hydrate formation/dissociation. The initial water to hydrate conversion (before the dissociation cycle) was 0.91.

Sample	PVA concentration (wt %)	H17 concentration (wt %)	Cycle numbers		
			1	2	3
Cryogel	5	8	5	22	-
-//-	7	5	1	11	17
“Dry water”	0	5	24	43	53

4. Conclusion

1. It was established that the PVA inclusions into the frozen water solutions might ten times increase the gas hydrate formation rate and the water to hydrate conversion.
2. It was shown that the stability of PVA cryogel dispersions to the processes of gas hydrate formation/dissociation might be ten times higher than that for “dry water”.

Acknowledgements
The present work was supported by the Basic Research Program of RAS No. IX.135.2.3. 502 One of us, N S Molokitina, expresses her grateful thanks to the Council of Russian President grant for President's scholarship SP-3113.2018.1.

References

[1] Horiguchi K, Watanabe S, Moriya H, Nakai S, Yoshimitsu A and Taoda A 2011 *Proc. Int. Conf. on Gas Hydrates* (Edinburgh: Scotland) P5.053
[2] Rehder G, Eckl R, Elfgen M, Falenty A, Hamann R, Kähler N, Kuhs W F, Osterkamp H and Windmeier C 2012 *Energies* **5** 2499
[3] Wang W X, Bray C L, Adams D J and Cooper A I 2008 *J. Am. Chem. Soc.* **130** 11608
[4] Carter B O, Wang W X, Bray C L, Adams D J and Cooper A I 2010 *Langmuir* **26** 3186
[5] Mel'nikov V P, Podenko L S, Nesterov A N, Drachuk A O, Molokitina N S and Reshetnikov A M 2016 *Doklady Chemistry* 466 53
[6] Binks B P and Murakami R 2006 *Nature Mater.* **5** 865
[7] Lozinsky V I 1998 *Rus. Chem. Rew.* 67 573
[8] Mel'nikov V P, Podenko L S, Nesterov A N, Drachuk A O and Molokitina N S 2016 RU Patent 2574403
[9] Podenko L S, Nesterov A N, Drachuk A O, Molokitina N S and Reshetnikov A M 2014 *Rus. J. Phys. Chem.* **88** 1257
[10] Podenko L S, Drachuk A O, Molokitina N S and Nesterov A N 2017 *Earth's Cryosphere* **21** 43
[11] Peng D and Robinson D B 1976 *Ind. & Eng. Chem. Fund.* **15** 59
[12] Circone S, Kirby S H and Stern L A 2005 *J. Phys. Chem.* **109** 9468

IOP Conf. Series: Earth and Environmental Science **193** (2018) 012020 doi:10.1088/1755-1315/193/1/012020

Laboratory study of the acoustic properties of sand samples containing ice

G A Dugarov, A A Duchkov and M I Fokin

Trofimuk Institute of Petroleum Geology and Geophysics, Novosibirsk, Russia

E-mail: dugarovga@ipgg.sbras.ru

Abstract. The paper considers a series of experiments with sand samples containing ice. The acoustic measurements were performed in specialized laboratory setup which allows also to form synthetic hydrate-bearing samples in order to find differences between ice and methane hydrate. We find that the velocities of compressional and shear waves depend on the rate of temperature drop below 0°C. During the long-time maintaining of the constant negative temperature the decreasing of the velocities was observed. These dependences can possibly be explained by that in a case of quick freezing formed ice is not energetically favourable, after formation ice particles reconstruct into the pore space with less mechanical stress.

1. Introduction

Natural gas hydrates are considered as a promising new source of natural gas [1, 2]. Also, the presence of gas hydrates in the area of deposit development could cause complications such as accidents due to gas dynamic phenomena during hydrate decomposition and hydrate formation in the bottomhole zone [3]. For stability of gas hydrates high pressure and low temperature are needed [4]. These conditions exists in deep-water sediments and in permafrost. But in the permafrost regions it is hard to distinguish gas hydrate from ice.

The physical properties of samples containing hydrate and ice depending on many factors, such as material of rock matrix, type of pore fluid, saturation by hydrate or ice, etc. Specialized laboratory setups are used for study of the properties of synthetic hydrate-bearing samples [5-7]. In previous works we used our laboratory setup [8] for investigation of acoustic properties of sand samples containing methane hydrate and ice [9]. In this work the results of additional experiments with samples containing ice are presented.

2. Method

For forming sand samples, we place the mixture of quartz sand with a grain size of 0.2-0.25 mm and distilled water in special caprolone cells with ends closed by grids. The internal diameter of the cells is 26 mm; the height varies in the range from 35 mm to 45 mm for compensation of axial compression. We place the cells in the high pressure chamber with temperature maintaining system. The external axial and lateral pressure is about 300 atm.

The acoustic measurements were done by piezoceramic elements mounted in the upper and lower punches. Two discs with a diameter of 18 mm and thickness of 2 mm are mounted in each punch. One disk is used for generating or receiving compressional (P) wave, the second one for generating or receiving shear (S) wave. The signal frequency range is 300-700 kHz. Acoustic measurements were performed every 5 min during the whole experiment. The height of the samples was also measured during the whole experiment for correct velocity estimation and accounting samples compaction.

The saturation of the samples is controlled by the amount of initial added water. The water saturation of the samples is about 0.3 (except one dry sample), the porosity of the samples is about 0.45.

For forming methane hydrate in sand samples in previous works we applied temperature cycles within methane hydrate stability region or maintain a constant temperature low enough for forming hydrate. Therefore, in this work we form the samples containing ice by maintaining of a constant negative temperature or reformed it many times with temperature cycles.

3. Results

The variations in velocities of P- and S-waves at temperature cycles are presented on figure 1. The temperature varies from 5°C to -15°C. The duration of temperature cycles increases from 1 hour to 24 hours. We can see fast velocity increase when the temperature drops below 0°C due to cementation of the sand by formed ice. But we see that a slower decrease of the temperature leads to a lower increase of the velocities.

For checking that this dependence is determined by the rate of temperature drop and not by the number of temperature cycles we did an experiment with reverse sequence of temperature cycles (figure 2). We can see that a faster decrease of the temperature leads to a stronger increase of the velocities, which means that the rate of temperature decrease is the determining factor.

On figure 3 we show the values of P- and S-waves at -10°C for temperature decreasing at different rate for both experiments: with increasing cycle duration (experiment from figure 1) and decreasing cycle duration (experiment from figure 2). The differences between velocity values for shorter and longer temperature cycles are about 0.21 km/s (6.1%) for P-wave and 0.12 km/s (5.6%) for S-wave.

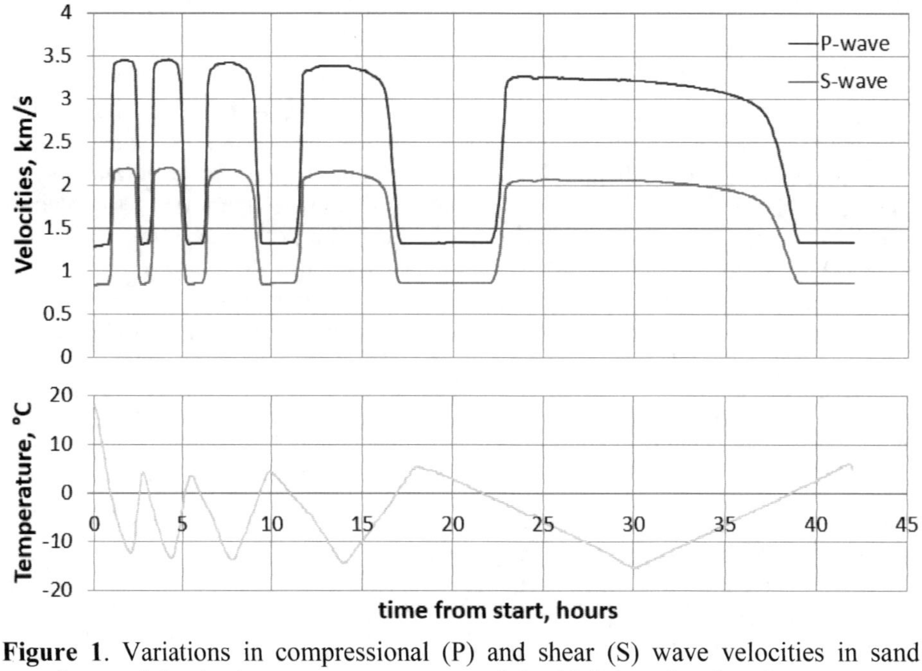

Figure 1. Variations in compressional (P) and shear (S) wave velocities in sand sample containing ice/water at temperature cycles with increasing duration.

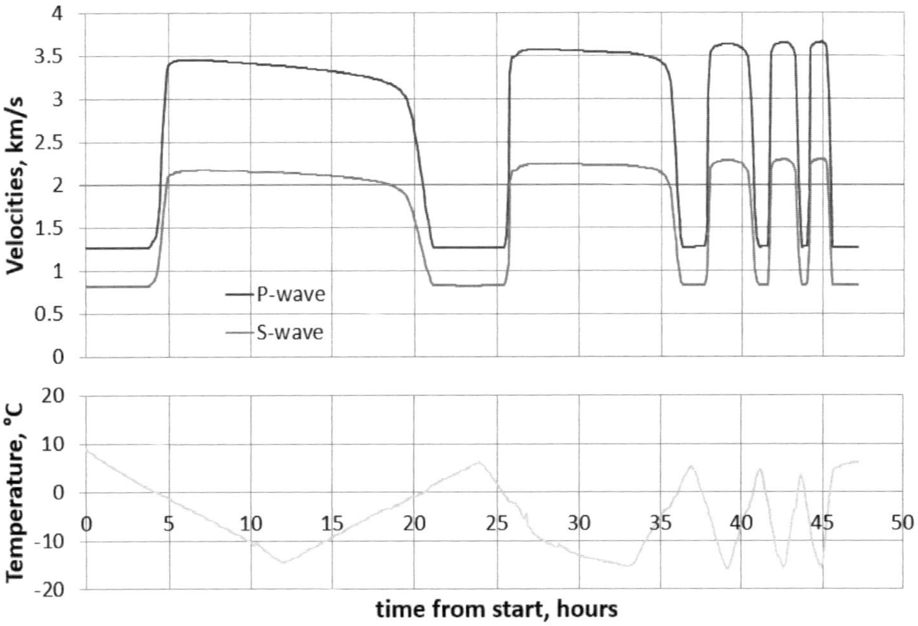

Figure 2. Variations in velocities in sand sample containing ice/water at temperature cycles with decreasing duration.

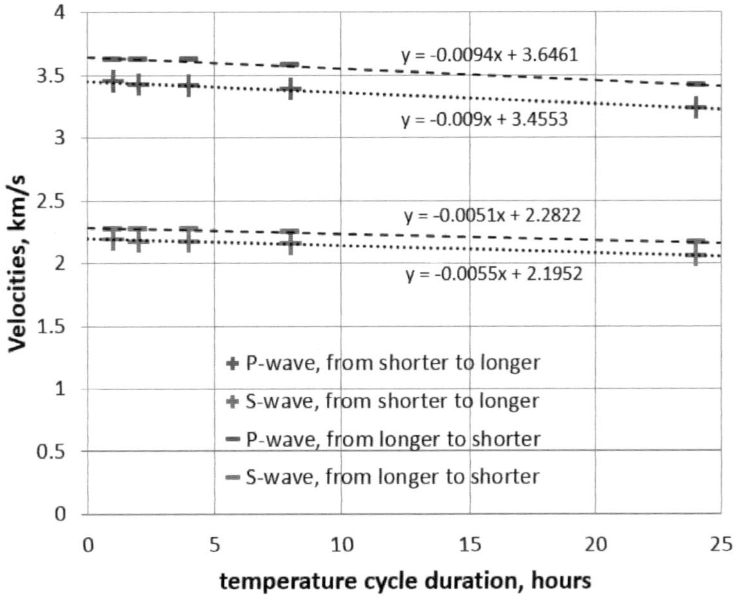

Figure 3. Velocities at -10°C for temperature decreasing at different rate (from figure 1 and 2), lines show linear approximation.

We see that the velocity dependences from the rate of temperature drop can be approximated by linear functions. The slopes of the approximation functions for both experiments are the same (figure 3), but they are different for different type of waves. For P-wave the slope (0.009) is almost twice larger than for S-wave (0.005). The vertical shift of the approximation functions can be explained by the small differences in the initial water saturation of the samples.

IOP Conf. Series: Earth and Environmental Science **193** (2018) 012020 doi:10.1088/1755-1315/193/1/012020

On figures 1 and 2 we can see also the decrease of the velocities during the time for long-time cycles. For checking this we made an experiment with long maintaining of the constant temperature (figure 4). We can see permanent decrease of both velocities during 120 hours. After 120 hours P-wave velocity decreased by 0.51 km/s (14.6%) and S-wave velocity decreased by 0.3 km/s (13.3%). The values of the velocity drop of every 10 hours presented on figure 5. After 90 hours the decreasing of the velocity drops is so small that it is almost constant, 0.15 km/s every 10 hours for P-wave and 0.1 km/s every 10 hours for S-wave. Similar experiment was made with dry sample (figure 6). After 80 hours of maintaining of the constant negative temperature we see no drops in P- and S-wave velocities.

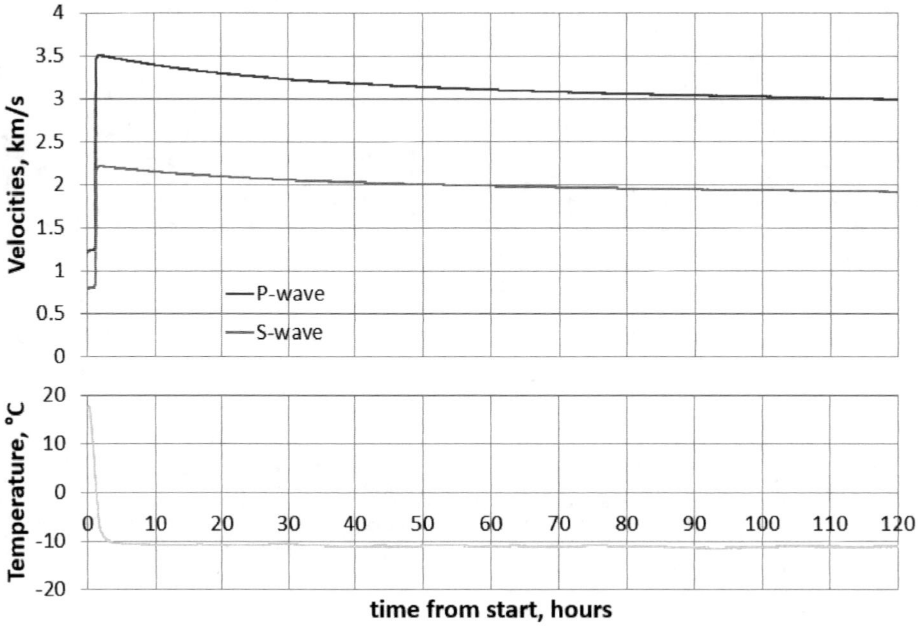

Figure 4. Decreasing of velocities with time at the constant negative temperature.

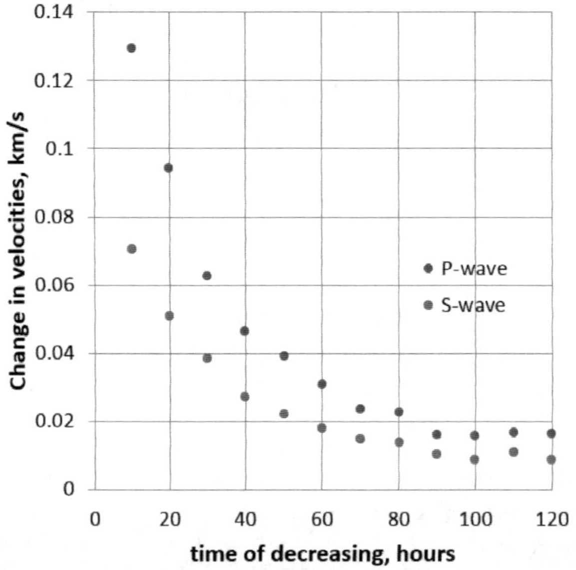

Figure 5. The velocity drops values for every 10 hours.

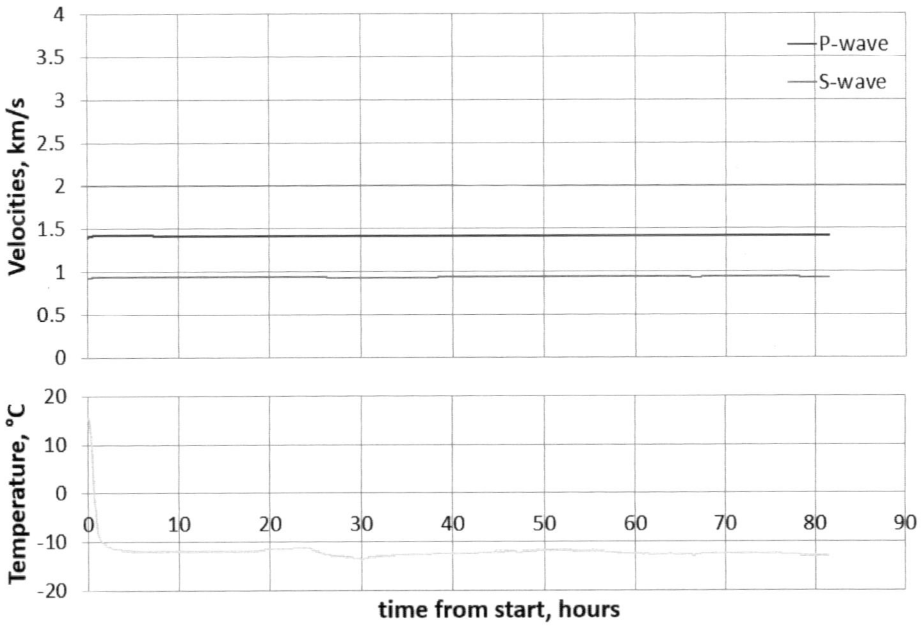

Figure 6. Variation in velocities with time at the constant negative temperature for dry sample.

4. Conclusions

The results of the first two experiments show that the values of velocities after temperature drops below 0°C depend on the rate of the temperature decreasing. A slower decrease of the temperature leads to a lower increase of the velocities. These dependencies could be approximated by linear functions. The slopes of the approximation functions for both experiments are the same but they are different for different type of waves. For P-wave the slope is almost twice larger than for S-wave.

The decreasing of the velocities is also observed during long-time maintaining of the constant negative temperature. The decrease of the velocities for the sample containing ice can be explained by reduction of cementation degree. Both dependences can possibly be explained by that in a case of quick freezing formed ice is not energetically favorable, ice takes over a part of the load on the host matrix. When the water-saturated sample is quickly frozen and the negative temperature is maintained the crystal of ice can slowly pass into an equilibrium state, ice particles possibly reconstruct into the pore space with less mechanical stress.

Acknowledgements

This research has been carried out according to the Complex Program of Basic Research of the Siberian Branch of the Russian Academy of Sciences. The work was supported by Russian Foundation of Basic Research and "National Intellectual Development" foundation [17-35-80023].

References

[1] Chong Z R, Yang S H B, Babu P, Linga P and Li X S 2016 *Applied Energy* **162** 1633-52
[2] Makogon Y F and Omelchenko R Y 2013 *Journal of Natural Gas Science and Engineering* **11** 1-6
[3] Istomin V A, Moiseykin P A, Abrashov V N, Fedulov D M, Chernykh V V, Medvedev S G and Sopnev T V 2013 *Vesti Gazovoy Nauki* 99-104
[4] Manakov A Yu and Duchkov A D 2017 *Russian Geology and Geophysics* **58** 240-52
[5] Winters W J, Dillon W P, Pecher I A and Mason D H 2000 *Natural Gas Hydrate in Oceanic and Permafrost Environments* ed M D Max (Springer) Chapter 24 311–22

[6] Kulenkampff J and Spangenberg E 2005 *Scientific Results from the Mallik 2002 Gas Hydrate Production Research Well Program, Mackenzie Delta, Northwest Territories, Canada* ed S R Dallimore, T S Collett (Canada: Natural Resources Canada) p 16

[7] Priegnitz M, Thaler J, Spangenberg E, Rücker C and Schicks J M 2013 *Review of Scientific Instruments* **84** 104502

[8] Duchkov A D, Golikov N A, Duchkov A A, Manakov A Yu, Permyakov M E and Drobchik A N 2016 *Seismic Instruments* **52** 70–8

[9] Duchkov A D, Duchkov A A, Dugarov G A and Drobchik A N 2018 *Doklady Earth Sciences* **478** 74–8

Double layer composite material based on elastomer and ultra-high molecular weight polyethylene

A A Dyakonov[1,2], **N V Shadrinov**[1,2], **M D Sokolova**[1] **and A A Okhlopkova**[1,2]

[1]Institute of Oil and Gas Problems, Yakutsk, Russia
[2]North-Eastern Federal University, Yakutsk, Russia

E-mail: afonya71185@mail.ru

Abstract: Some physical and mechanical properties of a double layer composite material with ultra-high molecular weight polyethylene on one side and elastomer on the other side were evaluated. As elastomer, nitrile butadiene rubber (NBR), isoprene rubber (IR) and styrene butadiene rubber (SBR) were used, and dephynilguanidine (DPG) were used, as adhesive. In this paper we present a progress of our investigations.

1. Introduction

Despite a high variety of the polymer composite materials with unique characteristics, new materials possessing improved operational characteristics are still important [1-3]. One distinctive approach to improve physical and mechanical properties is double layer composite material with a polymer on one side and elastomer on the other side which has been investigated in [4, 5]. In this paper the influence of dephynilguanidine (DPG) to adhesion bond between ultra-high molecular weight polyethylene (UHMWPE) and nitrile butadiene rubber (NBR), isoprene rubber (IR) and styrene butadiene rubber (SBR) as elastomer is investigated.

2. Materials and methods of investigation

Ultra-high molecular weight polyethylene (UHMWPE, GUR-4113) [6], nitrile butadiene rubber (NBR, BNKS-18, Russia), isoprene rubber (IR, SKI-3, Russia), styrene butadiene rubber (SBR, SKS-30, Russia) [7], and dephynilguanidine (DPG, VEKTON) were used. Combination of elastomers with a UHMWPE is possible owing to their close temperature processing windows and a production method –hot pressing technique [8, 9]. Double layer composite material (figure 1) processing is made in 4 stages [10]:

1st stage. Cold molding of UHMWPE in a mould with a pressure 90 kgf/cm^2 within 5 minutes at room temperature. UHMWPE powder is poured into a mould and there we mold it;

2nd stage. Homogenous laying of the rubber mixture layer upon UHMWPE. Non homogenous laying may cause crack formation of the UHMWPE;

3rd stage. Mould with a UHMWPE and a rubber is placed into hot press for 30 minutes at temperature 155°C and a pressure 90 kgf/cm^2.With a hot molding UHMWPE is pressed into a rubber mixture and with a heat development is passed into a high elastic state, and it enables macromolecule diffusion;

4th stage. Hot pressed sample is cooled to 60°C in a mould under pressure. Shock cooling of a UHMWPE may be a cause of its deformation.

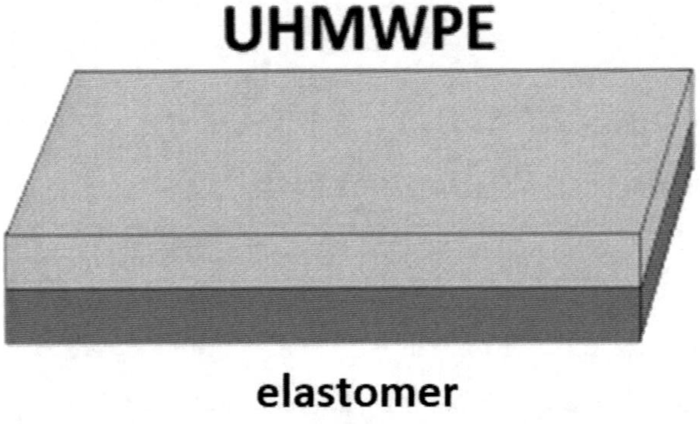

Figure 1. Double layer composite material based on elastomer and UHMWPE.

Investigations of mechanical and physical properties, include resistance to aggressive environments of the rubber with a different content of a DPG. Stress related characteristics were investigated at rupture machine Shimadzu Autograph according to GOST 270-75. Corrosion resistance IRM-901 in a stress-free state is investigated according to GOST 9.030-74. Investigation of the adhesion bond between elastomers containing different concentration of the DPG and UHMWPE has been made according to GOST 6768-75. Investigation of the structure of interface elastomer boundary layer and UHMWPE has been made at scanning electron microscope JEOL JSM-7800F.

3. Results and discussion

Investigation results of elastomer features based on NBR, IR and SBR are presented in the table 1. Concentrations of DPG as – 0.1, 0.2 and 0.3 phr for 100 phr of the rubber.

Table 1. Features based on nitrile butadiene rubber, isoprene rubber and styrene butadiene rubber depending dephynilguanidine concentration.

N	Sample	f, MPa	ε p, %	$f_{100\%}$, MPa	H, SH A	ρ, g/cm^3	ΔQ, %
1	NBR	10.6	410	2.4	64	1.16	8.3
2	NBR+0,1 phr DPG	9.3	313	2.5	62	1.16	8.3
3	NBR+0,2 phr DPG	9.1	303	2.9	63	1.15	7.9
4	NBR+0,3 phr DPG	8.9	285	3.0	67	1.15	6.8
5	IR	7.2	707	0.7	41	1.13	201.6
6	IR+0,1 phr DPG	14.4	724	1.3	62	1.12	129.0
7	IR+0,2 phr DPG	14.1	689	1.3	67	1.12	139.6
8	IR+0,3 phr DPG	14.1	666	1.3	69	1.12	116.2
9	SBR	15.9	688	1.9	62	1.17	41.2
10	SBR+0,1 phr DPG	12.3	417	2.3	66	1.16	38.5
11	SBR+0,2 phr DPG	11.9	414	2.4	66	1.16	34.1
12	SBR+0,3 phr DPG	10.9	343	2.5	67	1.15	35.7

f, MPa-tensile stress at rupture; ε_p, %- rupture strain; $f_{100\%}$, MPa –conventional stress at strain 100%; SH A-shore hardness A; ΔQ, %- swelling rating IRM-901.

Table 1 shows that injection of the DPG to the NBR and SBR will lead to the reduction of the extension coefficient and tear resistance as compared to the initial sample. Nevertheless, we may see slight module elevation at 100% tensile strain. Injection of the DPG has rather different impact on the mechanical and physical properties of the IR: its rupture strain is changing slightly, nevertheless its tear resistance is doubled at 100% tensile strain.

Injection of the DPG has a different affect on the rubbers according to the Shore hardness. Quintessential effect of the injection has on IR where Shore hardness rises up to 1.6 times. Shore hardness A of the NBR and SBR is slightly changing with injection of the DPG. There has been observed a rise of a Shore hardness A of a SBR by 4-6 durometers. Rubbers density is not changed by the injection of the DPG. Corrosion resistance of all samples IRM-901 rises up. Most significant rise has been observed (up to 1.7 times) with IR.

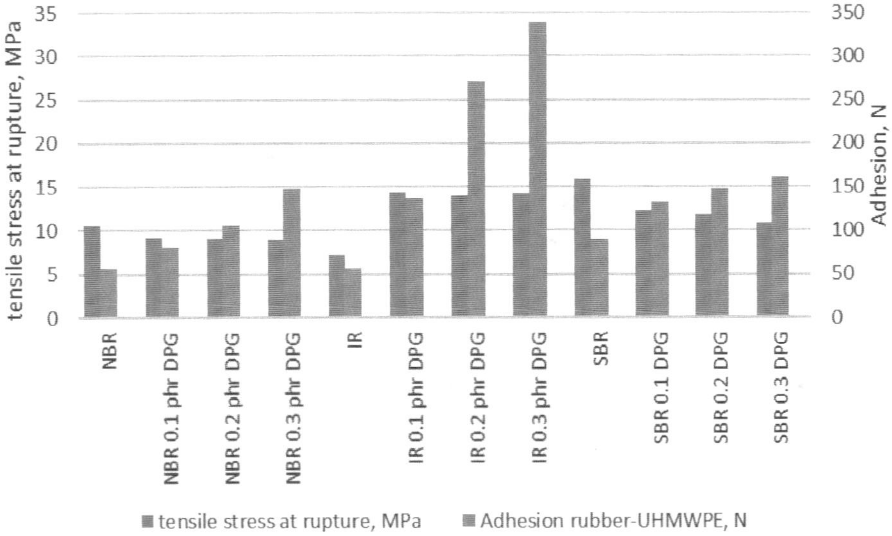

Figure 2. Double layer composite material adhesion dependence and failure resistance of elastomers on dephynilguanidine.

Figure 2 shows dependence diagrams of elastomers, failure resistance, and adhesion force between the rubbers and UHMWPE depending on DPG content. With increase of DPG concentration in the rubber mixture, supposedly owing to formation of sulphide entities between macromolecules of elastomer and UHMWPE, adhesion build appears.

It is clear that injection of DPG into rubber mixture, and also its increasing concentration will lead to adhesion. However, it should be noted that adhesion measurement at the rupture machine shows that in some cases there is a layer separation across phase boundary between the rubber and UHMWPE, and in some cases across the rubber. Rubber layer separation indicates excess of adhesion force between the rubber and UHMWPE over cohesion strength inside the rubber. In such cases, observed adhesion force in N, is an indication of rubber strength properties but not of adhesion.

For the NBR separation of specimens containing 0.1 and 0.2 phr of DPG as UHMWPE is passing in integrated way, i.e. wearing takes part partially by rubber and partially by interfacial interaction rubber/UHMWPE. Wearing has a cohesion character if there is a separation of specimens containing 0.3 phr.

If there is a rise of concentration of DPG in IR one may observe as hardening at elastomer rupture, as adhesion bond with UHMWPE. Hardening at materials structure violation was equal 370% regarding the rubber without DPG. Still, rubber separation with 0.3 phr of DPG is going on across elastomer.

After injection of 0.1 phr of DPG into SBR, separation is taking part partially on boundary and partially on rubber. After injection of 0.2 and 0.3 phr of DPG into the rubber mixture, separation is taking part across elastomer.

Rise of adhesion bond between the rubbers containing DPG mixed with UHMWPE is probably explained by the fact that DPG is a vulcanization accelerator where sulphide generation process between macromolecules is accelerated [8, 11]. It is determined in papers [12-14] that injection of DPG into polyethylene sulfonation will lead to formation of sulphides between polyethylene macromolecules that may point at formation of sulphides between the rubber and polyethylene. Usage of DPG as an elastomer adhesion bond catalyst with UHMWPE leads to the formation of sulphides between them, it is shown at figure 3.

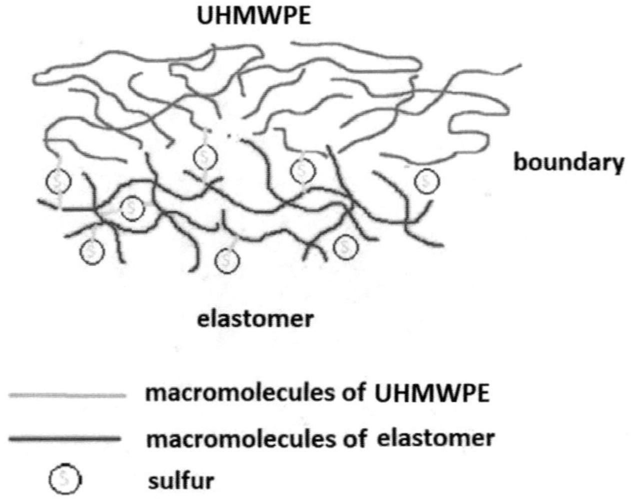

Figure 3. Formation of sulphides between elastomer and UHMWPE.

Research of supramolecular structure of the double layer composite materials of the interphase boundary revealed that injection of the DPG into the rubber mixture leads to the significant changes by way of formation vesicular structure of the UHMWPE. Figure 4 (a, c, e) shows micrographs of UHMWPE boundary with the rubber without adding DPG. Figure 4 (b, d, f) corresponds to injection of 0.3 phr of DPG into the rubber mixture. Observed structural changes may indicate formation of sulphides between UHMWPE and elastomers, which lead to rise of adhesion bond.

4. Conclusions

It is shown that hardening of adhesion bond between elastomers and UHMWPE mainly depends on vulcanizing system. Injection of the DPG into the rubber mixture leads to an increase of the adhesion bond between the elastomer and UHMWPE and influences vesicular structure near interface boundaries.

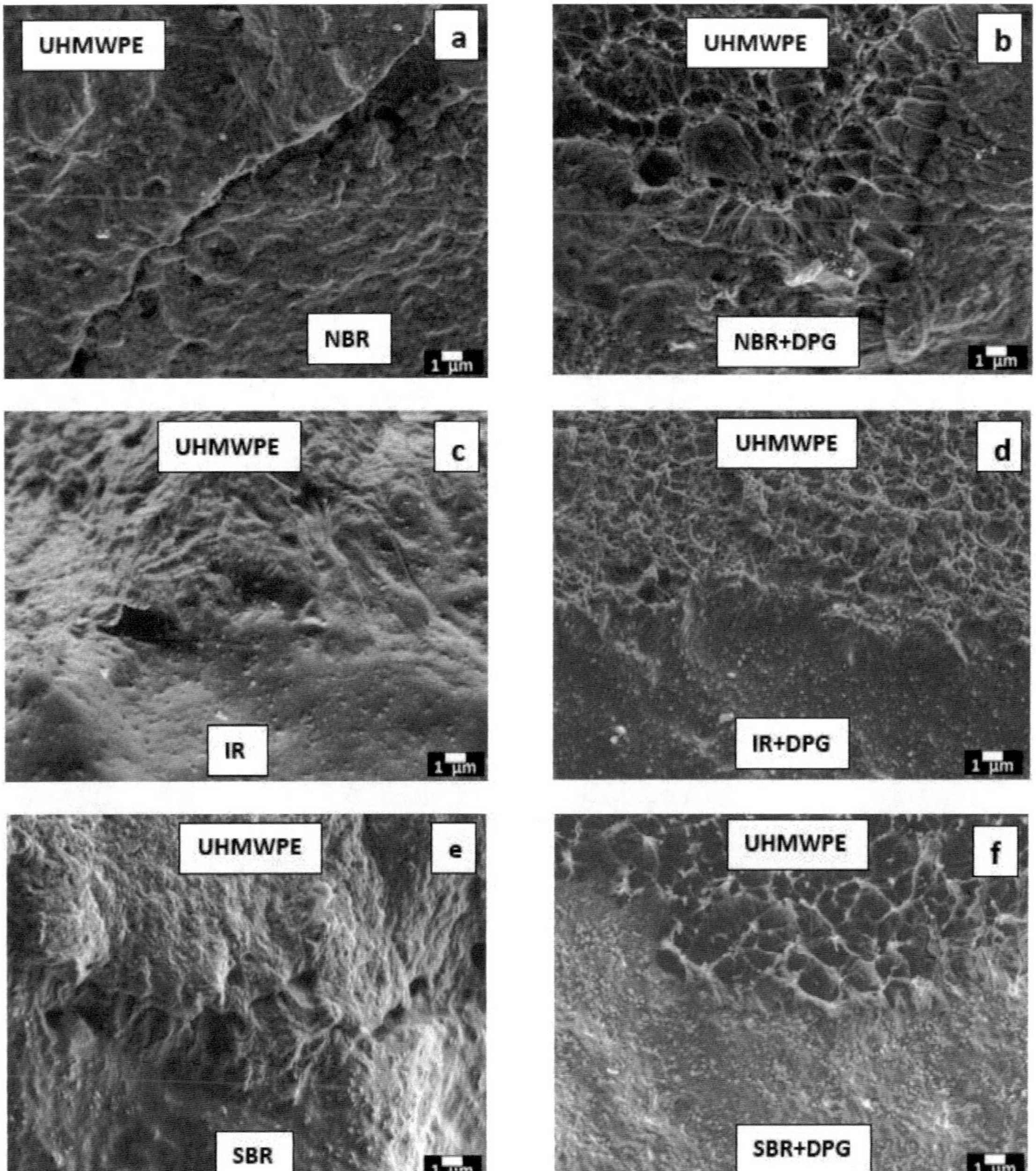

Figure 4. Micrographs of UHMWPE transitional layer with elastomers a) NBR b) NBR + 0.3 phr DPG c) IR d) IR + 0.3 phr DPG e) SBR f) SBR + 0.3 phr DPG.

Acknowledgments
This paper has been done with a financial support of a National assignment Ministry of education and science of the Russian Federation N 11.1557/4.6, N 11.7221.2017/6.7 and Russian foundation of basic research N 16-33-60070 "Pigment effect research for tensiles of the composite materials in nanometer range".

References

[1] Carraher C E 2018 *Carraher's Polymer Chemistry* (Boca Raton: CRC Press, Taylor & Francis Group)
[2] Kickelbick G 2007 *Hybrid Materials: Synthesis, Characterization, and Applications* (Weinheim: Wiley - VCH)
[3] Zyss J 1994 *Molecular Nonlinear Optics: Materials, Physics, and Devices* (Boston: Academic Press)
[4] Dyakonov A A, Sokolova M D, Shadrinov N V and Sleptsova S A 2017 *AIP Conference Proceedings* 020037
[5] Jin M, Zhang X, Emeline A V, Numata T, Murakami T and Fujishima A 2008 *Surface and Coatings Technology* **202** 1364–70
[6] Kelly J M 2002 *Journal of Macromolecular Science, Part C: Polymer Reviews* **42** 355–71
[7] Chaikun A M, Eliseev O A, Naumov I S and Venediktova M A 2013 *Trudi VIAM* **12**
[8] Kornev A E, Bukanov A M and Sheverdyaev O N 2009 *Elastomer Technology* (Moscow: NPPA Istek)
[9] Galibeev S S, Hairullin R Z and Arhireev V P 2008 *Kazan Technology University Bulletin* **2**
[10] Sokolova M D, Dyakonov A A, Popov S N, Shadrinov N V, Morova L Ya, Okhlopkova A A and Sleptsova S A 2017 Surface modified composite material (*Patent of the Russian federation N 2615416*)
[11] Simpson R B 2002 *Rubber Basics* (Shrewsbury, Shropshire: Rapra Technology)
[12] Karaseva Y S, Bashkatova T V, Cherezova E N and Husainov A D 2006 *Kazan Technology University Bulletin*
[13] Shwartz A G and Dinzburg B N 1972 *Matching of the Rubber with a Plastics and Synthetic Resins* (Moscow: Chemistry)
[14] Zaborski M and Slusarski L 1996 The Effect of Network Structure of Elastomers on Their Interactions with plastomers *Polimery (In Poland)*

The development of conceptual options to the construction of ice-resistant floating wind power plants

V V Elistratov, A S Bolshev, S A Frolov and A A Panfilov

Peter the Great Saint-Petersburg Polytechnic University, St.Petersburg, Russia

Abstract. This paper addresses the conceptual engineering solutions for a Marine Floating Power Plant which uses the Wind energy and is designed to supply electrical power to Russian Arctic coastal and island territory. The analysis of different types current Wind Turbines (WT) made a basis for a construction configuration for sea based Arctic Marine Floating Wind Power Plants and helped define the main wind, wave and ice loads influencing Marine Floating Wind Power Plants in Arctic conditions. Also, different types of Marine Floating Wind Power Plant substructures were analyzed. Numerical modelling of the behavior of the considered conceptual options of Marine Floating Wind Power Plants was carried out. Strength analysis of substructures for Arctic offshore wind power plant options in question has been performed.

Introduction

Russian grid energy is traditionally based on major hydraulic, nuclear and thermal power plants. However, offshore, island, and coastal areas of the Arctic Ocean have significant Wind Energy Resources [1, 2]. Using them through building Autonomous Offshore Marine Floating Wind Power Plants, Coastal Wind Power Plants, and Wind-Diesel Power Plants can significantly decrease deliveries of diesel fuel to the Northern Territories and improve power supply [1]. The work examines Marine Floating Wind Power Plants in more detail, possessing high mobility, can be quickly transported to a particular water area and provide the electrical power to the development of industrial projects on these remote territories.

Commercial exploitation of sea based Wind Turbines (WTs), designed for electric power generation, started in the 1990s. The depths allowing WT to be set become much deeper. Shallow depth WT are usually placed on fixed substructures. However, the cost of WP fixed substructures, set deeper than 50 meters underwater, is inadequately higher than the profit from the WT. There is already the current global experience in carrying out Marine Floating Wind Power Plant projects. Except that the known projects were carried out for ice-free seas, which means almost no experience in carrying out Marine Floating Wind Power Plants in ice conditions. The following problems are of great current interest and in need of scientific evidence: the problems related to finding reliable and energetic types and WT parameters adapted to Arctic conditions; the problems related to the development of optimal design shapes of Marine Floating Wind Power Plants substructure receiving wind, wave and ice loads; the problems related to the minimizing metal consumption and Marine Floating Wind Power Plants costs; the problems related to drilling of the towing technologies and schematics, positioning and anchorage of Marine Floating Wind Power Plants as well as the problems related to the transportation of the electrical power to the shore [3].

In the Marine Floating Wind Power Plants, already constructed or being designed, it is usually suggested to use three main types of substructures of floating WTs: SPAR type (*Hywind* project [4]); TLP type (*GICON* project [5]); and semi-submersible substructures (*WindFloat* project [6]).

The present work discusses some conceptual Marine Floating Wind Power Plant options for relatively shallow sea depths so the options of exploiting substructures of SPAR type, which are usually designed for depths over 100 meters, haven't been analyzed.

The choice of a floating substructure for a Marine Floating Wind Power Plant was based on the following requirements:

- Stability of the floating Marine Floating Wind Power Plant substructure.
- Acceptable Marine Floating Wind Power Plant oscillation parameters (linear and angular displacements) under the action of environmental loads while producing electrical power or a standstill due to storms.
- Acceptable linear and angular displacement of Marine Floating Wind Power Plant under the influence of ice features in the extreme ice impact mode.
- Acceptable metal consumption of the structure options and the minimum cost of their production.
- Maximum constructability.

1. Conceptual options of marine floating wind power plant construction

The present work offers conceptual options of Marine Floating Wind Power Plant with a capacity of 1.5 MW having different types of substructures and the exploitation of current WT types:

– Concept 1 – single-column MFWPP-1 with one horizontal axial WT and ice-resistant TLP substructure (Figure 1).

– Concept 2 – three-column MFWPP with three vertical axial WT with ice-resistant semisubmersible substructure (Figure 2).

From a structural point of view, a single-column MFWPP-1 substructure consists of a supporting column with the diameter of 5 m and three main buoyancy volumes interconnected with the column by tubular braces with the diameter of 2,5 m. To reduce ice loads on the facility in the ice belt area the diameter of the support column in reduced to 3 m. The overall height of the substructure is 47 m.

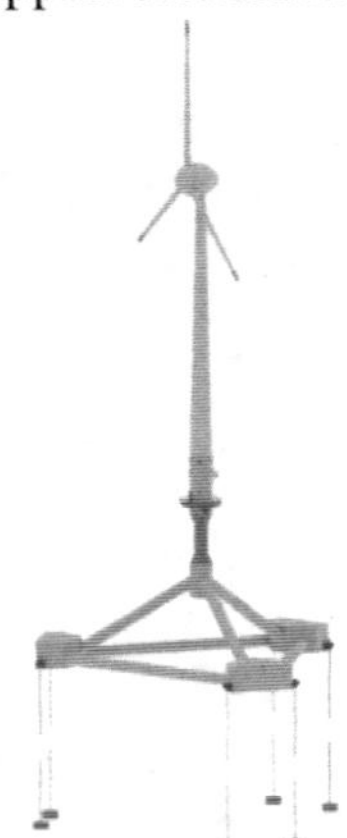

Figure 1. MFWPP-1.

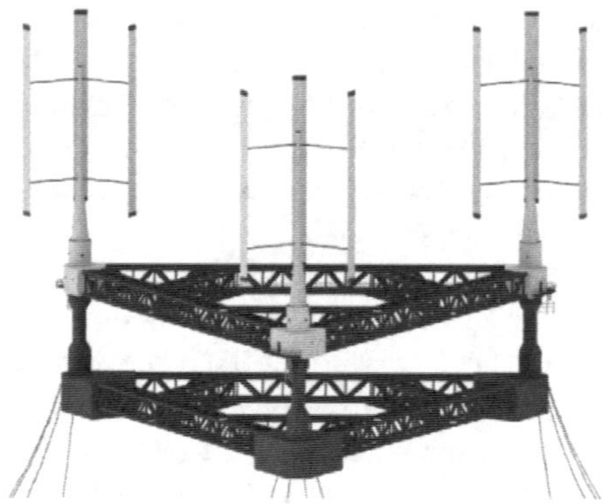

Figure 2. MFWPP -2.

MFWPP-1 hull can be conditionally divided into two main components:

— Bottom frame of Marine Floating Wind Power Plant hull being a triangle substructure with the overall size equal to or below 60 m. The main buoyancy blocks are placed in the construction corners; each corner has ballast tanks for receiving and pumping liquid ballast while Marine Floating Wind Power Plant positioning and displacement from the base. At each side wall of the Marine Floating Wind Power Plant main buoyancy blocks, there are cantilevered hull structures for installation of mooring chain tension devises and one tow pad eye. The main buoyancy volumes are interconnected by non-watertight structures.

— The WT tower foundation is cylinder-shaped. The WT substructure's lower part is connected with the main buoyancy blocks by watertight cylinder-shaped constructions. WT Tower foundation of the single-column Marine Floating Wind Power Plant with one horizontal axial wind-powered generator can be conditionally vertically divided into 3 belts:

a) lower belt of the foundation is cylinder-shaped. The lower belt is divided into two levels. There is a sealed cable termination for electric power supply to shore in the flooring of the lower level;

b) ice breaking belt is cylinder-shaped with conical parts at the top and bottom parts designed to take ice impacts;

c) the top of the WT tower foundation is cylinder-shaped. The main Marine Floating Wind Power Plant electrical equipment is placed in the top belt of the WT tower foundation: switchgears, a transformer, a WT control and state monitoring cabinet and converters;

For a single-column MFWPP-1, a 1.5 MW capacity horizontal axial WT is set. All main equipment, i.e. generator set, gear, system of swinging to the wind, blade rotation system, control system, auxiliary service board, etc. is located inside the cradle.

WT Tower consists of meta tapered tubes, with cradle rotation gear attached to the top part and the bottom part attached to the substructure body.

The Marine Floating Wind Power Plant anchor mooring system comprises tensioned mooring lines. These mooring lines are fixed to cantilevered chain tension gears located at side walls of the Marine Floating Wind Power Plant main buoyancy blocks. Mooring lines are mooring chains. Gravity or pile-type anchors may be applied depending on soil type.

Structurally, the MFWPP-2 substructure consists of three cylindrical support columns with diameter of 5 m, placed within 100 m apart from each other and interconnected by spatial horizontal trusses and braces which, in their turn, consist of built-up double T-beams. Both top and bottom parts of each column are provided with three-dimensional structures to connect the trusses and the column. And these bottom three-dimensional structures are used as buoyancy volumes. In order to decrease ice loads on the facility, the diameter of the support columns in the ice belt area was reduced.

The MFWPP-2 hull with three wind turbines is a triangle support structure; in its corners there are three WT Towers support foundations. Three-column Marine Floating Wind Power Plant hull may be conditionally divided into three main components:

– Bottom frame of the hull that is a triangle structure with overall dimension of about 100 m. Ballast tanks are located in each of three frame fixation areas. Mooring system pad eyes are located on side panels of truss fixation areas at the Marine Floating Wind Power Plant hull bottom frame;

– Top frame of the hull that is a triangle structure. Each of frame truss fixation areas accommodates rooms for Marine Floating Wind Power Plant main electrical equipment;

– Three WT Tower foundations are cylinder-shaped. WT Tower foundations are integrated into truss fixation volumes both at the top and bottom frame. WT Tower foundation of the three-column MFWPP may be conditionally divided into three vertical belts as follows:

1) WT foundation bottom belt is cylinder-shaped. This belt is divided into two levels, with the lower level integrated into the truss fixation volumes at the Marine Floating Wind Power Plant hull bottom frame. Pump room serving the ballast tanks is located at the lower level. The pump room flooring is equipped with a sealed subsea cable termination for electric power supply to shore. The second level accommodates ballast system manifold room.

2) ice breaking belt is cylinder-shaped with conical parts in the top and bottom parts designed to take ice impacts;

3) WT Tower foundation top belt is cylinder-shaped. Bottom part of this belt is integrated into the truss fixation volumes at the floating Marine Floating Wind Power Plant hull top frame.

The WT Tower foundation top belt houses the electrical equipment. For the three-column MFWPP-2 the vertical axial WTs of Darrieus type are set; their capacity is 0,5 MW and they possess an H-shaped wind wheel with swept surface diameter of 30 meters and of blade width of 42 meters. WT Towers consist of three tapered tubes that have main WT bearings inside.

Mooring system of the three-column MFWPP-2 comprises slugged mooring lines. The bundles are located at the substructure corners at the area of bottom column frame. Mooring lines are mooring chains. Also, gravity or pile-type anchors may be applied depending on soil type.

Mass of MFWPP-2 substructure is almost 4 times greater than the mass of the MFWPP-1 substructure.

2. Mathematical modelling of marine floating wind power plant behavior under effects of environmental loads

Buoyancy and stability of the Marine Floating Wind Power Plant was ensured based on its structural analysis. In order to analyze Marine Floating Wind Power Plant behavior under the environmental loads and to select Marine Floating Wind Power Plant mooring system parameters, mathematical modelling of their behavior under effects of environmental loads has been performed.

The environmental conditions that Marine Floating Wind Power Plant was designed for included varying parameters of current and wind velocity, wave and ice parameters typical for expected conditions of Marine Floating Wind Power Plant deployment. The Marine Floating Wind Power Plant is expected to operate in working mode with extreme wind velocity up to 25 m/s. Wind velocity is combined with current velocity (up to 1.5 m/s) and moderate storm (the height of the waves with 3% probability – 6.0 m, the average period of 9.0 s) or ice effects (the thickness of the level ice being 1.0 m and a compressive strength bring 1.5 MPa). Thus, such combination of factors describes the extreme conditions for operational mode.

Generation and calculation of wind stream induced loads and moments for horizontal axial WT and vertical axial WT considering static and dynamic components have been determined according to the proprietary methodology [7]. During design of WT as a part of the Marine Floating Wind Power Plant, all the most typical design cases and load combinations that may occur from a floater-based WT operating in severe climatic conditions have been studied. Design combinations of static and dynamic loads used for checking static strength and stability, fatigue strength and stiffness, both from WT structure and acting on the already vibrating under wave effects floater, have been studied for the following design cases: WT working mode; WT emergency operation; transportation, installation and maintenance. In addition, loads related to WT operation in severe climatic conditions (icing of structures and blades) have been taken into account.

In all WT design cases, the following requirements are applied to civil-engineering structures of WT: static strength and stability when WT members are affected by ultimate loads; WT member fatigue strength where loads are specified versus time; WT structure stiffness, i.e. provision of its members' vibrations where natural frequencies of the WT and its members are beyond the structural resonance zone.

Two factor combinations were taken as an extreme mode of survival with respect to wind and wave effects. The first combination included extreme wind velocity with 100-year return period (40 m/s), extreme current velocity (up to 1.5 m/s), and moderate storm (the height of the waves with 3% probability – 6.0 m, the average period of 9.0 s). The second combination included extreme storm with 100-year return period (the height of the waves with 3% probability – 12.3 m, the average period of 11.0 s), extreme wind (up to 40 m/s), and moderate current (0.5 m/s). In this survival mode, Marine Floating Wind Power Plant does not generate electrical power but shall withstand environmental loads and keep the mooring system undamaged.

In extreme mode with respect to ice load, extreme wind (up to 40 m/s) was combined with effect of ice with maximum thickness (1.0 m) and strength (1.5 MPa). The selected ice movement speed was moderate speed since maximum forces could be expected at very low ice drift speed, but very low drift speeds are unlikely due to high wind velocity that partially affects this speed.

For analysis and generalization of environmental conditions affecting installation and operation of Marine Floating Wind Power Plants, coastal areas of the Kara Sea, the Barents Sea, and the Okhotsk

Sea have been selected, where energy infrastructure development is especially relevant in view of execution of major projects related to development of the Russian continental shelf mineral resources.

Design loads from level ice acting on a vertical cylindrical barrier were calculated according to procedures given in Russian SP 38.13330.2012 [8] and ISO 19906 [9] for the specified ice thickness and compression strength.

Ice and wave load calculation approaches have been integrated into *Anchored Structures* [10] software package. The same software package was used for calculation of current- and wave-induced loads, and hydrodynamic characteristics of Marine Floating Wind Power Plants. The software package was also used for time domain simulation of Marine Floating Wind Power Plant behavior under environmental loads.

For the systematic study of Marine Floating Wind Power Plant options' behavior, a design case matrix has been developed that allows, by means of subsequent mathematical modelling, selecting mooring systems for Marine Floating Wind Power Plant options in question, studying their behavior under various combinations of environmental conditions, estimating viability of structural designs with various directions of environmental impacts and various water depth at site.

Some cases of mathematical modelling of Marine Floating Wind Power Plant behavior under environmental impacts are shown in Figures 3–6.

Analysis of mathematical modelling results for a single-column MFWPP-1 reveals the following: behavior of Marine Floating Wind Power Plant with vertically tensioned lines under wind, current, waves, and ice effects meets regulatory requirements of the Russian Maritime Register of Shipping (RMRS) at typical water depths.

Analysis of mathematical modelling results for a three-column MFWPP-2 reveals the following:

1. In operational mode at waves, the facility's motions do not exceed design values and the three-column Marine Floating Wind Power Plant maintains its operability at typical water depth.
2. At minimum installation water depth, the facility losses its stability during heavy storms and its bottom touches the seabed.

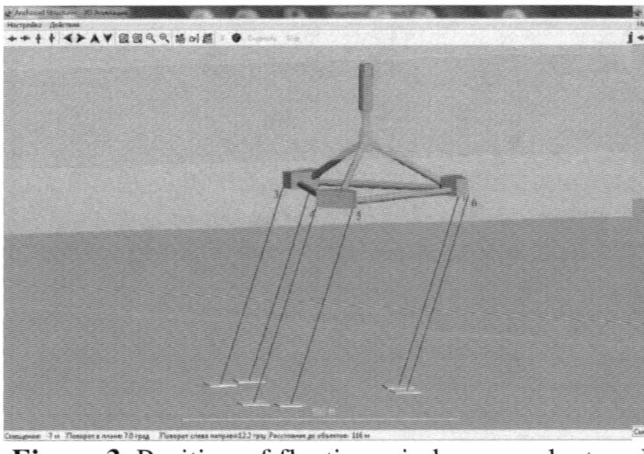

Figure 3. Position of floating wind power plant with vertically tensioned lines under ice load impact.

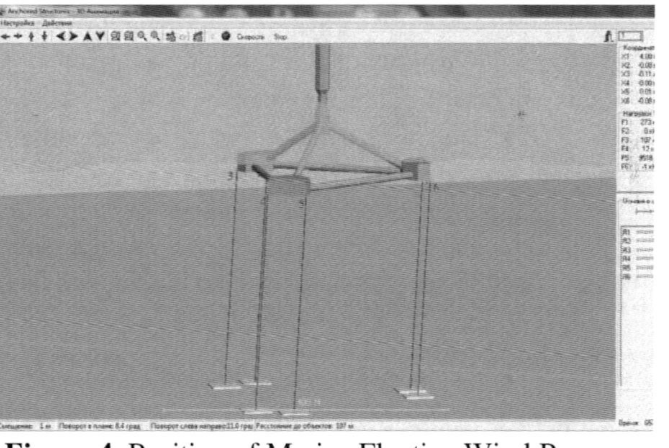

Figure 4. Position of Marine Floating Wind Power Plant with vertically tensioned lines under wave load impact.

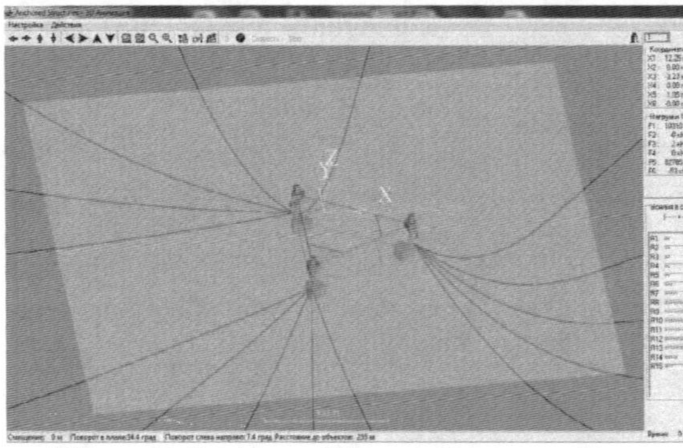

Figure 5. Position of three-column marine floating wind power plant under ice load impact.

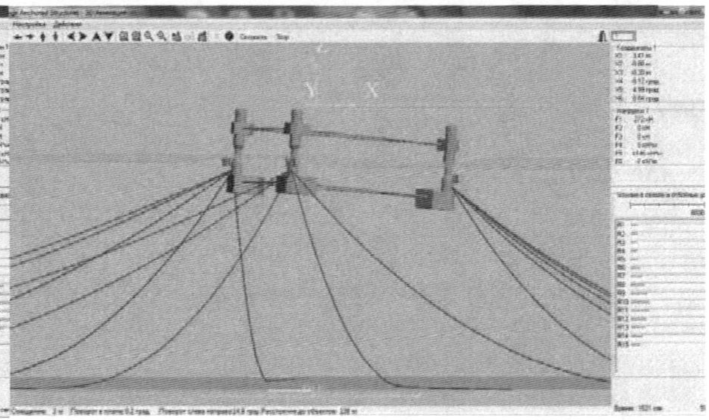

Figure 6. Position of three-column marine floating wind power plant under wave load impact.

3. Estimations of marine floating wind power plant structural member strength

Strength estimations considered the worst environmental load combinations that may cause maximum structural stresses [10]. Structural member ultimate strength criterion for the operating mode complies with regulations of RMRS [11] and was determined using the following equations:

$$\sigma \leq \eta_1 \cdot R_d; \quad \tau \leq 0{,}57\ \eta_1 \cdot R_d; \qquad (1)$$

where σ and τ are the highest normal and tangential stresses in structural members; η_1 is safety factor; R_d is material design yield strength.

The studied substructure members are referred to the main criticality category according to the RMRS Rules for MODU/FOR (η_1=0.68 in working operation mode). During design stage, steel grade and dimensions of Marine Floating Wind Power Plant members were selected based on this criterion.

Strength estimations for the developed substructure designs were performed considering the following three design cases:

- Design case 1. Structural dead weight effects;
- Design case 2. Structural dead weight effects and loads from WT on the floater mating surface;
- Design case 3. Structural dead weight and ice field effects.

Stressed and strained state of the substructure for various load cases was determined by means of FE analysis using ANSYS software package.

Maximum values of effective structural stresses in the substructures obtained from the above estimations do not exceed the allowable limits.

Conclusions

The obtained study results allow to conclude the following:

1. WT types and parameters preferable for application as an energy component of Marine Floating Wind Power Plants have been analysed.
2. Main combinations of static and dynamic loads and moments acting on Marine Floating Wind Power Plants from WT of various configurations under normal operational and extreme working conditions have been determined.
3. Developed Marine Floating Wind Power Plants conceptual options maintain their operability under wind-wave and ice effects typical for the Russian Northern seas.
4. Design values of stresses in structural members under extreme conditions do not exceed allowable limits. Marine Floating Wind Power Plants substructures presented in this paper meet the required stability factors.
5. Both studied substructure options are able to operate at water depths from 50 to 100 meters.

IOP Conf. Series: Earth and Environmental Science **193** (2018) 012022 doi:10.1088/1755-1315/193/1/012022

6. Floating substructures at water depth less than 50 meters are not reasonable. These water depths require fixed substructures.

References

[1] Elistratov V V 2016 *Renewable energy* 3rd edition, amended (St. Petersburg: St.Petersburg Polytechnical University Publish House) p 424

[2] Elistratov V V, Dyuldin M V, Slivkanich M A 2016 Specific Features of Estimation of Wind Power Resources in Russian Polar Regions. *Polar Mechanics: Proc. of the 3rd Int. Conf.* (electronic source) (Vladivostok, Russia) 807–15

[3] Bolshev A S, Frolov S A, Chernetsov V A and Kupreev V V 2007 The design of substructures for offshore wind turbines *Hydraulic engineering Journal* (Moscow, Russia) **5** 31–6

[4] Rune Yttervik 2015 Hywind Scotland – status and plans. *EERA DeepWind'2015* [electronic source]. URL: https://www.sintef.no/globalassets/project/eera-deepwind 2015/ presentations/ closing/ rune-yttervik_statoil.pdf

[5] Grobmann J, Dahlhaus F, Adam F and Schuldt B Experimental Studies and numerical Modelling of structural Behavior of a Scaled Modular TLP Structure for Offshore Wind Turbines. [electronic source]
URL: http://www.sintef.no/globalassets/project/deepwind2014/presentations/e/adam-f_gicon.pdf

[6] Principle Power [electronic source]. URL: http://www.principlepowerinc.com/en/key-markets-projects?location=8

[7] Elistratov V V 2011 Design and Operation of Unconventional and Renewable Energy Units. *Wind Power Plants: textbook*/Elistratov V V and Panphilov A A (St. Petersburg: Polytechnical University Publish House, Russia) p 115

[8] SP 38.13330.2012 Loads and Effects on Hydro-technical Facilities (Wave, Ice and Ship-induced) Updated Revision of SNiP 2.06.04-82* *Analytik* (Moscow, Russia) 111–2

[9] ISO 19906 2010 Petroleum and natural gas industry. Arctic offshore structures p 428

[10] Bolshev A S, Frolov S A and Kuteynikov M A 2013 Mathematical Modelling of Offshore Floating Facilities Using *Anchored Structures* Software Package // scient. and engineer. *Book of the Russian Maritime Register of Shipping* **36** 68–90

[11] Rules for the classification, construction and equipment of Mobile Offshore Drilling Units and Fixed Offshore Platforms 2014 *RMRS* (St. Petersburg, Russia) p 479

Polar compression of planets of solar system and ways of development of geophysical processes

B W Levin[1, 2] and E V Sasorova[2]

[1]Institute of Marine Geology and Geophysics, Yuzhno- Sakhalinsk, Russia
[2]Shirshov Institute of Oceanology, Moscow, Russia

E-mail: levinbw@mail.ru, sasorova_lena@mail.ru

Abstract. The global parameters of the planets of the solar system are considered taking into account periodic variations of the main parameters, and it is shown that the cyclic variation of the polar compression for some planets should lead to pulsations in the shape of the planets and the intensification / weakening of the intensity of geophysical processes.

The planets of the solar system in the first approximation are an ellipsoid of revolution, which on the velocity of rotation determines the magnitude of the polar compression (ε) of the celestial body. This parameter is defined as (R-H) / R, i.e. - the difference between the equatorial (R) and the polar radius (H) of the body, referred to the magnitude of the equatorial radius.

Table 1. Characteristics of the planets needed to calculate the centrifugal and gravitational forces. The flatness column indicates the observed and calculated values.

Planets	Mean radius, m	Mass, kg	Rotation period, s	Angular velocity, sec^{-1}	The flatness, (compression)	Distance to the Sun, million km
Venus	$6.05*10^6$	$5.72*10^{24}$	243 days	$3*10^{-7}$	$5.20\ 10^{-8}$	108
Mercury	$2.43*10^6$	$3.30*10^{23}$	58.65 days	$1.24\ 10^{-6}$	$0.999\ 10^{-6}$	58
Moon	$1.737*10^6$	$7.35*10^{22}$	29.5 days	$2.47*10^{-6}$	$6.51*10^{-6}$	
Earth	$6.37*10^6$	$6*10^{24}$	24 h	$7.3*10^{-5}$	$3.36*10^{-3}$ calc $3.45*10^{-3}$	146
Mars	$3.4*10^6$	$6.4*10^{23}$	24h 37min 22.7sec	$7.1*10^{-5}$	$4.61*10^{-3}$ calc 0.0046	228
Jupiter	$69.2*10^6$	$1.9*10^{27}$	9h 55 min	$1.76*10^{-4}$	$8.1*10^{-2}$ calc. 0.081 (1:12.3)	778
Saturn	$57.3*10^6$	$5.7*10^{26}$	10h14min on equator	$1.7*10^{-4}$	$1.43*10^{-1}$ calc 0.14285 (1:7)	$143*10^3$

It should be noted that the polar compression of the planets is one of the most important parameters of the planet, characterizing its movements. A number of global parameters of the planets of the solar system, such as: the mean radius, the mass of the body, rotation period around the axis, the compression or flattening of the planet and the distance to the Sun are given in table 1.

Theoretically, the compression of planets can be calculated from the expression $\varepsilon = \omega^2 R^3 / GM$, where ω is the angular velocity of rotation of the planet around the axis, R is the mean radius of the body, M is the planet mass, G is the gravitational constant ($G = 6.673 * 10^{-11}$ m^3 $kg^{-1}sec^{-2}$). In addition to computational methods, the value of ε can be determined by observational methods, by measuring the equatorial and polar radii. Figure 1 shows a scatter diagram for the angular velocity of rotation of planets and the degree of their flatness.

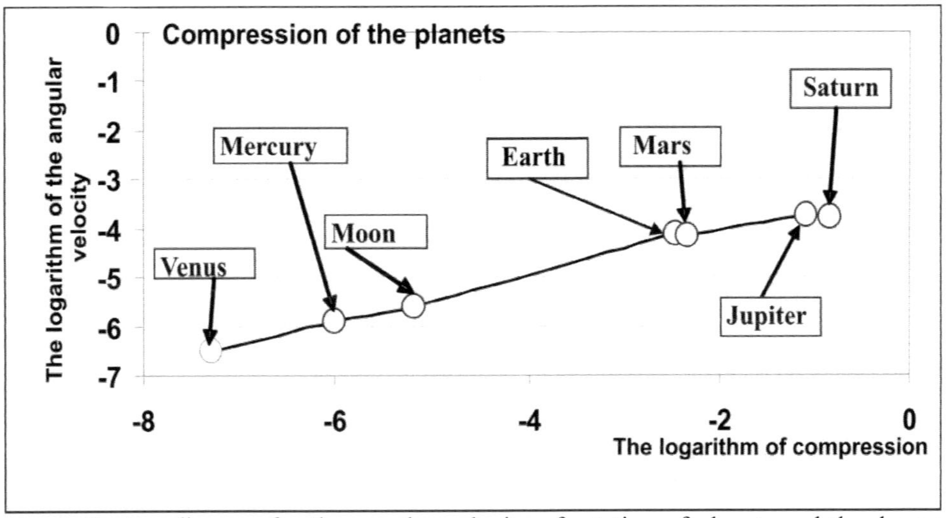

Figure 1. Scatter diagram for the angular velocity of rotation of planets and the degree of their flatness (along the X axis - the decimal logarithm of the flatness value, and along the Y axis - the decimal logarithm from the value of the angular velocity of rotation of the planets).

A comparative analysis of the global parameters of the planets and the scattering diagram shows the existence of a group of slow-moving planets (Venus, Mercury) with weak polar compression. For example, the period of revolution of Venus around the axis of rotation is 243 terrestrial days, and the period of its rotation around the Sun is 224 days. Paraphrasing the words of the famous Russian poet, we can say: “longer than a year, lasts a day”. For Mercury, the period of revolution around the rotation axis is 58.65 earth days and the period of its revolution around the Sun is 88 days. The compression ratio of these planets is insignificant: $5.20*10^{-8}$ for Venus and $0.999*10^{-6}$ for Mercury. This group of planets is located relatively close to the Sun.

The second group includes planets: Earth, Mars, and the Moon. For these celestial bodies the velocity of rotation and associated with it centrifugal forces are quite significant. This is manifested in the appearance of lunar seismicity, which was recorded in the experiments of 1969-1977. And it is described in a number of papers on the geophysics of the Moon. Unfortunately, on other planets, experiments on observational seismology were not performed. The compression of the Earth and Mars is approximately 1/300 and 1/217, respectively.

The third group consists of the planets Jupiter and Saturn, distant from the Sun, whose period of rotation around the axis is almost 2.5 times smaller than that of the Earth (about 10 hours). In connection with this, the polar compression of the planets is 1/10 and 1/15, which strongly distinguishes them from other planets. It is known that the satellites of the giant planets are characterized by high activity of their subsoil. Numerous craters are located on the surface of the satellites; the ejections of liquid heated jets to a large height have been repeatedly recorded.

The second and third groups of planets are located much further from the Sun than the first group. It is interesting to note that as the distance from the Sun increases, the angular velocity of rotation of the planets around the axis increases.

Observations show that the angular velocity of rotation of the planets does not remain constant, that it varies with time. For the Earth, the duration of observations of variations in the velocity of rotation (or the period of the earth's day) is more than 300 years. The long-period variations lasting several years, or several ten years and medium-period variations associated with the moving of the planet around the Sun. are noted. Regular changes in the velocity of rotation of the planet cause regular changes in the degree of its compression and shape (pulsation). If the planets would rotate at a constant velocity the values of ε, R and H remain constant.

However, when the rotational velocity changes, the degree of its compression and the surface area of the planet change. As the velocity of rotation increases, its equatorial radius increases, and the polar radius decreases, as the velocity of rotation decreases, the equatorial radius decreases and the polar radius increases. That is, in response to changes in the velocity of rotation, there is a pulsation of the Earth's shape, a change in its surface while maintaining and retaining the volume. The scheme of such pulsations is shown in figure 2.

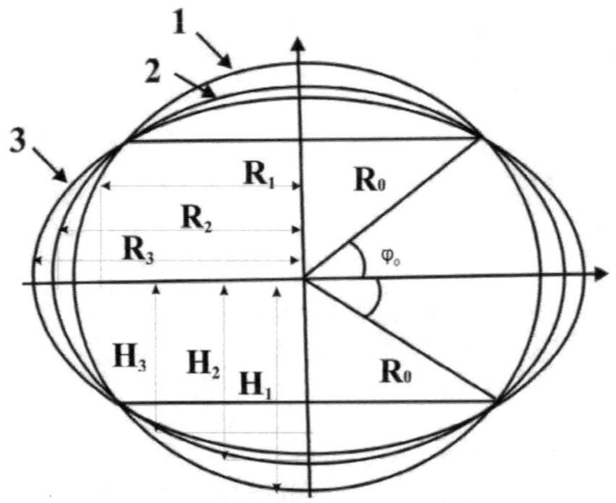

Figure 2. Diagram of the pulsation of the ellipsoid surface as a function of variations in the angular velocity of rotation - dω. The compression ratio is ε, H and R are the polar and equatorial radii, R_0 is the radius of the sphere that is equal in volume to the ellipsoid, 1) dω1 = 0 (H_1, R_1) is the sphere, ε = 0; 2) dω2> 0, (H_2, R_2), ε2> 0; 3) dω3> dω2, (H_3, R_3), ε3> ε2.

Consider the effect of variations in the value of the angular velocity of rotation of the planet on the total energy of the rotating body, which is described as the product of the moment of inertia (I) by the square of the angular velocity ν. The energy of the body as a result of the instability of rotation should increase as a result of adding the angular velocity and the value of the moment of inertia of the body:

$$dE / E = 2 * d\nu / \nu + 1/2 * dI / I.$$

The relative value of the variations in the angular velocity of rotation for the Earth is estimated from the observations of [1] as $d\nu/\nu \approx 10^{-8}$. According to the estimates given in [2, 3] for the Earth, the value of dE is equal to 10^{21} J. This value exceeds by two orders of magnitude the annually allocated energy from earthquakes on the whole Earth. Even a seemingly small change in the velocity of rotation causes a significant energy release due to the huge mass of the planet. Such variations in the rotational velocity can generate amplification or attenuation of a whole series of catastrophic geophysical phenomena. In recent years, scientific literature has published papers showing the relationship between the variations in the angular velocity of the Earth's rotation and the increase in seismic and volcanic activity [3-6].

In the works cited above, based on a joint analysis of time series of seismic event density and variations of the Earth's rotation velocity of about 300 years, it is shown that the maxima of seismic activity (SA) correspond to the final stages of the angular rotation velocity reduction (i.e., deceleration stages). Intervals of time with a minimum CA correspond to the final stages of the stage of increasing the angular velocity. The local minima of the angular velocity increments ΔVn / Δt practically coincide in time with the maxima of the SA. Thus the maximum increase of the CA occurs at the maximum of braking stage. A comparative analysis of earthquake density in five-year intervals for the period from 1895 to 2016 and variations in the Earth's rotation velocity is shown in figure 3.

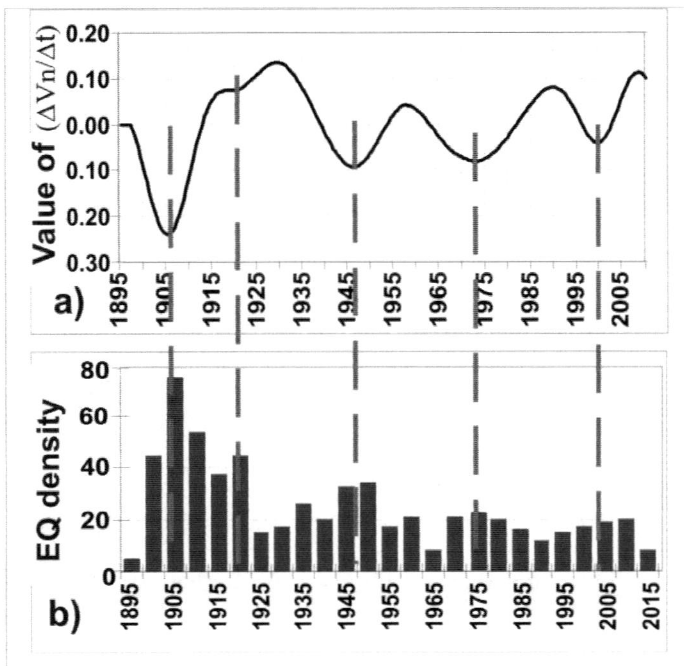

Figure 3. Comparative analysis for the period 1895 -2016.Fragments:
a) - is the increment of the angular velocity;
b) density distribution of strong earthquakes (with M≥7.5) over five-year intervals, along the vertical axis - the number of earthquakes in each five-year interval. On the horizontal axes - the upper boundaries of the five-year intervals (year).

Unfortunately, at the present time we have practically no observational material for seismic, volcanic activity and other geophysical manifestations on other celestial bodies, except the Earth and the Moon.

However, in recent years works have appeared in the scientific literature showing the relationship between the variations in the angular velocity of the Earth's rotation and the increase in seismic and volcanic activity. In the work [7] considers the relationship between variations in angular velocity and volcanic activity of the Earth. Decreasing of the angular velocity of rotation of the planet (the braking stage) leads to an increase in volcanic activity, and vice versa, an increase in the rotation velocity of the Earth leads to a decrease in it (figure 4).

The presence of a negative correlation between volcanic activity and the relative angular velocity of the Earth's rotation is shown. All catastrophic eruptions with the volcanic explosivity index (VEI)> 5 and 88% of large eruptions (with VEI> = 5) begin during the braking period.

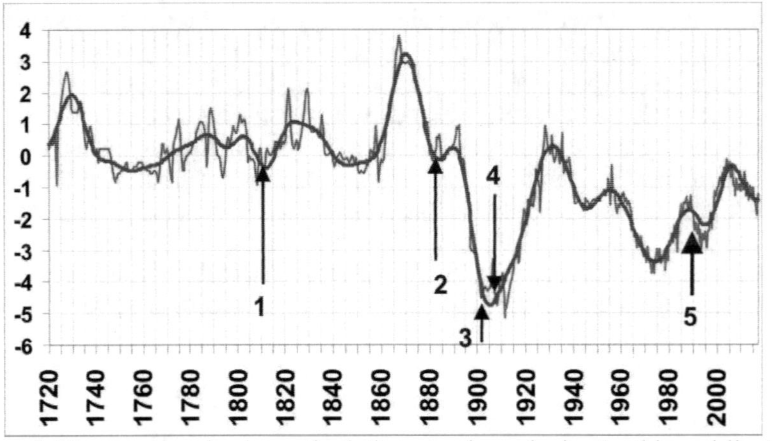

Figure 4. Time series of values of relative angular velocity (v, thin red line) and its low-frequency component (Vn, thick blue line) - for the period 1720 - 2017. Black vertical lines with arrows - the initial stages of the strongest eruptions in the last 300 years (1 - the volcano Tambora (1812), 2 - Krakatau (1883), 3 - Santa Maria (1902), 4 - Novarupta (1912), 5 - Pinatubo 1991)). On the horizontal axis - time (year), along the vertical axis the values of $v * 10^8$ and $Vn * 10^8$.

At present, the Earth enters the initial phase of the new braking process, which has already led to an increase in global seismic and volcanic activity in recent years. Unfortunately, now we practically have no observational material behind seismic, volcanic activity and other geophysical manifestations on other celestial bodies, except the Earth and the Moon. A comparative analysis of lunar and terrestrial seismicity was carried out in [8].

All planets of the solar system move along elliptical orbits around the Sun and periodically experience acceleration in the region of perihelion and braking in the aphelion region. At the same time, the planets rotate around their own rotation axes. Variations in the angular velocity of rotation cause constant periodic changes in the degree of polar compression (oblateness) and, accordingly, the pulsation of the shape of the planets.

Until now, the assumptions about the pulsation of the Earth's shape have been based only on theoretical positions. At present, 814 GPS observation stations are located on the Earth's surface, and we have geodetic observations of changes in the vertical strain components at the observation points [9]. The duration of these observations is from 6.5 to 10 years. Now we have observational data on the changes in the vertical components of the GPS station coordinates, which confirm our theoretical proposals about the periodic pulsations of the Earth's shape. The amplitudes of these pulsations (about 20 mm) depend on the latitudinal locations of GPS stations and the characteristic periods of a year or sometimes a year and a half a year. This confirms the medium-period pulsations of the Earth's shape [10].

At present, the duration of the GPS observation period is not sufficient to confirm the changes in the vertical components of the Earth's surface deformation in the low-frequency range.

Acknowledgments

The work was performed within a state task No. 0149-2018-0015 (IORAN) and No. 0285-2018-0014 (IMGiG of the FEB RAS) Russia. The work is partially supported by RFBR grants 16-05-00089.

References

[1] Sidorenkov N S 2002 *Physics of the Earth's Rotation Instabilities* (Moscow: Nauka)
[2] Levin B W et al 2014 *Adv. Geosci.* **35** 137-44
[3] Sasorova E V and Levin B W 2018 *Geography and Geology* **10** 43-55

[4] Levin B W and Sasorova E V 2015 *Doklady Earth Sciences* **461** 254-9
[5] Levin B W and Sasorova E V 2015 *Doklady Earth Sciences* **464** 351-5
[6] Bendick R and Bilham R 2017 *Geophys. Res. Lett.* **44** 8320-7
[7] Levin B et al 2018 *Geophysical Research Abstracts. EGU General Assembly 2018* **20** p 3814
[8] Levin B W and Sasorova E V 2010 *Doklady Earth Sciences* **434** 1249-52
[9] Scripps Orbit and Permanent Array Center, http://sopac.ucsd.edu/
[10] Levin B W et al 2017 *Geodesy and Geodynamics* **8** 206-12

IOP Conf. Series: Earth and Environmental Science **193** (2018) 012024 doi:10.1088/1755-1315/193/1/012024

Influence of peculiarities of the form of a submerged body on the parameters of generated waves in the ice motion

V L Zemlyak[1], V M Kozin[1,2] and N O Baurin[1]

[1]Sholom-Aleichem Priamursky State University, Birobidzhan, Russia
[2]Institute of Machining and Metallurgy, FEB RAS, Komsomolsk-on-Amur, Russia

E-mail: vellkom@list.ru

Abstract. It is known that when a submerged body moves under the ice surface in an ice cover, a system of flexural-gravity waves (FGW) appears, with a certain intensity of which it is possible to achieve partial or complete destruction of the ice. There are many theoretical works in which the submerged body is modelled in various ways, the influence of physicomechanical properties, the depth of immersion and the length of the body on the stress-strain state of the ice cover is analysed. Similar works of an experimental nature are practically not encountered. An experimental study of the influence of the relative elongation of a submerged body and the depth of its immersion in under-ice motion on the deflection and curvature of the generated flexural-gravitational waves is performed. With the help of the proposed criterion of ice destruction, the ice-breaking capacity of the FGW was assessed.

1. Introduction

Today the process of development of flexural-gravity waves by the motion of a solid body under the ice cover near its surface is studied little. Normally, when solving similar tasks at the theoretical level an immersed body of a simplified form is used. Kheisin [1] studied the plane steady-state problem of the motion of vortex under a layer of broken ice. Bukatov and Zharkov [2] studied the steady motion of a point source of mass under a floating elastic plate and analyzed the effect of the velocity of motion, the depth of the source and the thickness of the plate on its deflection. In Pogorelova's work [3] a SV was modelled by a source-sink system, moving in a fluid close to the surface of an elastic plate. In Sturova's work [4] the solution of the linear steady problem of the flow of nonviscous liquid around a sphere under the ice cover, broken ice, a membrane, and also under the free surface is performed. Korobkin [5] considers the nonlinear problem of non-stationary waves caused by the submerged elliptic cylinder, moving under the ice cover. In Pogorelova's work [6] the influence of the thickness of the plate, the size of a submarine and its velocity on the amplitude of the deflections of floating plates is analyzed numerically. It has been shown theoretically that the wave formation in the ice cover and the efficiency of its destruction depend significantly on the relative elongation and displacement of the immersed body, as well as the speed of its movement and the depth of immersion.

In work on the basis of experimental studies, the influence of relative elongation and depth of immersion of a submerged body on the parameters of waves generated by it under the ice sheet motion is analyzed.

2. Equipment and technique for conducting experiments

Experimental studies of a submerged body motion under the ice sheet carried out in the ice tank laboratory [7] “Ice technology” in Solom-Aleikhem Priamursky State University” (Birobidzhan, Russia). The main element of the tank is a bowl with the dimensions $L \times B \times H = 14 \times 3 \times 1$ m, equipped with a rail with a towing carriage which allows a model submerged body to be towed. The pool tank is a welded construction reinforced with ribs installed on the foundation and covered with a layer of insulation. To prevent ice on the inner surface of the sides of the tank heating elements are fixed on the outer wall surfaces. The ice tank is equipped with a system for collecting and model ice melting.

Figure 1. General view of the ice tank.

Due to the lack of refrigerating machines, experimental studies were carried out only in the winter period (December-January). The model of the ice cover was made in the ice basin by freezing natural freshwater without any dopant up to the set thickness (0.002 m) in natural conditions. The room where the basin is located was not heated for the entire period of the experiment. The experiments were carried out only at night time from 9.00 p.m. to 5.00 a.m. At night there were negligible fluctuations in the room temperature, within the difference of 1°C to 2°C, although the room temperature varied from -10°C to -18°C, depending on the outside temperature, varying from -25°C to -30°C. Relatively uniform ice sheets were thus obtained in 1-1.5 hours at night, except for the patches of thaw on the ice. Effects of snow and wind from the outside were minimized at the ice basin. Due to a layer of insulating material installed on the sides and under the bottom of the basin and compliance with constant temperature mode, the thickness of the ice at the walls of the basin and in the middle of the ice field did not exceed 0.0005 m. After carrying out of each model test, broken ice pieces were removed and the depth of the model was readjusted, and then ice sheet was formed. To conduct the experiments necessary equipment and measuring system to register vibrations of model ice was prepared. By towing the model vessel its velocity and profiles of FGW were determined. The towing system allowed carrying out model tests at velocities of up to u_m=2.4 m/s under steady motion conditions. The time of motion of the model after starting work at stationary regime did not exceed 4 sec.

To carry out model experiments models of submerged bodys were made at a scale of λ_l=1:120 was made as drop-shaped body of rotation, length-diameter *ratio* L_m/B_m=8.4-11.4 (where B_m=0.12 m). The parameters of the model were chosen considering previously executed experimental studies [8] in view of modeled ice conditions these parameters are optimal.

Harnessing of the model in the towing system was made through carabiners connecting the tow cable with attachment points located in the extremities of the body.

Simulation scale equal was selected basing on the size of the ice pool, with the length of the channel providing access to the steady motion of the model SV. The width was sufficient to eliminate the influence of the waves reflected from the channel sides, on the principle wave system [9]. Modelling of sea ice cover can be executed by using various ice models and for each of them there are corresponding conditions of similarity. The usual modelling is performed with partial satisfaction of conditions of similarity [10]:

$$\lambda_E = \lambda_w = \lambda_h = \lambda_l, \tag{1}$$

where λ_E is the model scale for Young's modulus; λ_w is the model scale for deflections; λ_h is model scale for ice thickness.

In this case, the requirements for similarity conditions with respect to Poisson's ratio μ and density ρ of the model ice are satisfied.

The model vessel must be geometrically similar to the real one and their displacements must be in proportion to the cubed modulus of the geometrical scale-ratio:

$$L_n/L_m=\lambda_l, \qquad D_n/D_n=\lambda_l^3, \tag{2}$$

where L_n is the length of the full-scale vessel; D_m is the model vessel displacement.

The model motion velocity u_m is determined by the condition of similarity:

$$u_n/u_m=\lambda_l^{1/2}, \tag{3}$$

where u_n is the full-scale vessel velocity.

The parameters of the model FGW are converted to natural ones in accordance with the modulus of geometrical similarity:

$$\lambda_n/\lambda_m= w_n/w_m=\lambda_l, \tag{4}$$

where λ_n is the length of the full-scale FGW; λ_m is the length of the model FGW; w_n is the full-scale FGW deflections; w_m is the model FGW deflections.

When natural ice cover is used as model one, the thickness of the modeling ice will be calculated when converted to natural one in accordance with the following relations:

$$h_n = h_m \lambda_l^{4/3} \left([\sigma_u]_n / [\sigma_u]_m \right)^{-\frac{1}{3}}, \tag{5}$$

where h_n is the natural ice thickness; h_m is the modeled ice thickness; $[\sigma_u]_n$ is the natural flexural stresses; $[\sigma_u]_m$ is the modeled flexural stresses.

Determination of flexural strength of the model ice was experimentally performed by testing beams and for that purpose the model ice cover was prepared by building up ice of the required thickness h_m=0.002 m [11]. The beams had a rectangular shape with the parameter of $l \times b$=0.15×0.45 m which were prepared in the model ice by cutting out according to the stamp by the rotating Dremel tool. The force required for beam destroying was determined by using the electronic dynamometer Mark - 10 (USA) (Fig. 4) (Kozin et al, 2016). The average value of the flexural strength for natural ice was the value of $[\sigma_u]_n$=0.7 MPa, value of the flexural strength for model ice was of $[\sigma_u]_m$ =1.75 MPa. The thickness of the modeling ice cover after conversion to natural one was h_n=1.6 m.

It is known that at modeling sea ice by using fresh-water model ice, modulus of elasticity of the natural ice should be less then modulus of elasticity of the model ice $E_n > E_m$, however the ratio E/σ_u for sea ice is practically the same as that for fresh-water ice. To meet this condition at modeling ice is very important [12]. To enhance the visibility of fractures of ice, the model ice surface was covered with a thin layer of snow, soon after the model field formation.

The coefficient α was used as a criterion for breaking model ice, that is it is equal to the maximum value of the tangent of the surface slope of the ice plate. The relationship between failure of ice cover and the coefficient α was already empirically determined in Kozin's experimental work [10]. It was stated, that for the load motion over the ice cover as well as for motion of a submarine vessel when the maximum value of the slope of the ice surface is greater than the value of 0.04, it leads to complete ice failure and cracks opening. The value of α in experimental studies was defined by the formula:

$$\alpha=2\pi w_m/\lambda_m. \quad (6)$$

Due to the complexity of the study, the duration of the preparation of the model ice field and limited cold period (December-January), aiming to identify optional parameters of the experiment such as model displacement velocity, relative deepening and water depth a series of preliminary test model runs was executed. According to the performed study the most significant results to be expected at velocity of the model vessel equal to u_m=1.42÷2.17 m/s, model deepening equal to h_{svm}=0.16÷0.25 m.

3. Results of the model experiments

Figs. 2-5 show the main results of model experiments. The maximum values of the deflection of ice were taken at a critical speed of the load, which amounted to $u_m\approx1.85$ m/s.

From Figs. 2 and 3, it follows that the critical velocity are great enough. The value of the critical velocity did not depend on the parameters of the models and the depth of immersion.

The maximum destruction rate of model ice was reached when moving at a critical speed (Figs. 4-5).

The nature of the destruction of model ice after passing through the load is shown in Figs. 6-8. Movement at subcritical speeds led to a significant decrease in the area of destruction of model ice (Fig. 6). At velocities above the critical value, the fracture area increased, but there was no intensive destruction of ice and grinding of ice floes. It can be seen from the graphs that an increase in the relative elongation of a submerged body leads to a decrease in the ice-breaking ability of flexural-gravitational waves, since their curvature drops (Fig. 7). With increasing depth, the deflections of model ice (Fig. 3) and the curvature of the FGW (Fig. 5) are also significantly reduced.

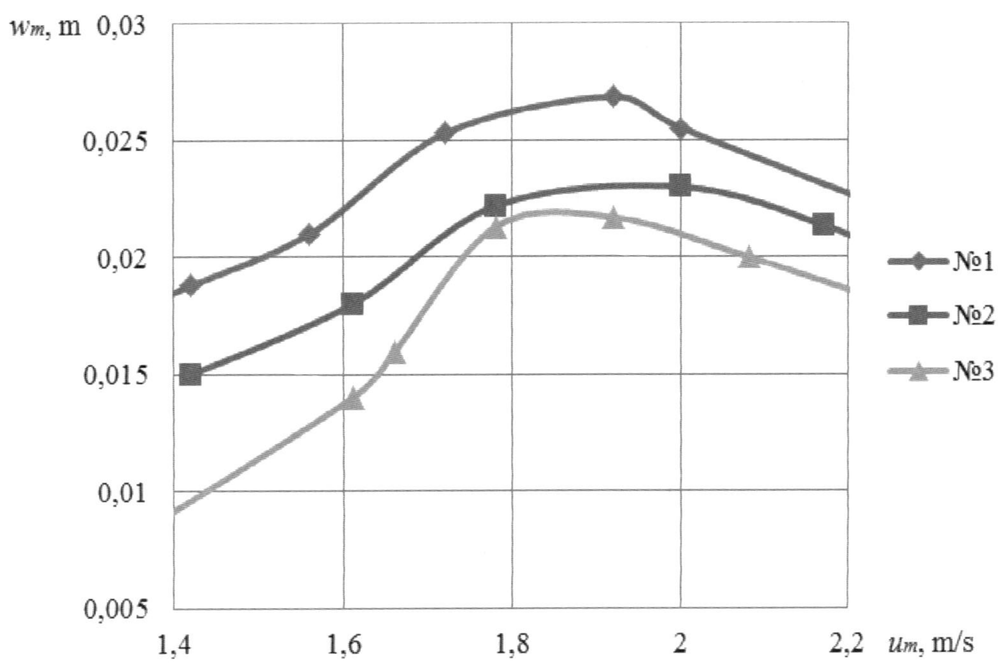

Figure 2. The relationship of the deflections w_m at different submerged body velocity: №1 L_m/B_m=8; №2 L_m/B_m=9.9; №3 L_m/B_m=11.4.

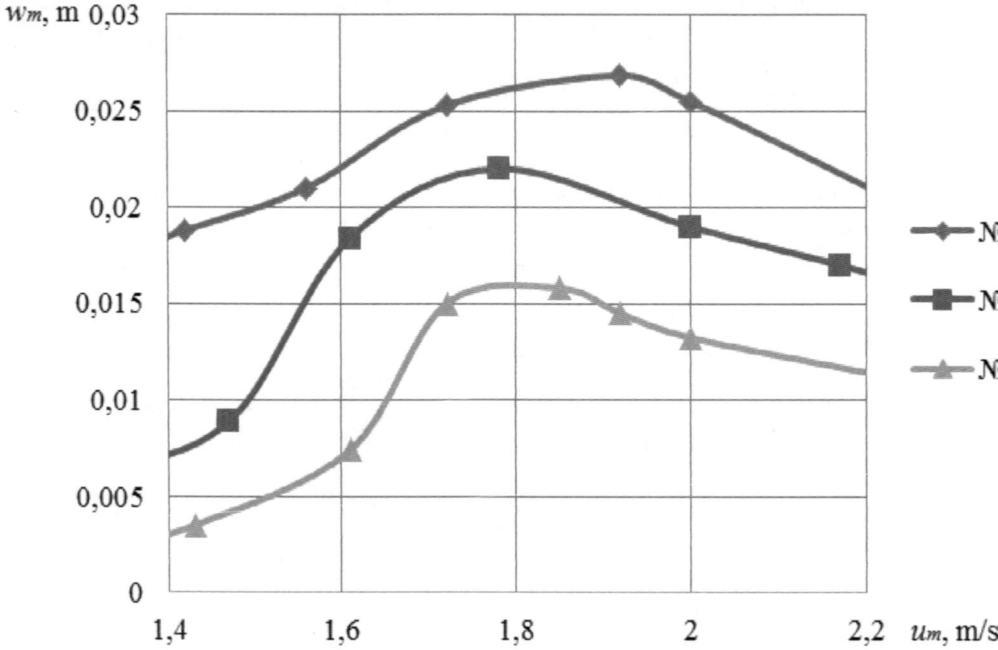

Figure 3. The relationship of the deflections w_m at different the depth of submerged body L_m/B_m=8: №1 h_{svm}=0.16; №2 h_{svm}=0.21; №3 h_{svm}=0.25.

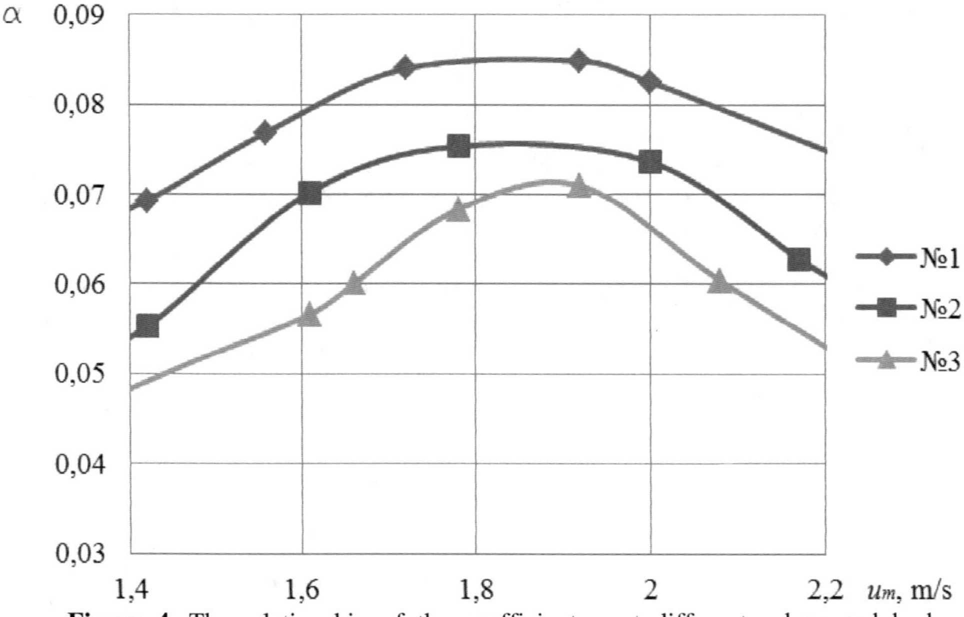

Figure 4. The relationship of the coefficient α at different submerged body velocity: №1 L_m/B_m=8; №2 L_m/B_m=9.9; №3 L_m/B_m=11.4.

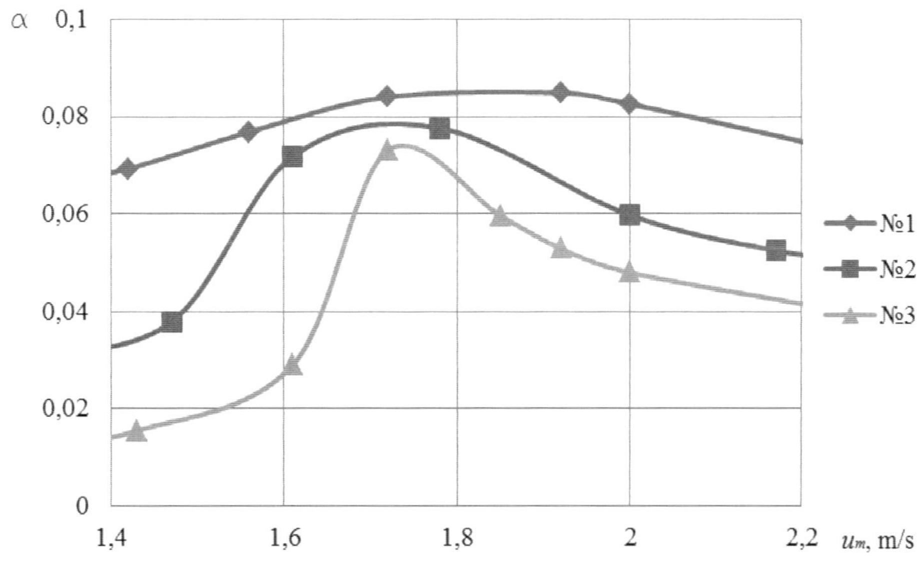

Figure 5. The relationship of the coefficient α at different the depth of submerged body L_m/B_m=8: №1 h_{svm}=0.16; №2 h_{svm}=0.21; №3 h_{svm}=0.25.

Figure 6. The nature of the destruction of model ice after passing through the load (L_m/B_m=8, h_{svm}=0.16 m, u_m=1.85 m/s, h_m=0.002 m).

Figure 7. The nature of the destruction of model ice after passing through the load (L_m/B_m=8, h_{svm}=0.16 m, u_m=1.42 m/s, h_m=0.002 m).

Figure 8. The nature of the destruction of model ice after passing through the load (L_m/B_m=11.4, h_{svm}=0.16 m, u_m=1.66 m/s, h_m=0.002 m).

4. Conclusion

The results of the studies show that the depth of submerged body and ratio length-diameter has a significant impact on the parameters of FGW. The maximum values of the deflection of ice were taken at a critical speed of the load, which amounted to um≈1.85 m/s. Change the depth and ratio length-diameter area of the body do not lead to a decrease in the critical velocity.

Acknowledgments
The reported study was funded by RSF (Russian Science Foundation) according to the research project No. 16-19-10097.

References

[1] Kheisin D E 1967 *Dynamics of Ice Cover* (Gidrometeoizdat: Leningrad) p 216
[2] Bukatov A E and Zharkov V V 1995 *Izv. Ross. Akad.Nauk, Mekh. Zhidk. Gaza* **2** 118–125
[3] Kozin V M and Pogorelova A V 2008 *Int. J. Offshore Polar Eng* **18** 271–276
[4] Sturova I V 2012 *Applied math and mechanics* **76** (3) 406-417
[5] Korobkin A A, Kostukov V K and Makarenko N I 2012 *Vestnik, Quart. J. of Novosibirsk State Univ., Series: Math., mech. and informatics* **12** (4) 76–81
[6] Pogorelova V, Kozin V M and Zemlyak VL 2012 *J. of Applied Mechanics and Technical Physics* **53** (1) 27-37
[7] Zemlyak V L, Baurin N O and Kurbackiy D A 2013 *Vestnik Priamurskogo gosudarstvennogo universiteta im. Sholom-Aleykhema* **12** (1) 75-84
[8] Zemlyak V L, Kozin V M, Baurin N O and Petrosyan GV 2014 *Proc 24th Int Offshore and Polar Eng Conf ISOPE* 1135-1140
[9] Voytkunsky J I 1985 *Handbook of ship theory* (Shipbuilding: St. Petersburg) p 584
[10] Kozin V M, Onishchuk A G and Mar'in B N 2005 *The Ice-Breaking Capacity of Flexural-Gravity Waves Produced by Motion of Objects* (Dal'nauka: Vladivostok) p 250
[11] Stepanjuk I A 2001 *Test Technology and sea ice modeling* (Gidrometeoizdat: St. Petersburg) p 78
[12] Lewis J W 1982 *Offshore Technol. Conf. Houston* **4** p 493-498

Peculiarities of hydrate formation in "natural gas/asphaltene-resin-paraffin deposits/water" systems

I K Ivanova[1,2], M Ye Semenov[2] and V V Koryakina[2]
[1] M. Ammosov North-Eastern Federal University, Yakutsk
[2] Institute of Oil and Gas Problems, Siberian Branch, Russian Academy of Sciences, Yakutsk, Russia

E-mail: iva-izabella@yandex.ru

Abstract. The results of research into kinetic parameters of formation and decomposition of hydrates synthesized in "natural gas/asphaltene-resin-paraffin deposits (ARPD)/water" model systems and in a "natural gas/water" system have been considered. The synthesis and decomposition of hydrates have been conducted in a calorimetric cell of the high-pressure differential scanning calorimetry (DSC). Calculations of kinetic parameters of the processes under study: the rate constants and order of reaction of formation/decomposition of a hydrate – containing phase in paraffin deposits and water, have been performed applying the Avrami equation. The rate of decomposition of hydrate - paraffins has been identified to be half that of hydrates formed in distilled water, thus indicating the stability of hydrate - paraffins. The degree of water conversion into the hydrate has been determined to decrease with an increase of the water content in ARPD emulsions.

1. Introduction

Hydrate plug formation is a major problem of exploitation of oil and gas condensate fields in the Arctic shelf, Far North, Western and Eastern Siberia. Hydrates are formed during the reaction of associated petroleum gas dissolved in oil with water emulsified in oil [1-6], and during condensation of water vapors on paraffin deposits in the form of a drop or film [7]. Low climatic temperatures and specific thermobaric conditions of occurrence of oil deposits under the effects of permafrost thickness favor this process. The above-mentioned factors initiate either hydrate formation, or deposition from petroleum of high molecular weight paraffins on the surface of oilfield equipment including formation of paraffin plugs in wellbores. In their turn, paraffin deposits on the walls of drill tubing form neckings that lead to an even greater intensification of hydrate formation. Hence, formation of gas hydrates in oil, as well as paraffin and hydrate plugs in wellbore complicate oil production technology, transportation, and storage on the shelf and in areas with low average annual temperatures.

As the review of scientific literature has shown, there is a considerable quantity of works on research and a thermodynamic description of phase equilibrium of gas hydrates in water-in-oil emulsions applying the method of high-pressure differential scanning calorimetry (HP DSC) [8-17]. A special attention is paid to kinetics of hydrate formation in these systems, since detailed knowledge of formation, accumulation, and decomposition of hydrates is a key factor in development of new technologies to prevent problems in hydrate formation. In [8], the classical crystal growth theory helps describe kinetics of methane hydrate growth in water droplets. Four parameters have been used in this model: the rate constant, supersaturation of methane in the oil phase, and the mean and standard deviation of the induction period. The induction period of hydrate formation has been shown in [10] to be strongly dependent on the degree of overcooling of the system, affecting the symmetry of signals of the thermal curves. It has also been concluded that each water droplet in water-oil emulsions behaves like a micro reactor where, irrespective of other droplets, the process of hydrate nucleation occurs. The Avrami model has been shown in [17] to be applicable to description of kinetics of hydrate formation in water-in-oil emulsions. The rate of hydrate formation has been defined to depend on the water quantity in the emulsion [16]: as the water content increases, the rate of hydrate formation decreases, while kinetic parameters of hydrate formation in emulsions defined by this model can help to assess the risks of hydrate formation in oil pipelines.

The analysis of the above-mentioned literature has shown that research into gas hydrates in oil dispersed systems is focused on theoretical and experimental study of thermodynamics and kinetics of growth/disassociation of methane hydrates in water-in-oil emulsions with various compositions, but the process of hydrate formation in natural gas mixtures in emulsions of paraffin deposits has not been thoroughly studied yet.

In this paper, we present the data of the research of kinetics of formation/decomposition of natural gas hydrates applying the method of high-pressure differential scanning calorimetry in emulsions of asphaltene-resin-paraffin deposits of oil at the Irelyakhskoye oil and gas field (Sakha Republic (Yakutia).

2. Experimental part

ARPD samples have been taken from the tubing surface at the Irelyakhskoye oil and gas field (Sakha Republic (Yakutia) with the following composition: 59.6 wt.% of hydrocarbons; 20.0 wt.% of paraffins, 4.1 wt.% of asphaltenes, 12.6 wt.% of resins; 3.7 wt.% of mechanical impurities with 0.882 g/ cm^3 density. Emulsion samples were prepared by mechanical mixing of ARPD and distilled water in the H_2O/ARPD weight ratio of: 20/80, 40/60, 60/40 and 80/20 without additon of any synthetic surfactants using a household mixer with blades at a free rotation speed of 11000 rpm during 30 minutes. Mixture of water and ARPD with a weight ratio of 80/20 revealed the maximum saturation of ARPD with water and the actual component ratio in this sample to be 60/40, therefore sample 80/20 was excluded from the study. Natural gas of the Srednevilyuisk gas condensate field (GCF) (Sakha Republic (Yakutia) with a high content of methane has been used as model of a gas - hydrate-forming agent. The choice of this gas as a model has been predetermined by the similarity of its composition to that of reservoir gases of the Irelyakhskoye oil and gas field [18].

Hydrates of this gas synthesized in ARPD emulsions were selected as subjects of the investigation. Their phase transitions were studied in a calorimetric cell of the high-pressure differential scanning calorimetry DSC 204 HP Phoenix HP DSC (Netzsch, Germany). The relative error in the enthalpy measurements makes ±3%, while the error in the temperature measurements makes ±0.1° C. Steel crucibles closed with perforated aluminum caps were used in the experiments. The weight of the ARPD emulsion loaded into the crucible was equal to 35 mg. The discharge pressure in the cell was 50 bar, and it was dynamically maintained up to + 6 ° C, then was switched to a stationary mode. The shooting mode of thermograms consisted of two segments: cooling to -12 ÷ -22 ° C with a rate of a temperature decrease of 0.1 ° C/min with the crystallization peak of the hydrate-containing phase consisting of ice and hydrates; and heating to + 20 ° C with a rate of the temperature increase of 0.2 ° C/min with the peaks of ice and hydrate melting. For each of the samples under study, no less than 2 thermograms were obtained within the coordinates "signal [mW/mg] - time (temperature) [min, (° C)]".

Experimental curves of the conversion degree of water into the hydrate-containing phase (α) and hydrate decomposition (β) are obtained in the coordinates (α or β - t, min) in the automatic mode according to the procedure given in the original software of the DSC. The obtained curves of the degree of conversion of the water contained in emulsions into hydrate, and the curves of the degree of hydrate decomposition were analyzed using the Avrami equation [19] for topochemical reactions:

$$-\ln\left(1-\frac{q}{100}\right)=kt^{n},$$

where q is a degree of crystallization (ice+hydrate) (α) upon cooling or a degree of hydrate decomposition (β) upon heating; n is a constant that determines the character of crystallization/decomposition (with n <1– the diffusion process, n=1 – reaction of the first order where the rate of chemical interaction is comparable with the diffusion rate); k is a constant determining the rate constant of crystallization/decomposition; t is time, min.

The rate constants of crystallization/ decomposition (min^{-1}) were calculated using the Sakovich equation [20]:

$$K = nk^{1/n}.$$

On the basis of the enthalpy of ice melting, the quantitative composition of hydrates in paraffin deposit emulsions was calculated according to the following relations:

$$\alpha(hydrate) = \frac{\Delta H_{melt}}{\Delta H_{CH_4}} * 100\%,$$

where α (hydrate) is the hydrate content, wt.%, ΔH_{melt} is the specific heat of ice melting, J/g, 334 is the enthalpy of ice melting, J/g, ΔH_{CH4} is the enthalpy of hydrate methane melting 438,54 J/g [9].

3. Results and discussions

One or two peaks of crystallization of the water phase are observed within the temperature range of -1.87÷-21.8°C in the process of cooling of all samples in the systems under study; while one or two peaks are observed upon heating: the peak of ice melting falls within the temperature range of - 1.10÷-2.06°C, and the second peak of hydrate melting is within the temperature range of 7.87÷10.70°C. It is worth noting here that melting endotherms are charcaterized by the absence or presence of a peak of ice melting, and the peak of hydrate melting has been recorded in all the samples under study.

Figure 1 shows curves of the degree of phase transformations: crystallization of water droplets in the composition of emulsions into the hydrate-containing phase upon cooling and a curve of hydrate melting upon heating.

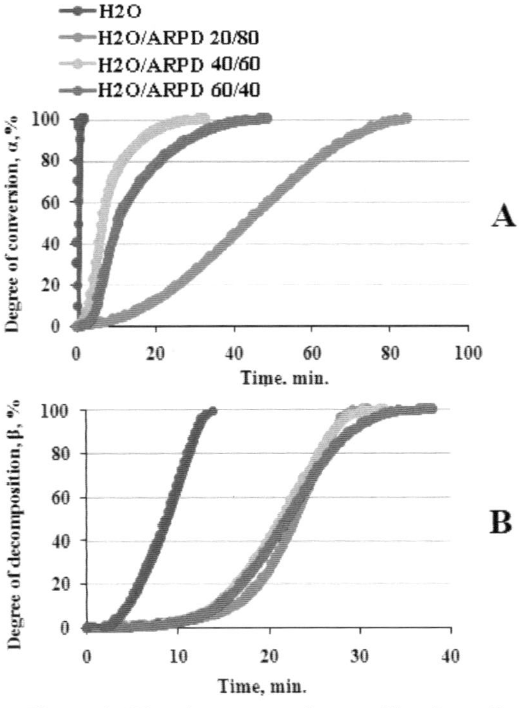

Figure1. Kinetic curves of crystallization of hydrate – containing phases (A) and hydrate melting (B), synthesized in distilled water and «H_2O/ARPD» systems.

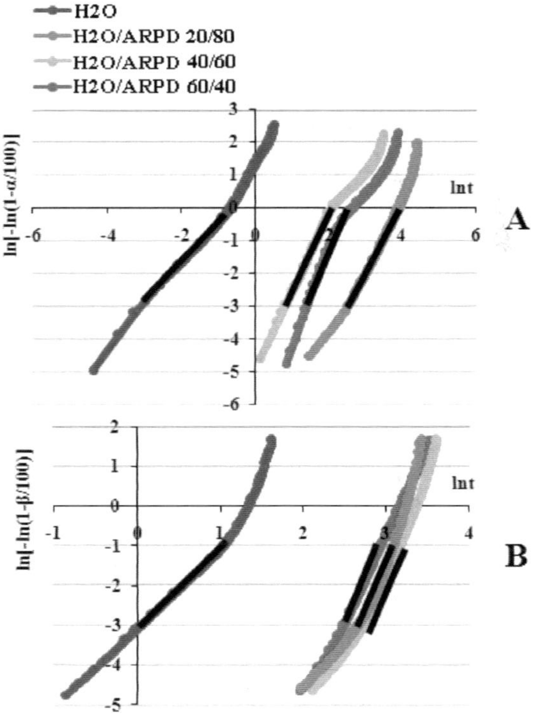

Figure 2. Logarithmic anamorphoses of kinetic curves of crystallization of hydrate-containing phases (A) and hydrate melting (B), synthesized in distilled water and « H_2O / ARPD» systems.

Data on crystallization and melting of hydrates synthesized in distilled water are also compared in this figure. The kinetic curves of crystallization of hydrate – containing phases of ARPD emulsion samples are evident to be S-shaped (Fig. 1 A), while the curves differ among themselves in duration of the induction period. The curve of crystallization of a hydrate-containing phase in distilled water is characterized by the maximal initial speed, since the acceleration period immediately follows the initial period on this curve, i.e. the crystallization process starts immediately over the entire reaction surface or throughout the whole liquid phase volume. The acceleration period for all crystallization curves covers the range of transformation degree values from 10 to 70%.

The decomposition curves of hydrate emulsions of ARPD and water (Fig.1 B) are also S-shaped, while their induction periods are clearly pronounced, as compared with the crystallization curves, which is apparent to relate with a slow release of gas as a result of hydrate melting. During acceleration period, mass hydrate melting and enhanced gas evolution occur within the range of the curve from 10 to 70%. Periods of acceleration are known to be more informative on the ongoing process; therefore, this very section of the curves is more interesting for kinetics analysis. The values of the Avrami equation parameter n, the half-lives $\tau_{0,5}$, and the rate constants K of crystallization/decomposition have been calculated for the sections of crystallization and melting curves in accordance with the periods of acceleration (Fig.2 A and B), and also the values of approximation accuracy r^2 have been given (Table 1). Based on the obtained results, the quantitative content of hydrate $\alpha(hyd)$ in the emulsions and distilled water under study has been calculated (Fig.3).

Table 1. Kinetic parameters of crystallization of hydrate-containing phases and hydrate decomposition in distilled water and systems «H_2O/ARPD»

Sample H_2O/ARPD	n	K, min^{-1}	r^2	$\tau_{0,5}$(graph), min
Crystallization				
20/80	2.1	0.04	0.999	43.4
40/60	2.5	0.32	0.996	6.6
60/40	2.8	0.24	0.992	10.5
Distilled Water	1.5	3.40	0.991	0.4
Melting				
20/80	4.6	0.16	0.998	22.5
40/60	4.3	0.18	0.999	21.1
60/40	4.3	0.18	0.999	22.0
Distilled Water	2.0	0.46	0.992	3.4

High values of reliability approximation demonstrate the validity of the Avrami equation for description of kinetics of hydrate formation and decomposition in ARPD emulsions. It is evident that, as compared with ARPD emulsions, crystallization of the hydrate-containing phase in bulk-distilled water proceeds an order of magnitude faster, values of the half-life are minimal, and the content of the hydrate phase makes only 3% (Fig. 3). The hydrate-containing phase in ARPD emulsions is crystallized by an order of magnitude slower, while the half-life varies from 6.6 to 43.4 min. The slow rate of crystallization could be determined by the high viscosity of emulsions [13], as well as by inhibitory effects of the asphaltenes in the ARPD content on the hydrate formation process [21]. Resulting from slow crystallization, the process proceeds with a high degree of water conversion into hydrate, and the hydrate content is within the range of 49 - 82%, while with an increase of the water mass fraction in the emulsion, the content of the hydrated phase in the samples decreases (Fig.3).

The mechanism of crystallization of hydrate-containing phases in emulsions, regardless of the water cut, proceeds in the kinetic area ($n>1$), i.e. the growth of the hydrate is directed towards the dispersed water droplets. The process of crystallization in distilled water occurs in the diffusion-kinetic area ($n=1.5$), i.e. the hydrate is formed at gas – water interface. The obtained values n correlate well with sample images (Fig.4), it is evident that in distilled water, hydrates are formed on its surface (Fig. 4 (A) in the form of white inclusions, while in ARPD process of hydrate formation occurs in droplets, dispersed all over the ARPD volume (Fig. 4 B).

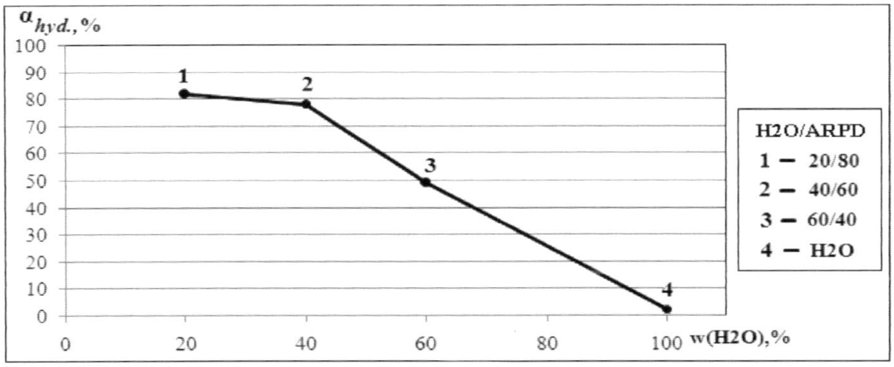

Figure 3. Dependence of hydrate quantity $\alpha(hyd)$ on water content (w(H_2O)) in ARPD emulsion.

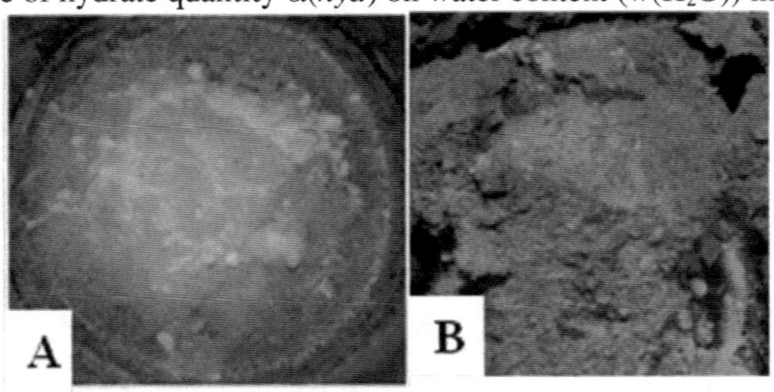

Figure 4. Images: (A) – hydrates, synthesized in distilled water and (B) – H_2O/ARPD sample with component ratio of 40/60 in the cross-sectional view: the arrows indicate inclusions of ice/hydrate.

The values of n>1 of hydrate decomposition indicate the complexity of this mechanism, and in cases of melting of hydrate - paraffins, this index has the maximum values and lays within the range from 4.3 to 4.6. The rate of decomposition of the hydrate synthesized in distilled water is two times higher, while its period of semi-decomposition is six times less, compared with hydrate - paraffins, which indicates the stability of the latter. The stability of hydrate - paraffins is believed to be due to their high sorption ability, resulting to coating of hydrates in these systems with a film of liquid and solid hydrocarbons such as: resins, asphaltenes, and paraffins that strengthen hydrates and make them more resistant to destruction.

4. Conclusions

The Avrami equation has been shown to be applicable to description of kinetic processes of formation of hydrate containing phases and hydrate decomposition, synthesized in ARPD emulsions. The crystallization rate of the hydrate – containing phase in bulk distilled water has been identified to be an order of magnitude faster than that of hydrate crystallization in ARPD emulsions. It has been demonstrated that the degree of conversion of water into hydrate in ARPD emulsions is in agreement with the water content and crystallization rate: the lower the rate constant and the quantity of water phase in emulsion are, the higher degree of water conversion into hydrate is. Hydrate growth in dispersed water droplets of ARPD emulsion is directed inwards the droplets, while hydrate formation in bulk water occurs at the "water-gas" interface. The analysis of the melting curves has shown that hydrated paraffins are more resistant to decomposition as compared with hydrates synthesized in distilled water. With the ARPD water cut of 20 to 80 wt.%, stable hydrate paraffins have been defined to be formed, causing blockage of the wellbore and creating an emergency situation in gas-oil wells of the Nepa-Botuoba anteclise.

Acknowledgments

The research has been performed with the financial support provided by the Russian Federation Ministry of Education and Science, within the framework of the basic part of the State task, Project

10.7697.2017/ВУ "Organization of Performance of Scientific Research" and State Task of the Federal Agency for Scientific Organizations №0377-2016-0003 and by Russian Foundation for Basic Research (RFBR) and the Government of the Sakha Republic the Scientific Project № 18-45-140035.

References

[1] Sloan E D and Koh C A 2008 *Clathrate hydrates of natural gases* (CRC Press) p 720
[2] Sum A K, Koh C A and Sloan E D 2014 *Proc. Int. Conf. on Gas Hydrates (Beijing, China)*
[3] Sloan E D 2000 *Hydrate engineering* ed by J B Bloys (Richardson/Texas) vol 21 p 89
[4] Davies S R, Boxall J A, Koh C A, Sloan E D, Hemmingsen P V, Kinnari K J and Xu Z G 2009 *SPE Prod. Oper.* **24** 573
[5] Turner D, Miller K and Sloan E 2009 *Chem. Eng. Sci.* **64** 3996
[6] Akhfash M, Aman Z M, Ahn S Y, Johns M L and May E F *2016 Chem. Eng. Sci.* **140** 337
[7] Sayakhov F L and Bagautdinov N Ya 2003 *Electrothermal methods of action on hydrate-paraffin sediments* (Moscow: OOO Nedra-Business) p 119
[8] Dalmazzone D, Hamed N and Dalmazzone C 2009 *Chem. Eng. Sci.* **64** (9) 2020
[9] Gupta A, Lachance J, Sloan E D and Koh C A 2008 *Chem. Eng. Sci.* **63** 5848
[10] Dalmazzone D, Hamed N, Dalmazzone C and Rosseau L 2006 *J. Therm. Anal. Calorim.* **85** 361
[11] Delgado-Linares J G, Majid Ahmad A A, Sloan E D, Koh C A and Sum A K 2013 *Energy Fuels* **27** (8) 4564
[12] Greaves D, Boxall J, Mulligan J, Sloan E D and Koh C A 2008 *Chem. Eng. Sci.* **63** 4570
[13] Stoporev A S, Manakov A Y, Aladko E Y, Altunina L K, Bogoslovskii A V and Strelets L A 2014 *Petrol. Chem.* **54** (3) 171
[14] Semenov M E, Manakov A Yu, Shitz E Yu, Stoporev A S, Altunina L K, Strelets L K, Misyura S Ya and Nakoryakov V E 2015 *J. Therm. Anal. Calorim.* **119** (1) 757
[15] Dalmazzone D, Dalmazzone C, Hamed N, Herzhaft B and Rousseau L 2005 *Proc. Int. Conf. on Gas Hydrates* (Trondheim, Norway)
[16] Talatori S and Barth T 2012 J. *J. Petroleum Sci. Eng.* **80** 32
[17] Talatori S, Fotland P and Barth T 2008 *Proc. Int. Conf. on Gas Hydrates* (Vancouver, Canada)
[18] Kalacheva L P and Fedorova A F 2017 *Baza dannyh sostav gidratov prirodnyh gazov mestorozhdenij Leno-Tungusskoj neftegazonosnoj provincii* (svidetelstvo o gos registracii bazy dannyh 2017620891 ot 11 08 2017 po zayavke 2017620612 ot 21 06 2017)
[19] Avrami M 1939 *J. Chem. Phys.* **7** 1103
[20] Rozovskiy A Ya 1974 *Kinetika topokhimicheskikh reaktsiy* (Moscow: Khimiya) p 224
[21] Zi M, Chen D, Ji H and Wu G 2016 *Energy Fuels* **30** (7) 5643

Study of hydrates formation in mineralized solutions and kinetic regularity of their decomposition

L P Kalacheva, I I Rozhin and A S Portnyagin

Institute of Oil & Gas Problems, Yakutsk, Russia

E-mail: lpko@mail.ru

Abstract. Processes of hydrate formation in water solutions of electrolytes that imitating the composition of reservoir water of eastern Siberian platform were studied. The investigation of hydrates properties, obtained from mineralized solutions showed notable influence of electrolytes on the processes of their growth and decomposition, as well as texture and component composition of gases in hydrates. Equilibrium conditions of hydrate formations in solutions of electrolytes shifted to the field of higher pressures and lower temperatures. The texture of the hydrates depends on the type of cation in the electrolyte solutions. If filiform crystals of hydrates are formed in the water, sodium cations promote to the formation of granular forms and calcium does to the formation of laminate hydrates. The research of the component composition of the gas obtained by the decomposition of hydrates showed that the mineralization of solutions leads to the concentration of hydrocarbons C_2-C_4 in clathrate phase and to increasing of the fatty coefficient. The shape of the kinetic curves of hydrate decomposition depends on the type and mineralization degree of the model reservoir waters.

1. Introduction

In the terms of stratification of oil and gas fields, which referred to the permafrost area, the low stratum temperature promotes a transition of hydrocarbon gases to the hydrate form [1–3]. For the fields of Eastern Siberia, high mineralization of formation water is typical, which obstructs processes of hydrate formation both in free volume and in porous media. However despite of the high antihydrate activity of the reservoir waters, the possibility of hydrate formation in the bottom hole of the reservoir zone due to abnormally low formation temperatures and in the borehole in the result of thermal cooperation with surrounding frozen rocks is not excepted [4]. From the other side, for the commercial production of natural gas hydrates with the purpose of storage and transportation it is possible to use water from natural sources that contain different quantity of dissolved substances so research of the influence of the composition of electrolyte solution on the kinetics of formation and decomposition of natural gas hydrates is an actual problem [5,6].

The oil and gas fields of the east of the Siberian platform are placed within the Leno-Tunguska and the Lena-Vilyui oil and gas provinces. The fields of the Leno-Tunguska oil and gas provinces are characterized by abnormally low reservoir temperature and reservoir pressure [7], and its reservoir waters are referred to the chloride-calcium type, and on type of the dominated cation either to the calcium or sodium subgroup of waters. The mineralization of reservoir water is able to reach 400 g/l and higher [8]. For the fields of the Lena-Vilyui oil and gas provinces an abnormal high reservoir pressure is 36.3 MPa, the temperatures reach +66°C [7]. The mineralization of reservoir waters is from 40 up to 180 g/l, there can be waters of chloride-calcium and bicarbonate-sodium types [8].

2. Experimental part

For hydrate formation aqua solutions of calcium and sodium chlorides, sodium hydrogen carbonate, imitating the composition of a stratum waters of the East Siberian platform were used. The concentration of solutions of chlorides varied from 50 up to 180 g/l, and of bicarbonates from 2.5 up to 20 g/l. The composition of natural gas is: methane – 92.7; ethane – 5.24; propane – 1.21; isobutane – 0.10; *n*-butane – 0.12; carbon dioxide – 0.05; nitrogen – 0.58% mol. The molar mass of the gas is 17.25 g/mol, the relative density on air is 0.596.

The conditions of natural gas hydrate formation in the absence of salts that dissolved in reservoir water were calculated by Sloan's method [9], where the equation of state of Redlich-Kwong is used. Then these conditions for mineralized reservoir water of the chloride-calcium type for the sodium and calcium subgroups were recalculated on the methods proposed in the work [10]. The results of calculations of the equilibrium conditions of natural gas hydrates formation are presented on figure 1, where the dot-dash graph matches to an absence of dissolved salts in the reservoir water. With increasing of mineralization of electrolyte solutions the equilibrium pressures increase at given temperature and a hydrate formation temperatures decrease at a given pressure.

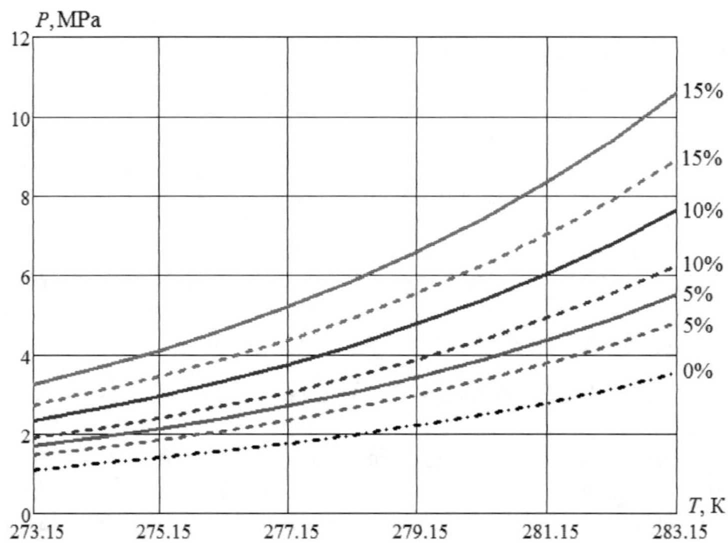

Figure 1. The equilibrium conditions of the natural gas hydrate formation depending on the concentration of calcium chloride (solid graph) and sodium chloride (dashed graph) solutions.

In the laboratory conditions hydrates of natural gas in salt solutions obtained at a temperature of 278 K. The experiments were carried out in a closed type facility in static conditions at the same time in 4 high pressure chambers. The gas pressure, which exceeded the equilibrium pressure of hydrate formation in advance, was pressurized to avoid refueling the chambers. Each chamber (figure 2) is a steel glass (1) with a volume of 1000 cm^3, closed with a steel cap (2) with a standard gage (3) of the type MO-160 (precision is 0.4) and a gas valve (4). In the hydrate chamber, 100 ml of distilled water (or electrolyte solution of the specified concentration) is introduced (5), then natural gas is introduced from the balloon (6) up to a pressure of 8 MPa and an experiment is conducted. The end of hydrate formation process in aqueous solutions is fixed by the constant of pressure in the chambers.

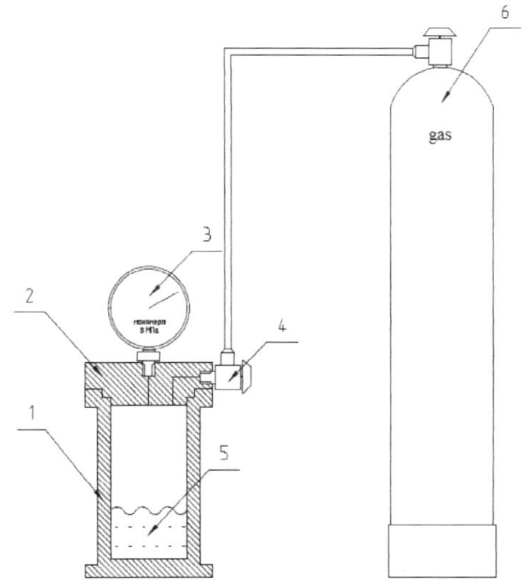

Figure 2. Scheme of experimental facility for gas hydrates generation: 1 – steel glass; 2 – cover; 3 – standard pressure gage; 4 – refueling faucet; 5 – solution; 6 – balloon with natural gas.

The texture of synthesized hydrates depends on the type of solutions. In the absence of mineralization, white filiform crystals are formed, which occupy the entire volume of the chamber (figure 3a). Hydrates obtained in solutions of chloride and sodium bicarbonate have a granular structure (figure 3b and 3c), and in solutions of calcium chloride a stratified structure (figure 3d). With increasing of mineralization of solutions the volumes of synthesized hydrates decreased. This regularity is observed both for hydrates obtained in solutions of chloride and sodium bicarbonate, and in solutions of calcium chloride.

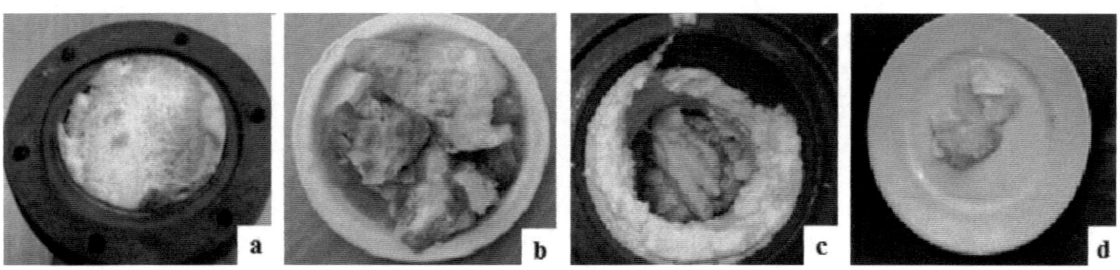

Figure 3. The texture of synthesized hydrates, depending on the type of solution used for hydrate formation: a – water; b – sodium chloride solution; c – sodium bicarbonate solution; d – calcium chloride solution.

To determine the volume of gas contained in hydrates, an experimental facility was assembled (figure 4). The experiment was carried out as follows. At the end of the hydrate formation process the pressure in the chamber (1) decreased to atmospheric one. Then the chamber was placed in a thermostat (2), connected to a gasometer of the tympanic type (3). If gas sampling was required for research of the component composition a gas burette (4) with an expansion capacitance (5) was connected to the facility. The volume of gas made during the decomposition of hydrates was measured every minute.

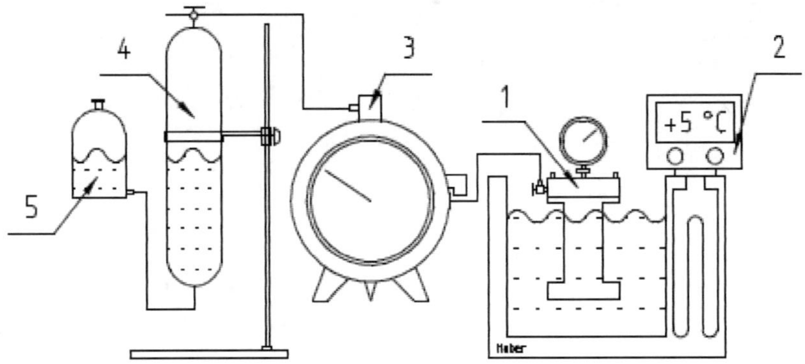

Figure 4. The scheme of the facility for natural gas hydrates decomposition: 1 – high pressure chamber; 2 – thermostat; 3 – gasometer of the tympanic type; 4 – gas burette; 5 – an expansion capacitance of the gas burette.

It is established that the maximum volume of gas contains hydrates obtained in solutions of sodium bicarbonate (table 1). The volume of gas in the hydrates obtained in solutions of chlorides always decreases with increasing of a concentration.

Table 1. The volume of gas in hydrates obtained in solutions of salts.

ω, %	The volume of gas, *l*		ω, %	The volume of gas, *l*
	NaCl	$CaCl_2$		$NaHCO_3$
0	14.3	14.3	0.25	17.77
5	11.7	6.8	0.5	18.86
10	5.4	2.1	1.0	18.43
15	1.7	3.9	2.0	16.20

Analysis of the gas composition in hydrates obtained in solutions of chlorides indicated that under its formation in mineralized solutions reallocating of the components of the sourced gas mixture occurs (table 2). The reallocating coefficient *B* is calculated as the ratio of the content of hydrocarbons in the hydrated gas to its content in natural gas.

Thus, the composition of the gas in hydrates depends on the composition and concentration of model reservoir water. The maximum concentration by the formation of natural gas hydrates in salt solutions is observed for isobutane and propane. With the increase in the concentration of solutions their content in hydrates increases. In the presence of calcium cations the reallocating coefficients of specified components of natural gas are higher than in solutions containing sodium cations.

Table 2. Coefficients of the natural gas components reallocating *B* at the formation of its hydrates depending on the type and concentration of solutions.

Component	*B*						
	H_2O	NaCl			$CaCl_2$		
		5	10	15	5	10	15
iso-C_4H_{10}	3.75	5.3	10.7	15.4	10.0	12.45	17.85
C_3H_8	5.12	5.79	9.60	12.58	9.43	11.74	13.29
C_2H_6	2.63	2.91	3.11	2.57	2.90	3.81	2.77
n-C_4H_{10}	2.04	1.75	2.08	1.75	1.71	5.17	1.88
CH_4	0.85	0.83	0.76	0.75	0.77	0.69	0.72

A composition of gases in hydrates lets to estimate fatty coefficients. A fatty coefficient (the ratio of the sum of homologues of methane to methane content) can be dry (0.3 – 8%), bold (8 – 20%), fat (20 – 30%) and high-fat gases (>30%). Fat and high-fat gases are a valuable raw material for the oil and gas chemical industry. The sourced natural gas according to Vysotsky's classification [11] refers to dry (table 3). However in the process of hydrate formation in electrolyte solutions when the sodium-type changes to calcium one the fatty coefficient of the gas in the hydrate increases due to an increase of the C_2-C_4 alkane content.

Table 3. Fatty coefficient of natural gas and gases in hydrates.

	NG	A gas in hydrates obtained in different liquid phases						
		H_2O	Mineralization NaCl, g/l			Mineralization $CaCl_2$, g/l		
			51.48	106.47	168.48	52.17	107.67	175.38
C_{2+}/CH_4	7.07	26.26	29.95	41.45	43.83	40.51	56.37	49.02

To study the particularities of hydrate decomposition reactions obtained in mineralized solutions kinetic parameters were found by the Erofeev equation, which is more often used for the analysis of isothermal solid-phase reactions [12]:

$$\alpha = 1 - \exp(-kt^n) \tag{1}$$

where α is the degree of conversion, k is the generalized speed constant, t is the decomposition time, and n is the kinetic parameter. The rate of conversion is defined as the ratio of the gas volume emitted by the time moment t to the volume of gas emitted upon complete decomposition of the hydrate: $\alpha = V_t / V_\infty$.

The equation (1) in twice logarithmic type is a straight equation, that lets to calculate the values of the parameters n and k. The first of them characterizes the main driving force of the reaction: when $n > 1$ the process is determined by the kinetics, when $n < 1$ is determined by the diffusion. The parameters k and n let us to find the speed constant of reaction K according to Sakovich's formula: $K = nk^{1/n}$ (12).

The parameters of the hydrate decomposition process are summarized in table 4, where you can see that the values of n for the decomposition reactions of all hydrate samples are more than one. It means that these reactions occur in kinetic region, in other words, the speed of decomposition of hydrates is determined by the speed of physical desorption of the gas from the clathrate lattice.

Table 4. Kinetic characteristics of decomposition reactions of hydrates synthesized in different liquid phases.

Solution	H_2O	NaCl			$CaCl_2$			$NaHCO_3$			
		5%	10%	15%	5%	10%	15%	0.25%	0.5%	1%	2%
A type of graph	Sigmoidal with induction period	Sigmoidal without induction period									
n	1.499	2.264	1.791	1.330	1.698	1.664	1.551	1.213	1.026	1.015	1.159
A region of reaction	Kinetic: the speed of the process is determined as the speed of physical desorption										
k	0.029	0.004	0.047	0.127	0.037	0.045	0.092	0.074	0.066	0.076	0.063
K, min^{-1}	0.141	0.191	0.324	0.282	0.245	0.257	0.333	0.142	0.072	0.080	0.107

Note that the hydrates decomposition speed obtained in solutions of sodium bicarbonate at the maximum gas content has minimal decomposition speed. It should be considered in the industrial production of hydrates with the purpose of the natural gas storage and transportation.

3. Conclusions

The processes of growth and decomposition of natural gas hydrates in model solutions of various mineralizations, imitating the reservoir waters of the East Siberian platform are researched. It is established that equilibrium hydrate formation curves in electrolyte solutions are shifted to the field of high pressures and low temperatures. The texture of the hydrates depends on the type of cation in the electrolyte solutions. If filiform crystals of hydrates are formed in the water sodium cations promote the formation of granular hydrates and calcium-layered hydrates. The mineralization of solutions leads to the concentration of C_2-C_4 hydrocarbons in the clathrate phase, and to an increase of the fatty coefficient and caloric power of the gas. The shape of the kinetic curves of hydrate decomposition depends on the type and degree of mineralization of model reservoir waters. A conversion of natural gas to hydrates for storage and transportation is expediently carried out in diluted solutions of electrolytes, because it increases their thermodynamic stability.

Acknowledgments
The work is carried out within the framework of State Order №0377-2016-003.
The reported research was partly funded by Russian Foundation for Basic Research and the Government of the Republic of Sakha (Yakutia) of the Russian Federation, grant №18-45-140035.

References

[1] Makogon Yu F 1974 *Hydrates of Natural Gases* (Moscow: Nedra) p 208 (in Russian)
[2] Yakushev V S 2009 *Natural Gas and Gas Hydrates in the Permafrost Zone* (Moscow: VNIIGAZ) p 192 (in Russian)
[3] Yakutceni V P 2013 Gas hydrates – unconventional gas sources, their formation, properties, distribution and geological resources *Petroleum Geology – Theoretical and Applied Studies* **8**(4) 1-24 (in Russian)
[4] Bulejko V M, Vovchuk G A, Grigor'ev B A and Istomin V A 2014 Phase action of hydrocarbon systems in a water-saturated sand reservoir under conditions of hydrate formation *Vesti Gazovoy Nauki Scientific-Technical Collection Book* **4**(20) 156-63 (in Russian)
[5] Mimachi H, Takeya S, Gotoh Y, Yoneyama A, Hyodo K, Takeda T and Murayama T 2015 Dissociation behaviors of methane hydrate formed from NaCl solutions *Fluid Phase Equilibria* **413** 22-7
[6] Sato H, Tsuji T, Nakamura T, Uesugi K, Kinoshita T, Takahashi M, Mimachi H, Iwasaki T and Ohgaki K 2009 Preservation of methane hydrates prepared from dilute electrolyte solutions *International Journal of Chemical Engineering* 843274
[7] Safronov A F and Safronov T A 2008 *Geological and Economic Aspects of the Development of the Oil and Gas Complex of the Republic of Sakha (Yakutia)* (Yakutsk: Yakutsk Scientific Center of the Siberian Branch of the Russian Academy of Sciences) p 184 (in Russian)
[8] Zorkin L M 1989 *The Waters of the Oil and Gas Fields of the USSR* (Moscow: Nedra) p 382 (in Russian)
[9] Sloan E D and Koh C A 2008 *Clathrate Hydrates of Natural Gases* (Boca Raton USA: Taylor&Francis Group/CRC Press) p 720
[10] Istomin V A and Kvon V G 2004 *Prevention and Elimination of Gas Hydrates in Gas Production Systems* (Moscow: IRTs Gazprom) p 508 (in Russian)
[11] Pirogov S Yu, Akulov L A, Vedernikov M V, Kirillov N G, Naumchik I V, Sokolova I V and Sof'in A P 2006 *Natural Gas Methane: handbook* (St. Petersburg: NGO Professional) p 848 (in Russian)
[12] Rozovskij A Ya 1974 *Kinetics of Topochemical Reactions* (Moscow: Khimiya) p 224 (in Russian)

IOP Conf. Series: Earth and Environmental Science **193** (2018) 012027
doi:10.1088/1755-1315/193/1/012027

Method for estimation of icebreaker propulsion performance in ice

G I Kanevsky [1], A M Klubnichkin [1], A V Ryzhkov[2] and K E Sazonov[1]

[1] Ice laborotary, Krylov State Research Centre, St.Petersburg, Russia
[2] Central Design Bureau «Iceberg», St.Petersburg, Russia

E-mail: K_Sazonov@ksrc.ru

Abstract. Classical methods of ship propulsion estimations based on the traditionally applied system of hull and propeller interaction coefficients are not suitable for prediction of icebreaker performance in ice. This is because the wake fraction tends to assume negative values in operating conditions typical of navigation in ice. This paper suggests an alternative system of hull/propeller interaction coefficients to overcome the problem. The alternative system comprises the thrust deduction factor t and the factors allowing for hull effect on thrust i_{TB} and hull effect on torque i_{QB}. An alternative method of ship propulsion performance estimations is developed based on this system of coefficients to take into account all characteristics of icebreaker sailing in ice. The method was used to analyze the full-scale data obtained from sea trials of the icebreakers *Vladivostok* and *Novorosiisk*. The initial data inputs for the analysis were ship speed in ice and consumed power measured during the sea trials. The number of propeller revolutions and ice resistance were calculated. The estimated propeller revolutions were compared with the full-scale data gathered during the sea trials in ice. The ice resistance was compared with the predictions based on model tests in ice basin. It is shown that at slow speeds in ice, ranging from 0.5 to 5.0 knots, the predicted number of propeller revolutions is in good agreement with the data measured during the sea trials. Propulsion performance of the *Vladivostok* icebreaker in 1.5-m thick ice was estimated

1. Introduction

In view of the modern developments in Arctic shipping it is required to increase the operational efficiency of icebreakers and ice-going vessels and, in this connection, to drastically raise the accuracy of predictions regarding the propulsion performance of ships in ice. The propulsion performance in ice has two components: ice resistance and thrust of propulsion system. Recently the most vivid progress has been achieved in development of experimental [1, 2] and computational [3–7] methods for ice resistance evaluations. All these references as well as many other studies provide for ice resistance predictions with acceptable accuracy for design purposes, in particular for the cases of close-to-limit ice conditions.

A vast majority of studies on icebreaker propulsion systems have focused only on the interaction between ice and propellers or propulsion pods with an emphasis on strength considerations [8-12]. However, little attention has been paid to the hydrodynamic aspects of icebreaker propulsions. Only one reference can be cited [13] where some specific hydrodynamic issues of icebreaker propulsors have been addressed, e.g. cavitation phenomena.

Traditional approaches used today enable us to confidently predict only the limit ice-going capability of icebreakers and vessels, i.e. the maximum thickness of ice field where the ship is able to sail without stopping at a given power of main engines. However, it is impossible to predict the thrust, torque, speed of revolutions and efficiency of icebreaker propulsors in a broader range of ice navigation conditions, which is necessary for optimizing operation of icebreakers and ice-class vessels. In this context it appears that prediction of ship propulsion in ice for the entire spectrum of operating conditions is very much relevant today.

2. Traditional and "bollard pull" system of hull/propeller interaction coefficients

The traditional system of hull/propeller interaction coefficients is made up of three factors: t – thrust deduction coefficient, w – wake fraction and i_Q – factor allowing for non-uniform wake effect on torque. The values of these coefficients and their dependence on propeller load are determined experimentally by self-propelled model tests, comparing propeller curves obtained in open water and behind model hull. This comparison is done at equal thrust coefficients in open water K_{T0} and behind model hull K_T. Figure 1 gives a typical curve of wake fraction versus effective thrust of propulsors K_{DE}, showing that the wake fraction turns negative at K_{DE} variations in the range from 0 to 1.0, which is typical for ice sailing. The traditional approach to propulsion performance prediction in this case is no longer applicable because of negative wave fraction values.

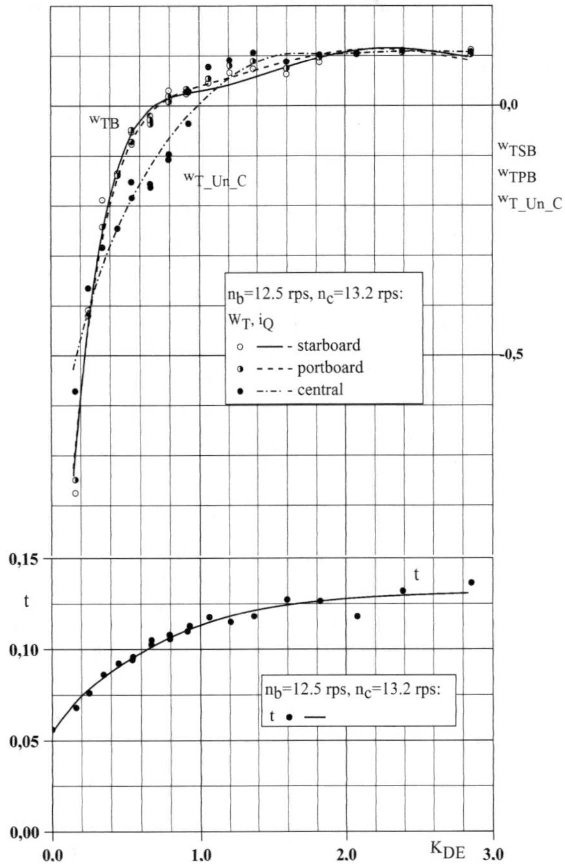

Figure 1. Typical results obtained using traditional hull/propeller interaction coefficients

The alternative system of interaction coefficients is based on the data traditionally obtained from towing tank experiments [14]. Coefficients are determined at a given constant advance ratio of propeller J. For this purpose the test data plots are used to pick up the values of thrust and torque in open water K_{T0}, K_{Q0} and behind model hull K_T, K_Q, as well as effective thrust coefficient K_E. The interaction coefficients are defined by the following equations:

IOP Conf. Series: Earth and Environmental Science **193** (2018) 012027 doi:10.1088/1755-1315/193/1/012027

- factor allowing for hull effect on thrust $i_T = \dfrac{K_T(J)}{K_{T0}(J)}$

- factor allowing for hull effect on torque $i_Q = \dfrac{K_Q(J)}{K_{Q0}(J)}$

- thrust deduction factor $t = 1 - \dfrac{K_E(J)}{K_T(J)}$.

The thrust deduction factor is estimated in the same way as it is done under the traditional approach.

A specific feature of the "bollard pull" approach is that flow velocities in open water and behind model hull are equal, hence the advance ratios are also equal, however, the propeller thrusts in this case are different. The propulsive efficiency η in the "bollard pull" approach is defined as:

$$\eta = \eta_{sh}\eta_0 \frac{i_T(1-t)}{i_Q}$$

where η_{sh} - shafting efficiency, η_0 - propeller efficiency in open water.

Figure 2 shows values of alternative interaction coefficients for a four-shaft icebreaker sailing ahead.

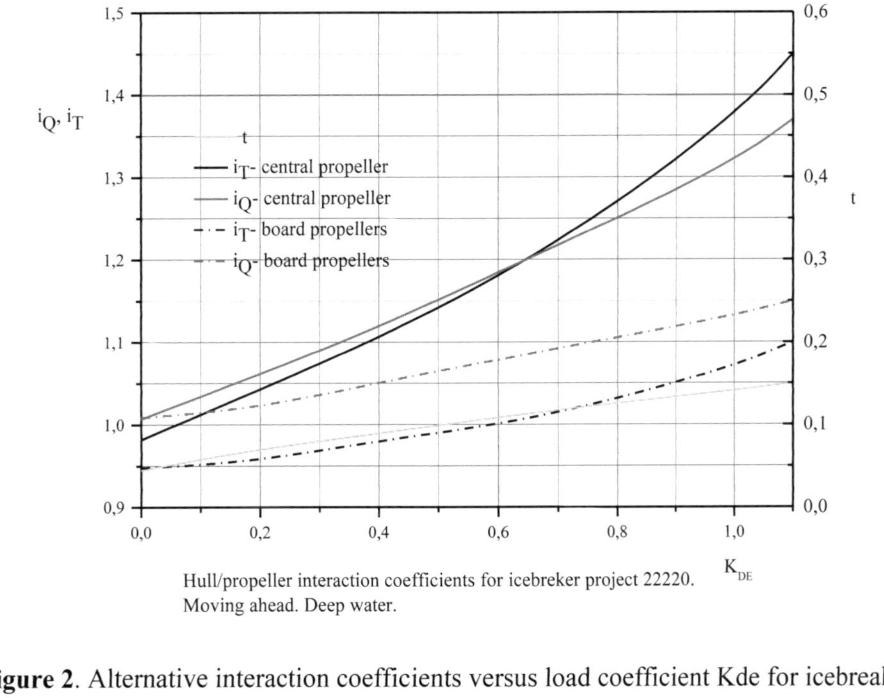

Figure 2. Alternative interaction coefficients versus load coefficient Kde for icebreaker.

3. Propulsion performance estimations for icebreaker

Introduction of the "bollard pull" coefficients system enabled us to work out a method for estimation of icebreaker propulsion performance in ice environment. The main assumption is that the hull/propeller interaction coefficients estimated based the "bollard pull" approach can be extended to the icebreaker operation mode in ice. It is a quite conventional assumption that was used before to study ship sailing in ice. The efforts to develop a new method required certain correction of the traditional performance prediction technique as well as generalization of results to the case of a multi-shaft vessel whose propulsion system comprises different type propulsive units, e.g. conventional screw propellers and podded thrusters.

The new approach makes it possible to estimate the revolutions and thrust of propulsors. Figure 3 shows a diagram relating all main characteristics of icebreaker performance in ice: speed, thrust and number of revolutions.

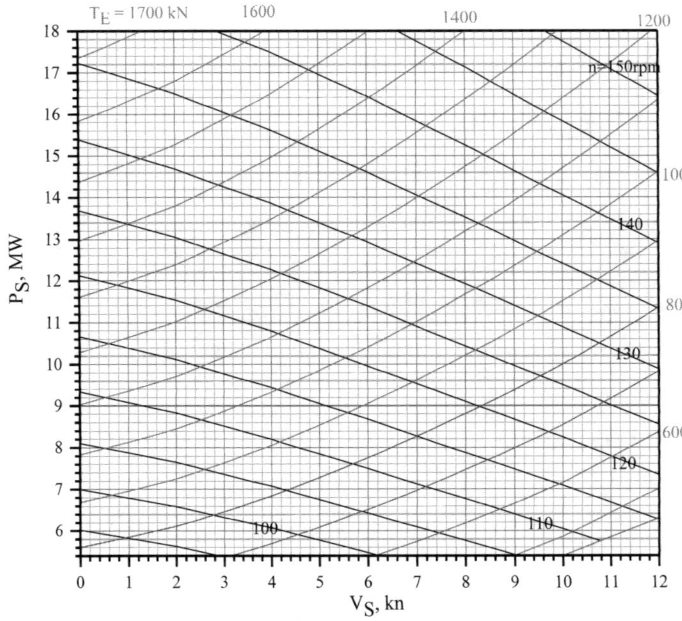

Figure 3. Propulsion performance diagram for icebreaker Pr.21900m

4. Application of results

The new method is applicable to a range of practical tasks in the design of icebreakers. The main advantages are new opportunities in the analysis of full-scale data from icebreaker sea trials, insights into the ice influence on hydrodynamic characteristics of icebreakers, more accurate test in ice basins with self-propelled models of icebreakers.

Sea trials. As it is seen from the diagram of Figure 3 it is possible to find the pulling thrust developed by propulsors based on the power and speed of the icebreaker measured in sea trials. The number of propulsor revolutions also determined from the above diagram serves as an indicator to confirm the procedure correctness.

If we pick the records of ship speed, power and propeller revolutions at the sections where propellers do not show any significant interaction with ice, it is possible to estimate the thrust of propulsion system and, therefore, ice resistance for these sections. Actually, the suggested method makes it possible to estimate the ice resistance in full scale conditions. With this capability of estimating full-scale ice resistance one can address a range of issues. In the analysis of full-scale data the information about ship ice resistance enables us, first, to make full use of the measurements taken at partial power outputs of ship power plant. Secondly, the ice resistance data enable extrapolation of full-scale test results to different ice strength and thickness values using well-known correction techniques employed in model experiments. Figure 4 illustrates results of full-scale data processing. It shows the speed of *Vladivostok* icebreaker versus ice thickness referred to a standard ice bending strength of 500 kPa, as per full-scale trial data [15]. Squares on the graph refer to the results obtained by traditional processing method with extrapolation of power and speed data. Circles refer to the results of ice resistance extrapolations by formulas used in model experiments. The results are in good agreement.

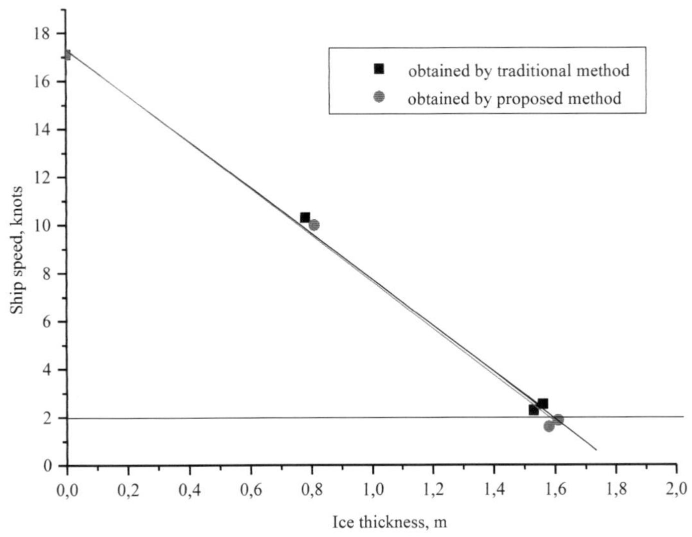

Figure 4. Speed vs ice thickness obtained by different methods for the *Vladivostok* icebreaker

Investigation of interaction coefficients in ice conditions. Investigation of hull/propeller interaction coefficients for a ship sailing in ice is a very challenging task that has not been resolved yet. One of the reasons is negative wake fractions mentioned above. However, our alternative system of interaction coefficients brings us closer to the solution of this problem. The new system enables us to examine ice effects on the interaction coefficients both in model experiments and in the analysis of full-scale trials.

The values of interaction coefficients under the alternative method can be obtained by testing self-propelled ship models towed in ice basin by towing carriage at constant speed in ice field. During these tests the frequency of propeller revolutions is varied to change K_{DE}. Model experiments in ice field with variation of model speed at constant frequency of propeller revolutions cannot be used for evaluation of hull/propeller interaction coefficients because in this case ice pieces adjacent to the underwater hull change their size [16]. Obviously, it may have significant influence on the coefficient values. The functions $K_T = f(J)$ and $K_Q = f(J)$ obtained from ice tests can be used to determine the alternative system coefficients. Then their comparison with the coefficients obtained in open water enables us to judge the ice effect.

Sea trial data can also provide information about hull/propeller interaction coefficients in ice. If propeller revolutions measured in full scale are different from predictions, e.g. based on the diagram of Figure 3, then it is always possible to choose an appropriate correction factor to adjust the assumed coefficient values and remove discrepancies in the frequency of propeller revolutions. With accumulation of statistical data from full-scale trials of various ships it would be possible to generalize the results and derive empirical relationships for interaction coefficients in ice.

Calculation of propulsion performance in ice for model conditions (with stock propellers) make it possible to set the frequency of propeller revolutions more accurately in model tests to better simulate real conditions. This selection of propeller revolutions is most important when slipstreams of propellers may have effect on the ice on the underwater hull of model (sailing astern, bow propellers, etc.)

5. Conclusions

The alternative system of hull/propeller interaction coefficients is helpful in addressing a broad range of issues related to ship propulsion performance in ice. In addition to estimation of critical icebreaker parameters in ice: power, propeller revolutions and pulling thrust, the proposed system of coefficients paves the way for systematic studies on hull/propeller interaction coefficients in ice, providing requisite "initial conditions" for this purpose.

The alternative system of interaction coefficients enables rather accurate predictions of ice resistance for full-scale sea trials, increasing the efficiency of such trials. Also, the ice resistance calculations may well be used by designers of ship engines for their improvement.

References

[1] Resistance Test in Ice 7.5-02-04-02.1 *International Towing Tank Conference 2017 (Recommended Procedures and Guidelines)* P 8

[2] Resistance Test in Ice 7.5-02-04-02.2 *International Towing Tank Conference 2017 (Recommended Procedures and Guidelines)* P 8

[3] Lindqvist G 1989 A straightforward method for calculation of ice resistance of ships *Proceedings of POAC* 722–35

[4] Su B, Riska K, Moan T 2010 A numerical method for the prediction of ship performance in level ice *Cold Reg. Sci. Technol* **60** 177–88

[5] Su B, Riska K, Moan, T, 2011 Numerical simulation of local ice loads in uniform and randomly varying ice conditions *Cold Reg. Sci. Technol* **65** 145–59

[6] Tan X, Riska K, Moan T, 2014 Effect of dynamic bending of level ice on ship's continuous-mode icebreaking *Cold Reg. Sci. Technol* **106 107** 82–95

[7] Valanto P 2009 On Computed Ice Load Distributions, Magnitudes and Lengths on Ship Hulls moving in Level Ice *Proc. of the 20th Int. Conf. on Port Engineering under Arctic Conditions POAC* (Luleå)

[8] Andryushin A V, Hänninen S, Heideman T 2013 Azipod Azimuth Thruster for large capacity arctic transport ship with high ice category Arc7. Ensuring of operability and operating strength under severe ice conditions *Proc. of the 22rd Int. Conf. on Port Engineering under Arctic Conditions POAC'13 (*Espoo)

[9] Appolonov E M, Nesterov A B and Sazonov K E Regulation of extreme ice loads acting on hull of azimuth propulsion systems for ice ships *Ships and Offshores Structures* vol 6 239–47

[10] Dobrodeev A A, Sazonov K E, Andryushin A V, Fedoseev S S, Gavrilov S V, 2017 Experimental Studies of Ice Loads on Pod Propulsors of Ice-Going Support Ships *Proceedings of the 24th International Conference on Port and Ocean Engineering under Arctic Conditions POAC* (Busan)

[11] Ikonen T, Peltokorpi O, Karhunen J, 2015 Inverse ice-induced moment determination on the propeller of an ice-going vessel *Cold Regions Science and Technology* **112** 1–13

[12] Sampson R, M. Atlar M, Sasaki N, 2009 Propeller ice interaction – effect of milling *Proc. of the 20th Int. Conf. on Port Engineering under Arctic Conditions POAC'09* (Luleå)

[13] Alekseev Y N, Beljashov V A, Sazonov K E, 1993 Hydrodynamic problems of propellers for icebreaking ships *Proc. 12th Int. Conf. POAC'93* (Hamburg) **1** 351–8
[14] Kanevskii G I, Klubnichkin A M, 2017 Propeller-hull interaction coefficients: classic vc alternative system *Proceedings of 5th Int. Symp. on Marine Propulsors* **3**
[15] Kostylev A I, Sazonov K E, Timofeev O Ya, Egiazarov G E, Soloviev A S, Egorov D N, Shtrambrant V I, 2016 Sea trials of "Vladivostok" *Sudostroenie* **6** 9–12
[16] Alekseev Y N, Sazonov K E, 1993 A Method for Ship Level Ice Resistance Computation *Proc. 12th Int. Conf. on Port and Ocean Eng. POAC 93* (Hamburg) **2** 755–62

IOP Conf. Series: Earth and Environmental Science **193** (2018) 012028 doi:10.1088/1755-1315/193/1/012028

The method of calculation the pressure gradient in multiphase flow in the pipe segment based on the machine learning algorithms

E Kanin[1], A Vainshtein[1], A Osiptsov[1] and E Burnaev[1]

[1]Skolkovo Institute of Science and Technology (Skoltech), 3 Nobel Street, 143026, Moscow, Russian Federation

E-mail: Evgenii.Kanin@skoltech.ru

Abstract. Engineering tools allowing pressure gradient calculation in the pipe segment commonly use stationary correlation and mechanistic models such as Beggs and Brill, Ansary, etc. This is well known and convenient way which gives rough estimate of pressure gradient due to friction losses and liquid phase interference. It avoids solving complex dynamic pressure equation and derives quick results with comfortable precision margin for large scale systems, such as horizontal pipes and wells. In order to enlarge the applicability zone and accuracy of existing methods, a new method of pressure gradient definition is evolved. It is included three surrogate models that are based on Machine Learning (ML) algorithms. The first model predicts liquid holdup in the segment, the second defines flow pattern and the third predicts pressure gradient. In order to create these models, several ML algorithms are applied such as Random Forest, Gradient Boosting, Support Vector Machine and Artificial Neuron Network.

Involvement of the latest machine learning algorithms will allow applying this method to wider range of input data compared with standard multiphase flow correlations and mechanistic models. The proposed method demonstrates high accuracy – on the collected experimental data set it gives $R^2 = 0.985$ for pressure gradient prediction. That is why it could help to carry out correct calculation of bottom hole pressure and pressure distribution along the length of the pipeline.

1. Introduction

Multiphase flow is simultaneous flow in pipes of two or more phases (liquid, gas or solid). It could be characterized by flow pattern according to physical distribution of phases in the pipe. During multiphase flow, the flow regime depends on magnitudes of forces that act on the fluid from other fluids or from pipe wall. Pressure gradient depends on flow pattern significantly, that is why it is necessary to identify it correctly before calculating pressure gradient. Several articles are devoted to the application of ML algorithms in the identification of flow pattern. In these papers, authors plotted experimental data points in suitable coordinates and tried to construct models in order to match these points. In paper [1] author created Artificial Neuron Network, in [2] authors applied Support Vector Machine algorithm and in [3] fuzzy inference system was used.

The other very important parameter of multiphase flow is liquid holdup. It is a fraction of pipe volume that occupied by liquid phase. This characteristic undoubtedly influences on pressure gradient of the flow that is why it should be correctly calculated. Liquid holdup is also important in planning design of separation equipment. Several papers were devoted to the identification of this parameter by machine learning tools. For example, in [4] and [1] authors have applied Artificial Neuron Network.

These two parameters – flow pattern and liquid holdup – are calculated in different multiphase flow correlation for the pressure gradient definition. There are plenty of correlations that were developed using laboratory experiments. The most popular are Beggs and Brill [5], Mukherjee and Brill [6] correlations and others. Many studies have been done in order to find out the applicability of the correlations and authors of these articles considered that single correlation couldn't be used for any ranges of input parameters because each correlation works correctly under its own ranges of data. There are also several semi-empirical mechanistic models that are used for prediction of different multiphase flow characteristics. The most popular ones are Hasan and Kabir [7], Ansari [8] and others. Mechanistic models have advantages in specific flow pattern prediction but they aren't more accurate in pressure drop predictions compared to empirical correlations.

Researchers also tried to apply ML in order to predict pressure gradient or output pressure directly. In [9] article authors construct ANN for prediction the bottomhole pressure, in [10] authors also predict bottomhole pressure with the use of ANN but they suggested to divide well into segments and to define flow pattern in each segment.

In this paper, three machine learning models are constructed for prediction the following parameters of multiphase flow pipe segment: liquid holdup, flow regime and pressure gradient consistently. Pipe segment to be a part of the pipe which has homogeneous flow type and approximately constant pressure gradient. In the process of creating models, four machine learning algorithms are tried out such as Random Forest Regressor, Gradient Boosting Regressor, Support Vector Regressor and ANN in order to compare their predictive capability.

2. The method description

2.1. Method of calculation the pressure gradient in the pipe segment

In this part of the article, the process of calculation the pressure gradient in the segment will be described. The proposed method allows estimating pressure gradient in the segment that could be oriented to angles from $-90°$ to $90°$. Positive and negative angles describe uphill and downhill flow respectively and angle equals to zero is related to horizontal flow.

In order to decrease the number of input parameters for ML algorithms and to make the model similar for different types of liquids, the set of dimensionless variables is used according to paper [11]. These parameters are called velocity number of gas, velocity number of liquid, diameter number, viscosity number and defined by the following equations:

$$N_{vg} = v_{sg}\sqrt[4]{\frac{\rho_l}{g\sigma_{lg}}}, N_{vl} = v_{sl}\sqrt[4]{\frac{\rho_l}{g\sigma_{lg}}}, N_d = d\sqrt{\frac{\rho_l g}{\sigma_{lg}}}, N_\mu = \mu_l\sqrt[4]{\frac{g}{\rho_L\sigma_{lg}^3}} \qquad (1)$$

where v_{sg} and v_{sl} are gas and liquid superficial velocities, ρ_l - liquid density and σ_{lg} is surface tension between liquid and gas phases, d - diameter of the tube and μ_l - liquid viscosity.

In order to predict pressure gradient in the segment, the regression model is constructed. Since pressure gradient depends significantly on the flow pattern and liquid holdup in the segment, these features are included into input parameters of the model. These two characteristics are calculated by other two ML models which will be described later. Apart flow pattern and liquid holdup, the following parameters are among input features of this regression model: inclination angle, dimensionless parameters (eq. 1), average pressure and temperature.

The second ML model that is a multi-class classifier for prediction flow regime. In this paper, the following four flow patterns are distinguished: bubble, slug, annular mist and stratified flows. The input features of this classifier are inclination angle, dimensionless parameters (eq. 1), liquid holdup, average pressure and temperature. In order to increase the predictive power of the model, the dataset is divided into three physically homogeneous parts: data points related to horizontal flow, to up-flow and to down-flow and, consequently, besides the general model that uses all data, three ML models were constructed and trained on the appropriate data set.

In both previous models, the liquid holdup parameter is also included in input variables. As a result, the final regressive model is devoted to liquid holdup prediction. This model is trained on the following input parameters: inclination angle, dimensionless parameters (eq. 1), average pressure and temperature. Similar to the flow pattern prediction model, in this case three ML models for horizontal flow, upstream and downstream are constructed.

2.2. Sources of datasets

In order to compose dataset for training ML models, the data points are collected from the articles, books and PhD dissertations published in open source. From paper [12] 111 data point for horizontal flow are taken. Author of this paper carried out experiments using kerosene and water as liquid phase and air as the gas phase. Next 88 data points are from the article [13] in which author performed an experiment for horizontal flow using kerosene and air. Further, 1400-points dataset is taken from [14] which consists of uphill, downhill and horizontal flows in pipes, with inclination angle varying from $-90°$ to $90°$. The author uses kerosene and lube oil as liquid phase and air as the gas phase. From [15] 238 data points of horizontal multiphase flow of water and natural gas are used. The final 188 data points of water and air multiphase flow are taken from [16]. Among this data we also find flow in pipes oriented to angles $-10°$ and $10°$ except horizontal flow.

As a result, the total number of data points for construction ML model for liquid holdup prediction and flow pattern identification is approximately 2000. Among these points, about 1100 points are applied for training, validation and testing ML model for pressure gradient prediction. In the composed dataset the flow pattern is provided for about 1400 data point. In order to fill remaining 400 points the flow pattern map created by Mukherjee [6] is used.

2.3. Applied Machine Learning algorithms, tuning and evaluation scores

Four Machine Learning algorithms are considered in this paper: Random Forest, Gradient Boosting, Support Vector Machines and Artificial Neuron Networks. All these algorithms are tried out in each of aforesaid models for liquid holdup prediction, flow pattern identification and pressure gradient calculation in order to compare its predictive capability.

Firstly, let's introduce some designations. $\mathbf{X}$ is a matrix with arguments with size $m \times d$ (m is a number of samples and d is number of features). y is a matrix that contains interested real values. It has size $m \times r$ where $r = 1$ when single output in the problem and $r > 1$ when multiple output task. Value $\hat{y}$ is also introduced that is a matrix with predicted values and has the same size as matrix y.

The first ML algorithm that is applied – **Random Forest** [17]. It is an ensemble machine learning method which is used for solving regression and classification problems. This algorithm builds several independent decision trees and average them in order to obtain final result:

$$h_N(\mathbf{X}) = \frac{1}{N}\sum_{i=1}^{N} h_i(\mathbf{X}) \tag{2}$$

where $h_i(\mathbf{X})-$ is decision tree with number i, N - number of decision trees.

In this article, Random Forest algorithm is used in the form that is implemented on Python language in the Scikit-learn library [18].

The second method that is used in the construction of ML models is **Gradient Boosting** [19]. This method also refers to ensemble methods and is used in both regression and classification problems but has another structure. Algorithm constructs several decision trees and results as the weighted sum of them:

$$h_N(\mathbf{X}) = \sum_{i=1}^{N} \alpha_i h_i(\mathbf{X}) \tag{3}$$

Each decision tree $h_i(X)$ tries to fit anti-gradient of loss function (logistic, exponential loss fucntions):

$$-\frac{\partial L(f(x_j), y_j)}{\partial f(x_j)}\Bigg|_{f(\mathbf{X})=h_i(\mathbf{X})}, j = 1, ..., m$$

where $f(x_j) = \hat{y}_j$ is predicted value for sample with number j.

Weights in the sum of decision trees (α_i) are found from the minimization of loss function:

$$\alpha_i = \text{argmin}_{\alpha>0} \sum_{j=1}^{m} L(f_j + \alpha h_i(\mathbf{x}_j), y_j)$$

Similar to Random Forest, Gradient Boosting algorithm is applied in the version of the Scikit-learn library [18].

Further **Support Vector Machine** algorithm [20] is considered. SVM is a machine learning algorithm that is used in classification and regression analysis. Very often matrix of arguments $\mathbf{X}$ is transformed into high-dimensional feature space by non-linear mapping before application of SVM algorithm F. Dot product of transformed vectors $\mathbf{x}$ and $\mathbf{x'}$ is a kernel of transformation $K(\mathbf{x}, \mathbf{x})$: $K(\mathbf{x}, \mathbf{x}) = F(\mathbf{x}) \cdot F(\mathbf{x'})$. In the construction of ML models using SVM in this paper Gaussian kernel with width σ is taken: $K(\mathbf{x'}, \mathbf{x}) = \exp\left(\frac{||\mathbf{x}-\mathbf{x'}||^2}{2\sigma^2}\right)$. This transformation is applied in order to make the data linearly separable (in the classification task) or to make the transformed data fit with the line function: $\mathbf{w} \cdot F(\mathbf{X}) + b$.

SVM algorithm solves the optimization problem of maximizing margin defined as $\frac{1}{||\mathbf{w}||}$ which is the same to minimize the norm $||\mathbf{w}||$. This optimization problem is solved in dual formulation with the use of Lagrangian.

SVM algorithm is also exploited in the form of the Scikit-learn [18].

The final ML technique used in this paper is **Artificial Neuron Networks** [21]. This is a mathematical representation of neurons in the brain. The mathematical or computer model of representation of the biological system is also called multi-layer perceptron. This ML algorithm is also applied in classification and regression problems. ANN consists of several layers and in each layer have a certain number of nodes. The first layer contains input parameters, so it has p nodes. The last one contains output parameters: in the case of regression problems or binary classification there is only one node in the case of multiclass classification the number of nodes which is equal to the number of classes. The algorithm consists of forward and backward propagation. In the forward propagation process, algorithm calculates values in each node in the following way: in order to obtain value in node k in layer $h + 1$ algorithm performs linear combination of values in nodes in layer h with definite weights and apply to this linear combination activation function $g(\cdot)$:

$$\mathbf{y}_{k,h+1} = g(\theta^T_{k,h\to h+1}\mathbf{y}_h) \tag{4}$$

In the backward process, the algorithm adjusts weights by using gradient descent optimization algorithm in order to decrease the value of loss function $L(f_i, y_i)$.

In classification problems sigmoid $\left(f(x) = \frac{1}{\exp(-x)+1}\right)$ or hyperbolic tangent ($f(x) = \tanh(x)$) activation functions are used. In the case of regression - ReLu ($f(x) = \max(0, x)$).

Artificial Neuron Network is presented in many libraries on Python language. In this paper, the functions from Scikit-learn library [18] are used.

For tuning models hyperparameters M cross-validation procedure is used. In this technique dataset is divided into M equally sized parts and $M - 1$ partitions are exploited as training dataset and remaining partition as validation dataset This process is repeated M times and on each iteration different validation partition is used. The model that gives the best score is the model with optimal hyperparameters. In order to evaluate the score of the model with the best hyperparameters $N \times M$ cross-validation procedure is applied. In this method, dataset is divided into M equally sized parts N times and in each N-step the partitions are different. Thanks to this technique the calculated score of the model ($mean$) is more correct compared to M cross-validation and also allows to build confident intervals of such a score ($\pm 2 \cdot std$). In the case of M cross-validation $M = 5$ is used and in the case of $N \times M$ cross-validation $M = 5$ and $N = 20$ are exercised.

During construction of ML models the following set of evaluation metrics is applied. In the case of multi-classification problem, following accuracy score is exploited:

$$\text{Accuracy} = \frac{1}{m}\sum_{i=1}^{m} 1_{f(x_i)=y_i} = \sum_{y} \frac{TP_y + TN_y}{TP_y + FN_y + TP_y + FN_y} \tag{5}$$

where TP_y - true positives of class y, FP_y - false positives of class y, FN_y - false negatives of class y and TN_y - true negatives of class y.

In the regression problems the coefficient of determination (R^2 score) is used:

$$R^2 = 1 - \frac{\sum_{i=1}^{m}(y_i - \hat{y}_i)^2}{\sum_{i=1}^{m}(y_i - \bar{y})^2} \tag{6}$$

where $\bar{y} = \frac{1}{m}\sum_{i=1}^{m} y_i$

3. Results

In this part of the paper, the obtained results are represented, namely, scores of the algorithms with the best predictive capability, cross-plots.

Let's start with the first model that predicts liquid holdup. In all considered cases – horizontal, uphill, downhill flows and in the case when all dataset is used in model construction the Gradient Boosting algorithm has the best predictive capability. When all dataset is used for training, validation and testing model, Gradient Boosting has $R^2 = 0.925 \pm 0.033$. In model for horizontal flow $R^2 = 0.974 \pm 0.015$, for uphill flow $R^2 = 0.965 \pm 0.013$ and for downhill flow $R^2 = 0.83 \pm 0.102$. From these results, the following conclusion could be made: models for horizontal and uphill flows demonstrate well predictive capability with high coefficient of determination, while the model for downhill flow reveals the relatively low result.

Further, let's move on to results of the second model that predicts flow pattern in the segment. Gradient Boosting algorithm has the best predictive capability when all dataset is used for model creation with accuracy score 0.85 ± 0.037. In the model for horizontal flow also Gradient boosting has the best accuracy score that is equal to 0.924 ± 0.043, for uphill flow neuron network has the best predictive capability with accuracy 0.875 ± 0.052. Finally, in the case of uphill flow ANN performs the best result with an accuracy score 0.769 ± 0.063. From these results, one could make a conclusion that models for horizontal and up flows have good accuracy score, while the model for downhill flow reveals a poor result.

Now, let's consider the third model which predicts pressure gradient in the segment. In this case division on horizontal, uphill and downhill doesn't apply because of small data points that belong to each class. So, the only one model for pressure gradient prediction is built which uses all dataset in training, validation and testing stages. In this model, flow pattern feature is used among input parameters. It is included via one hot encoding method in which four new columns are created (their number is equal to the number of flow regimes in this problem) instead of one column which contains names of flow pattern. These new features consist of 0 and 1. When sample belongs to slug flow pattern it will have 1 value in the column that is responsible for indication of an appurtenance to the slug flow class. In this model, SVM algorithm gives the best score – $R^2 = 0.985 \pm 0.011$ whereas Gradient Boosting demonstrates slightly worse result – $R^2 = 0.971 \pm 0.014$.

Since model for the pressure gradient estimation is the most important in the proposed method for pressure drop calculation, it is necessary to construct a cross-plot. In this graph on X-axis real values are plotted while predicted values are on Y-axis. At Figure 1 such cross-plot is demonstrated.

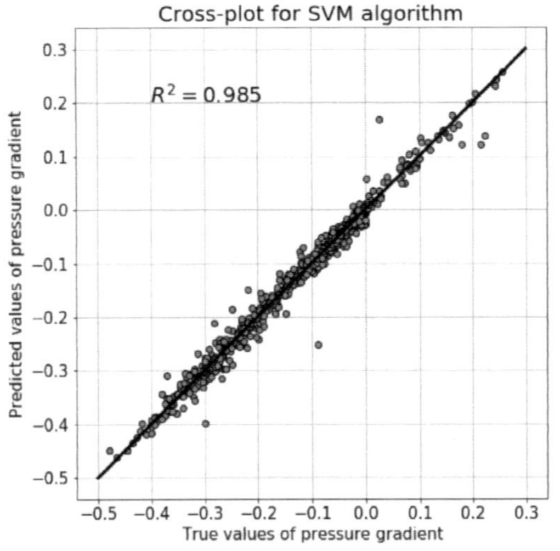

Figure 1. Cross-plot for SVM algorithm in the pressure gradient model.

In order to summarize results of created models, the Table 1 with scores is written. In this table results of all applied machine learning algorithms in the case where all data set is used are represented.

Models	Model 1	Model 2	Model 3
Algorithm \ Metric	R^2	Accuracy	R^2
Random Forest	0.917 ± 0.04	0.848 ± 0.036	0.955 ± 0.02
Gradient Boosting	0.925 ± 0.033	0.85 ± 0.037	0.971 ± 0.014
SVM	0.858 ± 0.043	0.846 ± 0.039	0.985 ± 0.011
ANN	0.884 ± 0.04	0.833 ± 0.037	0.98 ± 0.011

Table 1. ML models results of all applied machine learning algorithms in the case of usage of all dataset.

4. Conclusions and future directions

In the present paper, the new method of pressure gradient calculation in the pipe segment is developed. It consists of three surrogate models that are based on the Machine Learning algorithms such as Random Forest, Support Vector Machine, Gradient Boosting and Artificial Neuron Network. The first model predicts liquid holdup parameter. The best R^2-score in this model achieved by using Gradient Boosting algorithm and it is equal to 0.925. The second one is focused on flow pattern identification. The best accuracy score in this model also belongs to Gradient Boosting algorithm and equal to 85.5%. The last one calculated pressure gradient. In this case SVM method performs best of all and shows R^2-score equal to 0.985. The distinguishing feature of the proposed method is high accuracy. As a result, this method calculates bottomhole pressures, pressure distributions along the pipelines and other very important characteristics in petroleum engineering domain more correctly allowing to solve different optimization (production maximization e.g.) and planning tasks (oil transportation system enchancements e.g.) more effectively.

There are also some possible ways to improve the model, its predictive capability and expand boundaries of applicability. Firstly, it is necessary to add data points to the collected dataset: it could be experimental data points from other researchers, synthetic data calculated by using multiphase flow simulators such as OLGA, ANSYS Fluent and others. Secondly, temperature effects should be introduced. In the present paper, temperature is linearly interpolated between boundaries but it is more correct to calculate entropy, heat transfer along the tube.

References

[1] Osman E S 2004 *Oil Production & Facilities* **19** 33–40
[2] Li X, Miskimins J L, Sutton R P, Hoffman B T *et al.* 2014 *SPE Annual Technical Conference and Exhibition* (Society of Petroleum Engineers)
[3] Popa F, Dursun S, Houchens B *et al.* 2015 *SPE Annual Technical Conference and Exhibition* (Society of Petroleum Engineers)
[4] Shippen M E, Scott S L *et al.* 2002 *SPE Annual Technical Conference and Exhibition* (Society of Petroleum Engineers)
[5] Beggs D H, Brill J P *et al.* 1973 *Journal of Petroleum technology* **25** 607–617
[6] Mukherjee H, Brill J P *et al.* 1983 *Journal of Petroleum Technology* **35** 1–003
[7] Hasan A, Kabir C *et al.* 1986 *SPE California Regional Meeting* (Society of Petroleum Engineers)
[8] Ansari A, Sylvester N, Shoham O, Brill J *et al.* 1990 *SPE Annual Technical Conference and Exhibition* (Society of Petroleum Engineers)
[9] Osman E S A, Ayoub M A, Aggour M A *et al.* 2005 *SPE Middle East Oil and Gas Show and Conference* (Society of Petroleum Engineers)
[10] Li X, Miskimins J, Hoffman B T *et al.* 2014 *SPE Annual Technical Conference and Exhibition* (Society of Petroleum Engineers)
[11] Duns Jr H, Ros N *et al.* 1963 *6th World Petroleum Congress* (World Petroleum Congress)
[12] Minami K, Brill J *et al.* 1987 *SPE Production Engineering* **2** 36–44
[13] Abdul-Majeed G 1996 *Journal of Petroleum Science and Engineering* **15** 271–280
[14] Mukherjee H 1979 *An Experimental Study of Inclined Two-Phase Flow* Ph.D. thesis U. of Tulsa
[15] Eaton B 1966 *The prediction of flow patterns, liquid holdup and pressure losses occurring during continuous two-phase flow in horizontal pipelines* Ph.D. thesis U. of Texas at Austin
[16] Beggs H 1973 *An Experimental Study of Two-Phase Flow in Inclined Pipes* Ph.D. thesis U. of Tulsa
[17] Breiman L 2001 *Machine learning* **45** 5–32
[18] Pedregosa F, Varoquaux G, Gramfort A, Michel V, Thirion B, Grisel O, Blondel M, Prettenhofer P, Weiss R, Dubourg V, Vanderplas J, Passos A, Cournapeau D, Brucher M, Perrot M and Duchesnay E 2011 *Journal of Machine Learning Research* **12** 2825–2830
[19] Friedman J 1999 *GreedyFuncApproxSS. pdf*
[20] Cortes C and Vapnik V 1995 *Machine learning* **20** 273–297
[21] Rosenblatt F 1958 *Psychological review* **65** 386

Satellite monitoring of atmospheric sulphur dioxide pollution in polar latitudes

V B Kashkin[1], D V Zuev[1], M A Kurako[1], A A Romanov[1], T V Rubleva[1] and K V Simonov[2]

[1] Siberian Federal University, 79 Svobodny pr., 660041, Krasnoyarsk, Russia
[2] Institute of Computational Modeling SB RAS, 50/44 Akademgorodok, 660036, Krasnoyarsk, Russia

E-mail: mkurako@sfu-kras.ru

Abstract. The study is devoted to the mapping of areas of environmental risk. Risk assessment is based on a satellite monitoring data and an evaluation of the impact of air pollution on the vegetation cover of urban areas. Ecological risk zones mapping is shown on the example of Norilsk industrial zone impact on vegetation cover. There is a brief description of the existing system of satellite monitoring data of sulphur dioxide emission for the Norilsk industrial zone. Our approach for risk assessment mapping uses elastic grid method for generating elastic net and visualizing spatial satellite data from OMI and OMPS databases. The results of calculations based on the methodology of mapping zones of environmental risk are presented.

1. Introduction

The work of Norilsk industrial zone factories leads to technogenic pollution of atmospheric air, reservoirs, and destruction of the fertile layer of the earth and vegetation cover. The share of sulphur dioxide emission by Norilsk industrial zone (NIZ) factories is 25% of Russian industrial emission [1]. High concentration of sulphur dioxide in the atmosphere causes serious damage to the leaves, which leads to chronic plant damage. Emission of sulphur dioxide (SO_2) causes changes in soil composition and damage to forests around the factories of the northern territories of the Krasnoyarsk region.

Sulphur dioxide is a pollutant that has a significant impact on the environment and climate at the global, regional, and local levels. It is oxidized to sulphate aerosols that reduce visibility, influence on the cloud formation, and cause acid rains. Remote satellite monitoring enables new possibilities in studying the gas composition of the atmosphere, monitoring the state of the environment, forecasting technogenic and natural emergencies, as well as simplify the study of climate change.

Evaluation of the ecological situation allows to obtain objective information on the state of the environment and population health; establish scientifically-based boundaries of the territory or its individual sections; organize a system of monitoring by priority indicators of environmental change and population health; rationale emergency and rescue activities for the rehabilitation of public health and environment.

It should be noted that it is important not to be limited only to the total value of environment pollution, when considering the degree of ecological disadvantage of urban territory. It is necessary to consider the indicators of the public health because of the total technogenic load on a person. Any industrial cluster is characterized by almost complete change of environmental parameters: soil pollution, atmosphere, reservoirs, climate, etc. The urgency of these problems is very high and will increase in future.

There are sanitary and hygienic standards that are used for assessing an impact on plant ecosystem, but these standards are not adapted to that purpose in view of biodiversity [1-2]. The ecological risk, as a probability of occurrence of environment adverse situations because of economic activity of people, allows more reasonably estimate a degree of influence of negative factors on components of environment. Therefore, the problem of estimating an environment risk of impact on urban territory vegetation cover is very important. However, environmental risk assessment in the Arctic zone is not sufficient. In this regard, the aim of the work is to outline the methodological issues of assessing and mapping environmental risks on plant ecosystem using remote sensing data.

2. Data description and evaluation

Monitoring of sulphur dioxide (SO_2) emission is performed by OMI and OMPS spectrophotometers. The Ozone Monitoring Instrument (OMI) data [3-14] is a key source for sulphur dioxide emission levels. The OMI measures SO_2 emission concentration in Dobson units (DU) for four altitudes above the sea level: 0.9 km, 2.5 km, 7.5 km and 17 km. 1 DU equals 0.01 mm of sulphur dioxide deposited layer thickness at 0 °C and atmospheric pressure equals 1013 GPA. Also, OMI provides graphical database of SO_2 emission for the NIZ territory. Database incorporates data from 2004 to 2018 and includes data about SO_2 emission mass, distribution area, maximum values of SO_2 concentration and time of measurement.

OMPS is a new instrument for SO_2 emission monitoring [15, 16]. The OMPS spectrometer, like its predecessor, carries out the control of SO_2 emission on a global scope. The OMPS graphical database has the same format as the OMI graphical database. It contains detailed information about SO_2 emission and incorporates data from April 2013 to 2018.

Analysis of SO_2 emission is performed by MATLAB script using data from graphical database for Norilsk industrial zone. Script generates daily and monthly average concentration values and allows to build SO_2 emission concentration map for the studied day. Based on the obtained evaluation data set, the possibilities of using OMI–OMPS data for the NIZ territory was studied [11-13, 17].

Firstly, the correlation of data was evaluated according to the method described in [15, 16]. There are two types of correlation. Spatial correlation is the correlation between OMI and OMPS average annual SO_2 concentration values within the NIZ. Temporal correlation is the correlation between the OMI and OMPS daily masses of SO_2 for the studied area. Correlation calculation encompasses period from 2013 to 2017.

The spatial correlation of OMI-OMPS annual averages is high and equal to 0.8. The relative difference equals 8.2%. It can be explained by the difference in observation time, since the SO_2 column can move relatively quickly in space. The maximum value (0.91) of spatial correlation of average annual values was observed in 2017. Average annual concentrations of SO_2 are 3.56 DU and 2.64 DU for OMI and OMPS, respectively. Average relative difference OMPS-OMI equals 26%. The minimum value (0.79) was observed in 2014, while the relative difference is comparable to other years and equals 11.7%. Temporal correlation of OMI-OMPS data is based on daily SO_2 emission masses in 2013–2017. The calculation involves only days, when both values were registered. The lowest value of correlation coefficient is 0.62. The average daily SO_2 emission mass equals 3.56 kt for OMI and 2.65 kt for OMPS data, respectively. The highest value of correlation corresponds to May, July and November.

In general, SO_2 emission mass from OMI is slightly higher than that of OMPS. A comparison of OMI and OMPS data shows that there is a well enough correlation for the NIZ region. OMPS-based metrics are slightly smaller than OMI metrics, probably due to lower spatial resolution for the studied region. However, despite these differences the OMPS spectrometer is capable to perform monitoring with given accuracy.

3. Monitoring data mapping by elastic grids method

In this section, we demonstrate our approach for mapping and visualizing OMI and OMPS data. These data are currently being used to provide daily information about regional air pollution and average annual information about point sources of pollution. Consider, following [18], the data visualization algorithm based on elastic grids method.

Let us define a grid with $p \cdot q$ nodes and enumerate nodes y_{ij} with indexes $i = \overline{1,p}$, $j = \overline{1,q}$. The grid should satisfy to the following properties:

- Closeness to the data points. Grid should be in some sense like the plane of the first two principal components.
- Elasticity. This property provides a uniformity and smoothness of the grid.

Let us divide data set X into $p \cdot q$ subsets $K_{ij}, i = \overline{1,p}, j = \overline{1,q}$, where each subset contains only points, which are closer to the y_{ij} node than any other node and denote this condition as follows

$$K_{ij} = \left\{ x \in X \mid \left\|y_{ij} - x\right\|^2 \leq \|y_{kl} - x\|^2, \forall k,l \right\}. \tag{1}$$

Measure of closeness between nodes and data points is defined as Root Mean Square (RMS) distance. Each node (except the boundary nodes) is connected to four neighbors. The greater the average edge length, the stronger grid is stretched. The degree of curvature is determined by a point estimate of the second difference derivative. As a result, we obtain the following functional:

$$D = \frac{D_1}{|X|} + \lambda \frac{D_2}{pq} + \mu \frac{D_3}{pq}, \tag{2}$$

where λ, μ are the elastic coefficients, responsible for stretching and bending, respectively, D_1 is the measure of closeness, D_2 is the measure of grid stretching, D_3 is the measure of grid curvature. Terms D_1, D_2, D_3 are defined as follows:

$$D_1 = \sum_{i,j} \sum_{x \in K_{ij}} \left\|x - y_{ij}\right\|^2, \tag{3}$$

$$D_2 = \sum_{i=1}^{p} \sum_{j=1}^{q-1} \left\|y_{ij} - y_{i,j+1}\right\|^2 + \sum_{i=1}^{p-1} \sum_{j=1}^{q} \left\|y_{ij} - y_{i+1,j}\right\|^2, \tag{4}$$

$$D_3 = \sum_{i=1}^{p} \sum_{j=2}^{q-1} \left\|2y_{ij} - y_{i,j-1} - y_{i,j+1}\right\|^2 + \sum_{i=2}^{p-1} \sum_{j=1}^{q} \left\|2y_{ij} - y_{i-1,j} - y_{i+1,j}\right\|^2. \tag{5}$$

Emission concentration maps based on the elastic grids method and emission and concentration values for NIZ in 2015–2017 were constructed. Figure 1 depicts SO_2 concentration mapping for 2015–2017 and spatial distribution of emission, where colors indicate the SO_2 concentration. Note that almost the entire territory of the NIZ near Norilsk has an increased value of SO_2 concentration.

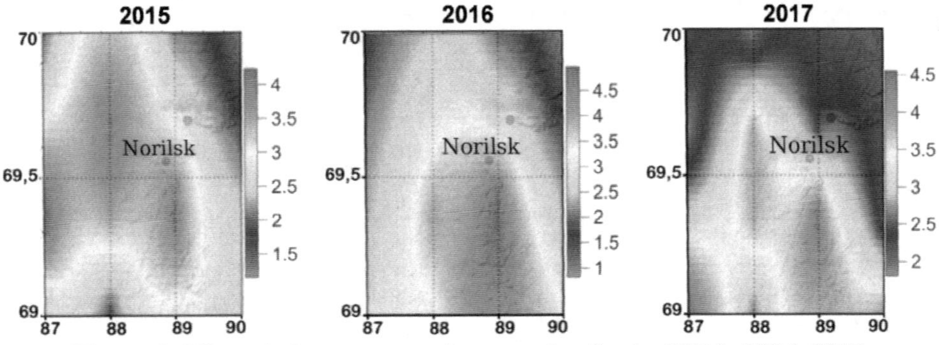

Figure 1. SO_2 emission concentration mapping for the NIZ in 2015–2017.

4. Environmental risk assessment

Environmental risk is a quantitative or qualitative assessment of environment hazard of adverse impact on environment [2, 19-20]. Environmental risk is considered as probability of adverse situation occurrence, ecosystem destruction or animal population death because of economic activity of people. There are two the most studied risk assessment areas: assessment of the possibility of death or public health damage and determination of the average risk of damage caused by atmosphere pollution.

The constant presence of SO_2 in the environment increases a degree of environmental risk. There are three levels of environmental risk: acceptable, high, and unacceptable [2, 20]. An important issue is the determination of the concentrations of pollutants corresponding to different levels of risk.

The level of atmospheric pollution in the fractions of Maximum Permissible Concentration (MPC) corresponding to the zone of unacceptable risk is determined by data and methodological materials [2, 19]. Air pollution degree is calculated considering the excess of the average annual value of MPC, pollutant hazard class, permissible repeatability of concentrations of a given level, and the mass of substances.

The annual average MPC_{a} is associated with an average daily value of MPC_{d} as follows

$$MPC_{\mathrm{a}} = a \cdot MPC_{\mathrm{d}}, \tag{6}$$

where a is the coefficient in the range 0.1 to 1.0 depending on the pollutant hazard class.

Unacceptable risk zone is defined as the territory, where the value of average annual air pollution index P corresponds to the environmental emergency and is equal 8, the value of a for SO_2 is 1:

$$8MPC_{\mathrm{a}} = 8 \cdot (1 \cdot MPC_{\mathrm{d}}) = 8MPC_{\mathrm{d}}. \tag{7}$$

MPC_{d} for SO_2 equals 0.05 mg / m^3. Therefore, the boundary of unacceptable level of risk equals 1.28 DU [20]. If the atmosphere is contaminated with substances with different hazard classes, complex variant of air pollution index is calculated:

$$P = \sqrt{\sum_j K_j^2}, \tag{8}$$

where K_j is the MPC for jth pollutant converted to the concentration of third hazard class pollutant.

Consider the proposed approach for building risk maps. Existing methodological developments to risk assessment are outlined in [2]. The risk assessment is carried out by combining data for all significant natural hazards (air quality, water quality, vegetation cover quality) typical for the studied region. Construction of risk maps is based on geodatabase with topographic and thematic layers, as well as template of cartographic material (with color-coded information).

Most plants are more sensitive to harmful factors than people. Taxonomic groups of plants according to the degree of sensitivity to phytotoxic gases are arranged in the following order [19]: mosses, lichens and mushrooms, softwood trees, hardwood trees, and grass vegetation. It should be noted [2, 19] that environmental standards of air quality on the level of chemical pollution are recommendatory in nature and are not currently defined for all pollutants. The sensitivity of groups of plants was determined by analysis of the environmental standards [2, 19].

Environmental risks mapping is based on the following sensitivity coefficients: 1 for grass vegetation, 0.75 for small-leaved forest, and 0.5 for softwood forest. The higher the sensitivity, the lower concentrations of pollutants cause damage to the relevant vegetation types. Levels of environmental risk are selected in accordance with the sensitivity coefficient for each group of plants (Table 1).

Table 1. Levels of air pollution (DU).

Environment risk level	Softwood forest	Small-leaved forest	Grass vegetation
Unacceptable	>0.625	>0.94	>1.25
High	0.625…0.50	0.94…0.75	1.25…1
Acceptable	<0.50	<0.75	<1

Figure 2 shows an example of mapping environmental risk zones for the Northern territories of the Krasnoyarsk region (April, 2017).

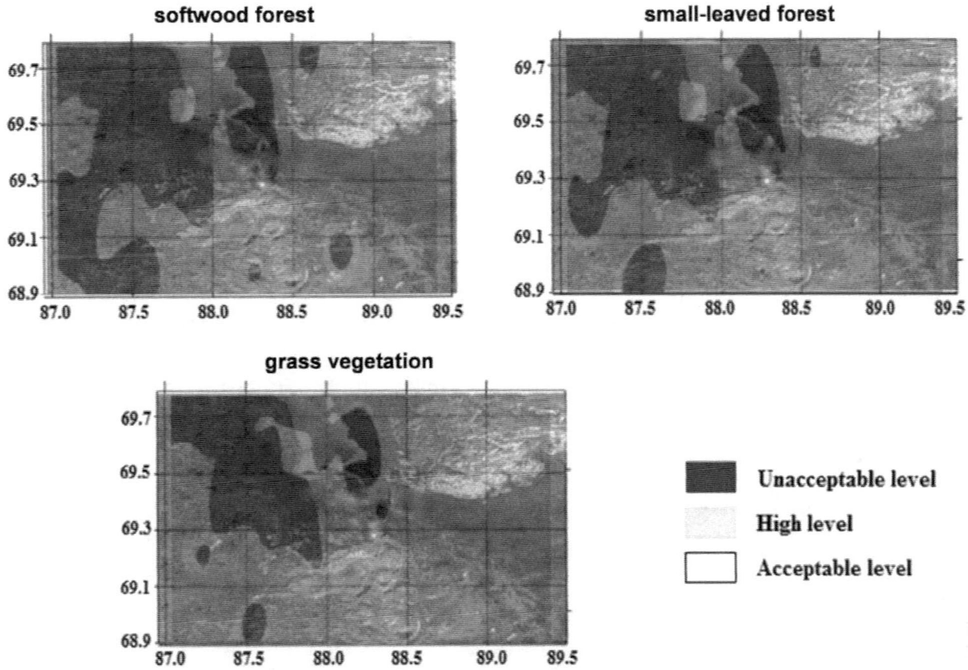

Figure 2. Environmental risk zones mapping for the Northern territories of the Krasnoyarsk territory.

There is a significant air pollution caused by SO_2 emission for the studied area, leading to an increased risk of negative impact to the human body and plants. These maps can serve as a signal for the factories of the Northern territories of the Krasnoyarsk region to reduction of production capacity. Thus, the satellite monitoring makes it possible to carry out environmental monitoring of the atmosphere, while being one of the most cost-effective methods of observation. OMI and OMPS instruments provide information on SO_2 emission at different altitudes and an opportunity to determine the future direction of emission. The satellite monitoring is an independent and objective source of information, which allows to draw conclusions about changes in the composition of the atmosphere for the Northern territories of the Krasnoyarsk region.

References

[1] State Report on the state and protection of the environment in 2016 [Internet]. 2017 [cited 12.02.2018]. Available from: http://www.mpr.krskstate.ru/dat/bin/art_attach/8804_2017.10.11_doklad_2016_gotovij_variant.pdf

[2] Criteria for assessing the ecological situation in the territories for identifying zones of an environment emergency and zones of ecological disaster. Methodology of the Ministry of Natural Resources of the Russian Federation [Internet]. 1992 [cited 12.02.2018]. Available from: http://docs.cntd.ru/document/901797511

[3] Krotkov N A, Carn S A, Krueger A J, Bhartia P K and Yang K 2006 Band residual difference algorithm for retrieval of SO2 from the Aura Ozone Monitoring Instrument (OMI) *IEEE Transactions on Geoscience and Remote Sensing* **44**(**5**) 1259–66

[4] Carn S A, Krueger A J, Krotkov N A, Yang K and Levelt P F 2007 Sulfur dioxide emissions from Peruvian copper smelters detected by the Ozone Monitoring Instrument *Geophys. Res. Lett.* **34** L09801

[5] McLinden C A, Fioletov V E, Krotkov N A, Li C, Boersma K F and Adams C 2016 A decade of change in NO2 and SO2 over the Canadian oil sands as seen from space *Environ. Sci. Technol.* **50** 331–7

[6] Krotkov N A et al 2008 Validation of SO2 retrievals from the Ozone Monitoring Instrument

over NE China *J. Geophys. Res. Atmos.* **113** D16S40
[7] Krotkov N A, Schoeberl M R, Morris G A, Carn S A and Yang K 2010 Dispersion and lifetime of the SO2 cloud from the August 2008 Kasatochi eruption *J. Geophys. Res.* **115** D00L20
[8] Krotkov N A et al 2016 Aura OMI observations of regional SO2 and NO2 pollution changes from 2005 to 2015 *Atmos. Chem. Phys.* **16** 4605–29
[9] Fioletov V E, McLinden C A, Krotkov N A and Li C 2015 Lifetimes and emissions of SO2 from point sources estimated from OMI *Geophys. Res. Lett.* **42** 1969–76
[10] Fioletov V E, McLinden C A, Krotkov N A, Li C, Joiner J, Theys N, Carn S and Moran M D 2016 A global catalogue of large SO2 sources and emissions derived from the Ozone Monitoring Instrument *Atmos. Chem. Phys.* **16** 11497–519
[11] Official site of NASA for global monitoring of sulfur dioxide [Internet]. 2018 [cited 12.02.2018]. Available from: https://so2.gsfc.nasa.gov/index.html
[12] NASA Goddard Earth Sciences (GES) Data and Information Services Center (DISC): The OMI PBL SO2 product (OMSO2v1.2.0) [Internet]. 2018 [cited 12.02.2018]. Available from: http://disc.sci.gsfc.nasa.gov/Aura/data-holdings/OMI/omso2_v003.shtml
[13] NASA/GSFC Aura Validation Data Center (AVDC): The OMPS PBL monthly SO2 [Internet]. 2018 [cited 12.02.2018]. Available from: http://avdc.gsfc.nasa.gov/index.php?site=1868800100
[14] Spinei E, Carn S A, Krotkov N A, Mount G H, Yang K and Krueger A 2010 Validation of ozone monitoring instrument SO2 measurements in the Okmok volcanic cloud over Pullman WA, July 2008 *J. Geophys. Res* **115** D00L08
[15] Yang K, Dickerson R R, Carn S A, Ge C and Wang J 2013 First observations of SO2 from the satellite Suomi NPP OMPS: Widespread air pollution events over China *Geophys. Res. Lett.* **40** 4957–62
[16] Zhang Y, Li C, Krotkov N A, Joiner J, Fioletov V and McLinden C 2017 Continuation of long-term global SO2 pollution monitoring from OMI to OMPS *Atmos. Meas. Tech.* **10** 1495–509
[17] Zuev D V and Kashkin V B 2013 Analysis of sulfur dioxide emissions from the OMI instrument (AURA satellite) for the Norilsk industrial zone *Optics of the Atmosphere and Ocean* **26(9)** 793–7
[18] Gorban A N, Zinoviev A Yu and Pitenko A A 2000 Visualization of data by the method of elastic maps *Information Technology* **6** 26–35
[19] Koplan-Dix V A and Alekhova M V. To the issue of developing environmental quality standards for atmospheric air [Internet]. 2009 [cited 12.11.2018]. Available from: https://distant.msu.ru/mod/resource/view.php?id=14264
[20] Polishchuk Yu M and Tokareva O S 2011 Methodological issues of mapping zones of ecological risk of the impact of oil production on vegetation cover *Izvestiya Tomsk Polytechnic University* **318(1)** 166–9

Research of the physical-mechanical properties of sea ice at the scientific research station «Ice base the cape of Baranov»

S M Kovalev, V N Smirnov, V A Borodkin and N V Kolabutin

Arctic and Antarctic Research Institute, Bering str., 38, 199397 Sankt-Petersburg, Russia

E-mail: skovalev@aari.ru

Abstract. In 2013, on the island of Bolshevik, the Northern Earth archipelago, the scientific research station "Ice base the cape of Baranov" resumed a permanent work. The composition of complex studies of sea ice cover is continuously expanding and now it includes: standard visual ice observations; researches of morphological characteristics of level ice; researches of the physical-mechanical properties of ice; researches of the dynamics of the ice cover; the study of spatial and temporal non-homogeneity of the fast ice structure and physical properties.

1. Introduction

In present time the research station "Ice base the cape of Baranov" is the only station in all western part of the Laptev Sea where it is possible to make regular complex observations for a status of a sea snow-ice cover and land ice. The main task of ice research works is expansion of a full complex of cryospheral observations in accordance with standards of World Meteorological Organization. For the solution of this task the following ice research works are carried out at the station: standard ice observations, measurements of morphometric characteristics of level ice, ridges and icebergs, research of physical-mechanical properties of sea ice, creation and introduction of new techniques of ice experiments and observations. In 2015 by the decision 17 of the Congress World Meteorological Organization the "Ice base the cape of Baranov" along with observatory of Tiksi is included in the list of 36 ground stations for observing the Global Cryosphere Service network.

2. Main Part

Visual observations over a status of the ice cover on a visible space in Shokalsky Strait in the water area of the "Ice base the cape of Baranov" are made for the purpose of mapping and prediction of an ice situation in an adjacent section of the Laptev Sea. The main objective of observations is systematic and complete fixing of all characteristics of a status of an ice cover in the studied territory. Operations are carried out by the observer according to Manual, 1984 [1] and the Order of 12.02.2015 No. 64 of Federal Service on the hydrometeorology and monitoring of the environment defining a transmission order of sea coastal hydro-meteorological observations and offshore ice observations in high latitudes [2].

Observations are performed from the observation post organized according to normative documents in a point with coordinates 79° 16'49"N and 101° 37'20"E., located on the steep coast (height above sea level 24 m) in the territory of the station. Estimated distance of the visible horizon from observation point under favorable weather conditions is 18 km. Observations are performed on the main and additional object of ice observations (Figure 1).

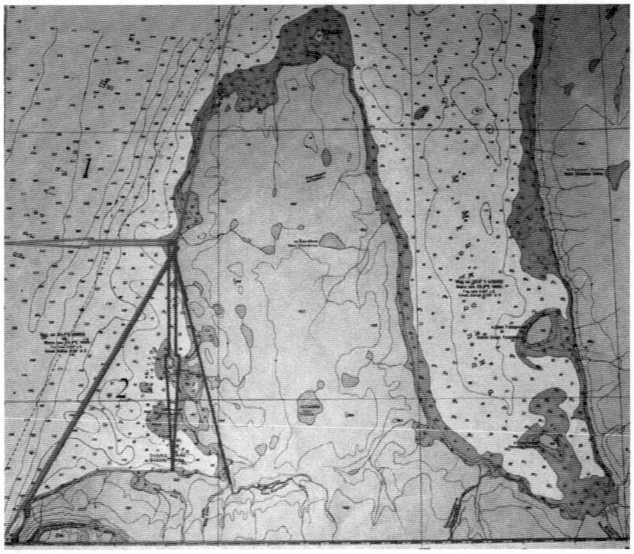

Figure 1. The layout of the main and additional areas for ice observations at the "Ice base the cape of Baranov". 1 – main object (red sector), 2 – additional object (blue sector).

As the main water object for ice observations (the sector of the review of a sea surface) the northern part of the Shokalsky Strait dividing islands the Bolshevik and the October Revolution of the archipelago of Northern Earth adjacent to the western part of the Laptev Sea is used. The greatest width of the passage around observations is 45 km, depth is up to 350 m. The coastal territory, adjacent to an object, is characterized by existence of powerful (up to 800 m) dome-shaped glaciers. The object of observation is the wide range of natural ice: sea fast ice and drifting ice, icebergs. The sector of observations from ice point and its alignment is designated in Figure 1 (the red sector). As additional subject to ice observations the bay Amba, which is an internal bay of Bolshevik island. The additional sector of observations from ice point and its alignment are designated in Figure 1 (the blue sector).

A task of the observer with perhaps bigger accuracy is to note all characteristics of the ice cover and to fix all changes occurring in it. In the case of suitable light conditions, the photo fixing of an ice situation is carried out. As a standard terminology at the description of results of observations the terms and symbols defined in [3, 4] are used.

The research of morphological characteristics of unstrained ice and snow in space and in time allows to monitor increase and melting of the ice cover, change of its physical properties during all annual cycle. Besides, regular measurements of thickness of ice of different age and its physical properties are the one of key parameters of estimation climate state of the Arctic basin.

The purpose of work is the research of evolution of morphometric characteristics of level ice and snow cover during an annual cycle taking into account an ice structure near Shokalsky Strait, at the following tasks are solved:

– collection of data about change of the main morphometric characteristics of unstrained ice and snow on the ice ground near a station by method of contact drilling;

– collection of data about thickness of level ice by a contactless method by means of the EM31 Ice device;

– performance of a complex of temperature measurements of a snow surface, temperature on snow/ice border; construction of vertical profiles of temperature, salinity and density of ice; performance of a textural and structural analysis;
– test of strength of thin plates on a flexure.

Morphometric measurements by the contact way are carried out on the polygon by the size 80×100 m with internal profiles through each 10 m for the main (central) profile and 20 m for other profiles. Ice thickness, snow height (or depth of a meltwater pool), excess of the top surface of the ice cover over water level are measured in each point of the polygon. Repeatability of works on the ground of 5-10 days during the periods of intensive processes of thawing or increase (October-November and June-July) and 10-15 days during other periods. Measurements are taken by ice auger drilling. The error of measurement is up to 1 cm.

Measurement of physical properties of ice in a control point of the polygon is carried out by the established techniques within the time limits close to carrying out metric measurements on the polygon (with a difference of 0-2 days).

Measurements are taken in the contactless way on the same polygon by means of EM31Ice within the time limits as close as possible to date of contact measurements.
The measurements of ice thickness with EM31Ice are carried out with use of a snowmobile on long profiles on level ice. After the end of measurements in several points of the profile the control measurements of ice and snow thickness with a contact method are made.

On the control point a core is taken for textural and structural analysis in laboratory conditions. Before taking the core is necessarily oriented by sides of the world. In the laboratory a description of ice texture and photographing of vertical and horizontal sections taking to account their orientation is carried out, making (preparation) of thin sections for structure analysis is performed.

The main tasks of researches of ice cover properties are:
– the study of physical-mechanical characteristics of level ice and ridges by traditional methods on samples (texture, density, temperature, salinity, ice strength characteristics);
– the evaluation of spatial and seasonal changeability of total-scale strength of the fast ice on total thickness by method of borehole probe-indenter and comparative test of ice samples strength;
– obtaining of new data for solution of engineer tasks of bearing capacity of ice at load operations on the fast ice by a method of control of parameters of deflection under load and registration of flexural stresses in the ice field;
– obtaining of nature data of large-scale dynamic processes in the system "ice-water" for estimation ice impacts onto shores and sea bottom.

For solution of these tasks in a character place of the fast ice the polygon is marked. In the nodes of the grid by a auger motodrill holes are drilled at total ice thickness. In the holes by a borehole probe-indenter ice strength tests to compression (local strength) are carried out.
On the polygon in one point a taking of ice samples as cylinders for evaluation of physical properties and performing of strength tests of ice samples is made. The test is carried out on hydraulic press and field test machine.

The samples at uniaxial compression are made of the same horizons on which local strength was evaluated.

On the fast ice the following tasks are solved:
– influence of ice construction (structure and texture) onto local strength on several polygons, different by dominating ice types [5];
– influence of azimuthal anisotropy onto local strength caused by stable flow and formation of azimuthal anisotropy. The most part of crystals has the main optical axis directed along the flow. Therefore, change of strength of "horizontal" ice samples in dependence of their orientation occurs. The strength of ice samples at the uniaxial compression has 2 maximums and one minimum. Maximums of strength of ice samples are achieved at sawing out the samples parallel and perpendicular C-axis, the minimum with angle of 45°. Since local strength is a strength of ice in the conditions of uniform compression the azimuthal anisotropy of local strength is expressed less than

one of samples of ice at uniaxial compression;
– detection (reveal) of changeability of ice structure at penetration of the indenter into wall of ice borehole. The core with failure zone is drilled out and transported in the ice laboratory where the failure zone, changes of texture and structure of ice are studied. The ice plates are photographed with high resolution;
– influence of indenter diameter onto character of destruction. For investigation of "scale effect" at evaluation of local strength experiments of penetration of indenters of different diameters (6, 9 and 12 cm) are performed.
– influence of penetration velocity of the indenter to ice.
In the place of tests works of measurement of ice physical properties similar to measurements on the control point of the ice polygon are made.

During studies of ice properties and ice cover a technology of monitoring of physical-mechanical condition of the fast ice at impact on it both natural and anthropogenic factors was developed. As a rule, the fast ice during total period of existence is undergone natural oscillation and wave processes causing flexural stresses in the ice. Data about appearance in the fast ice trains of waves with duration to one hour with a period of oscillations 30–50 seconds and amplitude more than one centimeter. The most probable cause of appearance of these oscillations is swell waves from storms in open sea during iceberg formation. On the fast ice several tens such events were fixed. In the Figure 2 features of propagation of flexural-gravity waves with horizontal polarization of wave process are marked. It is obvious that in this case a process of remote storm sea on the open water is registered. Physics of wave propagation from source is characterized not only long-period oscillations of fast ice but and clearly distinguished components of horizontally polarized waves.

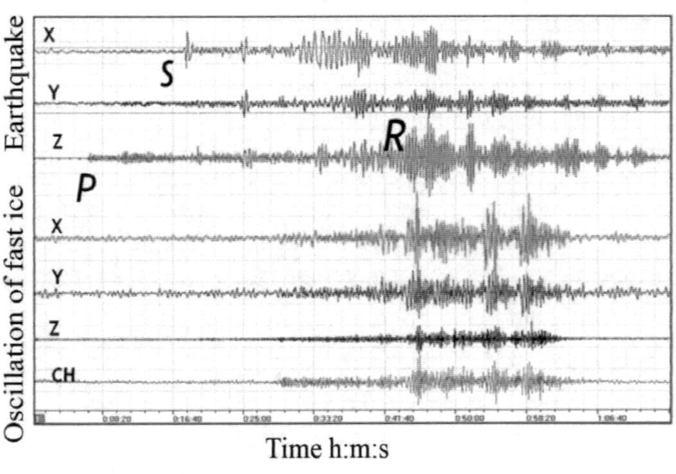

Figure 2. Initial record of long-period oscillations in the fast ice with three-component seismometer SME. The scales are same. Ice base the cape of Baranov, 01.07.2015.

Records of ground mechanical oscillations on the coast of Laptev Sea revealed two important moments: oscillations occur both from local and remote earthquakes and from processes caused by "icequakes" (impact of fast ice onto sea bottom and shore, dynamics of glaciers, plowing by icebergs the sea bottom …). Earthquakes are registered with clear detection of volume and surface seismic waves. Events of impact of seismic waves onto fast ice and glaciers as triggers of ice destruction are considered.

In the first time experiments for registration of wave processes at movement of atomic ice-breaker "Yamal" in the fast ice are performed. For a few kilometers devices fix ice breaking signals accompanied by flexural-gravitational waves with a period of up to 15 seconds. Total process of ice-breaker movement gives the extend information for large-scale experiments. Applicable (suitable) approaches for determination of elastic and strength properties of fast ice, study of mechanism of deforming and destruction of level ice field and control of bearing capacity of ice during operations at the fast ice are developed.

The main tasks of study of dynamic properties of ice formations are:
– dynamics and mechanics of the fast ice deforming as an indicator of local and large-scale processes of sea ices interaction in the system ice-water in the Laptev Sea;
– large-scale physical-mechanical processes of deforming and destruction of ice with purpose to obtain data for development and introduction of a method of short-term forecast of fast ice destruction and break-off the with advance time of several hours;
– parameters of mechanics of the fast ice deforming and dynamics to determine relative deformations of the fast ice at compression, shearing and hummocking;
– spectra of flexural-gravity waves and self-excited oscillations in the ice cover caused by compression and hummocking;
– characteristics of swell waves in the ice cover from storms in open water as precursors of the fast ice destruction;
– horizontal and vertical motions of the fast ice at tidal and wind impacts;
– parameters of free waves at calvings of outlet glaciers;
– parameters of free oscillations of icebergs in the fast ice;
– characteristics of plowing processes by ridges and icebergs of the sea bottom.

Performing of instrumental monitoring of the fast ice dynamics, processes of fracture of outlet glaciers, collapse of glacier fragments and formation of icebergs, plowing by ridge keels and icebergs of the sea bottom is provided by observation methods. For this purpose, three-component seismometers and two-component tiltmeters are used. Sensibility of seismometers CME 4 kV/m/s in the frequency range 0.01-100 Hz, tiltmeters is 10^{-6} V/rad. Seismometers are installed on both the fast ice and on coast ground in specially equipped bunker. Then oscillation velocities of displacement of three components horizontal and vertical are registered. On the fast ice the devices are installed at distance not less than 300 m from tide crack; on the coast at 500 m from place of the minimum anthropogenic influence. Frequency of signal for each channel is 100 Hz.

With a purpose to obtain the fullest information about seismicity of coast of Laptev Sea since 2016 year on the station with Federal State Budgetary Institution of Science "Federal Research Center for Comprehensive Arctic Studies of the Russian Academy of Sciences (FRC CAS RAS) observations of oscillations of coast ground are performed. Seismic observation point of "Ice base the cape of Baranov" was organized in accordance with methodological directions [6]. The selected place with coordinates 79,276°N и 101,657°E is indicated in the Figure 3 (number 4).

In October, 2016 in the seismic bunker CMG-6TDc seismometer which at the same time measures three components of oscillations of an earth surface in the range of frequencies from 0.033 to 50 Hz was set with GPS antenna. Data are transferred with using of the software package of Guralp Systems directly to AARI and FRC CAS RAS.

Results of the described techniques and further tasks are necessary for development and deployment of the short-term forecast of the fast ice destruction. Appearance of long-period ripple waves, rise of cyclic offsets of ice, signals of crack formation in ice, etc. can serve as warning. In combination with rising tide and squeezing wind these natural phenomena supplement an overall picture of ice destruction and release of the fast ice in the sea. The received results can be used in case of the decision of engineering tasks of hydrotechnical construction and exploitation of sea constructions on the Arctic shelf and also in case of support of the security regulation by operations on ice.

One of the main results of work is improvement of the forecast of a dynamic state of the sea ice which is based on contact and remote measurements of large-scale mechanics of ice (swell, flexural and gravitational waves, large-scale strength of ice, interaction of the drifting ice and the fast ice, seismic waves from earthquakes and "icequakes", estimate of large-scale parameters of physical-mechanical properties of ice). For this purpose, installation on the coast of ice base the remote measuring system allowing to obtain information about ice processes coming on removal up to 50 km from base in real time is planned. This technology is used for studying of dynamics of glaciers and sea ice on the Arctic shelf and in Arctic Ocean (at the drifting stations "North Pole" and in coastal zones).

Measuring devices are installed on ice, electric signals are given by a radio channel on the registering complex. For preliminary processing special programs in statistics, the spectral analysis, etc. are used. At this stage testers of remote system directly from the coast of station "Ice base the cape of Baranov" in 2019 are prepared.

Figure 3. Location of seismic equipment at Ice base the cape of Baranov in 2016. 0 – Recording equipment, 1 – ground seismometer, 2 – seismometer on the fast ice, 3 – seismometer with the "Baikal 7 HR" logger, 4 – « seismic bunker», 5 – location of recording equipment, 6 – seismometer with the "Baikal 7 HR" logger. Lines 0-1-2 and 4-5 – cable communication lines.

The task of spatial heterogeneity of a structure and physical properties of fast ice research is monitoring of the spatial and temporary heterogeneity of a structure and physical properties of equal fast ice arising at various hydro-meteorological conditions during its formation begun in 2014 and also development of a technique of mapping of the level ice cover on the prevailing ice types.
In the water area of station "Ice base the cape of Baranov" places are revealed according to their geographic location or ice formation conditions that have differences from the place of an arrangement or to site ice formations conditions where the main ice ground is chosen. It can be bays, available on technical capabilities, passages, feuds, zones of the frozen leads or sites close to the ice hole behind fast ice, a coastal zone with an intensive snow accumulation, ice around icebergs and at hummocks.

The choice of the place of works is defined according to technical capabilities of base, especially with a possibility of delivery of people and the equipment to the chosen place.

On the chosen site are carried out ice lines from several points, ice polygons or certain ice stations where are carried out works similar to measurements on a control point of the main ice ground. The quantity of points is chosen taking into account the fullest illumination of the studied ice object.

Acknowlegments
Work is performed within planned scientific thematic of AARI on the project of Roshydromet "Scientific and methodical developments of monitoring of a physical-mechanical state of ice in the Arctic in real time for short-term forecasting of the critical and dangerous ice phenomena".

Reference

[1] Instruction to hydrological stations and posts. Issue 9 Hydrometeorological observations at sea stations and posts. Part 1 Hydrometeorological observations at coastal stations and posts 1984 (Leningrad: Gidrometeoizdat) p 312
[2] The code for the operative transfer of marine coastal hydrometeorological observations data (KN-02 SEA) 2014 (Moscow: Roshydromet) p 44
[3] Nomenclature of sea ice. *Conventional signs for ice cards 1974* (Leningrad: Gidrometeoizdat) p 77

IOP Conf. Series: Earth and Environmental Science **193** (2018) 012030 doi:10.1088/1755-1315/193/1/012030

[4] Bushuev A V and Volkov N A *Atlas of ice formations* 1974 (Leningrad: Gidrometeoizdat) p 141
[5] Cherepanov N V 1976 Classification of ice of natural reservoirs. *Proc. of the AARI* vol 331 p. 77–99
[6] Recommendations for the organization of seismic observation points in the Arctic 2016 (Arkhangelsk: Federal Research Center for Comprehensive Studies of the Arctic) p 34

Experimental and numerical research of the stressed-deformed state of ice beams reinforced by surface reinforcement

V M Kozin[1,2], V L Zemlyak[1], A S Vasilyev[1] and K I Ipatov[1]

[1]Sholom-Aleichem Priamursky State University, Birobidzhan, Russia
[2]Institute of Machining and Metallurgy, FEB RAS, Komsomolsk-on-Amur, Russia

E-mail: vellkom@list.ru, vasil-grunt@mail.ru

Abstract. It is known, that ice crossings are actively used in the Russian Federation, mainly in its eastern part, during the winter period. If the thickness of the ice cover is not sufficient for the safe exploitation of the crossing, it can be used the traditional methods of increasing the bearing capacity of ice. But traditional methods are not always effective, and the physical and mechanical properties of ice can strongly depend on various external factors. In this connection, the task of increasing the bearing capacity of ice by using alternative methods, for example, the introduction of reinforcing elements into the ice is becoming very relevant. The aim of the work was experimental and numerical study of stress-strain state of ice samples reinforced by surface reinforcing frame. The results of experiments on loading the samples were compared to the numerical results of calculations of ANSYS.

1. Introduction

In winter period in the case of absence of bridge structures or when the arrangement of ferry crossings is impossible the ice crossings arrange when the ice cover of required thickness is formed on the water barriers. If the thickness of the ice cover is not sufficient for the safe operation of the crossing, traditional methods of increasing of ice bearing capacity can be used, such as ice freezing from below, ice frosting from above, or ice strengthening with a wooden cover [1]. Practical experience reveals that the physical and mechanical properties of the ice cover, strengthened by these methods can strongly depend on various external factors (ambient temperature and presence of snow and wind when freezing). Ice, produced by the accelerated sprinkling method is often a firn mass capable of adhering to the wheels of vehicles and breaking away from the natural ice. This method is effective only for the ice cover of a certain thicknesses, and it increases the probability of formation of deep cracks in the ice. In this connection, the task of increasing the bearing capacity of ice by using alternative methods, for example, the introduction of reinforcing elements into the ice is becoming very relevant.

Yakimenko [2] describes experimental studies on the "surface reinforcement" of ice crossings by geosynthetic materials. Surface reinforcement by freezing the steel meshes was proposed by Nikitin [3]. A number of solutions are known in which steel elements are frozen to increase the bearing capacity in the ice cover [4, 5, 6, 7]. The method of surface reinforcement by introducing the welded steel frames into the relatively thin ice cover of 0.3-0.4 m thick can actually be sufficiently promising.

In paper [8], was present model experiments on testing the strength of ice reinforced with steel reinforcement, as well as numerical experiments on strengthening of ice samples by steel reinforcement and reinforcement of various types of composite materials. This paper presents a similar research, but other samples reinforcement schemes.

2. Preparations for the experimental and numerical research

To carry out the model experiments, a universal loading installation (Figure 1) was designed and assembled, this universal loading installation consisted of power frame, loading device and measuring module. The power frame consisted of pillars, frames, top and bottom beams. The loading device consisted of a hydraulic cylinder 3 with a nominal pressure of 9 atm. and distributing power beam 2. The force of the loading device was transmitted to the sample 1 through the hinge supports 6. The loading system was arranged so that it provided a pure flexure in the middle part of the span of the ice sample. The vertical displacement of the section of the sample in the middle of the span was measured by using non-contact laser sensor LAS-Z by Way Con (Germany) 5 fixed to an independent pillar. The load experienced by the sample was recorded by using an electronic weighing indicator SH-20 of TOKVES (Russian Federation) 4. The loading velocity for all samples was 135 kPa/s constantly.

A wooden framework made of boards of thickness 40 mm was arrange to manufacture the ice samples. The framework made it possible to prepare ice beams with dimensions L × B × H = 2000 × 200 × 200 mm. Two-layer polyethylene film with a thickness of 0.03 mm and a reinforcing frame were placed in the assembled framework. After that, the framework was filled with water. The liquid was exposed to low atmospheric temperatures (t < 0 ° C) until its complete freezing. The time of preparation of the sample, depending on the weather conditions, was from 5 to 7 days at an ambient temperature of -16 ° C to -28 ° C.

An all-welded frame of steel reinforcement with a diameter of 6 mm was used to reinforce the ice beams. The frame scheme is shown in Figure 2.

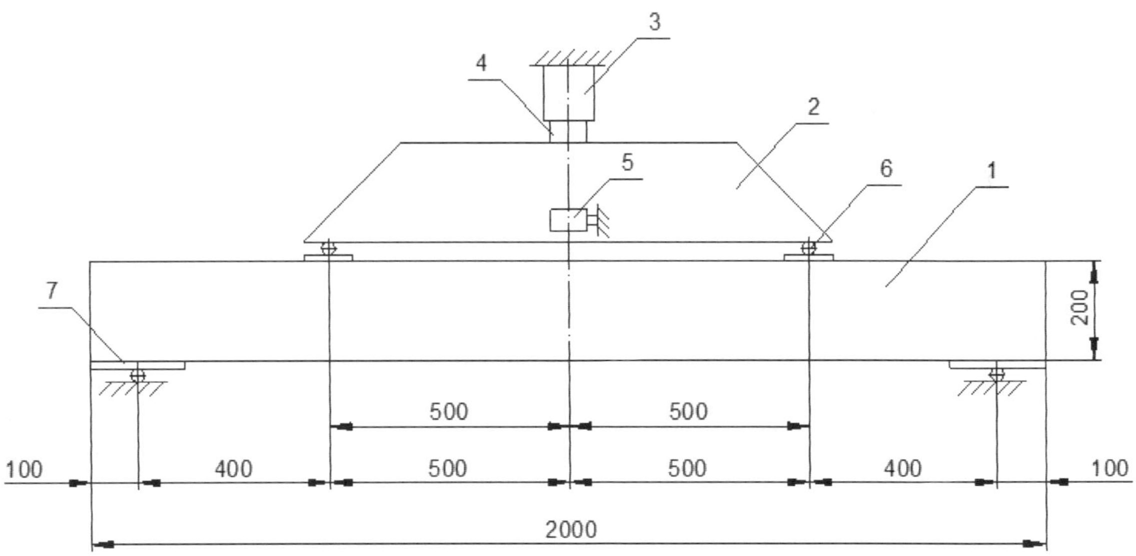

Figure 1. Scheme for experimental installation: 1 - ice beam; 2 - distribution beam; 3 - hydraulic cylinder; 4 - weighing indicator SH-20; 5 - sensor of vertical displacements LAS-Z; 6 - hinged supports of a distribution beam; 7 - hinged supports of the ice beam.

A wooden framework made of boards of thickness 40 mm was arrange to manufacture the ice samples. The framework made it possible to prepare ice beams with dimensions L × B × H = 2000 ×

200 × 200 mm. Two-layer polyethylene film with a thickness of 0.03 mm and a reinforcing frame were placed in the assembled framework. After that, the framework was filled with water. The liquid was exposed to low atmospheric temperatures (t < 0 ° C) until its complete freezing. The time of preparation of the sample, depending on the weather conditions, was from 5 to 7 days at an ambient temperature of -16 ° C to -28 ° C.

An all-welded frame of steel reinforcement with a diameter of 6 mm was used to reinforce the ice beams. The frame scheme is shown in Figure 2.

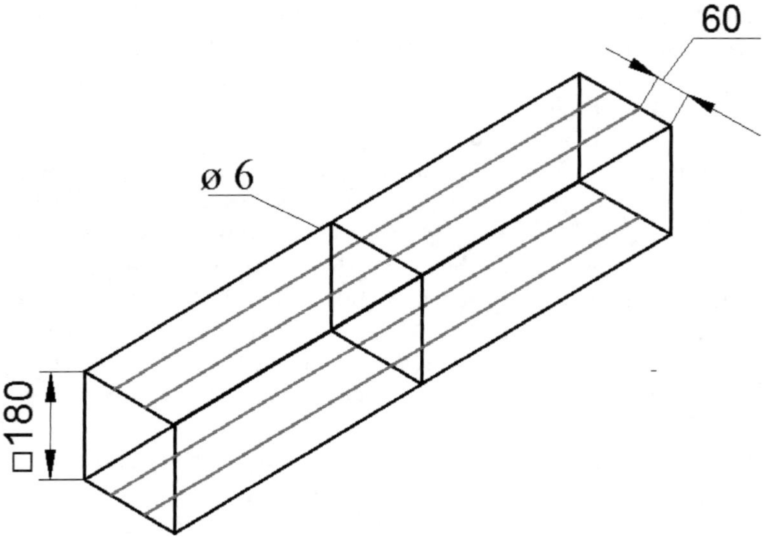

Figure 2. The scheme of reinforcement of ice beam.

Experiments on ice beams were carried out to evaluate the impact of the surface reinforcement of the stretched area strengthened with the reinforcing framework on carrying capacity of beams in the conditions of pure flexure.

The numerical calculation of stress-strain state of ice samples was carried out in the software system ANSYS Workbench, with using ANSYS Mechanical module. The model consists of 80160 finite elements and 89150 nodes. For ice, nonlinear finite element elements SOLID 65 in the form of a hexahedron were used to simulate elements that allow cracking in tension, and also allow calculations based on the [9].

The finite element BEAM188 was used to simulate the work of reinforcing materials. This is a beam element with flexural rigidity. Each longitudinal bar of the model was divided into 196 beam FE (Finite elements), each transverse bar into 18 beam.

The following mechanical characteristics of ice were used to calculate the reinforced ice beam: initial modulus of elasticity E = 200 MPa, uniaxial compression strength Rb = 0.55 MPa, uniaxial tensile strength Rbt = 0.4 MPa, density ρ = 930 kg/m3, Poisson ratio μ = 0.3.

The characteristics of the reinforcement: the initial modulus of elasticity E = 2 • 105 MPa, the design resistance Rs = 355 MPa.

The mechanical characteristics of materials used to enhance ice in numerical experiments presented in table 1: hot-rolled reinforcement A400 (composite number 1); reinforcement fiberglass composite (composite number 2); reinforcement carbon (composite number 3); reinforcement aramid composite fiber (composite number 4); reinforcement combined with a combination of glass and basalt (composite number 5).

Table 1. Calculated mechanical characteristics of steel and composite reinforcement.

Parameter name	№ 1	№ 2	№ 3	№ 4	№ 5
Maximum tensile stress, $σ_{bt,n}$, MPa	365	168	840	448	320
Compression resistance, $σ_{bt,n}$, MPa	365	63	180	96	96
Modulus of elasticity, E, MPa	$20 \cdot 10^4$	$50 \cdot 10^3$	$130 \cdot 10^3$	$70 \cdot 10^3$	$10 \cdot 10^4$

With account of the multiplication of the standard characteristics $σ_{bt}$ and $σ_{bc}$ by the corresponding coefficients, the calculated mechanical characteristics of used steel and composite reinforcement are shown in table 1.

3. The results of experimental and numerical research

A preliminary series of experiments on loading of unreinforced ice beams was carried out to evaluate the impact of surface reinforcement on the ultimate load-carrying capacity of samples. During the experiments, it was determined the beam deflection caused by the load applied to the test sample. According to the experimental data the destructive load was 3.6 kN, with a numerical calculation of about 3.9 kN [8].

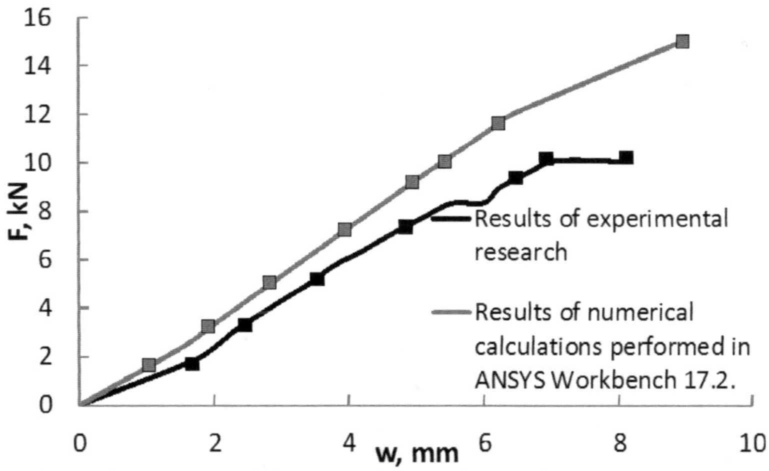

Figure 3. Diagram of the dependence of the deflection of reinforced A400 samples from the load (— - results of experimental research, — - results of numerical calculations performed in ANSYS Workbench 17.2.).

Figure 3 presents the results of a model experiment with reinforced A400 sample. It is apparent that the maximum load that the samples sustained in the experiment was about 10 kN, which significantly exceeded the maximum load that the unreinforced sample could withstand. At this the formation of through cracks in the tested samples occurred, and the maximum deflection reached 8.14 mm after which the complete destruction of the beams occurred.

As you can see, the results of model experiments and numerical calculations are in good agreement in the elastic zone. As a criterion for the destruction of the ice beam, a sharp increase in deformation was taken. The criterion characterized by a loss of bearing capacity and destruction of most of the section. In this case, the reinforcement did not reach the yield point, and as a result of the action of the bending moment in the middle of the beam span the destruction with the formation of extensive through cracks in the ice occurred.

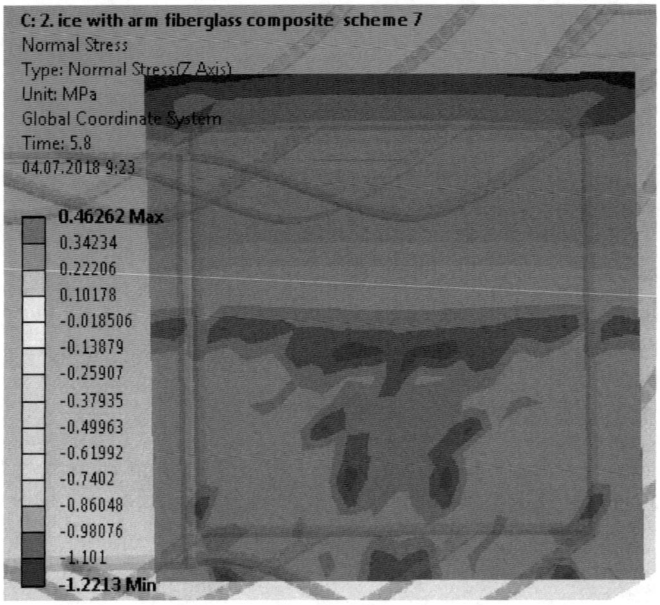

Figure 4. Normal stresses in the median section of the ice beam of sample №2 upon failure.

An example of calculating the stress-strain state of a reinforced sample №2 with an fiberglass composite reinforcement in the ANSYS Workbench 17.2 software package is shown in Figure 4-8.

Figure 4 shows the displacement of the neutral axis of the modeled sample (bottom section tensions are within zero) when the breaking load of about 9.6 kN. Figure 4 shows that ice no longer works at the bottom of the section, but the load is received by the reinforcing material in the stretching zone.

Figure 5. Scheme of cracks in sample №1 (A400) during the experimental research

Figure 5 shows the appearance of cracks in the ice beam during the experiment.

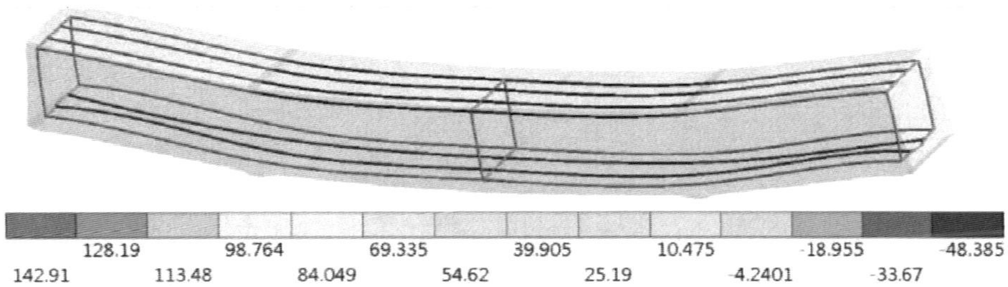

Figure 6. Stresses in the reinforcement when breaking the sample №2 (MPa).

As can be seen from Figure 6 the stresses in the reinforcement of the stretched part did not reach the yield point and was about 143 MPa.

The main results of numerical calculations presented as diagrams of dependence of load from deflection for samples strengthened with reinforcement of various materials are shown in Figure 7.

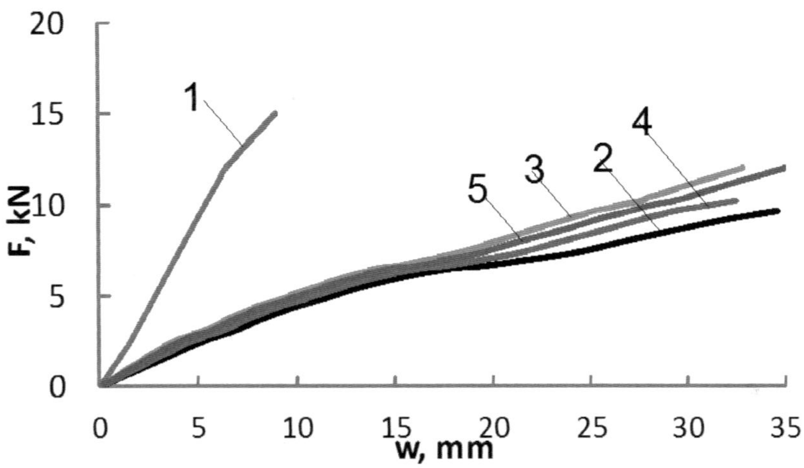

Figure 7. Results of load-deflection calculations.

Taking into account the load at which the unreinforced sample destructed was 3.6 kN [8], it can be judged, that the bearing capacity of reinforced sample is much higher.

The stresses in the middle sections of each of the reinforced samples presented in Figure 8.

The reinforced sample №. 2 breacking before the remaining samples at a load of 9.6 kN. Analysis of the data shows that the use of reinforcing carcass of various composite materials significantly increases the bearing capacity of ice. The maximum carrying capacity in comparison with the unreinforced was demonstrated by a sample reinforced with steel reinforcement A400.

The lowest bearing capacity was demonstrated by samples strengthened with reinforcement material № 2, 4. The maximum stresses at a load of 9.6 kN (The load at which the weakest of the samples (No. 2) lost its bearing capacity) for samples reinforced with material № 1, 2, 3, 4, 5, were -0.168 MPa; -1.094 MPa; -0.664 MPa; -0.778 MPa; -0.68 MPa respectively. The fastest growth of normal stresses with increasing the load was observed in samples of composite №2, the slowest growth observed in samples reinforced with №1.

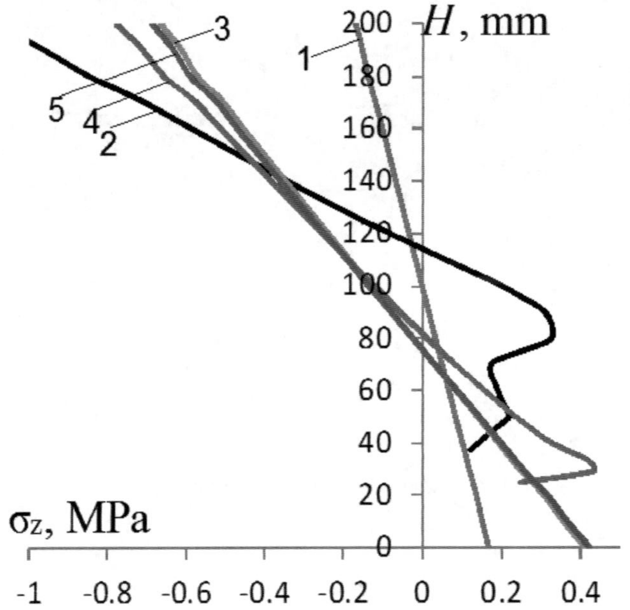

Figure 8. Distribution of normal stresses σz in the beam section at load of 9.6 kN.

4. Conclusion

Resulting from the study, the following conclusions can be drawn:

As a result of experimental and numerical research, it was established that the use of surface reinforcement of ice with various materials under this reinforcement scheme it possible to increase the bearing capacity from 166 to 233%.

Destruction of samples in all cases occurred as a result of formation of extensive through cracks in the ice caused by the action of bending moment in the middle of the beam span. At this the reinforcement did not reach the yield point. In numerical calculations at the stages of failure the normal stresses in the lower part of the section approached zero.

The results of model experiments and numerical calculations are in good agreement in the elastic zone. As for deflections and destructive loads, deviations in the comparison of data did not exceed 8.9% and 33% respectively.

Acknowledgments

The study is performed under the project 9.4934.2017 of the task for fulfillment of the government works in the field of science within the base part of the government task for higher education establishment.

References

[1] Common house needs 218.010-1998. *Instruction on the design, construction and operation of ice crossings*

[2] Yakimenko O V and Sirotyuk V V 2015 *Strengthening of ice crossings by geosynthetic materials* (*SibADI.* Omsk) p 168

[3] Nikitin P E, Nikitina M P 2015 *The way of creating a reinforced ice crossing for wide reservoirs* Patent for invention of the Russian Federation No. 2569694 published on 27.11.2015

[4] Kostenko A V, Serdecny A S and Serdecny A A 2005 *Ice crossing* [in Russian] Patent for the invention of the Russian Federation No. 2260648 published on 20.09.2005

[5] Kozin V M, Kustov A N and Morozov V S 2003 *Method for creating an ice crossing* Patent for

invention of the Russian Federation No. 2198355 published on 10.02.2003
[6] Kozin V M, Vidyakin A V and Popenko N V 2011 *Method for creating an ice crossing* Patent for invention of the Russian Federation No. 2431012 published on 10.10.2011
[7] Kozin V M, Vidyakin A V and Popenko N V 2012 *Method for creating an ice crossing* Patent for invention of the Russian Federation No. 2459900 published on 27.08.2012
[8] Kozin V M, Zemlyak V L, Vasilyev A S and Ipatov K I 2018 The Research of the Stressed Strain State of Ice Beams Reinforced by Surface Reinforcement, *Proceedings of the Twenty-eighth International Ocean and Polar Engineering Conference* (Sapporo, Japan) pp 1511-1515
[9] Willam K J and Warnke K J 1974 Constitutive model for the triaxial behavior of concrete *Seminar of concrete structures subjected to triaxial stresses* (Bergamo, Italy) pp 3-11

Seismogeological characteristics and oil-and-gas content of the Kara Sea shelf (South Kara, North Kara sedimentary basins)

V A Kontorovich[1,2], **A U Kalinin**[1,2], **L M Kalinina**[1,2], **M V Solovev**[1,2], **S M Guseva**[1]

[1]Trofimuk Institute of Petroleum Geology and Geophysics, Novosibirsk, Russia
[2]Novosibirsk State University, Novosibirsk, Russia

Abstract. On the basis of a comprehensive analysis of geological and geophysical materials, seismogeological models of sedimentary complexes in the South Kara oil-and-gas province (OGP) and in the North Kara potential oil-and-gas province (POGP) were constructed. The conclusion is drawn that the South-Kara OGA is the northern end of the West Siberian oil-and-gas province and the Jurassic and Cretaceous deposits are of the greatest interest here. North-Kara POGP is the continuation of the Barents Sea province and Triassic and Jurassic rocks have the biggest petroleum potential. On most of the POGP, the main prospects of oil production are associated with the Neoproterozoic-Paleozoic sedimentary complex.

1. Introduction

Two separate sedimentary basins are allocated in the Kara Sea. South, located to the south of the Novaya Zemlya Archipelago, part of the water area is identified as part of the South Kara Regional Depression, which is the northern end of the West Siberian sedimentary basin [1].

In the respect of oil and gas content, the southern part of the Kara Sea is separated as part of the South Kara oil and gas province (OGP), which is part of the West Siberian oil and gas province, the northern one is an independent North Kara prospective oil and gas province (POGP).

The South Kara and North Kara sedimentary basins are separated by the North Siberian sill, a large contrast erosion-tectonic basement high. The presence of this barrier does not allow comparison of seismogeological complexes in the South Kara regional depression and the North Kara POGP (figure 1).

2. South Kara OGP

At the north of Western Siberia the five seismogeological megacomplexes were identified in the Mesozoic-Cenozoic deposits: Triassic, Neocomian (Berrias-Lower-Aptian), Aptian-Albian-Cenomanian, Turonian-Maastricht and Cenozoic. All sedimentary seismogeological complexes that developed in the north of Western Siberia continue to the southern part of the Kara Sea and were developed in the South Kara OGP.

Triassic-Jurassic (T_{2-3}-J), Neocomian (Berrias-Lower-Aptian) (K1) and Aptian-Albian-Cenomanian (K_1-K_2) megacomplexes are the main oil and gas bearing complexes on the territory of the West Siberian oil and gas province.

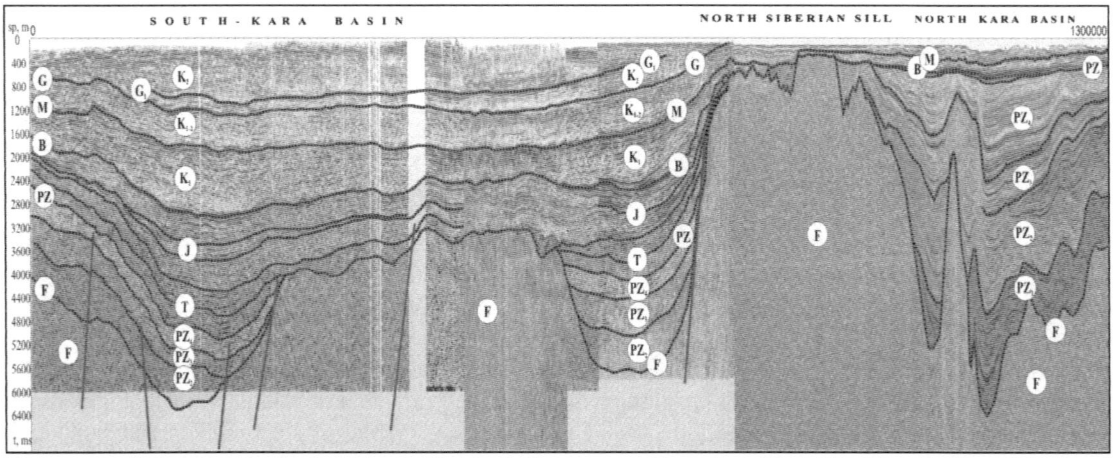

Figure 1. Seismogeological cross-section of the composit regional profile "South Kara Regional Depression – North Siberian sill – North Kara Basin".

All Mesozoic-Cenozoic sedimentary megacomplexes are controlled in the top by regionally developed marine clay packs - mega-regional impermeable beds which are characterized by abnormally low acoustic characteristics and the most energetically expressed reflecting seismic horizons are confined to them.

The Mesozoic-Cenozoic sedimentary cover lies on a Paleozoic basement. In most of Western Siberia Paleozoic deposits have undergone the processes of Hercynian folding and Early Triassic rifting. In these regions, the dislocated and metamorphosed Paleozoic rocks serve as the basement. In the east of the West Siberian basin and in the Yenisei-Khatanga regional trough, the Neoproterozoic-Paleozoic platform sediments, similar to the complexes of the Siberian Platform, lie beneath the Mesozoic-Cenozoic deposits.

The analysis of the time cross-sections allows to assume that the Vendian-Cambrian platform deposits of the Yenisei-Khatanga trough continue to the northeast of Western Siberia and are developed in the Gydan Peninsula and continue to the southern part of the Kara Sea as separate blocks. In the South Kara regional depression, two large submerged massifs are allocated, within which the thickness of platform sediments increases to 12000 m. These submerged blocks located in the western and eastern parts of the South Kara regional depression are separated by a large basement high, within which the thickness of platform deposits is reduced to 5000 m.

In the Arctic regions of Western Siberia, the most prospecting gas content is the Aptian-Albian-Cenomanian rock complex, which is associated with the unique gas fields of the Nadym-Pur, Yamal and Gydan oil and gas areas (OGA). In the South Kara OGP three oil and gas fields were discovered on the shelf: Rusanovskoe, Leningradskoe and Pobeda. The main reserves of the Leningradskoe and Rusanovskoe fields are concentrated in the Aptian-Albian reservoirs, at the Pobeda field the gas is concentrated in the Cenomanian reservoir, yje light oil is in the Upper Jurassic reservoir.

Based on the results of the researches carried out at IPGG SB RAS in the southern part of the Kara Sea, 35 structures have been identified - potential traps for hydrocarbon reservoirs. The most perspective for liquid hydrocarbons in Western Siberia and in the southern part of the Kara Sea are the Neocomian and Middle-Upper Jurassic reservoirs. In particular, the light oil reservoir is discovered in the Upper Jurassic sandstones of the Pobeda field. The Neocomian complex of rocks with more than 80% of the oil reserves in the territory of Western Siberia has a clinoform structure in the South Kara OGP as well as in Western Siberia, and hydrocarbon deposits, controlled both by anticlinal structures, and lithological traps.

3. North Kara POGP

The North Kara basin has not been studied by drilling and the model of the geological structure of this region can be constructed only from the analysis of seismic materials, potential fields and geological survey data for the islands. At the present day, a network of regional seismic profiles has been worked out in the northern part of the Kara Sea, the interpretation of which has formed the basis for the present studies.

In the northwest, the North Kara basin borders on the Barents Sea basin, which has been well studied by seismic exploration and deep drilling. Comparison of wave fields at the regional composite time cross-section along the profile passing through the Barents Sea-Kara Sea line makes it possible to note that the Barents Sea sedimentary basin has been extended in the northwestern part of the Kara Sea.

In this zone, a series of Paleozoic, Triassic, Jurassic and Cretaceous sedimentary seismic complexes are identified on the cross-sections, which are well joined to the Barents Sea complexes. In the southern and southeastern directions, the Mesozoic seismogeological complexes are regionally rising, their thicknesses are sharply reduced, and they are wedged to the erosion surface of the Paleozoic top.

In most of the North Kara basin, the erosion surface separating the Paleozoic and low-thickness Mesozoic deposits, with which a large break in sedimentation is associated in this region, is clearly recorded on the cross-sections. The analysis of time seismic cross-sections makes it possible to identify in the Paleozoic platform deposits of the North Kara POGP 4-5 seismogeological complexes, according to the overlaying, controlled by energetically expressed reflecting horizons.

In the Arctic sector, the Paleozoic deposits are located everywhere at big depths and are almost not drilled. As a result, when stratifying seismogeological complexes, as a rule, data on islands are used [2,3].

The composite section of the Novaya Zemlya archipelago includes a complete sequence of Paleozoic deposits from the Cambrian to Permian inclusive. If we operate with the average thickness of the sediments composing the various complexes, then in the presence of a complete sequence in the section, the average thickness of the Paleozoic deposits on the Novaya Zemlya archipelago will be about 10,000 m. On the islands of the Severnaya Zemlya, the composite section of the Paleozoic deposits includes the Cambrian-Devonian interval, the average thickness of which is about 8500 m. Considering that in the most submerged parts of the North Kara basin the maximum thickness of Paleozoic sediments estimated according to seismic data is 13,000 m, it can be assumed that a complete Paleozoic section was developed in these zones. In this case, the basal seismic complex can be conditionally dated by the Cambrian, and the seismic complexes overlapping it Ordovician-Silurian, Devonian-Carboniferous and Permian, respectively.

The analysis of the time cross-sections allows us to conclude that in the central part of the basin under the erosion surface overlapped by the Mesozoic, the youngest deposits appear, while in the direction to the side parts under the Mesozoic, more ancient seismic complexes are emerging. Thus, in different parts of the basin in the top of the Paleozoic, there will be layers of different ages, probably from the Ordovician to the Permian. The basal complex, composed of supposedly Cambrian deposits, is developed only in contrasting depression zones and does not come out under the erosion surface, but wedges out onto the contiguous basement highs.

Within the researches, a complex interpretation of the CDP seismic data for the South Kara Regional Depression and the North Kara basin was carried out, the set of the structural maps were constructed, including united structural maps for the Paleozoic top and the basement of platform deposits-the basement top (figure 2). On the cross-sections the Paleozoic top, to which the reflecting horizon A is confined, identified quite confidently, the correlation of the basement top (horizon F) was carried out along the envelope of the reflecting horizons, which emphasize the platform appearance of the sediments. Below this horizon, on the cross-sections are fixed chaotic seismic images.

With the use of these materials, maps of thicknesses of Paleozoic and Mesozoic deposits were also constructed on the Kara Sea, and a structural map on the top of the basement served as the basis for

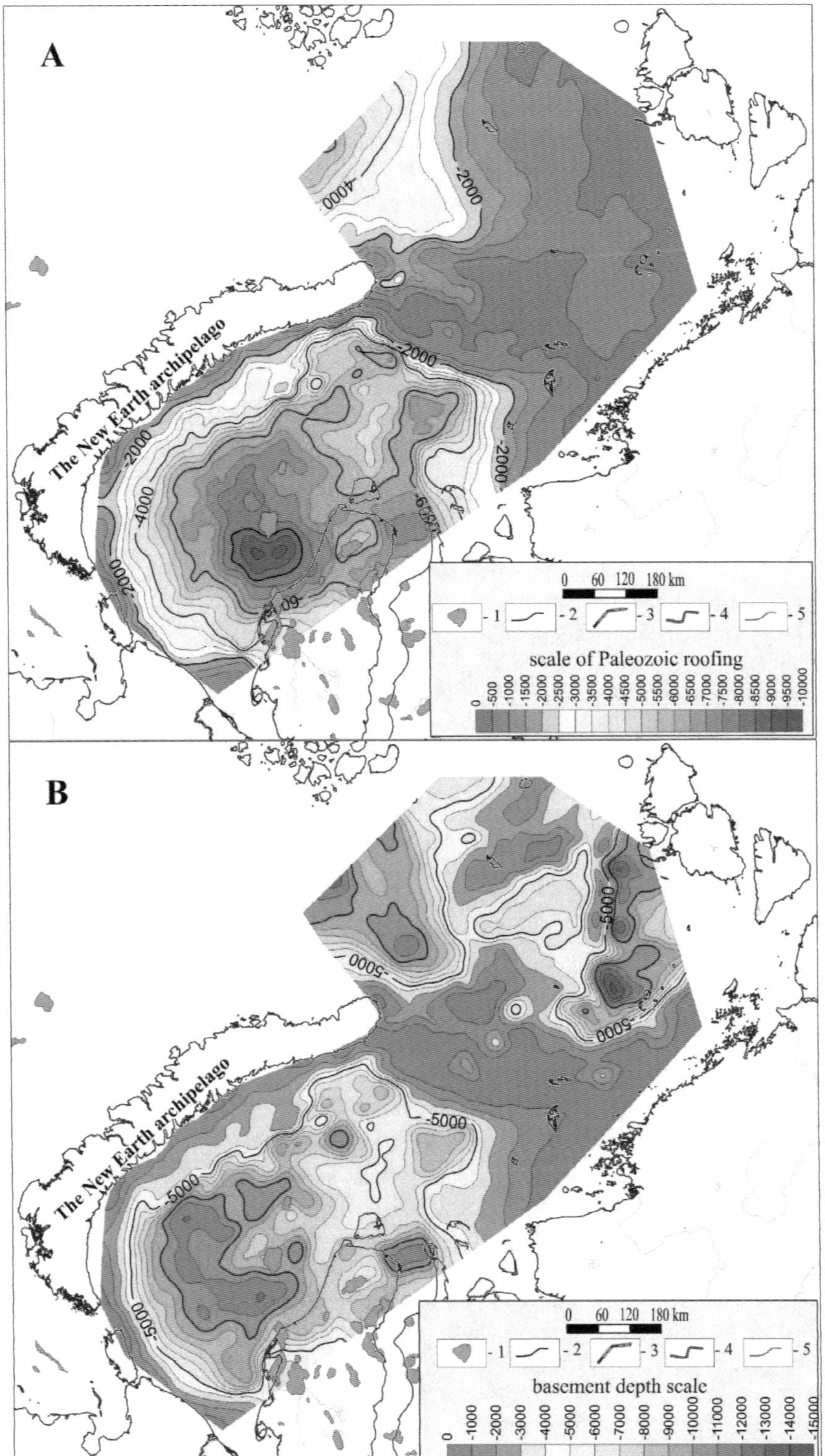

Figure 2. Structural maps of the top of the Paleozoic (A) and the basement (B). Legend: 1 - oil and gas fields; 2 - isohypses; 3 - administrative boundaries; 4 - coastline; 5 - rivers.

constructing a scheme for tectonic zoning of the Kara Sea (the South Kara OGP, the North Kara POGP).The analysis of structural maps on the tops of the Paleozoic and the basement showed that in the South Kara Regional Depression these surfaces are largely similar. In the western part, there is a large depression zone - the South Kara megasineclise, in the east there is a large, semi-closed positive structure in the latitudinal direction - the Northern megascarp, complicated by smaller closed uplifts. In the North Kara basin, structural plans for the tops of the Paleozoic and the basement differ fundamentally. In this part of the shelf in a relief of the Paleozoic top, a monocline is plunged in the northwestern direction, practically uncomplicated by positive and negative structures. The relief of the basement top, on the contrary, is strongly dissected and a series of large positive and negative structures is identified here.

The analysis of the thickness maps of the Paleozoic-Mesozoic platform deposits in general, and also of the thickness maps of the Paleozoic and Mesozoic deposits, showed that in the South Kara and North Kara sedimentary basins, the areas of which are 410000 km2 and 335,000 km2, the sedimentation volume is 2300000 km3 and 1550000 km3, respectively. At the same time, Paleozoic platform deposits in the South Kara OGP account for 23%, in the North Kara POGP - 71%.

4. Conclusions

The analysis of geological and geophysical materials in the Kara Sea suggests that the sedimentary basins in the South Kara and North Kara sedimentary basins are of undoubted interest in relation to oil and gas potential.

The oil and gas potential of the South Kara OGP is proved by the discovery of the Leningradskoe, Rusanovskoe and Pobeda oil and gas fields. In this part of the shelf there are 35 uplifts - potential traps for hydrocarbon reservoirs, Neocomian clinoform complex is developed, with more than 80% of oil fields in Western Siberia, and the oil and gas content of the Upper Jurassic is proved.

The analysis of time cross-sections and the results of structural constructions allows to draw a conclusion that from the seismostratigraphic and structural-tectonic positions the North Kara basin is also perspective in terms of oil and gas potential. Here, anticline, structural-tectonic, structural-stratigraphic, structural-lithological, tectonically screened traps are identified in sedimentary complexes. According to a number of researchers, salt-dome structures can also be developed in the Ordovician-Silurian complex of the North Kara POGP [4].

Acknowledgements

The reported study was funded by RFBR according to the research project № 18-05-70105 (Arctic resources).

References

[1] Suprunenko O I, Ustrickij V I and Zujkov O N 2009 Geologic-geophysical zoning of northern Barents-Cara shelf by seismic prospecting data *Geology of oil and gas* **4** 17-25

[2] Dolgunov K A, Martirosyan V N, Vasil'eva E A and Sapozhnikov B G 2011 Structural and tectonic peculiarities of structure and prospects of oil and gas potential of the northern part of Barents-Kara region *Geology of oil and gas* **6** 70-83

[3] Stupakova A V 2011 Structure and petroleum potential of the Barents-Kara shelf and adjacent territories *Geology of oil and gas* **6** 99-115

[4] Martirosyan V N, Vasil'eva E A, Ustrickij V I, Suprunenko O I and Vinokurov I YU 2011 The North of Kara Sea – highly promising area for oil and gas of the Acrtic shelf of Russia *Geology oil and gas* **6** 59-69

IOP Conf. Series: Earth and Environmental Science **193** (2018) 012033 doi:10.1088/1755-1315/193/1/012033

Seismogeological characteristics and oil and gas potential of the Arctic regions of the Siberian platform and the Laptev Sea

V A Kontorovich[1,2], I A Gubin[1], A Yu Kalinin[1,2], L M Kalinina[1,2], M V Solovev[1,2] and E S Surikova[1,2]

[1]Trofimuk Institute of Petroleum Geology and Geophysics, Novosibirsk, Russia
[2]Novosibirsk State University, Novosibirsk, Russia

Abstract. During the research seismogeological models of the Neoproterozoic-Phanerozoic sedimentary complexes of the Siberian Platform (Anabar-Khatanga and Lena-Anabar oil and gas regions) northern part were constructed. It is concluded that in the Anabar-Khatanga and Lena-Anabar oil-and-gas areas Permian anticlinal traps complexly related to the salt diapirs have the best prospects of petroleum production. The carbonate deposits of Riphean-Carbon are also of interest. The existing geological and geophysical materials do not allow precisely stratify the Laptev Sea basin sedimentary cover. The thickness of platform deposits in which reaches 14-15 kilometers. From the structural-tectonic and general-scientific positions, this sedimentary basin is located in favorable conditions for the formation of hydrocarbon fields.

1. Introduction

According to modern schemes of oil and gas zoning, the Arctic regions of the Siberian platform and the Laptev Sea shelf cover the Anabar-Khatanga and Lena-Anabar oil and gas areas (OGA) of the Lena-Tunguska oil and gas province (OGP) and the Laptev Sea prospective oil and gas province (POGP).

At present, 16850 km of seismic CDP profiles have been worked out in this territory and deep wells have been drilled in the continental part, the complex analysis of which allows to build regional models of the geological structure on a new level and to specify oil and gas potential of this region.

2. Geological structure and oil and gas content of the Anabar-Khatanga and Lena-Anabar OGA

The Anabar-Khatanga and Lena-Anabar OGA, tectonically located within the Anabar-Khatanga saddle and the Leno-Anabar regional depression, are in the extreme north of the Siberian Platform and administratively located in the north-east of the Krasnoyarsk Territory and the north-west of the Sakha Republic (Yakutia), on the coast of the Laptev Sea. The analysis of geological and geophysical data allows to draw a conclusion that a Neoproterozoic-Mesozoic sedimentary basin has developed in this territory, within which the thickness of platform sediments reaches 14-16 km [1-3].

2.1. Seismogeological characteristics.

The analysis of seismic cross-sections and drilling data allows to identify 5 regional-developed seismogeological megacomplexes in the sedimentary cover of this region: Riphean, Vendian, Paleozoic, Permian and Mesozoic, which are controlled by seismic marker horizons in the top and bottom [4] (figure 1).

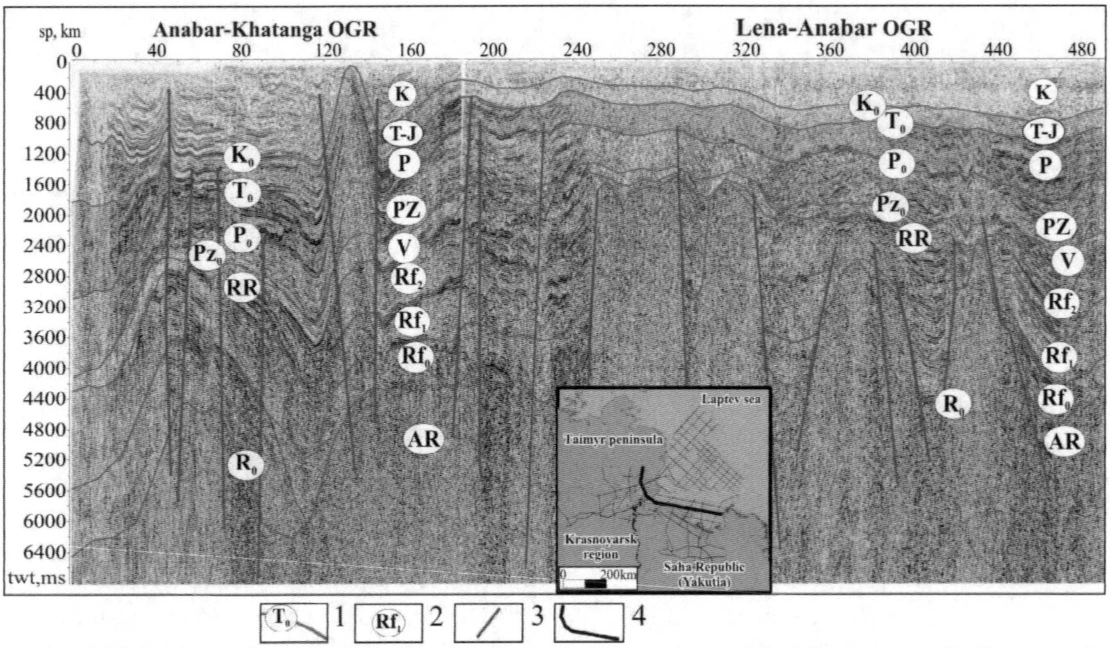

Figure 1. Seismic section showing seismogeological characteristic of the Neoproterozoic-Phanerozoic deposits of the Siberian platform Arctic regions. Legend:1 - reflector, 2 - seismic complex, 3 – faults, 4 – seismic line.

Mesozoic seismogeological megacomplex. It lies in the upper part of the sedimentary cover and includes terrigenous deposits of the Triassic, Jurassic and Cretaceous. At the time cross-sectionss, the Mesozoic seismocomplex in the bottom is controlled by the reflecting horizon T_0, in the top by the surface. In the north of the Siberian platform, the thickness of the Mesozoic region decreases regionally from 3000 m to 0 m in the south. Reducing the thickness of these deposits also occurs over the Beregovaya ridge, the Solanocupole mezoswell and the contrasting anticlinal structures.

In the north of the Lena-Anabar depression and in the Anabar-Khatanga saddle, in areas where the thickness of the Mesozoic is more than 1000 m, a reflecting horizon K_0, characterized by a high energy level, is identified in this megacomplex, associated with the top of Jurassic.

Permian seismogeological megacomplex. It is controlled by the reflecting horizons T_0 at the top and P_0 at the bottom. The reflector P_0 is identified practically on the entire studied territory and is absent only in the southeast, where the Cambrian emerges on the surface. The reflecting horizon P_0 controlling the bottom of the Permian complex, is at the same time associated with a large pre-Permian (early Permian) sedimentation break. In the Lena-Anabar OGA, this break is clearly fixed on time cross-sections and in various parts of this region the underlying Riphean-Cambrian reflecting horizons are cut off by this erosion surface.

In the Permian megacomplex two stable reflection horizons are identified, associated with the tops of the Lower Kozhevnikov and the Tustakh formations. The thickness of Permian deposits also regionally increases in the north direction; the maximum thicknesses, reaching 2500-3000 m, are fixed in the northern part of the Khatanga Bay.

Paleozoic seismogeological megacomplex. It is controlled at the top by reflector P_0, the reflection horizons PZ_0 in the Vendian development zone and RR in the areas of its absence is associated with the bottom of the complex. The reflecting horizons PZ_0 and RR, associated, respectively, with the tops of the Vendian and Riphean, are characterized by a high energy level and reliably traceable on time cross-sections.

In the Lena-Anabar and Anabar-Khatanga OGA, the volume of the Lower-Middle Paleozoic deposits is significantly different.

In the Lena-Anabar OGA, the Paleozoic megacomplex of rocks is represented mainly by Cambrian carbonates, which are overlapped with terrigenous Permian sediments with nonconformity; the deposits of Ordovician-Silurian and Devonian-Carboniferous in most of this area are eroded and are not present in the section. In the southeast of the Lena-Anabar OGA, on the seismic cross-sections in a Paleozoic megacomplex it is identified two stable reflecting horizons, K_1 and K_2, which are formed at the geological boundaries, associated respectively with the basal and upper parts of the Cambrian deposits. In the south-west direction, to the Anabar dome, the thickness of the Cambrian deposits is first significantly reduced, and then the upper part of the Cambrian and the horizon K_1 are cut off by the horizon P_0, associated with the pre-Permian regional sedimentation break. On the seismic cross-sections of this region in the Paleozoic seismocomplex also the cross-bedded (clinoform) pattern of the seismic record is clearly fixed (figure 1). Reflecting horizon K_2, which is associated with the geological boundary lying near the Cambrian top, in the southern direction is submerged first to the middle, and then to the bottom of the complex, forming the clinoforms of the southern fall.

In the Anabar-Khatanga OGA located in the west of the study area, at the base of the Paleozoic megacomplex lies the Cambrian mainly carbonate deposits, which, with a break, is covered with limestones of the Devonian-Carboniferous, or terrigenous Permian deposits, which are characterized by a series of subparallel reflective horizons at the time cross-sections. A feature of the Paleozoic deposits geological structure of the Anabar-Khatanga saddle is the presence in the Early Middle Devonian salt layer and the associated series of salt domes, which largely determined the structural-tectonic construction of this region [1,4]. Located on this territory, the Nordvik and Kozhevnikov diapires break through the upper Devonian-Carboniferous, Permian and Mesozoic deposits and come to a surface. The analysis of geological and geophysical data allows to consider that the most of local high-amplitude positive structures identified in the Anabar-Khatanga OGA are associated with the coming to a surface and buried salt diapires which on seismic cross-sections are characterized by a columnar outphasing and decreasing of the wave fields amplitude-energy characteristics – a chaotic seismic pattern recording (see figure 1).

The Ordovician-Silurian deposits in the Anabar-Khatanga OGA are also eroded and absent in the section.

Vendian seismogeological megacomplex. It is controlled at the seismic cross-sections by the reflecting horizon PZ_0 at the top and by the reflector RR at the bottom. The megamplex is characterized by a two-membered structure. The lower part of the Vendian is represented by carbonates and has a thickness of 650-700 m, the upper part is composed of sandstones, siltstones and argillites of the Kessyusin formation. Lying at the top of Vendian-the lower parts of Cambrian, the terrigenous deposits of the Kessyusin formation is enriched with organic matter and is considered as an oil producing one.

The fundamental difference between the Vendian megacomplex and the overlying deposits is that it is not cut off by the erosion surface, but is wedged out to the elevated blocks of the Riphean.

Riphean seismogeological megacomplex. It is controlled at the bottom by the reflection horizon R_0 and at the top by the reflector RR. Associated to the base of the Riphean megacomplex, the reflecting horizon R_0 separates at the seismic cross-sections high-amplitude Riphean wave fields, emphasizing the platform shape of the sediments, from the rocks of the Archaean-Proterozoic basement characterized by a chaotic seismic record.

In the most complete sections of the Riphean, developed in the southeast of the Lena-Anabar depressionh and in the Anabar-Khatanga saddle, the thickness of the Riphean megacomplex reaches 8 km, and in it are identified from four to five conformable seismogeological complexes for which tops the energetically expressed reflecting seismic horizons are confined. In the southern direction, towards the Anabar dome, the Riphean reflecting horizons regionally rise and are successively cut off by the reflecting horizon P_0, confined to the pre-Permian sedimentation break. At the same time within the most contrasting basement highs the thickness of the "cut off" Riphean deposits is 3-4 km.

2.2. Regional structural-tectonic characteristics.

Within the studied region, the Neoproterozoic-Paleozoic reflecting horizons regionally plunge to the north. The lowest depths are fixed in the southern part of the study area, within the North-Siberian megamonoclysis and in the northeast, in the Pritaymyr zone; the largest - in the most submerged parts of major depressions - the Yenisei-Khatanga regional depression, the South Laptev and Lena-Anabar syneclises (figure 2).

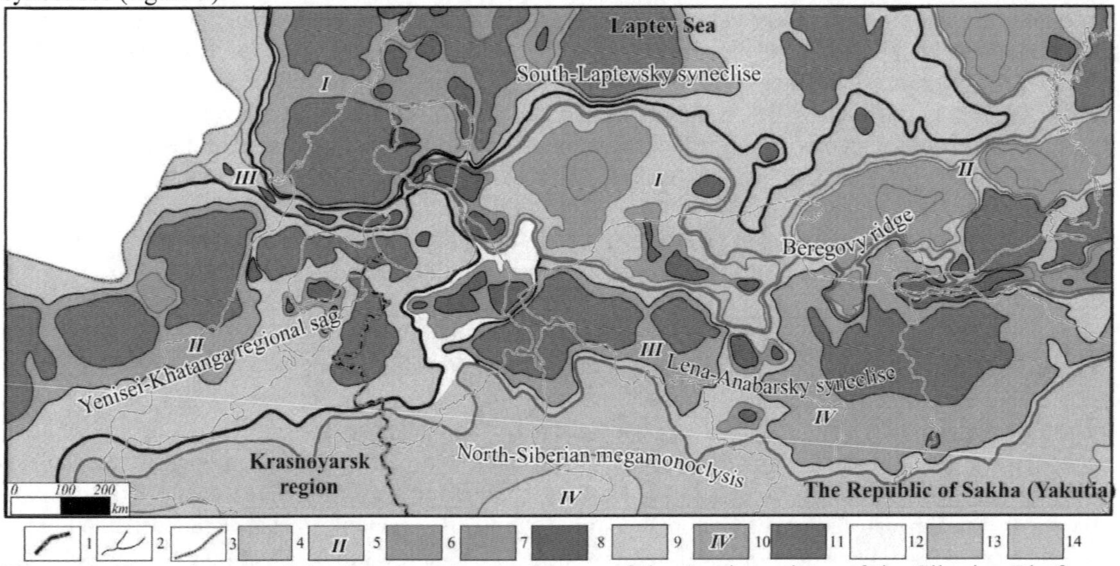

Figure 2. Tectonic map of the Permian structural stage of the Arctic regions of the Siberian Platform (Lena-Anabar and Anabaro-Khatanga OGA). Legend: 1 - administrative boundaries, 2-coastal line, rivers, 3 - zone of absence of platform sediments.Tectonic elements: positive structures: 4 - superorder, 5 - I rank (: I - Yuryangsky megaswell; II - Pogranishny megaswell; III - Salt-dome mezaswell; IV - Druokhansky megaswell), 6 - II rank, 7 - III rank, 8 - IV rank, negative: 9 - superorder, 10 - I rank (I - Begichevsky megadepression; II - Hetsky megasag; III - Yuryang-Khayansky megasag; IV - Taymylarsky megasag), 11 - II-III rank; intermediate structures: 12 - saddles, 13 - monoclases, 14 – monoclines.

In the northern part of the study area the South Laptev syneclise is located, in the western part of which the Begichev megadepression is allocated, separated from the Yenisei-Khatanga regional depression by the Solanocupole mesoswell.

To the west of the Solanocupole mesoswell there is a large superorder positive structure, stretched latitudinal direction – the Beregovaya ridge. The ridge is complicated by a series of positive structures of I-II orders and extends along the entire coast from the Khatanga Bay to the estuary of the Lena River, separating the South Laptev syneclise from the Anabar-Lena syneclise.

Controlled by the Beregovaya ridge in the north and the North-Siberian monocline in the south, the Lena-Anabar syneclise is a large depression, stretched out in the latitudinal direction, complicated by the Yuryang-Khayan and Taimylyr megadepressions.

The most of the III-IV orders uplifts of the Lena-Anabar and Anabar-Khatanga, which are traditional anticlinal oil and gas perspective objects, are located within larger positive structures, or in saddles that separate the 0-II orders depressions. For example, the Solanocupole mesoswell is complicated by the Nordvick and West Nordvick local uplifts, in the southern part of the Beregovaya ridge, the Ulakhan-Yuryakh and Ust-Olenek uplifts are identified, the group of Kozhevnikov structures and the Chidakh uplift are identified in the junction zone of the Yenisei-Khatanga regional depression and the Lena-Anabar syneclise.

The analysis of geological and geophysical data also allows us to conclude that the dominant influence of the modern structural and tectonic structure of the northern areas of the Siberian Platform has had three fundamentally important tectonic processes:

1. The pre-Permian (early Permian) sedimentation break, as a result of which the paleorelief of the studied territory was leveled and a poorly dislocated surface - peneplain was formed. Given that in the Anabar-Khatanga OGA under the terrigenous sediments of Perm there is the Devonian-Carboniferous, and in the Lena-Anabar OGA – Cambrian, it can be assumed that to the beginning of the pre-Permian sedimentation break the relief of the studied territory was plunging regionally in the west direction, towards the Yenisei-Khatanga regional depression.
2. A salt tectogenesis, as a result of which in the Anabar-Khatanga OGA have been created coming to the surface and buried salt diapires that formed in the structural plans of the Upper Devonian, Carboniferous, Permian and Mesozoic anticlinal structures.
3. The intensive young Late Cretaceous and Cenozoic tectonic movements, which have predetermined the differentiation of the relief and the formation of positive and negative structures of various orders.

These tectonic processes have predetermined the similarity of structural plans and the modern relief of the marker stratigraphic levels in the lower and upper parts of the sedimentary cover of the Lena-Anabar and Anabar-Khatanga OGA.

2.3. Oil and gas prospects.

The main prospects for the oil and gas potential of the Anabar-Khatanga OGA are associated with Permian terrigenous deposits, to a lesser extent with the carbonates of the Devonian-Carboniferous and Cambrian. In the Perm rock complex in the Lower Kozhevnikov and the Tustakh formations, clayey packs enriched with organic matter are capable of generating significant volumes of hydrocarbons are identified.

In the Anabar-Khatanga region, the most interesting in relation to oil and gas are the anticlinal structures associated with the salt domes. Near the known diapires intersecting the entire of the Upper Devonian, Carboniferous and Permian deposits, structural-tectonic traps associated with the wedging out of terrigenous and carbonate reservoirs onto salt stocks can form; over the subsurface salt domes, in the Permian and Mesozoic deposits classical anticlinal traps can be formed. The analysis of the seismic cross-sections also allows to identify in the Cambrian, Devonian and Carboniferous carbonates the wave fields characteristic of organogenic structures, which may also be of interest in the oil and gas potential.

The deposits of Perm, Cambrian, Vendian and Riphean represent the interest in the Lena-Anabar OGA in relation to oil and gas.

Oil and gas perspective objects in the Permian sediments can be associated with anticlinal structures located on the coast of the Laptev Sea and in the saddle separating the Yuryang-Khayan and Taimylyr megadepressions.

The Cambrian and Vendian are perspective, mainly in the southern part of the Lena-Anabar OGA, where hydrocarbon reservoirs can be controlled by stratigraphic, structural-lithological and structural-tectonic traps in the zones wedging, cutting and lithologic substitution of reservoir rocks. The erosion-tectonic ridges of Riphean located in the depths of 1400-1500 m and overlapped with Permian deposits are also of interest in the oil and gas content in the southern part of the Lena-Anabar OGA.

3. Geological structure and oil and gas potential of the Laptev Sea POGP.

3.1. Seismogeological characteristics.

The analysis of time cross-sections indicates that a thick mass of normally sedimentary platform deposits has developed in the Laptev Sea POGP, within which it is possible to identify 6 seismogeological complexes controlled by reliably traced reflecting seismic horizons – seismic markers indexed (bottom-up) S_1-S_5. At the base of the platform sediments, a reflecting horizon F is identified, which is associated with the base of the platform deposits – the top of the basement (figure 3).

At present, the views on the geological structure of the Laptev Sea area, in which no wells have been drilled, differ cardinally. Many researchers consider that the western and central parts of the Laptev Sea shelf are a continuation of the ancient Siberian platform, and both ancient and young sedimentary complexes from the Riphean to the Cenozoic are developed on this territory [5,6].

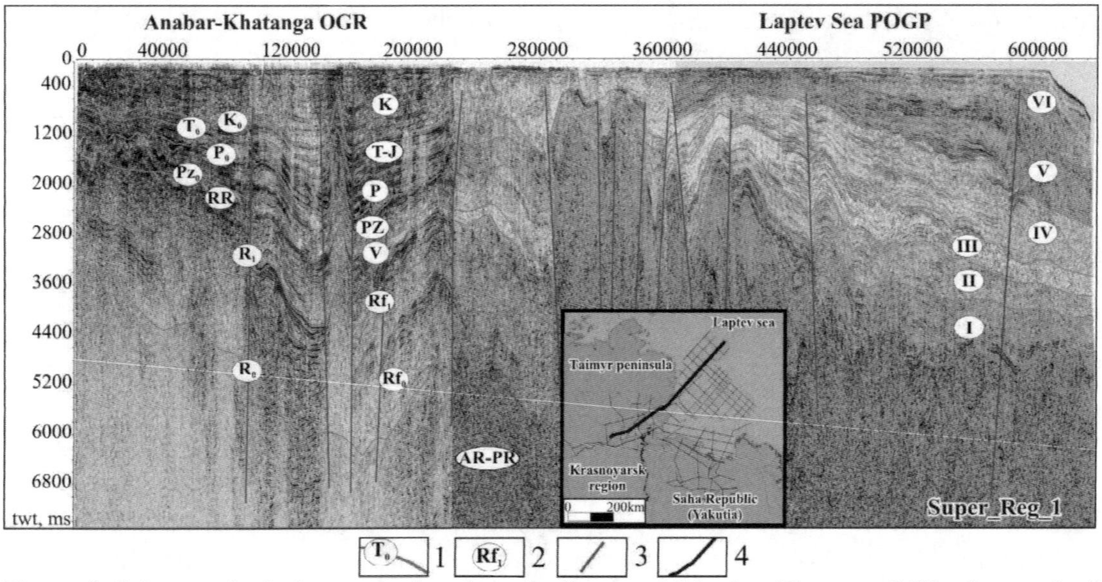

Figure 3. Seismogeological cross-section along the line reg_1 "Anabar-Khatanga OGP – Laptev Sea". Legend: 1 - reflector, 2 - seismic complex, 3 – faults, 4 – seismic line.

Another group of experts adheres to a fundamentally different point of view and believes that in the water area the sedimentary cover is composed of Upper Cretaceous-Cenozoic terrigenous deposits lying on the Early Cimmerian basement [7,8].

The analysis of available seismic materials does not allow answering this question. The analysis of wave fields allows us to note that the Anabar-Khatanga and Laptev Sea basins are separated by a contrasting basement high, characterized by a chaotic wave field character that does not allow stratifying seismic complexes identified in sea cross-sections.

3.2. Structural-tectonic characteristics.

Within the researches, structural constructions were made for all marker seismic horizons. The analysis of the structural map on the reflecting horizon F associated with a bottom of platform deposits indicates that in the studied part of the Laptev Sea basin the thickness of platform deposits varies in the range of 1500 up to 16000 m. The minimum thickness of the platform sediments is fixed in the western part of the basin, in the Pritaymyr region, where a large positive structure extended in the northwestern direction is identified – a ridge complicated by a series of uplifts of I-III orders. A large positive structure, unfolding in the northeastern direction, is also identified in the southeastern part of the basin. The submerged part of the Laptev Sea basin, within which the Central-Laptev megadepression is allocated, is also complicated by a series of large closed positive structures. At the regional level, the reliefs of the overlying reflecting horizons have a similar structure.

The subparallel occurrence of reflecting horizons, the similarity of structural surfaces, and minor change of seismic complexes thicknesses make it possible to conclude that the modern relief of the Laptev Sea POGP is formed due to intensive young tectonic movements. The analysis of seismic materials also allows to note that these processes were followed by the formation of faults, most of which intersect almost the entire platform deposits.

3.3. Oil and gas prospects.
From the structural-tectonic and general geological positions, the Laptev Sea POGP is in favorable conditions for the formation of hydrocarbon reservoirs. In this basin:

- a large sedimentary cover is allocated, the thickness of the platform sediments reaches 15-16 km;
- modern reliefs of the sedimentary complexes are largely differentiated and they contain large, deep depressions that can be associated with oil and gas formation zones and large positive structures located on their periphery – oil and gas accumulation zones;
- large positive and negative structures are complicated by closed uplifts of II-III orders – potential anticlinal traps for oil and gas reservoirs;
- the sedimentary cover is broken by a large number of faults that intersect almost the entire platform deposits.

4. Conclusion
Within the researches on the basis of complex interpretation of seismic data, well-logging and drilling results regional models of the Neoproterozoic-Phanerozoic deposits geological structure of the Arctic regions of the Siberian Platform and the Laptev Sea were constructed.

The analysis of the results has shown:

1. In the Anabar-Khatanga and Lena-Anabar OGA, the structural surfaces of various sedimentary complexes are largely similar. The main influence on the modern structure of these regions was rendered by: the pre-Permian sedimentation break, as a result of which the relief of the territory has been leveled; processes of salt tectogenesis, which predetermined the formation of salt domes and associated anticline structures and young Upper Cretaceous and Cenozoic tectonic movements, which predetermined the modern differentiation of the reliefs of various sedimentary complexes. Prospects for the oil and gas potential of the Anabar-Khatanga OGA are related to the Permian terrigenous deposits and Cambrian-Devonian carbonates, and the platform deposits of Riphean and Vendian are also of interest in the Lena-Anabar OGA.
2. Modern geological and geophysical data do not allow to stratify seismogeological complexes in the Laptev Sea shelf. At the same time, from the structural-tectonic and general geological positions of the Laptev Sea basin is in favorable conditions for the discovery of hydrocarbon reservoirs. There is a large sedimentary basin, which consists of six seismogeological complexes; the relief of the territory is largely differentiated and in the structural plans of the marker reflecting horizons large positive and negative structures of 0-I orders are identified - oil and gas formation zones and oil and gas accumulation zones.

Acknowledgments
The reported study was funded by RFBR according to the research project № 18-05-70105.

References
[1] Kalinko M K 1959 *History of geological development and oil and gas potential prospects of the Khatanga depression* (Leningrad) p 360
[2] Kontorovich V A, Kontorovich A E, Gubin I A, Zoteev A M, Lapkovsky V V, Malyshev N A, Soloviev M V and Fradkin G S 2013 Structural-tectonic characteristics and model of the geological structure of the Neoproterozoic-Phanerozoic deposits of the Anabar-Lena zone *Geology and Geophysics* **54**(8) 1253-74
[3] Stupakova A V, Bordunov S I, Sautkin R S, Suslova A A, Perotolchin K O and Sidorenko S A 2013 Oil and gas basins of the Russian Arctic *Geology of oil and gas* **3** 30-47
[4] Kosygin Yu A 1973 *Salt tectonics of the Siberian platform* (Novosibirsk: Science) p 162
[5] Vinogradov V A, Gaponenko G I, Gramberg I S and Shimaraev V N 1976 Structural-formation

complexes of the Arctic shelf of Eastern Siberia *Soviet Geology* **9** 23-38
[6] Kim B I, Evdokimova N K and Kharitonova L Ya 2012 Sedimentary cover of the Laptev Sea and its oil and gas potential *Geology of oil and gas* **6** 116-32
[7] Daragan-Suhsova L A, Petrov O V, Daragan-Suhsov Yu I and Rukavishnikova D D 2010 A new look at the geological structure of the sedimentary cover of the Laptev Sea *Regional geology and metallogeny* **41** 5-16
[8] Khoroshilova . M A, Franke D, Kirillov T, Muli B and Nikishin A M 2014 Detection and correlation of marker seismic horizons in the Laptev Sea basin *MSU Bulletin* **5** 3-12

IOP Conf. Series: Earth and Environmental Science **193** (2018) 012034 doi:10.1088/1755-1315/193/1/012034

Seismogeological criteria for gas deposits forecasting in the north of Western Siberia

V A Kontorovich[1,2], E S Surikova[1,2] and D V Ayunova[1]

[1]Trofimuk Institute of Petroleum Geology and Geophysics, Novosibirsk, Russia
[2]Novosibirsk State University, Novosibirsk, Russia

Abstract. Seismogeological criteria for forecasting the gas content of the Aptian-Cenomanian sediments of Western Siberia are identified and substantiated in the study based on a comprehensive interpretation of seismic data, well log data, test results and petrophysical studies results. As a reference object we use the Jubilee oil and gas condensate field, which is located in the Nadym-Pur interfluve. The field contain the unique gas deposit in the Cenomanian reservoir horizon PK_1, the Albian sand layer PK_{18} is productive too. Based on the results of the research, seismogeological criteria for the gas-bearing forecast are formulated. If there is a giant gas deposit in Cenomanian reservoir: 1) there is a reflectors formed on gas-water contacts (GWC) in seismic cross sections, amplitude characteristics of the reflector G, which is confined to the top of the Cenomanian complex, decrease; 2) time thickness (ΔT) increases, the interval velocities (Vint) decrease and amplitude-energy characteristics of the seismic record in the Aptian-Cenomanian complex decrease. The Aptian-Albian sheet gas reservoirs are displayed in seismic cross sections a jump in amplitudes of the seismic record, forming the "bright spot" anomalies in the time sections.

1. Introduction

The northern and arctic regions of Western Siberia, including the Yamal-Nenets Autonomous District and Kara Sea southern part, are one of the world's largest gas-bearing region. In terms of oil and gas zoning within this territory five oil and gas bearing areas (OGA) have been allocated: Nadym-Pur, Pur-Taz, Yamal, Gydan and South- Kara (figure 1).

The main gas reserves in the north of the West Siberian oil and gas province are controlled by high-amplitude anticline structures and concentrated in the Aptian-Cenomanian sedimentary megacomplex [1-4]. In the Aptian-Cenomanian megacomplex two regionally gas-bearing macro reservoirs are distinguished: Cenomanian and Aptian, which are covered by transgressive clay beds that act as reservoir cap rock for hydrocarbon deposits.

In the southern part of the Yamal-Nenets Autonomous District, Nadym-Taz interfluve, there are such unique gas giant fields as Medvezhye, Urengoy, Yamburg, Yamsovey and others. In this area there are no regionally developed reservoir cap rocks within the Aptian-Cenomanian complex, and in this region fields the main gas reserves are concentrated in Cenomanian sand body PK_1, which is covered by a thick clay Turonian regional reservoir cap rock (Kuznetsov formation). The Cenomanian deposits are massive and they are controlled by large, high-amplitude anticline structures. Substantially smaller gas, condensate and oil deposits are concentrated in the Aptian-Albian, Neocomian and Middle-Upper Jurassic sand beds.

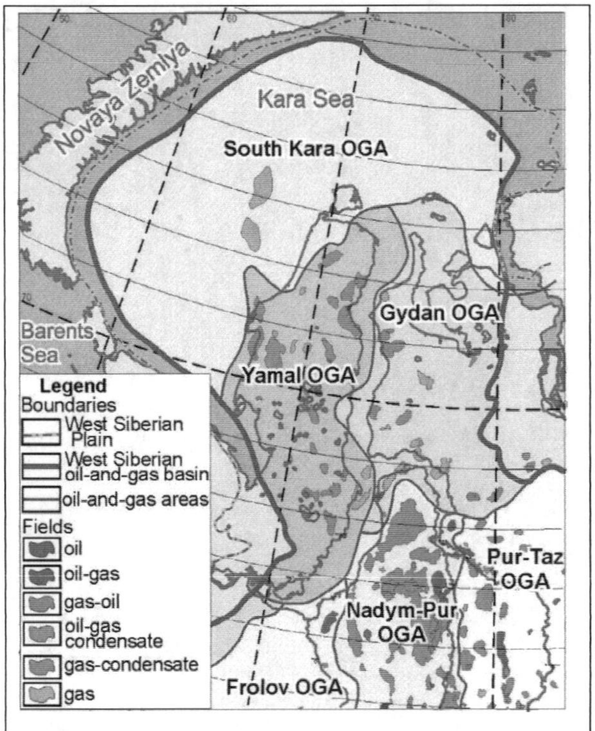

Figure 1. Northern part of the West Siberian oil-and-gas basin

In Far North regions of West Siberia and in the southern part of the Kara Sea, in the section of the Aptian-Cenomanian megacomplex there is a thick clay Khanty-Mansiisk (Yarongsky) reservoir cap rock lying in the lower part of Albian sediments. As a result, in the fields of the Yamal, Gydan and South Kara OGA, there are traditional Cenomanian deposits, but the main gas reserves are concentrated in the Aptian sandstones of the TP group that underlay the Khanty-Mansiisk reservoir cap rock. This situation takes place in Kharasavey, Bovanenkovo, South-Tambey, Nurminskoye, Arctic, Srednyamalsk and other fields. As a rule, Aptian gas deposits are multi-layered, and are localized in a series of closely located, hydrodynamically unconnected sand beds lying in the upper part of the Tanopchin formation; the type is Layer-uplifted deposit.

Cenomanian and Aptian deposits differ not only in structure, but also in gas composition. The Cenomanian deposits contain only dry gas - methane, whose density by air is 0.554; Aptian deposits - rich gas, the density of which is 1.03 - 2.97.

2. Seismogeological criteria for gas deposits forecasting in the Cenomanian sediments

In the north of West Siberia, the Aptian-Cenomanian megacomplex in seismic time cross-sections is limited by the reflectors M and G, which are associated to the top of the Neitin clay unit (top of Barremian) and the bottom of the Kuznetsov formation (top of the Cenomanian) , respectively [5-7]. The Neocomian (Berriasian-Lower-Aptian) sedimentary complex underlies Aptian-Cenomanian complex, it`s controlled by the horizon M in the bottom and the reflector B in top, which forms on the top of the Jurassic sediments.

2.1. Physical nature of seismic level reflector confined to «GWC»

In this region, gas-water contacts (GWC) can be frequently identified in seismic cross-sections on large positive structures, to which the unique Cenomanian deposits are confined, because an intense reflected wave forms on the contact of natural gas and water-saturated sandstones of PK_1 (figure 2). On such objects, there are locally developed seismic reflectors under anticline structures in the reflector G relief, which merge with G horizon in the direction of the positive structures slopes. The reflectors associated with the GWC are usually traced quasi-horizontally or have a convex downward form.

Reflected seismic waves are formed on the boundaries of environments characterized by different physical properties. To form reflected waves, such characteristics of rocks as the density and velocity model for the sediments are determinative. The product of these parameters is the acoustic impedance and the differences in the acoustic impedance at the geological boundaries define the reflection coefficients and, as a consequence, the energy level of the seismic horizons formed on the boundaries.

The cross-section of the Cenomanian complex and the productive horizon PK_1 (as a part of the complex) is represented by interbedding of siltstones, argillites and sandstones with a sandstones predominance. The sand layers of the sand body PK_1 are good reservoir rock which porosity coefficient

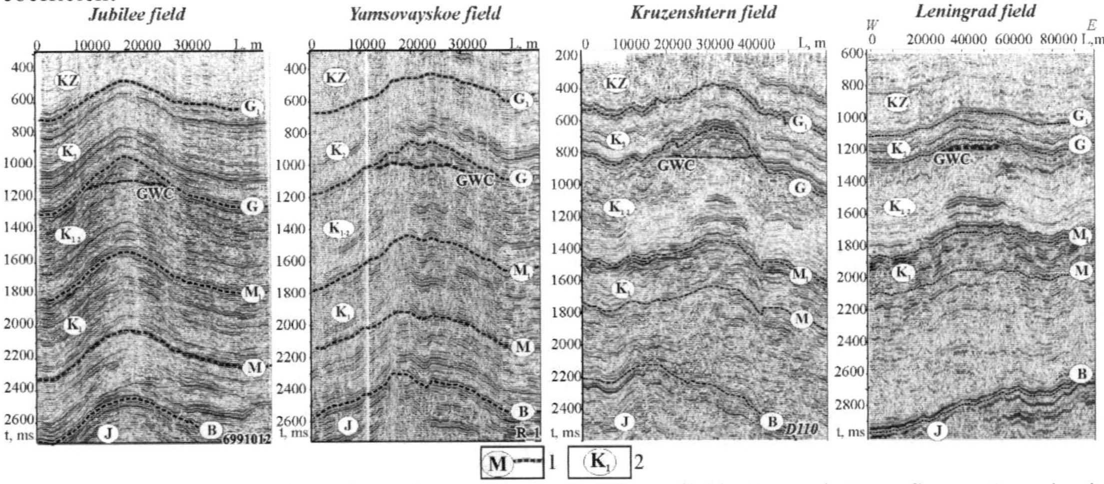

Figure 2. Time cross-sections along the profiles, crossed gas fields. Legend: 1 – reflector, 2 - seismic complex.

is about 30-35%, but separating them silty clay beds, on the one hand, are characterized by low filtration and capacitance properties, on the other - are not effective impermeable rock. As a result, the Cenomanian gas reservoirs are massive and have bottom water-drive.

Analysis of acoustic logging data for hydrocarbon fields in the north of Western Siberia showed that the Cenomanian section ,in general, is poorly acoustically differentiated - the p-wave velocity differences at the boundaries between different lithological rocks do not exceed 100-150 m/s , as a rule. At the same time, the presence of a reflector at the GWC level indicates that at this boundary there is a drop in the acoustic impedance. The reflector forming on the GWC is not isochronous and "cut" the different-age layers within the structure, which are parts of a single productive horizon.

It is known from published materials that in terrigenous rocks at depths of up to 1-1.5 km, p-wave velocities in water- and gas-saturated reservoirs differ by 15-25% [8]. The analysis of the results of tests and well-log data for a number of fields in the north of West Siberia which was carried out at IPGG of the SB RAS showed that the fluid in the sand beds of the TP, XM and PK bodies, separated in the Aptian-Cenomanian section, significantly affects the acoustic characteristics of rocks: the p-wave velocity in gas-saturated reservoirs are 2100-3000 m / s, in water-saturated reservoirs - 2500-3800 m / s [9]. On the average, the velocity difference at the boundary between gas and water-saturated sandstones is of the order of 500-600 m / s. Densities behave in a similar way. The density of methane is 0.00072 g / cm3, Cenomanian water is 1.01-1.03 g / cm3.

The synchronous decrease in the velocities and densities of the gas-saturated part of the section leads to a sharp drop in the acoustic impedance at the GWC and is the reason for the formation of an intense reflected wave on this physical boundary.

It should be noted that not all Cenomanian gas deposits generate reflectors associated with GWC in time cross-sections. This may be due to both the quality of the seismic material and the limited seismic resolution.

2.2. Interval velocities

The gas-saturated intervals of the Cenomanian section are characterized by underspeed of p-waves. When reservoir altitude is sufficiently high this factor will lead to interval velocities (Vint) decrease and, as a consequence, time thickness (ΔT) increase of the entire Aptian-Cenomanian seismic megacomplex.

IOP Conf. Series: Earth and Environmental Science **193** (2018) 012034 doi:10.1088/1755-1315/193/1/012034

Figure 3 shows the dependences of ΔT on ΔH for the Turonian-Cenozoic and Aptian-Cenomanian megacomplexes constructed from the Nadym-Pur interfluv wells data. We can see the standard distribution for West Siberia in the Turonian-Cenozoic complex, which is located in the upper part of the sedimentary cover and does not contain oil&gas pools. Figure 3A shows the observation times of the reflector G (the depth of occurrence of the Kuznetsov formation) are increased in the direction of the depression zones, as the thickness of the megacomplex increases. In the Aptian-Cenomanian inverse dependence takes place caused by the velocities drop in the gas-saturated part of the section.

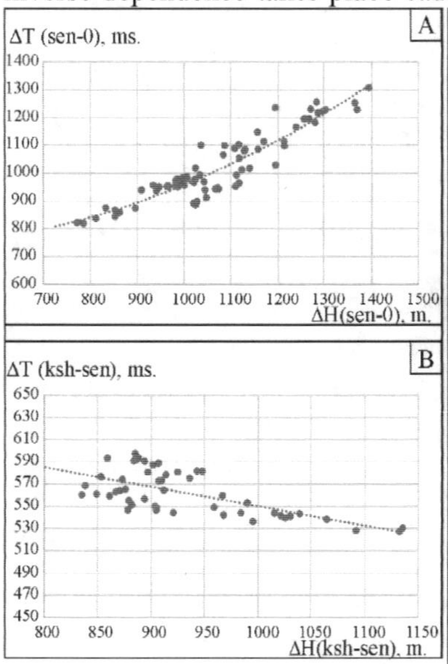

Figure 3. Dependence of time thickness (ΔT) on thickness (ΔH) for the Turonian-Cenozoic (A) and Aptian-Cenomanian (Б) megacomplexes (Nadym-Pur interfluv).

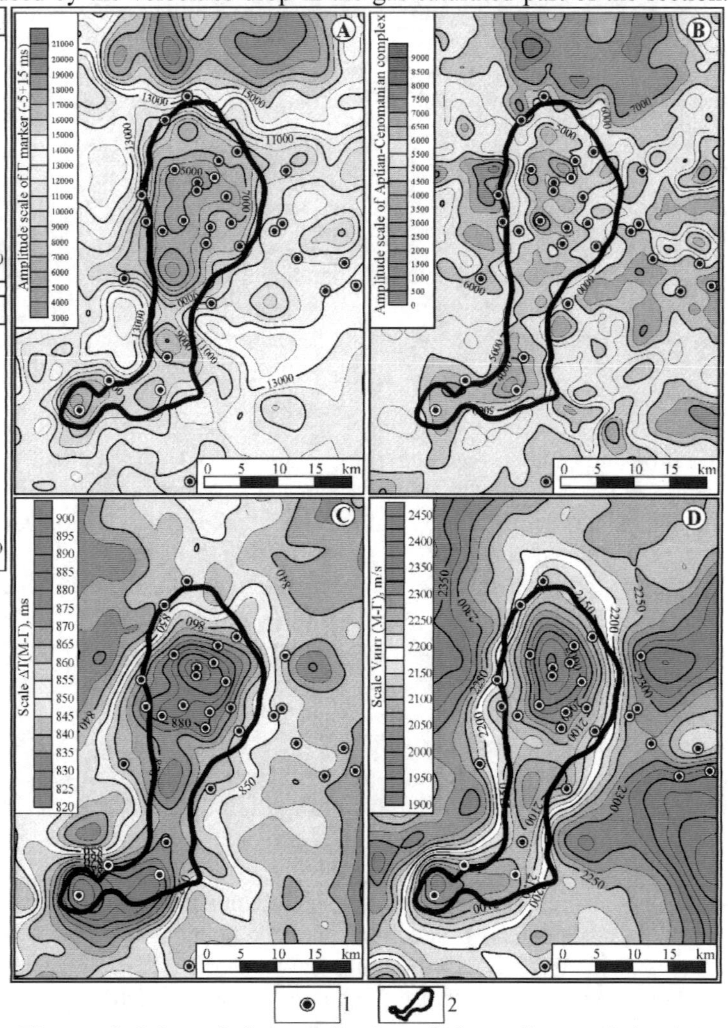

Figure 4. Map of the amplitudes of the reflector G in the interval -5 + 15 ms (A);map of modulus mean amplitudes of the seismic record of the Aptian-Cenomanian complex (B), the map of time thickness between reflectors M and G (C); map of the interval velocity in Albian-Cenomanian megacomplex (D) (Jubilee field). Legend: 1-wells, 2- Cenomanian deposit outline.

The effect of velocity drop is also reflected in paleoreconstructions. In seismic time paleosections flattened on reflector G local depressions in M and B reflectors paleoreliefs conform to the large anticlinal structures controlling huge Cenomanian gas deposits. Formation of these depressions is due the decrease of p-wave velocities in Cenomanian gas deposits, not to the geological structure of the objects (figure 3).

2.3. Amplitude attributes of the reflection record

1. Loss of densities and p-wave velocities leads to a decrease in acoustic impedance of the entire gas-saturated layer and, as a result, to a change in the reflection coefficients not only on its bottom but also on its top - at the boundary between the clays of Kuznetsov formation and the Cenomanian sandstones. The reflector G, which controls the top of the Cenomanian complex, is formed on the bottom of the Kuznetsov formation clays, p-wave velocity in which is about 2000 m/s. In the development zone of the hydrocarbon deposit at the "fluid-reservoir" boundary, the reflection coefficient and, as a consequence, the energy of the reflected wave, decrease since acoustic impedance decreases in the gas-saturated rocks.

2. As noted above the Cenomanian reservoirs filled with natural gas are characterized by reduced velocities and densities. At the same time, gas-saturated rocks have an increased absorptive capacity. Seismic waves passing through the gas-saturated part of the section, significantly lose energy, that leads to a drop in the amplitude waves characteristics within the entire Aptian-Cenomanian megacomplex and in the underlying Neocomian and Jurassic sediments.

The seismogeological criteria for gas deposits forecasting in the Cenomanian sediments were tested on seismic materials for a number of fields in the north of Western Siberia. As an example, in figure 4 the results of a comprehensive analysis of geological and geophysical materials for the Jubilee field, where a unique gas pool is concentrated in the Cenomanian reservoir, are given.

The results of these studies showed:

1. In the gas-saturated part of sand body PK_1, the interval velocity is 2200-2300 m/s, in the water-saturated part - 2800-2900 m/s; the difference in velocity characteristics of the section is about 500-600 m/s, that predetermines the formation of a reflector associated with the GWC at this boundary (figure 2).

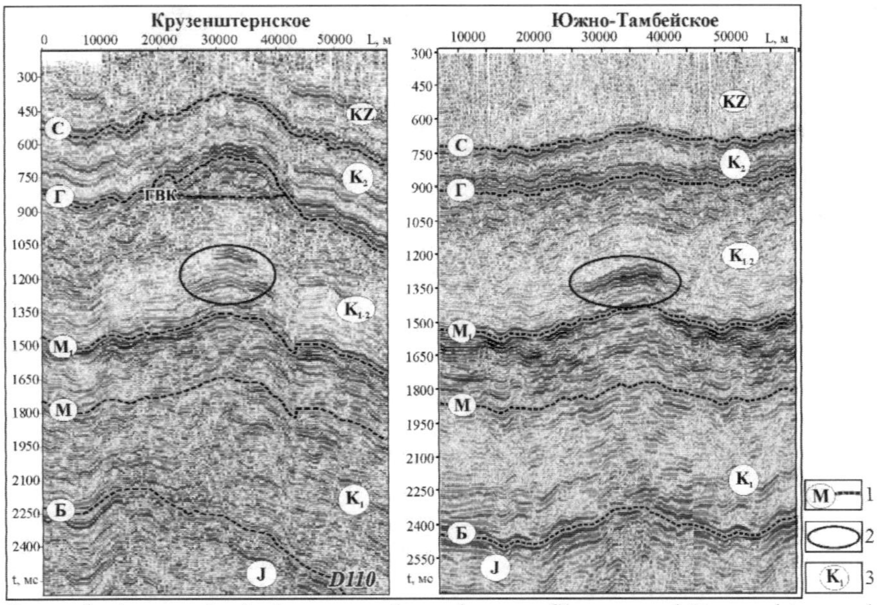

Figure 5. Parts of seismogeological cross-sections along profiles crossed Kruzenshtern and South-Tambey fields. Legend: 1 – reflector, 2 – "bright spot", 3 - seismic complex.

2. The contour of the Cenomanian gas reservoir is characterized by:

• lowered (less than 8000 cu) amplitudes of the reflector G (figure 4A) and low values of the modulus mean amplitudes of the seismic record (less than 5000 ms) inside the entire Aptian-Cenomanian megacomplex (figure 4B).

Increased (more than 860-870 ms) ΔT between the reflector M and G (figure 4C) and lower values (less than 2150-2200 m / s) of Vint (figure 4D).

3. Seismogeological criteria for gas deposits forecasting in the Aptian-Albian sediments

In principle, gas pools in Aptian-Albian sand layers, which are associated with the main gas reserves in the north of the West Siberian oil and gas province - Gydan, Yamal and South Kara OGA, affect the nature of the wave field [10]. In these regions, sand beds of the TP group, containing significant gas reserves, can be displayed in time cross-sections by jump in the amplitudes of the seismic record - a "bright spot" [6,7]. As an example, figure 5 shows the time sections crossing the Kruzenshtern and South-Tambeyskoye fields located in the Yamal OGA. In sections above the reflector M, confined to the Neitin clay unit, which lies in Aptian-Cenomanian megacomplex bottom part, there are local high-amplitude seismic records anomalies that characterize gas pools.

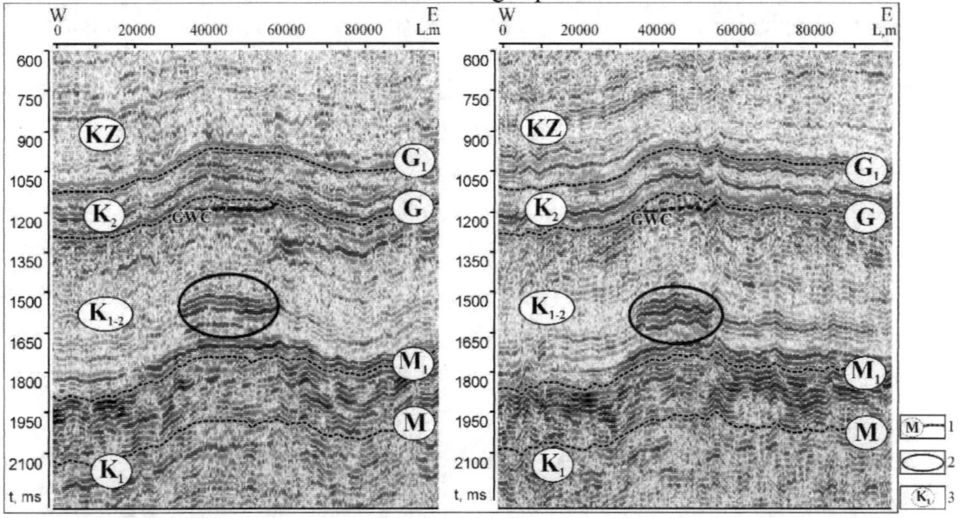

Figure 6. Seismogeological sections along Leningrad field profiles Legend: 1 – reflector, 2 – “bright spot”, 3 - seismic complex.

As noted earlier, in the West Siberian oil and gas province the Aptian-Cenomanian complex, composed of sandstones, siltstones and argillites, is characterized by weak acoustic differentiation. The Khanty-Mansiysk reservoir cap rocks developed in the northern part of the Yamal-Nenets Autonomous District and in the southern part of the Kara Sea, and the host rocks are characterized by the very close p-wave velocities, and reflector is not formed on it.

Under conditions of weak acoustic differentiation, only high- velocity thin calcareous sandstones and low-velocity gas-saturated sand beds have anomalous acoustic characteristics in this part of the section. Analysis of acoustic logging data showed that in productive wells where PK beds are gas-saturated, they are characterized by anomalously low p-wave velocities of the order of 2200-2400 m/s. In wells with water-saturated sandstones of PK beds, these strata do not differ in acoustic characteristics from overlying and underlying rocks, the p-wave velocity of which is 2800-3000 m/s. Thus, on the top and bottom of the Aptian-Albian gas-saturated sandstones, significant jumps of acoustic impedance take place. As a result high energy reflections form at these boundaries. The mathematical modeling results of wave fields show when the thickness of low-velocity gas-saturated sandstone exceeds 15-20 m, resonance summation of the waves from its top and bottom take place, which leads to an increase in energy of the interference signal and the effect "bright spot" generates. A similar effect is observed in the case of multi layered reservoir, when the formation of an interference wave happen on a series of closely located thin gas-saturated sandstone beds.

The anomalies of "bright spot" are reliably mapped over the area, it allows us to outline the contours of the gas pools from the interpretation of the seismic data. As an example, figure 6 shows two time cross-sections along profiles crossing the Leningrad field located in the Kara Sea, where seismic anomalies are not difficult to identify. Calculation of amplitude characteristics of seismic recording in the interval of productive layers allows contouring of gas pools. It should be borne in mind that in the case of multi layered reservoirы concentrated in close beds, this contour will be integral character.

Seismogeological criteria for forecasting gas deposits in the Aptian-Albian and Cenomanian reservoirs formulated in the paper are reliably confirmed by the results of two-dimensional mathematical modeling of wave fields.

4. Conclusion

The study aims to develop seismogeological criteria for the gas deposits forecasting in the Aptian-Cenomanian deposits in the north of Western Siberia. The research was carried out on the basis of a comprehensive scientific analysis of seismic data, well-log data, test results and lithologic-petrophysical studies.

As a result we determine a series of seismogeological criteria allowing to forecast of gas deposits in the Aptian-Cenomanian sediments of the north of Western Siberia.

If there is Cenomanian gas deposit, in wave seismic fields we can see:

• the presence in the relief of the reflector G anticlinal structures, in the base of which reflectors, associated with gas-water contacts, are identified in time sections;

• Increase in ΔT (time thickness) and decrease in interval p-wave velocity (Vint) in the Aptian-Albian-Cenomanian megacompex;

• the drop in amplitude characteristics the reflector G confined to the top of the Cenomanian reservoir , and the decrease in the amplitude-energy characteristics of the seismic record within the entire Aptian-Cenomanian megacomplex.

The Aptian-Albian bedded gas deposits are displayed in time cross-sections by a sharp increase in seismic record amplitudes and the seismic anomaly "bright spot" formation.

Acknowledgments

The reported study was funded by RFBR according to the research projects № 18-35-00492 and № 18-05-70105.

References

[1] Kontorovich A E, Nesterov I I, Salmanov F K, Surkov V S, Trofimuk A A and Erve Y G 1975 *Geology of oil and gas of Western Siberia* (Moscow: Nedra) p 679

[2] Ermilov O M, Karogodin Yu N, Kontorovich A E, Ter-Sahakyan, Yu G, Agalakov S E, Belyaev S Yu, Borisova L M, Bukreeva G F, Burstein L M, Gordeev V N, Dmitruk V V, Žilina I V, Kontorovich V A, Krasavchikov V O, Suprunenko O I, Chupova I M and Fursenko E A 2004 *Features of the geological structure and development of unique gas deposits of the far North of West Siberia* ed I I Nesterov (Novosibirsk: SB RAS) p 140

[3] Kazarinov V P 1963 Levels of petroleum potential in the Mesozoic sediments of West Siberia *Problems of Siberian oil* 46-57

[4] Nesterov I I, Salmanov F K, Shpilman K A 1971 *Oil and gas fields of Western Siberia* (Moscow: Nedra) p 463

[5] Kontorovich V A 2009 The Mesozoic-Cenozoic tectonics and petroleum potential of West Siberia *Geology and Geophysics* **50**(4) 461-74

[6] Kontorovich V A, Ajunova D V, Gubin I A, Kalinin A Yu, Kalinina L M, Kontorovich A E,

Malyshev N A, Skvortsov M B, Solov'ev M V and Surikova E S 2017 Tectonic evolution of the arctic areas and the water areas of the west Siberian petroliferous province *Geology and Geophysics* **58**(3-4) 423-44

[7] Kontorovich V A, Ajunova D V, Gubin I A, Yershov S V, Kalinin A Yu, Kalinina L M, Kanak M S., Solov'ev M V, Surikova E S and Shestakova N I 2016 Seismic stratigraphy formation history and gas potential of the Nadym-Pur interfluve area *Geology and Geophysics* **57**(8) 1583-95

[8] Bondarev V I 2007 *Exploration seismology* (Yekaterinburg: URSMU) p 690

[9] Gubin I A 2015 The influence of lithology and saturation on acoustic properties of the aptian-cenomanian sediments by the example of geophysical oil field *Proc. XI Int. scientific. Congr. Interexpo GEO-Siberia-2015* (13-25 April 2015, Novosibirsk,): Int. scientific. Conf. on "Subsurface managment. Mining. Directions and technologies of search, exploration and development of mineral deposits. Geoecology vol 1 (Novosibirsk: SSUGT) pp 33-7

[10] Kazarenkov V A, Ershov S V, Ryzhkova S V, Borisov E V, Ponomareva E V, Popova N I and Shaporina M N 2014 Geological structure and oil and gas potential of Jurassic and Cretaceous regional reservoirs in the Cara-Yamal region and prognosis of hydrocarbon distribution *Geology of oil and gas* **1** 27-49

IOP Conf. Series: Earth and Environmental Science **193** (2018) 012035 doi:10.1088/1755-1315/193/1/012035

Oil emulsions as medium of natural gas hydrate formation

V V Koryakina[1], I K Ivanova[1,2] and M E Semenov[1]

[1] Laboratory of Technogenic Gas Hydrates, Institute of Oil and Gas Problems, Siberian Division, Russian Academy of Sciences, Oktyabrskaya St. 1, Yakutsk, Russia, 677980

[2] Institute of Natural Sciences, Ammosov North-Eastern Federal University, Yakutsk, Russia

E-mail: KoryakinaVV@ipng.ysn.ru

Abstract. Results of research into kinetics and a mechanism in water-in-oil emulsions with various contents of the water phase have been given. The study was undertaken with application of the differential scanning calorimetry technique and statistical analysis of drop distribution according to their size prior and post hydrate formation. Rate constants of hydrate formation in emulsions have been identified to be in accordance with system overcooling, not with the ratio of water and oil phases. The anticipated mechanism of formation of natural gas hydrates in emulsions has been shown to depend on their water content and to occur according either to the micro reactor type with formation of weakly stable thinly dispersed hydrates, or the relay type with formation of coalescence of water drops and stable hydrate agglomerates. Effects of the water phase content of emulsions on processes of water drop enlargement have been determined.

1. Introduction

Today, about 80% of Russian oil is extracted with large volumes of water. This problem becomes even more pronounced at the stage of oil extraction from oil reservoirs heavily underlain by water and further – at the stage of separation of the extracted water-in-oil emulsion.

Water influx in crude oil in combination with dissolved natural gas would result in gas hydrate formation, notably in oil and gas fields located in the Extreme North and the Arctic with their severe climatic conditions and low reservoir temperatures. In this case, there is a threat of formation of gas hydrate deposits on oilfield equipment, promptly leading to a drop in production of hydrocarbon raw materials. Thus, one of the key factors of successful development of Arctic deposits is a solution of problems referring to the issue of hydrate formation in wet crude oil.

There have been a large number of studies on physical and chemical features of gas hydrate formation and decomposition in oil emulsions [1-8]. These works mainly focus on study of hydrates of simple gases in reverse oil emulsions containing synthetic surfactants (SAS). In addition, the hydrate synthesis process is carried out by mixing organic and gas phases, that distorts the true nature of the hydrate formation process in water-oil emulsions. However, in practice, the formation of oil emulsions occur in the presence of natural emulsifiers, that are part of their composition: asphaltenes, naphthenes, resins, and paraffins.

The research by [1] has showed that the rate of hydrate formation in oil emulsions depends on the water quantity in emulsions. It has been identified that with an increase of water content in 0-50 wt.%

emulsions, the rate of methane hydrate formation increases, while in emulsions with the water content of 80 wt%, hydrate formation proceeds according to a two-step mechanism with the presence of an inversion phenomenon. Therefore, study of hydrate formation in oil with large volumes of water has practically never been conducted.

The objective of this paper is research of effects of water cuts in the range 20-80 wt. % in "the water- in- oil" emulsion on kinetic features of formation of natural gas hydrates.

2. Research methods

The subjects of the research were emulsions of paraffineous oil of the Irelyakskoe gas and oil field (Sakha Republic (Yakutia) and water produced by mechanical mixing of oil and water with ratio 20/80, 40/60, 60/40 and 80/20 without adding synthetic SAS. Natural gas from the Srednevilyuisky gas condensate deposit (Sakha Republic (Yakutia) containing 92.88 vol.% of methane, 5.25 vol.% of ethane, 1.21 vol.% of propane, and 0.24 vol.%. of other homologs has been used as hydrate forming gas.

Synthesis of natural gas hydrates in oil emulsions and research into its kinetics have been performed on the differential high-pressure scanning calorimeter DSC 204 HP Phoenix by Netzsch (Germany). The obtained DSC-grams have been applied to construct kinetic curves of transformation α_{micro} of the water phase into the hydrate one in oil. Then, the kinetic curves have been analyzed applying the Avrami equation [9]. The Avrami index characterizing the mechanism of the crystallization process and rate constants of hydrate formation in oil have been obtained.

In order to define peculiarities of the crystallization mechanism of the water phase into hydrate in emulsions according to digital micro-images obtained with application of the optical microscope Olympus BX 41, we have performed the statistical analysis of distribution of water droplets by size prior to and post hydrate formation.

3. Results and discussion

Real kinetic parameters of formation of natural gas hydrates in oil-in-water emulsions have been studied by cooling and heating of DSC thermograms.

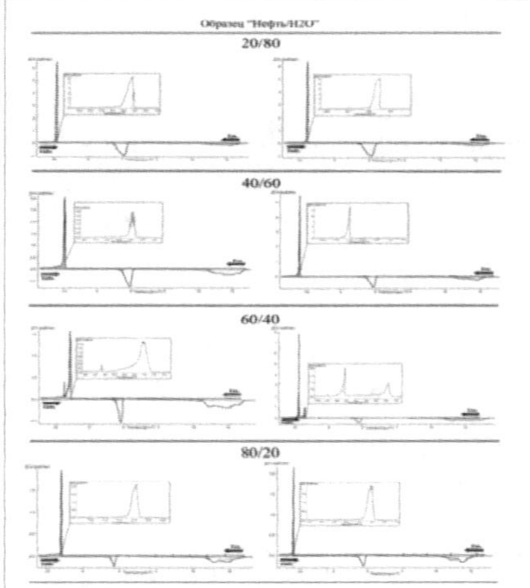

Figure 1. DSC – thermograms of cooling (segment Exo.) and heating (segment Endo) of water-oil emulsions in the presence of natural gas.

The segment of cooling of DSC thermograms of all emulsions reveals only one registered asymmetric exothermal signal, while in the segment of heating, there are two consecutive and independent endothermal effects, those of ice melting and decomposition of natural gas hydrate (Figure 1). Thus, formation of a hydrate-containing mixture has been identified in oil-in water - emulsions, and it has been defined to consist of phases of ice and natural gas hydrate.

The shape of the crystallization peak of the emulsion with composition 80/20 resembles the classical Gaussian curve, thus evidencing freezing of thinly dispersed emulsions. The crystallization peak of the emulsion with the 20/80 composition is asymmetrical and has the abrupt commence and extended termination, which is characteristic of the thermal effect of freezing of the continuous phase. These peculiarities of peak shapes are likely to predetermine the mechanism of crystallization of emulsions in the presence of natural gas. Crystallization of the emulsion containing 20 wt .% of water is likely to proceed according to the micro-reactor mechanism, while, in case of the 40-60 wt.% water content, the mechanism of crystal formation is mixed, and, with the 80 wt.% water content, droplets of the water phase are crystallized according to the relay mechanism. Transfer from the micro-reactor mechanism of water phase crystallization to the relay mechanism with an increase of its quantity in the emulsion is related with an increase of the degrees of freedom of the crystallization process which is characterized by the increase of the parameter n_{cr} (Table.1).

Table 1. Values of Avrami parameter of crystallization $n_к$ of water phase in water-in-oil emulsions with various contents of water phase in presence of natural gas (R^2 - approximation accuracy).

Water quantity in emulsion, wt %	n_{cr}	R^2
20	2.5	0.998
40	2.5	0.999
60	2.7	0.997
80	3.9	0.984

Water droplets of emulsions containing 60 and 80 wt.% of water where the relay mechanism of crystal formation prevails, have been defined to increase in size by 2-6 μm (Table 2) following hydrate formation. Thus, the process of hydrate formation in emulsions with the water phase content more than 60 wt. % leads to coalescence of water droplets with their final enlargement in size.

Table 2. Average diameters of water droplets (D_n) prior and post hydrate formation in oil emulsions with various water contents.

Water content in emulsion wt.%	D_n prior to hydrate formation μm	D_n post hydrate formation , μm
20	15.30±7.65	14.65±8.48
40	15.42±9.17	14.98±7.84
60	16.17±8.74	18.55±8.97
80	17.74±11.39	23.02±10.70

The processes of formation of methane hydrates in oil emulsions accelerate with cooling of the system, which is defined as a difference between an equilibrium temperature of hydrate formation and the DSC temperature of crystallization experimentally observed.

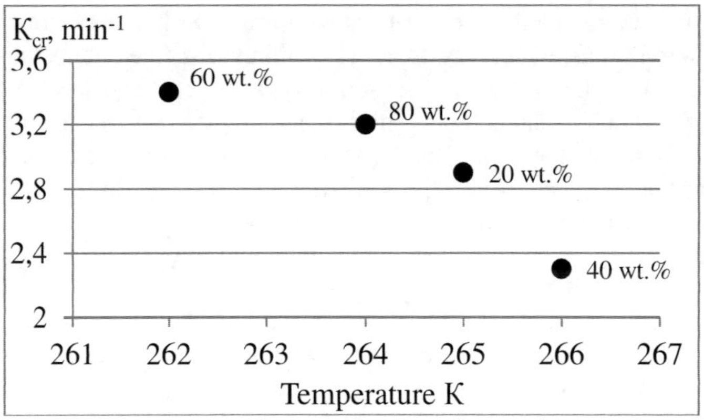

Figure 2. Dependence of rate constants of crystallization of natural gas hydrate K_{cr} in oil emulsions with different water cuts on temperature.

Values of rate constants of crystallization K_{cr} of oil emulsions are not dependent on its water content, and they increase with a temperature decrease (Figure 2). Thus, the higher overcooling of the "water and oil emulsion- natural gas" system, the higher the rate of hydrate formation.

As a rule, values of the melting constant evidence stability of gas hydrates. The hydrates of natural gas produced in emulsions are shown to be 2-2.5 times more stable than those synthesized from distilled water (Figure 3). Hydrates synthesized in emulsions with 80 wt. % of the water content have been identified most stable. Thus, the lower the melting constant of natural gas hydrates produced in oil emulsion, the higher their stability.

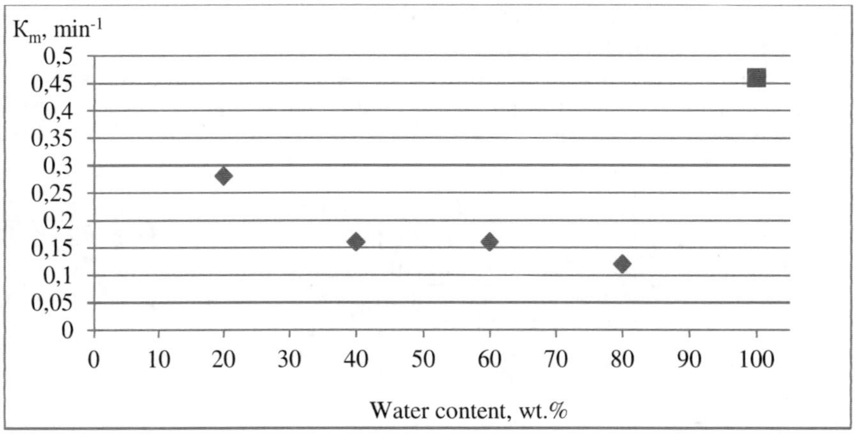

Figure 3. Rate constants of melting of natural gas hydrates K_m synthesized in oil emulsions (diamonds) and in distilled water (square).

The increase of stability of natural gas hydrates is likely to be predetermined by changes in dispersity of hydrates formed from emulsions with various water contents. So, thinly dispersed easily melting hydrates form in emulsions with 20 wt. % of water, while slowly discomposing hydrate agglomerates grow in emulsions with 80 wt.% of water.

Thus, the general scheme of the mechanism of hydrate formation in oil emulsions in accordance with their water content is presented as follows (Table 3):

Table 3. Mechanism of natural gas hydrate formation in oil emulsions with various contents of water phase.

1Water content in oil emulsions (less than 40 wt.%)	
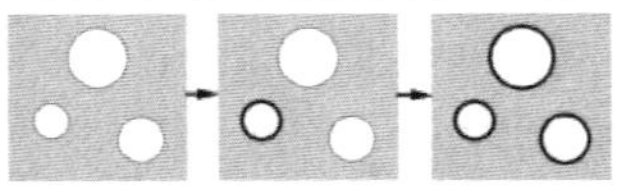	"Micro-reactor" mechanism: -low degree of crystallization freedom (n_{cr}=2.5); -without enlargement of water droplets; - formation of thinly dispersed weakly stable gas hydrates (K_{squ}= 0.28 min^{-1})
Water quantity in oil emulsions (40-60 wt.%)	
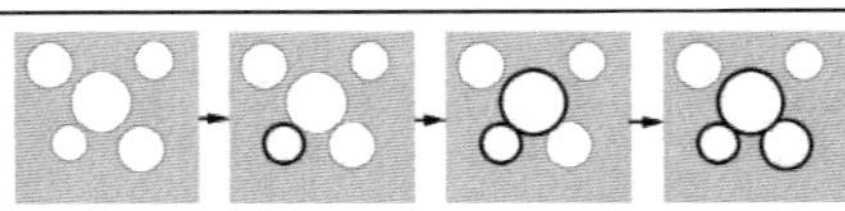	Mixed mechanism: - n_{cr}=2.7; - enlargement of droplets by 0-2 mcm; - K_{cr} = 0.16 min^{-1}
Water quantity in oil emulsion (more than 60 wt.%)	
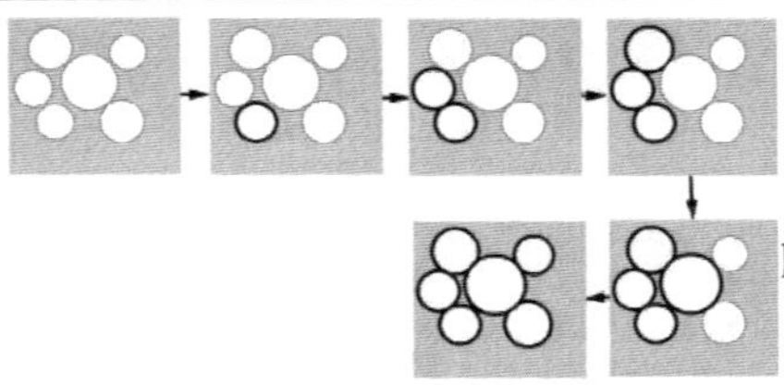	"Relay" mechanism: - high degree of crystallization freedom (n_{cr}=3.9); - enlargement of water droplets up to 6 mcm; - formation of stable agglomerates of natural gas hydrates (K_{squ}= 0.12 min^{-1})

4. Conclusion

Thus, the research has shown that the mechanism of formation of natural gas hydrates in emulsions is defined by the water quantity and proceeds according to the micro reactor type with formation of weakly stable thinly dispersed hydrates or according to the relay type with coalescence of water droplets and formation of stable hydrate agglomerates. Here, the rate of crystallization of water droplets is not affected by the water content in emulsions and it depends on the super-cooling of the system.

Acknowledgments
The research has been performed within the framework of the State task of the Federal Agency for Scientific Organizations #0377-2016-0003 and with the financial support provided by the Russian Federation Ministry of Education and Science, within the framework of the basic part of the State task, Project 10.7697.2017/BY "Organization of Performance of Scientific Research".

References

[1] Talatori S and Barth T 2012 *J. Petroleum Sci.&Eng.* **80** 32
[2] Sum A K, Koh C A and Sloan E D 2012 *Energy&Fuels* **26** 4046
[3] Semenov M E, Manakov A Yu, Shitz E Yu et al 2015 *J. Therm. Anal. Calorim.* **119** 757
[4] Ivanova I K, Semenov M E, Koryakina V V and Rozhin I I 2016 *Acta Chemica Iasi* **24** (2) 139
[5] Dalmazzone D, Hamed N and Dalmazzone C 2009 *Chem. Eng. Sci.* **64** 2020
[6] Akhfash M, Aman Z M, Ahn S Y et al. 2016 *Chem. Eng. Sci.* **140** 337
[7] Turner D, Miller K and Sloan E 2009 *Chem. Eng. Sci.* **64** 3996
[8] Lachance J W, Sloan E D and Koh C A 2008 *Chem. Eng. Sci.* **63** 3942
[9] Avrami M 1939 *J. Chem. Phys.* **7** 1103

IOP Conf. Series: Earth and Environmental Science **193** (2018) 012036 doi:10.1088/1755-1315/193/1/012036

Unsteady motion of circular cylinder under ice cover

V K Kostikov[1,2], **N I Makarenko**[1,2] **and A A Korobkin**[3]
[1] Lavrentyev Institute of Hydrodynamics, Novosibirsk, Russia
[2] Novosibirsk State University, Novosibirsk, Russia
[2] University of East Anglia, Norwich, UK

E-mail: makarenko@hydro.nsc.ru

Abstract. An interaction of the moving submerged circular cylinder with elastic solid ice cover is investigated. The method of reduction to integral-differential system of equations is extended to the case of elastic ice sheet floating on the fluid surface. The elastic properties of the ice are accounted for by the Euler's beam theory. A special iterative procedure is developed to construct small-time asymptotic solution for the cylinder moving with constant acceleration from rest. Elastic-gravity waves in the ice cover are studied in a wide range of parameters.

1. Introduction

In this paper we study analytically the deflection of an elastic ice-plate floating on the surface of ideal fluid caused by unsteady motion of submerged circular cylinder. The most recent theories and data related to the problem of external loading on the surface of the ice were discussed in [1]. The problem of unsteady motion of submerged body under ice cover in homogenous or stratified fluids was considered in linear statement in [2]. The linearised problem on hydroelastic behaviour of the floating plate of finite size was investigated in [3]. Our analysis uses reduction of the Euler equations to the nonlinear boundary integral-differential system of equations for the wave elevation together with normal and tangential fluid velocities [4]-[6]. In this method the key role is played by the integral equation for a normal fluid velocity that discribes interaction of the cylinder with the floating ice-cover.

2. Formulation of the problem

The plane irrotational flow of a heavy inviscid deep fluid is considered in the coordinate system Oxy with a vertical y-axis. The circular cylinder moves totally submerged under ice cover. Trajectory of the cylinder center $(x_c(t), y_c(t))$ is known at any time moment $t > 0$. The ice cover is modeled by an elastic solid plate with constant small thickness h and density ρ_{ice} floating freely on the free surface. The shape of the plate $y = \eta(x,t)$ is unknown and to be determined during solution of the problem. Initially, the fluid, ice plate and cylinder are at rest. The line $y = 0$ corresponds to undisturbed contact surface between ice and fluid (Figure 1). Let h_0, u_0, and ρ be initial submergence depth of the cylinder center, characteristic speed of the cylinder, and fluid density, respectively. The dimensionless variables use the depth h_0 as a length scale, the speed u_0 as a velocity scale, the quantity ρu_0^2 as a pressure unit, and the ratio h_0/u_0 as a time unit. Unsteady flow of incompressible fluid under ice cover is described by the Euler equations for the velocity of the flow $\mathbf{u} = (U, V)$ and hydrodynamic pressure p

$$\begin{cases} U_t + UU_x + VU_y + p_x = 0, \\ V_t + UV_x + VV_y + p_y = -\lambda, \\ U_x + V_y = 0, \quad U_y - V_x = 0, \end{cases} \tag{1}$$

where $\lambda = gh_0/u_0^2$ is the square of the inverse Froude number and g is the gravity acceleration.

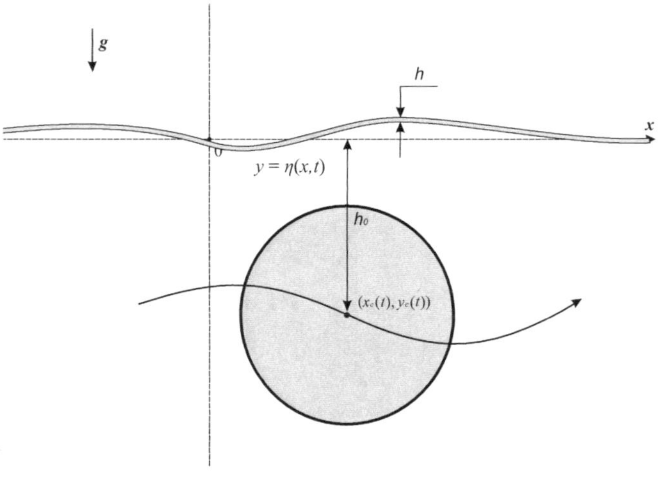

Figure 1. Scheme of motion.

We assume in our considerations that fluid remains to be in contact with an ice cover. Moreover, in contrast to the free surface flow, the pressure at the ice-covered surface depend on the deflection of the ice-plate. Therefore the following boundary kinematic and dynamic conditions should be satisfied

$$\eta_t + U\eta_x = V, \quad p = \alpha\eta_{tt} + \beta\eta_{xxxx} \qquad (y = \eta(x,t)), \tag{2}$$

where $\alpha = \rho_{ice}h/(\rho h_0)$ and $\beta = Eh^3/(12\rho h_0^3 u_0^2)$ are dimensionless parameters responsible for inertia and elasticity of the ice cover respectively. Here E is Young modulus. The boundary condition at the moving cylinder surface has the form

$$(\mathbf{u} - \mathbf{u_{cyl}}) \cdot \mathbf{n} = 0 \qquad \Big((x - x_c(t))^2 + (y - y_c(t))^2 = r^2\Big), \tag{3}$$

where $\mathbf{n}$ is the unit normal and r is non-dimensional radius of the cylinder. We assume that the fluid is at rest at infinity, so we have $U, V, \eta \to 0$ as $|\mathbf{x}| \to \infty$. The initial velocity field $\mathbf{u}_{|t=0} = \mathbf{u}_0$ should be irrotational

$$U_{0x} + V_{0y} = 0, \quad U_{0y} - V_{0x} = 0,$$

and also it should satisfy the compatibility condition

$$(\mathbf{u}_0 - \mathbf{u_{cyl}}(0)) \cdot \mathbf{n}_0 = 0 \qquad \Big((x - x_c(0))^2 + (y - y_c(0))^2 = r^2\Big).$$

These relations are satisfied, in particular, when the cylinder starts moving smoothly from rest.

3. Integro-differential equations on the free surface

We introduce the tangential velocity $u(x,t) = (U + \eta_x V)_{|y=\eta}$ and normal velocity $v(x,t) = (V - \eta_x U)_{|y=\eta}$ of the fluid particles at the interface $y = \eta(x,t)$ between ice and fluid, and reduce the basic equations (1)-(2) to an equivalent system of boundary integral-differential equations for the unknown functions u, v, η

$$\eta_t = v, \qquad u_t + \frac{1}{2}\frac{\partial}{\partial x}\left(\frac{u^2 - 2\eta_x uv - v^2}{1+\eta_x^2}\right) + \lambda\eta_x + \alpha\eta_{ttx} + \beta\eta_{xxxxx} = 0. \tag{4}$$

$$[I + A(\eta) + r^2 A_r(\eta)]v = [B(\eta) + r^2 B_r(\eta)]u + v_d(\eta). \tag{5}$$

The integral operators A and B are given by

$$\begin{aligned} A(\eta)u(x) &= \frac{1}{\pi}\int_{-\infty}^{+\infty} \frac{\eta(x) - \eta(s) - (x-s)\eta'(x)}{(x-s)^2 + [\eta(x) - \eta(s)]^2} u(s)ds, \\ B(\eta)u(x) &= \frac{1}{\pi}\int_{-\infty}^{+\infty} \frac{x - s + [\eta(x) - \eta(s)]\eta'(x)}{(x-s)^2 + [\eta(x) - \eta(s)]^2} u(s)ds, \end{aligned} \tag{6}$$

The operators A_r and B_r act by the formulae

$$\begin{aligned} A_r(\eta)u(x) &= \frac{1}{\pi}\int_{-\infty}^{+\infty} \frac{[\eta(s) - y_c - r^2P(x)]Q'(x) - [s - x_c - r^2Q(x)]P'(x)}{[\eta(s) - y_c - r^2P(x)]^2 + [s - x_c - r^2Q(x)]^2} u(s)ds, \\ B_r(\eta)u(x) &= \frac{1}{\pi}\int_{-\infty}^{+\infty} \frac{[\eta(s) - y_c - r^2P(x)]P'(x) + [s - x_c - r^2Q(x)]Q'(x)}{[\eta(s) - y_c - r^2P(x)]^2 + [s - x_c - r^2Q(x)]^2} u(s)ds, \end{aligned} \tag{7}$$

with Poisson kernels

$$P(x) = \frac{\eta(x) - y_c}{(x - x_c)^2 + (\eta(x) - y_c)^2}, \quad Q(x) = \frac{x - x_c}{(x - x_c)^2 + (\eta(x) - y_c)^2}.$$

Functions v_d in (5) is defined by relation

$$v_d(\eta) = 2r^2\Big(\dot{y}_c Q(x) - \dot{x}_c P(x)\Big)'. \tag{8}$$

Prime superscript $(\dots)'$ in formulae (6)–(8) denotes partial differentiation with respect to the spatial variable x. Respectively, the time variable t was omitted from the functions η, u, v in formulae (5)–(8) for simplicity since it appears there only as a parameter.

4. Small-time asymptotic solution

We consider the unsteady flow which starts from rest, $\eta(x,0) = u(x,0) = v(x,0) = 0$, and is caused by two-dimensional motion of the cylinder with constant acceleration. The trajectory of the cylinder can be taken in dimensionless form as $x_c(t) = t^2\cos\theta$, $y_c(t) = -1 + t^2\sin\theta$. The solution of system (4)-(5) is derived by the small-time expansion method in the form

$$\begin{aligned} \eta(x,t) &= t^2\eta_2(x) + t^3\eta_3(x) + t^4\eta_4(x) + \dots, \\ u(x,t) &= tu_1(x) + t^2u_2(x) + t^3u_3(x)\dots, \\ v(x,t) &= tv_1(x) + t^2v_2(x) + t^3v_3(x) + \dots. \end{aligned}$$

where the coefficients η_n and u_n may be evaluated via v_n by formulae following from differential equations (4)

$$\begin{aligned}
&\eta_{n+1} = \frac{v_n}{n+1} \quad (n \geq 1),\\
&u_1 = -\alpha v_1', \quad u_2 = 0, \quad u_3 = \frac{1}{6}[v_1^2 - \lambda v_1 - \beta v_1^{(4)}]' - \alpha v_3',\\
&u_4 = 0, \quad u_5 = \frac{1}{20}[4v_1 v_3 - \lambda v_3 - \beta v_3^{(4)}]' - \alpha v_5', \quad \ldots
\end{aligned} \tag{9}$$

Using the power expansion of the free surface elevation η we can determine similar power series for the integral operators defined by formulae (6) and (7)

$$\begin{aligned}
&A(\eta) = t^2 A^{(2)} + t^3 A^{(3)} + \ldots, \quad B(\eta) = B^{(0)} + t^2 B^{(2)} + \ldots,\\
&A_r(\eta) = A_r^{(0)} + t^2 A_r^{(2)} + t^3 A_r^{(3)} + \ldots, \quad B_r(\eta) = B_r^{(0)} + t^2 B_r^{(2)} + \ldots
\end{aligned}$$

Here the leading-order coefficients of the operators A and B have the form:

$$A^{(2)}u(x) = \frac{1}{\pi}\int_{-\infty}^{+\infty} \frac{\eta_2(x) - \eta_2(s)}{(x-s)^2} u(s)ds - \eta_2'(x)\frac{1}{\pi}\int_{-\infty}^{+\infty} \frac{u(s)ds}{x-s}, \quad B^{(0)}u(x) = Hu(x) = \frac{1}{\pi}\int_{-\infty}^{+\infty} \frac{u(s)ds}{x-s},$$

where H denotes the Hilbert transform. Operators $A_r^{(0)}$ and $B_r^{(0)}$ can be evaluated as follows

$$A_r^{(0)}v(x) = \frac{1}{\pi}\int_{-\infty}^{+\infty} \frac{(1 - r^2 p(x))q'(x) - (s - r^2 q(x))p'(x)}{(1 - r^2 p(x))^2 + (r^2 q(x) - s)^2} v(s)ds,$$

$$B_r^{(0)}u(x) = \frac{1}{\pi}\int_{-\infty}^{+\infty} \frac{(1 - r^2 p(x))p'(x) + (s - r^2 q(x))q'(x)}{(1 - r^2 p(x))^2 + (r^2 q(x) - s)^2} u(s)ds,$$

with functions

$$p(x) = P(x)|_{\{\eta=0,\, x_c=0,\, y_c=-1\}} = \frac{1}{1+x^2}, \quad q(x) = Q(x)|_{\{\eta=0,\, x_c=0,\, y_c=-1\}} = \frac{x}{1+x^2}. \tag{10}$$

The dipole term v_d has the power expansion $v_d(x,t) = t v_d^{(1)}(x) + t^3 v_d^{(3)}(x) + \ldots$ with coefficients

$$\begin{aligned}
&v_d^{(1)}(x) = 4r^2(q'(x)\sin\theta - p'(x)\cos\theta),\\
&v_d^{(3)}(x) = 4r^2(p''(x)\cos 2\theta - q''(x)\sin 2\theta) + 4r^2(\eta_2(x)[p'(x)\sin\theta + q'(x)\cos\theta])'.
\end{aligned} \tag{11}$$

Thus, integral equation (5) for normal fluid velocity v leads to the set of equations for coefficients v_n

$$[I + r^2 A_r(0)]v_n = \varphi_n, \quad (n \geqslant 1)$$

where functions φ_n can be evaluated explicitly due to relations (9) via the coefficients v_1, v_2, $\ldots$, v_n by the formulae

$$\begin{aligned}
&\varphi_1 = v_d^{(1)} + Hu_1 + r^2 B_r^{(0)} u_1, \quad \varphi_2 = 0,\\
&\varphi_3 = v_d^{(3)} + Hu_3 + r^2\left(B_r^{(0)} u_3 - A_r^{(2)} v_1\right) - A^{(2)} v_1, \quad \varphi_4 = 0, \quad \ldots
\end{aligned} \tag{12}$$

The main difficulty created by the presence of an ice cover is that small-time expansion for the tangential velocity $u(x,t)$ starts with the first order of time variable t, which is due to the presence of the terms η_{ttx} in differential equation (4). This fact prevents us from formulating explicit formulae expressing v_n via the previous coefficients $v_1, \dots, v_{n-1}$ as it was made in [6] in case of the free surface problem. Instead of that we have series of integro-differential equations

$$\begin{aligned} & v_1 + \alpha(H + r^2 B_r^{(0)})\partial_x v_1 = v_d^{(1)}, \quad v_2 = 0, \\ & v_3 + \alpha(H + r^2 B_r^{0})\partial_x v_3 = v_d^{(3)} + \frac{1}{6} H \partial_x (v_1^2 - \lambda v_1) + \\ & + r^2 \left(\frac{1}{6} B_r^{(0)} \partial_x (v_1^2 - \lambda v_1)' - A_r^{(2)} v_1 \right) - \frac{\beta}{6} \left(H + r^2 B_r^{(0)} \right) \partial_x^5 v_1, \end{aligned} \tag{13}$$

To solve these equations we apply the method of successive approximations. By this approach we suppose $\alpha = \beta = 0$ (i.e. we ignore the presence of ice cover in the first order) and calculate v_i from equations (13). Coefficients obtained by this way correspond to the problem on free surface waves caused by circular cylinder. Expansion for the wave elevation $\eta(x,t)$ can be found from the first relation in formulae (9) as follows

$$I_0(\eta_2(x)) = 2r^2 \Big(1 - \frac{r^2}{4}\Big) \Big(q'(x)\sin\theta - p'(x)\cos\theta\Big) + O(r^6), \quad \eta_3(x) = 0,$$

$$\begin{aligned} I_0(\eta_4(x)) = r^2 \Big(1 - \frac{r^2}{4}\Big) & \Big(p''(x)\Big[\cos 2\theta + \frac{\lambda}{6}\sin\theta\Big] + \Big[\frac{\lambda}{6}\cos\theta - \sin 2\theta\Big] q''(x)\Big) + \\ & + \frac{r^4}{4}\Big(p'(x)\Big[\sin 2\theta - \frac{\lambda}{3}\cos\theta\Big] + \Big[\frac{\lambda}{3}\sin\theta + \cos 2\theta\Big] q'(x)\Big) + \\ & + \frac{r^4}{9}\Big(p''''(x)\cos 2\theta - q''''(x)\sin 2\theta\Big) + \frac{r^4}{3}\Big(p''(x) - \frac{7}{4} q'(x)\Big) + O(r^6). \end{aligned} \tag{14}$$

Operator I_0 means that η_2 and η_4 were found at the zeroth step. Then we substitute obtained coefficients (14) into equations (13), which get the following form

$$\begin{aligned} & I_1(v_1) + \alpha(H + r^2 B_r^{(0)})\partial_x I_0(v_1) = I_0(v_1), \quad v_2 = 0, \\ & I_1(v_3) + \alpha(H + r^2 B_r^{(0)})\partial_x I_0(v_3) = I_0(v_3) - \frac{\beta}{6}\left(H + r^2 B_r^{(0)}\right) \partial_x^5 I_0(v_1), \end{aligned} \tag{15}$$

The values of coefficients η_2 and η_4 at the next itaration step can be evaluated directly from equations (15):

$$\begin{aligned} I_1(\eta_2(x)) = \frac{r^2}{2} & \Big(4 + r^2(\alpha - 1)\Big)\Big(q'(x)\sin\theta - p'(x)\cos\theta\Big) + \\ & + \frac{\alpha r^2}{2}(4 - r^2)\left(q''(x)\cos\theta + p''(x)\sin\theta\right), \end{aligned} \tag{16}$$

$$\begin{aligned} I_1(\eta_4(x)) = r^2 \Big(2 + \frac{\alpha r^2}{16}[\lambda + 8] - \frac{r^2}{12}[\lambda + 6 + 45\beta]\Big) & \left(p'(x)\cos\theta - q'(x)\sin\theta\right) + \\ & + 3r^4\left(1 - \frac{3\alpha}{2}\right)\left(p'(x)\sin 2\theta + q'(x)\cos 2\theta\right) + \\ & + r^2\left(\frac{\lambda}{6}\left[1 - \frac{r^2}{4}\right] + \alpha\left[\frac{r^2}{2}\left(1 + \frac{\lambda}{6}\right) - 1\right]\right)(p''(x)\sin\theta + q''(x)\cos\theta) + \\ & + r^2\left(1 + \frac{r^2}{4}[\alpha - 1]\right)(p''(x)\cos 2\theta - q''(x)\sin 2\theta) + \end{aligned} \tag{17}$$

$$+\frac{\lambda\alpha r^2}{6}\left(1-\frac{r^2}{4}\right)(p'''(x)\cos\theta - q'''(x)\sin\theta) - \alpha r^2\left(1-\frac{r^2}{4}\right)(p'''(x)\sin 2\theta - q'''(x)\cos 2\theta)+$$
$$+\frac{r^4}{9}\Big([p^{(4)}(x)-\alpha q^{(5)}(x)]\cos 2\theta - [q^{(4)}(x)+\alpha p^{(5)}(x)]\sin 2\theta\Big)+$$
$$+\frac{2\beta r^2}{3}\left(1-\frac{r^2}{4}\right)(p^{(6)}(x)\sin\theta + q^{(6)}(x)\cos\theta) - \frac{r^4}{12}\Big(7q'(x)+[7\alpha-4]p''(x)+\frac{\alpha}{4}q'''(x)\Big)$$

Consequently, repeating this procedure we can calculate ice deflection profile coefficients $I_n(\eta_2)$ and $I_n(\eta_4)$ with any demanded accuracy with respect to parameters α and β from the following formulae

$$\begin{aligned} I_{n+1}(\eta_2) &= I_n(\eta_2) - \alpha(H + r^2 B_r^{(0)})\partial_x I_n(\eta_2), \\ I_{n+1}(\eta_4) &= I_n(\eta_4) - \alpha(H + r^2 B_r^{(0)})\partial_x I_n(\eta_4) - \frac{\beta}{3}(H + r^2 B_r^{(0)})\partial_x^5 I_n(\eta_2). \end{aligned} \tag{18}$$

Here $I_n(\eta)$ denotes the value of η calculated at the n-th step of iterative procedure.

Having the explicit expression for ice deflection $\eta(x,t)$ we can determine pressure p in the fluid under the ice plate from dynamic boundary condition (2). If we look for a solution in the form of power series then the pressure p at the n-th step of iterative procedure takes the form

$$I_n(p(x,t)) = 2\alpha I_n(\eta_2(x)) + \beta\partial_x^2 I_n(\eta_2(x))t^2 + (12\alpha I_n(\eta_4(x)) + \beta\partial_x^4 I_n(\eta_4(x)))t^4, \tag{19}$$

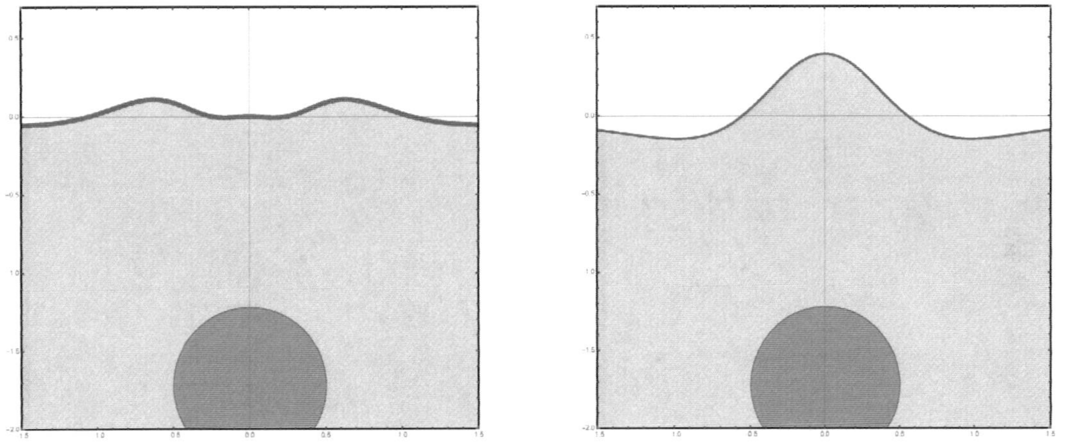

a. Ice plate ($\alpha = 0.4$, $\beta = 0.001$) *b. Free surfac.* ($\alpha = 0$, $\beta = 0$)

Figure 2. Deformation of the boundary $y = \eta(x,t)$ at the time $t = 0.85$ caused by vertical sumbersion of the cylinder of radius $r = 0.5$ with parameter $\lambda = 10$.

5. Analysis of the solution

Coefficients (16) and (17) determine the deformation of the boundary $y = \eta(x,t)$ in wide range of paramters: acceleration rate of the cylinder (parameter λ), direction of motion (angle θ), size of the cylinder r, mass and elastic constant of the ice cover (parameters α and β). In particularly, when $\alpha = \beta = 0$ we have the case of the free surface flow; when $\beta = 0$ we have the case of the flow under broken ice.

Figures 2 and 3 demonstrate the differences in the flow patterns generated by the cylinder moving under ice plate and free surface. Indeed, in case of the free surface flow, the vertically submerging cylinder ($\theta = -\frac{\pi}{2}$) causes inertial deflection of free boundary with subsequent formation of intense vertical splash jet. In contrast, two additional deflection zones appear symetrically above the cylinder submerging under elastic ice plate. Similarly, for horizontal motion of the cylinder ($\theta = 0$), the floating ice cover supress essentially impulsive response of the fluid, but small-amplitude elastic waves are generated in this case. Figure 4 shows the pressure distribution in the fluid under the ice plate. In accordance with this picture, the elastic plate experiences the highest loads in two symetric zones of intense deflection above the cylinder. Therefore we can predict that ice is more likely to break in these zones.

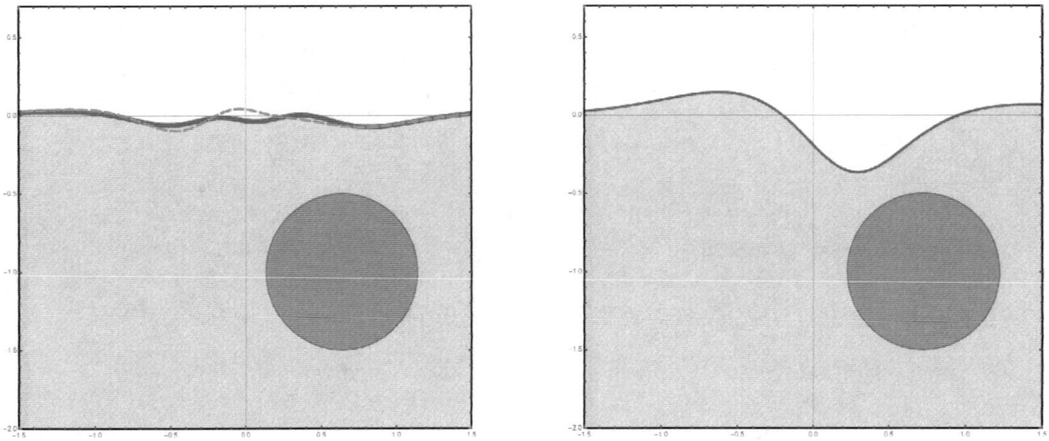

a. Ice plate ($\alpha = 0.3$, $\beta = 0.002$) *b. Free surface* ($\alpha = 0$, $\beta = 0$)

Figure 3. Deformation of the boundary $y = \eta(x,t)$ at the time $t = 0.85$ caused by horizontal motion of the cylinder of radius $r = 0.5$ with parameter $\lambda = 10$.

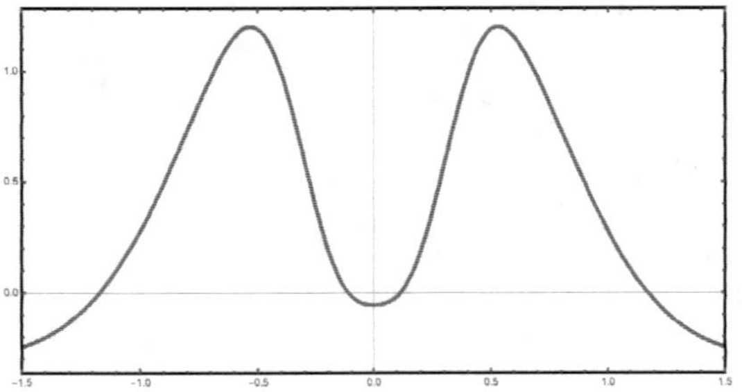

Figure 4. Pressure distribution in the fluid under the ice plate at the time $t = 0.85$ for submerging cylinder of radius $r = 0.5$; $\lambda = 10$, $\alpha = 0.4$, $\beta = 0.001$.

Acknowledgments

This work was supported by RFBR grants (grants No. 16-08-00291, 18-01-00648).

References

[1] Squire V A, Hosking R J, Kerr A D and Langhorne P J 1996 *Moving Loads on Ice Plates* (Dordrecht: Kluwer)
[2] Sturova I V 2012 *J. Appl. Mech. Tech. Phys.* **12** 114–22
[3] Korobkin A A 2000 *J. Appl. Mech. Tech. Phys.* **41** 286–92
[4] Makarenko N I 2003 *JOMAE* **125** 72–5
[5] Kostikov V K, Makarenko N I and Korobkin A A 2012 *Proc. 27th Int. Workshop on Water Waves and Floating Bodies*
[6] Kostikov V K and Makarenko N I 2018 *J. Eng. Math.* https://doi.org/10.1007/s10665-018-9962-x

Genuinely nonlinear impulsive ultra-parabolic equations and convective heat transfer on a shock wave front

I V Kuznetsov[1,2] **and S A Sazhenkov**[1,2]

[1] Novosibirsk State University, Pirogova str. 1, Novosibirsk, Russia
[2] Lavrentyev Institute of Hydrodynamics, ac. Lavrentyev av. 15, Novosibirsk, Russia

E-mail: sazhenkovs@yandex.ru

Abstract. In the present paper, we derive the kinetic equation and impulsive condition and formulate a class of kinetic solutions to impulsive ultra-parabolic equations. Here ultra-parabolic equations are linked with diffusion processes with inertia (convective heat transfer) on a shock wave front.

Key words: ultra-parabolic equation, genuine nonlinearity condition, kinetic solution, convective heat transfer, impulsive condition

1. Introduction

In this paper, we aim to give a description of heat transfer in continuous medium in presence of shock waves, taking into account radiation on a shock wave front [1] in the form of a convective heat transfer. It is known that the maximum entropy production principle leads to the radiative heat transfer on a shock wave front [2]. Moreover, outside of a shock wave front, the heat transfer is conductive. In view of possible application, our research is directed towards study of heat evolution in glaciers. It is important to note that convective heat transfer in a glacier is not a novelty, see [3]. In this paper, we apply the ultra-parabolic equation. Instead of inertia linked with the unidirectional flow [4, 5], we deal with inertia because of the maximum entropy production principle on a shock wave front. Also, it is important to mention impact-induced thermal convection phenomenon [6, Equations (1)–(7)], [7]. Here we assume that a shock wave is propagating in the direction s. The dissipation term in the direction s can be ignored. Therefore heat flux $a(\boldsymbol{x}, t, s, u)$ in the s direction depends also on other physical directions $\boldsymbol{x} = (x_1, \dots, x_d)$, time variable t, and the sought specific heat distribution u.

We assume that, in a glacier, a shock wave [8] is responsible for convective heat transfer. In this model we include the Bridgman effect in ice [9, 10] which is simulated with the help of an impulsive condition. Some references on impulsive partial differential equations are given in [11]. Considering, in particular, the process of the crater formation at the Yamal Peninsula, we assume that it is linked with ice-melting following degasation on a shock wave [12].

The following report is devoted to kinetic solutions of Problem Π_0, see sections 2 and 5. We derive the kinetic equation and justify that initial (in t) and impulsive conditions hold true. Strong traces of kinetic solutions would be justified in our future research. The questions linked with the uniqueness are still open.

IOP Conf. Series: Earth and Environmental Science **193** (2018) 012037 doi:10.1088/1755-1315/193/1/012037

2. Problem Π_0 – the basic formulation

Let Ω be a bounded domain of spacial variables $\boldsymbol{x} \in \mathbb{R}^d$ with a smooth boundary $\partial\Omega$ ($\partial\Omega \in C^2$). Let $t \in [0,T]$ and $s \in [0,S]$ be two independent time-like variables. Here T and S are given positive constants. Let $\tau \in (0,T)$ be a fixed time moment.

Denote $G_{T,S} := (\Omega \times (0,\tau) \times (0,S)) \cup (\Omega \times (\tau,T) \times (0,S))$, $\Xi^1 := \overline{\Omega} \times [0,S]$, $\Xi^2 := \overline{\Omega} \times [0,T]$, $\Gamma_0^1 := \overline{\Omega} \times \{t=0\} \times [0,S]$, $\Gamma_\tau^1 := \overline{\Omega} \times \{t=\tau\} \times [0,S]$, $\Gamma_T^1 := \overline{\Omega} \times \{t=T\} \times [0,S]$, $\Gamma_0^2 := \overline{\Omega} \times [0,T] \times \{s=0\}$, $\Gamma_S^2 := \overline{\Omega} \times [0,T] \times \{s=S\}$, $\Gamma_l := \partial\Omega \times [0,T] \times [0,S]$.

In this paper we study the following Cauchy-Dirichlet problem.

Problem Π_0. *For arbitrary initial and final data $u_0^{(1)} \in C_0^{0,\alpha}(\Xi^1)$, $u_0^{(2)}, u_S^{(2)} \in C_0^{0,\alpha}(\Xi^2)$, and impulsive data $\beta \in C_0^{0,\alpha}(\Xi^1)$ ($\alpha \in (0,1)$), it is necessary to find a function $u : G_{T,S} \mapsto \mathbb{R}$ satisfying*

- *quasilinear ultra-parabolic equation*

$$\partial_t u + \partial_s a(\boldsymbol{x},t,s,u) + \mathrm{div}_x \boldsymbol{\varphi}(\boldsymbol{x},t,s,u) = \Delta_x u + g(\boldsymbol{x},t,s,u), \quad (\boldsymbol{x},t,s) \in G_{T,S}, \tag{2.1a}$$

- *initial condition with respect to time-like variable t*

$$u|_{t=0} = u_0^{(1)}(\boldsymbol{x},s), \ (\boldsymbol{x},s) \in \Xi^1, \tag{2.1b}$$

- *initial and final conditions with respect to time-like variable s*

$$u|_{s=0} \approx u_0^{(2)}(\boldsymbol{x},t), \ u|_{s=S} \approx u_S^{(2)}(\boldsymbol{x},t), \ (\boldsymbol{x},t) \in \Xi^2, \tag{2.1c}$$

- *homogeneous boundary condition*

$$u|_{\Gamma_l} = 0, \tag{2.1d}$$

- *impulsive condition*

$$u(\boldsymbol{x},\tau+0,s) = u(\boldsymbol{x},\tau-0,s) + \beta(\boldsymbol{x},s), \ (\boldsymbol{x},s) \in \Xi^1. \tag{2.1e}$$

By $C_0^{0,\alpha}(\overline{\mathcal{O}})$ we standardly denote the space of Hölder continuous with exponent $\alpha \in (0,1)$ finite functions on a closed set $\overline{\mathcal{O}} \in \mathbb{R}^N$ supplemented with the norm

$$\|\Phi\|_{C^{0,\alpha}(\overline{\mathcal{O}})} = \max_{\zeta \in \overline{\mathcal{O}}} |\Phi(\zeta)| + \sup_{\substack{\zeta,\eta \in \mathcal{O} \\ \zeta \neq \eta}} \frac{|\Phi(\zeta) - \Phi(\eta)|}{|\zeta - \eta|^\alpha}.$$

In (2.1c) the relation sign $\approx$ means that $u_0^{(2)}$ and $u_S^{(2)}$ may be unattained by a solution u on some parts of the sets Γ_0^2 and Γ_S^2, respectively. The fact whether $\approx$ becomes equality ($=$), or not, is figured out ***a posteriori***, i.e., after a solution of Problem Π_0 is constructed somehow.

In the formulation of Problem Π_0 the functions $a = a(\boldsymbol{x},t,s,\lambda)$ and $g = g(\boldsymbol{x},t,s,\lambda)$, and the nonlinear vector-function $\boldsymbol{\varphi} = (\varphi_1(\boldsymbol{x},t,s,\lambda), \ldots, \varphi_d(\boldsymbol{x},t,s,\lambda))$ are given and satisfy the following conditions.

Conditions on $a\&\boldsymbol{\varphi}\&g$.

(i) *There exist constants $b_1, b_2 > 0$ such that for all $(\boldsymbol{x},t,s) \in G_{T,S}$ and $\lambda \in \mathbb{R}$ the inequality*

$$(\partial_s a(\boldsymbol{x},t,s,\lambda) + \mathrm{div}_x \boldsymbol{\varphi}(\boldsymbol{x},t,s,\lambda) - g(\boldsymbol{x},t,s,\lambda))\lambda \geqslant -b_1\lambda^2 - b_2 \tag{2.2a}$$

is valid.

(ii) *Functions* $a = a(\boldsymbol{x}, t, s, \lambda)$, $\varphi_i = \varphi_i(\boldsymbol{x}, t, s, \lambda)$, $g = g(\boldsymbol{x}, t, s, \lambda)$, *and their derivatives* $\partial_s a$, $\partial_{x_j} a$, $\partial_\lambda a$, $\partial_s \varphi_i$, $\partial_{x_j} \varphi_i$, *and* $\partial_\lambda \varphi_i$ $(i, j = 1, \ldots, d)$ *are Hölder continuous in* $\boldsymbol{x}$, t, s, *and* λ *with continuity exponents* γ, $\frac{\gamma}{2}$, γ, *and* γ $(\gamma \in (0,1))$, *respectively, on* $\overline{\Omega} \times [0, \tau] \times [0, S] \times [-M, M]$ *and on* $\overline{\Omega} \times [\tau, T] \times [0, S] \times [-M, M]$ *for any finite* $M > 0$.

(iii) *Function* $a = a(\boldsymbol{x}, t, s, \lambda)$ *satisfies* ***the genuine nonlinearity condition***

$$\operatorname{meas}\{\lambda \in \mathbb{R} : \ \xi_1 + \partial_\lambda a(\boldsymbol{x}, t, s, \lambda)\xi_2 = 0\} = 0 \tag{2.2b}$$

for every $(\boldsymbol{x}, t, s, \boldsymbol{\xi}) \in \overline{\Omega} \times [0, T] \times [0, S] \times \mathbb{S}^1$.

By $\mathbb{S}^1$ we denote the unit circle in $\mathbb{R}^2$ centered at the origin, $\mathbb{S}^1 := \{\boldsymbol{\xi} = (\xi_1, \xi_2) \in \mathbb{R}^2 : \xi_1^2 + \xi_2^2 = 1\}$. In (2.1e) and further in the article we standardly denote $u(\boldsymbol{x}, \tau+0, s) := \lim\limits_{t \to \tau+0} u(\boldsymbol{x}, t, s)$, $u(\boldsymbol{x}, \tau - 0, s) := \lim\limits_{t \to \tau - 0} u(\boldsymbol{x}, t, s)$.

Remark 2.1. *Note that, with* $u\big|_{s=0} = u_0^{(2)}$ *and* $u\big|_{s=S} = u_S^{(2)}$ *on* Ξ^2 *on the place of (2.1c), Problem* Π_0 *becomes ill-posed. Indeed, since for a.e.* $(\boldsymbol{x}, t, s) \in G_{T,S}$ *the function* $\lambda \mapsto a(\boldsymbol{x}, t, s, \lambda)$ *is nonlinear and, in general, non-monotonous, it may be impossible to equate a solution* u *of (2.1a) to* $u_0^{(2)}$ *and* $u_S^{(2)}$ *on the* ***entire*** *sets* Γ_0^2 *and* Γ_S^2. *Therefore we set up a more loose non-classical condition (2.1c) following the original idea presented in [13–15], [16, Sec. 2.6–2.8], [17, 18].*

In order to deal with the above described sophisticated non-classical features of Problem Π_0, we introduce the notion of a kinetic solution. To this end, we systematically study a strictly parabolic regularized formulation.

3. Parabolic regularization. Problem Π_ε

We construct a kinetic solution of Problem Π_0 as a singular limit of weak solutions u_ε of the following strictly parabolic model, as $\varepsilon \to 0+$.

Problem Π_ε. *For arbitrary initial and final data* $u_0^{(1)} \in C_0^{0,\alpha}(\Xi^1)$, $u_0^{(2)}$, $u_S^{(2)} \in C_0^{0,\alpha}(\Xi^2)$, *and impulsive data* $\beta \in C_0^{0,\alpha}(\Xi^1)$ $(\alpha \in (0,1)$ *is given), it is necessary to find a function* $u_\varepsilon : G_{T,S} \mapsto \mathbb{R}$ *satisfying*

- *quasilinear parabolic equation*

$$\partial_t u_\varepsilon + \partial_s a(\boldsymbol{x}, t, s, u_\varepsilon) + \operatorname{div}_x \boldsymbol{\varphi}(\boldsymbol{x}, t, s, u_\varepsilon) = \Delta_x u_\varepsilon + \varepsilon \partial_s^2 u_\varepsilon + g(\boldsymbol{x}, t, s, u_\varepsilon),$$
$$(\boldsymbol{x}, t, s) \in G_{T,S}, \tag{3.1a}$$

- *initial condition*

$$u_\varepsilon\big|_{t=0} = u_0^{(1)}(\boldsymbol{x}, s), \quad (\boldsymbol{x}, s) \in \Xi^1, \tag{3.1b}$$

- *boundary conditions*

$$u_\varepsilon\big|_{s=0} = u_0^{(2)}(\boldsymbol{x}, t), \ u_\varepsilon\big|_{s=S} = u_S^{(2)}(\boldsymbol{x}, t), \quad (\boldsymbol{x}, t) \in \Xi^2, \tag{3.1c}$$

$$u_\varepsilon\big|_{\Gamma_l} = 0, \tag{3.1d}$$

- *impulsive condition*

$$u_\varepsilon(\boldsymbol{x}, \tau + 0, s) = u_\varepsilon(\boldsymbol{x}, \tau - 0, s) + \beta(\boldsymbol{x}, s), \quad (\boldsymbol{x}, s) \in \Xi^1. \tag{3.1e}$$

Here $\varepsilon \in (0, 1]$ is an arbitrarily fixed small parameter.

Let us formulate the notion of a weak solution to Problem Π_ε. Let $\hat{u} \in L^\infty(G_{T,S}) \cap W_2^1(G_{T,S})$ be an arbitrary extension of $u_0^{(1)}$, $u_0^{(2)}$ and $u_S^{(2)}$ into $G_{T,S}$ such that $\hat{u}\big|_{\Gamma_l} = 0$.

Remark 3.1. *Such $\hat{u}$ may be constructed as the solution of the Dirichlet problem for Laplace's equation $\Delta_{x,t,s}\hat{u} = 0$ (in $G_{T,S} \cup \Gamma_\tau^1$) with boundary data $\hat{u}\big|_{\Gamma_l} = 0$, $\hat{u}\big|_{\Gamma_0^1} = u_0^{(1)}$, $\hat{u}\big|_{\Gamma_T^1} = 0$, $\hat{u}\big|_{\Gamma_0^2} = u_0^{(2)}$, $\hat{u}\big|_{\Gamma_S^2} = u_S^{(2)}$, for example. This $\hat{u}$ belongs, in fact, to the space $C^2(\overline{G}_{T,S})$ and is represented in terms of Green's function [19, Sec. 2.2.4, Th. 12].*

Definition 3.1. *Function $u_\varepsilon \in L^\infty(G_{T,S}) \cap L^2(0, T; W_2^1(\Xi^1))$ is called a weak solution of Problem Π_ε, if it satisfies the following demands:*

1. *The equality $u_\varepsilon - \hat{u} = 0$ holds on $\Gamma_0^1 \cup \Gamma_0^2 \cup \Gamma_S^2 \cup \Gamma_l$ in the trace sense.*
2. *One-sided traces $u_\varepsilon(\boldsymbol{x}, \tau \pm 0, s)$ exist on Γ_τ^1 and satisfy the impulsive condition (3.1e) for all $(\boldsymbol{x}, s) \in \Xi^1$.*
3. *The integral equality*

$$\int\limits_{G_{T,S}} \big(- u_\varepsilon \partial_t \phi - a(\boldsymbol{x}, t, s, u_\varepsilon)\partial_s \phi - \boldsymbol{\varphi}(\boldsymbol{x}, t, s, u_\varepsilon) \cdot \nabla_x \phi + \nabla_x u_\varepsilon \cdot \nabla_x \phi + \varepsilon \partial_s u_\varepsilon \partial_s \phi - g(\boldsymbol{x}, t, s, u_\varepsilon)\phi \big)\, d\boldsymbol{x}dtds = 0 \quad (3.2)$$

holds for every $\phi \in L^\infty(G_{T,S}) \cap \overset{o}{W}{}_2^1(G_{T,S})$.

Proposition 3.1. *Whenever $u_0^{(1)}$, $\beta \in C_0^{0,\alpha}(\Xi^1)$, $u_0^{(2)}$, $u_S^{(2)} \in C_0^{0,\alpha}(\Xi^2)$ ($\alpha \in (0, 1)$), and Conditions on a&$\boldsymbol{\varphi}$&g hold, for any small fixed $\varepsilon > 0$ there is a unique weak generalized solution $u_\varepsilon = u_\varepsilon(\boldsymbol{x}, t, s)$ of Problem Π_ε such that $u_\varepsilon \in H^{\alpha', \frac{\alpha'}{2}}(\overline{\Omega} \times [0, \tau] \times [0, S]) \cap H^{\alpha', \frac{\alpha'}{2}}(\overline{\Omega} \times [\tau, T] \times [0, S]) \cap W_2^{2,1}(G')$, $\nabla_x u_\varepsilon$, $\partial_s u_\varepsilon \in H^{\gamma', \frac{\gamma'}{2}}(G_{T,S})$, where $\alpha', \gamma' \in (0, 1)$ depend on α and ε, and G' is an arbitrary strictly interior subdomain of $G_{T,S}$.*

Moreover, the maximum principle

$$\|u_\varepsilon\|_{L^\infty(\Xi^1 \times (0,\tau))} \leqslant \inf_{\xi > b_1} \Big\{ e^{\xi\tau} \max\big\{ \|u_0^{(1)}\|_{L^\infty(\Xi^1)}, \|u_0^{(2)}\|_{L^\infty(\Omega \times (0,\tau))}, \|u_S^{(2)}\|_{L^\infty(\Omega \times (0,\tau))} \big\}, e^{\xi\tau}\sqrt{\frac{b_2}{\xi - b_1}} \Big\}, \quad (3.3a)$$

$$\|u_\varepsilon\|_{L^\infty(\Xi^1 \times (\tau,T))} \leqslant \inf_{\xi > b_1} \Big\{ e^{\xi\tau} \max\big\{ \|u_\varepsilon(\cdot, \tau - 0, \cdot) + \beta\|_{L^\infty(\Xi^1)}, \|u_0^{(2)}\|_{L^\infty(\Omega \times (\tau,T))}, \|u_S^{(2)}\|_{L^\infty(\Omega \times (\tau,T))} \big\}, e^{\xi\tau}\sqrt{\frac{b_2}{\xi - b_1}} \Big\}, \quad (3.3b)$$

and the energy estimate

$$\int\limits_{G_{T,S}} \Big(|\nabla_x u_\varepsilon|^2 + \varepsilon\, |\partial_s u_\varepsilon|^2 \Big)\, d\boldsymbol{x}dtds < C \quad (3.4)$$

hold. The constant C does not depend on ε.

Remark 3.2. *Bounds* (3.3a) *and* (3.3b) *immediately imply that*

$$\|u_\varepsilon\|_{L^\infty(G_{T,S})} \leqslant M, \tag{3.5}$$

where the constant M *depends on* $\|u_0^{(1)}\|_{L^\infty(\Xi^1)}$, $\|u_0^{(2)}\|_{L^\infty(\Xi^2)}$, $\|u_S^{(2)}\|_{L^\infty(\Xi^2)}$, $\|\beta\|_{L^\infty(\Xi^1)}$, b_1 *and* b_2, *and does not depend on* ε.

In the formulation of Proposition 3.1, by $H^{\lambda',\frac{\lambda'}{2}}(G_{T,S})$ we standardly denote the spaces of Hölder continuous functions with respect to variables $(\boldsymbol{x}, s)$ and t with exponents λ', λ', and $\frac{\lambda'}{2}$, respectively ($\lambda' \in (0,1)$). By $W_2^{2,1}(G')$ we denote the Sobolev space of measurable integrable functions v on G' such that $\partial_t v$, $\partial_{x_i} v$, $\partial_s v$, $\partial^2_{x_i x_j} v$, $\partial^2_{ss} v$, $\partial^2_{s x_i} v \in L^2(G')$, $i, j = 1, \dots, d$.

Proof of Proposition 3.1 is based on the well-known theory of quasilinear parabolic equations of the second order [20], [21, Chapter 1, §2, Chapter 5, §6]. On the first step we notice that there is a unique weak generalized solution of Problem Π_ε for $t \in (0, \tau)$. On the second step we notice that there exists a weak generalized solution for $t \in (\tau, T)$ with the Cauchy data $u_\varepsilon(\boldsymbol{x}, \tau, s) = u_\varepsilon(\boldsymbol{x}, \tau - 0, s) + \beta(\boldsymbol{x}, s)$ on Ξ^1.

4. Relative compactness of $\{u_\varepsilon\}_{\varepsilon>0}$ in $L^1(G_{T,S})$

The first main result of this paper is the following.

Theorem K. *(1) The family of weak generalized solutions* $\{u_\varepsilon\}_{\varepsilon>0}$ *of Problem* Π_ε *(in the sense of Definition 3.1) is relatively compact in* $L^1(G_{T,S})$, *as* $\varepsilon \to 0$.

(2) Function $u = \lim\limits_{\varepsilon\to 0} u_\varepsilon$ *solves kinetic equation* (5.1a) *along with nonnegative Radon measures* m *and* n, *and satisfies kinetic initial condition* (5.1b) *and kinetic impulsive condition* (5.1c).

Proof. Relative compactness of the family of weak solutions to Problem Π_ε is proved with the help of the Perthame-Souganidis and Lazar-Mitrovic averaging compactness theorems [22, Theorem 6], [23, Theorem 7]. On the strength of relative compactness, we derive (5.1a) from (3.2) by means of usual routine. Relations (5.1b) and (5.1c) are proved with the help of Aleksic-Mitrovic strong trace existence theorem [24, Theorem 4]. □

5. Notion of a kinetic solution to Problem Π_0

Theorem K and the results of [25] motivate us to introduce the notion of kinetic solutions to Problem Π_0. This formulation is the second main result of the paper.

Set

$$\chi(\lambda; v) = \begin{cases} +1, & \text{for } 0 < \lambda < v, \\ -1, & \text{for } v < \lambda < 0, \\ 0, & \text{elsewhere.} \end{cases}$$

Definition 5.1. *Let* $N \in \mathbb{N}$, $L > 0$, $\mathcal{O}$ *be an open set of* $\mathbb{R}^N$ *and a function* $h \in L^\infty(\mathcal{O}\times(-L, L))$ *satisfy the inequality* $0 \leqslant h(\boldsymbol{z}, \lambda)\mathrm{sgn}(\lambda) \leqslant 1$ *for a.e.* $(\boldsymbol{z}, \lambda) \in \mathbb{R}^{N+1}$. *We say that* h *is a* χ*-function if there exists a function* $v \in L^\infty(\mathcal{O})$ *such that* $h(\boldsymbol{z}, \lambda) = \chi(\lambda; v(\boldsymbol{z}))$ *for a.e.* $\boldsymbol{z} \in \mathcal{O}$.

Lemma 5.1. *[26, Lemma 2.1.1]. The following two identities hold true (say for* Ψ *measurable on* $\mathcal{O} \times \mathcal{R}_\lambda$ *and locally Lipschitz continuous in* λ, *i.e.* $\partial_\lambda \Psi(\boldsymbol{z}, \cdot) \in L^\infty_{\mathrm{loc}}(\mathbb{R})$*):*

(i) $\int\limits_{\mathbb{R}_\lambda} \partial_\lambda \Psi(\boldsymbol{z}, \lambda)\chi(\lambda; v)\, d\lambda = \Psi(\boldsymbol{z}, v) - \Psi(\boldsymbol{z}, 0)$, *in particular,* $\int\limits_{\mathbb{R}_\lambda} \chi(\lambda; v)\, d\lambda = v$;

(ii) $\int\limits_{\mathbb{R}_\lambda} \partial_\lambda \Psi(\boldsymbol{z}, \lambda)\, |\chi(\lambda; v) - \chi(\lambda; w)|\, d\lambda = \mathrm{sgn}(v - w)(\Psi(\boldsymbol{z}, v) - \Psi(\boldsymbol{z}, w))$, *in particular,* $\int\limits_{\mathbb{R}_\lambda} |\chi(\lambda; v) - \chi(\lambda; w)|\, d\lambda = |v - w|$.

Additionally,

(iii) $|\chi(\lambda; v) - \chi(\lambda; w)| = |\chi(\lambda; v) - \chi(\lambda; w)|^2$.

The following lemma establishes the link between sequences of χ-functions and their limits.

Lemma 5.2. *[27]. Let $\mathcal{O}$ be an open set of $\mathbb{R}^N$ and $h_n \in L^\infty(\mathcal{O} \times (-L, L))$ be a sequence of χ-functions converging weakly to $h \in L^\infty(\mathcal{O} \times (-L, L))$. Set $v_n(\cdot) = \int_{-L}^{L} h_n(\cdot, \lambda)\, d\lambda$ and $v(\cdot) = \int_{-L}^{L} h(\cdot, \lambda)\, d\lambda$.*

Then the three assertions are equivalent:

- *h_n converges strongly to h in $L^1_{\text{loc}}(\mathcal{O} \times (-L, L))$,*
- *v_n converges strongly to v in $L^1_{\text{loc}}(\mathcal{O})$,*
- *h is a χ-function.*

Definition 5.2. *Function $u \in L^\infty(G_{T,S}) \cap L^2((0, T) \times (0, S); \overset{o}{W}{}^1_2(\Omega))$ is called a kinetic solution of Problem Π_0 if it satisfies the following kinetic equation and initial, boundary, and impulsive conditions:*

- *kinetic equation*

$$\begin{aligned}
&\partial_t \chi(\lambda; u(\boldsymbol{x}, t, s)) + \partial_s((\partial_\lambda a(\boldsymbol{x}, t, s, \lambda))\chi(\lambda; u(\boldsymbol{x}, t, s))) + \operatorname{div}_x\big((\partial_\lambda \boldsymbol{\varphi}(\boldsymbol{x}, t, s, \lambda))\chi(\lambda; u(\boldsymbol{x}, t, s)) \\
&- \nabla_x \chi(\lambda; u(\boldsymbol{x}, t, s))\big) + \partial_\lambda\big((g(\boldsymbol{x}, t, s, \lambda) - \partial_s a(\boldsymbol{x}, t, s, \lambda) - \operatorname{div}_x \boldsymbol{\varphi}(\boldsymbol{x}, t, s, \lambda))\chi(\lambda; u(\boldsymbol{x}, t, s))\big) \\
&+ \big(\partial^2_{\lambda s} a(\boldsymbol{x}, t, s, \lambda) + \partial_\lambda \operatorname{div}_x \boldsymbol{\varphi}(\boldsymbol{x}, t, s, \lambda) - \partial_\lambda g(\boldsymbol{x}, t, s, \lambda)\big)\chi(\lambda; u(\boldsymbol{x}, t, s)) \\
&- \delta_{(\lambda=0)} g(\boldsymbol{x}, t, s, \lambda) - \partial_\lambda\big(m(\boldsymbol{x}, t, s, \lambda) + n(\boldsymbol{x}, t, s, \lambda)\big) = 0, \quad (5.1a)
\end{aligned}$$

- *kinetic initial condition*

$$\operatorname*{esslim}_{t \to 0+} \int_{-M}^{M} \int_{\Xi^1} \left|\chi(\lambda; u(\boldsymbol{x}, s, t)) - \chi(\lambda; u_0^{(1)}(\boldsymbol{x}, s))\right| d\boldsymbol{x} ds d\lambda = 0, \quad (5.1b)$$

- *kinetic impulsive condition*

$$\int_{-M}^{M} \chi(\lambda; u(\boldsymbol{x}, \tau + 0, s))\, d\lambda = \int_{-M}^{M} \chi(\lambda; u(\boldsymbol{x}, \tau - 0, s))\, d\lambda + \beta(\boldsymbol{x}, s), \quad (\boldsymbol{x}, s) \in \mathbb{R}^{d+1}, \quad (5.1c)$$

- *kinetic boundary conditions*

$$\begin{aligned}
&\partial_\lambda a(\boldsymbol{x}, t, 0, \lambda)\big(\chi(\lambda; u_0^{\text{tr},(2)}(\boldsymbol{x}, t)) - \chi(\lambda; u_0^{(2)}(\boldsymbol{x}, t))\big) \\
&- \delta_{(\lambda = u_0^{(2)}(\boldsymbol{x}, t))}\big(a(\boldsymbol{x}, t, 0, u_0^{\text{tr},(2)}(\boldsymbol{x}, t)) - a(\boldsymbol{x}, t, 0, u_0^{(2)}(\boldsymbol{x}, t))\big) = \partial_\lambda \mu_0^{(2)}(\boldsymbol{x}, t, \lambda), \quad (5.1d)
\end{aligned}$$

$$\begin{aligned}
&\partial_\lambda a(\boldsymbol{x}, t, S, \lambda)\big(\chi(\lambda; u_S^{\text{tr},(2)}(\boldsymbol{x}, t)) - \chi(\lambda; u_S^{(2)}(\boldsymbol{x}, t))\big) \\
&- \delta_{(\lambda = u_S^{(2)}(\boldsymbol{x}, t))}\big(a(\boldsymbol{x}, t, S, u_S^{\text{tr},(2)}(\boldsymbol{x}, t)) - a(\boldsymbol{x}, t, S, u_S^{(2)}(\boldsymbol{x}, t))\big) = -\partial_\lambda \mu_S^{(2)}(\boldsymbol{x}, t, \lambda), \quad (5.1e)
\end{aligned}$$

where $m, n \in \mathcal{M}^+(G_{T,S} \times (-M, M))$, $n = \delta_{(\lambda=u)} |\nabla_x u|^2$; $\mu_0^{(2)}$, $\mu_S^{(2)} \in \mathcal{M}(\Xi^2 \times (-M, M))$. *Here* $\mathcal{M}^+$ *denotes the space of finite positive Radon measures and* $\mathcal{M}$ *denotes the space of finite Radon measures. Functions* $u_0^{\mathrm{tr},(2)}$ *and* $u_S^{\mathrm{tr},(2)}$ *are strong traces of a solution* $u = u(\boldsymbol{x}, t, s)$ *(if any) of the kinetic equation* (5.1a) *on the planes* $\{s = 0\}$ *and* $\{s = S\}$, *respectively:*

$$\operatorname*{esslim}_{s\to 0+} \int\limits_{\Xi^2} \left|u(\boldsymbol{x}, t, s) - u_0^{\mathrm{tr},(2)}(\boldsymbol{x}, t)\right| d\boldsymbol{x}dt = 0, \quad \operatorname*{esslim}_{s\to S-0} \int\limits_{\Xi^2} \left|u(\boldsymbol{x}, t, s) - u_S^{\mathrm{tr},(2)}(\boldsymbol{x}, t)\right| d\boldsymbol{x}dt = 0. \quad (5.1f)$$

The full proof of existence of a kinetic solution under Conditions on $a\&\varphi\&g$ is yet incomplete. In the stationary case ($a = a(\lambda)$, $\varphi = \varphi(\lambda)$, $g \equiv 0$), the existence and uniqueness were proved, see [25]. Existence of strong traces on Γ_0^2 and Γ_S^2 is not obvious. Even if strong traces do exist, we cannot assume that measures $\mu_0^{(2)}$ and $\mu_S^{(2)}$ are positive measures. Also, the question linked with the uniqueness is open.

6. Conclusion

In this paper we have formulated Theorem K and have given a brief scheme of its proof. As well, we have set up the definition of a kinetic solution to problem Π_0.

Acknowledgments

The work was supported by the Federal Agency for Scientific Organizations of the Russian Federation (project no. III.22.4.2) and by the Russian Foundation for Basic Research (grant no. 18-01-00649).

References

[1] Zel'dovich Y B and Raizer Y P 1969 *Annu. Rev. Fluid Mech.* **1** 385–412
[2] Rebhan E 1990 *Phys. Rev. A* **42**(2) 781–8
[3] Cuffey K M and Paterson W S B 2010 *The Physics of Glaciers* (Amsterdam: Elsevier)
[4] Escobedo M, Vázquez J L and Zuazua E 1994 *Trans. Amer. Math. Soc.* **343**(2) 829–42
[5] Tersenov Al S 2005 *J. Evol. Equ.* **5**(2) 277–89
[6] Ghods A and Arkani-Hamed J 2007 *J. Geophys. Res.* **112** E03005 1–17
[7] Senshu H, Kuramoto K and Matsui T 2002 *J. Geophys. Res.* **107** E12,5118 1–13
[8] Stewart S T and Ahrens T J 2005 *J. Geophys. Res.* **110** E03005 1–23
[9] Fateev E G 2001 *JETP Lett.* **73**(8) 432–5
[10] Fateev E G 2008 *Tech. Phys. Lett.* **34**(5) 363–6
[11] Kuznetsov I 2018 *Math. Model. Nat. Phenom.* **13**(2):20 1–7.
[12] Kuznetsov V V 2018 *Bulletin KRASEC. Physical & Mathematical Sciences* **2018**(1) 117–30 (In Russian. English abstract.)
[13] Bardos C, Leroux A Y and Nedelec Y C 1979 *Commun. Part. Differ. Equ.* **4**(9) 1017–34
[14] Dubois F and Le Floch P 1988 *J. Differ. Equ.* **71**(1) 93–122
[15] Otto F 1996 *C. R. Acad. Sci. Paris Ser. I Math.* **322**(8) 729–34
[16] Málek J, Nečas J, Rokuta J and Růžička M 1996 *Weak and Measure-Valued Solutions to Evolution Partial Differential Equations* (London: Chapman & Hall)
[17] Kwon Y-S 2008 *J. Math. Anal. Appl.* **340**(1) 543–9
[18] Coclite G M, Karlsen K H and Kwon Y-S 2009 *J. Func. Anal.* **257**(12) 3823–57
[19] Evans L C 2010 *Partial Differential Equations* (Providence, RI: Amer. Math. Soc.)
[20] Carrillo J 1999 *Arch. Ration. Mech. Anal.* **147**(4) 269–361
[21] Ladyženskaja O A, Solonnikov V A and Uralceva N N 1968 *Linear and Quasi-Linear Equations of Parabolic Type* (Providence, RI: Amer. Math. Soc.)
[22] Perthame B and Souganidis P E 1998 *Annales Scientifiques de l'É.N.S. 4e ser.* **31**(4) 591–8
[23] Lazar M and Mitrović D 2012 *Dynam. Part. Differ. Equ.* **9**(3) 239–60
[24] Aleksić J and Mitrović D 2013 *J. Hyper. Differ. Equ.* **10**(4) 659–76
[25] Kuznetsov I V 2017 *Sib. Elect. Math. Rep.* **14** 710–31
[26] Perthame B 2002 *Kinetic Formulation of Conservation Laws* (Oxford: Oxford Univ. Press)
[27] Vasseur A 2001 *Arch. Ration. Mech. Anal.* **160**(3) 181–93

Simulation of fracture of elasto-platic solids with cracks under conditions of out-of plane deformation

V D Kurguzov

Lavrentyev Institute of Hydrodynamics SB RAS, Novosibirsk, Russia

E-mail: kurguzov@hydro.nsc.ru

Abstract. The paper considers a tearing mode crack (mode III fracture) propagation in elasto-plastic materials having ultimate strain. The fracture process of such material has been described in the modified Leonov-Panasyuk-Dugdale model with using an extra parameter such as the plastic zone width (the pre-fracture zone width. The two-parameter criterion of quasi-brittle fracture for mode III cracks in elasto-plastic material has been derived for conditions of the low-scale yielding in the presence of a stress field singularity of the field in the vicinity of a crack. The deformation fracture criterion has been deduced for the initial crack tip, whereas the force criterion has been deduced for tangential stresses with account for their averaging at the tip of a model crack. The lengths of initial and model cracks differ by the length of the pre-fracture zone. The sequential analysis of application possibility of the proposed strength criterion has been performed when determining fracture loads for solids with tearing mode cracks. Fracture diagrams of quasi-brittle fracture under conditions of antiplane deformation have been plotted for a band with an edge crack.

1. Introduction

In studies [1–3], the two-parameter (coupled) discrete-integral fracture criterion has been proposed. This criterion may be used when constructing fracture diagrams of plane specimens with opening mode cracks (mode I fracture). In the plane representing stress versus crack length, curves are plotted, which divide this plane into three subareas corresponding to the absence of fracture, damage accumulation in a pre-fracture zone, and fragmentation of a specimen. Fitting constants applied for analytical description of fracture diagrams of materials in the presence of cracks is by the way of approximation of the classical stress-strain diagram of starting material and the critical stress intensity factor (SIF).

The coupled fracture criterion is applied in the present paper for determination of critical loads for an edge mode III crack in a band made from elasto-plasic material.

2. Quasi-brittle fracture diagrams tearing mode crack

The proposed model [1–3] uses the non-classical scheme of material failure.

Suppose that a ($\tau-\gamma$)-diagram is obtained during laboratory experiment when testing macro-specimen. We take the simplest approximation of the real ($\tau-\gamma$)-diagram of material under study when this diagram is approximated by a two-link broken line. When approximating, original material is substituted for ideally plastic material having the ultimate strain. Given in figure 1, *a* are the original ($\tau-\gamma$) - diagram (curve 1) and its two-link approximation (curve 2). Parameters of this approximation are selected so that the areas under the curves should be coincident. The approximation parameters are:

G is shear modulus, τ_Y is shear yield point of the material and constant stresses acting according to the Leonov-Panasyuk-Dugdal (LPD) model [4, 5], γ_0 is the maximum elastic strain of material $(\tau_Y = G\gamma_0)$ and γ_1 is the maximum material strain. The approximation of the ($\tau - \gamma$)-diagram within the segment $\gamma_0 < \gamma < \gamma_1$ can be treated as ideal plasticity.

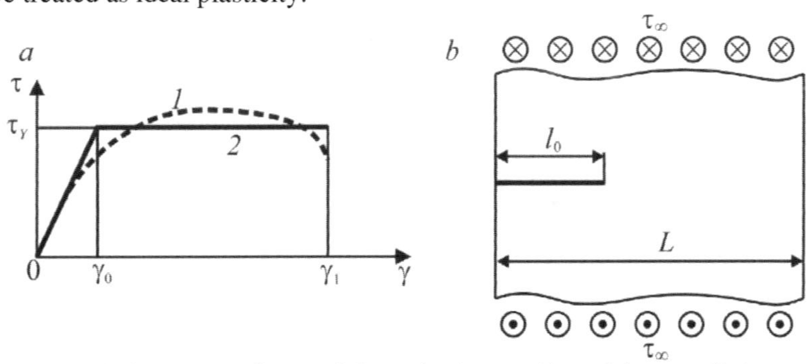

Figure 1. The original diagrams of material strain (curve 1) and its two-link approximation (curve 2) (*a*); the loading scheme of a band with edge tearing mode crack (*b*).

Suppose that r for material with the regular structure is the grain diameter, more exactly, it is the effective diameter of breakdown structures [3]. The Neuber-Novozhlov approach [6, 7] allows solutions having the singular term with integrable singularity to be used for structured media.

Consider an edge tearing mode crack in a band with the width L (Figure 1, *b*). Let this plane tearing mode crack be extended rectilinearly. Except for the length l_0 of a real crack-cut, we introduce into consideration the length of a model crack-cut. The length of the model crack-cut is $l = l_0 + \Delta$, the pre-fracture zone of length Δ being located on the real crack continuation. The fracture problem has two linear scales: if a grain diameter r is governed by the material structure, then the second linear scale is produced by the system itself. This second linear scale is the pre-fracture zone length Δ that changes in accordance with how the following values change: 1) the real crack length and 2) the load intensity. Emphasize that critical pre-fracture zone length Δ^* under single loading is the well-defined parameter ($l^* = l_0 + \Delta^*$ is the critical macro-crack length).

When plotting diagrams of quasi-brittle fracture, sufficient fracture criteria are used [1–3] when short, long macro-cracks and those of medium length are considered. The sufficient (coupled) criterion can be represented in the form of two relations for short macro-cracks and macro-cracks of medium length

$$\frac{1}{r}\int_0^r \tau_{yz}(x,0)dx = \tau_Y , \quad 2w(-\Delta^*) = \delta^* . \tag{1}$$

Here $\tau_{yz}(x,0)$ are shear stresses on the continuation of cracks; $Oxyz$ is the Cartesian coordinate system provided that the coordinate origin is coincident with the model crack tip in the modified LPD model [4, 5], the axis Ox is directed along the crack plane; $2w = 2w(x)$ is the displacement of the model macro-crack flanks ($x < 0$); δ^* is the critical displacement of flanks of this crack, and Δ^* is the critical pre-fracture zone length (critical values obtained via the sufficient and necessary fracture criteria are marked with superscripts * and ^{0}). Attention should be given to the fact that the proposed criterion (1) and (2) is coupled. Figure 2 demonstrates compressive stresses acting in the LPD model on the crack continuation (*a*) and the approximation of a plastic zone by a rectangular pre-fracture zone (*b*). Under conditions of antiplane deformation, by “compressive” are meant stresses opposite in sign to stresses τ_∞, acting at remote boundaries. The pre-fracture zone occupies only a part of the plastic zone.

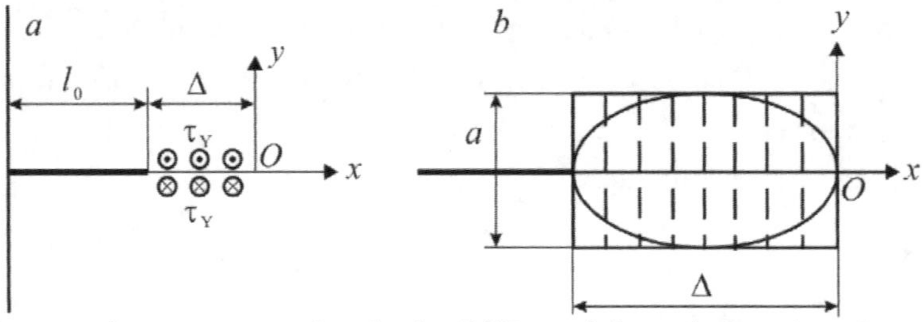

Figure 2. Compressive stresses acting in the LPD model on the continuation of a crack (*a*); approximation of the plastic zone (ellipse) by the rectangular pre-fracture zone (*b*).

The field of shear stresses $\tau_{yz}(x,0)$ on the model crack continuation $x>0$ may be represented as a sum of two terms

$$\tau_{yz}(x,0)=\frac{K_{\text{III}}}{\sqrt{2\pi x}}+\tau_{nom}\,. \tag{2}$$

Here $\tau_{nom}=Y_r\tau_\infty$ are nominal stresses, otherwise the regular part of the stress field in the vicinity of a model crack, $Y_r=Y_r(l/L)$ is the correction factor to be determined, $K_{\text{III}}=K_{\text{III}\infty}+K_{\text{III}\Delta}>0$ is the total stress intensity factor (SIF); $K_{\text{III}\infty}>0$ is the SIF generated by test conditions; $K_{\text{III}\Delta}<0$ is the SIF generated by the given constant stresses τ_Y, acting in the pre-fracture zone. The first and second summands in relation (2) are the singular and regular part of the solution, respectively. The first equality (1) in the coupled criterion keeps track of stresses acting on the continuation of a model crack and their attainment of yield point τ_Y after averaging, and the second equality of this criterion describes blunting of a real macro-crack tip.

We pass to assessment of singular terms of stress fields for edge cracks. Inasmuch as deformation of material under conditions of the low-scale yield is studied, then for specimens with sharp edge cracks, the following relations [8] are obtained for the total SIF $K_{\text{III}}=K_{\text{III}\infty}+K_{\text{III}\Delta}>0$:

$$K_{\text{III}\infty}=Y_s\tau_\infty\sqrt{\pi l}\,,\quad Y_s=\sqrt{\frac{2L}{\pi l}\text{tg}\frac{\pi l}{2L}}\,,\quad K_{\text{III}\Delta}=-\tau_Y\sqrt{\pi l}\left[1-\frac{2}{\pi}\arcsin\left(1-\frac{\Delta}{l}\right)\right]. \tag{3}$$

When considering quasi-brittle fracture under conditions of low-scale yield with account for inequality $\Delta^*/l_0\ll 1$ to the accuracy of higher order of infinitesimals for the summand $\arcsin(1-\Delta/l)$ in relation (3) the representation $\arcsin(1-\Delta/l)\approx\pi/2-\sqrt{2\Delta/l}$ is valid for $\Delta/l\ll 1$. Finally, the simplified expression for SIF $K_{\text{III}\Delta}$ has the form

$$K_{\text{III}\Delta}=-2\tau_Y\sqrt{\frac{2\Delta}{\pi}}\,. \tag{4}$$

The correction factor $Y_r=\dfrac{1}{1-l/L}-Y_s\sqrt{\dfrac{2l/L}{1-l/L}}$, that accounts for a band width L, is deduced from equilibrium condition of stresses applied to remote boundaries and stresses acting on the ligament $L-l$. As $l\to L$, then $\tau_{nom}\to\infty$ are nominal stresses that reflect the rise in stresses with vanishing the net cross-section under constant loading. Both coefficients Y_s and Y_r for an edge tearing mode crack in a semi-infinite solid are identically equal to unity.

When there exists the singular term of the solution under conditions of the low-scale yield, the model crack opening $2w$ in a band is given as

$$2w(-x,0)=\frac{4K_{\text{III}}}{G}\sqrt{\frac{-x}{2\pi}}. \tag{5}$$

The critical crack opening of a model crack δ^* in relation (1) depends on a plasticity margin $\gamma_1-\gamma_0$ of studied material and the plastic zone width a at the tip of a real crack. This critical opening of model cracks δ^* calculated for specimens with a sharp crack is given in such a way

$$\delta^*=(\gamma_1-\gamma_0)a. \tag{6}$$

Let the pre-fracture zone width a in relation (6) be proportional to the plastic zone diameter for specimens with a sharp crack manufactured from a homogeneous material [9]

$$a=\chi\frac{1}{\pi}\left(\frac{K_{\text{III}\infty 0}}{\tau_Y}\right)^2. \tag{7}$$

Here $\chi<1$ is some correction factor for determination of which it is necessary to use the data of the numerical or laboratory experiments. Emphasize that the plastic zone width in relation (7) depends on the length l_0 of an initial crack, i.e., for SIF $K_{\text{III}\infty 0}=K_{\text{III}\infty 0}\left(l_0/r,l_0/L\right)$ representation (3) is used. The model crack length l and parameter l/L in this representation are substituted by the initial crack length l_0 and parameter l_0/L, respectively. The critical opening of the model crack δ^* in relation (6) is fitted so that the material at the real crack tip fails when γ_1 is the maximum material strain.

Let us obtain estimates of the critical material state at the crack tip. All necessary analytical expressions are available in relations (2), (3), and (5) – (7) to make use of sufficient (coupled) criterion (1). After appropriate transformations, the original equalities of criterion (1) change into approximate equalities for a band with a crack:

$$K_{\text{III}\infty}\left(\tau_\infty^*,l^*,\frac{l^*}{L}\right)+K_{\text{III}\Delta}\left(\tau_\infty^*,l^*,\Delta^*\right)=\left(\tau_Y-Y_r\tau_\infty^*\right)\sqrt{\frac{\pi r}{2}}, \tag{8}$$

$$\frac{4}{G}\left[K_{\text{III}\infty}\left(\tau_\infty^*,l^*,\frac{l^*}{L}\right)+K_{\text{III}\Delta}(\tau_\infty^*,l^*,\Delta^*)\right]\sqrt{\frac{\Delta^*}{2\pi}}=\chi(\gamma_1-\gamma_0)\frac{2}{\pi}\left(\frac{K_{\text{III}\infty}\left(\tau_\infty^*,l_0,\frac{l_0}{L}\right)}{\tau_Y}\right)^2. \tag{9}$$

Draw attention to the fact that the initial crack length l_0 enters into equation (9). Since the quasi-brittle approximation is studied, then this length l_0 is substituted by the model crack length l^*, taking into consideration the inequality $\Delta^*/l_0\ll 1$. After transformation, we get the system of equations

$$\bar{\tau}_\infty^* Y_s\sqrt{\frac{2l^*}{r}}-\frac{2\sqrt{2}}{\pi}\sqrt{\frac{2l^*}{r}}\sqrt{\frac{\Delta^*}{l^*}}=1-Y_r\bar{\tau}_\infty^*, \tag{10}$$

$$\frac{\sqrt{2}\tau_Y}{G}\left[\bar{\tau}_\infty^* Y_s-\frac{2\sqrt{2}}{\pi}\sqrt{\frac{\Delta^*}{l^*}}\right]\sqrt{\frac{\Delta^*}{l^*}}=\chi\left(\gamma_1-\gamma_0\right)(Y_s\bar{\tau}_\infty^*)^2. \tag{11}$$

Here $\bar{\tau}_\infty^*=\tau_\infty^*/\tau_Y$ are critical dimensionless stresses in a band via the sufficient fracture criterion. After opening square brackets in the left-hand side of equation (11), the term with the factor Δ^*/l^* appears, which can be dropped as a value of the higher order of infinitesimals in comparison with $\sqrt{\Delta^*/l^*}$ in virtue of the inequality $\Delta^* / l_0 \ll 1$. As a result, only terms with factors $\sqrt{\Delta^* / l^*}$ remain in the system of equations (10) and (11). After solving the obtained system of equations, taking into account the equality $\tau_Y / G = \gamma_0$, we find analytical expressions of dimensionless critical stresses $\overline{\tau}_\infty^*$ and the dimensionless length $\overline{\Delta}^* = \Delta^* / l^*$ of the pre-fracture zone:

$$\overline{\tau}_\infty^* = \left[Y_r + \left(1 - \frac{2}{\pi} \chi \overline{\gamma}_{\text{III}} \right) Y_s \sqrt{2\overline{l}^*} \right]^{-1}, \tag{12}$$

$$\overline{\Delta}^* = \frac{1}{2} \left(\chi \overline{\gamma}_{\text{III}} Y_s \overline{\tau}_\infty^* \right)^2 \tag{13}$$

Here $\overline{l}^* = l^* / r^*$ is the critical dimensionless crack length, $\overline{\gamma}_{\text{III}} = (\gamma_1 - \gamma_0) / \gamma_0$ is the parameter characterizing plasticity margin in longitudinal shear. There is the restriction $2\chi\overline{\gamma}_{\text{III}} < \pi$, wherein the quasi-brittle fracture exists under condition of the low-scale yielding of a homogeneous material in the pre-fracture zone.

Within the limit for $\gamma_1 \to \gamma_0$ the following formula, corresponding to the sufficient fracture criterion, can be derived from relation (13)

$$\overline{\tau}_\infty^0 = \left(Y_r + Y_s \sqrt{2\overline{l}_0} \right)^{-1}. \tag{14}$$

Here $\overline{\tau}_\infty^0 = \tau_\infty^0 / \tau_Y$ is the dimensionless load via the sufficient criterion and $\overline{l}_0 = l_0 / r_0$ is the dimensionless crack length. Relation (14) describes brittle fracture of materials. It is obvious that $\overline{\tau}_\infty^0 < \overline{\tau}_\infty^*$ for $l_0 < l^*$.

If the critical SIF $K_{\text{III}c}$ and the classical $(\tau - \gamma)$ diagram (more precisely its approximation) are obtained in two laboratory experiments, then, using three parameters $r_0, \tau_Y, \chi\overline{\gamma}_{\text{III}}$ or $r^*, \tau_Y, \chi\overline{\gamma}_{\text{III}}$ it is possible to construct two critical curves $\overline{\tau}_\infty^0 = \overline{\tau}_\infty^0(\overline{l}_0, \overline{L}_0)$ and $\overline{\tau}_\infty^* = \overline{\tau}_\infty^*(\overline{l}^*, \overline{L}^*)$ within the wide range of variation in crack lengths. These curves depend on the crack lengths $\overline{l}_0$, $\overline{l}^*$ as well as on the width of a band $\overline{L}_0 = L / r_0$ and $\overline{L}^* = L / r^*$. Let the plane $(\overline{l}_0, \overline{\tau}_\infty^0)$ be compatible with the plane $(\overline{l}^*, \overline{\tau}_\infty^*)$. On the combined plane "crack length versus stresses" $(\overline{l}, \overline{\tau}_\infty)$, where $\overline{l} = l / r$ and $\overline{\tau}_\infty = \tau_\infty / \tau_Y$, we construct diagrams of quasi-brittle fracture for the specimen type being considered. The critical curves on the plane $(\overline{l}, \overline{\tau}_\infty)$ depend on the relation $\overline{L} = L / r$. Let the load intensity $\overline{\tau}_\infty$. be given. Then the diagram of the quasi-brittle fracture allows one to estimate the state of a solid with a crack. Two critical curves $\overline{\tau}_\infty^0$ and $\overline{\tau}_\infty^*$ divide the plane $(\overline{l}, \overline{\tau}_\infty)$ into three subareas: the subarea $\overline{\tau}_\infty < \overline{\tau}_\infty^0$, where there is no fracture; ; the subarea $\overline{\tau}_\infty^0 < \overline{\tau}_\infty < \overline{\tau}_\infty^*$, where damage accumulation takes place in material of the pre-fracture zone; the subarea $\overline{\tau}_\infty > \overline{\tau}_\infty^*$, where the specimen falls under monotonic loading.

Figure 3 demonstrates the dimensionless critical stresses $\overline{\tau}_\infty^0 = \overline{\tau}_\infty^0(\overline{l}_0, \overline{L}_0)$ (curves 1 – 5) and $\overline{\tau}_\infty^* = \overline{\tau}_\infty^*(\overline{l}^*, \overline{L}^*)$ (curves 1' – 5') for specimens with a sharp crack in the log-log coordinates When the concrete implementation of calculation was conducted, the parameters $\overline{L}_0 = 200, 400, 800, 1600, \infty$ for curves 1, 2, 3, 4, 5 and $\overline{L}^* = 200, 400, 800, 1600, \infty$ for curves 1', 2', 3', 4', 5', respectively, were chosen.

The pairs of curves 1 – 1', 2 – 2', 3 – 3', 4 – 4' and 5 – 5' are diagrams of quasi-brittle fracture for the type of a studied specimen manufactured from a homogeneous material.

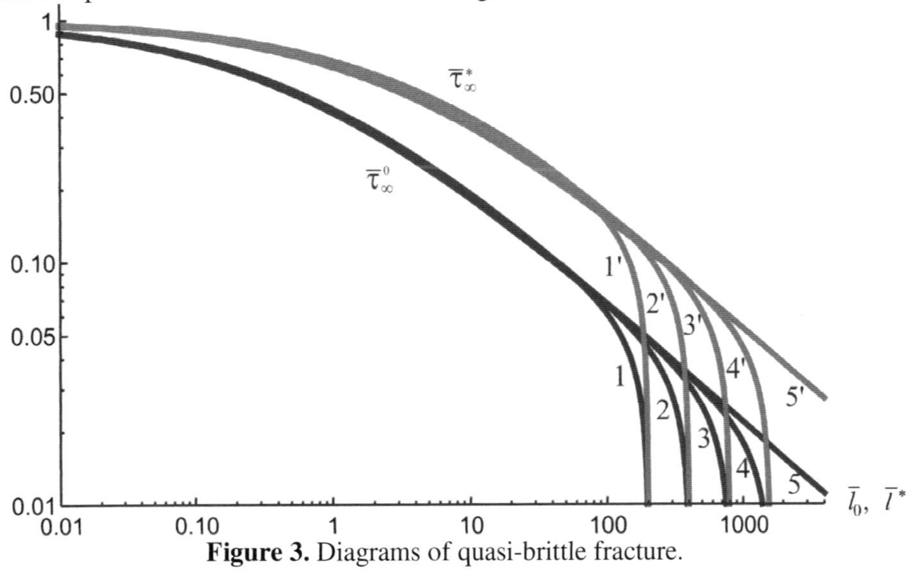

Figure 3. Diagrams of quasi-brittle fracture.

3. Conclusion

The quasi-brittle fracture diagram of an edge crack of the longitudinal shear in a band of the finite width was constructed in the plane "load versus crack length". The diagram consists of two curves, which divide the plane into domains. In the first domain, there is no fracture. Damage accumulation under repeated loading occurs in the second domain. Finally, fragmentation of the specimen under monotonic loading occurs in the third domain. The constants for the analytical description of the fracture diagrams of quasi-brittle materials with cracks were chosen making use of the approximation of the classical stress–strain diagram for the original material and the critical stress intensity factor. The derived structural relations (12) – (14) may be useful for prediction of a critical fracture loading and estimation of the pre-fracture zone length when specimens are loaded with respect to the mode III (antiplane strain loading) in structured materials. These relations express the value of critical loading and the length of a pre-fracture zone in terms of the crack length making use of the following four parameters: r is the characteristic linear parameter of material structure, γ_0 and γ_1 are parameters of the stress-strain diagram, χ is the correction coefficient. The first three parameters are found as a result of a laboratory experiment, and the last parameter is found during computer simulation.

Acknowledgments

This work was supported by the Russian Foundation for Basic Research (Grant Nos. 16-08-00483 and 18-08-00528).

References

[1] Kornev V M 2010 *Physical Mesomechanics* **13** 47
[2] Kornev V M and Demeshkin A G 2011 *J. Appl. Mech. Tech. Phys.* **52** 975
[3] Kornev V M 2013 *Physical Mesomechanics* **16** 25
[4] Leonov M I and Panasiuk V V 1959 *Prikladnaya mekhanika* **5** 391
[5] Dugdale D S 1960 *J. Mech. Phys. Solids* **8** 100
[6] Neuber G 1937 *Kerbspannunglehre: Grunglagen fur Genaue Spannungsrechnung* (Berlin: Springer-Verlag)
[7] Novozhilov V V 1969 *Prikl. Mat. Mekh* **33** 212.

IOP Conf. Series: Earth and Environmental Science **193** (2018) 012038 doi:10.1088/1755-1315/193/1/012038

[8] Murakami Y 1986 *Stress intensity factors handbook* (Oxford: Pergamon Press)
[9] Rice J R 1968 *Fracture* vol 2 (New York: Academic Press)

The using of the shaped charges for destruction of ice cover

V I Kolpakov[1], N A Kudyukov[1], S V Ladov[1], Ya M Nikolskaya[1] and S V Fedorov[1]

[1] Special Machine building department, Bauman Moscow State Technical University, Moscow, Russia

E-mail: sm4-2009@mail.ru

Abstract. The possibility of using cumulative charges for the destruction of the ice cover is considered. On the basis of experimental researches and numerical calculations, the features of the penetration of a cumulative jet into an ice obstacle, as well as the nature of the destruction of ice in the zone of the through hole formed, are analysed. In this case, the existence in the neighborhood of the penetration whole the zones of strength weakened has been established and quantitatively described: zones of fine-dispersed compacted snow-ice, zones of fine-crystalline crushed ice particles and cracking zones. This feature can be used in the development of an appropriate combined explosive device for the destruction of the ice cover

1. Introduction

At the present time, shaped charges (SC) with metal liners of conical or hemispherical shape find wide application both in the military field and in a number of civil explosive technologies [1–3]. At the same time, the most probable problem is the penetration of classical axisymmetric SC, forming by the high-speed shaped charge jets (SCJ), through the monolithic metal barriers [4, 5].

In engineering methods of estimating the depth of penetration from SCJ in this rigid-plastic barriers, as a rule, a hydrodynamic cumulation model with corresponding correction coefficients is used. The role of such amendments (correction coefficients) is especially increasing in the transition to non-traditional material of barriers, such as rock, concrete, ice, frozen ground, water, which have recently attracted interest because of the possibility of expanding the scope of use of SC.

As a rule, such materials have a low density and a sufficiently high compressibility in relation to, for example, steel. Some of them are characterized by the inertial motion of the environment and the fragile nature of the destruction; in addition to the triggering elements of SCJ, an important contribution to the formation of a cavity can be made by a pestle penetrating like a rigid body [6].

Let us consider the features of penetration of ice barriers under the action of SC on the basis of experimental and numerical researches.

1.1 Some information about ice barriers

It is known that an ice barrier of finite thickness has a substrate in the form of water, however the thickness of ice in itself can be very large and for small SCs it is an barrier of semi-infinite thickness. Ice is rather difficult environment with the characteristics changing depending on surrounding conditions [7]. Semicrystalline ice, which consists of randomly interlocked or intergrown individual crystals, is generally spread. Therefore, when assessing the penetrating power of SCJ with a sufficient degree of accuracy, ice can be considered isotropic.

We will provide data on some physicomechanical characteristics of ice [7]. Density of pure ice at 0 ° C is 916,8 kg/m3, sea ice because of existence of a time and impurity 800 … 916 kg/m3. The module of elasticity changes within 9000 … 9500 MPa, the shift module – 3000 … 3400 MPa, Poisson's coefficient – 0,31 … 0,37. Strength at compression changes in the wide range 0,98 … 7,85 MPa, at stretching – 0,66 … 2,75 MPa. Dynamic hardness of ice at decrease in temperature from 0 ° C to - 25 ° C changes from 5 to 8,3 MPa.

It must be said about the heterogeneity of the structure of the intruding ice, which can be layered. As a rule, the basis is crystalline (transparent) ice, which comes from its lower surface, from water, and above it - the so-called cloudy ice (the freezing of wet snow or the freezing of water on the upper surface of the ice cover). The strength of cloudy ice is about 2 times weaker than crystalline ice, but it breaks up under the effect of an explosion more difficult. It is even more difficult to break up layered ice (ice-snow-ice).

All this must be taken into account, first of all, in the calculation methods.

2. Experimental researches

In figure 1 shows the corresponding SC schematic diagrams with a typical explosive charge (EC) form and the shaped charge liners (SCL) that were used in the experiments and calculations.

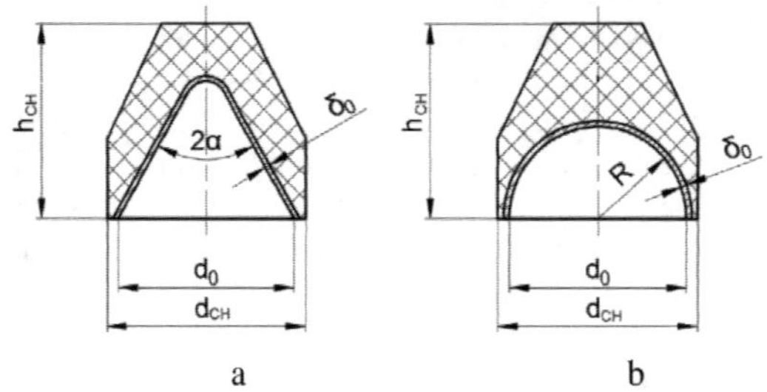

Figure 1. Constructive schemes SC: a - with conical SCL; b - with hemispherical SCL.

In laboratory experiments has been used SC with copper conical SCL of three scales: diameter 20, 36 and 50 mm, which were "shot" either in a rigid steel pipe, or in a steel ice-filled tank (box) by layer wise freezing of fresh ice by the method of slow crystallization of water under natural conditions at temperatures -5° ... -15°C. The result was a fine-grained hemicrystalline ice without significant contamination with an average density of 905 kg / m^3.

In the course of the experiments, the depth of penetration of SCJ into the ice Lice was measured at its installation thickness H_{ice}, equal to the height of the cage (tank, box), and the profile of the hole formed in the ice. At the same time in a case of penetration of the entire ice thickness, a characteristic hole profile is obtained that has an ejection funnel from the front surface (D_f, H_f), a spall funnel from the rear surface (D_S, H_S) and a through-hole of a cylinder-conical shape with an average hole diameter d (Figure 2). For more accurately determine the profile of the through hole and evaluate the state of the material (ice) around it, in some experiments, blocks of pierced ice were extracted from the pipe and cut along the cross-section along the resulting hole.

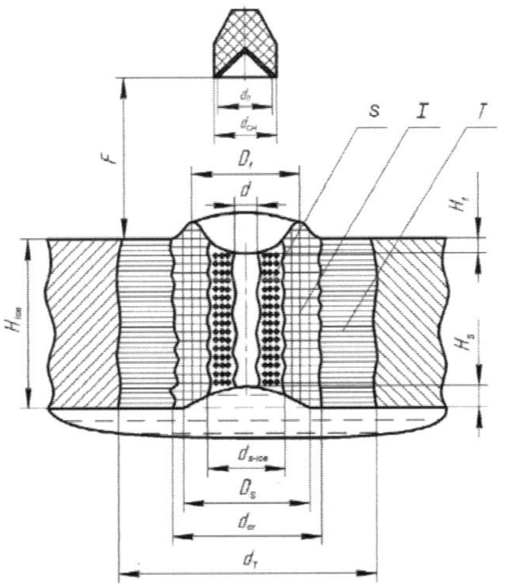

Figure 2. Scheme of interaction of SC with ice barrier.

It was found that in the neighborhood of the through hole three zones are distinctly distinguished: the zone of fine-dispersed compacted snow-ice (zone "S") of diameter ds-ice, which is adjacent directly to the hole and is easily removed, since it has almost no density (the density of the caked snow is 200 ... 400 kg / m^3); zone of fine-crystalline crushed ice particles (zone "I") of diameter d_{cr}, more solid, but unconnected with the bulk of the ice; zone of cracking (zone "T") in an array of diameters d_T. This feature is noted also in work [8].

In the experiments, the open-shell charges of the standard cumulative perforators PKOS-32 (d_{CH} = 20 mm, h_{CH} = 15 mm, d_0 = 18 mm, copper conical liner, 2α = 75 °, δ_0 = 0.6 mm, m_{EC} = 3.4 g) and ZPKO-73 (d_{CH} = 36.5 mm, h_{CH} = 38 mm, d0 = 33 mm, copper conical lining, 2α = 60 °, δ0 = 0.9 mm, m_{EC} = 28 g) equipped with phlegmatized hexogen (ρ_{EC} = 1, 65 g / cm^3, D = 8100 m / s) are used, as well as a laboratory charge of SC-50 from a mixture of Composition B (ρ_{EC} = 1.6 g / cm^3, D = 7600 m / s) having the following geometric parameters : d_{CH} = 50 mm, d0 = 45 mm, copper conical lining, 2α = 50°, δ_0 = 0,7 / 1,4 mm, m_{EC} = 275 g, where m_{EC} is the mass of the charge EC, ρ_{EC}, D - density and velocity of detonation EC. The charges were detonated by means of high-speed electric detonators 9VP through a thin layer of highly sensitive elastic EC.

The results of laboratory experiments on the penetration depth of Lice are given in Table. 1. In the same place, for comparison, the results of calculations based on the well-known engineering method of V.M. Marinin [5], stated in earlier work of authors [6], and also the known data from of natural experiments using SC with a diameter d_{CH} = 70 ... 215 mm with conical steel SCL for penetrating ice (frozen ground) barriers are given [3, 8].

The best relative results were obtained for an SC with a diameter d_{CH}= 50 mm (variant No. 3, Table 1), which were: Lice = (18 ... 20) d_{CH}, d = (0.5 ... 0.6) d_{CH}. In comparison with the known results of natural experiments (variants No. 4 and No. 5, Table 1), the data of laboratory experiments are maximally underestimated.

In general, SC with copper and steel conical SCL are capable of penetrating a semi-infinite barrier from solid ice (frozen ground) to a depth of at least (20 ... 25) d_{CH}, with an average diameter of the through-hole being at least (0.6 ... 0.7)d_{CH}, where d_{CH} is the diameter of the charge EC. When the barrier penetrates the final thickness, this value increases by 10 ... 15% and a pronounced spall funnel appears (Figure 2). This forecast is based on experiments made in middle latitudes on fresh ice. Obviously, in conditions of real Arctic sea ice and substantially low below-zero temperatures, the penetration capacity of SC can decrease

Table 1. The results of calculations for the penetration of ice by a SC

Mechanical and structural parameters			
Variant of calc.	Parameters of SC (Figure 1)		
	d_{CH}, мм	SCL's shape	SCL's material
1	20	cone 2α = 75°	copper
2	36	cone 2α = 60°	copper
3	50	cone 2α = 50°	copper
4	70	cone 2α = 50°	steel
5	215	cone 2α = 70°	steel

The main results of SC

Variant of calc.	F, mm (Figure 2)	Calculations by engineering methods		Experiment	
		L_{ice}, mm	d, mm	L_{ice}, mm	d, mm
1	30	218	8,8	225…250	8..10
2	50	314	15,9	325	15
3	150	915	23,2	925	30
4	190	1310	32,7	1300	38…44
5	340	2034	127,2	2000	160

As a result of the experimental researches, the sizes of the identified zones of weakened ice strength were determined during penetration of SC with copper conical liners with a cone angle of 2α = 50 ... 75°, which were: $d_{s\text{-}ice}$ = (2 ... 3)d_{CH}, d_{cr} = (4 ... 5) d_{CH}, $d_T \geq 20d_{CH}$ (Figure 2). The provided data indicate that in actual practice around the hole in ice from the action of SCJ a certain "fictitious" destruction zone with a diameter of approximately (2 ... 5)d_{CH} is formed for SC with high copper conical lining. The dimensions of this zone depend on the SCJ parameters: more massive and low-gradient jets (conical liners with large solution angles 2α = 90 ... 140° and hemispherical linings) should provide larger sizes, in particular, zones "S" and "I". The same tendency should be observed when using steel and aluminum (instead of copper) as SCL material. In practical terms, this can be important for the subsequent action on a weakened ice of a high explosive EC charge or a high-speed penetrating solid cylindrical element [9, 10].

3. Numerical researches

To estimate the cumulative effect on ice barriers, SC calculations with conical and hemispherical copper SCL (Figure 1) using numerical simulation methods [5, 11, 12] were performed.

The tasks of penetrating the ice barrier with cumulative charges with conical and hemispherical facing were solved in two stages.

At the first stage, the cumulative effect of charges on a steel barrier was investigated with the aim of justifying the applicability of the chosen mathematical model for solving this class of problems.

Wherein thickness of the steel barrier was 180 mm, the focal length F = 50 mm. The free space around the SC and the obstructions is filled with air. The dimensions of the calculated area were: 250 mm (≈ 6.9 d_{CH}) along the z axis, 100 mm (≈ 2.8 d_{CH}) along the r axis.

In the mathematical formulation of the problem for steel, an ideal elastic-plastic model with destruction was accepted. The criterion of destruction was taken in the form $\sigma_i > \sigma_p$, where σ_p is the spall strength of the material. For the liner material (copper), an ideal elastic-plastic model without destruction was accepted.

As the equations of state (EQS) of interacting media, the following ratios were used: for steel and copper – barotropic dependence; for detonation products - EQS in the form of JWL; for air - EQS of ideal gas. Parametrs in this equation were accepted: for steel: ρ_0 = 7,85 g/cm^3, K = 175 GPa; G = 80,8 GPa; Y = 0,8 GPa, σ_p = 1,65 GPa; for copper: ρ_0 = 8,96 g/cm^3, K = 152,8 GPa; G = 39,9 GPa; Y = 0,1 GPa; for air: k = 1,4; ρ_0 = 1,225·10-3 g/cm^3, p_0 = 1,013·10-4 GPa.

The characteristics for EC and the constants in EQS in the form of JWL were calculated using the procedure [13] and are presented in table. 2.

The boundary conditions were assigned as follows:

- on contact surfaces $\sigma_{n1} = \sigma_{n2}$.
- the parameters at the front of the detonation wave correspond to the parameters of the Chapmen-Juge.
- conditions for the flow of the medium outflow along the boundaries of the calculated area that are not in contact with the barrier material are established.

The results of the performed calculations are illustrated in figure 3.

Table 2. Characteristics and constants of pressed hexogen

Parameter	Value
ρ_0, g/cm^3	1,65
D, km/s	8,1
Q, kcal / kg	1250
Q, MJ / kg	5,23
pCJ, GPa	27,0
A, GPa	1200
B, GPa	44,1
r_1	5,91
r_2	1,82
ω	0,32
k	3,0
e_0, GJ / m^3 = GPa	8,64

As can be seen from figure 3a, the penetration depth of the steel barrier by the charge ZPKO-73 with a conical copper liner is approximately 87 mm, which differs from the experimental results (85 mm) by only 2%, with the diameter of the formed whole being approximately -12 mm. This allows us to state that the chosen mathematical model corresponds to reality. The penetration depth of steel SC with hemispherical liner (see figure 3b) is approximately 40 mm, the jet, as in the previous case, is clogged into a hole with a diameter of 20 mm.

Thus, based on the results of numerical calculations, the penetration depth of a steel semi-infinite SC with a conical copper SCL is at least 2 times greater than the depth of penetration of SC with a hemispherical SCL, however, the diameter of the hole formed when using hemispherical SCL is about 1.7 times larger.

At the second stage, the effect of the same SC on the ice barrier was investigated. The thickness of the ice barrier was 740 mm. The dimensions of the calculated area were: 750 mm ($\approx$20.8 d_{CH}) along the z axis, 100 mm ($\approx$ 2.8 d_{CH}) along the r axis.

IOP Conf. Series: Earth and Environmental Science **193** (2018) 012039 doi:10.1088/1755-1315/193/1/012039

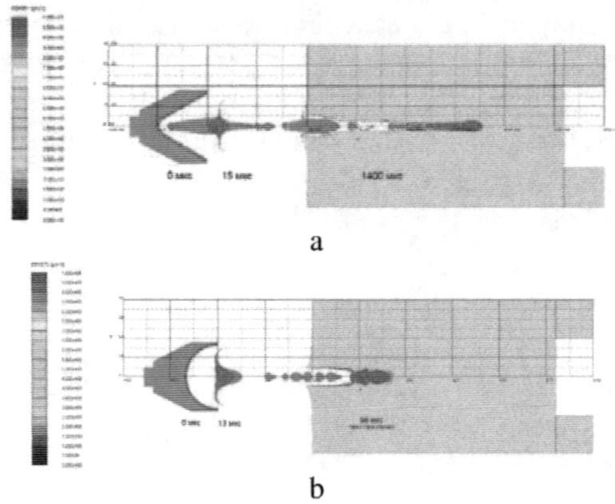

Figure 3. Results of calculations the penetration of the of steel barrier by SC (in different shades the density field is shown): a - with conical liner; b - with hemispherical liner.

Similar equations of state for ice in the mathematical formulation were given; the ice behavior model is an ideal elastoplastic with destruction. The corresponding characteristics of ice: $\rho_0 = 910$ kg / m^3, K = 9.9 GPa; G = 3.6 GPa; Y = 0.01 GPa, $\sigma_p = 0.003$ GPa [7].

The results of the calculations are shown in figure 4.

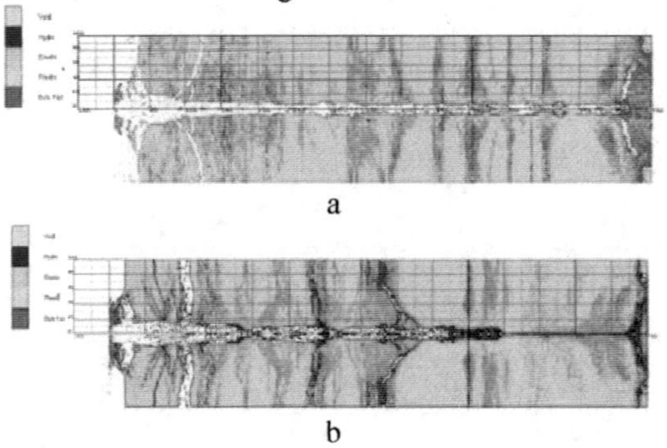

Figure 4. Results of calculations of penetration of the ice barrier for the moment of time t = 1400 μs (in different shades the material state field is shown):a - with conical liner, b - with hemispherical liner.

As can be seen in Figure 4a, the penetration depth of the ice SCJ formed from the conical liner is about 650 mm, and the diameter of the hole formed is approximately 18 mm. A jet of hemispherical liner provides a cavern in the ice with a depth of approximately 500 mm and a diameter of about 30 mm (figure 4b).

Thus, the depth of penetration of ice when using a charge with a conical copper SCL is approximately 1.5 times greater than when using a charge with a hemispherical liner, however, the diameter of the hole formed is 1.7 times smaller.

At the same time for both cases of SCJ penetration into the ice (see figure 4), it should be noted the following:

- along all width of an ice barrier of final thickness active crack propagation and strong destruction of the surface layer of ice are observed, followed by its outflow (the thickness of the ejected part is approximately 70 mm for both cases).

- in the finite thickness barrier, there is a splitting having parameters $D_S \approx 120$ mm, $H_S \approx 40$ mm for both cases (figure 2).
- SCJ pestle, formed from conical liner, does not get stuck in the punched hole, but continues to penetrate into the barrier, which was also observed in experimental studies (at the time of 2820 μs, pestle penetrates the ice barrier through and through), in the case of hemispherical liner such a pestle is not formed.

In comparison with the data of Table. 1 the obtained results of numerical calculations give overestimated results in relation to the laboratory experiment and calculation by the engineering method (variant No. 2). At the same time, the obtained results fit perfectly into the recommended range of penetration depths and hole diameters in the ice barrier under the action of SC with conical and hemispherical liners.

4. Conclusion

As a result of the laboratory experiments and theoretical researches, the features of the penetration of ice barriers by axisymmetric SC with conical and hemispherical SCL are studied. The obtained results can be used in the development of appropriate explosive devices for the destruction of the ice cover.

Acknowledgment

The work was financially supported by the Ministry of Education and Science of the Russian Federation within the framework of the basic part of the state task under the section "Initiative Scientific Projects" (code 9.5330.2017.BP).

References

[1] Babkin A V, Veldanov V A and Gryaznov E F 2016 *Ammunition* (Moscow: BMSTU Publishing) p 506
[2] Minin I V and Minin O V 2013 *Shaped charges* (Novosibirsk: SGGA) p 200
[3] Ladov S V and Kobylkin I F 1995 *Use of cumulative charges in explosive technologies* (Moscow: BMSTU Publishing) p 47
[4] Walters W P and Zukas J A 1989 *Fundamentals of shaped charges* (N.Y.: John Wiley and Sons Publishing) p 398
[5] Andreyev S G, Babkin A V and Baum F A 2004 *Physics of explosion* (Moscow: Fizmatlit Publishing) p 656
[6] Ladov S V, Kolpakov V I and Fedorov S V 1995 Features of penetration the ice and grunto-concrete barriers cumulative charges *Defensive technique* **4** 39–45
[7] Maeno N 1988 Science about ice (Moscow: Patterus Publishing) p 231
[8] Ozeretskovskiy O I 2007 Action of explosion on underwater objects (Moscow: Federal State Unitary Enterprise TsNIIHM Publishing) p 262
[9] Ladov S V, , Klimachkov S I, Okhitin V N and Fedorov S V 2018 About a possibility of use of the explosive device the combined type for destruction of an ice cover *Engineering Journal: Science and Innovation* **2** 74.
[10] Veldanov V A, Isaev A L, Ladov S V and Fedorov S V 2002 *Penetration of a solid cylindrical body into the continuous and weakened by an opening ice Defensive technique* **11** 46–51
[11] Kolpakov V I, Ladov S V and Rubshov A A 1998 *Numerical simulation of shaped charges* (Moscow: BMSTU Publishing) 87–92
[12] Fedorov S V, Ladov S V and Nikolskaya Ya. M 2018 Comparative analysis of formation of shaped charge jets from conical and hemispherical liners *Engineering Journal: Science and Innovation* **1** 73.
[13] Kolpakov V I 2016 Determination of the constants of the equation of state of detonation products in the form of Jones-Wilkins-Lee *Izvestiya RARAN* **4** 28 87–92

IOP Conf. Series: Earth and Environmental Science **193** (2018) 012040 doi:10.1088/1755-1315/193/1/012040

Numerical modeling of salt domes formation in Anabar-Khatanga oil and gas area (Siberian sector of the Russian Arctic)

B Lunev[1], T Abramov[2], A Kalinin[1], V Kontorovich[1] and V Lapkovskiy[1]

[1]Trofimuk Institute of Petroleum Geology and Geophysics, Novosibirsk, Russia
[2]Inversion Sensor, Novosibirsk, Russia

Abstract. The present work is devoted to modeling the processes of formation and analysis of specific features of the salt domes structure in the Anabar-Khatanga OGA. The performed numerical experiments showed that the specificity of the salt tectogenesis in this region is determined by the features of its geological structure and the history of tectonic development.

1. Introduction

The Anabar-Khatanga saddle, located in the northeast of the Krasnoyarsk Region and the northwestern part of the Sakha Republic (Yakutia), is one of the perspective areas for the Siberian sector of the Russian Arctic.

The analysis of drilling data and seismic cross-sections makes it possible to identify six regionally developed complexes in the sedimentary cover of the Anabar-Khatanga oil and gas area (OGA): Riphean, Vendian, Paleozoic, Permian, Triassic and Cretaceous (figure 1). Riphean limestones lying at the base of the platform deposits are covered with Vendian carbonates from above. Above there are the Cambrian mainly carbonate deposits, which, with a break, are covered by limestone of the Devonian-Carboniferous, or terrigenous Permian deposits. The upper part of the deposits is composed of terrigenous rocks of the Triassic, Jurassic and Cretaceous. A feature of the geological structure of this region is the presence in the Early Middle Devonian salt strata and the associated series of salt domes, which largely determined the structural-tectonic construction of this region. The Nordvik and Kozhevnikov diapirs that were studied on this territory break through the deposits of the Upper Devonian, Carboniferous, Permian and Mesozoic and come out to the surface. The analysis of geological and geophysical materials allows us to assume that the most of local high-amplitude positive structures identified in the Anabar-Khatanga OGA are associated with surface-coming and subsurface salt diapires which in the time cross-sections are characterized by a staggering out of phase and a reduction in the amplitude-energy characteristics of the wave fields – a chaotic pattern of seismic recording (figure 1).

In this region, the salt-dome anticlinal structures are of the greatest interest in relation to oil and gas potential. Near the known diapires intersecting the entire of the Upper Devonian, Carboniferous and Permian deposits, structural-tectonic traps associated with the wedging out of terrigenous and carbonate reservoirs onto salt stocks can form; over the subsurface salt domes, in the Permian and Mesozoic deposits classical anticlinal traps can be formed. These circumstances determine the interest in studying the salt tectonics of the Anabar-Khatanga region.

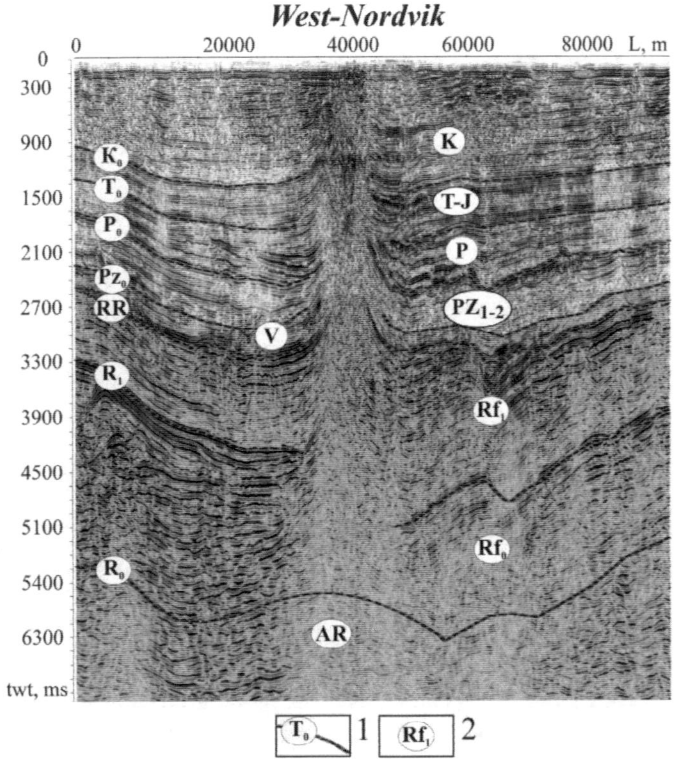

Figure 1. Seismogeological cross-section showing the West-Nordvic salt dome. Legend: 1 – reflector, 2 - seismic complex.

The goal of the present research: using the algorithms and software developed at IPGG SB RAS and taking into account the specific features of the geological structure of the Anabar-Khatanga OGA, to perform numerical experiments on the modeling of salt tectogenesis processes with the aim of establishing the basic regularities of the salt-dome structures formation - the main oil and gas objects of this region.

The mechanism of salt tectogenesis has been thoroughly studied in a century and a half of research. The development of salt tectonics is mainly determined by the fact that salt, unlike terrigenous and biogenic sediments, practically does not compact under the weight of overlying rocks. Therefore, after the burial of salt having a density of 2100-2200 kg/m3 to a depth of more than 1000-1500 m, density inversion occurs and the salt layer begins to float, forming the typical structures of Rayleigh-Taylor instability development. The process dynamics and the shape of the formed structures are determined by the ratio of the densities, thicknesses and (to a lesser extent) the rheological properties of the floating and overlapping layers, as well as the rate of accumulation of the latter and time.

The perturbing factors that distort the symmetry of the resulting structures are external, in relation to the system, movements, edge effects due to the initial form of the salt layer, and the systematic nature of the change in the weight of the overlapping deposits (increase in their thickness or/and density in a certain direction and time).

The highly effective methods and programs of numerical modeling of the described process developed at IPGG SB RAS made it possible to calculate various variants of the development of salt tectogenesis and to reveal the main regularities and specifics of the formation of the salt-dome structures of the Anabar-Khatanga OGA.

2. The numerical modeling methods, the basic geological model of the object

In accordance with the theory of simple fluids with fading memory [1], the first approximation of the rheological equation of state, to describe the irreversible deformation of almost any material, is the equation of the Newtonian fluid. In this case, the Newtonian viscosity is interpreted as the "natural viscosity" of a given material, defined as the upper asymptote of its viscometric viscosities. This approximation is adequate for deformation rates less than a certain critical value for this material. In the case of rocks, the critical deformation rate can be estimated to be of the order of 10^{-14} s^{-1} [2]. Available data [3] show:

- the rate of deformation in the processes of salt domes growth does not exceed this limit;
- in spite of a very large variation in the estimates of the effective viscosity of sedimentary rocks, especially salts, depending on the estimation method and deformation rate, the upper asymptotic for all is close to 10^{20} Pa s;
- the typical mushroom shape of mature salt diapires certainly indicates the proximity of viscosities of salt and host rocks in this process.

Proceeding from it, for the considered class of tasks, it is completely correct to represent the medium as a homogeneous viscous Newtonian fluid. The known smallness of the Reynolds number (of the order of 10^{-22}) determines the investigated flow as a "creeping", the evolution of which can be represented by a sequence of quasistationary states connected together.

Thus, the modeling of salt tectogenesis in this paper is reduced to the calculation of the gravity driven creeping flow of a Newtonian fluid with a inhomogeneous density, bounded from above by a free surface. In the rectangular Cartesian coordinates X_1, X_2, X_3 we consider the half-space $F\,(\boldsymbol{x},t) = x_3 - h\,(x_1,x_2,t) = 0$ $x_3 \le h\,(x_1,x_2,t)$; $\{\boldsymbol{x}\} = \{x_1,x_2,x_3\}$, t is the time, $\boldsymbol{n}$ is the vector of the unit normal to this surface. The half-space is composed by a set of immiscible layers (or closed bodies) D_k, separated by boundaries $S_{k(x,t)}$, whose configuration is varied by the calculated flow. The flow is driven by normal gravity force $\boldsymbol{g}$ applied to the density perturbation, which is related to the configuration of the boundaries $S_{k(x,t)}$. The initial conditions are determined by any given configuration of the boundaries $S_{k(x,t)}$. The density, stresses, and pressure are represented by the sums $\rho_{(x,t)} = \rho^0_{(x_3,t)} + \sigma_{(x,t)}$; $\boldsymbol{T}_{(x,t)} = \boldsymbol{T}^0_{(x_3,t)} + \boldsymbol{\tau}_{(x,t)}$; $P_{(x,t)} = P^0_{(x_3,t)} + p_{(x,t)}$, where $\rho^0_{(x_3,t)}$, $\boldsymbol{T}^0_{(x_3,t)}$, $P^0_{(x_3,t)}$ – are the characteristics of the hydrostatic state $T^0_{ij} = -\delta_{ij}P^0 = -\delta_{ij}g\int_0^{x_3}\rho^0_{(x_3)}dx_3$, δ_{ij} – is the Kronecker delta, and $\sigma_{(x,t)}$, $\boldsymbol{\tau}_{(x,t)}$, $p_{(x,t)}$ are their small perturbations. Flow $\boldsymbol{v}$ is associated with the perturbations.

The problem of calculating the creeping flow is divided into a quasistationary and evolutionary. In the case of a quasistationary problem $S_{k(x,t_n)}$ given at some time t_n, and the corresponding density perturbation $\sigma_{(x,t_n)}$, it is required to find the flow field $\boldsymbol{v}_{(x,t_n)}$, $p_{(x,t_n)}$ and the shape of the free boundary $F_{(x,t_n)}$. The smallness of the perturbation of the free boundary h (with respect to the horizontal dimensions) and the derivatives $\frac{\partial h}{\partial x_1}, \frac{\partial h}{\partial x_2}$, allows to linearize the boundary conditions of the quasistationary problem and write it with respect to perturbations in the form:

$$\begin{gathered} \sigma_{(x,t_n)} = \rho_{(x,t_n)} - \rho^0_{(x_3,t_n)},\ \rho_{(x,t_n)} = \rho_k \text{ для } \boldsymbol{x} \in W_k, \\ \mu\nabla^2\boldsymbol{v} - \boldsymbol{\nabla}p = -\sigma\boldsymbol{g},\ \mu - \text{viscosity}, \\ \boldsymbol{\nabla}\cdot\boldsymbol{v} = 0, \\ (v_3 = \tau_{31} = \tau_{32} = 0)_{x_3=0}, \end{gathered} \tag{1}$$

with the additional condition for determining the perturbation of the free surface $(\tau_{33})_{x_3=0} = -\rho^0|\boldsymbol{g}|h$, ($\boldsymbol{v}$ and p are continuous everywhere in the half-space).

The evolutionary problem consists in finding the motion of the boundaries $S_{k(x,t)}$ from equations:

$$\frac{\partial S_k}{\partial t} + \boldsymbol{v}\cdot\boldsymbol{\nabla}S_k = 0, \tag{2}$$

with certain initial conditions $S_{k(x,t_0)}$, $\boldsymbol{v}_{(x,t_0)} = 0$.

The evolution of the flow is calculated by solving the problem (1) with specified by (2) $S_{k(x)}$ and the corresponding density perturbation $\sigma_{(x)}$ and the subsequent integration of (2) (with the obtained $\boldsymbol{v}_{(x)}$), with respect to small time increments δt (2).

Equations (2) can be easily solved numerically. The main computational difficulties in the calculation of creeping flows of the studied type are related to the quasistationary problem (1). Its solution was obtained analytically in the form of a Green function [4], so that the calculation of the creeping flow at each instant of time reduces to calculating the convolution integrals:

$$v_{i(x)} = g \iiint \sigma_{(\xi)} \mathrm{V}_{i(x,\xi)} \, d\xi_1 d\xi_2 d\xi_3,$$

$$p_{(x)} = g \iiint \sigma_{(\xi)} \mathrm{p}_{(x,\xi)} \, d\xi_1 d\xi_2 d\xi_3,$$

where $\mathrm{V}_{i(x,\xi)}$, $\mathrm{p}_{(x,\xi)}$ – corresponding Green functions.

The use of the convolution integral gives greater computational advantages than the solution of the system of linear algebraic equations arising in the realization of difference methods [5]. Exact solution of the boundary value problem in the form of a Green function makes it possible to apply the convolution theorem with the help of the fast Fourier transform procedure [6]. It was realized in the CoreModuleFFT program developed in the laboratory of mathematical modeling of natural oil and gas systems of IPGG SB RAS, used in the paper,

The above method allows us to calculate the flow caused by the action of the Archimedean forces, creating structures of salt tectonics, in general. To model their formation under the specific geological conditions, it is necessary to take into account the regional vertical movements forming the sedimentary basin and the processes of accumulation, compaction and erosion of sediments for each moment of evolution.

The regional model that determines the specificity of the demonstration of salt tectogenesis are based on the following data on the development of the basin, derived from the results of the geological and geophysical study of the territory [7, 8, 9].

- 407-384 million years (Eifelian age) – the formation of salt evoparitic sediments. At the same time, according to most researchers, the salt basin was of limited dimensions and occupied the submerged part of the depression extended in the northeast direction. Therefore, the starting position of the salt pack was set in the form of a strip about 20 km wide (the length of the calculated model was 40 km). The thickness of the pack was set from 0 m at the edges to a maximum of 700 m in the three most powerful "depocenters" in the axial part of the depression, spaced about 10 km apart. Considering that in the cores of the diapires emerging on the surface, the salt formation is represented by practically pure rock salt - up to 99.7% NaCl, the density of these deposits was set at 2100 kg/m3. The density of the underlying rocks, mainly represented by limestones and dolomites with anhydrite lenses, is assumed to be 2800 kg/m3.
- 385-303 million years – the formation of mainly carbonate sediments of the Upper Devonian-Carboniferous with a thickness of 1500-1600 m.
- 303-286 million years – the depositional break, erosion of 500 m of sediments.
- 286-225 million years (Permian, Triassic) - the formation of terrigenous deposits with a thickness of 5200-5500 m.
- 225-208 million years (Triassic) – the raising of the territory, erosion of the 100 m of sediments.
- 208-97 million years (Jurassic-Lower Cretaceous) – the formation of terrigenous deposits with a thickness of 1500 m.
- 97-94 million years (Cretaceous) – the raising of the territory, erosion of the 300 m of sediments.
- < 94 million years (Upper Cretaceous, Cenozoic) – the formation of terrigenous deposits with a thickness of 300 m.

Based on these data, a piecewise constant rate of immersion and upwelling of the basin with accumulation or erosion of sediments is given for the territory. It was believed that the accumulated deposits were compacted as the immersion, so that the density of overlapping carbonate deposits of the Devonian and Carboniferous reaches 2700 kg/m^3, the density of the most submerged terrigenous rocks of the Permian age reaches 2,500 kg/m^3, and for the Upper Cretaceous 2300 kg/m^3 (with intermediate values in the Triassic and Lower Cretaceous).

3. Modeling results

In figure 2 the results of modeling are presented in the form of a series of sequence stages of system development. As shown by numerical experiments, the development of salt tectogenesis in the Anabar-Khatanga OGA is characterized by the following features.

The insignificant width of the basin, in which the accumulation of salt occurred, did not allow the development of parallel swells and ridges of domes growing from them, with a distance between swells (ridges) of the order of the main characteristic wavelength $\lambda{\sim}2d$, where d is the thickness of the convection layer (in this case – the thickness of the layer from the bottom of the salt pack to the surface). As a result, only one range of diapires was formed from an unstable pack.

A large deficit in the density of the salt pack, in relation to all overlapping rocks, including terrigenous deposits of the Upper Paleozoic and Mesozoic, has led to the fact that the most developed salt diapires rise all the thickness of the deposits, coming in some cases to a surface.

The considerable thickness of the deposits overlapping the unstable pack, combined with the high density of the underlying dolomites and anhydrite, led to the growth of diapires in the form of pillars - streams, crowned with a mushroom-shaped bulb, expanding as it approached the free surface. (When the density of rocks underlying the unstable pack is close to the density of the overlapping rocks, the compression of the diapiric base does not occur, and salt domes grows as the anticline smoothly narrowed to the top [10]). The high density of carbonate rocks, initially lying on the salt pack, promotes an additional compression of the growing domes base. As a result, an essential feature of mature diapires is their pipelike structure, with a core made by a substrate that rises, in a number of cases, to the entire height of the diapir (which suggests the possibility of finding the substrate rocks in the cap-rock of the domes that have come out to the surface and eroded). This effect, along with the pronounced 3- dimensional character of the structures, is strongly promoted by the extremely low density of the salt-pack rocks. An important (including, in practical terms) feature of the salt tectogenesis revealed by modeling for the study area is the formation of a significant number of small, buried, deep-lying diapires located relatively close to each other - at a distance of about 3-4 km (although the most large and mature, the diapires that have reached the surface are 10-15 km apart). This phenomenon is largely due to the fact that the salt pack is covered by high-density carbonates thicker than 1 km (and initially, up to erosion - more than 1.5 km), above which lie significantly less dense terrigenous rocks. As a result, there is a local instability that forms the growth of an additional perturbation mode with a wavelength $\lambda_m{\sim}2d_m$,, where d_m is the thickness of the layer from the bottom of the salt to the top of the carbonate formation, where a density contrast occurs.

According to the calculations, noticeable growth of the domes began in early Permian and by the beginning of the Triassic, their amplitude reached 1000 m. The main growth of these diapires occurs in the Mesozoic. By the end of the Triassic, the most intensively growing domes passed the "pillow" and "finger" stages, and during the Jurassic the amplitudes were significantly increased and classic mushroom-shaped diapires were formed. In the Cretaceous and Cenozoic the process of growth of the main domes continued and at the same time their transformation into funnel-shaped bodies took place. In the upper parts of the domes, typical overhangs continued to form, and in the axial parts of the diapires hats depressions were formed. This is due to the fact that against the backdrop of the continuous process of floating and spreading salt under a free (day) surface, saturated with low-density salt the diapires edge continued to float, while the inner parts of diapires enriched with heavy substrate began to plunge (figure 2, the final stages).

The main growth of the buried domes of the shorter-wave mode - cryptodiapires, forming in the sedimentary cover anticlinal traps perspective for oil and gas potential, occurred most intensively in the Late Cretaceous and Cenozoic.

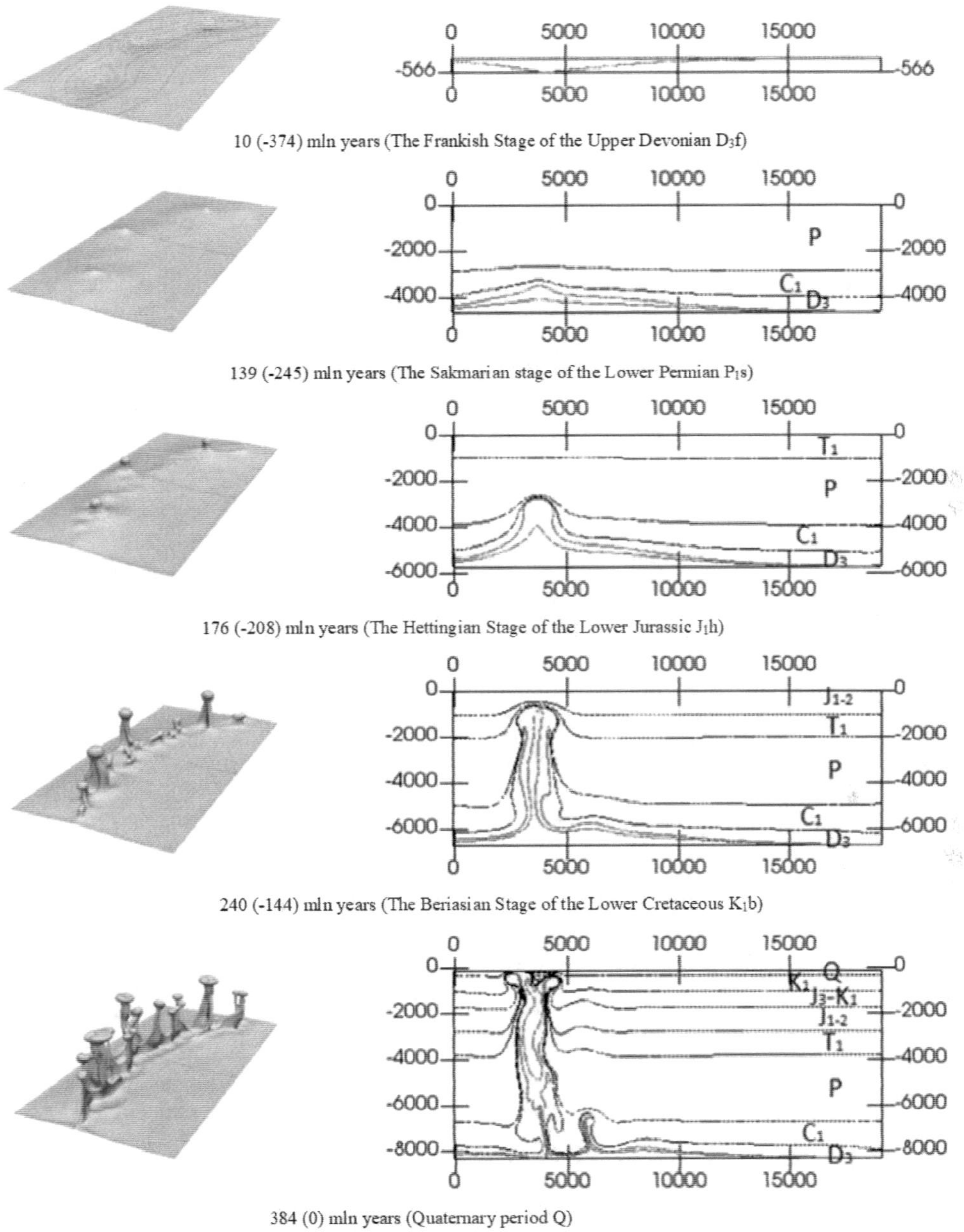

Figure 2. The calculated stages of the salt-dome structures evolution of the Anabar-Khatanga saddle. Above the section for each stage is the time of evolution of the model and the corresponding geological age (in brackets). Red color shows the bottom and top of the salt pack.

4. Conclusion

The present work is devoted to modeling the processes of formation and analysis of specific features of the structure of salt domes in the Anabar-Khatanga OGA. The performed numerical experiments showed that the specificity of the salt tectogenesis in this region is determined by the features of its geological structure and the history of tectonic development:

1. The accumulation of the Devonian salt pack is not universal, but in a separate, relatively narrow depression.
2. Significant depth of burial of salt deposits.
3. Extremely low density of salt deposits, composed of practically pure rock salt, with respect to all deposits, primarily to high-density carbonates underlying and directly overlapping rocks.

The results of numerical experiments, which are in good agreement with the available data of the geological structure of the Anabar-Khatanga saddle, have shown that in the region as a result of the salt tectogenesis processes, the diapires emerging to the surface and buried, with which most of the oil-and-gas-perspective objects are associated, were formed. With known salt domes, structural-tectonic traps can be associated, closed cryptodiapires in the structural plans of the Permian and Mesozoic stratigraphic levels have developed classical anticlinal structures.

The performed works had mainly regional character, at the same time the methods, algorithms and programs developed in the course of the conducted researches can be used at creation of detailed models of the perspective oil and gas objects connected with salt domes that cannot be deciphered in wave seismic fields.

Acknowledgements. The reported study was funded by RFBR according to the research project № 18-05-70105 (Arctic resources).

References

[1] Astarita G and Marucci G 1974 *Principles of Non-Newtonian Fluid Mechanics* (New York: McGraw-Hill) p 289
[2] Lunev B V 1996 *Rus. Geol. Geophys.* **37** 87-101
[3] Jackson M P A and Talbot C J 1986 *Geological Society of America Bulletin* **97** 305-23
[4] Lunev B V 1986 *Dokl. AN USSR, Earth Sci.* **290** 72-6
[5] Abramov T, Lavrentiev M and Lunev B 2016 *Proceedings of the 2nd International Conference on Applications in Information Technology* (Aizu-Wakamatsu: The University of Aizu Press) pp 121-4
[6] Abramov T V 2016 *Computational Technologies* **21** 3-11
[7] Konishchev V S 1982 *Tectonics of the Regions of the Halocanosis of the East European and Siberian Platforms* (Minsk: Science and Technology) p 258
[8] Kosygin Yu A 1960 *Types of Salt Structures of Platform and Geosynclinal Regions* (Moscow: Publishing House of the USSR AS) p 90
[9] Yu A Kosygin 1973 *Salt Tectonics of the Siberian Platform* Ed (Novosibirsk: Science) p 162
[10] Kontorovich V A, Lunev B V, Lapkovsky V V and Filippov Yu F 2014 *Geology and Mineral Resources of Siberia* **2C** 105-15

Nonlinear waves in marginally stable flows of Antarctic Bottom Water

N I Makarenko[1,2], J L Maltseva[1,2], E G Morozov[3], R Yu Tarakanov[3] and K A Ivanova[4]

[1] Lavrentyev Institute of Hydrodynamics, Lavrentyev pr. 15, Novosibirsk 630090, Russia
[2] Novosibirsk State University, Pirogova st. 2, Novosibirsk 630090, Russia
[3] Shirshov Institute of Oceanology, Nakhimovski pr. 36, Moscow 117997, Russia
[4] Aix–Marseille University, Marseille 06, France

E-mail: makarenko@hydro.nsc.ru

Abstract. In this paper, we study internal waves and stratified flows in the Romanche Fracture Zone of the Mid Atlantic Ridge. Theoretical analysis involves mathematical formulation of stratified flows which uses LADCP- and CTD-data obtained from field observation in the cruises of RV *Akademik Sergey Vavilov*. We discuss the 2.5-layer model of the fluid with exponential stratification of both layers. The long-wave approximate model describing travelling waves is constructed by means of scaling procedure with a small Boussinesq parameter. It is demonstrated that solitary wave regimes can be affected by the Kelvin — Helmholtz instability arising due to interfacial velocity shear in upstream flow.

1. Introduction

Deep basins of the Atlantic Ocean are filled with the water of an Antarctic origin which can spread to the mid-latitudes of the Northern Hemisphere [1]. This water is usually called Antarctic Bottom Water (AABW) which means the water mass limited from above by an isotherm of potential temperature 2.0°C. Abyssal water exchange between basins occurs through the fractures of underwater ridges [2]. Our paper is inspired by recent field observations [3] related to extremely long series of internal waves and Kelvin — Helmholtz billows at the equatorial Romanche Fracture Zone of Mid-Atlantic Ridge. This fracture presents an active zone in which a sharp increase in the intensity of the turbulent mixing occurs. Internal waves play significant role here due to the wave breaking which intensifies the mixing. We consider a theoretical model intended to describe internal solitary waves in a two-layer flow with the fluid density depending exponentially on the height in both the layers. The method of derivation involves analysis of the non-linear Dubreil-Jacotin — Long equation that results from stationary Euler equations of stratified fluid. Long-wave scaling procedure uses small Boussinesq parameter which characterizes small slope of the density profile in the layers and small density jump at their interface. This asymptotic procedure combines approaches, applied formerly to pure two-fluid system [4], with perturbation technique developed in the paper [5] for a continuous stratification. Parametric range of solitary wave is considered in the framework of constructed mathematical model. It is demonstrated that these wave regimes can realize to be close to parametric domain of the Kelvin — Helmholtz instability. Such a marginal stability of long internal waves could explain the formation mechanism of a very long billow trains which intensify the mixing of abyssal waters.

2. Basic equations

We consider a 2D motion of inviscid two-layered fluid which is weakly stratified under gravity in each layer. The fully nonlinear Euler equations describing stationary flow are

$$\rho(uu_x + vu_y) + p_x = 0, \quad \rho(uv_x + vv_y) + p_y = -\rho g, \quad u\rho_x + v\rho_y = 0, \quad u_x + v_y = 0, \tag{1}$$

where ρ is the fluid density, (u, v) is the fluid velocity, p is the pressure and g is the gravity acceleration. It is supposed that the flow is confined between the flat bottom $y = -h_1$ and the rigid lid $y = h_2$ (see Figure 1). The layers are separated by the interface $y = \eta(x)$ having the equilibrium level $y = 0$.

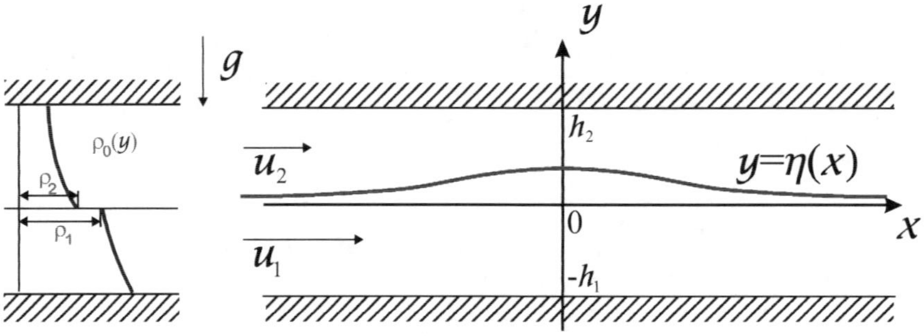

Figure 1. Scheme of the flow

Looking for a solitary-wave solutions, we require the fluid velocity (u, v) to attain the upstream velocity $(u_j, 0)$ as $x \to -\infty$. Non-disturbed parallel flow has no vertical velocity and elevation (i.e. $v = 0$, $\eta = 0$) but the horizontal velocity $u = u_0(y)$ may be piece-wise constant, so we have $u_0(y) = u_1$ by $-h_1 < y < 0$, and $u_0(y) = u_2$ by $0 < y < h_2$. Upstream fluid density $\rho = \rho_0(y)$ and the pressure $p = p_0(y)$ should be coupled by the hydrostatic equation $dp_0/dy = g\rho_0$. We consider the density profile depending exponentially on height,

$$\rho_0(y) = \begin{cases} \rho_1 \exp(-N_1^2 y/g) & (-h_1 < y < 0), \\ \rho_2 \exp(-N_2^2 y/g) & (0 < y < h_2), \end{cases} \tag{2}$$

where $N_j = \text{const}$ is the Brunt — Väisälä frequency in j-th layer, and the constants ρ_1 and ρ_2 are such that $\rho_2 < \rho_1$.

It is known [6] that stationary system (1) can be reduced to single scalar equation for a stream function ψ defined by formulae $u = \psi_y$, $v = -\psi_x$. Namely, the fluid density ρ can be determined from the upstream condition by the relation $\rho(\psi) = \rho_0(\psi/u_j)$ in the j-th layer. By that, excluding the pressure p from (1) leads to the non-linear Dubreil-Jacotin — Long (DJL) equation

$$\Delta\psi = \frac{N_j^2}{g u_j}\left\{ g\left(y - \frac{\psi}{u_j}\right) + \frac{1}{2}\left(|\nabla\psi|^2 - u_j^2\right)\right\}, \tag{3}$$

where $\Delta = \partial_x^2 + \partial_y^2$ is the Laplace's operator, $\nabla = (\partial_x, \partial_y)$; the indexes j should be taken $j = 1$ in lower layer, and $j = 2$ in upper layer. Kinematic boundary conditions at the bottom, at interface and at the lid have the form, respectively

$$\psi = -u_1 h_1 \ (y = -h_1), \qquad \psi = 0 \ (y = \eta), \qquad \psi = u_2 h_2 \ (y = h_2). \tag{4}$$

In addition, the continuity of the pressure p provides the nonlinear boundary condition at unknown interface

$$2g(\rho_1-\rho_2)\eta=\rho_2(|\nabla\psi|^2-u_2^2)|_{y=\eta(x)+0}-\rho_1(|\nabla\psi|^2-u_1^2)|_{y=\eta(x)-0}. \tag{5}$$

3. Non-linear long-wave model

We introduce non-dimensional independent variables $\bar{x}$, $\bar{y}$ and functions $\bar{\eta}$, $\bar{\psi}$ by the formulae

$$(x,y,\eta)=\frac{h_1}{\pi}(\bar{x},\bar{y},\bar{\eta}),\quad \psi=\frac{u_jh_j}{\pi}\bar{\psi}$$

considered separately in the lower layer ($j=1$) or in the upper layer ($j=2$). Using the ratio of undisturbed thicknesses of the layers $r=h_1/h_2$, we locate the bottom as $\bar{y}=-\pi$, and relation $\bar{y}=\pi/r$ defines the rigid lid. Scaling procedure involves dimensionless parameters

$$F_j=\frac{u_j}{\sqrt{g_jh_j}},\quad \sigma_j=\frac{N_j^2h_j}{\pi g},\quad \lambda_j=\frac{N_jh_j}{\pi u_j}\quad (j=1,2);\quad \mu=\frac{\rho_1-\rho_2}{\rho_2}.$$

Here, the densimetric (or internal) Froude number F_j presents scaled fluid velocity u_j in the j-th layer, defined with reduced gravity acceleration $g_j=(\rho_1-\rho_2)g/\rho_j$. The Boussinesq parameters σ_1, σ_2 characterize the slope of the density profile (3) in continuously stratified layers, and the Atwood number μ determines the density jump at interface. The Long's numbers λ_j are not independent, these constants are coupled with the parameters σ_1, σ_2, μ and F_j by the relations

$$\lambda_1^2=\frac{\pi\sigma_1(1+\mu)}{\mu F_1^2},\quad \lambda_2^2=\frac{\pi\sigma_2}{\mu F_2^2}. \tag{6}$$

Parameters σ_1, σ_2 and μ are small in the case of low stratification. However, the limit passage $\sigma_j\to 0$, $\mu\to 0$ is singular because the Long's numbers λ_j involve the ratios σ_j/μ in formulae (6). The 2.5-layer model uses the hypotheses that the Boussinesq parameters σ_1, σ_2 and the Atwood number μ are of the same order, so we can use a single small parameter σ by setting

$$\sigma=\sigma_1=\sigma_2=\mu. \tag{7}$$

In accordance with hypothesis (7), the derivation procedure of long-wave model should involve the slow horizontal variable $\xi=\sqrt{\sigma}\,\bar{x}$, as it was demonstrated originally by Benney & Ko (1978) in the case of slight linear stratification. Thus, we obtain the equations for scaled stream function $\bar{\psi}$ and non-dimensional wave elevation $\bar{\eta}$ as follows (bar is omitted throughout what follows):

$$\sigma\psi_{\xi\xi}+\psi_{yy}+\lambda_1^2(\psi-y)=\frac{1}{2}\sigma\left(\sigma\psi_\xi^2+\psi_y^2-1\right)\qquad(-\pi<y<\eta(\xi)) \tag{8}$$

$$\sigma\psi_{\xi\xi}+\psi_{yy}+\lambda_2^2r^2(\psi-ry)=\frac{1}{2}\sigma\left(\sigma\psi_\xi^2+\psi_y^2-r^2\right)\qquad(\eta(\xi)<y<\pi/r) \tag{9}$$

Kinematic boundary conditions (4) can be rewritten now as follows:

$$\psi(\xi,-\pi)=-\pi,\quad \psi(\xi,\eta(\xi))=0,\quad \psi(\xi,\pi/r)=\pi. \tag{10}$$

Correspondingly, Eq. (5) providing continuity of pressure at interface $y=\eta(x)$ takes the form

$$2\eta=F_2^2(\sigma\psi_\xi^2+\psi_y^2-r^2)|_{y=\eta+0}-F_1^2(\sigma\psi_\xi^2+\psi_y^2-1)|_{y=\eta-0}. \tag{11}$$

We find the stream function ψ to be expanded in a power series with respect to σ as

$$\psi = \psi^{(0)}(\xi, y) + \sigma\,\psi^{(1)}(\xi, y) + ... \tag{12}$$

where the leading-order term $\psi^{(0)}$ defines the hydrostatic mode, and the coefficient $\psi^{(1)}$ provides the correction due to non-linear dispersion. These coefficients are uniquely determined via the wave elevation η by the equations (8)–(10), so we obtain

$$\psi^{(0)}(\xi, y) = \begin{cases} y - \eta\,\dfrac{\sin\lambda_1(\pi + y)}{\sin\lambda_1(\pi+\eta))} & (-\pi < y < \eta), \\ ry - r\eta\,\dfrac{\sin\lambda_2(\pi - ry)}{\sin\lambda_2(\pi - r\eta)} & (\eta < y < \pi/r), \end{cases}$$

where $\lambda_j = 1/\sqrt{\pi}F_j$ $(j = 1, 2)$ should be taken at the leading order in σ in accordance with formulae (6) and (7). The dispersive term $\psi^{(1)}$ is much more complicated, but it can be calculated also in explicit form via the function η. Now we substitute power expansion (12) for function ψ into the condition (11) and truncate the terms with the powers higher than the first power of σ. By that, we obtain the first-order ordinary differential equation for the wave elevation $\eta(x)$ as

$$\left(\frac{d\eta}{dx}\right)^2 = \eta^2\,\frac{D(\eta; F_1, F_2)}{Q(\eta; F_1, F_2)} \tag{13}$$

where function D is given by the formula

$$D(\eta; F_1, F_2) = \sqrt{\pi}F_1 \,\mathrm{ctg}\left\{\frac{\pi + \eta}{\sqrt{\pi}F_1}\right\} + \sqrt{\pi}F_2 \,\mathrm{ctg}\left\{\frac{\pi - r\eta}{\sqrt{\pi}F_2}\right\} + \frac{1}{3}\,(1 - r)\eta - 1.$$

Denominator Q in (13) has one more complicated form, so we refer to the paper [7] for details. The model equation (13) generalizes the non-linear models early suggested in [4,8,9] for a system with constant densities in both layers, as well as the latest "2.5-layer" models considered in [10-12].

Solitary-wave solutions of Eq. (13) are given in the implicit form by the formula

$$x = \pm \int_a^\eta \sqrt{\frac{Q(s; F_1, F_2)}{D(s; F_1, F_2)}}\,\frac{ds}{s} \tag{14}$$

where parameter a determines non-dimensional amplitude of the wave. Parametric range of solitary waves described by Eq. (13) is formed by the domain in (F_1, F_2)-plane where the radical function Q/D in (14) is ensured to be non-negative. The function $Q(s; F_1, F_2)$ is positive in the vicinity of point $s = 0$; therefore, function D plays the determining role here. Depending on F_1 and F_2, this function can change the sign even by small s. As a consequence, the map of solitary-wave regimes is formed by the Froude numbers (F_1, F_2) such that inequality $D(0; F_1, F_2) > 0$ holds. This inequality defines the range of non-linear waves, which are supercritical with respect to the phase speed of linear harmonic wave-packets. Spectrum of normal modes, defined on the (F_1, F_2)-plane, is demonstrated on Figure 2 (left panel), where modal bounds are formed by separate branches of the curve

$$\sqrt{\pi}\,F_1 \,\mathrm{ctg}\frac{\sqrt{\pi}}{F_1} + \sqrt{\pi}\,F_2 \,\mathrm{ctg}\frac{\sqrt{\pi}}{F_2} = 1.$$

The domain covered only by the first mode is marked here with the blue color. Correspondingly, the embedded domain of the second mode is highlighted with the yellow, and the third mode is marked with the pink color. The range of internal solitary waves is shown in the right panel of Figure 2, this is the narrow band marked by the orange color. The narrowness of this range reveals anomalously low amplitude dispersion of solitary waves described in the framework of considered 2.5-layer model of a weakly stratified fluid.

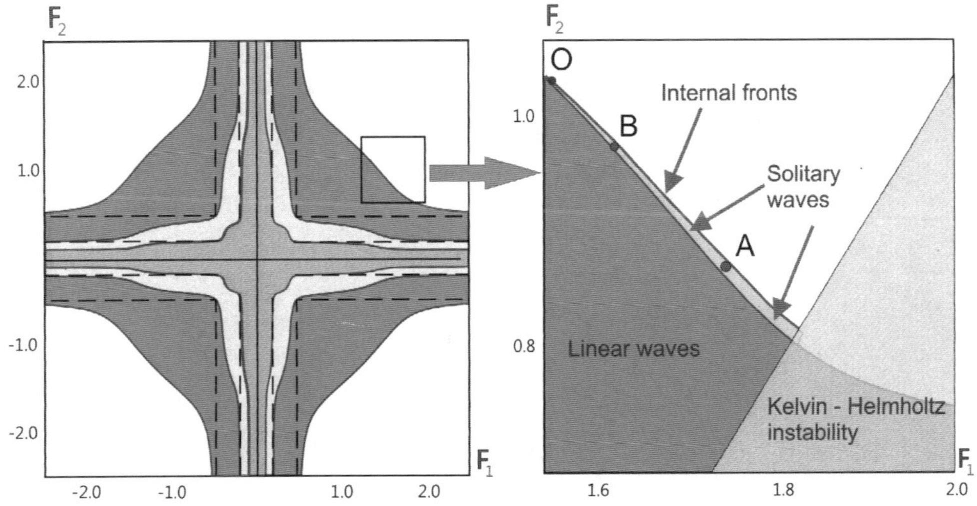

Figure 2. Spectrum of linear waves (colored modes 1-3) (left) and fragment of parametric domain of solitary waves (right).

4. Flows of Antarctic Bottom Water in the Romanche Fracture Zone

We present in this Section a comparison of solutions of Eq. (13) with the field data measured for internal solitary waves in weakly stratified abyssal currents. Figure 3 demonstrates fragments of temperature distribution in a quasi-steady shear bottom flow recorded from mooring station with a 350-m line of thermistors located over a depth of 4720 m at the entrance to the Romanche Fracture Zone in the equatorial Atlantic [3]. Trains of short-period (20÷30 min) internal waves modulated by tide propagate here along a sharp interface corresponding to the 0.85°C isotherm which separates the lower layer of cold Antarctic Bottom Water (AABW) from the overlying warmer AABW layer with temperature $\theta < 2.0$°C. Moored temperature and velocity data show permanently marginal stability of the flow with the Richardson numbers $0.25 < \mathrm{Ri} < 1$. In this context, internal Froude numbers F_1 and F_2 characterize the magnitude of the velocity jump at the interface in upstream flow. The shear $u_1 \neq u_2$ between the layers initiate the development of the Kelvin — Helmholtz instability which provides non-stationary formation of billow trains [13] Long-wave perturbations give the greatest contribution to this instability due to their increased power intensity. Constant two-layer flow is linearly stable in the long-wave limit if the inequality

$$|u_1 - u_2| < \sqrt{\frac{g(\rho_1 - \rho_2)(\rho_1 h_2 + \rho_2 h_1)}{\rho_1 \rho_2}} \tag{15}$$

holds, and this flow is unstable in the opposite case. Exactly the same bound (15) follows from the *non-linear* stability criteria predicted by the shallow water theory ([14]; see also [15]) for a *variable* difference $|u_1 - u_2|$ and *variable* layer thicknesses h_1 and h_2. Tidal amplification of the shear triggers the formation of small-scale overturns which create long trains of the Kelvin — Helmholtz billows.

Bold curves in Figure 3 show overlapped profiles of internal waves calculated from (14). The upstream parameters used in the calculation were chosen from CTD and LADCP data of density and currents measured immediately at the fronts of solitary waves. Upstream velocities u_1, u_2 were assumed as $u_1 = 35\div45$ cm/s and $u_2 = 10\div20$ cm/s. These values correspond to the mean velocities measured in the western part of the Romanche Fracture Zone by a velocity LADCP

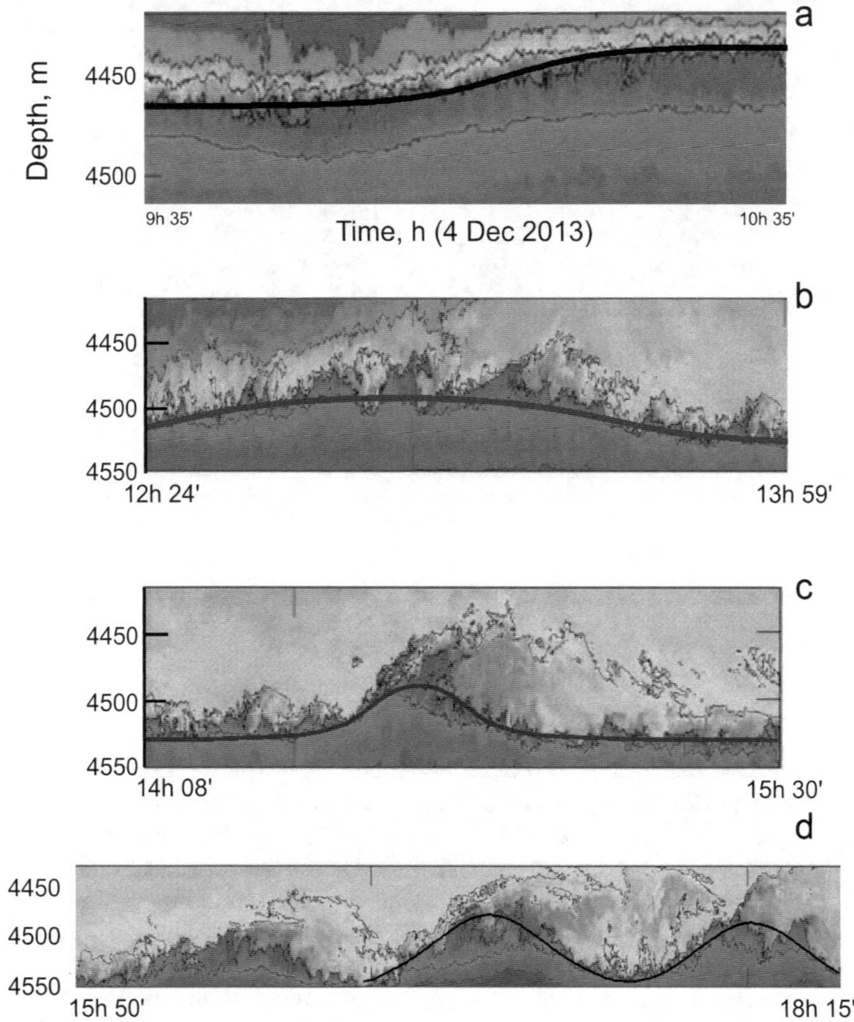

Figure 3. Internal waves in AABW flows affected by the Kelvin — Helmholtz instability

profiler in the cruises of the R/V "Akademik Sergey Vavilov" [16] and measured on the mooring station [3]. Maximal velocity of the flow in the lower layer, recorded by the mooring station within 6 months is as high as 65 cm/s. That intense shear is induced here by the permanent inflow of bottom waters to the fracture zone through a narrow gap in its southern wall.

Figure 3a demonstrates front-like wave (smooth internal bore) affected by moderate small-scale shear instability. The non-dimensional amplitude of this wave is $a = \eta_{max}/h_1 = 0.16$, and calculated solution corresponds to the Froude point A with coordinates $(F_1, F_2) = (1.719, 0.891)$ shown in Figure 2 (right panel). Point A belongs to the bore diagram which is tangential to the spectrum boundary at the point O. Figure 3b demonstrate a long series of overturns which are distributed uniformly along with gently sloped wave top. It is interesting that similar overturning near the middle part of broad solitary wave was observed in laboratory experiments [17]. In contrast, solitary wave shown on Figure 3c is relatively short, and the flow is apparently non-symmetric due to intensive breaking which localizes downstream the sharp wave-crest. Figure 3d demonstrates the train of three subsequent solitary waves with intense overturns in the lee side of these waves. The amplitude of the leading wave shown in Figure 3d is clearly lower than the amplitudes of the following two waves, which occur randomly and are almost identical. The solitary-wave type of these two waves corresponds to Froude point B with coordinates $(F_1, F_2) = (1.634, 0.959)$.

5. Acknowledgments

This work was supported by the grants of the Russian Foundation for Basic Research (project nos. 17-08-00085, 18-01-00648) and Interdisciplinary Program II.1 of SB RAS (project No 2). Roman Tarakanov was supported by the Russian Science Foundation (grant No 16-17-10149).

6. References

[1] Mantyla A W and Reid J L 1983 Abyssal characteristics of the World Ocean waters *Deep-Sea Res.* **30** 805-833
[2] Morozov E, Demidov A, Tarakanov R and Zenk W 2010 *Abyssal Channels in the Atlantic Ocean: Water Structure and Flows* (Dordrecht: Springer)
[3] Van Haren H, Gostiaux L, Morozov E and Tarakanov R 2014 Extremely long Kelvin — Helmholtz billow trains in the Romanche Fracture Zone *Geophys. Res. Lett.* **44**(23) 8445–8451
[4] Ovsyannikov L V, Makarenko N I, Nalimov V I et al 1985 Nonlinear Problems of the Theory of Surface and Internal Waves, (Novosibirsk: Nauka) [in Russian]
[5] Benney D J and Ko D R S 1978 The propagation of long large amplitude internal waves *Stud. Appl. Math.* **59** 187-199
[6] Yih C-S 1980 *Stratified Flows* (New York: Academic Press)
[7] Makarenko N, Maltseva J, Tarakanov R and Ivanova K 2018 Internal solitary waves in a layered weakly stratified flow *The Ocean in Motion* (Dordrecht: Springer) 55–66
[8] Miyata M 1985 An internal solitary wave of large amplitude *La Mer* **23** 43–48
[9] Choi W and Camassa R 1999 Fully nonlinear internal waves in a two-fluid system *J.Fluid Mech.* **396** 1–36
[10] Voronovich A G 2003 Strong solitary internal waves in a 2.5- layer model *J. Fluid Mech.* **474** 85–94
[11] Makarenko N I and Maltseva J L 2009 Phase velocity spectrum of internal waves in a weakly-stratified two-layer fluid *Fluid Dyn.* **44**(2) 278–294
[12] Makarenko N I and Maltseva J L 2009 Solitary waves in a weakly stratified two-layer fluid *J. Appl. Mech. Techn. Phys.* **50**(2) 229–234
[13] Thorpe S A 1985 Laboratory observations of secondary structures in Kelvin — Helmholtz billows and consequences for ocean mixing *Geophys. Astrophys. Fluid Dyn.* **34** 175–190
[14] Ovsyannikov L V 1979 Two-layer shallow water model *J. Appl. Mech. Tech. Phys.* **20** 127–135
[15] Gavrilyuk S L, Makarenko N I and Sukhinin S V 2017 *Waves in continuous media* (Cham, Switzerland: Birkhäuser/ Springer)
[16] Tarakanov R Yu, Makarenko N I and Morozov E G 2013 Antarctic bottom water flow in the western part of the Romanche Fracture Zone based on the measurements in October of 2011 *Oceanology* **53**(6) 655–667
[17] Grue J, Jensen A, Rusås P-O and Sveen J K 2000 Breaking and broadening of internal solitary waves *J.Fluid Mech.* **413** 181–217

Co-deposition of gas hydrate and oil wax from water-in-crude oil emulsion saturated with CO_2

A S Stoporev [1,2,3], A G Ogienko [2,3], L.K. Altunina[4] and A Y Manakov [2,3,5]

[1] Gubkin Oil and Gas University, Department of Physical and Colloid Chemistry, Leninsky ave. 65, Building 1, Moscow, 119991, Russian Federation
[2] Nikolaev Institute of Inorganic Chemistry SB RAS, Ac. Lavrentiev ave. 3, Novosibirsk, 630090, Russian Federation
[3] Novosibirsk State University, Pirogova Str. 2, Novosibirsk, 630090, Russian Federation
[4] Institute of Petroleum Chemistry SB RAS, Akademichesky ave., 4, Tomsk, 634021, Russian Federation
[5] Author to whom any correspondence should be addressed

E-mail: manakov@niic.nsc.ru

Abstract. Samples of water-hydrate-paraffin sediments with the regions of different colored were obtained by passing the emulsion of CO_2-saturated water into oil through a cooled tubular reactor Each of the regions was characterized by means of powder X-ray diffraction. It was shown that prevailing joint deposition of paraffins and hydrate occurs at some regions; complicated water-hydrate-paraffin agglomerates containing almost no admixture of the colored oil components. The possibility to form these agglomerates should be taken into account in the development of the methods preventing the deposition of hydrates and paraffins under real conditions.

1. Introduction

Gas hydrates are widespread natural clathrate compounds in which the host framework is composed of hydrogen-bonded water molecules, while the guest components occupy the cavities of this framework [1]. Technological attention to gas hydrates is due to the possibility of the formation of hydrate plugs in the equipment of gas and oil producing facilities and in pipelines. For this reason, hydrate prevention is an urgent subject for investigation in oil and gas industry [2-4]. Hydrate formation occurs during oil production when multiphase flows coming up from the wells are cooled (oil, water/brine and associated petroleum gas) directly in boreholes or in field pipelines. According to the most widespread model, hydrate formation occurs during the reaction of associated petroleum gas dissolved in oil with water emulsified in oil [1,2,5]. The mechanisms of hydrate formation in emulsions were considered in [6,7]. The major stages of this process include hydrate nucleation on the surface of water – oil contact, rapid growth of hydrate film on the entire surface of water drops, and slower ingrow of the hydrate inside the drop. The latter process is limited by gas diffusion through the ice layer which has already been formed. Most frequently thermodynamic inhibitors (methanol, glycols) are used to prevent hydrate formation. The addition of these inhibitors into the multiphase flow causes a shift of the equilibrium conditions of hydrate formation to the region of low temperatures/high pressure. Increasing application is won by so-called low-dose inhibitors of hydrate formation: their action is due to their adsorption on the surface of hydrate particles. This slows down the nucleation of the hydrate phase (kinetic inhibitors) and prevents the agglomeration of hydrate particles with each other (anti-agglomerating agents), which allows preventing pipeline blocking in many cases [8,9].

It is well known that, in addition to gas hydrates, paraffins present in oil can also form deposits on pipeline walls (for example, [10,11]). There are some data that evidence in favor of the mutual effect of hydrate particles formed in the multiphase flow and depositing paraffin species. It was shown in [12] that the formation of gas hydrate particles in the flow promotes the precipitation of paraffins. In turn, paraffin deposition promotes the agglomeration of the formed hydrate particles. The possibility of the formation of hydrate-paraffin agglomerates was demonstrated. The authors of [13,14] state that the presence of paraffin crystals in the oil phase decelerates hydrate nucleation. Paraffin deposition on the surface of water drops and hydrate particles was discovered, and these agglomerates were observed to stick together. In the opinion of the authors of [14], adhesion of the hydrophobic paraffin particles on the hydrophilic surfaces is provided by the presence of surface-active species in the system. There is a trend to an increase in the rate of hydrate growth with an increase in the concentration of paraffin crystals in oil. In addition, the presence of paraffin crystals causes an increase in the viscosity of hydrate suspensions, which points to higher agglomeration of the particles in the presence of paraffins. Paraffins promote hydrate formation in dispersions stabilized with solid particles with highly and moderately hydrophobic surface; the situation is inverse for the particles with low hydrophobicity [15]. So, by present the major regularities of the processes involved in the formation of hydrate-paraffin agglomerates in multiphase systems hydrate – water – hydrate-forming gas have been already studied. At the same time, we are unaware of any works considering the structure and phase composition of these agglomerates. In the present work we report on the first results on the formation of hydrate-paraffin agglomerates in a flow reactor and describe the results obtained in the investigation of the resulting hydrate-paraffin agglomerates.

2. Experimental section

A schematic of the experimental set-up is shown in figure 1. The inner diameter of the tubular reactor is 6 mm (the wide part) and 2 mm (the narrow part), the site in which the reactor width changes is marked as (3); the length of the reactor is 500 mm. All parts of the set-up except the tubular reactor (1) were kept at room temperature; the necessary low temperature in the reactor (1) was maintained with the help of the fluid cryostat pumping the coolant through jacket (2). The formation of paraffin-hydrate plug (9) occurred in the narrowing spot (3) of the tubular reactor. To simplify the extraction of plug (9) from the reactor, in some experiments teflon bushings (4) with the total length of 100 mm and inner diameter 4 mm were placed next to the insertion (3).

During the experiment, CO_2-saturated emulsion of water (50 wt%) in paraffin-containing oil (paraffin content 3.7 mass %, density 826 kg/m^3, the size of drops in the emulsion 72±14 μm) was pressed from the pressure cylinder (5) under the action of gas pressure created by the gas from gas vessel (6) through reactor (1), then the mixture was depressurized through a fine-adjusting regulating valve (10). Valve (10) allowed controlling the rate of emulsion flush through the reactor. Pressure at the reactor inlet and outlet, as well as in the pressure cylinder (5), was measured with the help of pressure sensors (7) with the accuracy of 0.03 MPa. Experiment management was performed with the help of cutoff valves (8). Before experiment, the emulsion was charged into the pressure cylinder (5) in which the pressure of CO_2 was maintained at a level of 5.5 MPa at room temperature. For saturation, the emulsion was kept before experiment under pressure for 24 h. The necessary reactor temperature (-1.5 or -5.0°C) and the rate of emulsion flush ~2.5 ml/min were set at the beginning of the experiment. Flow rate was decreased almost to zero during several hours, and then it was again set at the previous level. The pressure difference at the inlet and outlet of the tubular reactor was recorded during the experiment.

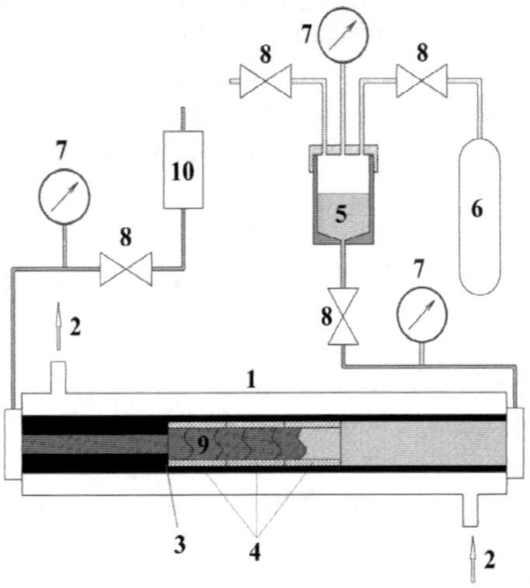

Figure 1. Schematic view of the experimental set-up:. (1) Tubular reactor, (2) cooling jacket, inlet and outlet of cooling liquid into the jacket of the reactor, (3) site in which the section of the reactor narrows, (4) tubular teflon inserts, (5) pressure reservoir with the emulsion saturated with gas, (6) vessel with gas, (7) pressure sensor, (8) valves, (9) the formed plug, (10) microadjusting valve to control the flow of the emulsion through the tubular reactor.

3. Results and discussion

Six independent experiments on the formation of paraffin-hydrate plugs were carried out. The appearance of the samples and the results of their investigation turned out to be similar in all experiments, so below they will be considered jointly. Typical pressure curves at the reactor inlet and outlet are shown in figure 2. In the case of the free flow of the emulsion through the reactor the pressure difference was close to zero. The flow of the emulsion through the reactor was hampered with the formation of the paraffin-hydrate plug, and thus the pressure difference occurred. The formation of agglomerates started within several hours after temperature was decreased to -1.5 - -5.0°C, and manifested itself as the pressure drop and the reactor outlet, while the pressure at the reactor inlet remained unchanged or decreased to a smaller extent (figure 2). At a temperature of +1°C, pressure change and plug formation were not observed during pumping for a long time (up to 24 h). It was stressed that, as a rule, plug formation occurred during several hours after the flow of emulsion through the reactor was stopped. To exclude the participation of ice in plug formation, reactor temperature was increased to +0.5 - +1°C for about 1 hour before the end of the experiment, and pressure difference either remained at the former level or increased (most likely due to additional growth of the plug). This points to the fact that ice, if any, was present in the reactor at that moment, it did not play the major role in the formation of the agglomerate. At the end of the experiment, temperature was decreased to -25°C, and after that the residual pressure was released, the resulting sample was taken out of the reactor and immediately frozen in liquid nitrogen.

The samples taken from the zone of teflon inserts ((4) in figure 1) and from the wide part of the reactor differed from each other substantially (figure. 3). In all cases, the samples from the wide part of the reactor were uniformly colored, with porosity visible for unaided eye (region (5) in figure 3). We suppose that these samples correspond to the frozen dispersion of hydrate particles and ice in oil. Gas bubbles could be formed during oil degassing at the stage of freezing / pressure release. The samples from the narrow part of the reactor, as a rule, consisted of several regions differing from each other in appearance; their porosity was not visible for unaided eye (figure 3). In the region which was nearest to the wide part of the reactor, as a rule, three types of regions were interchanging (figure 3): a light region, which was non-transparent, with white dull surface, sometimes contaminated with oil (III) in figure 3, a dark region – non-transparent, with nonuniform dark color on the surface, without light spots (II) in figure 3; spotty – with nonuniform dark color of the surface and with visible light spots on this surface (IV) in figure 3. The transparent part of the sample (region (I) in figure 3) was formed in the region which was the closest to the narrow part of the tubular reactor. ((3) in figure 1).

Light regions verged on the wide part of the reactor, further on the regions of different types were interchanging.

Each region of the sample was characterized by means of powder X-ray diffraction. Typical diffraction patterns are shown in figure 4. In all cases, three phases are revealed in the diffraction patterns: ice, hydrate of carbon dioxide, and crystalline oil paraffin. The reflections related to ice and hydrate were assigned relying on the data reported in [16] and [17], respectively. According to the data of [18,19], there are two reflections of oil paraffins in the angle range 2Θ (CuK_α radiation) shown in figure 4: 2Θ = 21÷22 and 24÷25°. Their positions depend on temperature and composition of the mixture of paraffins. The small amount of substance in the sample available for us does not allow us to apply the known procedures of quantitative X-ray analysis to the examination of our sample. For this reason, we only compared relative amounts of different phases in different regions of the sample. One can see in the powder X-ray diffraction patterns that the transparent regions (I) in figure 3 contain mainly ice, with small admixtures of hydrate and paraffin (weak reflections of paraffin and hydrate at the background of strong reflections of ice). For the samples from the wide part of the reactor and for spotty (IV) in figure 3 and dark (II) in figure 3 regions, the intensities of reflections from all phases are comparable; the most intensive reflection from ice is somewhat larger than the most intensive reflections of paraffin and hydrate. It may be assumed that these regions contain all three phases in comparable amounts. Finally, the situation is quite contrary for light regions (III). Here the most intensive reflection of ice is weaker than the most intensive reflections of paraffin and hydrate, while the relative intensity of the ice and hydrate reflections varies. So, the light regions contain mainly paraffin and hydrate, the transparent regions contain mainly water, while in other regions the amounts of these phases are comparable. According to the data of thermovolumetric analysis of averaged samples from the dark, light and spotty regions, the average hydrate content in the mixture varied within the range 6-9 mass %. More detailed analyses of the composition of different regions will be carried out in future.

We suppose that the most interesting result of the present work is experimental confirmation of the possibility of the joint deposition of pure agglomerates of water, hydrate and oil paraffins in the reactor channel. It is known [2] that the paraffin-hydrate plugs are formed most frequently in the points of channel narrowing or bending, which is observed in our case. It may be assumed that the deposition starts at the walls of the channel and then reaches its center. Alternating formation of light, dark and spotty regions may be connected with the different manners of the joint crystallization of paraffins and hydrate depending on the concentrations of components present in a given point of the reactor, and on the degree of thermodynamic supersaturation. We suppose that the crystallization of pure paraffins capturing hydrate and water into their volume (all components are not colored, that is why light regions are formed) takes place in the sites with the highest supersaturation of the emulsion

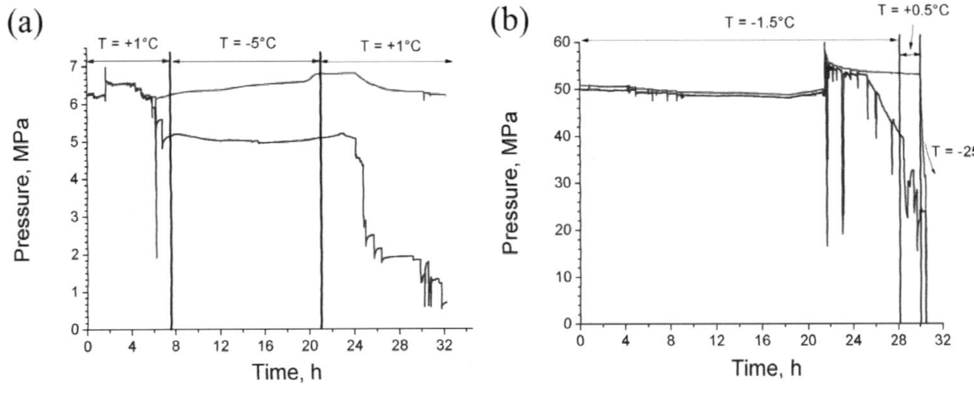

Figure 2. Typical time dependencies of pressure at the inlet of the tubular reactor (red line) and at the outlet of it (blue line). The intervals of time during which the corresponding temperature was kept in

the reactor are shown. Pressure drop in experiment (a) is jumpwise (connected with an increase in the rate of liquid phase discharge from the reactor), pressure decrease in experiment (b) is smooth. The regions of a sharp pressure increase on the curves correspond to gas pumping into the pressure cylinder.

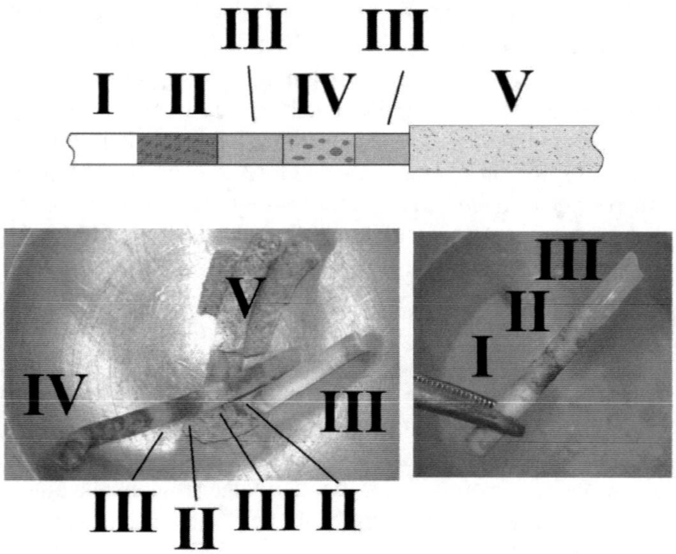

Figure 3. Different regions of the samples of paraffin-hydrate plugs obtained in the experiments. Regions: (I) transparent, (II) dark, (III) light, (IV) spotty, (V) from the wide part of the reactor. Corresponding regions are also marked in photographs.

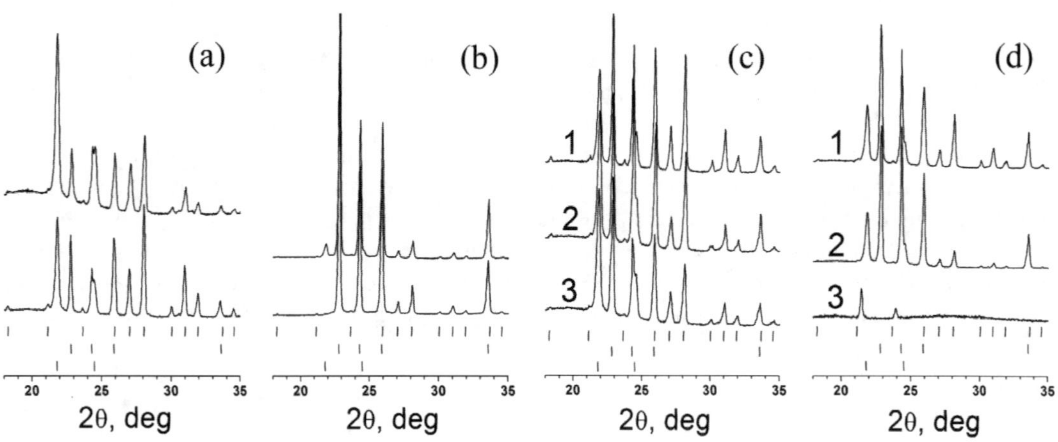

Figure 4. Typical powder diffraction patterns of different regions of frozen water-paraffin-hydrate agglomerates. (a) Light regions, (b) transparent regions, (c) 1 and 3 – spotty regions, 2 – dark region, (d) 1 – sample from the wide part of the reactor, 2 – spotty region at -100°C, 3 – spotty region at 0°C. Strokes mean the expected positions of the reflections from different phases: upper position - CO_2 hydrate, middle position - ice, lower position – paraffins.

with the hydrate-forming gas and paraffin. In this situation, water – hydrate – paraffin agglomerate is formed due to hydrophobization of the hydrophilic surfaces of water and hydrate by surfactants that are present in oil. Due to this, the surface of water drops and hydrate particles has good adhesion to paraffin, and they are capable of joint deposition. The mechanism is similar to that considered previously in [20]. Enrichment of the light regions with water and paraffin proceeds due to pushing the major part of oil further along the reactor. After the deposition of a substantial amount of water – hydrate – paraffin agglomerate, colored oil components start to get included into the deposit; so spotty and dark regions are formed. A detailed mechanism leading to the formation of differently colored regions is still unclear at present, and the amount of data is insufficient to discuss these questions. Finally, because of the deposition of emulsion-stabilizing components, water which remains not included into the agglomerates and the hydrate phase is separated from oil and pushed into the peripheral part of the sample where it forms a transparent region. In this region, hydrate is formed from dissolved CO_2 dissolved in water; paraffin particles are brought mechanically from preceding regions.

Thus, the possibility of spontaneous formation of water – hydrate – paraffin agglomerates in oil dispersed systems is demonstrated experimentally in the work. This possibility should be taken into account when developing the methods against the formation of hydrate-paraffin deposits during oil production in the regions with cold climate.

Acknowledgements
The work was supported by the Russian Scientific Foundation Project № 17-17-01085.

References

[1] Sloan E D and Koh C A 2008 *Clathrate hydrates of natural gases* 3rd ed. (London, New- York: CRC Press, Boca Rator) p 731
[2] Sloan E D 2000 *Hydrate engineering* In Henry L Doherty series (Book 21) ed J B Bloys (Richardson, Texas) p 89
[3] Mokhatab S, Wilkens R J and Leontaritis J 2007 *Energ. Source Part A* **29**(1) 39
[4] Creek J L 2012 *Energy&Fuels* **26**(7) 4112
[5] Sum A K, Koh C A and Sloan E D 2009 *Ind. Eng. Chem. Res.* **48**(16) 7457
[6] Greaves D, Boxall J, Mulligan J, Sloan E D and Koh C A 2008 *Chem. Eng. Sci.* **63**(18) 4570
[7] Turner D J, Miller K T and Sloan E D 2009 *Chem. Eng. Sci.* **64**(18) 3996
[8] Kelland M A 2006 *Energy&Fuels* **20**(3) 825
[9] Perrin A, Musa O M and Steed J W 2013 *Chem. Soc. Rev.* **42**(5) 1996
[10] Aiyejina A, Chakrabarti D P, Pilgrim A and Sastry M K S 2011 *Int. J. Multiphase Flow* **37**(7) 671
[11] Zheng S, Fogler H S and Haji-Akbari A 2017 *AICHE J.* **63**(9) 4201
[12] Gao S 2008 *Energy&Fuels* **22**(5) 3150
[13] Zheng H M, Huang Q Y, Wang W, Long Z and Kusalik P G 2017 *Ind. Eng. Chem. Res.* **56**(29) 8330
[14] Shi B H, Chai S, Ding L, Chen Y C, Liu Y, Song S S, Yao H Y, Wu H H, Wang W and Gong J 2018 *AICHE J.* **64**(9) 3502-18
[15] Raman A K Y and Aichele C P 2017 *Energy&Fuels* **31**(5) 4817
[16] Rottger K, Endriss A, Ihringer J, Doyle S and Kuhs W F 1994 *Acta Cryst. B* **50** 644
[17] Udachin K A, Ratcliffe C I and Ripmeester J A 2001 *J.Phys.Chem.B.* **105**(19) 4200
[18] Negoro K 1962 *Bull. Chem. Soc. Jap.* **35**(3) 375
[19] Heyding R D, Russell K E, Varty T L, St-Cyr D 1990 *Powder Diffraction* **5** 93
[20] Zheng S, Khrutphisit T and Fogler H S 2017 *Energy&Fuels* **31**(1) 34

IOP Conf. Series: Earth and Environmental Science **193** (2018) 012043 doi:10.1088/1755-1315/193/1/012043

A discontinuous Galerkin method for mathematical modeling of ice melting at the interaction with the environment

S I Markov[1,2]

[1]Trofimuk Institute of Petroleum Geology and Geophysics of SB RAS, Novosibirsk, Russia
[2]Novosibirsk State Technical University, Russia

E-mail: www.sim91@list.ru

Abstract. A Stefan problem with a moving boundary in a heterogeneous two-phase medium is considered. For solving the Stefan problem, a discontinuous Galerkin method in a nonsymmetrical Inner Penalty formulation is used. We propose to split the problem solution into a discontinuous component in the phase transition zone and a continuous component outside the phase transition zone. A multiscale method allows to reduce a number of freedom degrees by using the discontinuous Galerkin method in the phase transition zone only. Mathematical modeling results of the ice melting at the interaction with the environment are given.

1. Introduction

A study of the ice cover and permafrost layer state is an important scientific problem, since it is aimed at preventing man-caused and environmental disasters. According to [1, 2], two main factors of negative influence on the ice cover and permafrost layer are possible. The first factor is associated with natural phenomena, which can be seasonal in nature [1]. For example, anomalously warm and arid summer. The second factor is due to technogenic impacts [2] in the human activity process. From the physical point of view, both types of influence initialize the heat transfer process with different types of interface conditions and moving boundaries.

A solution of problems with a moving boundary is required in the study of physical processes with a phase transition or mixing of substances. Stefan's problem of the "solid-liquid" phase transition is a classic example of a problem with the moving boundary. This problem is of great practical importance in predicting a permafrost layer state and creating phase-change composite materials [1 – 3].

The solid surface interacts with a medium whose temperature is higher than the phase transition temperature. As a result, the boundary between the liquid and solid is moved toward the liquid phase.

A modeling procedure of the heat transfer process in media with the phase transition is complicated by the search for the current position of the moving boundary [3]. Therefore, a computational scheme must be adaptable to the specifics of the problem and maximally independent of computational domain meshes.

To solve such a class of problems, computational schemes of finite differences methods (FDM), finite volume methods (FVM) and finite element methods (FEM) are used usually. FDM and FVM are successfully applied to solve one-dimensional multiphase Stefan problems, since they allow the use of such popular methods for tracking the phase transition front as catching the front node [4], enthalpy and phase field methods [5]. However, their application for solving the three-dimensional problems is complicated, since the FDM is sensitive to the orientation of the mesh [4], and the FVM requires the construction of special multilevel interpolation schemes [5].

Modern computational schemes of FEM are implemented using explicit methods for tracking the phase transition front with fixed or moving meshes. An extended FEM (XFEM) uses the fixed meshes. XFEM is based on the introduction of additional freedom degrees in the front-moving zone [6]. The effectiveness of XFEM decreases when solving the multiphase Stefan problems, since the account of the phase transition boundaries becomes nontrivial at the functional level.

We propose to use a multiscale Discontinuous Galerkin Method (MDG) with a moving nonconformal mesh. In the MDG-method, the solution is represented as the sum of the discontinuous and continuous components. The discontinuous component exists in the phase transition zone only. The continuous component is defined in the entire computational domain. The strategy allows to construct a scheme of the discontinuous Galerkin method with the computational structure of the classical FEM [7, 8].

The multiscale discontinuous Galerkin method in the nonsymmetrical variational "Inner Penalty" formulation for the two-phase three-dimensional Stefan problem using nonconformal tetrahedral moving meshes are used. Results of solving the problem of melting ice with constant heat exchange with the surrounding medium are presented.

2. Problem Definition

Figure 1 shows a computational domain. The computational domain consists of two phases: ice and water regions. Let Ω_l is the liquid phase (water), and Ω_s is the solid phase (ice). At the initial time, the medium is completely filled with ice. On the upper and lower boundaries S_1 the heat exchange occurs with the medium with the temperature of 1 °C and –5 °C, respectively. The side faces of S_2 are assumed to be thermally insulated. A front of the phase transition (melting) is denoted as ξ(t). At the initial instant of time, front is located in the plane Z = 0. The phase transition temperature is 0° C. The length of the computational domain is 25 m, width is 16 m, and height is 3 m.

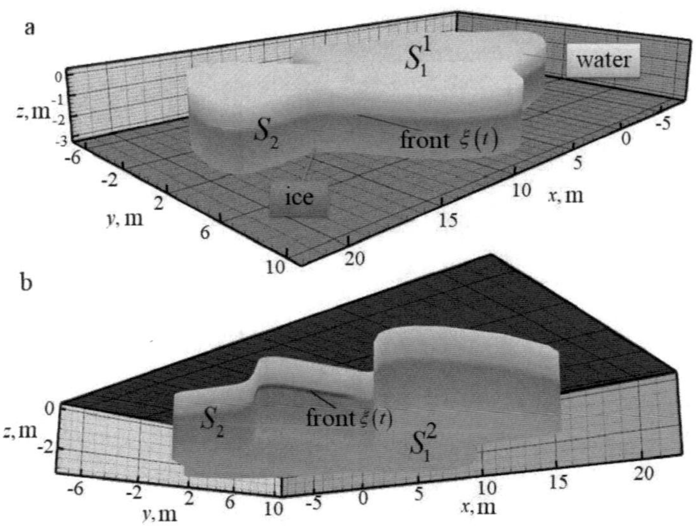

Figure 1. Computational domain: top view (a), bottom view (b).

It is necessary to determine the behavior of the "ice-water" phase system over time.

3. Mathematical Model of the Stefan Problem

The temperature distribution in the each phase is described by the heat conductivity equations [3]

$$\rho_l c_l \frac{\partial T_l}{\partial t} = \nabla \cdot (\lambda_l \nabla T_l) \text{ in } \Omega_l, \quad (1)$$

$$\rho_s c_s \frac{\partial T_s}{\partial t} = \nabla \cdot (\lambda_s \nabla T_s) \text{ in } \Omega_s, \quad (2)$$

with initial conditions

IOP Conf. Series: Earth and Environmental Science **193** (2018) 012043 doi:10.1088/1755-1315/193/1/012043

$$T_l\big|_{t=0} \equiv 1,\ T_s\big|_{t=0} \equiv -5, \tag{3}$$

ρ – density, kg/m^3; c – specific heat, J/(kg·°C); λ – thermal conductivity, W/(m·°C); T – temperature, °C; l – liquid phase, s – solid phase.

At the side faces S_2

$$\left.\frac{\partial T_l}{\partial \mathbf{n}}\right|_{S_2} \equiv 0,\ \left.\frac{\partial T_s}{\partial \mathbf{n}}\right|_{S_2} \equiv 0, \tag{4}$$

at the upper and lower boundaries, the process of heat exchange with the medium is given as

$$\lambda_l \left.\frac{\partial T_l}{\partial \mathbf{n}}\right|_{S_1^1} + \beta\left(T_l\big|_{S_1^1} - 1\right) \equiv 0,\ \lambda_s \left.\frac{\partial T_s}{\partial \mathbf{n}}\right|_{S_1^2} + \beta T_s\big|_{S_1^2} \equiv 0, \tag{5}$$

where $\beta = 10$ W/(m^2·°C) – heat transfer coefficient.

On the boundary of the phase transition, ideal contact conditions for the temperature distribution are determined as

$$[T]\big|_{\xi(t)} \equiv 0, \tag{6}$$

and conjugation conditions for the heat flux are satisfied

$$[\lambda \nabla T \cdot \mathbf{n}]\big|_{\xi(t)} := -\rho L \frac{\partial \xi(t)}{\partial t}, \tag{7}$$

where L – phase transition heat, J / kg.

4. Variational Formulation of the Discontinuous Galerkin Method for the Stefan Problem

Let $\Xi_h(\Omega)$ be a partition into disjoint sets Ω_k. Let $\Gamma := \cup \partial\Omega_k$ be the set of the boundaries, $\Gamma_0 := \Gamma \setminus \partial\Omega$ be the set of the inner boundaries, and $\mathrm{T}(\Gamma) := \prod_{\Omega_k \in \Xi_h(\Omega)} L^2(\partial\Omega_k)$ be the trace space. We introduce finite-dimensional subspaces on the set $\Xi_h(\Omega)$ as

$$P^h := \left\{q \middle| q \in L^2(\Omega) : q \in \Im_m(K)\, \forall K \in \Xi_h(\Omega)\right\}, \tag{8}$$

$$\mathbf{V}^h := \left\{\mathbf{w} \middle| \mathbf{w} \in \mathbf{H}(\mathrm{div}, \Omega) : \mathbf{w} \in \left[\Im_{m+1}(K)\right]^3 \forall K \in \Xi_h(\Omega)\right\}, \tag{9}$$

where $\Im_m(K)$ is a polynomial space of degree m.

For building the trace operator on the inter-element boundaries, we consider the average $\{\cdot\}$ and jump $[\cdot]$. For the functions $\mathbf{v} \in [\mathrm{T}(\Gamma)]^3$ and $p \in \mathrm{T}(\Gamma)$ at the external boundary $\partial\Omega$ we can write [7, 8]

$$\begin{gathered} \underline{[\mathbf{v}]}\big|_{\partial\Omega} = \mathbf{v} \otimes \mathbf{n},\ [\mathbf{v}]\big|_{\partial\Omega} = \mathbf{v} \cdot \mathbf{n}, \\ \{\mathbf{v}\}\big|_{\partial\Omega} = \mathbf{v},\quad [p]\big|_{\partial\Omega} = p\mathbf{n},\quad \{p\}\big|_{\partial\Omega} = p, \end{gathered} \tag{10}$$

At the inner boundary $\Gamma_0 := \partial\Omega_k \cap \partial\Omega_n$ between elements Ω_k and Ω_n we can write [7, 8]

$$\begin{gathered} \underline{[\mathbf{v}]}\big|_{\Gamma_0} = \mathbf{v}_k \otimes \mathbf{n}_k + \mathbf{v}_n \otimes \mathbf{n}_n,\ [\mathbf{v}]\big|_{\Gamma_0} = \mathbf{v}_k \cdot \mathbf{n}_k + \mathbf{v}_n \cdot \mathbf{n}_n, \\ \{\mathbf{v}\}\big|_{\Gamma_0} = (\mathbf{v}_k + \mathbf{v}_n)/2, \qquad [p]\big|_{\Gamma_0} = p_k\mathbf{n}_k + p_n\mathbf{n}_n, \\ \{p\}\big|_{\Gamma_0} = (p_k + p_n)/2, \end{gathered} \tag{11}$$

where the index indicates belonging to Ω_i

Let $\Omega := \Omega^C \oplus \Omega^D$ and $T^h := T_C^h + T_D^h$. Here T_C^h is the continuous solution component of the problem (1) – (7) determined over in Ω. T_D^h is discontinuous solution component determined in Ω^D only. This means that [8]

$$T_D^h\big|_{\partial\Omega^D} \equiv 0. \tag{12}$$

The finite element interpolants $T_C^h \in P^h$ and $T_D^h \in P^h$ are defined as [8]

$$T_D^h := \sum a_i p_i^D,\ p_i^D \in P^h \text{ in } \Omega^D, \tag{13}$$

$$T_C^h := \sum b_i p_i^C,\ p_i^C \in P^h \text{ in } \Omega, \tag{14}$$

where $[p^C]|_\Gamma \equiv 0,\ \{p^C\}|_\Gamma := p^C$.

The nonsymmetrical formulation of the discontinuous Galerkin method for the problem (1) – (7) [8]: to find such $T_C^h \in P^h \times [0,\mathrm{T}]$, $T_D^h \in P^h \times [0,\mathrm{T}]$, $\forall p_C^h \in P^h$ and $p_D^h \in P^h$

$$\begin{aligned} a\left(T_C^h, p_C^h\right) + a\left(T_D^h, p_C^h\right) &= \left(f^h, p_C^h\right), \\ a\left(T_C^h, p_D^h\right) + a\left(T_D^h, p_D^h\right) &= \left(f^h, p_D^h\right), \end{aligned} \tag{15}$$

$$a\left(T^h, p^h\right) := \int_\Omega c\rho T_t^h p^h d\Omega + \int_\Omega \lambda \nabla T^h \cdot \nabla p^h d\Omega + \int_\Gamma \lambda\left(\left[T^h\right]\cdot\left\{\nabla p^h\right\} - \left\{\nabla T^h\right\}\cdot[p^h] + \mu[T^h]\cdot[p^h]\right) dS, \tag{16}$$

$$\left(f^h, p^h\right) := -\int_{S_1} \beta T_g p^h dS, \tag{17}$$

where T_g – temperature on the S_I. The problem of choosing the stabilization parameter μ was solved in [9].

We can take the Stefan condition (7) into account as

$$\int_{\xi(t)} \lambda\left\{\nabla T^h\right\}\cdot[p^h] dS := \int_{\xi(t)} \frac{\lambda^s \nabla T^s + \lambda^l \nabla T^l}{2} \cdot p^h \mathbf{n} dS := -\int_{\xi(t)} \frac{1}{2} \rho V L p^h dS,$$

where V is the velocity of the front motion.

To estimate the velocity of the front motion and the approximation of the time derivative, we apply the implicit Euler scheme

$$\frac{\partial f}{\partial t} = \frac{f(t+\Delta t) - f(\Delta t)}{\Delta t}, \tag{18}$$

where Δt – discretization time step.

5. Experimental Results

To solve the problem, we choose adaptive nonconformal tetrahedral meshes. Figure 2 shows the 3D triangulation of the computational domain at the time moment t = 100 h.

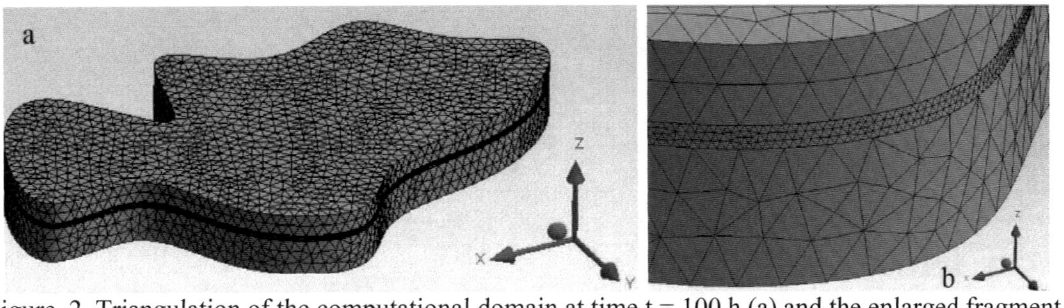

Figure. 2. Triangulation of the computational domain at time t = 100 h (a) and the enlarged fragment of the mesh: the average length of the tetrahedron edge is 2 cm near the front and 10 cm for the coarse mesh (b), 651114 tetrahedra with 2604456 freedom degrees.

The combined solver BiCG + GMRES (5) is used to solve SLAE. BiCG is the oscillating method and allows you to quickly approach an exact solution. The GMRES method is a smoothing agent and ensures uniform convergence to the solution. Physical parameters are presented in the Table 1.

Table 1. Media Physical Parameters.

Media	ρ, kg/m^3	c, J/(kg·°C)	λ, W/(m·°C)	β, W/(m^2·°C)	L, J / kg
Water	1000	4187	0.6	10	$42.3 \cdot 10^3$
Ice	900	2060	2.25		$334.1 \cdot 10^3$

Figure 3 shows the dynamics of the position of the phase transition front over time depending on the time discretization step (from 10 min to 40 min).

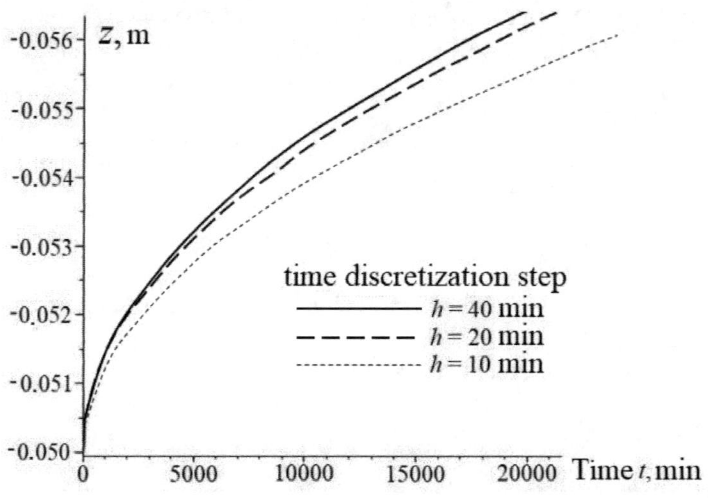

Figure 3. Dynamics of the phase transition front over time depending on the time discretization step.

Figure 4 shows the volume temperature distribution along the Oz axis after 6000 min of ice melting.

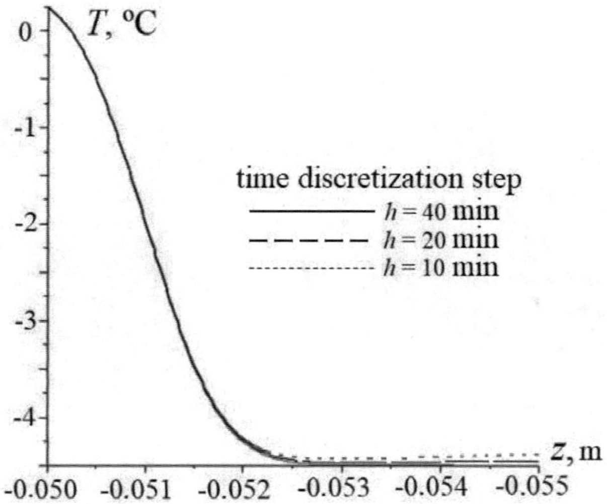

Figure 4. Temperature distribution after 6000 min of ice melting.

6. Discussion of Results

Analyzing the graphs in Figure 3 and Figure 4 allows us to conclude that the results of modeling the process of ice melting when using the proposed computational scheme is correct. At the time t = 3000 min, the position of the phase transition front corresponds to the coordinate z = – 0.052 m (see Figure 3), which is confirmed by the graphs in Figure 4. At the point z = – 0.052 m, a "bend" of the curve is observed (see Figure 4), since at this point the physical properties of the medium is changed (the boundary of the phase transition of the ice into water).

7. Conclusion

Ice melting at the Interaction with the Environment is described by the two-phase Stefan problem. For solving the two-phase Stefan problem, we used the computational scheme of the multiscale discontinuous Galerkin method using the nonconformal tetrahedral meshes. The Stefan condition is taken into account in the weak form in the IP (Inner Penalty) variational formulation of the DG-method. The mathematical modeling results of ice melting are obtained at the constant heat exchange with the media with the positive temperature depending on the time discretization step.

Acknowledgments

The reported research was funded by financial support of the President's scholarship (PS-3627.2016.5).

References

[1] Johnson K D, Harden J W, McGuire A D, Clark M, Yuan F and Finley A O 2013 Permafrost and organic layer interactions over a climate gradient in a discontinuous permafrost zone *Environ. Res. Lett.* **8** 1–12
[2] Markov E V, Pulnikov S A and Gerber A D 2015 Problems of defining boundary conditions for modeling the thermal interaction of "hot" pipelines with permafrost soil *Technical Sciences: Fundamental Research* **2** 2106–10
[3] Viscor M and Stynes M 2010 Numerical method for a Stefan-type problem with interior layers *Progress in Industrial Mathematics at ECMI* **17** 479–84
[4] Mitchell S L and Vynnycky M 2014 On the numerical solution of two-phase Stefan problems with heat-flux boundary conditions *Journal of Computational and Applied Mathematics* **264** 49–64
[5] Popov N, Tabakova S and Feuillebois F 2004 Numerical modelling of the one-phase Stefan problem by finite volume method *Numerical Analysis and Its Applications* **3401** 456–62
[6] Dave M, Hicham C, Jean-Loup R, Ziegler D and Fafard M 2016 A XFEM Lagrange multiplier technique for Stefan problems *Frontiers in Heat and Mass Transfer* **7(31)** 1–9
[7] Arnold D N, Brezzi F, Cockburn B and Marini L D 2002 Unified analysis of discontinuous Galerkin methods for elliptic problems *SIAM J. Numer. Anal.* **39** 1749–79
[8] Bochev P, Hughes T J R and Scovazzi G 2006 A multiscale discontinuous Galerkin method *Computer Methods in Applied Mechanics and Engineering* **195** 2761–87
[9] Itkina N B and Markov S I 2016 Discontinuous Galerkin method for solution of singularly perturbed problems *Computational Technologies* **21** 49–63

Accumulation of gases dissolved in water saturating a nonisothermal porous massif in the presence of water freezing zones

B S Maryshev and D S Goldobin

Institute of Cotinuous Media Mechanics UB RAS, 614013, 1 Akad. Korolev street, Perm, Russia

E-mail: bmaryshev@mail.ru

Abstract. A one-dimensional problem of gas transport through a bubbly medium under the effect of a temperature wave is considered, taking into account the freezing of the near-surface layer. The transport processes are described within the framework of the bubbly medium approximation. The solubility of gases in water is calculated on the basis of the scaled particle theory. It is shown that the average gas transfer to the surface increases under the action of a temperature wave, and the freezing effect leads to the slowdown of the gas transport near the surface and thus causes the accumulation of gases beneath it. Solving the heat transport equation with a phase transition we calculate the temperature field. With this time-dependent temperature field, from the solute transport equation, the distribution of the gas mass in the near-surface zone is obtained and the accumulation rate is evaluated. The effect of the parameters of the problem (the minimal annual temperature, the amplitude of the annual temperature oscillations and the intensity of gas generation) on the phenomenon of the gas accumulation is studied.

1. Introduction

The diffusion of dissolved gases in liquids is a well-studied problem (see, e.g., [1, 2]). Usually, the gas–liquid interface is presented as surface and the transport processes between gas and liquid are described by the boundary conditions. But the transport of gases through the liquid which saturates porous media is a specific process. The free gas is presented as bubbles in the porous media and the interface is distributed in the volume of media. Such media is called "bubbly" media [3, 4]. If the concentration of dissolved gas is greater than the solubility, the gas transits into bubbles. The small bubbles are trapped in porous matrix by the capillary forces. For the big enough bubbles, the buoyancy force can overcome the capillary one and the bubbles will move upwards. However, a bubble displacing in a porous medium is always unstable to splitting into smaller bubbles [5]; the splitting of bubbles occurs repetitively until they become small enough to be immobilized by the capillary force. The critical immobilisation gas volume fraction in pores varies from system to system, but remains within the range from 0.5–1% [6] to several percents [7]. The solubility depends on pressure and temperature. In the case of mechanical equilibrium, pressure linearly depends on the depth, which means that the gas bubbles are more inclined to formation in the upper layer of media where the solubility is lower. The low temperature at the boundary slows down the process of bubbles generation because the solubility increases with the decrease of temperature. Simultaneously, the water freezing blocks the gas transport; gases accumulate in the near-surface layer and transit into the bubbles. Thus, the solving of the heat transport problem with phase transition (Stefan problem [8]) is an important part of this work.

The paper is organised as follows. Sec. 2 is devoted to the derivation of the equations for the gas transport in nonisothermal porous media. In Sec. 3, the problem of one-dimensional heat transport with phase transition is formulated and solved. Sec. 4 presents the formulation of gas transport problem with freezing and discussion of the results of numerical simulation. In the conclusion we summarize all the results and discuss them.

2. Diffusion in bubbly media with heterogeneity of thermodynamic properties

The dissolved gas transits into bubbles if the solution concentration exceeds the solubility. In porous media the small bubbles are immobilized by the capillary forces. The solubility of gases at moderate pressure depends on temperature (T) and pressure (P) as follows (see [4], which corresponds to the scaled particle theory [9]):

$$X = X_0(T_0, P_0)\frac{T_0}{T}\frac{P}{P_0}\exp\left[q\left(\frac{1}{T}-\frac{1}{T_0}\right)\right], \tag{1}$$

where T_0 and P_0 are reference values, X_0 is the solubility at the reference temperature and pressure, $q = G_i/k_b$, G_i is the interaction energy between a solute molecule and the surrounding solvent molecules, $k_b \approx 1.38 \cdot 10^{-23}\,\mathrm{J/K}$ is the Boltzmann constant. The equation for transport of dissolved gas with the gas generation in the porous massif can be written in the following form:

$$\frac{\partial G}{\partial t} = -\frac{\partial J}{\partial z} + A\exp(-Kz)\,. \tag{2}$$

Here $G = C + Q$ is the net molar fraction of gas in the pore matter, Q is the molar fraction of the nondissolved gas, C is the molar fraction of the dissolved gas; for realistic $Q \ll 1$, the difference between C and the aqueous solution concentration can be neglected and one can treat C as the solute concentration. The molar concentration flux J of the solute is purely diffusive; $A\exp(-Kz)$ is the source function of the gas production (see, e.g., [10, 11]), where (A/K) is the gas production per the ground surface area and K is the attenuation coefficient. The diffusive flux in nonisothermal media with freezing zones can be written in form:

$$J = -D(T)\,C\left(\frac{1}{C}\frac{\partial C}{\partial z} + \frac{\alpha}{T}\frac{\partial T}{\partial z}\right); \qquad D(T) = \begin{cases} D_*, & T > T_*; \\ 0, & T \le T_*; \end{cases} \tag{3}$$

where α is the thermodiffusion constant, D_* is the effective diffusivity coefficient in the liquid-saturated porous medium, T_* is the freezing point. The gas molecules transport is solely due to the solution flux through the solvent phase [12, 13]. The gas in ice and in bubbles is immobilized. The net molar fraction G is $G = C$ for $G < X$ and $G = X + Q$ for $G > X$ (the gas excess over the solubility transits into the bubbles). Pressure is hydrostatic, $P = P_0 + \rho g z$, where g is the gravity acceleration and ρ is the pore fluid density. The transport problem (2)–(3) significantly depends on the temperature profile $T(z,t)$. The temporal evolution of this profile is to be determined from the heat conductivity problem with the account for the phase transition (freezing of water)—Stefan problem. The next section is devoted to the solution of this problem.

3. Temperature wave with phase transition

The main reason of a possible accumulation of gas bubbles is a nonlinear dependence of the diffusive flux and the solubility on temperature. The temperature profile in the massif with

IOP Conf. Series: Earth and Environmental Science **193** (2018) 012044 doi:10.1088/1755-1315/193/1/012044

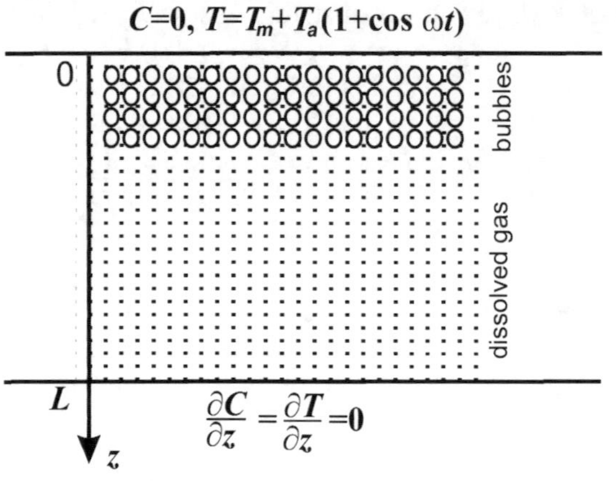

Figure 1. The sketch of the problem.

the account for the phase transition is given by the solution of a one-dimensional Stefan problem (see Figure 2). At the upper boundary of the massif, temperature depends on time as $T_m + T_a(1+\cos\omega t)$, where T_m is the minimal annual temperature, T_a and ω are the amplitude and the frequency of the annual temperature oscillation, respectively.

For solving the Stefan problem it is convenient to introduce the fraction of frozen water in pores, $h(z,t)$. In what follows, we neglect the frost heaving of soils, since it is small for regions, where the air temperature becomes negative for some part of the year. Then, the governing equations for the heat transfer read

$$
\begin{gathered}
(\rho c_P)_{\text{p.m.}}\frac{\partial T}{\partial t} = \frac{\partial}{\partial z}\left(k_{\text{p.m.}}\frac{\partial T}{\partial z}\right) + Q(z,t)\,;\\
Q(z,t) = -\rho_l\lambda\phi\frac{\partial h}{\partial t}\,;\\
\frac{\partial h}{\partial t} = \begin{cases} r_\lambda(T_* - T), & \text{for } T > T_*,\ h > 0\,;\\ r_\lambda(T - T_*), & \text{for } T < T_*,\ h < 1\,;\\ 0, & \text{otherwise}\,;\end{cases}\\
T|_{z=0} = T_m + T_a(1+\cos\omega t), \qquad \left.\frac{\partial T}{\partial z}\right|_{z=L} = 0\,.
\end{gathered}
\tag{4}
$$

Here the first equation describes the temperature field evolution, $(\rho c_P)_{\text{p.m.}}$ and $k_{\text{p.m.}}$ are the volumetric specific heat and the heat conductivity of the saturated porous medium (generally, these parameters depend on h), $Q(z,t)$ is the heat release (consumption) rate due to water phase change, ρ_l is the liquid density, λ is the specific latent heat of phase transition, ϕ is the porosity. The third equation of system (4) describes the ice melting, where it is present ($h > 0$) and temperature is high ($T > T_*$), and water freezing, where liquid water is present ($h < 1$) and temperature is low ($T > T_*$). The phase transition rate r_λ is practically so high that one can consider the transition to the thermodynamic equilibrium between ice and liquid water as instant. Accordingly, for numerical simulation the specific value of this parameter is also not important if it is large enough. In the last line of (4), the boundary conditions are specified. For a real porous massif, one can assume $L = \infty$; for numerical simulations, the value of L was chosen large enough to have no effect of its doubling on the results of calculations.

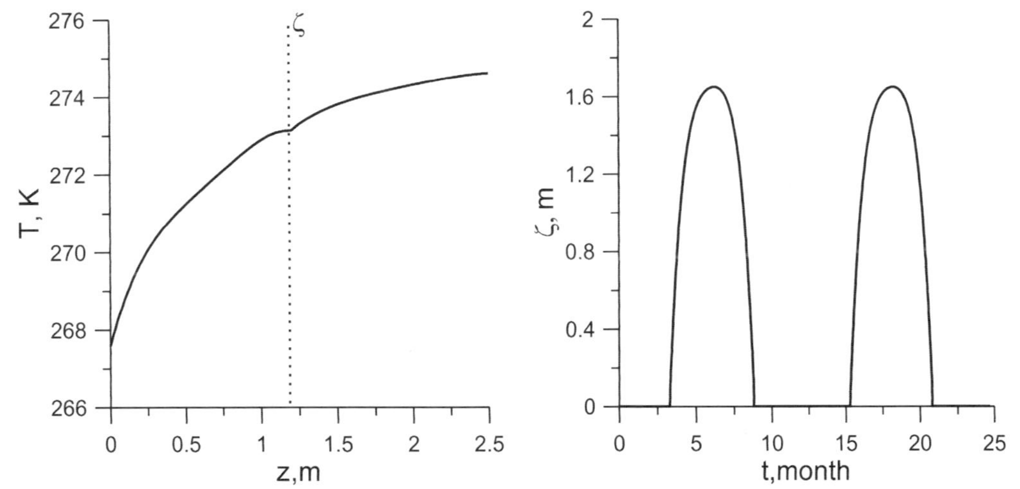

Figure 2. The temperature profile at time moment $t = 475$ months, the position of the ice–water interface $z = \zeta$ is indicated with dots (left panel). The position of interface ζ versus time is plotted in right panel. Parameters: $T_a = 15\,\mathrm{K}$, $T_m = 260\,\mathrm{K}$.

The problem (4) was solved numerically by the implicit finite differences scheme of the second order of accuracy in space and the first order of accuracy in time. Here we describe the annual oscillations of temperature; therefore, the convenient unit of time is month and all distances are measured in meters. The parameters are chosen for the system of water in a peat bog: the temperature diffusivity of the saturated massif is nearly the same for liquid water and ice and was adopted $\chi_{\mathrm{p.m.}} = 1.82\,\mathrm{m}^2/\mathrm{month}$, $k_{\mathrm{p.m.}} = 1.5\,\mathrm{W/(m\,K)}$, other parameters were $\lambda = 3.3\cdot 10^5\mathrm{J/kg}$, $\phi = 0.6$, $\omega = 2\pi/12\,\mathrm{month}^{-1}$ and $T_* = 273.15\,\mathrm{K}$ [4, 12, 13].

The depth of saturated layer in peat bogs varies from 20 to 150 m, but the characteristic depth of thermal wave penetration can be estimated as $d = \sqrt{2\chi/\omega} \sim 2.5\,\mathrm{m}$. The size of the calculation domain for temperature was $L = 50\,\mathrm{m} = 20\cdot d$. The example of the results for the temperature profile evolution problem are presented in Figure 2.

4. Diffusion in bubbly media under action of temperature wave with freezing

For a given temperature profile $T(z,t)$, Eqs. (1)–(3) yield the full gas transport problem:

$$
\begin{gathered}
\frac{\partial G}{\partial t} = \frac{\partial}{\partial z}\left[D(T)\left(\frac{\partial C}{\partial z} + \frac{\alpha C}{T}\frac{\partial T}{\partial z}\right)\right] + A\exp(-Kz)\,;\\
D(T) = \begin{cases} D_*, & T > T_*\,;\\ 0, & T \le T_*\,;\end{cases}\\
C = \begin{cases} G, & G \le X\,;\\ X, & G > X\,;\end{cases}\\
X = X_0\frac{T_0}{T}\left(1 + \frac{\rho g z}{P_0}\right)\exp\left[q\left(\frac{1}{T} - \frac{1}{T_0}\right)\right];\\
C|_{z=0} = 0\,, \qquad \left.\frac{\partial C}{\partial z}\right|_{z=L} = 0\,.
\end{gathered}
\tag{5}
$$

Here the condition at the atmosphere–porous-media interface corresponds to the release of the emitted gas into the atmosphere; the condition at $z = L$ means no gas flux from beneath. The problem (5) is solved numerically with an implicit finite differences scheme of the second

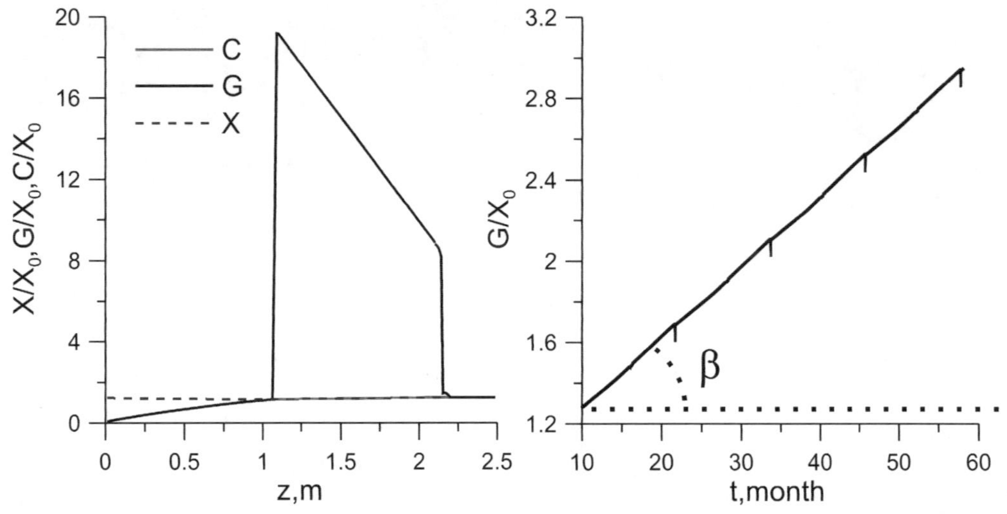

Figure 3. The profiles of the dissolved gas concentration C, the net molar fraction G and solubility of gas X for time moment $t = 475$ months (left panel). The dependence of net molar fraction on time at $z = 1.5$ m (right panel). The results are presented for the temperature profile shown in Figure 2; parameters: $T_a = 15$ K, $T_m = 260$ K.

order of accuracy in space and the first order of accuracy in time. The most noticeable implication here is the accumulation of methane in a bogged soil. Hence, all reference values of parameters are chosen for methane and water; specifically, $q = 1138$ K, $D_* = 2.8 \cdot 10^{-3}$ m^2/month, $X_0 = 2.6 \cdot 10^{-5}$, $\alpha = 0.3$, $P_0 = 1$ atm and $T_0 = 293$ K (see [14, 15, 16, 17, 18]).

The example of spatial distribution of gas (total amount and dissolved) is presented in Figure 3 (left panel). It can be seen that the zone of methane accumulation is located just below the zone of soil freezing (see Figure 2) and beneath ≈ 1 m, which is consistent with experimental data [19]. In our model, the accumulated methane has no ways for emission into the atmosphere. However, in nature, if the volumetric fraction of gas bubbles exceeds some threshold value the gas seepage occurs and an excessive amount of gas is released from the soil [6]. These events are recognized as emission of methane from peat bogs, and natural observations show that these events are periodic [20]. In our model, the amount of gas in the bubbly layer increases linearly with time (see Figure 3, right panel) and the gas seepage is not included in the model. In this case the most interesting characteristic is the rate of accumulation ($\gamma = \tan \beta$) because it allows to calculate the periods of emission for any specific soil and peat bog.

The dependencies of the accumulation rate γ on the minimal temperature (T_m) and the annual temperature amplitude (T_a) are presented in Figure 4. One can seen that the accumulation rate is fixed if the near-surface soil is frozen all over the year and becomes lower with the increase of temperature. The dependencies of γ on the parameters of gas generation (A and K) are plotted in Figure 5. Figure 5 shows that the accumulation rate decreases with the decrease of the intensity of gas generation.

5. Conclusion

The gas transport through a bubbly medium under the effect of a temperature wave has been investigated with the account for the freezing of the near-surface layer. The gas transport and accumulation have been described within the framework of the bubbly medium approximation. It has been shown that the average gas transfer to the surface increases under the action of a temperature wave, and the freezing effect leads to the blockage of the gas transport—the

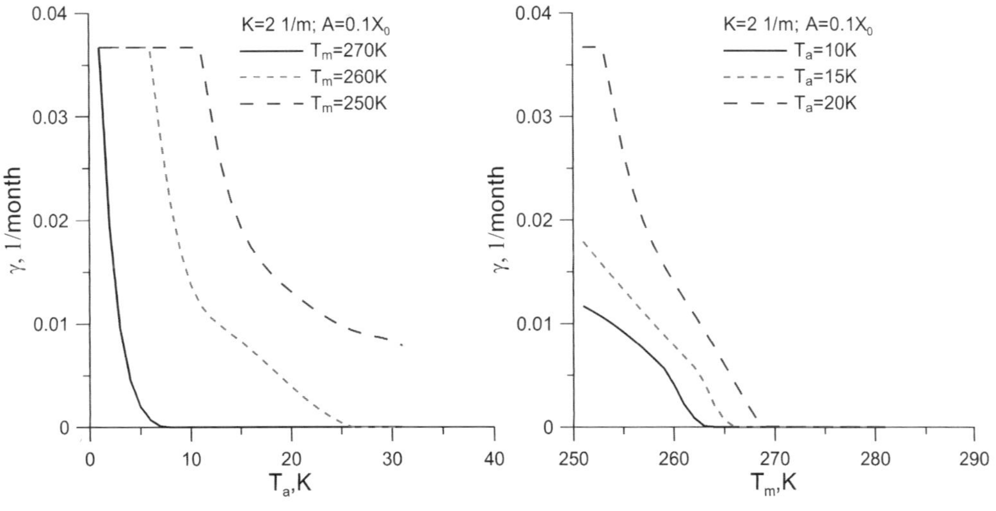

Figure 4. The dependencies of the gas accumulation rate on the annual temperature amplitude (T_a, left panel) and the minimal annual temperature (T_m, right panel). The values of parameters are indicated in legends.

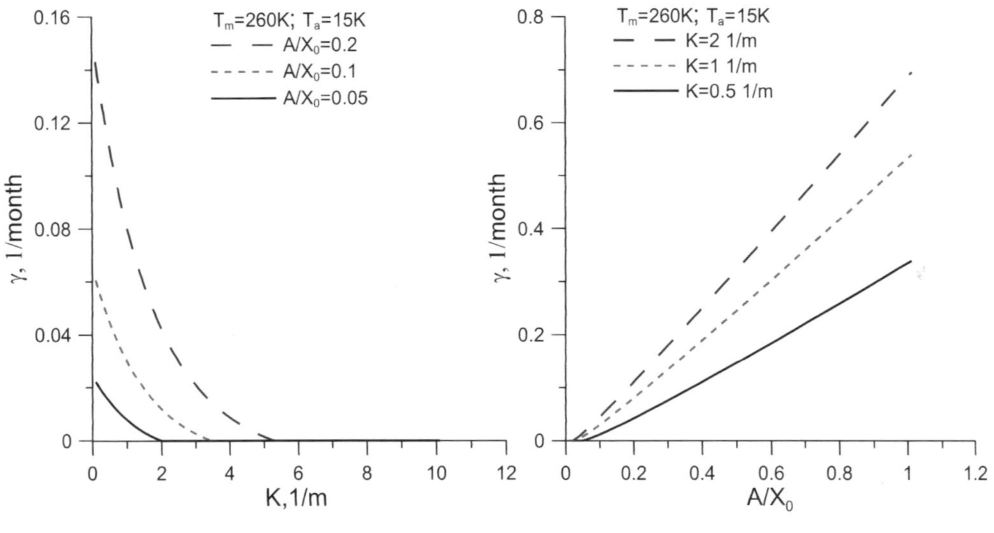

Figure 5. The dependencies of the gas accumulation rate on the attenuation coefficient (K, left panel) and the intensity of gas generation (A, right panel). The values of parameters are indicated in legends.

accumulation of gas immediately beneath the surface. We have obtained numerically the solution for the heat transport problem with a phase transition (Stefan problem) in form of temperature distribution. For this time-dependent distribution, from the transport equations, the distribution of the gas amount in the near-surface layer has been obtained numerically. The effect of gas accumulation has been revealed and the rate of the accumulation has been calculated. The effect of the parameters of the problem on the phenomenon of the gas accumulation has been analysed. The rate of accumulation has been shown to be constant if the surface of the massif is frozen all over the year and to decrease with the increase of the mean surface temperature. The rate of accumulation has been found to increase with the intensification of the gas generation in the porous massif.

Acknowledgments

This work was partially supported by a Grant of the President of Russian Federation (Grant No. MK-1447.2017.5).

References

[1] Bird R B, Hirschfelder J O and Curtiss C F 1954 *Molecular theory of gases and liquids* (New York: Wiley)
[2] Bird R B, Stewart W E and Lightfoot E 2007 *Transport Phenomena* (New York: Wiley)
[3] Yurkovetsky Y and Brady J F 1996 *Phys. Fluids* **8** 881
[4] Goldobin D S and Brilliantov N V 2011 *Phys. Rev.* E **84** 056328
[5] Lyubimov D V, Shklyaev S, Lyubimova T P and Zikanov O 2009 *Phys. Fluids* **21** 014105
[6] Firoozabadi A, Ottesen B, Mikkelsen M 1992 *SPE Form. Eval.* **7** 337
[7] Moulu J C 1989 *J. Petrol. Sci. Eng.* **2** 379
[8] Crank J 1984 *Free and moving boundary problems* (Oxford: Clarendon Press)
[9] Pierotti R A 1976 *Chem. Rev.* **76** 717
[10] Davie M K and Buffett B A 2001 *J. Geophys. Res.* B **106** 497–514
[11] Goldobin D S, Brilliantov N V, Levesley J, Lovell M A, Rochelle C A, Jackson P D, Haywood A M, Hunter S J and Rees J G 2014 *Eur. Phys. J.* E **37** 45
[12] Krauzin P V and Goldobin D S 2014 *Eur. Phys. J. Plus* **129** 221
[13] Goldobin D S and Krauzin P V 2015 *Phys. Rev.* E **92** 063032
[14] Verhallen P T H M, Oomen L J P, Elsen A, Kruger J and Fortuin J M H 1984 *Chem. Eng. Sci.* **39** 1535
[15] Sachs W 1998 *J. Petrol. Sci. Eng.* **21** 153
[16] Zeebe R E 2011 *Geochim. Cosmochim. Acta* **75** 2483
[17] Shvab I and Sadus R J 2014 *J. Chem. Phys.* **140** 104505
[18] Segers R 1998 *Biogeochemistry* **41** 23
[19] Hornibrook E R, Longstaffe F J and Fyfe W S 1997 *Geochim. Cosmochim. Acta* **61** 745
[20] Derwent R G, Simmonds P G, Manning A J, O'Doherty S and Spain G 2009 *Atmosph. Env.* **43** 2328

System for geophysical observation on the drifted ice floes for investigation of the lithosphere structure in the Arctic (Project)

V S Mogilatov[1,2], I Yu Koulakov[1,2] and V V Plotkin[2]

[1]Novosibirsk State University, Novosibirsk, Russia
[2]Institute of Petroleum Geology and Geophysics, Novosibirsk, Russia

E-mail: mvecs@ya.ru

Abstract. The Project aims at developing a geophysical observation system on drifting ice floes with the purpose of studying the lithosphere structure in the Arctic Regions. Exploration of the Arctic region is actual task that affects political and economic interests of several country. The knowledge of the lithosphere deep structure beneath the Arctic Ocean is necessary to reconstruct the stages of the regional geological evolution, which, in turn, provides important information for exploration of new mineral deposits and to understand their origin. Although the Arctic attracts much attention, the scientific investigations there advance quite slowly. The harsh climatic conditions and large expenses slow down realization of the fieldwork in high latitude areas. Therefore, scientists from over the world looks for new technologies, which could optimize and reduce the costs of the fieldworks that aimed at investigation of the geological structure beneath the Arctic Ocean. The main task of the Project is to develop the methodology of long-term multidisciplinary geophysical observations that are meant to study structure of the crust and offshore seismicity in the Polar Regions. To this proposes, the observation systems have to be adopted to the conditions of the deep sea covered by paleocrystic floes and should take into account drift of the ice floes. In the framework of the Project, the prototypes of such observation systems will be developed and tested in the natural conditions on ice of the Arctic Ocean and in other water reservoirs. Here, instead of standard methods, we propose performing complicated, but more efficient experiments with deployment of various geophysical equipment of the ice floes. The drift of the ice floes will provide the movement of the instruments above study area and thus will improve data coverage. The observation systems of the passive seismological and magnetotelluric sounding methods supplemented by the electromagnetic sounding method with innovative controlled source will be designed for installation on top of the drifted ice floes. The practical implementation of this set of the methods will be significantly less expensive than currently used technologies of active seismic survey, for example.

1. Introduction

The Project aims at developing a geophysical observation system on drifting ice floes with the purpose of studying the lithosphere structure in the Arctic Regions. Exploration of the Arctic region is actual task that affects political and economic interests of several country. The knowledge of the lithosphere deep structure beneath the Arctic Ocean is necessary to reconstruct the stages of the region geological evolution, which, in turn, provides important information for exploration of new mineral deposits and their origin. Although the Arctic attracts much attention, the scientific investigations there advance quite slowly. The harsh climatic conditions and large expenses slow down realization of the fieldwork in high latitude areas. Therefore, scientists from over the world looks for new technologies, which could optimize and reduce the costs of the fieldworks that aimed at investigation of the geological structure beneath the Arctic Ocean. The main task of the Project is to develop the methodology of long-term multidisciplinary geophysical observations that are meant to study structure of the crust and offshore seismicity in the Polar Regions. To this proposes the observation systems has to be adopted to the conditions of the deep sea covered by paleocrystic floes and should take into account drift of the ice floes. In the framework of the Project, the prototypes of such observation systems will be developed and tested in the natural conditions on ice of the Arctic Ocean and other water reservoirs.

Recent state of knowledge about structure of the Arctic Region is based on works of Russian (Soviet) and international polar scientists. Initially, the drifted research stations "North Pole" and hydrogeographic survey, which were supported by the military, have played an important role. The seismic methods and aeromagnetic measurements have been used at that time. However, the goal of that research was investigation of the ocean bathymetry. The geophysical investigation of the Arctic Region with the geological proposes gets real only during the last decades. Note that most of geophysical studies in the shelf zone are based on standard methods of marine geophysics in areas with lite ice conditions. Big progress was achieved within the "Marine Arctic Geological Expedition" (JSC MAGE) conducted in collaboration with several institutions in the framework of the project "Arctic-2014" (Kazanin et al., 2015). Unique results were obtained with the use of reflection and refraction seismic methods based on under ice observations. This very complicated and expensive field campaign involved special ships and atomic icebreaker making possible conducting observations in areas with thick ice reaching up to 240 centimetres thickness. These and other offshore campaigns requested enormous funds for performing standard marine geophysical methods, which in the polar conditions are very expensive. A cheaper alternative of polar studies is based on large and global scale investigations with the use of seismology, tectonic reconstructions, paleomagnetic analysis, magnetotelluric data, which provide indirect and generalized information about geological structure of the region.

The target geological structures in the Arctic are overlain by the deep ocean water layer and partly by the ice floes. This is the main problem, which has to be solved at a stage of the project development. This situation strongly limits the list of the technologies that could be applied. In particular, a number of ground-based technologies, methods of remote sensing with use of air and spacecraft, and even some approaches of marine geophysics could not be applied in areas covered by ice floes. What do we have? Standard methods of active seismic in the polar conditions are very expensive and logistically difficult. Satellite gravimetric and magnetometric investigations provide important information about regional structures, however, its resolution not enough to reveal local features that are the most important from the practical point of view. Electromagnetic methods with the use of natural fields could provide important information, but they also do not provide sufficient resolution. Thus, the technologies with active sources have to be used. Still, traditional methods of electromagnetic sounding with artificial sources aimed at providing the distribution of resistivity in geological media become inefficient in regions covered by thick high-conductivity sea water layer. On the other hand, the ice floes prevent the use of some marine technologies (for example CSEM), which are effective on the clear water. Instead of standard methods, we propose here another setup of the observation systems presuming deployment of measuring instruments on the ice floes. The drift of the ice floes will provide the movement of the instruments above study area and good data coverage. In the following, we propose several geophysical methods specially adapted for the condition of the deep water covered by ice floes.

The proposed set of the passive seismological and magnetotelluric sounding methods supplemented by the method of the electromagnetic sounding with innovative controlled source will be designed for installation on top of the drifted ice floes. The practical implementation of this set of the methods will be significantly less expensive than currently used technologies of active seismic, for example. The list is open and could be extended with other geophysical methods which are suitable for the polar condition. We propose a new approach that could be used for investigation of the geological structure of the ocean bottom in the areas covered by ice floes and does not required big expenses. At the same time, these investigations could provide detailed information about different structural features and discontinuities and point on potential mineral deposit sites.

Geophysical investigations with system installed on drifted ice floes is a novel direction and many aspects require consideration that is more detailed. The theory and field practice of marine MTS postulate that optimal results could be obtained if electro-magnetic measurements conducted close to the see bottom. In our case, measurements have to be conducted on a top of ice floes. We do not find any practical works of such kind. Theoretical studies based on mathematical modelling show that if measurements performed on the top of the ice floes then effectivity became lower but still sufficient [1]. We propose to use powerful controlled electro-magnetic source on the drifted ice floes in the Arctic, which is a novel task. Now scientists including the Project participant test this technology [2], [3].

2. Seismological monitoring

For the first time the idea to use the drifting ice floes for installation of the seismic network have been proposed several years ago. Since 2007 such kind of the experiments carried out by the Alfred Wegener Institute in Bremerhaven (AWI Bremerhaven) and directed by Vera Schlindwein [4]. The migration trajectories of stations from one of these networks have shown in the figure 1. This network were installed on the drifted ice floes in the Gakkel ridge region. During operation period a lot local earthquakes that occur in the spreading zone and several regional and teleseismic events were recorded. The examples of records of these events [5] are presented in figure 2. One can see that beside clear P-wave, the converted SgP-phase could be reliably detected. The SgP-phase could be easily converted to the S-wave.

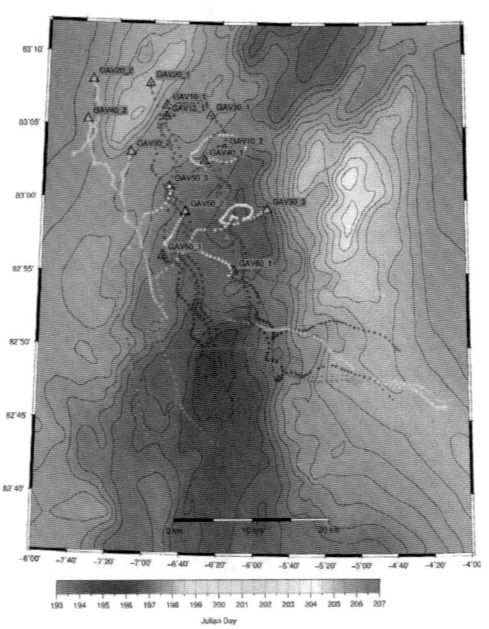

Figure 1. Trajectories of migration of seismic stations installed on drifting ice. The initial position of the stations is shown by triangles; the time of their functioning is marked by color. This network was installed on the drifted ice floes in the Gakkel ridge region.

This experiment has shown that seismic networks, installed on the ice floes provide valuable records of the P- and S-waves that could be used for the seismic tomography and other proposes. In spite of promising results obtained by the specialists from the AWI, we do not know any similar works performed by other institutions. Apparently, this approach should be developed and applied in Russian part of the Arctic. The aim of the Project is to develop technologies that will help to realize this proposal

In the framework of the Project, we propose to develop the method of seismological observation that used networks installed on the drifted ice floes. We will further develop and improve this technology on a base of theoretical modeling and field experiments. The main idea is to install a network of portable broadband seismometers on the ice floes. These seismometers will autonomously record the seismicity and GPS position during certain time period (for example one month). Analysis of the recorded data will include identification of local earthquakes and picking of the arrival times of different phases. Previous works show that such type of the observation system allows robust registration of the converted S-P phase. The 3D distribution of the P- and S- wave velocities beneath the sea bottom and location of the local earthquakes will be obtained after the iterative tomographic inversion of the arrival times. This inversion can be based on the existing tomography codes, such as LOTOS code [6]. However, the existing algorithms cannot be directly applied to the floating network data and will require some modifications. The updated version of the algorithm should take into the account the following points: (1) starting reference velocity model should include information about the water layer according to the available bathymetry; (2) the converted PS waves recorded at the drifting ice should be reduced to the S-rays ended at the sea bottom; (3) the structure of the input data has to be changed because of the constantly changed station positions due to the ice drift.

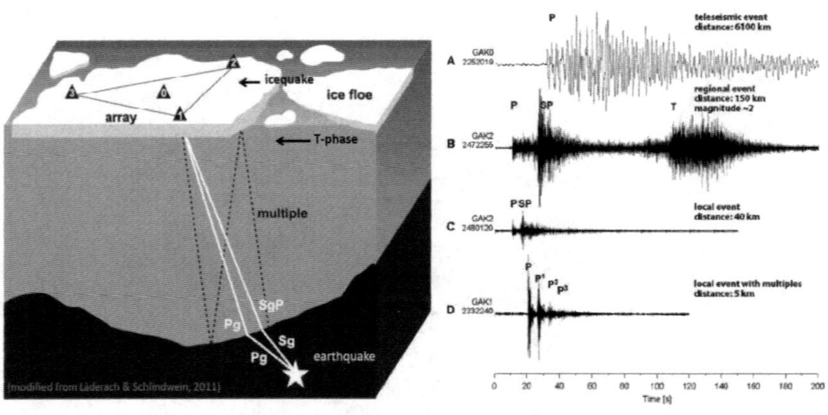

Figure 2. Scheme of arrival of seismic rays at the station, installed on ice. Right: examples of seismograms recorded by the station on a drifting ice floe, from teleseismic, regional and local events. Multiple waves reflected from the bottom and ice are also shown.

The theoretical modelling will include creation of a realistic synthetic model and its reconstruction with use of available methods. Parameters of the ice drift and distribution of seismicity obtained from previous field experiments [5] will be used for the creation of the synthetic model. We will create synthetic velocity model (with realistic shapes of anomalies or "Checkerboard" type of model). The created synthetic model and real distribution of the events will be used for the calculation of the travel times. In the next step, we will "forget" the model structure and events parameters and will reconstruct the model using absolutely the same steps of algorithm performance as in the case of the real data inversion. Thanks to the personal communications of the project participants with the German scientists, we have an access to the data obtained during several experiments conducted by the AWI Bremenhaven. This give us possibility to test developed technologies with the real data sets and to obtain tomographic models of real geological structures in the Gakkel region in the Arctic. The experimental part of the project will include installation of seismic stations on the ice in different water basins. In the end of the first year, to investigate various technological aspects the seismic stations will be installed on the ice of the Arctic Ocean at the vicinity of Tiksi. Except for the drift of the ice floes, this experiment will appropriately simulate the conditions of drifting ice in the Arctic. The registration period from December 2018 to April 2019 will be enough to register sufficient number of seismic events occurred at different distances. After analysis of the record of these events, we will be able to make a conclusion about effectivity of the proposed observation system. Beside body wave, we will try to extract the record of the Rayleigh waves. In addition, we will investigate if it is possible to make seismic noise correlation analysis and extract information about surface waves. According to our knowledge, there is not studies that used such analysis performed for the stations installed on the ice on water basins.

3. Electrical prospecting with controlled sources

We offer a unique electromagnetic sounding technique, based on the excitation of a certain (TM) polarization of electromagnetic field that technically is provided by using a specific field source (circular electric dipole - CED). This technique (Vertical Electric Currents Soundings – VECS, [7]) allows recording subtle anomalies in a response due to deep compensation caused by physical processes common background of the conductive geoelectric section, including a layer of sea water. The technique has already been widely tested at various sites onshore primarily with hydrocarbon deposits, and proved highly effective. Both theoretical aspects of the novel method and results of practical applications have already been extensively published. The VECS method is characterized by the use of specific source of field, which may seem very complicated from the traditional point of view. This is a fair price that we pay for high performance and which usually is indicated as a disadvantage, given the need to move the source. However, the issue is resolved in the most satisfactory way under this project - we employ the ice-drift to make it move. Within the project we propose a rationale for the following method of electromagnetic sounding in the geological environment of the ocean floor to a depth of several kilometers in order to determine resistivity distribution pattern, as well as other geoelectric parameters that reveal various regional and local geological objects by moving the entire experimental setup over the Arctic basin.

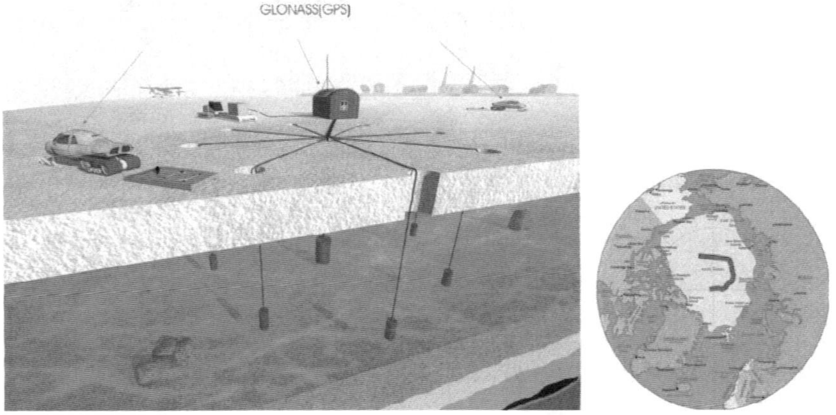

Figure 3. General view of the feed array on the Arctic ice and possible research area.

IOP Conf. Series: Earth and Environmental Science **193** (2018) 012045 doi:10.1088/1755-1315/193/1/012045

A pulse-feed array of circular electric dipoles is arranged on the ice (once and very carefully) (illustrated by figure 3) and is grounded at the center and at the ends of radial feeders through holes in the ice. CED radius (length of the radial line) can be up to 5-10 km (question to be studied). The observation system consists of electric receivers in the form of fixed, water-grounded, horizontal and vertical lines, as well as spatially distributed magnetic field measurement system, operated by the mobile inductive sensors. Alternatively, we have to decide whether to do continuous measurements (which is all but impossible), or to carry out measurement runs, tying it to a certain point, and repeat it over time (a day means 5-7 km of ice-drift). This issue is yet to be studied. Anyway, based on the experience of VECS application, the measurements may comprise the band along the source path with width equal to ten radii. So, given the CED radius is 10 km, we get total length of 4000 km of ice floe drift (which is consistent with the average for the Soviet NP stations), or 400 000 square kilometers of the surveyed area in the shape of, interestingly, very elongated winding corridor. The interpretation, which should include both 1D and 3D approaches, will extremely increase its confidence level due to the persistently accompanying seismic soundings, which will ultimately provide reliable a priori information on the boundaries. In the attached file, we provide a rationale for the CED- techniques applications. Moreover, CED-based techniques as applied in the project will address and other configurations of the receiver-transmitter array, which theoretically claim to be highly effective under the conditions considered.

4. Magnetotelluric investigations

In the addition to electromagnetic sounding with controlled source, we propose to use magnetotelluric (MT) sounding. The MT sounding could resolve deeper part of Earth than transient electromagnetic sounding (TEM), which provide better resolution in the upper part of the study volume. MT method uses natural fields and do not requires power sources in the addition to one used for the TEM. It is reasonably to perform geomagnetic-variation measurements in addition to the MT sounding. The theory and field practice of marine MT postulate that optimal results could be obtained if electro-magnetic measurements conducted close to the see bottom. In our case, measurements have to be conducted on a top of ice floes. Theoretical studies based on mathematical modelling show that if measurements performed on the top of the ice floes then effectivity became lower but still sufficient [1]. In the framework of the Project, we will investigate the applicability of the method in the polar conditions. Preliminary, to perform the MT sounding we propose to use the same system of electrical receivers, which have a form of vertical and horizontal lines, as one used for TEM. The CED source will be used for signal registration. That will provide possibility to register the vertical electrical component of the MT field, which is novel issue of the MT sounding. The standard method of MT sounding based on horizontally layered model, activated by the vertically incident plane wave. The influence of vertical component of the electric field is considered as neglect. On the border with weak conductive layer (atmosphere or ice), this component reduce to a zero value. However, if there are lateral inhomogeneity in the media or in the electromagnetic source field then vertical component of the electric field could became significant. Value of registered electrical component depend from length of the receiver line. Realization of the long vertical receiver line is technically difficult (even in the marine condition due to the sea undertows). The use of CED for registration of the vertical component is not applied in the actual practice. Therefore, it is of interest for scientist because it open a new possibility for interpretation of the vertical component of the electrical field. Presence of the vertical component could indicate the anisotropy of deep conductive layers. The use of CED with increased length of the radial lines will provide the ability to reveal such layers. Theoretical studies and mathematical modelling (including 3D) will be used for feasibility demonstration of each method. The field experiments in winter conditions on the ice of fresh and salty water basin are proposed. In the framework of the Project, we will study the different scientific aspects of the planned geophysical experiment in the Arctic. For example, we will calculate the necessary amount of the power sources. The planned geophysical works with fixed measurement configuration will provide high quality of the retrieved data. Thus, the necessary amount of repeated measures will significantly reduce. Another positive factor is absence of the industrial electromagnetic noise in the Polar Regions. However, a telluric noise and noise induced by local ice movements still could play an important role. Important part of the experiment concern the retrieving the registered data, its temporal storage and near real-time transfer to the mainland. Apparently, the data processing center that work in the near real-time regime have to be organized. This center should control the retrieving data and provide capability to identify possible defects in the observation systems (it will be impossible to return and repeat the measurements). Furthermore, it could be necessary to change parameters of measure instruments (for example, current impulse time, time of the signal registration for the TEM etc.) due to the ongoing results. Taking into account necessity for 3D analysis the use of a supercomputer may be necessary.

Acknowledgments
The study was performed with the financial support of the Russian Science Foundation (RSF), project no. 18-17-00095.

References
[1] Korotaev S M, Zhdanov M S, Orehova D A, Kruglyakov M S, Trofimov I L, Tshors Yu G and Shneer V S 2010 *Fizika Zemli* 35-47
[2] Mogilatov V, Goldman M, Persova M, Soloveichik Yu, Koshkina Yu, Trubacheva O and Zlobinskiy A 2016 *Journal of Applied Geophysics* **135** 17–31
[3] Mogilatov V S and Zlobinskiy A V 2016 *Geofizika* 75-80
[4] Schlindwein V, Müller C and Jokat W 2007 *Geophysical Journal International* **169** 100-12
[5] Läderach C and Schlindwein V 2011 *Seismological Research Letters* **82** 494-503
[6] Koulakov I 2009 *Bulletin of the Seismological Society of America* **99** 194-214
[7] Mogilatov V S, Zlobinskiy A V and Balashov B P 2017 *Geophysical Prospecting* **65** 1380-97

Numerical research of gas-hydrate deposit development in the conditions of negative temperatures

N G Musakaev[1,2,3], M K Khasanov[1] and S L Borodin[1,2]

[1]Sterlitamak Branch of Bashkir State University, Lenina Str. 37, Sterlitamak 453103, Russia
[2]Tyumen Branch of Khristianovich Institute of Theoretical and Applied Mechanics of SB RAS, Taymirskaya Str. 74, Tyumen 625026, Russia
[3]Industrial University of Tyumen, Volodarskogo Str. 38, Tyumen 625000, Russia

E-mail: musakaev@ikz.ru

Abstract. Numerical research of non-isothermal filtration flow in a hydrate saturated reservoir taking into account the decomposition of methane hydrate is carried out. It is established that the gas extraction in the case of the negative initial reservoir temperature (the gas hydrate decomposition occurs to gas and ice) is more beneficial than the methane extraction from the reservoir with the positive initial temperature (the methane hydrate decomposition occurs to gas and water). Calculations show that, in the first case, the time for the complete hydrate decomposition in a reservoir decreases with an increase in the initial reservoir temperature and an increase in the bottomhole depression.

1. Introduction

At present, there is an increase in the cost of natural gas production, which occurs against the background of growing demand for this raw material [1]. Accordingly, there is growing interest in the industrial development of non-traditional sources of hydrocarbons. Significant potential resources of natural gas (primarily methane) are in the deposits of gas hydrates [2, 3]. To date, over two hundred gas hydrate deposits have been discovered in the world. On the continents these deposits are confined mainly to the areas of permafrost [1, 4]. In the gas hydrates a huge volume of natural gas is concentrated, according to some estimates this volume several times greater than the volume of natural gas in free state [2, 3]. Industrial extraction of only 15% of methane from hydrate-containing deposits will provide the world with energy for 200 years at the current level of the gas consumption [3].

When developing deposits of gas hydrates it is necessary to transfer gas in a reservoir from the solid hydrate state to the free gas state. Then, methane is removed from the reservoir using traditional gas production technologies [1, 5]. At present, several methods are proposed for extracting methane from gas hydrate deposits: pressure reduction, heating, inhibitor input, injection of carbon dioxide into a reservoir. With the depressive method of the gas hydrate deposits development it is necessary that the pressure in a reservoir becomes lower than the equilibrium hydrate decomposition pressure. In this case, gas hydrate begins to dissociate to gas and water (or ice) with the heat absorption. It should be noted that any technological ideas and appropriate schemes for the hydrocarbon deposits development should be supported by appropriate theoretical calculations [6, 7].

Theoretical researches of the gas extraction from a hydrate saturated reservoir are presented in a number of works, for example [8-11]. In this paper, we present the numerical research results of the process of the gas extraction from a reservoir initially saturated with methane and its hydrate under conditions of negative (lower than 0°C) initial temperature of the reservoir.

2. Main equations

Let a horizontal porous reservoir of constant thickness be filled with methane and its hydrate in the initial state, the reservoir pressure p_0 and temperature T_0 correspond to the thermodynamic conditions of gas and hydrate existence in free state. The initial hydrate saturation of S_{h0} is known. Let the roof and bottom of the reservoir be impenetrable. We assume that the reservoir is homogeneous and isotropic, and also we neglect the influence of the upper and lower boundaries. In this connection, we can assume that the problem is one-dimensional and the process parameters depend only on the spatial coordinate x and time t [12].

We accept the following assumptions: methane hydrate is a two-component system with a constant mass concentration of gas G; the temperature of porous medium, gas, methane hydrate and ice (or water) at each point of the reservoir coincide (single-temperature process); the reservoir porosity m is constant; the porous medium skeleton, gas hydrate and ice (or water) are incompressible and immobile.

Initial and boundary conditions can be written in the following form:

$$t=0,\ x\in[0,L]:\ T=T_0,\ p=p_0,\ S_h=S_{h0},\ S_g=1-S_{h0},$$

$$t>0,\ x=0:\ p=p_e<p_0,\ \frac{\partial T}{\partial x}=0,$$

$$t>0,\ x=L:\ \frac{\partial p}{\partial x}=0,\ \frac{\partial T}{\partial x}=0.$$

Here L is the reservoir length; S_j ($j=i,h,g$) is the pore saturation with j-th phase (g – gas, i – ice, w – water, h – gas hydrate).

The system of main equations consists of the equations of conservation of mass and energy, the Darcy law and the equation of gas state. This system in the rectilinear-parallel approximation has the form [10, 12, 13]:

$$\frac{\partial}{\partial t}\left(\rho_g S_g+\rho_h S_h G\right)+\frac{\partial}{\partial x}\left(\rho_g S_g v_g\right)=0,$$

$$\frac{\partial}{\partial t}\left(\rho_i S_i+(1-G)\rho_h S_h\right)=0,$$

$$mS_g v_g=-\frac{k_g}{\mu_g}\frac{\partial p}{\partial x},\qquad k_g=kS_g^3,$$

$$\rho c\frac{\partial T}{\partial t}=\frac{\partial}{\partial x}\left(\lambda\frac{\partial T}{\partial x}\right)+\frac{\partial}{\partial t}\left(m\rho_h L_h S_h\right)-\rho_g c_g m S_g v_g\left(\frac{\partial T}{\partial x}+\varepsilon\frac{\partial p}{\partial x}\right)+\rho_g c_g m S_g\eta\frac{\partial p}{\partial t},$$

$$p=z\rho_g R_g T,\qquad z=\left[0{,}17376\cdot\ln(T/T_c)+0{,}73\right]^{p/p_c}+0{,}1\cdot p/p_c,$$

where ρ_j and c_j ($j=g,i,h$) are the density and specific heat of the j-th phase; ρ_{sk}, λ_{sk} and c_{sk} are the density, thermal conductivity and specific heat of the porous medium skeleton; v_g and μ_g are the velocity and dynamic viscosity of gas phase; k is the absolute reservoir permeability; ε and η are the throttling factor and adiabatic coefficient; L_h is the heat of hydrate decomposition; ρc and λ are the specific volume heat capacity and coefficient of thermal conductivity of the system "reservoir – saturating substance"; T_c and p_c are the empirical critical parameters of gas.

The values of temperature and pressure in the region of the hydrate decomposition are related by the condition of phase equilibrium [14, 15].

IOP Conf. Series: Earth and Environmental Science **193** (2018) 012046 doi:10.1088/1755-1315/193/1/012046

3. Calculation results

In the work, a numerical solution of the system of equations was carried out on the basis of the algorithm previously proposed by the authors [16]. During the calculations the following values of the used parameters were adopted: $L = 500$ m; $m = 0.1$; $k = 10^{-14}$ m^2; $S_{h0} = 0.2$; $p_c = 4.6$ MPa; $T_c = 190.56$ K; $c_{sk} = 1000$ J/(kg·K); $c_h = 2100$ J/(kg·K); $c_i = 2060$ J/(kg·K); $c_w = 4200$ J/(kg·K); $\lambda_{sk} = 1.5$ W/(m·K); $\lambda_h = 0.5$ W/(m·K); $\lambda_i = 2.2$ W/(m·K); $\lambda_w = 0.58$ W/(m·K); $R_g = 518.3$ J/(kg·K); $\rho_{sk} = 2000$ kg/m^3; $\rho_h = 900$ kg/m^3; $\rho_i = 900$ kg/m^3; $\rho_w = 1000$ kg/m^3; $G = 0{,}12$; the space coordinate step $\Delta r = 1$ m; the time step $\Delta t = 600$ s; the values c_g, λ_g, μ_g were determined by interpolating table data for methane [2, 16-18]. Assume that the width of the reservoir is 50 m, and its thickness is 20 m. Initial reservoir parameters when the gas hydrate decomposes into gas and ice – $p_0 = 2.8$ MPa, $T_0 = 270$ K; into gas and water – $p_0 = 3.8$ MPa, $T_0 = 276$ K.

Figure 1 shows the time distribution of the value of the accumulated gas production at the different values of the bottomhole depression ($\Delta p = p_0 - p_e$). Two cases are considered: 1) the initial reservoir temperature $T_0 < 0$°C, and the decomposition of the gas hydrate occurs to gas and ice ($L_{hi} = 166$ kJ/kg); 2) $T_0 > 0$°C, and the decomposition of the gas hydrate occurs to gas and water ($L_{hw} = 430$ kJ/kg). The solid lines correspond to the first case; the dashed lines correspond to the second case. For the second case ("gas hydrate → gas + water") the temperature in the reservoir did not drop below 0°C.

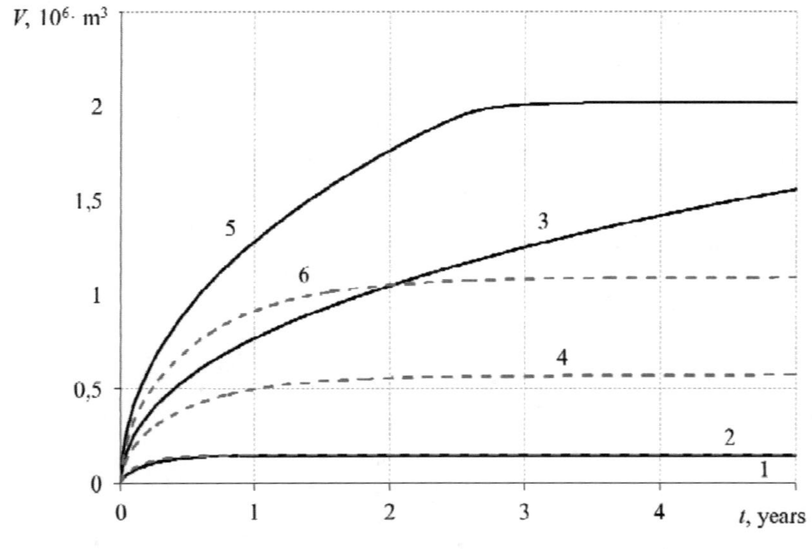

Figure 1. The change in time of the accumulated gas production. Lines 1, 3 and 5 – decomposition of gas hydrate to gas and ice at the bottomhole depressions $\Delta p = 0.3$, 0.6 and 0.9 MPa, respectively. Lines 2, 4 and 6 – decomposition of gas hydrate to gas and water at the bottomhole depressions $\Delta p = 0.3$, 0.6 and 0.9 MPa, respectively.

Figure 1 shows that with a small depression (lines 1 and 2) the amount of the accumulated gas production practically coincides for both cases. With this value of Δp there is no decomposition of gas hydrate and only free gas is taken. For large values of the reservoir depression (lines 3-6), the gas hydrate decomposes, and the amount of the accumulated gas production for the case "gas hydrate → gas + ice" being 2-2.5 times larger than the accumulated gas production for the case "gas hydrate → gas + water". Thus, in the realization of the first case, due to the lesser heat of the phase transition, more intensive gas hydrate decomposition takes place and, accordingly, a larger amount of gas is extracted from the reservoir.

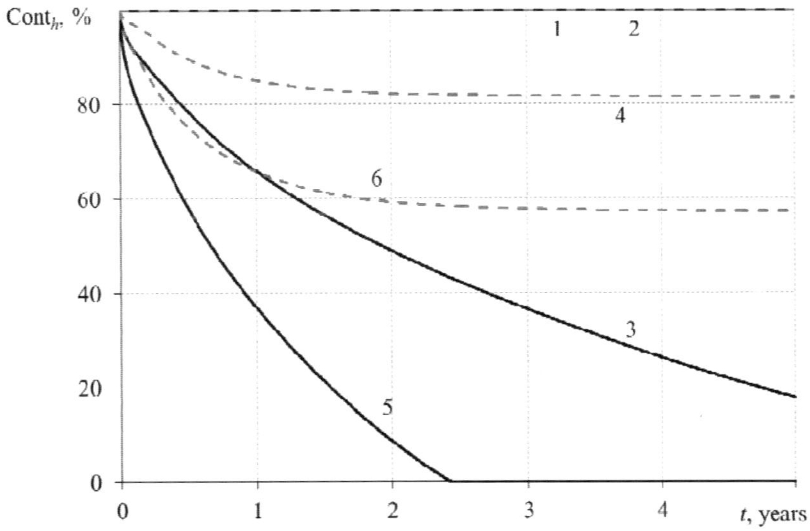

Figure 2. The distributions in time of the percentage of undecomposed gas hydrate in the whole reservoir at the different values of depression Δp. Lines 1, 3 and 5 correspond to $\Delta p = 0.3$, 0.6 and 0.9 MPa, respectively (phase transition "gas hydrate → gas + ice"); lines 2, 4 and 6 – $\Delta p = 0.3$, 0.6 and 0.9 MPa, respectively ("gas hydrate → gas + water").

A similar conclusion can be drawn by analyzing Figure 2. It can be seen that for small values of depression (lines 1 and 2) there is no decomposition of gas hydrate. In this case, the pressure in the reservoir does not fall below the equilibrium hydrate decomposition pressure. As the value of Δp increases (lines 3 and 4), the gas hydrate decomposes, but this process is rather slow (especially in the case "gas hydrate → gas + water"). So, for 5 years at $T_0 < 0°C$ (line 3) about 80% of the gas hydrate present in the reservoir has been decomposed, and this process continues. When decomposing gas hydrate to gas and water (line 4) only about 20% of the hydrate has been decomposed, and this process practically ceased after 2 years, which indicates a lack of the reservoir energy for the hydrate decomposition. The same lack of energy at $T_0 > 0°C$ is also observed for the large values of depression Δp (line 6). For the case of decomposition to gas and ice at $\Delta p = 0.9$ MPa (line 5) gas hydrate completely decomposed in about 2.5 years.

Figure 3 shows the dependence of the time for the complete decomposition of gas hydrate in the whole reservoir to gas and ice on the magnitude of the depression at different values of the initial reservoir temperature. It is seen that initially an increase in the depression Δp and an increase in the initial reservoir temperature T_0 ($T_0 < 0°C$) lead to a significant decrease in the time of full hydrate decomposition for the case "gas hydrate → gas + ice". Further increase in the depression does not lead to significant increase in the rate of the hydrate decomposition.

4. Conclusions

Mathematical model is proposed and the features of the methane hydrate decomposition in the hydrate saturated reservoir are studied in the cases of negative and positive initial reservoir temperature. It has been established that the mode of the gas extraction with decomposition of gas hydrate to gas and ice is more profitable, because with the same depressions more gas is extracted from the gas hydrate deposit. Calculations show that the lower the initial reservoir temperature is, the longer process of the hydrate decomposition to gas and ice becomes. At the same time, an increase in depression first leads to a significant increase in the gas hydrate decomposition rate (correspondingly to large volumes of produced gas), but with a further increase in the depression, the influence of this factor on the methane hydrate decomposition rate becomes insignificant.

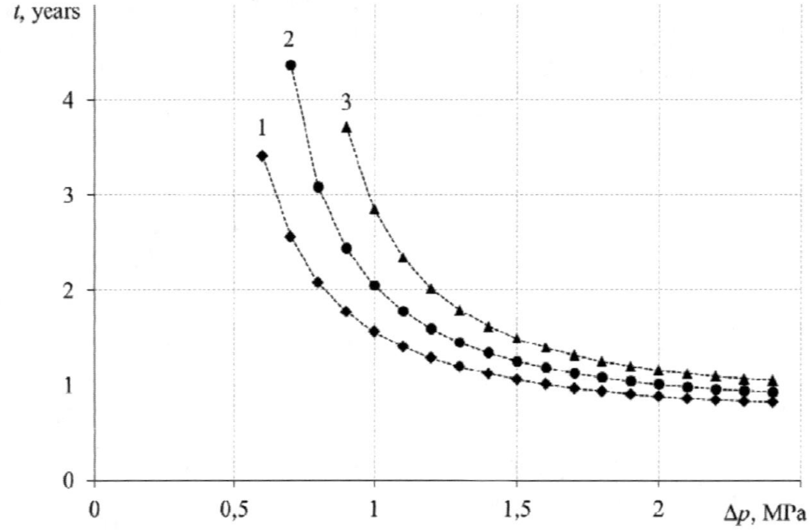

Figure 3. The time of the complete methane hydrate decomposition in the whole reservoir depending on the depression magnitude. Lines 1, 2 and 3 correspond to the initial reservoir temperature $T_0 = 272$, 270, 268 K.

Acknowledgments
The reported study was funded by Russian Science Foundation (project number 17-79-20001).

References

[1] Makogon Y F 2010 *J. of Natural Gas Science and Engineering* **2(1)** 49–59
[2] Istomin V A and Yakushev V S 1992 *Gas Hydrates under Natural Conditions* (Moscow: Nedra)
[3] Makogon Y F, Holditch S A, and Makogon T Y 2007 *J. of Petroleum Science and Engineering* **56** 14–31
[4] Koh A C, Sum A K and Sloan E D 2012 *J. of Natural Gas Science and Engineering* **8** 132–8
[5] Yang J, Okwananke A, Tohidi B, Chuvilin E, Maerle K, Istomin V, Bukhanov B and Cheremisin A 2017 *Energy Conversion and Management* **136** 431–8
[6] Shagapov V Sh, Musakaev N G, Khabeev N S and Bailey S S 2004 *Int. J. of Heat and Mass Transfer* **47(4)** 843–51
[7] Chernov A A and Pil'nik A A 2018 *J. of Crystal Growth* **483** 291–6
[8] Shagapov V Sh, Khasanov M K and Musakaev N G 2016 *J. of Applied Mechanics and Technical Physics* **57(6)** 1083–92
[9] Shagapov V Sh, Chiglintseva A S and Rusinov A A 2016 *Theoretical Foundations of Chemical Engineering* **50(4)** 449–58
[10] Vasil'ev V I, Popov V V and Tsypkin G G 2006 *Fluid Dynamics* **41(4)** 599–605
[11] Basniev K S, Kochina I N and Maksimov V M 1993 *Underground Fluid Mechanics* (Moscow: Nedra)
[12] Nigmatulin R I 1991 *Dynamics of Multiphase Media* (New York: Hemisphere Publ. Corp.)

[13] Shagapov V Sh, Urazov R R and Musakaev N G 2012 *Heat and Mass Transfer* **48(9)** 1589–600
[14] Musakaev N G and Borodin S L 2017 *MATEC Web of Conferences* **115** 05002
[15] Shagapov V Sh, Yumagulova Yu A and Musakaev N G 2017 *J. of Applied Mechanics and Technical Physics* **58(2)** 189–99
[16] Musakaev N G, Borodin S L and Khasanov M K 2018 *Int. J. of Heat and Mass Transfer* **118** 455–61
[17] Khasanov M K, Musakaev N G and Gimaltdinov I K 2015 *J. of Engineering Physics and Thermophysics* **88(5)** 1052–61
[18] Musakaev N G and Borodin S L 2016 *Heat and Mass Transfer* **52(5)** 981–91

IOP Conf. Series: Earth and Environmental Science **193** (2018) 012047 doi:10.1088/1755-1315/193/1/012047

Numerical simulation of hydrological and ecological regime of Siberian rivers' estuaries in the Russian sector of the Arctic

A F Voevodin[1], V S Nikiforovskaya[1] and N A Lapteva[2]

[1] Lavrentyev Institute of Hydrodynamics SB RAS, Novosibirsk, Russia
[2] FBSI SRC VB "Vector" Rospotrebnadzor, Koltsovo, Novosibirsk Region, Russia

E-mail: lapteva@sscc.ru; voevodin@hydro.nsc.ru

Abstract. Numerical simulation of hydrological and hydro-chemical regimes is performed using the mathematical model of water quality, developed at Lavrentyev Institute of Hydrodynamics SB RAS. This method is based on the one-dimensional Saint-Venant equations and the transfer equation of heat, oxygen and other impurities. The paper presents examples of numerical simulation.

1. Introduction

Global warming and intense development of the Arctic requires a thorough investigation (study) of the changes taking place in this region. In this regard, the problems of the water quality, protection of water bodies against pollution entering during the construction and operation of industrial enterprises, water resources utilization systems, hydraulic structures, etc., built on the waterways and adjacent territories are of particular relevance.

In some cases, there may be thermal pollution of water bodies caused, caused, for example, by emergency discharges of hot water by industrial enterprises, thermal and nuclear power plants. Assessment of the severity of exposure of a pollution source on the aquatic environment requires considering the resulting hydrodynamic, physical, chemical and biological processes.

2. Material and methods

Numerical modeling of possible changes in hydrological regimes in open water flows and their systems, as well as occurring hydrochemical processes in the case of an unsteady flow, is carried out using mathematical models and numerical methods developed at Lavrentyev Institute of Hydrodynamics SB RAS. To calculate the hydraulic characteristics, we used a system of one-dimensional shallow water equations (Saint-Venant equations) with various alternative options for taking into account surface runoff from the areas adjacent to the watercourse [1, 2]

$$B\frac{\partial Z}{\partial t}+\frac{\partial Q}{\partial x}=q\,,$$

$$\frac{\partial Q}{\partial t}+\frac{\partial}{\partial x}\left(\frac{Q^2}{\omega}\right)=-g\omega\left(\frac{\partial Z}{\partial x}+\frac{Q|Q|}{K^2}\right)+R_w$$

where t is the time; x is the coordinate, counted along the axis of the watercourse; $Z(x,t)$ is the level of the free surface; $Q(x,t)$ is water flow rate; $q(x,t)$ is the lateral inflow distributed along the length

IOP Conf. Series: Earth and Environmental Science **193** (2018) 012047 doi:10.1088/1755-1315/193/1/012047

of the watercourse; , $B(h,x)$, $\omega(h,x)$, $K(h,x)$ are the width of the free surface, the cross-sectional area, and the module of flow rate at a depth h; R_w are external factors (for example, wind upsurge (negative surge), the barometric pressure gradient); and g is the acceleration of gravity.

To calculate the water quality characteristics we used single-type transfer equations of heat, oxygen and other substances, namely: the heat balance equation, the equation of biochemical oxygen demand, and the equation of dissolved oxygen. The latter can be written in a general form

$$\frac{\partial}{\partial t}(\omega Y)+\frac{\partial}{\partial x}(QY)=\frac{\partial}{\partial x}\left(\omega E\frac{\partial Y}{\partial x}\right)+qy+F,$$

Using the continuity equation, this equation is reduced to:

$$\omega\frac{\partial Y}{\partial t}+Q\frac{\partial Y}{\partial x}-\frac{\partial}{\partial x}\left(\omega E\frac{\partial Y}{\partial x}\right)=F+q(y-Y). \qquad (1)$$

Here $Y(x,t)$ means either water temperature $T\,(x,t)$, or biochemical oxygen demand (BOD) per unit volume of water $L(x,t)$, or the concentration of dissolved oxygen $c\,(x,t)$ or salt $S(x,t)$ (or concentration of any dissolved substance); y is the characteristic of water quality: $T_q(x,t), L_q(x,t), c_q(x,t)$ or $S(x,t)$, coming together with the distributed inflow (outflow) of water per unit length of the watercourse; at that, $q \le 0$, then $y=Y$, otherwise y should be set; $E(h,x)$ is the longitudinal dispersion coefficient calculated by the Harleman formula [3]:

$$E=20.2\cdot\sqrt{g}\cdot|v|\cdot R/C,$$

where R is the hydraulic radius, C is the Chezy coefficient; depending on Y the value F is determined as follows:

$$-k_T B(T-T_E),\ \text{if}\quad Y\equiv T,$$
$$-k_1\omega L,\ \text{if}\quad Y\equiv L,$$
$$k_2\omega(c^o-c)-k_1\omega L,\ \text{if}\quad Y\equiv c,$$
$$0,\ \text{if}\quad Y\equiv S.$$

Here, k_T is the heat transfer coefficient with the environment; T_e is the equilibrium (or so called "natural") temperature, i.e. the water temperature at which, in given quasi-stationary meteorological conditions, the heat exchange between the water mass and the atmosphere would cease; c^o is the concentration of oxygen at saturation; $k_1(T)$ is the coefficient of mineralization, characterizing the rate of oxygen consumption due to the oxidation of organic contaminants (coefficient de-aeration); $k_2(T)$ is the coefficient of re-aeration, characterizing the rate of oxygen income from the atmosphere [4].

The GIDR software package, created on the basis of the developed numerical methods and algorithms, allows implementing these computer based algorithms for solving a broad range of research and practical (applied) tasks on the hydrological and environmental justification of various water resources utilization projects, assessment of possible negative impacts on the environment, and the development of effective protective measures in case of catastrophic phenomena on water bodies [5].

The numerical simulation was carried out using the DSIGMA software package included in the "GIDR" software package for calculation of unsteady water movements in branched systems of watercourses and channels taking into account water quality [6]. The package programs are universal, because they allow automatic adjustment for operation of any of the six modes mentioned below. In addition, they have a flexible modular structure, which allows making adjustments to blocks or replacement of some blocks with others and, if necessary, supplement the missing ones.

By means of the package programs, a wide range of one-dimensional hydraulic problems of open watercourses (channels) can be solved, for example, described in [7, 8]. These problems include the following;

- floods and releases, high water, daily and weekly regulation in the pool of hydro power plants;
- thermal pollution of water caused, for example, by emergency discharges of hot water by industrial enterprises, thermal power plants, and nuclear power plants;
- biochemical oxygen demand (BOD) of water;
- the concentration of oxygen contained in the water of rivers and reservoirs, necessary to the flora and fauna for their satisfactory habitat in the aquatic environment;
- water pollution by various impurities (water salinity or any other water-dissolved conservative substance).

3. Results.

A stream with a real morphometry consisting of 2 segments with a total length $L_{общ} = 156\,km$ is considered. At the junction of the segments (x=85 km) is the source of the contamination, occasionally discharging into a watercourse wastewater at a certain flow rate $Q^*(t)$ with characteristics of contamination $T^*(t), L^*(t), c^*(t), S^*(t)$ varying time. When calculating we used the following boundary conditions for the hydraulic equations. In the entrance range of the 1st segment: the water flow rate $Q(t)$ changing in time (3-hour rise of the flow rate from 1000 to 1500 m^3/s followed by further (after 9 hours) 6-hour descent to a constant value of 1000 m^3/s; in the main-stream station (outlet) of the system (the end of the 2nd segment) water level was constant in time z(t)=78,75=const. In Figure 1 at the top of the beginning of the 1st segment (I_n), characteristics of water quality of flow incoming through this cross section, needed for the substance transfer equations, are given. In the closing part of the 2nd segment, the boundary condition is not required, since it is produced by the transport equations themselves, assuming that at a great distance from the place of discharge of impurities (heat), the role of longitudinal dispersion becomes negligible, (i.e. $E = 0$), and the impurity concentration (temperature) at this boundary is determined mainly by advection transport. Thus, the equation (1) is reduced to the form:

$$\frac{\partial}{\partial t}(\omega Y) + \frac{\partial}{\partial x}(QY) = F,$$

The initial conditions for the hydraulic characteristics assumed a steady state condition, obtained by calculating for establishment of flow rate Q^0=1000 m^3/s and quality characteristics constant in time: temperature T^0=200C, BOD=0, oxygen concentration C^0=3 mg/l and concentration of impurities S^0=5 mg/l. The calculation was conducted with a time step dt=600 s, which was constant throughout the estimated period of time t=0–120 h. At that, the length increment on the 1st segment was equal to 2900 m, while on the 2nd segment – 3050 m. The Chezy coefficient was calculated according to the full formula of N.N. Pavlovsky, while the dispersion coefficient – according to the formula of Harleman.

The calculation results, shown in Figure 1 for the ends of the calculation segments, clearly reflect the changes in water quality characteristics, and allow tracing visually the process of distribution (restoration) of the watercourse depending on pollution (from top to bottom): water temperature (T), oxygen concentration (C), BOD (L), and the concentration of other impurities in water (S).

The analysis of the obtained results shows that in consequence of the provided operation mode of the treatment plant located at the junction of the segments, the consequences of emergency wastewater discharge can be eliminated only by the end of the 5th day, i.e. the achievement of water quality characteristics in the watercourse typical for its initial natural conditions.

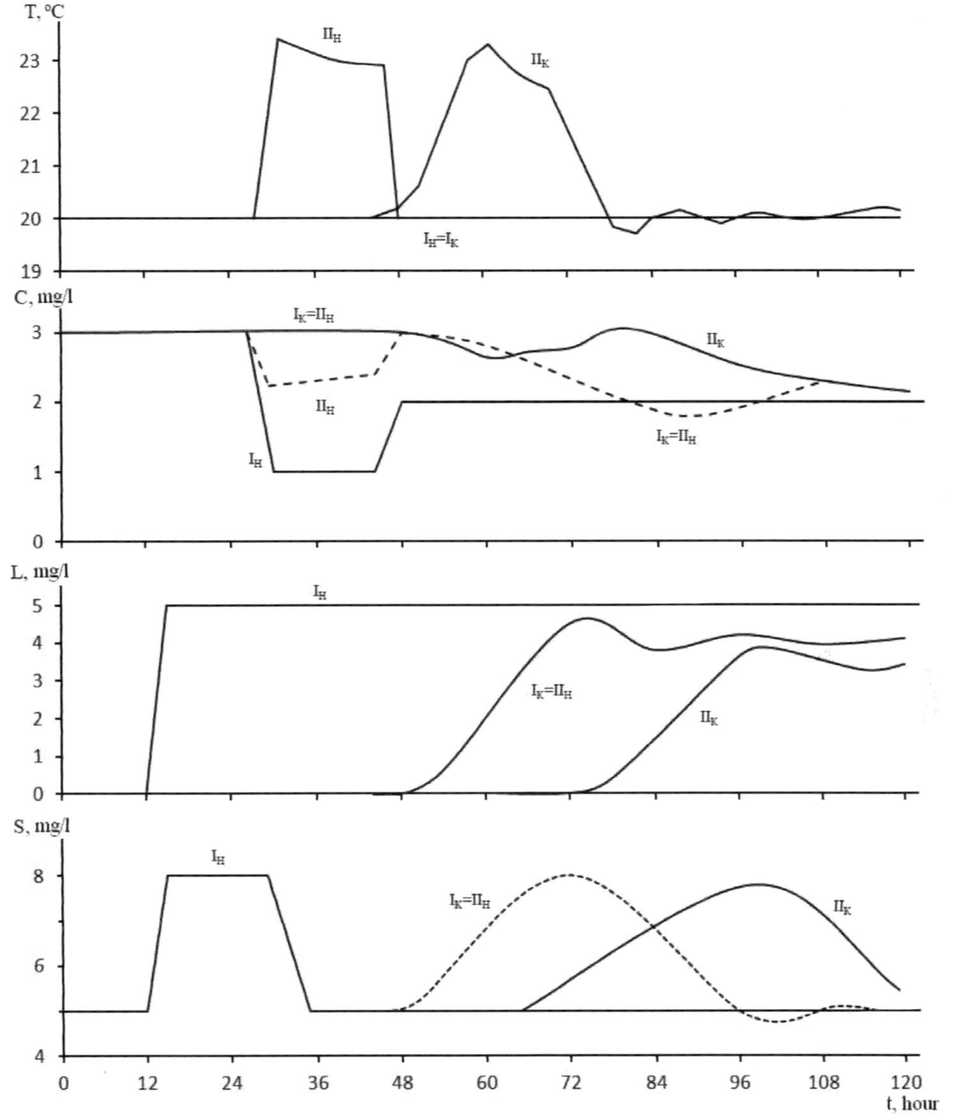

Figure 1. Distribution of temperature, oxygen, BOD and impurities in time at the ends of the segments.

4. Conclusion

The results obtained show the possibility of using this mathematical model for the study and prediction of water quality in the watercourse: the restoration of water quality performance, which was worsened due to the pollutants discharged into the watercourse to environmentally friendly standards required for flora and fauna.

Acknowledgments
The work was supported by the program of the Presidium of the Russian Academy of Sciences No. 51 "Climate change: causes, risks, consequences, adaptation and regulation problems".

References

[1] Voevodin A F. and Nikiforovskaya V S 1978 *Reports on mathematical modeling issues of water quality of reservoirs* Moscow
[2] Kuzin V I, Nikiforovskaya V S and Lapteva N A 2017 *Proceedings of III Russian scientific conference with international participation "Water and environmental problems of Siberia and Central Asia"* Barnaul **2** 142
[3] Ippen A T 1970 *Gigrometeoizdat* Leningrad.
[4] Streeter H W 1967 *Proc. ASCE. Journ. of Hyd. Div.* **93** 6
[5] Complex of hydraulic models for solving water resources management problems 1989 *Advertising circular Vneshtorgizdat* № 6120 MV.
[6] Voevodin A.F. Nikiforovskaya V.S. and Danilova Z.V. 1989 *Algorithms and Programs, Moscow, All-Russian Scientific and Technical Information Center* №50880000681 **2** 1
[7] Nikiforovskaya V.S. 1985 *In bulletin: Hydrophysical processes in rivers and reservoirs, Moscow Science* 211
[8] Nikiforovskaya V.S. 2013 *Meteorology and Hydrology* № 4 79

Theoretical substantiation of application of the hydrocarbon accumulation prospecting technique in Western Siberia based on the study of water-gas equilibria

D A Novikov[1,2]

[1] Trofimuk Institute of Petroleum Geology and Geophysics, Novosibirsk, Russia
[2] Novosibirsk State University, Novosibirsk, Russia

E-mail: NovikovDA@ipgg.sbras.ru

Abstract. Computer simulation of physicochemical equilibria and evasion-invasion processes in the water-gas system enables modeling gas saturation degree variability and the physical nature of gaseous diffusion redistribution at the hydrocarbon accumulation – formation waters interface and in the transition zone from hydrocarbon accumulation to the peripheral waters. Aquifers with waters characterized by low (< 0.2) to maximum saturation (0.8-1.0) with gases have been established within Cretaceous, Jurassic and Paleozoic deposits of the West Siberian sedimentary basin (WSSB), along with an increase in the saturation degree of formation waters with gases as the occurrence depth of reservoir intervals increases, and a relationship between gas saturation of formation waters and their total gas saturation value. All waters with total gas saturation exceeding 1.8 L/L become ultimately saturated with gases ($Kg = 1.0$), providing thereby theoretical prerequisites for the formation of hydrocarbon accumulations. The zone of Kg values spanning from 0.8 to 1.0 is associated with major gas condensate accumulations, while less saturated waters – with oil accumulations. The simulation results allow to assess the current state of the petroleum system of the West Siberian sedimentary basin, its part or a separate horizon within its limits, and to substantiate forecasts of its oil and gas potential at regional, zonal or local levels. The application of modeling techniques to water-gas equilibrium enables significant improvement of the reliability of oil and gas reserves assessment and largely contributes to the solution of problems with respect to searches of skipped accumulations in multi- reservoir fields during their further exploration.

1. Introduction

The fundamental task of studying geochemical processes and physicochemical equilibria in the water – gas system, being closely related to the backbone geological problem of the formation of oil and gas accumulations and degradation, is presently a topical research direction [1-4]. It is generally recognized that the present-day discoveries of new oil and gas fields in the West Siberian sedimentary basin (WSSB) can no longer rely on structural imaging and analysis of general geological and geochemical indicators derived predominantly from geophysical data. In this context, the study of physical and chemical equilibria in the water-gas system, providing information about the region-specific conditions for the oil and gas fields formation and conservation, as well as mass exchange processes with the surrounding groundwater, opens a new promising area of extensive research conjugated to geochemical prediction of oil and gas potential and evaluation of resrvoirs in local structures [5, 6].

Because of their extraordinary mobility, gases generate the largest scattering halos within sedimentary basins, during diffusion processes between hydrocarbon accumulations and surrounding reservoir waters, representing therefore one of the most reliable criteria for hydrocarbons prospecting [4, 7-12]. The factual knowledge resulted from the calculations performed by Yu.P. Gattenberger, V. M. Matusevich and others have provided a compelling evidence that "...at the present stage of the WSSB development, the water percolation rate in its submerged part is marked by negligible values even from the perspective of geological time, so that water percolation, as it is, does not take place. Under these conditions, the largest extents of mass transfer and, in particular, the appearance of water dispersion halos near oil and gas accumulations are associated with the diffusion processes» [13]. In this context, the role of inherent processes within the water-gas system increases far and away.

The study of water dissolved gases (WDG) in Western Siberia given a was start to more than 50 years ago by M. S. Gurevich and N. N. Rostovtseva. In the early 1950ies of the 20th century they were the first to define the nature of gas zonation of groundwater within the WSSB and revealed the prospecting value of WDG, which is largely determined by extent of their tension rather than qualitative composition alone. N. M. Kruglikov made a discovery about the diffusion and dispersion of gas being responsible for a decrease in dissolved gas tension as the distance from the gas-water contact increases.

L. M. Zorkin investigated different conditions potential for gas generation, their extraction from groundwaters and formation of gas accumulations. The study of groundwater and water-dissolved gases in WSSB attracted interest of many researchers who have greatly contributed to its advancement: A. E. Kontorovich, B. P. Stavitskii, A. A. Rozin, N. M. Kruglikov, V. V. Nelyubin, O. N. Yakovlev, V. M. Matusevich, A. A. Kartsev, A. D. Nazarov, S. L. Shvartsev and D. A. Novikov and many others [3, 4, 9, 11].

At this, conditions of physicochemical equilibria of gases remain largely understudied despite their representing a problem of great scientific and practical importance, which allows to solve more purposefully many genetic issues inherent in crude petroleum generation [14], and to reveal hitherto unknown patterns of the formation of groundwater [5, 6].

The above discussed equilibria allow revealing the direction of geochemical processes in the "water-gas" system both at the present stage of the petroleum system development, and in its geological past. These also address the problems of the deposits' preservation and enable prediction of their phase type. Their solution will result in providing a substantiation of a new oil and gas prospecting technology applicable to the conditions of WSSB.

2. Materials and Methods

In itself, the water-gas system is very complex, which is accounted for by the multi-component and multidirectional processes involved therein, which makes problematic the use of previously employed simplified calculus methods for gas saturation levels in formation waters, individual gases fugacities etc. The calculations in this research were made with the use of HG-32 (Hydrogeo) software package, currently having no analogs in the world, which was developed by M. B. Bukaty at the Trofimuk Institute of Petroleum Geology and Geophysics SB RAS. The simulations based on it allow considering all the parameters of the investigated system (density, TDS and the composition of groundwater, gas saturation, WDG composition, temperature and pressure conditions and other indicators) and determining the composition and a number of characteristics of the equilibrium free gas phase from the composition of dissolved gas and water, or conversely, the composition and other parameters of dissolved gas from the composition of free gas and solution, as well as modeling evasion-invasion of gases under variable pressure, temperature and composition of the solution [1, 2, 5].

The theory is underlaid by dependences of the mass-action law and the activity method. For each i component from n components of the gas mixture dissolved in groundwater, in this case, a reversible reaction of the solution-to- free phase transition is considered:

$$i_p \Leftrightarrow i_r. \quad (1)$$

The mass-action law can be written for it as

$$K_i = f_i / a_i^0, \tag{2}$$

where K_i -is thermodynamic equilibrium constant, f_i –is gas fugacity (volatility) in free phase газа, a_i^0 - is its activity in the solution in equilibrium with free phase (saturated with this gas) in the standard state.

The degree of water solution saturation with gas i is estimated using undersaturated index

$$L_i = \frac{a_i K_i}{f_i}, \tag{3}$$

where ai - is activity of gas in the investigated water.

The simplest approach consists in direct identification of fugacity F_i of each gas i in free gas phase in hypothetical equilibrium with the solution, employing semi-empirical methods developed by A.Yu. Namiot, E.S. Barkan and others, while

$$F_i \equiv a_i K_i \quad \text{и} \quad L_i = \frac{F_i}{f_i}. \tag{4}$$

The fi and Ki were calculated using a number of correction factors and a series of regression equations obtained M. B. Bukaty as a result of processing of experimental data [1, 2], which describe required thermodynamic parameters, solubility and distribution of gas between the phases depending on the composition of each, and real temperature and pressure conditions.

This paper summarizes the results of numerical physicochemical modeling of water-gas equilibria for more than 700 hydrocarbon accumulations localized within Cretaceous, Jurassic and Paleozoic sedimentary strata and delineated 132 oil and gas fields of WSSB.

3. Results and Discussion

Within the scope of the considered problem of interaction processes in the water-gas system, there are two main issues to be solved:

1. assessment of the groundwater saturation with gases (K*g*),
2. interaction in the water – gas system (formation water – hydrocarbon accumulation).

In the context of WSSB, the former aspect of the problem has been thoroughly discussed in papers authored by N.M. Kruglikov, V.N. Kortsenshtein, V.V. Nelyubin, O.N.Yakovlev, M.Ya. Rudkevich, L.S. Ozeranskaya, N.F. Chistyakova and others [15, 16, 17]. Their works provide data on the variability of groundwater saturation with gases at the regional level within different aquifers of the lower hydrogeological stage. Specifically, it is indicated that the gas saturation coefficient values for Jurassic complexes vary from 0.03 in the marginal part of the basin to 1.00 in its center.

Within the Neocomian complex, K*g* values tend to increase from the margins to the center of the basin (to 0.77 in the Middle Ob district), with the maximum values up to 0.94 identified in the Taz area (within the Nadym-Taz interfluve).While the Aptian-Albian-Cenomanian complex is characterized by a regular increase in Kg values from 0.02 to 0.87 in the direction from south to north of the WSSB. Calculations performed by D. A. Novikov showed that the saturation of groundwaters with gases appears of more complex nature at zonal and local levels (within a specific field) [4-6, 10, 18]. Within Paleozoic, Jurassic and Cretaceous water-bearing complexes, gas saturation of waters in the aquifers established from K*g* values varies from low (< 0.2) to maximum (0.8-1.0) (figure 1). Gas saturation levels of the detected formation waters vary from the unsaturated to extremely saturated (from K*g* values).

An increase in the gas saturation degree of formation waters with deeper occurrence of reservoirs and its dependence on the value of their total gas saturation is established. All waters whose value of total gas saturation is greater than 1.8 L/L tend to be extremely saturated with gases (K*g* = 1.0), providing thereby theoretical prerequisites for the formation of hydrocarbon accumulations. Whereas formation waters not saturated with gases are capable of dissolving the previously formed oil and gas accumulations. A direct relationship between the degree of saturation of formation waters with gases (Kg) and the phase composition of deposits is established. The zone with Kg values varying from 0.8 to 1.0 is associated with major gas condensate accumulations, and less saturated waters – with crude oil accumulations.

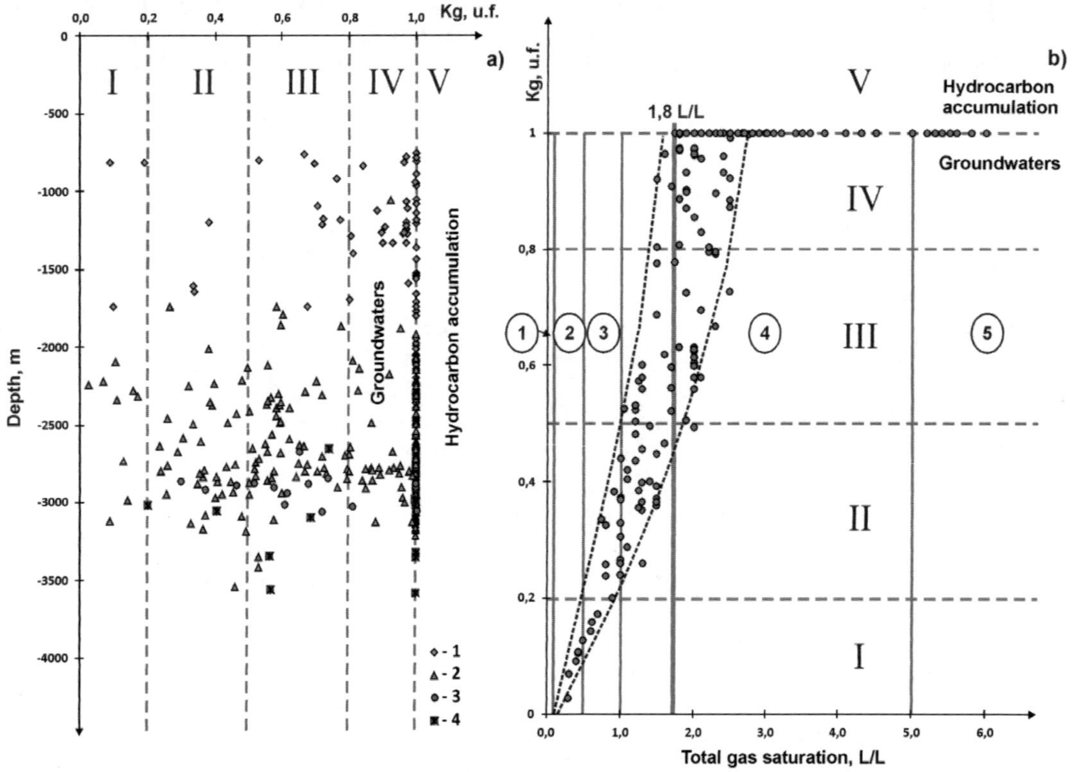

Figure 1. Variation of gas saturation coefficient of waters (K*g*) with depth (a) and versus total gas saturation value of groundwaters (b).

a) Zones by gas saturation coefficient of water: I - low, II - medium, III - high, IV – very rich, V – a hydrocarbon accumulation; gas saturation coefficient (K*g*) in the objects of major hydrogeological complexes: 1 – Aptian-Albian-Cenomanian, 2 – Neocomian, 3 – upper Jurassic, 4 – Lower-Middle Jurassic.

b) Zones by total gas saturation of ground waters (L/L): 1 – fairly low (< 0,1), 2 – low (0,1-0,5), 3 – medium (0,5-1), 4 – high (1-5), 5 – fairly high (> 5).

Gas saturation degree of water there shows a gradual decreasing trend from beds the with gas condensate accumulations localized within the K*g* values spanning from 0.8 to 1.0, to beds holding mostly oil accumulations confined to the Kg values in the range from 0.4 to 0.8. The total gas saturation K*g* therefore varies from 0.95 to 1.00in bed BP_{12}, indicating a maximum saturation of waters with gases.

K*g* values constitute 0.80-0.86 in bed BP_{14} while in the lower lying bed BP $_{15}$ they range from 0.74 to 0.83. The lower part the section is dominated by beds holding purely oil accumulations characterized by a lower degree of gas saturation. At this, the major oil-bearing bed BP_{16} is confined to the remarkably unsaturated zone, located within the Kg values range from 0.56 to 0.75. Bed BP $_{17}$ is marked by the minimum Kg values, within 0.32-0.42. As such, a rather complicated pattern of the revealed dependences indicates diverse chemical and gas composition of groundwaters, as well as the presence of different genetic types of water in the horizons within the oil-and-gas-bearing portion of the section.

In their works [15, 16, 17], N. M. Kruglikov and V. N. Kortsenshtein have analyzed data on interaction in the water – gas system (formation water – hydrocarbon accumulation) for PK_1 bed from Medvezhiye, Yamburgskoe, Yety-Purovskoe, Urengoyskoe fields and beds BU_{10}, BU_{12} of the Urengoyskoe field. In their opinion, the investigated hydrocarbon accumulations are in unstable (non-equilibrium) position with respect to the surrounding waters, which allowed them to deduce that the processes of degradation of the existing HC accumulations are currently taking place in Upper and Lower Cretaceous horizons.

In this paper, the redistribution of gases between HC accumulations and formation waters was estimated from the ratios of individual fugacities of gases calculated in the HC accumulation-groundwater and groundwater- HC accumulation systems. In the first case, the composition of the free gas phase was calculated from hypothetical equilibrium composition of WDG, while in the second – by the WDG calculated from the composition of hypothetical gas accumulation. The physicochemical calculations for more than 700 oil and gas accumulations performed using with this method allowed to estimate the direction of their interaction with the surrounding groundwaters. It has been established that the HC accumulations are mainly in an unstable position in relation to the surrounding waters. The analysis of methane, ethane and propane behaviors shown in figure 2 indicates that methane and carbon dioxide and argon dissipate almost in all of the investigated accumulations, which is compensated by the intake of helium and nitrogen, and to varying degrees, methane homologues. Quite numerous accumulations are found to be experiencing their phase transformation towards weighting of structure.

This suggests some specific features of the processes of redistribution of gases between HC accumulations and the surrounding formation waters, depending on the phase composition of HC accumulation. The exchange and retransformation processes occur most intensely within oil-gas-condensate and gas-condensate accumulations, and the least intensely – in oil accumulations. Analysis of changes in the ratios between individual fugitives of gases in formation waters and hydrocarbon accumulation has revealed some regularities. Thus the observed increase is in the methane, hydrogen, ethane, propane and carbon dioxide ratios versus their depth confirms inferences made earlier by B. N. Ryzhenko and V. P. Volkov about an increase in gas volatility within a wide range of PT conditions [19]. In this respect, butane, pentane and hexane appear much more complex.

Results of interactions in the water – gas system modeling showed that the zonality of the predicted-hypothetical composition of free gas phase, except hydrocarbon and non-hydrocarbon gases, is also manifested in the depth-dependent distribution of the helium-argon ratio which is directly related to the absolute age of hydrocarbon accumulations.

Unlike the widely applicable methods for determining the age of gases from the helium-argon ratio in groundwater and giving fairly problematic results, the algorithm used in the HG-32 (Hydrogeo) software package is based on the empirical equation by V. P. Savchenko for free gases, obtained by him from the data generalization for a multitude of hydrocarbon fields around the globe [20]. Thus, projected age of probable deposits, estimated from water-soluble gases, increases from 20-23 MA (Upper Oligocene) at the top of the Aptian-Albian-Cenomanian complex to 40-87 MA (Eocene-Upper Cretaceous) in the lower Neocomian [4]. The results obtained are in good agreement with the data on the isotopic composition of natural gases in giant deposits of the northern West Siberia obtained by N. N. Nemchenko, A. S. Rovenskaya and M. Shoell [21].

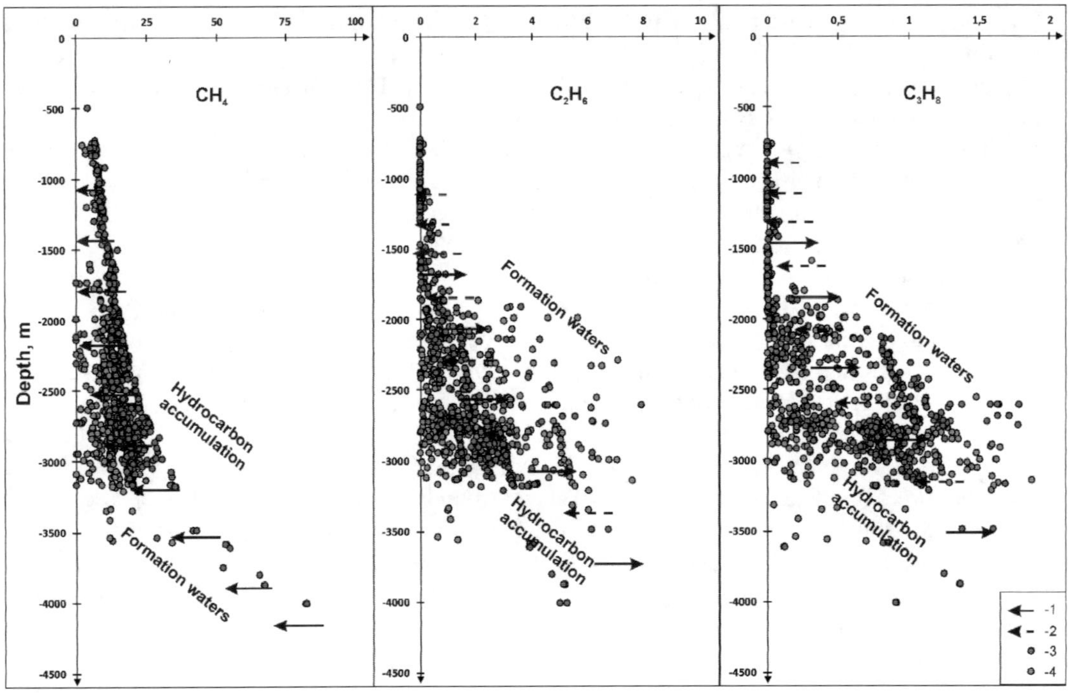

Figure 2. A relationship between CH_4, C_2H_6 and C_3H_8 fugacities in formation waters and hydrocarbon accumulations and their occurrence depths.
Direction of diffusion flow: 1 – from HC accumulation to formation waters, 2 – from formation waters to HC accumulation; gas fugacity: 3 – in HC accumulation, 4 – in formation waters.

4. Conclusions

The application of numerical modeling of water-gas balances thus allows to most reliably solve the problems on prediction of oil and gas content in sedimentary basins. However, the final stage of the calculations within the water-gas system additionally requires conducting extensive complex hydrogeological studies. In a nutshell, the proposed technology can be presented as the algorithm laid down below:

1. Collection and generalization of all available materials on the investigated object in the database within one of the common GIS packages.
2. GIS package-based generation of a set of electronic maps of the chemical and gas composition of groundwater, as well as hydrodynamic, geothermal, paleogeographic and lithological-facies maps of the main complexes with the obligatory allocation of the region-specific characteristics of gas-hydrogeochemical background.
3. Detailed paleohydrogeological reconstructions aiming to clarify the spatial boundaries of the main zones of oil and gas accumulations.
4. Construction of hydrogeological and hydrodynamic models of potential oil and gas accumulation zones, along with specifying the zonal gas-hydrogeochemical background.
5. Creation of hydrogeological models of reference fields, "empty" and promising structures.
6. Development of gas-hydrogeochemical field models within the reference structures in order to identify actual concentrations of oil and gas accumulations.
7. Computer simulation of physicochemical equilibria and processes in the water-gas system.
8. Modeling variability of the degree of gas saturation and the nature of diffusion redistribution of gases at the of hydrocarbon accumulation – formation water interface and in the transition zone from the HC accumulation to the peripheral waters.

9. Substantiation of recommendations on the direction of geological prospecting works and drilling the first and subsequent exploration wells.

In the context of the West Siberian sedimentary basin, the established nature of the water-gas equilibria allows to ascertain that the oil and gas deposits held in Mesozoic and Paleozoic sedimentary strata act as a conservation element of the lithosphere, representing "relics" of previous stages of its geological and geochemical evolution; while the surrounding groundwaters represent a more active component of the system, and appear to be significantly more advanced in their geochemical development. This is evidenced in the difference in the fugacity ratio of individual gases both in groundwater and HC accumulations. As a result, the composition of the latter undergoes a slow directional change towards establishing balance, which corresponds to the qualitatively new state of the water – gas geochemical system. The results of regional, hydrogeochemical and gas-hydrogeochemical studies suggest favorable conditions for the formation and conservation of hydrocarbon accumulations in most of the study area, with the exception of near edge zones of the sedimentary basin.

Acknowledgments
The research was supported by the Russian Foundation for Basic Research and the Ministry of Education, Science and Innovation Policy of Novosibirsk region within the scope of Research Project № 17-45-540086.

References
[1] Bukaty M 1992 *Geologiya Nefti i Gaza* **1** 17-9
[2] Bukaty M 1999 *Advertizing Technical Specification on a Program Complex HydrGeo* (Moscow: All-Russian scientific and technical information center) p 5
[3] Novikov D A 2017 *Geodynamics and tectonophysics* **8** 881-901
[4] Novikov D A 2018 *Neftyanoe Khozyaystvo - Oil Industry* **4** 16-21
[5] Novikov D A 2015 *Gas Industry* **3** 12-7
[6] Novikov D 2015 *Otechestvennaya Geologiya* **3** 7-15
[7] Namiot A 1958 *Gas industry* **12** 1-10
[8] Antonov P 1963 *Gas Industry* **9** 1-6
[9] Kartsev A et al. 1992 Theoretical fundamentals of oil and gas hydrogeology (Moscow: Nedra) p 208
[10] Novikov D A 2000 *Geologiya nefti i gaza* **3** 51-6
[11] Novikov D A and Lepokurov A 2005 *Geologiya Nefti i Gaza* **5** 24-33
[12] Novikov D A 2017 *Petroleum Exploration and Development* **44** 780-8
[13] Matusevich V, Rylkov A and Ushatinsky I 2005 *Geofluidal Systems and Problems of Oil-and-gas Content of the West Siberian Megabasin* (Tyumen: Tyumen State Oil and Gas University Press) p 225
[14] Sokolov V 1971 *Geochemistry of Natural Gases* (Moscow: Nedra) p 336
[15] Kruglikov N, Nelyubin V and Yakovlev O 1985 *Hydrogeology of the West Siberian Oil and Gas Bearing Basin and Feature of Formation of Deposits of Hydrocarbons* (St. Petersburg: Nedra) p 279
[16] Kruglikov N 1964 *Hydrogeology of a Northwest Board of the West Siberian Artesian Basin* (St. Petersburg: Nedra) p 166
[17] Kortsenstein V 1997 *Water-pressure Systems of the Largest Gas and Gas-condensate Fields of the USSR Hydrogeology* (Moscow: Nedra) p 247
[18] Novikov D A 2001 *Ob messenger* **1** 99-102
[19] Ryzhenko B and Volkov V 1971 **7** 760-73
[20] Shvartsev S L 1996 *General Hydrogeology* (Moscow: Nedra) p 423
[21] Nemchenko N, Rovenskaya A and Shoyell M 1999 *Geologiya Nefti i Gaza* **1** 45–56

Genetic classification of subsurface waters and brines of Arctic regions of Siberia

D A Novikov[1,2]

[1] Trofimuk Institute of Petroleum Geology and Geophysics SB RAS, Novosibirsk, Russia,
[2] Novosibirsk State University, Novosibirsk, Russia

E-mail: NovikovDA@ipgg.sbras.ru

Abstract. Specificity of hydrogeological aspects of the Siberian sedimentary basins is most pronounced in their Arctic part where the saltless type the cross section grades into saline type in the direction from the Yamal Peninsula (as part of the West Siberian sedimentary basin), passing through the Yenisei-Khatanga basin as far as the Anabar-Khatanga basin with salt domes in its structure. It has been established that the modern chemistry of groundwaters and brines of petroleum-bearing deposits is a product of long-term geological evolution affected by many geological and hydrogeological factors. Slightly metamorphosed groundwaters and weak brines of sodium chloride, chloride-bicarbonate sodium and sodium bicarbonate-chloride composition, with the TDS value varying from 2 to 63.3 g/dm^3 (after S. A. Shchukarev) are widely developed within the West Siberian and the Yenisei-Khatanga basins. Subsurface brines from the areas adjacent to the Anabar-Khatanga basin are found to be dominantly sodium chloride in composition, with TDS value varying from 52.3 to 312.3 g/dm^3, while those identified in the adjacent areas of the Siberian platform are strongly metamorphosed, of sodium chloride, sodium-calcium and calcium composition, with the TDS value up to 480 g/dm^3. Results of the detailed hydrogeochemistry analysis of petroleum reservoirs and paleohydrological history of Arctic parts of the Siberian sedimentary basins have prompted the presence of three genetic groups of subsurface waters in their section: infiltrogenic, sedimentogenic and condensatogenic. The first group includes modern and ancient infiltrogenic waters and brines resulting from salt domes leaching; the second group is composed of sedimentogenic waters (including brines), lithogenic waters and cryopegs. Comprising condensatogenic waters formed syngenetically with the hydrocarbon pools, the third group represents a special type, since the "water-rock-gas-organic matter" system has largely affected the processes of formation of their chemical composition.

1. Introduction

Given that the Arctic is a region with unique potential hydrocarbon resources, its sedimentary basins has justifiably attracted great scientific interest. The fact that more than 90% of Russia's gas and about 10% of oil is currently produced from the discovered oil and gas fields there [1] attests to the presence of commercial hydrocarbon accumulations in many of its sedimentary basins: the West Siberian, Yenisei-Khatanga, Anabar-Khatanga and others. The uniqueness of Arctic reserves consists in the structural level of hydrocarbon potential comprising deposits with the age spanning from the ancient Precambrian through the Quaternary, across the studied basins [1–6]. The hydrogeological structure of Arctic parts of the sedimentary basins of Siberia is also very complex and is controlled by many factors. It stands to reason that the exploration extent of their hydrogeology is extremely varied, with

the West Siberian sedimentary basin (WSSB) being thus far the best studied [7] (Figure 1). The significant role played by water in the formation of sedimentary rocks and the processes of migration, accumulation and degradation of hydrocarbons is distinctly inferred from the fundamental works of V. I. Vernadskii, V. A. Sulin, G. M. Sukharev, A. A. Kartsev and other scholars.
An enormous pool of factual material reflecting the chemical and gaseous composition of subsurface waters and brines has been accumulated since early days of exploration and prospecting for oil and gas in Arctic parts of the Siberian sedimentary basins. The numerous hydrogeochemistry research were conducted by A.S. Antsiferov, M. B. Bukaty, V.I. Vozhov, G.D. Ginsburg, Yu.G. Zimin, G.A Ivanova, M.K. Kalinko, A. A. Kartsev, A. E. Kontorovich, N. M. Kruglikov, A. R. Kurchikov, V. M. Matusevich, D. A. Novikov, B. P. Stavitskii, O. N. Yakovlev and others. The specific features of hydrogeological structure and conditions of the North-Alaska, Beaufort-Mackenzie, Sverdrup, West Greenland, East-Canadian, Alberta, Athabasca and the other basins are discussed in the works of H. Baadsgaard, S. Bachu, Z. Chen, C.A. Connolly, K. Dewing, M. Gascoyne, S.E. Grasby, J.S. Hanor, B. Hitchon, Y. Lee, F.J. Longstaffe, J.A. Nunn, C.M. Sauveplane, R.L. Stotler, J.R. Underschultz, L.M. Walter, and others. However, despite the lengthy period of study of the Siberian sedimentary basins, the genetic classification of subsurface waters and brines within them has still been largely overlooked.

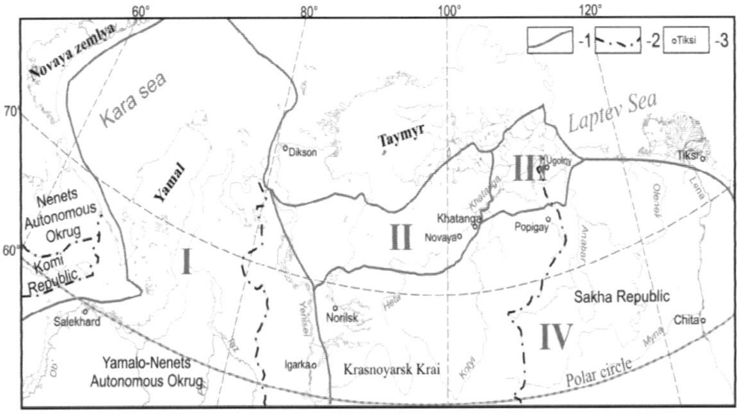

Figure 1. Location of the studied area. Boundaries of: 1 – basins limits, 2 – administrative; 3 – settlements. Sedimentary basins: I – West Siberian, II – Yenisei-Khatanga, III – Anabar-Khatanga, IV – Siberian platform.

In the first half of the last century, hydrogeochemichal studies involved a number of classifications of subsurface waters and brines. Suffice it to recall the chemical classification after S. Palmer (1911), M. G. Kurlov (1928), E. Karstens (1932), V. A. Aleksandrov (1932), V. I. Vernadskii (1933), S. A. Shchukarev (1934), V. A. Sulin (1935, 1946), I. I. Chebotarev (1955), H. Schoeller (1955), O. S. Dzhikia (1963), as well as the modified classification by V. A. Sulin L. Bojarski (1970) and others [8–18]. A number of chemical classifications with elements of genetic bias was developed by N. S. Kurnakov-M. G. Valyashko (1917, 1935, 1954), A. M. Ovchinnikov (1954), E. E. Belyakova (1968) and others. Whatever the Earth's hydrosphere origin and evolution, the water was involved in a variety of cycles, which led to the formation of different genetic types of subsurface waters generally understood as types, unified by single source of groundwater recharge, the mechanism of penetration into the subsurface strata and the general direction of water circulation processes [25]. Developing a genetic classification of subsurface waters was dealt with by many scholars, among them E. Suess (Deutsch: Eduard Sueß), R. A. Daly, A. A. Kozyrev, G. N. Kamensky, N. I. Tolstikhin, A. M. Ovchinnikov, E. T., Degens, H. Schoeller, A. A. Kartsev, D. E. White (D. E. White), E. V. Pinneker and others [19–24]. Previously, G.N. Kamenskii [19] identified three genetic cycles of subsurface waters: 1) infiltration, or continental, associated with the infiltration of atmospheric water and a complex of geochemical processes occurring in the uppermost earth's crust. 2) marine, or sedimentational, associated with the burial of marine waters during sedimentation and with their further metamorphization (metamorphic changes). 3) Metamorphic or magmatic type is associated with the formation of deep waters. Given that it is by far almost impossible to distinguish between magmatic and metamorphic waters, it makes sense to combine the two cycles into one – metamorphogenic-magmatic. Material manifestations of a cycle may not always be distinguished in the pure form, because of the cycles interacting with each other, which leads to the mixing of different genetic types of waters. As such, the interaction between cycles can be illustrated by the sedimentogenic water being displaced by infiltrogenic. It should be noted that the distinguished by many classifications genetic types of waters include such as metamorphic, volcanogenic and others. However, these notions actually imply not genetic types, but mixed waters of different origin.

2. Materials and Methods

Identification of the origin of subsurface waters is one of the key issues in modern hydrogeochemistry. Since the beginning of the 20th century and until now, several classification schemes for genetic types of subsurface waters of sedimentary basins and brines have been developed and used in hydrogeology (V. A. Sulin, A. A. Kartsev, E. V. Pinneker, S. L. Shvartsev and others). According to the classical works authored by S. A. Shchukarev, V. A. Sulin, S. B. Vagin, A. A. Rosin, M. I. Subbota, Ya.A. Khodzhakuliev, A. A. Kartsev, N. M. Kruglikov, V. V. Nelyubin, O. N. Yakovlev, V. M. Matusevich and other researchers, the genetic type of water is established using "genetic" coefficients, reflecting the relationship between different macro-and micro-components in its composition: rNa/rCl, Cl/Br, Ca/Cl, $r(HCO_3+CO_3)/r(Ca+Mg)$, B/Br, $Br/Cl \cdot 10^{-3}$, $Sr/Cl \cdot 10^{-3}$, rNa/(rCa+rMg), rNa+rMg/rCa, rCa/rMg, $rSO_4 \cdot 100/rCl$, $rHCO_3 \cdot 100/rCl$, $Br \cdot 10^3/M$, $I \cdot 10^3/M$, $NH_4 \cdot 10^3/M$, Br/I, HCO_3/SO_4, $(M/H) \cdot 100$, $(rCa/rNa) \cdot 100$, etc. The pool of data that served as the basis for this study is represented by the test results of more than 6200 objects from 437 search area, including 7286 samples of subsurface waters and brines. In this paper, we used the terminology and definitions of genetic types of subsurface waters from the classifications by A. A. Kartsev and S. L. Shvartsev [22, 23, 25], as they are described below.

Sedimentogenetic waters (salty waters and brines) imply aqueous solutions arriving to the sedimentary basin during sediment deposition in the marine environments. The same type should include aqueous solutions originating from lakes and lagoons of marine origin, usually saline (salt), sometimes desalinated [22]. Aqueous solutions (usually fresh, brackish, more rarely saline) of atmospheric origin, which penetrated into the sedimentary rocks by infiltration during hypergenesis should be called *ancient (fossil) infiltratogenic* waters. A. A. Kartsev, S. B. Vagin, V. M. Matusevich [22] and other researchers distinguished *lithogenic* type of waters released during thermodehydration of various minerals of sedimentary rocks. In the conditions of Western Siberia, they are found to be developed in Cretaceous and Jurassic reservoirs at depths of more than 2 km [7, 22, 26]. The author supports the viewpoint expressed by V. V. Nelyubin and O. N. Yakovlev [27] elucidating on the mechanism of the formation of *condensatogenic* waters and understands them as formation waters of the background composition and salinity, admixed with fresh (non-saline) condensation water released from the water-carbon mixture during its vertical migration, under the changing thermobaric conditions. The formation and preservation of condensatogenic water fringes are also favored by the formation of hydrocarbon pools through rapid vertical migration of hydrocarbons from petroleum kitchens into traps [28].

3. Results and Discussion

It has thus far been established that Arctic parts of the WSSB are characterized by relatively low TDS of subsurface waters and brines, which are < 63.3 g/dm^3 in the areas of Izvestinskoye, Komsomolskoye and Gubkinskoye fields in the Nadym-Taz interfluve area [26, 29] and is not greater than 51 g/dm^3 within the Malyginskoye group of fields on the Yamal Peninsula [30]; by sulfate-free, rich in dissolved OM and gas of hydrocarbon composition in subsurface waters of Mesozoic structural level, as well as by variable hydrogeochemical types and salinity of subsurface waters, both over area and in the cross section. At this, each of the Mesozoic confined aquifers (with pressurized waters) characterized by a number of hydrogeochemical signatures were influenced by the paleohydrogeological history [31] and the subsequent processes of metamorphization of subsurface waters composition, as well as by the waters' vertical migration from the Paleozoic basement to the sedimentary cover in the areas of thinned or missing Triassic volcanic-sedimentary strata [26]. The Yamal Peninsula is features by unique formations developing within the cryogenic strata – cryopegs consisting of high-salinity water and brines with negative temperature up to – 8,5° C, which formed as a result of cryogenic metamorphization of seawater and are of chloride sodium-magnesium composition and a TDS value up to 150 g/dm^3 [32].

Within Mesozoic complexes of the Yenisei-Khatanga basin, the waters with highest salinity and TDS more than 20 g/dm^3 were revealed in the Pelyatka area (24.8 g/dm^3, Aptian-Albian-Cenomanian complex), Sukhaya Dudinka area (23.3 g/dm^3, Lower-Middle Jurassic complex) and the Vankor area (23,0 g/dm^3, Upper Jurassic complex) [33]. Given the specific hydro-geothermal and hydrodynamic conditions in the aquifers within Jurassic and Cretaceous sediments [33], these helped establish general hydrogeochemical patterns: sodium- bicarbonate and sodium-chloride-bicarbonate waters with TDS up to 5 g/dm^3 are commonly distributed along the northeastern border, which are succeeded by sodium-chloride subsurface waters with salinity up to 15-20 g/dm^3 when approaching the central and southwestern parts. The hydrogeochemical picture of subsurface waters indicates the aquifers differentiation within Jurassic and Cretaceous sediments. In the cross section, various types of the vertical hydrogeochemical zonality can be distinguished – from direct, which is dominant within Messoyakha inclined ridge, to the inversed type. The diversity of the subsurface waters composition generally observed in the lower parts of the sedimentary cover declines up the section, while in Aptian-Albian-Cenomanian complex the hydrogeochemical conditions tend to become more even, with levelled off TDS concentrations in underground waters [7].

Further eastwards, the saltless type of the section characteristic of the WSSB and Yenisei-Khatangs basin is replaced by saline type within the Anabar-Khatanga basin, which is immediately reflected in the hydrogeochemical field, affecting also the nature of the vertical hydrogeochemical zoning [7]. This was established earlier by the characterization of the hydrogeological features in the transition type section (from the ancient Siberian platform to the young West Siberian plate), by the example of the Pre-Yenisei sedimentary basin located more southward [34, 35], where salt domes and stock structures typical for many world's salt basins are widely developed [36]. Presently, the brines of the Anabar-Khatanga basin have been identified within the seven aquifer complexes: in Lower-Middle Jurassic, Triassic formations, and in Permian, Carboniferous, Devonian, Vendian-Cambrian and Riphean deposits. The studied brines are of sodium chloride composition with the TDS value varying in a wide range from 52.3 to 312.3 g/dm^3. The chemical type of high-calcium brines and TDS reaching 54.6 – 144.6 g/dm^3 was established within the aquifer complexes of Riphean deposits in the Kostrominka area, within Cambrian deposits in the North-Soulem area and within Permian deposits in the Ulakhan and North-Soulem areas. The two types of vertical hydrogeochemical zoning established in the region include inversion type which is confined to the areas of salt domes distribution on the Yuryung-Tumus Peninsula, within the South Tigyan, Ilyin and other structures. The highest-salinity brines with TDS up to 312.3 g/dm^3 were revealed in aquifer aquifers of Triassic formations [7], which exhibits a significantly declining trend further down the section, though. High-salinity brines identified in the boundary zones of salt diapirs and appear to result from leaching of salts largely facilitated by modern and ancient infiltration waters. The direct (normal) type of vertical hydrogeochemical zoning is established within the structures showing the absence of salt domes. Whereas Cambrian, Vendian and Riphean deposits in the boundary areas of the Siberian platform, adjacent to the study area from the south, contain high-grade metamorphic brines of sodium chloride, sodium-calcium and calcium composition with TDS value up to 480 g/dm^3.

It has been established that with the increasing degree of metamorphization of subsurface waters and brines, which is basically growing after burial of calcium, bromine and strontium, significantly changing their composition. A comprehensive analysis of hydrogeochemical materials showed that among all the studied subsurface waters and brines, the most metamorphic have proven to be brines distributed within the Anabar crystalline shield of the Archean-Proterozoic age, which are characterized by a calcium chloride composition and TDS concentrations more than 300 g/dm^3. This was corroborated by the analysis of distribution of the main genetic coefficients. The amount of rNa/rCl is 0.2-0.3, Cl/Br ranges from 50 to 150, and the ratio of Ca / Cl exceeds 0.5 (Figure 2). Moderate metamorphization of brines is characteristic of Cambrian, Vendian and Riphean aquifers of the Siberian platform [37, 38]. Subsurface waters and weak brines of Western Siberia [39, 40] and the Yenisei-Khatanga [41] basin with halogen formations absent from their structures are at the earliest stage of metamorphization of their composition, the ratio of Ca/Cl usually does not exceed 0.20, and higher values are correlated with waters of active gas exchange within their limits. The infiltrogenic brines promoting leaching within the salt domes of the Anabar-Khatanga basin show genetic affinity with the modern leaching brines of the salt domes of the Iran-Pakistan basin and others. A natural chain of chemical and genetic types of subsurface waters and brines of the Siberian Arctic is basically built as indicated below. The earliest stage implies waters of variegated chemical composition coming from the zone of active gas-water exchange. This is followed by the subsurface waters of sedimentary basins characterized by the absence of salts in the hydrogeological section, which in turn are replaced by brines of sedimentary basins with halogen formations. The chain is culminated with the maximum degree of metamorphization of brines from crystalline shields of the Archean-Proterozoic age.

Ultimately, we can distinguish three genetic groups of waters: 1) infiltrogenic, 2) sedimentogenetic and 3) condensatogenic. The first includes modern and fossil infiltration waters, brines leaching salt dome structures, while sedimentogenetic waters (including brines), lithogenic waters and cryopegs belong to the second group. The third group should be a special type of condensatogenic waters developed syngenetically with hydrocarbon deposits. Modern and fossil infiltratogenic waters are aqueous solutions of atmospheric origin, penetrated into rocks by infiltration during hypergenesis. These are characterized by TDS concentrations up to 20 g/dm^3 and are chloride-bicarbonate and bicarbonate- sodium chloride in composition.

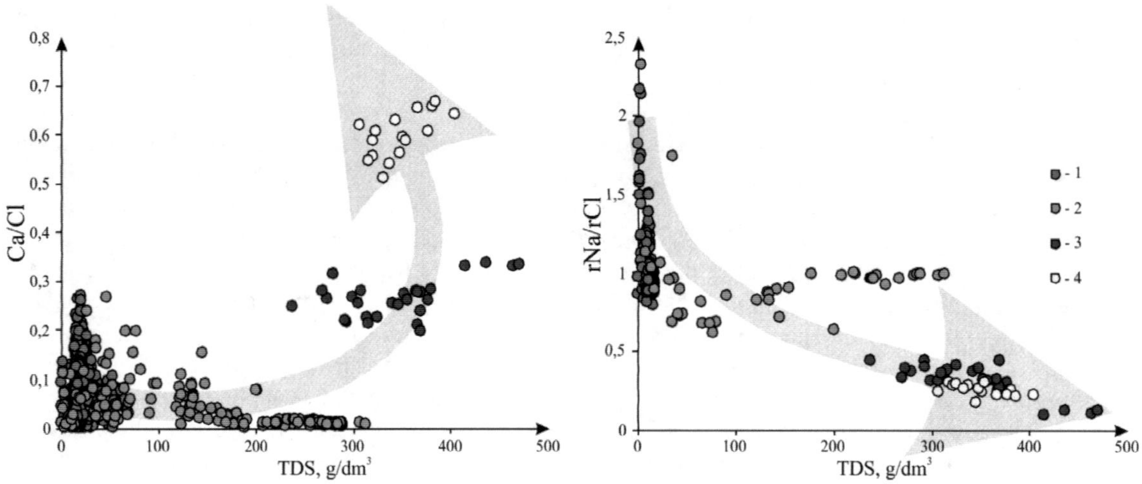

Figure 2. Change in the degree of metamorphism of groundwaters and brines of Arctic regions of Siberian sedimentary basins.

Sedimentary basins: 1 – West Siberian and Yenisei-Khatanga, 2 – Anabar-Khatanga, 3 – IV – Siberian platform, 4 – Anabar shield (southern slope). The arrow shows the direction of metamorphism of groundwaters and brines.

The fossil infiltratogenic waters should also include sodium chloride brines of rock salt leaching, with TDS at levels of 200-315 g / dm^3, developed within the salt domes of the Anabar-Khatanga basin. In the Arctic regions of Siberia, sedimentogenic waters and weak brines are developed within the Jurassic and Cretaceous aquifer complexes of the WSSB and Yenisei-Khatanga basin and are of sodium chloride composition and TDS concentrations from 25 to 63.3 g/dm^3. Lithogenic water released as a result of thermodehydration of mostly clay minerals are widely developed in the Cretaceous and Jurassic horizons at depths greater than 2 km, are characterized by a TDS levels between 5 and 15 g/dm^3, and are of chloride-bidrocarbonate and bidrocarbonate- sodium chloride composition. Sedimentogenic brines widely within the aquifers of the Cambrian, Vendian and Riphean of the Siberian platform with the TDS concentrations up to 480 g/dm^3 , are primarily of sodium chloride, sodium-calcium and calcium composition. cryopegs are widely spread within the Yamal and the Taimyr Peninsulas and found in the sediments of laidas, islands, floodplains in lower reaches of rivers characterized by the marine type of salinization. Cryopegs developed in these areas are characterized by high salinity up to 150 g/dm^3 and shallow occurrence depth, from 1 meter. Condensatogenic waters are characterized by the greatest diversity of chemical composition and variation of all genetic coefficients. Besides their low TDS, which is usually 2-10 g/dm^3, they are characterized by high gas saturation – to 2.0 L/L and more.

4. Conclusion

A detailed analysis of hydrogeochemical materials resulted in the following inferences: 1) subsurface waters and brines of chloride-bicarbonate sodium, sodium chloride, calcium chloride and other chemical types, with TDS ranging from 0.2 to 480 g/dm^3 are commonly developed in Arctic parts of the Siberian sedimentary basins. 2) Hydrogeochemical materials revealed that the subsurface waters of Mesozoic deposits within the West Siberian, Yenisei-Khatanga, Anabar-Khatanga basins are in the early stages of metamorphization of their chemical composition. The brines of the Siberian platform have reached the middle stage of metamorphizations. Most changed are the brines of the Anabar crystalline shield, on account of the processes of interactions between the host rocks and dispersed organic matter. 3) Specific traits of the geochemistry of subsurface waters and brines are largely determined by the composition of the host rocks (the presence or absence of salt-bearing formations), hydrodynamics of pressurized water systems, hydraulic interconnection of aquifers, and paleohydrogeological history. Their modern chemistry is a product of long-term geological evolution of fossil syngenetic waters. The initial composition is formed in the depositional environments. Later, during the stages of diagenesis and catagenesis, the captured waters experience significant changes, since they become involved in a variety of ongoing processes such as sediment compaction, elision water exchange, vertical and lateral migration, thermodehydration of clay minerals and interactions in the "water-rock-gas-OM" system. In this regard, the correct identification of the genetic appearance of subsurface waters and brines along with the composition of the dissolved substances requires a comprehensive study of their isotopic composition. The final classification of aqueous solution to a particular genetic type makes sense only after studying their isotopic composition, primarily distribution of δD, $\delta^{18}O$, ^3H, ^{37}Cl, ^{81}Br, $^{87}Sr/^{86}Sr$. 4) The three genetic groups of subsurface waters in the studied area are: infiltrogenic, sedimentogenetic and condensatogenic. The first group includes modern and ancient infiltration waters and salt domes leaching brines, the second comprises sedimentogenetic waters (including brines), lithogenic waters and cryopegs. The third group is designated as specific type of condensatogenic waters formed syngenetically with hydrocarbon deposits.

Acknowledgments

The research was supported by the Russian Foundation for Basic Research within the scope of scientific project № 18-05-70074 «The Arctic Resources».

References

[1] Kontorovich A E 2015 *Herald of the Russian Academy of Sciences* **85(3)** 213–22
[2] Bell J S, Bird T D, Hillier T L and Greener P L 1995 *Geological Survey Of Canada* p 541
[3] Bird K J and Magoon L B 1987 *U S Geological Survey Bull* **1778–80** p 50
[4] Franke D, Hinz K and Reichert C 2004 *Geomagnetism and Paleomagnetism/Marine Geology and Geophysics* **109(7)** B07106 p19

[5] Nemcok M et al. 2016 *Geological Society* **431** 1–38
[6] Vorren T O et al. *Norwegian Petroleum Society* **2** p 751
[7] Novikov D A 2017 *Petroleum Exploration and Development* **44** 780–8
[8] Palmer C **1911** *U S Geol Sum Bull* **749** 5–31
[9] Kurlov MG 1928 *Classification of Siberian medicinal mineral waters* (Tomsk: Physiotherapeutic Institute in Tomsk) p 73
[10] Karstens E E 1932 *Physics and Chemistry of Mineral Waters* (Moscow: Gosmedizdat) p 214
[11] Aleksandrov V A *Classification of mineral waters* (Moscow: Gosmedizdat) p 214
[12] Vernadsky V I 1935 *History of minerals of the earth's crust* (Moscow:Gostoptekhizdat) p 202
[13] Sulin V A 1953 *Waters of oil deposits of the USSR* (Moscow:Main edition of mining and fuel literature) p 367
[14] Sulin V A 1946 *Waters of Petroleum Formations in the System of Nature Waters* (Moscow:Gostoptekhizdat) p 96
[15] Chebotarev I I 1955 *Geochimica Cosmochim Acta* **8** 22–48, 137–170 and 198–212
[16] Schoeller H 1955 *Rev Znst Fr Pet* **10** 181–213, 219–246 and 507–52
[17] Bojarski L 1970 Die *2 Angew Geol* **16** 123–5
[18] Collins A G 1975 *Developments in Petroleum Science* p 496
[19] Kamensky G N 1947 *Search for and exploration of groundwater* (Leningrad: Gosgeolizdat) p 313
[20] White D E 1971 *Texas: Texas Water Development Board Report* **125** p 219
[21] Pinneker E V 1979 *Basic regularities of groundwater composition formation Problems of theoretical and regional hydrogeology* (Moscow: MSU) 10–5
[22] Kartsev A, Vagin S and Matusevich V 1986 *Hydrogeology of oil and gas bearing basins* (Moscow: Nedra) p 224
[23] Kartsev A et al. 1992 *Theoretical fundamentals of oil and gas hydrogeology* (Moscow: Nedra) p 208
[24] Zektser I S and Dzhamalov R G 2007 *Submarine Groundwater* (Taylor & Francis Group, LLC) p 466
[25] Shvartsev S L 1996 *General hydrogeology* (Moscow: Nedra) p 423
[26] Shvartsev S L and Novikov D A 2004 *Geologiya i Geofizika* **45** 1008–20
[27] Nelyubin V V and Yakovlev O N 1981 *The origin of oil and gas, their migration and patterns of formation and location of oil and gas deposits* 228-230
[28] Kolodiy V V 1975 *Underground condensation and salt water of oil, gas-condensate and gas fields* (Kiev: Naukova Dumka) p 122
[29] Novikov D A and Lepokurov A V 2005 *Geologiya Nefti i Gaza* **5** 24–33
[30] Novikov D A 2018 *Higher Educational Institutions News. Oil and Gas* **1** 35–42
[31] Sadykova Ya V and Dultseva M G *Water resources* **44** 168–81
[32] Aksenova V I, Bubnovb N G, Klinovaa G I, Iospaa A V and S G Gevorkyana 2011 *Water Resources* **38(7)** 934–43
[33] Kokh A A and Novikov D A 2014 *Water resources* **41** 375–86
[34] Novikov D A and Shvartsev S L 2009 *Russian Geology and Geophysics.* **50** 1131–43
[35] Dultsev F F and Novikov D A 2017 *Bulletin of the Tomsk Polytechnic University, Geo Assets Engineering* **328(11)** 6-15
[36] Hudec M R and Jackson M P 2007 *Earth-Science Reviews* **82(1/2)** 1–28
[37] Trifonov N S, Novikov D A and Yamskikh A A 2015 *Water Resources* **42** 909-921
[38] Novikov D A and Trifonov N S 2016 *Arabian Journal of Geosciences* **9** 1-14
[39] Novikov D A 2018 *Neftyanoe Khozyaystvo - Oil Industry* **4** 16-21
[40] Novikov D A and Sukhorukova A F 2015 *Arabian Journal of Geosciences* **8** 8703-8719
[41] Novikov D A *Geodynamics and tectonophysics* **8** 881-901

Abnormally high formation pressures in jurassic-cretaceous reservoirs of Arctic regions of Western Siberia

D A Novikov[1,2], F F Dultsev[1] and A V Chernykh[1,2]

[1] Trofimuk Institute of Petroleum Geology and Geophysics SB RAS, Novosibirsk, Russia,
[2] Novosibirsk State University, Novosibirsk, Russia

E-mail: NovikovDA@ipgg.sbras.ru

Abstract. Results of the study of the structure of hydrodynamic fields in Arctic part of the West Siberian sedimentary basin have designated the main region-specific hydrodynamic feature consisting in the phenomenon of abnormally high formation pressures widely spread at depths of 2.8-6.0 km within Jurassic-Cretaceous reservoirs. The two types of natural pressurized water systems which have developed in the region are: 1) elision (geostatic and geodynamic) system most common in the interior regions, and 2) infiltration aquifer system inherent in the external, near edge zones. As the depth increases, two hydrodynamic zones are distinctly distinguished (from top downwards): hydrostatic and enhanced evolving into AHFD (abnormally high formation pressure) . Most of the Aptian-Albian-Cenomanian aquifers belong to the first type. Hydrodynamic field stresses tend to increase in the lower lying Neocomian complex where the already elevated formation pressures gradually build up, to the extent of AHFP in the lower horizons. The Jurassic aquifer complexes in the central parts of the Yamal-Kara depression exhibit either enhanced or abnormally high formation pressures, which decline to hydrostatic levels towards the basin periphery. The anomaly coefficients of formation pressure reach the value of 2.2. The established piezo-minima zones extending along major petroleum charge zones, extending along the major hydrocarbon generation zones (Bolshaya Kheta and Kara megasyneclises) associated with the largest areas of petroleum accumulation (Vankor-Suzun, Bovanenkovo, Urengoy, etc.).

1. Introduction

Given the complexity of hydrogeological conditions of Arctic part of the West Siberian sedimentary basin (WSSB) it is critical to have reliable information on the hydrodynamic parameters of petroleum reservoirs. However, the quality of data obtained from exploration wells for most of the studied objects particularly those tested using drill-stem tester is very low. Analysis of the factual materials has shown that less than half of exploration wells that penetrated Mesozoic aquifers, provide reliable information on the hydrodynamic aspects. Notably, despite the thousands of deep exploration wells drilled in the studied region (Fig. 1), even the major hydrodynamic patterns have thus far been studied only in broad terms. The state of knowledge about the region's hydrogeology varies greatly, depending on many factors. Thus, Neocomian complex has received the most extensive coverage due to its high prospects for oil and gas and because of the many actual discoveries of hydrocarbon deposits. Numerous hydrodynamic research with the focus on Western Siberia are represented by the

IOP Conf. Series: Earth and Environmental Science **193** (2018) 012050 doi:10.1088/1755-1315/193/1/012050

works of B. L. Aleksandrov, G. D. Ginsburg, A. E. Gurevich, V. I. Dyunin, P. A. Kamenev, V. N. Kortsenshtein, N. M. Kruglikov, B. F. Mavritskii, V. V. Nelyubin, O. V. Ravdonikas, A. D. Reznik, O. N. Yakovlev and many other researchers [1-13].

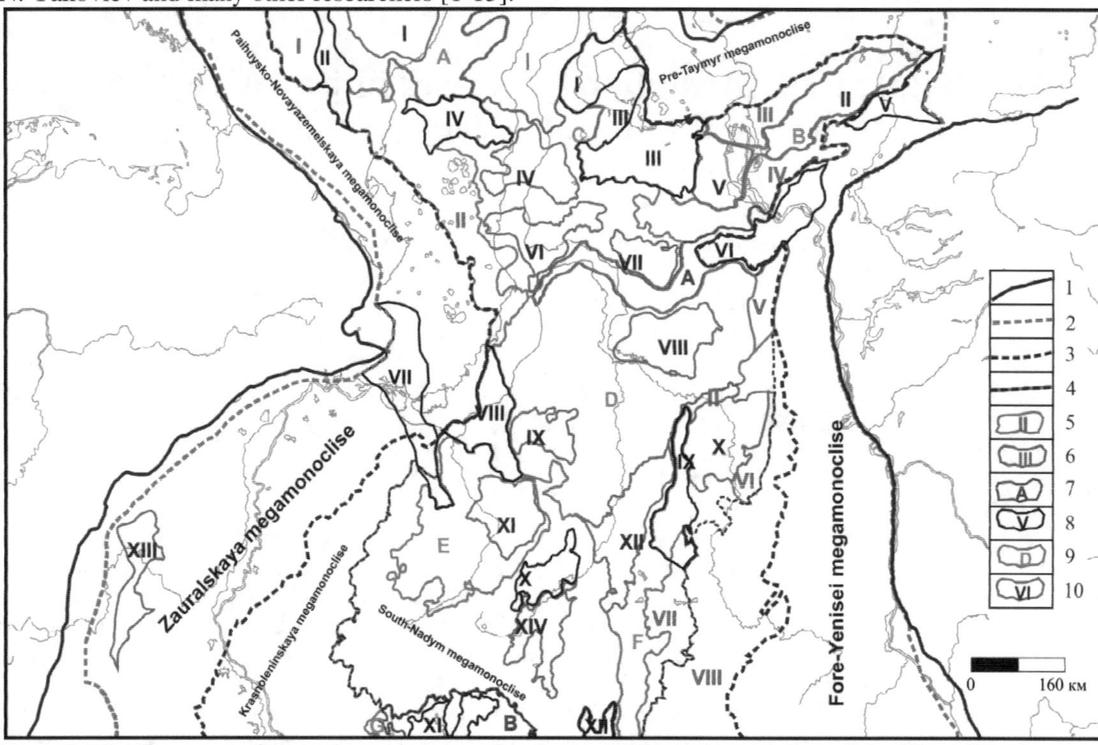

Figure 1. The fragment of tectonic zoning map for the top of Jurassic complex of the West-Siberian geosyneclise (simplified after [14])

Boundaries of: 1-West Siberian sedimentary basin; 2-Jurassic sedimentary basin; 3-Internal region and External belt; 4-Yamal-Kara depression; Intermediate tectonic elements: 5 – mega-, meso-, monocline; 6 – mega-, meso-, saddle. Positive tectonic elements: 7- zero-order structures; 8 – I-order structures. Negative tectonic elements: 9- 0-order structures; 10-structures of order I.

Intermediate tectonic elements: *1* – mega-, meso-, monoclim: I –North-Kara monoclise, II – East-Pakhoy monoclise, III –South-Taimyr megamonocline, IV –North-Messoyakha megamonocline, V – Dolgon mesomonocline, VI – East-Taz mesomonocline, VII – East-Pur megamonocline, VIII – Krasnosel'kup monoclise; *2* – mega-, meso-, saddles: I – South-Kara megasaddle, II – North-Chaselka saddle; Positive tectonic elements: *3* – zero-order structures: A – Messoyakha inclined ridge, B – Khantey hemianteclise; *4* – I-order structures: I – North-Gydan megaheadland, II –Pre-Paikhoy megaheadland, III – Tundra megaheadland, IV – Bovanenkovo-Nurminsky inclined megaswell, V – Gydan megaheadland, VI – Ust'-Port megaheadland, VII – Yarudey megaheadland, VIII – Medvezhiye –Nuginsky inclined megaswell, IX – Chaselka inclined megaswell, X –Northern Arch, XI – Surgut Arch, XII –Var'yegan-Tagrin megaheadland; Negative tectonic elements: *5* – zero-order structures: A – Kara megasyneclise, B – Agapa –Yenisei trench, C – Antipayuta-Tadebeyakha megasyneclise, D – Bolshaya Kheta megasyneclise, E – Nadym hemisyneclise, F –Middle-Pur inclined megatrench, G –Mansi syneclise; *6* – I-order structures: I –South-Kara megadepression , II – Agapa megathrough, III – Middle-Gydan mega-gully, VI – Yaptiksale megadepression, V – Yenisei megadepression, VII – East-Antipayuta megadepression, VIII – North-Taz megadepression, IX – Nerutinsky megadepression, X – Taz structural megabay XI – Upper-Tanlova megadepression, XII – Middle-Pur inclined megatrough, XIII – Visim inclined megatrough, XIV – Pyakupur-Amputa inclined megatrough.

The many years of fundamental research into the processes of filtration, and experimental studies of hydrodynamic systems of petroleum basins around the world evidence that the even systematic observations are limited to providing data on water flow direction and rate only for groundwaters or shallow artesian waters, largely ignoring the problems faced in the development of hydrocarbon deposits. Therefore, the main approaches applicable to determining directions, velocities and discharge rates of subsurface waters in deep horizons of petroleum basins are basically calculus-based.

2. Materials and Methods

In most natural "pressurized" aquifer systems, the density of subsurface waters is subject to variation both within its areal extent and throughout the section, whereas the aquifers' placement is non-horizontal. In this case, the differences in their water -table levels are largely controlled by variations in water density and the horizon occurrence depths, rather than by water motion alone. Ruling out the impact from these factors in order to focus on the dynamics effects alone, requires using calculated pressure heads (reduced, or corrected pressures) of subsurface waters, instead. It is only a differential of pressure head (i.e. reduced pressure) that can serve an indicator of the pressurized subsurface water flow. A reference (datum) plane is usually lined up with the bottom-hole of the deepest well. For the purpose of generalization and analysis of the hydrodynamic parameters of Jurassic-Cretaceous reservoirs, the following reference planes were accepted for major hydrogeological complexes: Aptian-Albian-Cenomanian – 2500 m, Neocomian – 4000 m, Upper Jurassic – 3800 m and Lower-Middle Jurassic – 5500 m. This study is largely based on the results of testing more than 4000 objects of 251 prospecting areas, as well as the interpretation of well-logs and drillcore analysis materials (more than 6200 samples). The best studied appear to be Cretaceous reservoirs. The structure of a hydrodynamic field of petroleum basins is known to evolve over a long time and is closely connected with the geological history through the processes of sedimentary rocks compaction and their post-sedimentation transformations from the ooze stage in early diagenesis and through the stage of metamorphism.

3. Results and Discussion

The revealed structural features and available information on the subsurface geology enabled identification of the Mesozoic-Cenozoic aquifer system in the cross section with characteristic pressurized (confined aquifers) and impermeable (aquicludes) complexes. The main hydrodynamic signature of this part of the WSSB is represented by a wide distribution of high formation pressures (AHFP) within the 2.8-6.0 km depth interval both in Jurassic aquifers and in the overlying horizons, up to Neocomian ones (Fig. 2). As the depth increases, two hydrodynamic zones become distinctly identifiable (from top downwards): 1) hydrostatic and 2) elevated pressures gradually developing into AHFP. Notably, most of the aquifers of the Aptian-Albian-Cenomanian aquifer complex belong to the first group. Given that the hydrodynamic field stress tends to increase in the lower lying Neocomian complex, the already elevated formation pressures gradually become AHFP in its lower horizons. Jurassic complexes in the central parts of the study area are dominated by elevated pressures and AHFPs, which anyway appear to be declining to the level of hydrostatic when approaching the basin periphery [15]. This pattern is distinctly observable on the example of the Yamal Peninsula, where in the direction from south to north from the Novoportovskoye oil and gas condensate field to the Malyginskoye field, the subsurface hydrogeological closure is tending to increase, which is explicitly manifested in the growth of Ka values from normal (0.95-1.00) to AHFP (1.60-1.92), respectively [13,16].

The Aptian-Albian-Cenomanian aquifer complex is characterized by formation pressures varying in the range between 3.5 and 28.4 Mpa and controlled by the occurrence depth of aquifers [17,18]. As is seen from Fig. 2, this complex is for the most part characterized by normal (hydrostatic) pressures. Notably, hydrodynamic stress analysis at a regional scale has shown that the study area falls almost entirely within the zone of normal pressures. In this context, quite extensive areas are confined to the elevated pressure zones in the areas of NW strike (the southern parts of the Kara megasyneclise) and in the central areas (the central regions of the Atipayuta-Tadebeyakha megasyneclise and the southern slope of the North-Taz megadepression complicating the structural plan of the Bolshaya Kheta megasyneclise). The sporadically distributed low-pressure hydrodynamic anomalies are associated with the existence of HC accumulations within the sediments of the complex, in itself, being a major gas-producing area of Western Siberia. Arctic regions hold unique hydrocarbon reserves, which is illustrated below by gas fields classification according to their largeness, trln m^3: Urengoyskoe – 9.5, Medvezhiye – 1.0, Yuzhno-Russkoe – 1.0, Yamsoveyskoe – 1.0, Komsomolskoe – 0.7, etc. [14,15]. The analysis of the low pressure anomalies propagation extents revealed a direct relationship between their existence and the presence of the unique and large gas deposits within the complex, which is, specifically, characteristic of the Cenomanian gas accumulations. As such, the anomalies were found within the Bovanenkovskoye, Messoyakhskoye, Tagulskoye, Utrenneye, Kharampurskoye, Yamsoveyskoye and other gas fields, where the Ka values vary in the range from 0.83 to 0.91.

Analysis of distribution of pressures reduced to the 2500 m plane within the Aptian-Albian-Cenomanian complex revealed extensive piezo-maxima zones acting as internal areas of groundwater recharge confined to the most plunging areas of the Middle Pur inclined megatrench, and the Bolshaya Kheta, Antipayuta-Tadebetakha and Kara megasyneclises. A general pressure differential of reduced pressures reaches 3.4 MPa. Whilst the piezo-minima zones delineate the contours of large zones of oil accumulation, for example, the Kharasaveyskoye field in the north, Vankor-Suzunskoye in the east, and Urengoiskoye in the central part, etc. In the Neocomian aquifer complex cross section, formation pressures vary between 6.3 and 77.9 MPa. Aquifers confined to the lowermost parts of the complex exhibit the presence of AHFP zones, and elevated pressures are remarkably observed throughout the cross section (Fig. 2). Analysis of the Ka values distributions has shown the presence of three zones characterizing the formation pressure as: normal, elevated and abnormally high. Remarkably, likewise the overlying Aptian-Albian-Cenomanian hydrogeological complex, it is dominated by the normal pressure zone. Against this backdrop, two major regions of elevated reservoir pressures are distinguished, with one of them confined to the central regions and delimited by the areal extent of the Bolshaya Kheta megasyneclise, while the second is established in the north, partly comprising the structures of the Kara and Antipayuta-Tadebeyakha megasyneclise, and encompassing the eastern part Agapa-Yenisei trench. Within these areas, we have identified seven hydrodynamic zones with IHFP. In the north, these are the areas of the South Kara megadepression, Tadebeyakha megatrough, and Yenisei megadepression.

Within the Bolshaya Kheta megasyneclise, the developing AHFP areas are confined to the Neruta megadepression structures. Likewise within the Aptian-Albian-Cenomanian complex, a number of hydrodynamic anomalies with the lower background formation pressures (Ka values are in range from 0.86 to 0.91) were established, also confined to large fields (Salmanovskoye, Kharasaveyskoye, Vankorskoye, etc.). The total pressure differentials reduced to the 4,000 m plane reaches 28.0 MPa within Neocomian complex. Analysis of their distribution revealed extensive high-gradient piezo-maxima zones acting as internal areas of subsurface recharge and confined to the most plunged areas of the Bolshaya Kheta, Antipayuta-Tadebeyakha, Kara megasyneclises and Agapa-Yenisei trench. The maximum lateral pressure decline gradients down to 0.8-1.2 MPa/km were established in the north-east of the study area in the transition region from the West Siberian artesian basin to the Khatanga basin, which is structurally confined to the Yenisei-Khatanga regional trough. What actually drives the values to be that large? A detailed consideration of the structural features of aquifers has thrown much light on the situation [9,11]. The identified localities on the North-Messoyakha megamonocline slope within the Nizhnyaya-Malaya Kheta structures cfault zones, have thus served as subsurface waters discharge areas. Westwardly, the top of Neocomian sediments has plunged deeper from 300 to 2000 m or more, while the reduced pressures have increased from 36 to 52 MPa. The revealed there extensive area with AHFP is delimited by the Paikhoy, Middle-Yarovskoye, Turka and Tanama areas. The highest Ka values are observed in the wells: Turkovskaya 1 – 2.45; Turkovskaya 2 – 1.46; Paiyakhskaya 1 – 1.45, and Paiyakhskaya 2 – 1.49.

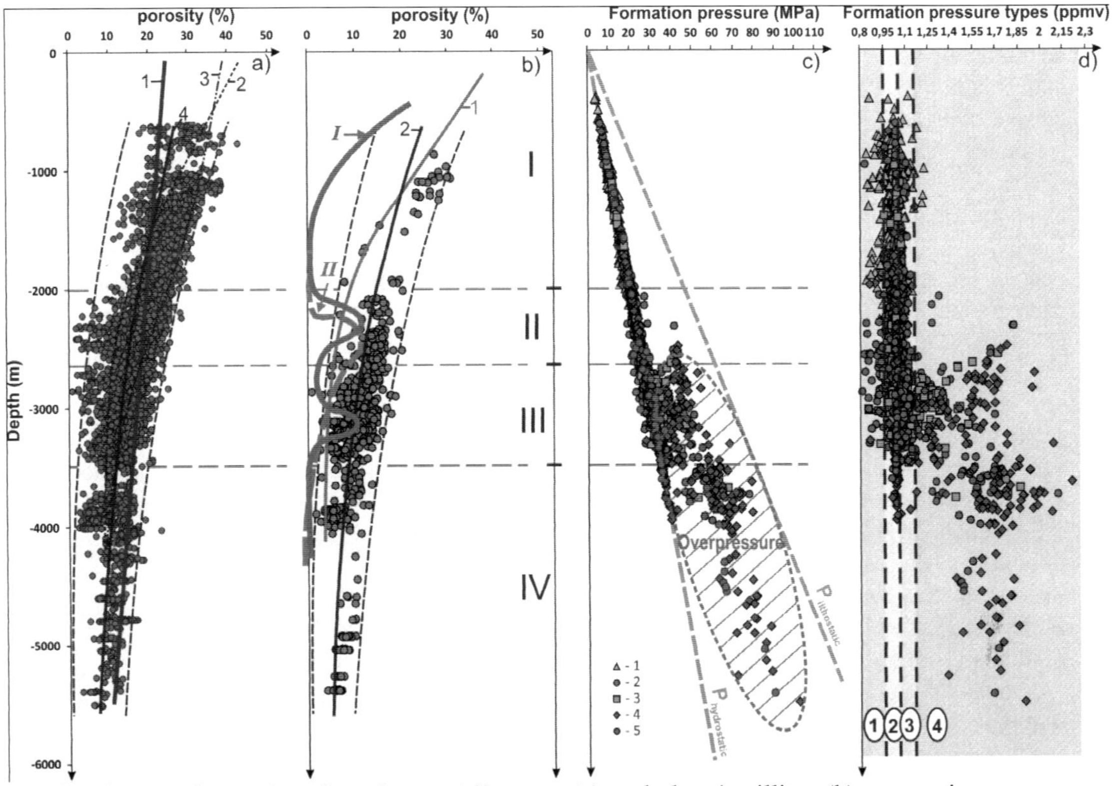

Figure 2. Change of porosity of sandstones/siltstones (a) and clays/argillites (b), reservoir pressures (c) and Ka (d) with depth within the Arctic regions of the West Siberian sedimentary basin [19].
a) trends in the porosity of sandstones/siltstones with depth: 1 – global [20]; 2 – in the Central regions of Western Siberia (alevrolites) [21]; 3 – in the Central regions of Western Siberia (sandstones[21]; 4 – in the Arctic regions of Western Siberia.
b) Trends in porosity change of clay / argillites with depth: 1 - G. I. Alekseev et al. [21]; 2 – in the Arctic regions of Western Siberia. Curves of dehydration of argillaceous sediments: I – after Burst [22]; II – after Perry and Hower [23]. Sediment Dehydration stages: I – free-water elision, II – the initial stage. 50% elision of interlayer water at the depths of 2.0-2.7 km, III – the next stage of 25% elission of interlayer water at the depths of 2.7-3.5 km, IV – the rest 25% elision of interlayer water at a depth of 3.5 km.
c) Hydrogeological complexes: 1 – Aptian-Albian-Cenomanian, 2 – Neocom, 3 – upper Jurassic, 4 – lower-middle Jurassic, 5 – pre-Jurassic.
d) zones of formation pressures in size Ka: 1 – lowered (0,8-0,95); 2 – normal (0,95-1,05); 3 – increased (1,05-1,15); 4 – anomalous-high (more than 1,15).

The AHFP zone is bordered on the zone of elevated pressures, which is expanding from the south-west and north-east, heading for the northwestern regions. The values of reduced pressures less than 35-36 MPa are associated with distribution of low pressures within the Ozernaya, Lower-Kheta, Malaya Kheta, Sukhaya Dudinka, Tochino, and Dolgan areas, which are interpreted to be influenced by the gas accumulations identified therein. These patterns also generally characterize the main piezo-minima proper to Arctic regions of Western Siberia, and account for their replicating the contours of major zones of petroleum accumulation. At this, the alignment of a number of oil and gas accumulation zones (e.g., the Urengoy area) with high pressure zones is remarkable, which may evidence the processes of vertical migration within them, from the deeper buried Jurassic complexes.

In terms of hydrodynamics, the upper Jurassic hydrogeological complex appears the most interesting, inasmuch as formation pressures vary considerably within its bounds, ranging from 9.9 to 67.7 MPa. The water dynamics specificity of the complex is represented by AHFP with anomaly coefficient reaching the value of 1.87 (Fig. 2). The analysis of the hydrodynamic field stress revealed the dominance of two pressure zones in the study area described as: normal (eastern parts) and abnormally high (western and central parts), with the zone of elevated pressures (Ka = 1.05-1.15) in between, stretching from south to north. In eastern parts, the dominance of the normal pressures is sporadically interfered by hydrodynamic anomalies with elevated pressures confined to the structures of the Yustyrmalskoye, Tekto-Kharampurskoye and Ravninnoye fields [17,11]. The presence of such an extensive AHFP zone in the central and western parts of the Nadym-Taz interfluve (near the Oxfordian regional clayey reservoir boundary) can be explained by the elision water regime in this part of the region [17,18]. V.M. Matusevich and co-authors [24,25] have identified the following pressurized aquifer systems striking from west to east: elisian- lithostatic, elisian-geodynamic (controlled by the -Koltogor -Urengoy graben-rift) and infiltration.

Analysis of distribution of pressures reduced to the 3800 m plane within Upper Jurassic complex has revealed extensive piezo-maxima zones confined to the boundary of argillization of Oxfordian regional reservoir in the west of the Nadym-Taz interfluve and to the most plunged parts of the Middle-Pur inclined megatrough and the Bolshaya Kheta megasyneclise. The total pressure differential across the complex reaches 34.0 MPa. The growth of reduced pressures at the boundary of the argillization of permeable sediments speaks of a reliable lithological screen and of possible lateral migration of fluids from west to east governed by the conditions of the elisian lithostatic pressurized water system. The zones of the main piezo-minima also delineate the areas of large oil and gas accumulations (e.g. the Kharampur within the Middle-Pur inclined megatrench, etc.). The lowermost of the thoroughly studied, Lower-Middle Jurassic aquifer complex is characterized by formation pressures 6.6-101.9 MPa (Fig. 2). The dominant controls in the central and northern regions are represented by elevated pressures and AHFPs; and by the zone of normal pressures further to the east and west, expanding as far as the sedimentary basin boundaries. The AHFP zone is aligned with the main super-order tectonic elements: the Kara, Antipayta-Tadebeyakha and the Bolshaya Kheta megasyneclises and the Arapa-Yenisei trench, where Ka values often reaches 1.5 or more. For example, in the Lower-Middle Jurassic reservoirs of the Bovanenkovskoye field they reach 1.54-1.77, and 1.79 in the West-Tambeyskoye field; vary between 1.97 and 2.21in the Zapolyarnoye field; and between 1.65 and 1.98 in the Urengoiskoye field [13].

Analysis of distribution of pressures reduced to the 5500 m plane within the Lower-Middle Jurassic aquifer complex has revealed extensive zones of piezo-maxima, delineating the areas of the developing AHFPs. The total pressure differential reaches 50 MPa, and the maximum pressure were identified within the Kara (>70 MPa) and the Bolshaya Kheta (> 80 MPa) megasyneclises. This ultimately allows an inference that the pressurized water system of the Lower-Middle Jurassic aquifer complex has preserved the relics of the exfiltration hydrodynamic regime, since most zones of elevated pressures are confined to large negative tectonic elements. The major piezo-minima zones replicate the contours of large zones of oil and gas accumulation (e.g. Novy Pert area in the north, the Kharampur area in the central parts, etc.).

4. Conclusion

It follows from the above that the pressurized water systems of Cretaceous and Jurassic complexes in Arctic parts of the West Siberian sedimentary basin, including petroleum-producing reservoirs identifiable within the extensive area, are isolated from each other by reliable aquicludes (impermeable beds), which is confirmed by the results obtained. Their isolation is impaired only locally, at sites marked either by the presence of numerous faults and tectonic dislocations, or lithological windows. Even within one complex, hydrodynamic conditions change significantly, and isolated hydrodynamic blocks have been established within their bounds. The low hydrodynamic pressure anomalies within the complex of Aptian-Albian-Cenomanian and Neocomian aquifers are associated with the presence of unique and large reserves of gas and gas-condensate deposits in their cross section, especially in Cenomanian reservoirs. Manifestations of elevated and abnormally high pressures within them are interpreted as a sign of a high level of hydrogeological closure of the subsurface, which characterizes the elision type of pressurized water system. The currently observed manifestation thereof in the study region belong to two types of natural pressurized water systems (after A. A. Kartsev, V. M. Matusevich): elision waters (geostatic and geodynamic) in the interior regions, and infiltration waters in the exterior parts, in near edge zones.

Especially widespread areas of AHFP are found to be within the Jurassic aquifer systems, dominating in the western and central parts (Upper Jurassic reservoirs), and in the northern and central regions (Lower-Middle Jurassic reservoirs). The Neocomian, Upper Jurassic and Lower-Middle Jurassic aquifer complexes are marked by the presence of extensive areas of piezo-minima, extending along the main zones of oil and gas generation (the Bolshaya Kheta and Kara megasyneclises), which are associated with the largest areas of petroleum accumulation (Bovanenkovo, Vankor-Suzun, Urengoy, and others).

Acknowledgments

The research was supported by the Russian Foundation for Basic Research within the scope of scientific project № 18-05-70074 «The Arctic Resources».

References

[1] Ginsburg G D, Gurevich A E and Reznik A D 1971 About the causes of low reservoir pressures in northern Siberia *Soviet geology* **9** 45-58

[2] Kortsenshtein V N 1977 *Water-supply systems of the largest gas and gas-condensate fields of the USSR* (Moscow: Nedra) p 247

[3] Kruglikov N M and Yakovlev O N 1981 Problems of the dynamics of groundwater in the north of Western Siberia *Hydrogeological conditions of oil and gas potential in some regions of the USSR* (Leningrad: Nedra) 78-100

[4] Kruglikov N M, Nelyubin V V and Yakovlev O N 1985 *Hydrogeology of the West Siberian oil and gas bearing basin and features of formation of hydrocarbon deposits* (Leningrad: Nedra) p 279

[5] Alexandrov B L 1987 *Abnormally high reservoir pressures in oil and gas basins* (Moscow: Nedra) p 216

[6] Matusevich V M and Ushatinsky I N 1998 Features of composition and formation of geofluidic systems of the West Siberian oil and gas-bearing megabasin *proc. of universities Oil and gas* **4** 28-35

[7] Dyunin V I and Korzun V I 2005 *Hydrogeodynamics of oil and gas bearing basins* (Moscow: science world) p 524

[8] Novikov D A 2013 Possibilities of oil-and-gas content in Middle-Jurassic deposits of Yamal peninsula by hydrogeological data *Geology of oil and gas* **6** 65-74

[9] Kokh A A and Novikov D A 2014 Hydrodynamic Conditions and Vertical Hydrogeochemical Zonality of Groundwater in the Western Khatanga Artesian Basin *Water Resources* **41** 396-405
[10] Novikov D A 2014 Hydrodynamics of neocomian oiland gas-bearing deposits of the transient territory from the Western-Siberian artesian basin to Khatangsky basin *Geology, geophysics and development of oil and gas fields* **2** 24–33
[11] Novikov D A 2017 Hydrogeological conditions for the presence of oil and gas in the western segment of the Yenisei-Khatanga regional trough *Geodynamics and tectonophysics* **8** 881-901
[12] Novikov D A 2017 Hydrogeochemistry of the Arctic areas of Siberian petroleum basins *Petroleum Exploration and Development* **44** 780-88
[13] Novikov D A 2018 Oil and gas fields exploration in the Jurassic-Cretaceous deposits of Yamal Peninsula based on the water - gas equilibrium *Oil Industry* **4** 16-21
[14] Kontorovich V A et al. 2001 Tectonic structure and history of development of the West Siberian geosyneclise in the Mesozoic and Cenozoic *Geology and Geophysics* **42** 1832-45
[15] Novikov D and Sukhorukova A 2015 Hydrogeology of the northwestern margin of the West Siberian Artesian Basin *Arabian Journal of Geosciences* **8** 8703-19
[16] Ses K V and Novikov D A 2017 Hydrodynamic features of hydrocarbon reservoirs of the central Yamal peninsula *Bureniye i Neft' (Drilling and Oil)* **5** 30-35
[17] Shvartsev S L and Novikov D A 2004 The nature of vertical hydrogeochemical zoning of petroleum deposits (exemplified by the Nadym-Taz interfluve, West Siberia) *Geology and Geophysics* **45** 1008-20
[18] Novikov D A and Lepokurov A V 2005 Hydrogeological conditions of petroleum potential deposits on the structures in the southern part of Yamalo-Karskoye depression *Geology of oil and gas* **5** 24-33
[19] Novikov D A 2018 Vertical hydrodynamic zoning of the Yamal-Kara depression (northern areas of Western Siberia) *Higher Educational Institutions News. Oil and Gas* **1** 35-42
[20] Ehrenberg S N and Nadeau P H 2005 Sandstone versus carbonate petroleum reservoirs: A global perspective on porosity-depth and porosity-permeability relationships *AAPG Bulletin* **89** 435–45.
[21] Alekseyev G I, Andreyev V N, Gorelov A A and Kaz'min L L 1982 *Methods of studying the compaction of terrigenous rocks in paleogeological reconstructions* (Moscow: Nedra) p 144
[22] Burst J F 1969 Diagenesis of Gulf Coast Clayey Sediments and Its Possible Relation to Petroleum Migration *AAPG Bulletin* **53** 73–93
[23] Perry E A and Hower J 1972 Late-stage dehydration in deeply buried politic sediments *AAPG Bulletin* **56**(10) 2013–21
[24] Matusevich V M and Bakuev O V 1986 Geodynamics of water-pressure systems of the West Siberian oil and gas-bearing megabasin *Soviet geology* **2** 117-22
[25] Matusevich V M and Ushatinsky I N 1998 Features of composition and formation of geofluidic systems of the West Siberian oil and gas-bearing megabasin *Higher Educational Institutions News. Oil and Gas* **4** 28-35

IOP Conf. Series: Earth and Environmental Science **193** (2018) 012051 doi:10.1088/1755-1315/193/1/012051

Paleohydrochemistry of Jurasic and Cretaceous deposits in arctic regions of Western Siberia

D A Novikov[1,2], Y V Sadykova[1], A V Chernykh[1,2], F F Dultsev[1] and A F Sukhorukova[1]

[1]Trofimuk Institute of Petroleum Geology and Geophysics, Novosibirsk, Russia
[2]Novosibirsk State University, Novosibirsk, Russia

E-mail: NovikovDA@ipgg.sbras.ru

Abstract. The Jurassic and Cretaceous paleohydrogeology of the Arctic regions of Western Siberia includes three hydrogeological cycles: Induan-Sinemurian, Pliensbachian-Cenomanian, Turonian-Serravallian. The paper presents results of the paleohydrological reconstructions of Jurassic and Cretaceous deposits. In the Triassic-Early Jurassic and Berriman-Cenomanian, the continental regime of sedimentation was prevalent within the study area, along with the processes of infiltration of fresh meteoric waters mainly hydrocarbonate calcium in composition with the TDS value not exceeding 5 g/dm^3. Marine sedimentation conditions were dominant during the Pliensbachian-Hauterivian and Turonian-Bartonian elision stages, when the processes of sedimentation were accompanied by syngenetic burial of sodium chloride waters, locally rich in magnesium, with the TDS value reaching 35 g/dm^3 in the deepest parts of the marine basin existing in the Volgian time. Modern chemistry of groundwaters of Jurassic and Cretaceous deposits in the Arctic regions of Western Siberia is a product of long-term geological evolution of the "water-rock-gas-OM" system.

1. Introduction

As a separate science, paleohydrogeology emerged early in the 20th century and studies the past stages in the development of sedimentary basins, aiming to clarify the role played by ancient groundwaters in the formation and preservation of mineral deposits, which was extensively discussed in the pioneer works of P.N. Chirvinskii. Later, this research line was largely contributed by many prominent scientists, among them L.A. Abukova, E.A. Baskov, E.A. Bars, S.B. Vagin, A.A. Kartsev, K.I., Makov, A.M. Nikanorov, E.V. Pinniker, A.N. Semikhatov, Ya. A. Semikhatov, Ya.A. Khodzhakuliev, S.A. Shagoyants, G.P. Yakobson and others. [1-3]. Given that paleohydrogeological studies are scant both in Russia and abroad, only few works devoted to sedimentary basins of Siberia, Eastern Pre-Caucasian, and the Pre-Kama Perm regions have been published recently [4-6].

Modern advanced-level scientific research requires aggregation of the latest data from many areas of the Earth Sciences, such as hydrogeology, lithology, stratigraphy, tectonics, etc. whose results and findings are helpful in solving many problems of modern hydrogeochemistry of oil and gas accumulations, including the composition of groundwaters, their origin, type of vertical hydrogeochemical zonality, the processes of petroleum generation and accumulation, and other aspects.

2. Materials and Methods

Paleohydrogeological studies of petroleum basins focus primarily on periodization of their hydrogeological history, consisting in determination of hydrogeological cycles and stages. According to A. A. Kartsev, S. B. Vagin and E. A. Baskov, a hydrogeological cycle begins with transgression and includes depositional processes with concomitant burial of sedimentary waters, spanning the period of subsequent elevation and regression, and culminates in a new submergence and transgression. The first part of hydrogelogocal cycle (from the start of transgression through the beginning of regression) is a sedimentary (elision) stage during which the formation of sedimentary waters and the elision water exchange take place. Its second part accounts for infiltration stage characterized by impoundment of waters of infiltration over the sedimentary basin progressively expulsing and replacing sedimentary waters [1].

Discrimination of elision cycles was largely based on the study of paleogeographic maps and lithofacies schemes, which provided the data on the predominantly marine regime of sedimentation. Infiltration cycles were determined from the data on the continental sedimentation regime existing in most of the study area, on the presence of hiatuses in sedimentation, inconformities, scours and basal layers of the overlying deposits [7-8]. Due to the inconsistencies between the boundaries of some cycles and those of large stratigraphic units (erathems, systems, divisions), their names were derived from their assignment to the generally accepted stages of the stratigraphic scale.

The restoration technique for the salt composition of waters from the ancient Mesozoic marine and lacustrine-alluvial basins is based on paleogeographic reconstructions and comparative-lithological analysis using natural-historical approach, which, in itself, is reduced to the fact that the formation of the ion-salt composition of waters in the continental and marine environments had common controls and are the product of identical processes occurring both in the past geological epochs and at the present time [9]. The paleohydrological reconstructions were based on the method for restoration of salt composition of waters of ancient Mesozoic marine and lacustrine-alluvial basins; by studies of paleogeographic maps of the Jurassic and Cretaceous periods compiled by the IPGG SB RAS researchers [6, 9], hydrogeological databank available at the IPGG SB RAS Laboratory for Hydrogeology of sedimentary basins of Siberia for the Arctic regions of Western Siberia, resulted from testing more than 3000 objects of 211 target areas, including the results of full chemical analysis of 3205 groundwater samples.

3. Hydroceochemical signatures

Groundwaters with the TDS value varying from 2 to 63.3 g/dm^3 and having different chemical composition were revealed within Jurassic and Cretaceous deposits of the Arctic regions of Western Siberia. The prevailing are Na Cl, Na Cl-HCO_3 and Na HCO_3-Cl water types (after Shchukarev S. A.), with each having its own patterns of main salt - forming macro-and micro components distribution, whose concentrations are dictated by their salinity levels: its growth directly entails an increase in Cl, Na, Mg, Ca, K amounts, as well as in micro components (Br, I, B, NH4 and Sr). The amounts of hydrogencarbonate (bicarbonate) ions tend to decrease starting from water salinity 15-20 g/dm^3 and more. The sulfate-ion concentrations do not exceed on average 20 – 60 mg/dm^3 , which is associated with the remarkable process of its reduction to hydrogen sulfide: $SO_4^{2-}+2C_{org}+2H_2O=2HCO_3^-+H_2S$.

The aquifer complexes (Aptian-Albian-Cenomanian, Neocomian, Upper Jurassic and Lower-Middle Jurassic) are marked by a number of hydrogeochemical signatures derived from the paleohydrogeological data [6], by subsequent processes of the groundwater chemistry metamorphization, as well as by the vertical migration of brines from Paleozoic basement to the overlying sedimentary cover in the areas of missing thick Triassic volcanic-sedimentary strata, acting as a desalinated "hydrogeochemical cushion" [10]. The anomalies in the hydrogeochemical field revealed in a number of areas in central areas of the study area accompany the process of inter-strata crossflows from the underlying sediments [11]. The diversity of groundwater composition observed at the base of the sedimentary cover generally decreases up the section, exhibiting significant smoothing of hydrogeochemical conditions and leveling of the groundwater salinity in the range from 15 to 20 g/dm^3 in the Aptian-Albian-Cenomanian complex [10-13]. This can be exemplified by the considered below features of hydrogeochemistry of the upper Jurassic aquifer complex whose water salinity was found to be the highest.

Groundwaters of Upper Jurassic sediments are characterized by varied chemical composition with dominance of Na Cl, Na Cl-HCO_3 and Na HCO_3-Cl water types whose TDS values vary in a wide range-from 5 to 63.3 g/dm^3 [10-16]. With increasing salinity, the chemical type of water changes from sodium bicarbonate chloride to sodium chloride (figure 1).

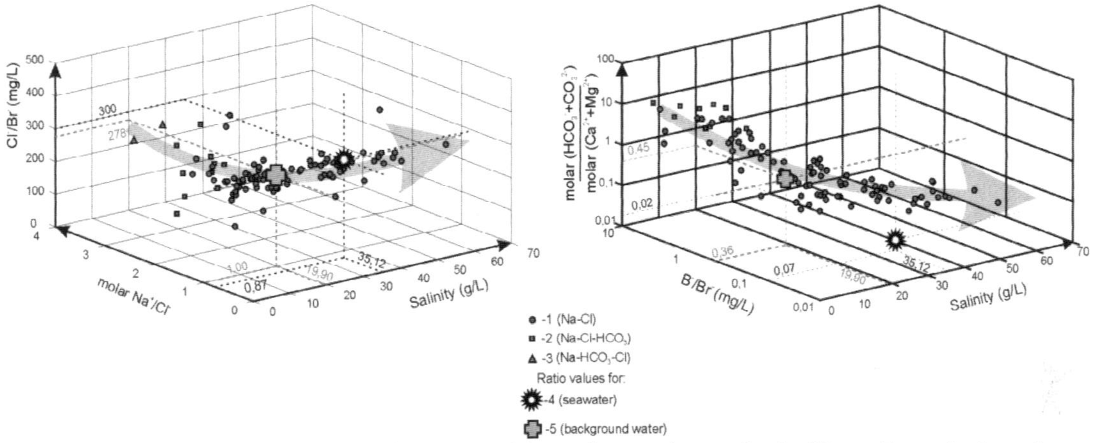

Figure 1. Change in the degree of metamorphism of groundwater in the Upper Jurassic deposits.

The diversity of waters with respect to chemical composition tends to decrease, as their salinity increases. The seven types allocated in water salinities ranging from 2 to 5 g/dm^3 are distributed among the following groups: 10-15 g/dm^3 – five, at 20-25 g/dm^3 – two, with TDS value more than 25 g/dm^3 – only one (sodium chloride) type. The background salinity value for groundwaters of the upper Jurassic aquifer complex in the studied region averages 19.9 g/dm^3. The area comprising the Severny (Northern) and Nizhnevartovsk arches, East Pur megamonocline and South Nadym megamonoclise is characterized distribution of groundwaters with salinities exceeding 25 g/dm^3. The highest values of the zonal hydrogeochemical background are accounted for the Severny arch structures where TDS of the background formation waters reaches 33.5 g/dm^3

Jurassic and Cretaceous paleohydrogeology and the existing paleohydrogeocheical conditions of the study area are considered below in detail.

4. Results and Discussion

The periodization of hydrogeological history revealed three hydrological cycles in the Jurassic and Cretaceous: Induan-Sinemurian, Pliensbachian-Cenomanian, Turonian-Serravallian. Hydrogeological history of the sedimentary cover starts with the Induan-Sinemurian cycle. The onset of sedimentation is associated with the active volcanic activity and the formation of volcanic-sedimentary (Turin series) strata [17]. In the Triassic-Early Jurassic, Western Siberia was an uplifted, slightly dissected, gently elevated land area with the dominance of the processes of infiltration of fresh atmospheric water.

In the earliest Pliensbachian-Cenomanian hydrogeological cycle, the sea moved inland from the north-west and north-east. The marine basin was shallow (25 m) with desalinated water, while the Gydan and Yamal Peninsulas were represented by the shoal zone, where significant fluctuations of water salinity from 2 to 15 g/dm^3 resulted from the unstable marine regime and river discharge. This is also corroborated by a paucity of the organic world (ammonites and brachiopods were not abundant, while pelecypods proved to be numerous, which is characteristic of the coastal areas), to some extent absorbed by clay complex (low K and Mg and high Na and Ca contents) [2, 18]. The southern part of the area was represented by a coastal plain, occasionally flooded by the sea and low-lying alluvial plain located on the site of present-day Bolshaya Kheta megasyneclise. Owing to the "mobility conditions" of the moving coastline there, coastal-marine and continental sediments were formed with the remains of leaf flora, plant detritus, accompanied by burial of predominantly Na Cl waters with high content HCO_3^- and Ca^{2+} and salinity values of 2-5 g/dm^3.

As the basin gradually shallowed in the southern direction, and the type of depositional settings changed, accordingly, from coastal-marine to continental (low-lying plain of accumulation), where processes of infiltration of fresh atmospheric water (from 0.5 to 2 g/dm^3) were predominant. The basin boards represented by themselves an elevated denudation plain, grading into the low-mountain relief (the Taimyr, Novaya Zemlya, Ural and Central Siberian uplands), acting as the provenance areas of clastic material and exterior source-areas of infiltration, within which Ca and HCO_3 waters salinity not exceeding 0.5 g/dm^3 were buried. The largest transgression of the sea basin over the time spanning the Early and Middle Jurassic is assigned to the Toarcian. The marine sedimentation zone has significantly extended within the study area, causing the basin deepening to a depth of 100 meters in the central part, with water salinity probably reaching 15-20 g/dm^3. This area encompassed the territory of the contemporary Kara, Antipayuta, Tadebeyakha and Bolshaya Kheta megasyneclises and the Agapa- Yenisei trench. The fine-dispersed dark grey and black clayey silts accumulated in the deep-water part of the Kiterbyut Formation, which evolved into the source-rock. The finds of marine fauna (rostra of belemnites, bivalve, foraminifera) indicate the established normal marine sedimentation regime in the region [7].

The shallow-water area (< 25 m) rimmed the marine basin along its entire periphery. Less saline (2-15 g/dm^3) NaCl waters with elevated amounts of HCO_3^- и Ca^{2+} ions were buried within its limits, in parallel with the processes of sedimentation. The second half of the Toarcian – the beginning of Aalenian is characterized by short-term, but extensive regression of the boreal seas, which led to the shallowing and desalination of the West Siberian marine basin. The area of marine sedimentation (25-100 m) in the north-west has declined due to the expansion of the shallow (< 25 m) water zone. In the second half of Aalen, the climate gradually turned to cool and humid, causing vegetation to perish, along with the wide distribution of carbonaceous strata and the appearance of Arctic genera of ammonites and belemnites in the basin [18].

The beginning of Baiosian was marked by short-term sea-level rises leaving no significant impact on the position of the paleogeographic regions. Despite the fact that in the Bathonian, the subsidence of the Arctic regions of Western Siberia continued, the position of zones of marine sedimentation practically remained unchanged, however the periods of sea ingression extended. In the north of the region, the extent of shoaling has increased due to the flooding of the low plain of accumulation rimming the basin boards. Callovian age is characterized by a continuation of the extensive marine transgression which resulted in further deepening of the marine basin occupying almost the entire area of the West Siberian geosyneclise, except its southern margins. At that time, clayey beds of the Golchikha and Yakovlevka Formations accumulated in the deep-sea environments, which subsequently evolved into the regional seals. The sea depth in the most submerged part reached 400 m; the deep-water area located in the central part of the region encompassed the area of the present-day Kara and Antipayuta-Tadebeyakha megasyneclises.

In this part of the marine basin, water salinity might reach 40 g/dm^3, which corresponds to average ocean salinity and is evidenced by the composition of authigenic minerals and a wide variety of fauna (bivalves, foraminifera, cephalopods and gastropods. This area was rimmed by shallower waters, where accumulated fine, substantially clayey sediments with layers of calcareous clays and glauconite with abundant fauna [18].This portion of the basin is featured by salinity was in the range 20-30 g/dm^3 with the increasing role of Mg in the progressively buried seawater. A small area of the shoal zone with a sea depth up to 25 meters and up to 80 km in width, where developed different sediments facies and Na Cl waters with high concentration of HCO_3^- и Ca^{2+} ions and salinity in the range from 5 to 15 g/dm^3 were buried. Accordingly, the area of continental sedimentation zones has significantly reduced. At the turn of the Callovian and Oxfordian, the transgression was succeeded by regression which reached its maximum in the middle of Oxfordian. Given that the basin was shoaling, thick sand layers

IOP Conf. Series: Earth and Environmental Science **193** (2018) 012051 doi:10.1088/1755-1315/193/1/012051

began to accumulate in its central and southern parts, forming the known today Oxfordian regional reservoir. The northern areas were still dominated by the marine environments, providing for the formation of predominantly clayey sediments of the Golchikha Formation, rich in organic matter and containing abundant remains of diverse marine fauna. Since almost the entire study area was represented by a zone with depths between 25 and 200 m, salinity levels of the waters that experienced burial were, accordingly, close to 30 g/dm^3 (figure 2). Given that the Jurassic transgression was the lengthiest event, the extent of the marine sedimentation have widened, and the depth in most of the area reached 200 m in the Late Oxfordian-Kimmeridgian time [7]. In the Volgian time (Tethonian-Early Berriasian) the dipping of the basin proceeded intensely, with the transgression reaching its maximum in the middle of the century. Climate in the Volgian was close to semiarid, causing thereby the dominance of chemical weathering conditions in the onshore areas and the accumulation of mainly clay and organogenic rocks on the continental shelf. In view of relief peneplanation, biogenic sedimentation was dominating, with clay sediments rich in organic matter, calcium, and silicon forming ubiquitously in the area (Bazhenov, Golchikha, Yanov Stan, Danilovka, Tutleima Formations) [18].

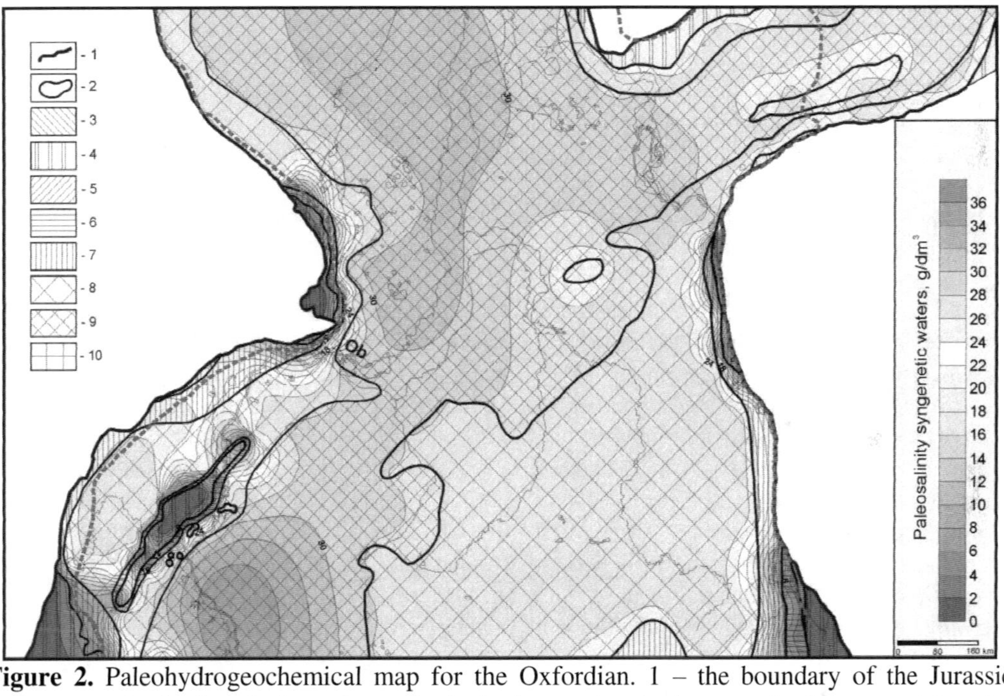

Figure 2. Paleohydrogeochemical map for the Oxfordian. 1 – the boundary of the Jurassic sedimentary basin; 2 – the boundary of paleohydrogeochemical zones: 3 – low mountains, the processes of infiltration of fresh meteorogenic waters with mineralization up to 0.5 g/dm^3 and with the predominance of HCO_3^- and Ca^{2+} ions predominate;3 – elevated plain, fresh and brackish infiltrogenic waters with a mineralization of 0.5-1.5 g/dm^3 with predominance in the composition of HCO_3^- and Ca^{2+} ions with an increased content of Cl^- and Na^+; 4 – denudation-accumulative plain, fresh and brackish infiltrogenic waters with a mineralization of 1.5-2.0 g/dm^3 with predominance in the composition of HCO_3^- and Ca^{2+} ions with an increased content of Cl^- and Na^+; 5 – low accumulative plain, burial of brackish water with mineralization 2-5 g / dm3 and predominance in the composition of Cl^- and Na^+ ions of increased content of HCO_3^- and Ca^{2+}; 6 – coastal plain, syngenetic brackish waters with a mineralization of 5-15 g/dm^3 with a predominance in the Cl^- and Na^+ ions with an increased content of Mg^{2+} and Ca^{2+} ions; 7 – shallow part of the shelf, coastal zone and island part of the shelf up to 25 m, saline thalassogenic waters with mineralization of 15.0-20.0 g/dm^3 with a predominance of Cl^- and

Na^+ ions with an increased content of Mg^{2+} ions; 8 – sea, 25-200 m, salty thalassogenic waters with mineralization 20.0-30.0 g/dm^3 with a predominance of Cl^- and Na^+ ions with an increased content of Mg^{2+} ions; 9 – deep sea 200-400 m, salty thalassogenic waters with salinity of 35.0 g/dm^3 with a predominance in the Cl^- and Na^+ ions with an increased content of Mg^{2+} ions.

Thus, by the end of the Volgian time, a deep basin with signs of hydrogen sulfide exposure had formed in Arctic region of Western Siberia. Almost the entire study area was flooded by the sea with depths of 200-400 m, where water salinity reached 35 g/dm^3. In addition, two deep depressions were distinguished within the borders of the Kara and the Bolshaya Kheta megasyneclises, where depths exceeded 400 meters, assumingly reaching 600 m where water salinity, hypothetically, could be 35 g/dm^3. The zone was surrounded by a narrow strip of shallower part of the basin (100-200 m), where marine sedimentation was complexed with the burial of saline thalassogenic waters featured by salinity 20-30 g/dm^3 and predominance of Cl^- и Na^+ ions in the composition, and enhanced Mg^{2+} ion concentrations. Delineation of the coastal zones has practically not changed: the coastal plain was almost entirely flooded by the sea and marked only by small narrow strip along the eastern board of the basin, while the alluvial plain existed only in the coastal part adjacent to the Taimyr upland. The basin coasts became flatter and peneplenized, representing by themselves elevated denudation land area. In addition, a series of peninsulas was discriminated along the eastern board. The burial of fresh atmogenic, slightly saline Ca HCO_3 waters with salinity not exceeding 1-2 g/dm^3 was taking place within the continental sedimentation zones.

A small in size deep-water zone was preserved in western Gydan Peninsula. Eastwardly, the shallow water area, which occupied almost half of the territory, has increased significantly. An extensive coastal plain formed in the south, where flood-plain, deltaic and beach depositional settings alternated depending on the sea level, and brackish waters with salinity 2-5 g/dm^3, Na and Cl in composition, with high concentration of HCO_3^- and Ca^{2+}ions were buried. In the period spanning from the Barremian to the end of the Cenomanian, the study area was characterized by alternation of sedimentation conditions between low-lying accumulation plane and coastal plain (Kontorovich et al., 2014), where infiltration of meteogenic Ca HCO_3 water with salinity from 0.5 to 2 g/dm^3 were mixed with sedimentary waters in the area of active water exchange, partially diluting and desalinating them. Later transgression, which began in the Turonian period, resulted in the formation of a vast shoal zone, covering the entire study area through the Late Campanian. Salinity of sea waters could reach 15 g/dm^3 in the deepest parts of the South-Western sea. In the Late Campanian, regression of the marine basin was accompanied by warming of the West Siberian sea, resulting in the deposition of calcareous sediments and causing an increase in the extent of the coastal plain. At this time, the composition of syngenetic waters was dominated by the infiltration-driven Ca HCO_3 with high concentration of Cl^- and Na^+ ions and salinity reaching 2-5 g/dm^3 in the coastal-continental environments. The following – Turonian-Serravallian – hydrogeological cycle commenced in the earliest Turonian. In Arctic regions of the West Siberian basin, shallow marine sedimentation conditions (up to 100 m) prevailed, the deepest part of the basin was situated in the western half of the study area. The shallow-water portion of the basin was rimmed by the low-lying accumulation and coastal plains from the east, and elevated – from the west. In the Late Campanian-Danish, the regression resulted in the marine basin shoaling (depth: < 25 m). Most of the region still comprises two shallow water zones, while a coastal plain, occasionally flooded by the sea developed within the eastern margin.

5. Conclusions

It follows from the above that the Jurassic-Cretaceous marine sedimentation regime dominated in Arctic regions of Western Siberia until the end of the Hauterivian and throughout almost entire Upper Cretaceous. Regular sea level fluctuations caused minor transgressions and regressions, the basin is found to have been the deepest in the Volgian time. The chemical composition of the Arctic basin bears no indications of any significant changes at the Triassic-Jurassic boundary. Systematic decreases were reported for potassium and sodium concentrations while the levels of calcium and magnesium tended to increase. Driven by the basin deepening, a significant increase in salinity occurred at the turn of the Middle - Late Jurassic, and desalination – at the end of the Early Cretaceous. At that time, salinity of syngenetic waters might reach 35 g/dm^3. A detailed analysis of modern hydrogeochemistry of petroleum-bearing deposits with account of paleohydrochemical reconstructions allowed to establish the dominance of sedimentary waters in the section, sometimes diluted by ancient infiltrates that penetrated during the marine basin regressions, which agrees well with the results of the paleogeographic constructions for the Jurassic and Cretaceous [7-8]. Local distribution was reported from condensate-prone waters of common origin with hydrocarbon accumulations. The "water-rock-gas-organic matter" system plays a critical role in the processes of the formation of chemical composition of groundwaters within petroleum-bearing deposits.

Acknowledgments
The research was supported by the Russian Foundation for Basic Research within Scientific Project No. 18-05-70074 «The Arctic Resources».

References

[1] Kartsev A, Vagin S and Baskov E 1969 *Paleohydrogeology* (Moscow: Nedra) p 152
[2] Gramberg I 1973 *Paleohydrogeochemistry of Terrigenous Rock Mass* (Leningrad: Nedra) p 172
[3] Khodzhakuliev Y and Abukova L 1985 *Paleohydrogeological Research in the Search for Oil and Gas* (Moscow: Nedra) p 209
[4] Shestov I, Tyurina I M and Shuvalov V M 2008 *Bulletin of Perm State University* **10** 16-9
[5] Hendry J M, Barbour S L, Novakowski K and Wassenaar L I 2013 *Water Resources Research* **49** 4580-92
[6] Sadykova Ya V and Dultseva M G 2017 *Water Resources* **44** 168-81
[7] Kontorovich A E et al. 2013 *Russian Geology and Geophysics* **54** 972-1012
[8] Kontorovich A E et al. 2014 *Russian Geology and Geophysics* **55** 745-76
[9] Baskov E 1983 *Fundamentals of Paleohydrogeology of Ore Deposits* (Leningrad: Nedra) p 263
[10] Shvartsev S L and Novikov D A 2004 *Geologiya i Geofizika* **45** 1008-20
[11] Novikov D A and Lepokurov A V 2005 *Geologiya Nefti i Gaza* **5** 24-33
[12] Novikov D A 2017 *Geodynamics and Tectonophysics* **8** 881-901
[13] Novikov D A 2018 *Neftyanoe Khozyaystvo - Oil Industry* **4** 16-21
[14] Kokh A A and Novikov D A 2014 *Water Resources* **41** 375-86
[15] Novikov D A and Sukhorukova A F 2015 *Arabian Journal of Geosciences* **8** 8703-19
[16] Novikov D A 2017 *Petroleum Exploration and Development* **44** 780-8
[17] Kontorovich V A, Belyaev S Yu, Kontorovich A E, Krasavchikov V O, Kontorovich A A and Suprunenko O I 2001 *Geologiya i Geofizika* **42** 1832-45
[18] Shurygin B N et al. 2000 *Stratigraphy of Oil and Gas Bearing Basins of Siberia. The Jurassic System* (Novosibirsk: SB RAS Publishing House, GEO branch) p 480

IOP Conf. Series: Earth and Environmental Science **193** (2018) 012052 doi:10.1088/1755-1315/193/1/012052

Geochemistry of brines in Vendian deposits of the Siberian platform

D A Novikov[1,2], A V Chernykh[1,2] and F F Dultsev[1]

[1] Trofimuk Institute of Petroleum Geology and Geophysics SB RAS, Novosibirsk, Russia,
[2] Novosibirsk State University, Novosibirsk, Russia

E-mail: NovikovDA@ipgg.sbras.ru

Abstract. Results of the geochemical studies on natural brines from regional Vendian reservoirs (Lower-Danilov, Tira, Nepa and Vilyuchan) of the Siberian platform are presented. TDS values of the studied groundwaters and brines range from 17.3 to 583.1 g/dm^3, where brines with salinity 280 – 400 g/dm^3 are predominant. Brines identified within the Ayan, Bratsk, Balaganka, Upper Chona, Shamanka and other areas, have the highest TDS values.The most widely spread brines belong to five chemical types of Cl Na to Cl Ca composition. TDS concentrations vary widely from 197 to 580 g/dm^3 in the Lower Danilov and Tira horizons, and from 170 to 470 g/dm^3 in the lower-sitting Nepa and Vilyuchan horizons. Na Cl brines whose TDS concentration reaches the halite precipitation phase should be assigned to infiltration-formed brines on account of salt rock leaching, while strongly metamorphosed brines (S > 300) mostly of Cl and Ca Cl Ca-Mg composition – to sedimentogenic type accounting for those descended from Cambrian salt-bearing deposits, thereby largely explaining the phenomenon of inverse hydrogeochemical zonality in subsalt horizons. The third genetic type, of Cl Ca-Na composition, is rooted in the processes of mixing sodium and calcium types of brines.

1. Introduction

Arctic sedimentary basins are of great interest as a region with high hydrocarbon potential. At the present, commercial petroleum pools have been found in many basins worldwide: Siberian platform, Athabasca, Beaufort-Mackenzie, West Siberian, Sverdrup and others [1-5]. The Siberian platform is unique in terms of its hydrogeological structure and geothermal, hydrodynamic and hydrogeochemical conditions, which have been amply discussed in the works of A. S. Antsiferov, E. A. Baskov, M. B. Bukaty, M. G. Valyashko, V. I. Vozhov, G. D. Ginsburg, V. V. Pavlenko, E. V. Pinneker, E. V. Stadnik, N. I. Tolstikhin, and many others [5-17].

Subsurface brines of the Siberian platform are characterized by high levels of salinity (up to 650 g/dm^3) and metamorphization, as well as by unique concentrations of numerous micro-constituents (Li, Rb, Sr, Br, B, I etc.). In recent years, the interest to the study area has been fueled by Gazprom company's works seeking for justification of the discoveries of giant gas fields in the Republic of

Sakha (Yakutia) area, and by the construction of the largest "Power of Siberia" (Sila Sibiri) and "ESPO - Pacific" gas- and oil-pipeline systems, and ongoing prospecting works in order to keep these facilities in operation ensured by sufficient hydrocarbon resource base.

Notably, the extent of hydrogeological study of the sedimentary cover of the Siberian platform is extremely varied, with terrigenous Vendian deposits being no exception, although they account for more than 50% of the total in-place gas resources and about 18% of total reserves and resources of crude oil [18]. The existing literature overviews fail to provide a comprehensive analysis of the currently available drilling data and results of geochemical testing of the regional horizons (top – down): Lower- Danilov, Tira, Nepa, and Vilyuchan.

2. Materials and Methods

The hydrogeochemical materials, which served as the basis for this work are the results of numerous dedicated research conducted over a long period of time. The electronic database includes results of a comprehensive chemical analysis (including micro constituents) of 809 conditioned samples, results of testing more than 1,237 objects within 101 prospecting areas. Based on the methodological approaches and chemical classifications after V.A. Sulin, A.A. Kartsev, S.L. Shvartsev, V.E. Pinneker, A.S. Antsiferov, E.A. Baskov, M.B. Bukaty, M.G., Valyashko, V.I. Vozhov and other researchers, geochemical signatures of subsurface brines have been thoroughly investigated.

The chemical types of brines were determined according to the S. A. Shchukarev's classification proposed in 1934 using the Kurlov formula. Unlike the latter, the component which is present in the amount of at least 25 equiv.% is considered to be predominant. Of the six main ions, by grouping them by two, three, four, five and six, S. A. Shchukarev identified as many as 49 classes of natural waters. These classes are represented in the form of a square whose sides are divided into seven equal parts. Of them, Shchukarev interprets Ca HCO_3, Na Cl и Ca SO_4 as staple, since their waters formed as a result of direct leaching of rocks. All other classes are a product of water mixing, metamorphization of their salt composition and cationic exchange [19]. Using the rNa/rCl, Cl/Br, Ca/Cl, $r(HCO_3+CO_3)/r(Ca+Mg)$, B/Br, $Br/Cl \cdot 10^{-3}$, $Sr/Cl \cdot 10^{-3}$ relationships, other indications of the origin of subsurface brines have been inferred. The degree of their metamorphization was determined from the integrated index of brine metamorphization (S) after S. L. Shvartsev [9] and the Ca/Cl, $Br/Cl \cdot 10^{-3}$, $Sr/Cl \cdot 10^{-3}$ relationships.

3. Results and Discussion

Subsurface waters and brines with TDS from 17.3 to 583.1 g/dm^3 are established in the Vendian reservoirs (table 1). In terms of the cationic composition, the investigated brines vary widely from sodium chloride to chloride with a predominance of calcium and magnesium cations (figure 1).

Despite the hydrogeochemical field varying significantly locally within Vendian deposits, the lateral zoning of the brine composition is clearly observed, particularly, in the zone of the Siberian platform joint with adjacent folded structures, where distribution of sodium brines is assumed to have resulted from leaching of rock salt. As the distance from the platform boundary increases, this zone is replaced by a strip of brines of mixed cationic, sodium-dominated composition, which are subsequently replaced by calcium-dominated brines. The calcium-magnesium brines are distributed locally, in the Nepa and Vilyuchan horizons within the Kupskaya area in the north of the Angara-Lena structural step, southern part of the Siberian platform. Brines from the Tira and Lower Danilov horizons tend to be of chloride type with their composition predominated by magnesium and calcium cations, revealed in brines within the Kupskaya Gruznovka, Bolshaya Tira and Ozernaya areas. The Nepa and Vilyuchan horizons are characterized by brines with the TDS concentrations from 170 g/dm^3 to 470 g/dm^3 (Figure 2). The Preobrazhenskaya, Ozernaya, Dulisma, Balagankino and Verkhne-Chona areas the TDS values reaches its maximum 430-470 g/dm^3. Brines with salinity up to 200 g/dm^3 are established on Omorinsk and South Solnechnaya areas. The overlying Tira and Lower-Danilov horizons are featured by distribution of brines with mineralization from 197 to 583 g/dm^3. The ultra-

IOP Conf. Series: Earth and Environmental Science **193** (2018) 012052 doi:10.1088/1755-1315/193/1/012052

high salinity brines are assigned to Ayan, Bratsk, Upper-Chona, Lower Uda, Ognevka, Ozerny, Preobrazhenka, Soba, Talakan and Shamanovka areas (table 1).

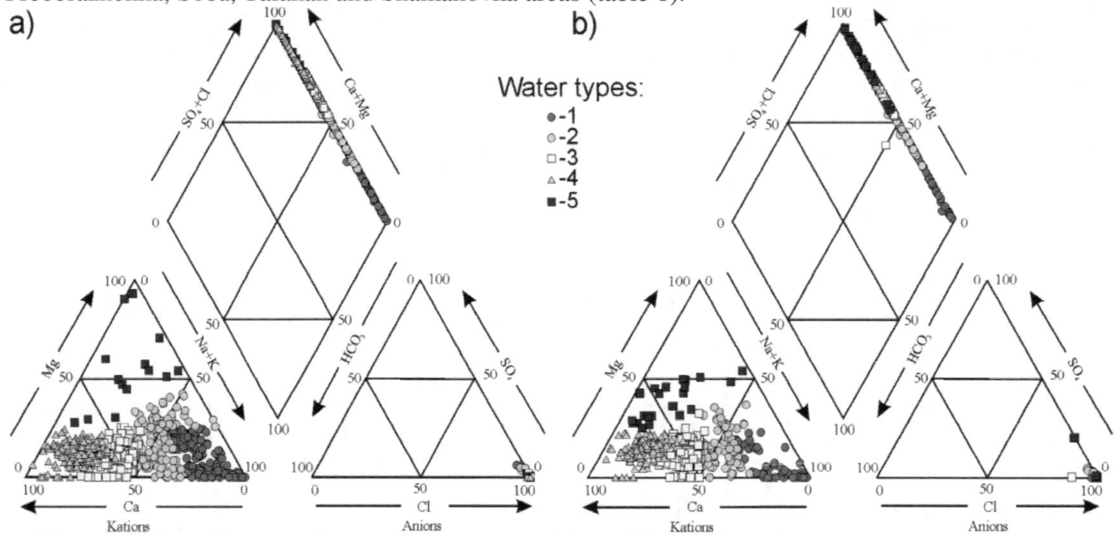

Figure 1. Piper diagram of groundwater composition of the regional Vendian reservoirs: a) Nepa and Vilyuchan and b) Lower-Danilov and Tira.

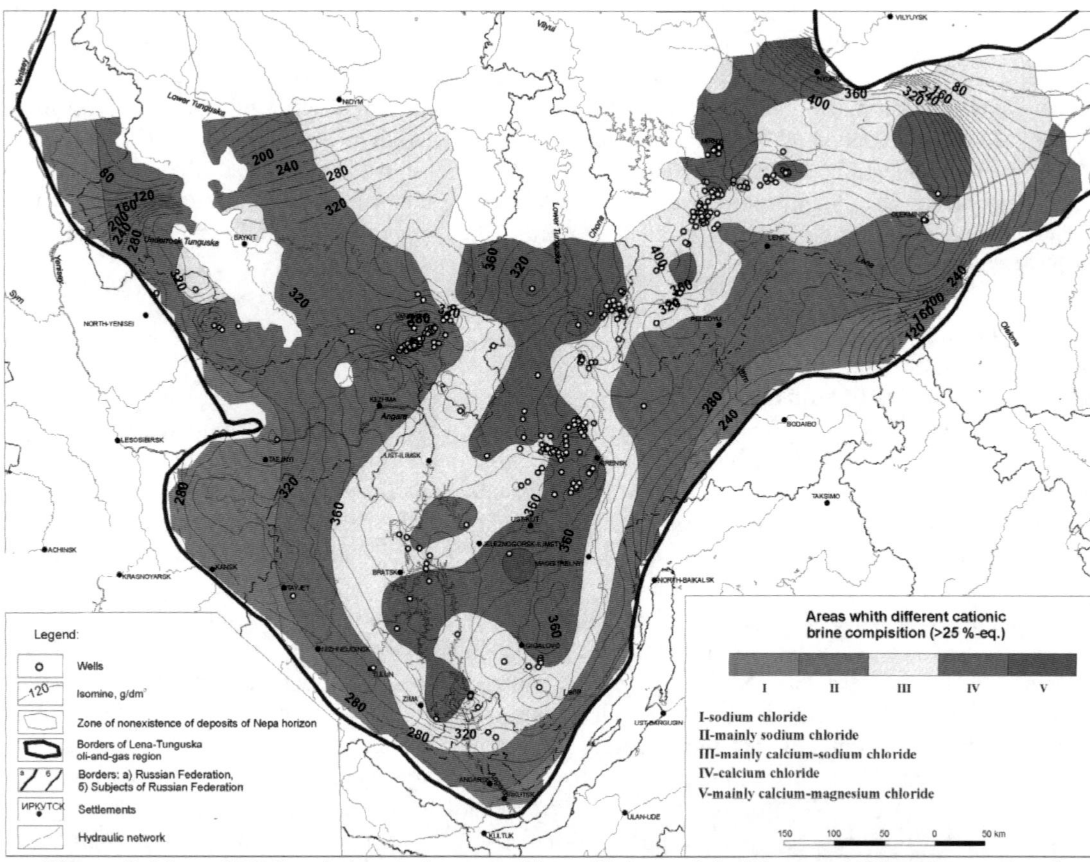

Figure 2. Map of TDS and chemical composition of brines of the Nepa and Vilyuchan regional Vendian reservoirs.

Table 1. Hydrogeological characteristics of regional reservoirs of Vendian deposits.

Index	Unit	Regional reservoirs			
		Lower Danilov	Tira	Nepa	Vilyuchan
Constitution of underground water[a]					
HCO_3^-	mg/dm^3	4-1819 (351,2)	2-1830 (229,3)	6-2440 (262,3)	12-290 (101,5)
SO_4^{2-}	- " -	87-3827 (964,8)	0,1-8723 (572,8)	5-16000 (648,5)	0,4-415 (141,8)
Cl^-	g/dm^3	14-267 (194,8)	10,2-410,6 (203,3)	5,9-291,2 (199,1)	173,7-257,7 (220,
Br^-	- " -	0,09-10,73 (2,7)	0,004-11,1 (3,6)	0,003-10,7 (3,6)	0,2-6,2 (3,1)
I^-	mg/dm^3	0,2-63 (12,3)	0,1-316 (29)	0,1-2200 (40,8)	2,5-105,5 (29,8)
F^-	- " -	1,9-11 (6,4)	0,4-19 (9,1)	0,3-125 (18,2)	-
Na^+	g/dm^3	8,3-125,4 (60,7)	2-125,4 (47,8)	1-146,4 (45,7)	12,2-85,6 (36,9)
Ca^{2+}	- " -	0,7-120,2 (45,9)	0,7-120,2 (55,7)	0,6-130,4 (56,4)	22-104 (82,9)
Mg^{2+}	- " -	0,1-41,5 (6,5)	0,1-66,2 (10,5)	0,1-42,3 (9,9)	0,6-15,8 (6,4)
K^+	- " -	0,1-21,3 (7,2)	0,1-45,4 (6,7)	0,1-25 (5,9)	0,2-42 (4,6)
Sr^{2+}	- " -	0,2-3,8 (1,2)	0,01-6,8 (2,1)	0,01-7,2 (1,7)	0,5-4 (2,2)
Li^+	mg/dm^3	1,4-72 (28,6)	0,5-439 (57,2)	1-168 (42,1)	0,9-38 (19,4)
Rb^+	- " -	0,4-59 (7,1)	0,1-98 (1,3)	0,05-130 (7,8)	1,6-10 (2,9)
Cs^+	- " -	0,3-0,6 (0,4)	0,3-2 (0,7)	0,01-3 (0,2)	-
Cu^{2+}	- " -	0,2-2 (0,8)	0,08-12,9 (1,4)	0,03-42 (1,1)	1-2 (1,7)
Zn^{2+}	- " -	0,87-5,8 (1,3)	0,2-1100 (55,1)	0,002-400 (29,3)	2,7-98 (30)
Pb^{2+}	- " -	16	0,5-16 (1,1)	0,06-11,5 (1,5)	2
Mn^{2+}	- " -	6-286 (72,3)	2-994 (127,2)	0,01-1000 (100)	72-210 (130,3)
NH_4^+	- " -	0,04-1025 (255,4)	0,03-1183 (249,8)	0,03-7500 (231,7)	24-313 (118,2)
SiO_2	- " -	2,6-20 (14,7)	0,2-66 (25,1)	0,3-102 (23,2)	95
Al^+	- " -	0,8-10 (3,3)	0,02-120 (7,9)	0,02-23 (4,3)	640
B^+	- " -	6-160 (41,1)	0,03-200 (41,9)	0,01-583 (39)	4,9-36 (12,5)
Fe общ.	- " -	0,9-450 (60,1)	0,1-1746 (198,8)	0,1-3000 (155,4)	0,3-380 (131,5)
TDS	g/dm^3	196-424,5 (316,5)	24,0-583,1 (325,4)	17,3-470,1 (318,6)	278,9-398,8 (350,8)
rNa/rCl	decimal quantity	0,07-1,07 (0,52)	0,01-1,01 (0,39)	0,01-1,55 (0,39)	0,08-0,72 (0,27)
Cl/Br	- " -	30-1176 (194)	23-910 (202)	18-939 (659)	39-755 (114)
Ca/Cl	- " -	0,01-0,53 (0,22)	0,01-0,74 (0,26)	0,01-0,74 (0,26)	0,12-0,44 (0,37)
$Sr/Cl \cdot 10^3$	- " -	1,1-15,7 (5,7)	0,01-137,4 (11,1)	0,01-137,2 (8,8)	2,9-17,6 (10,3)
$Br/Cl \cdot 10^3$	- " -	0,02-32,5 (12,7)	0,01-43,1 (16,5)	0,01-993,3 (18,9)	0,9-25,2 (13,9)
S	- " -	6-527 (189)	6-842 (249)	7-842 (236)	117-501 (270)
Composition of waters (S.A. Shchukarev)		Na Cl, Na-Ca Cl, Ca Cl, Ca-Na Cl, Mg-Ca Cl	Na Cl, Na-Ca Cl, Na-Ca-Mg Cl, Na-Mg Cl, Na-Mg-Ca Cl, Ca Cl, Ca-Na Cl, Ca-Na-Mg Cl, Ca-Mg Cl, Mg Cl, Mg-Ca Cl, Mg-Na Cl		Ca Cl, Ca-Na Cl

[a] min-max (avarage).

Statistical analysis of hydrogeochemical data allowed to reveal specific characteristics of both hydrogeochemical background and anomalies. The background brines of the Vendian horizons are predominantly of chloride calcium-sodium composition with the TDS concentration averaging 323 g/dm^3. The amount of HCO_3^- does not exceed 0.25 g/dm^3, SO_4^{2-} is 0.6 g/dm^3, Na^+ – 45.1 g/dm^3, Ca^{2+} - 58.3 g/dm^3, Mg^{2+} – 10.1 g/dm^3, K^+ – 6.1 g/dm^3; Sr^{2+} is not more than 1.84 g/dm^3, Br^- is not more than 3.65 g/dm^3, NH_4^+ is not more than 0.23 g/dm^3, etc.

Thirteen types of hydrogeochemical anomalies in the amounts of I, Br, B, SiO_2, Rb, Sr, Zn, Cu, Mn, Li, NH_4, Al, Cs were identified and delineated within the Vendian reservoirs. Thus, among the "traditional" micro-components, the highest concentrations of iodine > 0.03 g/dm^3 are established at Elokhta, Payga, Soba, Middle Botuoba, Taas-Yurakh, Byuk-Tanar and other areas; bromine (Br) in amounts > 3,6 g/dm^3 within the Ayan, Bratsk, Upper Dzhalinda, Upper Chona, Dulisma, Krivaya Luka, Markovo and other areas; boron (B) in amounts > 0,04 g/dm^3 was encountered in brines within the Ayan, Bratsk, Vakunayka, Upper Lena, Imbinsky, Payga, Sedanka, Middle-Botuoba, Yurubchen and other areas.

With respect to the alkaline constituents, the concentrations of rubidium in excess of 8.7 mg/dm^3 were established within the Bratsk, Vanavar, Imbinsky, Peleduy, Sedanka, Shamanovka and other areas; and that of lithium >46,1 mg/dm^3 - within the Ayan, Balagankino, Upper Lena, Engida, Kovykta, Sedanka, Shamanovka and other areas. Among alkaline earth elements, anomalously high are found to be Sr concentrations (> 1.84 g/dm^3) observed in the brines within the Ayan, Bolsheokinskoye, Bratsk, Upper-Vilyuchan, Kasatkino, Krivaya Luka, Markovo, Shamanovka, Yarakta and other areas. The zinc-dominated zones (>43.0 mg/dm^3) subsume brines from the Bratsk, Elochta, Soba, Chamba and other areas. Manganese concentrations are found to be remarkable >112.1 mg/dm^3 in brines from the Balagankino, Bratsk, Buyk-Tanar, Upper Tira, Dulisma, Kiya, Krivaya Luka, Middle Botuoba, Yarakta and other areas. Anomalous values of ammonium cation (in excess of 232.0 mg/dm^3) were revealed at Ayan, Bratsk, Upper Katanga, Dulisma, Katanga, Krivaya Luka, Soba, Tutura, Yarskaya and other areas. The hydrogeological retrospective of the study area and comprehensive analysis of the hydrogeochemical data suggest the presence of three major genetic types of brines within the Vendian reservoirs: 1) predominantly infiltration-driven sodium brines, 2) sedimentogenic-ancient infiltration-formed calcium-sodium and, less commonly, magnesium-sodium brines, 3) predominantly sedimentogenic chloride sodium-calcium and calcium brines (figure 3, table 1).

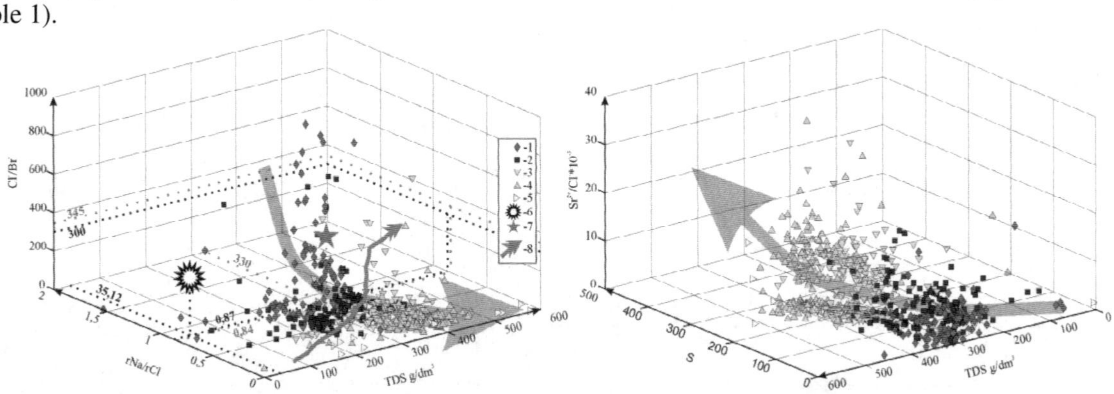

Figure 3. Variation of the metamorphization degree in subsurface brines of Vendian deposits within the Siberian platform, depending on the TDS concentrations.
The cationic composition of chloride brines: 1 – sodium, 2-with a predominance of sodium cations, 3-with a predominance of calcium and sodium cations, 4-calcium, 5-with predominance of calcium and magnesium cations; coefficient values for: 6-sea water, 7 – brines of the halite precipitation phase (after M. G. Valyashko); 8-McCaffrey curve [20].

The most metamorphosed, primarily, sedimentogenic chloride sodium-potassium and calcium brines were revealed near the southern boundary of the basin in the junction zone of the Pre-Sayany-Yenisei syneclise and the Angara-Lena structural step. They are represented by members of a uniformly continuous series featured by the increasing concentration of calcium and depletion of sodium. The TDS values of these brines vary from 344 to 583.1 g/dm^3.

The sodium-calcium to calcium brines ratio (both along the lateral and throughout the section) is governed by an increase in relative content of calcium and in mineralization of the enclosing rocks. Calcium brines with salinity in the range from 350 to 410 g/dm^3 appear to be dominating in the lower part of the sedimentary cover within the south-eastern margin of the Tunguska basin, delimited by the north-western slope of the Nepa-Botuoba anteclise. Brines of only sodium-calcium chloride composition are most common in the western part of the study region.

The distribution pattern of the second type of sodium chloride brines has a form of a narrow strip extending along the southwestern margin of the platform. Its composition and main genetic coefficients values show affinity to the modern evaporate-formed brines, however, at depths greater than a kilometer they bear evidence of the chemical composition metamorphization, specifically, a decrease in sulfate constituent, while relative concentrations of alkaline earth metals, bromine and other halogenic and halophilic micro constituents tends to increase. The studied sodium brines of Vendian deposits are therefore interpreted as transitional from leaching to metamorphosed brines whose TDS concentration does not exceed 440 g/dm^3, varying from 24 g/dm^3 within the Upper Chona area to 439.1 g/dm^3 within the Delikuton area. The salinity levels of the mixed sedimentogenic and ancient infiltration-formed calcium and sodium and, less commonly, magnesium-sodium brines, studied within the Kuyumba, Yurubchen, Taiga, Soba and other areas in the sediments of the Nepa horizon (southern part of the Siberian platform) do not exceed 455,8 g/dm^3. The formation of their chemical composition is associated with the processes occurring in the flushed paleo-karst reservoir within Riphean sandstones and the Vendian horizons of terrigenous complex characterized by largely hindered relationship between the overlying sulfate-terrigenous and sulfate-carbonate rocks of the Oskoba, Katanga, and Soba Formations.

At this, calcium-sodium type of brines is associated with a substantially desalinated, predominantly carbonate composition of brine-bearing rocks. This makes them significantly distinguished from the brines of sodium-calcium and calcium composition formed in the rocks of sulfate-bearing carbonate-terrigenous and sulfate-carbonate composition of the Ushakovka and Nepa Formations in the southeastern regions of the Yenisei syneclise and northwestern parts of the Angara-Lena step. Calcium and sodium brines occupying intermediate position between sodium- calcium and calcium levels by the metamorphization ratios and the presence of micro-constituents in their composition. As a distance from the boundaries of the platform increases, their salinity becomes lower and the macro-constituent composition is subject to alteration, which makes them in many aspects almost identical to the calcium type, allowing to separate them into a special genetic type; while the origin of brines from the marginal areas (the Taiga area) are most likely to result from the processes of mixing sodium and calcium types.

4. Conclusion

Summarizing the discussion above, the following inferences can be made. The brines of the Vendian horizons is mainly of sodium-calcium chloride and calcium chloride composition. Their salinity varies from 17.3 to 583.1 g/dm^3. Vendian deposits tend show a gradually decreasing trend in mineralization and concentrations of most macro-components from top downward, with the increasing distance from the salt beds of the Usol'ye Formation. Despite the local variability of the hydrogeochemical field of the Vendian horizons, there is a distinct regional zonality of brines of the Siberian platform, especially in the junction zone with adjacent folded structures, suggesting distribution of sodium- dominated leaching brines. With the increasing distance from the platform boundary, this zone is replaced by brines of mixed cationic composition with predominance of sodium, which become subsequently replaced by brines of mixed composition dominated by calcium (and sometimes magnesium), being dominant in the interior parts of the Nepa-Botuoba anteclise and other major structures in the southern part of the Siberian platform. As such, the hydrogeochemical zoning is influenced by many factors, among them is that the major salt-bearing strata is dated Lower Cambrian, allowing the Tira regional horizon to receive physically associated intercrystalline brines squeezed from the salt beds, and largely explaining the existence of the inverse hydrogeochemical zonality in the subsalt horizons.

Acknowledgments

The research was supported by the Russian Foundation for Basic Research within the scope of scientific project № 18-05-70074 «The Arctic Resources».

References

[1] Connolly C, Walter L, Baadsgaard H Longstaffe F 1990 Origin and evolution of formation waters, Alberta Basin, Western Canada Sedimentary Basin. II. Isotope systematics and water mixing *Applied Geochemistry* **5** 397-413

[2] Grasby S, Zhuoheng C and Dewing K 2012 Formation water geochemistry of the Sverdrup Basin: Implications for hydrocarbon development in the High Arctic *Applied Geochemistry* **27** 1623-32

[3] Hitchon B, Underschultz J R, Bachu S and Sauveplane C M 1990 Hydrogeology, geopressures and hydrocarbon occurrences, Beaufort-Mackenzie Basin *Bulletin of Canadian Petroleum Geology* **38** 215-35

[4] Kokh A A and Novikov D A 2014 Hydrodynamic Conditions and Vertical Hydrogeochemical Zonality of Groundwater in the Western Khatanga Artesian Basin *Water Resources* **41** 396-405

[5] Novikov D A 2017 Hydrogeological conditions for the presence of oil and gas in the western segment of the Yenisei-Khatanga regional trough *Geodynamics and Tectonophysics* **8** 881-901

[6] Antsiferov A 1989 *Hydrogeology of the most ancient layers of the Siberian platform* (Moscow: Nedra) p 176

[7] Baskov E 1976 Siberian platform: the main features of distribution and creation of basic types of underground brines *Works VSEGEI* **246** 61-75

[8] Valyashko M 1962 *Geochemical regularities in the formation of deposits of potassium salts* (Moscow: Moscow State University Press) p 403

[9] Valyashko M, Polivanova A, Zherebtsova I and Metkih B 1965 *Geochemistry and genesis of brines of the Irkutsk amphitheater* (Moscow: Science) p 159

[10] Vozhov V 2006 *Underground waters and hydromineral feedstock of the Leno-Tungussky oil-and-gas province* (Novosibirsk: Siberian Scientific Research Institute of Geology, Geophysics and Mineral Raw Materials) p 209

[11] Novikov D 2014 Hydrogeochemistry and mechanisms of groundwater formation in siberian arctic *GEO-Siberia-2014* Novosibirsk **1** 109-14

[12] Pinneker E 1966 *Brines of the Angaro-Lensky pool regularities of placement, structure, dynamics, formation and use* (Moscow: Science) p 332

[13] Sidkina E S, Novikov D A and Shvartsev S L 2012 Equilibrium of underground brines of western part of Tunguska artesian basin and minerals of enclosing rocks *Tomsk State University Journal* **11** 187-92

[14] Shvartsev S L 2000 The chemical composition and isotopes of strontium brines of the Tunguska basin in connection with the problem of their formation *Geologiya i Geofizika* **11** 1170-84

[15] Shvartsev S L and Bukaty M B 1995 On the role of rocks in the formation of strong brines of chloride-calcium type *Doklady of Academy of Sciences* **342** 530-3

[16] Novikov D A 2017 Hydrogeochemistry of the Arctic areas of Siberian petroleum basins *Petroleum Exploration and Development* **44** 780-8

[17] Novikov D A and Trifonov N S 2016 Hydrogeologic Implications of Industrial Effluent Disposal of the Yurubcheno-Tokhomo Field (Siberian Craton, Russia) *Arabian Journal of Geosciences* **9** 1-14

[18] Kushmar I, Grigorenko Yu, Belinkin V, Ananyev V and Gubina E 2006 Oil and gas of Eastern Siberia *Oil and gas of Eastern Siberia* (St. Petersburg: Nedra) p 100

[19] Samarina V 1977 *Hydrogeochemistry* (St. Petersburg: St. Petersburg State University Press) p 360

[20] McCaffrey M, Lazar and Holland H 1986 he evaporation path of seawater and the coprecipitation of Br- and K+ with the halite *Journal of Sedimentary Petrology* **57** 928-37

IOP Conf. Series: Earth and Environmental Science **193** (2018) 012053 doi:10.1088/1755-1315/193/1/012053

Experimental and theoretical basis for the creation of frost-resistant polymeric composite materials

E M Nurullayev and N Yu Lyubimova

Perm National Research Polytechnic University, Perm, Russia

Abstract. The structural-mechanical behavior of a three-dimensional cross-linked elastomer, filled with solid particles under uniaxial tension were considered. This paper presents a theoretical and experimental basis for the development frost-resistant (glass transition temperature of ~170K), elastic (deformation of at least 70% at T~223K) polymer composite material. The equations of mechanical stresses dependence on deformation were used. Based on the results of numerical experiments, a frost-resistant elastic polymer composite material is simulated. The material was tested with a bursting machine and the experimental results were compared with the calculated data.

1. Introduction

Polymer composite materials with an elastomeric matrix filled with solid particles are widely used in parts and units of various structures in the automotive, aircraft and shipbuilding industries. The development of frost-resistant structural material for parts and components of various types of road and air transport operated in the Far North and the Arctic of Russia is an important engineering task. One of the problems is the creation of an elastic material that would be capable to operate at temperatures down to 170-223 *K*, would exhibit enhanced wear resistance, and could be prepared from cheap components.

The most promising way to solve this problem is the use of high-molecular-mass synthetic hydrocarbon rubbers in the form of copolymers that do not crystallize in a wide temperature range, as a polymer base of composites [1, 2]. The theory of structural-mechanical behavior of filled elastomer, developed in our previous studies [2], can be recommended for accelerating the development of such materials. In these studies we used the relationship between the mechanical failure energy and the composition of the three-dimensionally cross-linked plastized elastomer.

2. Theoretical foundations and characteristics used in calculations

The structural-mechanical behavior of a three-dimensional cross-linked plasticized elastomer filled with solid particles under uniaxial tension has been considered. The elastic potential of the spatialpolymer network has the form:

$$U = C_1(I_1 - 3) + C_2(I_2 - 3).$$

The parameters ($2C_1$ and $2C_2$) allowed Mooney and Rivlin [2] to write the equation:

$$\sigma = (2C_1 + 2C_2\alpha^{-1})(\alpha - \alpha^{-2}).$$

Then as follows from [3]:

$$\sigma = 2C_1(\alpha - \alpha^{-2}) = \nu_{ch} k N_A T_\infty (\alpha - \alpha^{-2}) = \nu_{ch} R T_\infty (\alpha - \alpha^{-2}),$$

where σ is the conditional stress (tensile force related to the initial cross section). Mooney-Rivlin equation in part of the influence of physical (intermolecular interactions) component cross-linking in a polymer binder was clarified (AS Ermilov and E Nurullaev [4]). In the final version the equation has the form:

$$\sigma = \nu_{ch}\varphi_r^{1/3} R T_\infty \left\{1 + 29\exp\left[-0.225\cdot 10^{-3}(T - T_g)^2\right]\alpha^{-1}\alpha_{\dot{\alpha}}^{-1}\right\}\left(1 + 1.25\frac{\varphi/\varphi_m}{1 - \varphi/\varphi_m}\right)\left(\alpha - \alpha^{-2}\right).$$

The mechanical fracture energy of the filled elastomer is determined by the expression [4]:

$$W = \left\{\left(\frac{\alpha_b^3 - 3\alpha_b + 2}{2\alpha_b}\right) + 29\exp\left[-0.225\cdot 10^{-3}(T - T_g)^2\right]\right\}\alpha_{\dot{\alpha}}^{-1}\left(\frac{2\alpha_b^3 - 3\alpha_b^2 + 1}{2\alpha_b^2}\right)\times$$

$$\times\left(\nu_{ch}\varphi_r^{1/3} R T_\infty\right)\left(1 + 1.25\frac{\varphi/\varphi_m}{1 - \varphi/\varphi_m}\right)^2$$

where: α_b - tensile elongation; ν_{ch} - molar concentration of chemical transverse bonds in the polymer base of the binder; φ_r - volume fraction of three-dimensional cross-linked polymer in the binder, R - universal gas constant; $T\infty$ - equilibrium temperature at which the concentration of "physical" (intermolecular) cross-links tends to zero; T - the test temperature of the sample; Tg - the temperature of structural glass transition of the polymer binder; $a_{\dot{\alpha}}^{-1}$ - the coefficient of speed equals to 1 at the standard relative speed of stretching; φ - volume fraction of solid components; φ_m – ultimate degree of volume filling depending on the particle shape and their fractional composition [4, 5].

Polymer composite material is created on the basis of high molecular mass three dimensional cross-linked elastomer polyisoprene divinyl. Synthetic rubber butadiene of isoprene with the catalyst compounds of lithium (SCDI-L) and plasticized with dioctylsebacinate has the following characteristics:

- molecular mass of 286000;
- structural glass transition temperature of 178 K;
- density of 900 kg/m^3 and a volume fraction of 0.4;
- a plasticizer with a glass transition temperature of 169 K, a density of 910 kg/m^3 and a volume fraction of 0.6;
- glass transition temperature of the elastomer Tg = 170 K.

Value of fractional composition of silicon dioxide parameters are given in table 1.

Table 1. Parameters of fractional composition of silicon dioxide.

Fraction number	Weight-average particle size, μm	Limiting volume filling for fractions, φ_{mi}	Optimum volume of fractions	Limiting volume filling of the mixture, φ_m
1	1	0.535	0.2	
2	8	0.574	0.30	0.84
3	240	0.633	0.50	

The calculated values were compared with experimental data using the program [6].

IOP Conf. Series: Earth and Environmental Science **193** (2018) 012053 doi:10.1088/1755-1315/193/1/012053

3. Results and discussion

The experimental and calculated diagrams of uniaxial extension at different temperatures for composite samples based on SLDI-L high-molecular-mass copolymer are shown in figure 1.

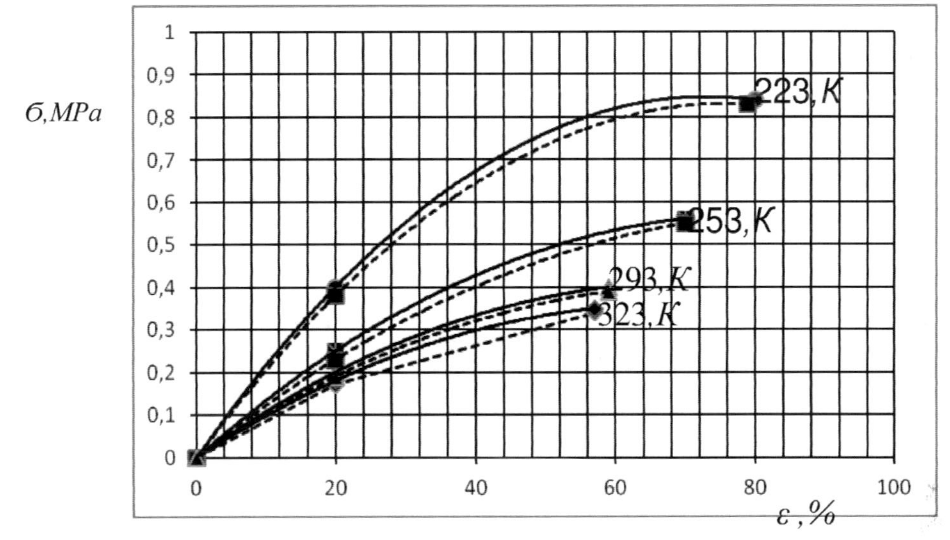

Figure 1. Stress σ as a function of strain ε for the composite based on SKDI-L at different temperatures of the experiment. Standard extension rate is 1.2×10^{-3} s^{-1}. Solid lines: experimental data; dashed lines: data of numerical calculation.

Both the breaking stress and breaking strain increase with decreasing temperature. Such frost resistance of the elastomer composite can be associated with the extremely low structural glass transition point of the polymer binder. Such thermo-mechanical behavior of the composite based on SKDI-L copolymer is also influenced by the quinol ether as a cross-linking agent. In contrast to the quinol ether, sulfur-based rubber-vulcanizing systems form less strong chemical cross-links [4]. The corresponding envelopes, plotted in the extension diagrams of both kinds of composites according to T Smith, show that the breaking strain of the composite based on SKDI-L high molecular-mass copolymer considerably exceeds that of the composites based on a blend of SKD-KTR and PDF-3B low molecular-mass rubbers, whereas the breaking stresses for these composites differ only slightly as shown on figure 2.

The experimental and calculated dependences of the mechanical failure energy on the breaking stress are shown in figure 3. As can be seen, with a decrease in the test temperature and, naturally, with an increase in the intermolecular interaction, the mechanical failure energy determining the service life of a material regularly increases.

In all types of experimental verification of the developed theory of structural and mechanical behaviour of a filled three-dimensional cross-linked elastomer based on high-molecular divinyl isoprene rubber is shown to have a satisfactory convergence of the experimental data and the calculated characteristics (an error of the order of 5%, which is satisfactory for such problems).

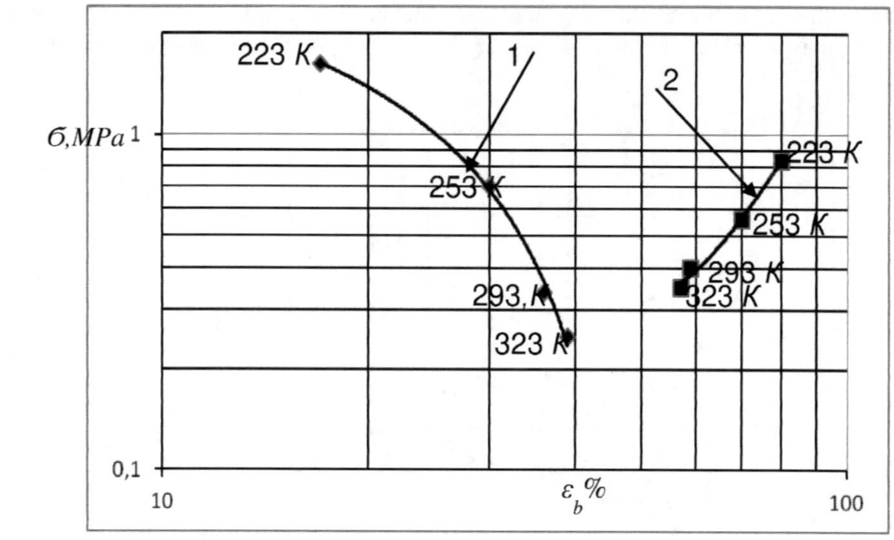

Figure 2. Envelopes of the break points (according to T Smith) of elastomer composite materials based on (1) a blend SKD-KTR and PDI - 3B low-molecular-mass rubbers and (2) SKDI-L high- molecular-mass copolymer. σ – stress and ε_b – strain.

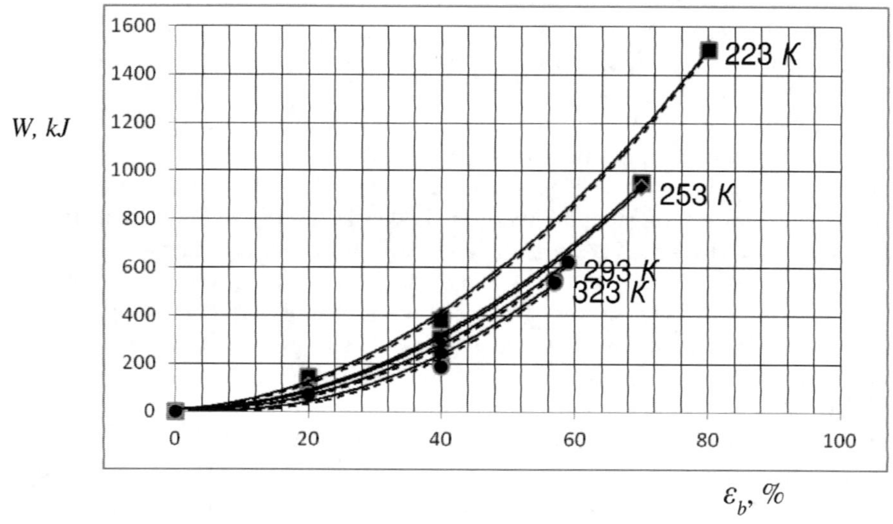

Figure 3. Mechanical failure energy *W* is as a function of breaking strain ε_b for the composite based on SKDI-L at different temperatures. Solid lines: experimental data; dashed lines: calculated data.

4. Conclusion

The theoretical basis of the structural and mechanical behaviour of filled three-dimensional cross-linked elastomer has been improved. The corresponding numerical calculations are carried out. Based on the results of the numerical experiment, a frost-resistant, three-dimensional cross-linked, filled with three-fraction silicon dioxide plasticized elastomer is proposed. A satisfactory correlation between theoretical and experimental data is shown. The developed elastomer is proposed by the authors for use as gaskets in equipment, automotive and aircraft tires, operated in the Arctic and the Far North.

References

[1] Ermilov A S and Nurullaev E M 2011 *J. Appl. Chem.* **84** 1558-61
[2] Smith T L 1964 *J. Appl. Phys.* **35** 27-32
[3] Treloar I R G 1974 *Rubber Rev.* **47** 625-96
[4] Smith T L and Chy W H 1972 *J. Polymer Sci.* **10** 133-50
[5] Ermilov A S and Nurullaev E M 2014 *Russian J. Appl. Chem.* **87** 500-8
[6] Ermilov A S and Nurullaev E M 2016 *Mech. Composite Materials* **52** 325-34

IOP Conf. Series: Earth and Environmental Science **193** (2018) 012054 doi:10.1088/1755-1315/193/1/012054

Proxy Pseudo3D model: the optimum of speed and accuracy in hydraulic fracturing simulation

Grigory V Paderin

LLC Gazpromneft Science & Technology Center, St. Petersburg, Russia

E-mail: Paderin.GV@gazpromneft-ntc.ru

Abstract. The paper describes an efficient approach to simulating hydraulic fracturing in Pseudo3D setting for reservoirs with a low permeable matrix. By reformulating the problem, the authors succeeded in reducing it to two ordinary first-order differential equations instead of a nonlinear second-order partial differential equation, which fundamentally affects the computational complexity. Also, this approach allows a more correct computation of a fracture front and fracturing fluids interface versus a traditional Cell-Based Pseudo3D model. Nowadays the main limitation of the developed software module is the absence of matrix leak-off.

1. Introduction

The paper describes the development of an effective method to solve the fracturing simulation problem in a Pseudo3D setting versus a traditional Cell-Based Pseudo3D model.

Hydraulic fracturing (frac) is the main approach to extract hydrocarbons from low-permeable reservoirs. The design and implementation of these operations is supported by calculations based on physical and mathematical models of fracture propagation. Nowadays there are a lot of models of hydraulic fracturing which are roughly divided [1] into four main types: Analytical 2D, Pseudo3D, Planar3D, and Full3D. A Pseudo3D model is considered to be the most popular class of models for practical calculations. This model evolved from a PKN model (Perkins-Kern-Nordgren) [2-3] developed in the 1960-70s. The need to develop the Pseudo3D model is explained by the inability to apply the PKN model for realistic cases because it leaves out the effect of distribution of vertical rock stresses and vertical rock properties on the hydraulic fracture geometry. Although the Pseudo3D model was introduced back in the 1980s [4-9], its development still continues [10-14].

Despite the fact that the Pseudo3D model is the fastest among the models used for practical calculations, the speedup task while maintaining satisfactory accuracy remains critical, and that is a focus of this work. The need for higher speed becomes even more demanding when solving problems of frac design optimization with multiple serial runs and a search of various treatment strategies for a huge number of wells. It can also be recommended to implement a Pseudo3D proxy model in cases where the tasks of simultaneous propagation of a significant number of hydraulic fractures are to addressed, for example, fracture network propagation in unconventional reservoirs or in other fractured structures.

Besides our previous paper [14] also focuses on building a simplified semi-analytical Pseudo3D model, the main idea being the cross-linking of analytical solutions in a simplified setting: a three-layer model and single hydraulic fracturing fluid. In contrast, this paper describes multilayered lithology with a more complex fracture height distribution lengthwise, including the effect of injecting a number of fracturing fluids with differing rheologies.

2. Formulation of Pseudo3D Model

The basic assumptions of the Pseudo3D model are as follows:

- The fracture propagates vertically
- The fracture length by far exceeds its height
- The fluid flow along the fracture is quasi-one-dimensional and horizontally-oriented
- The vertical propagation of the fracture is determined by the ratio of fracture pressure and the contrast of the minimum horizontal stress.
- Leak-off process is controlled by the Carter leak-off law.

All these assumptions are further expressed in the equations describing the three basic physical processes of hydraulic fracture development: rock destruction, rock deformation, and fluid flow. The strongest assumption related to a significant excess of length over height allows to consider a vertical mechanical problem using a 2D piecewise-homogeneous material and a 1D horizontal flow of fluid.

Let us introduce the following symbols. We assume that a hydraulic fracture intersects n geological layers, and $i = 0, ..., (n-1)$ is the layer number upward.; σ_i is the minimum rock stress in the i-th layer, K_{IC_i} is the fracture toughness of the i-th layer; h_i is the distance from the lower tip of the fracture to the upper boundary of the i-th layer at $i = 0, ..., (n-2)$; $h_{n-1} = h$, where h is the fracture height; h_{cp} is the distance from the lower tip of the fracture to the middle of the fracture initiation interval corresponding to the minimum σ_i value.

2.1. *Vertical Fracture Propagation*

The vertical fracture propagation is determined by the two-dimensional brittle fracture mechanism at each horizontal point independently. The Irwin criterion is used here which states the equality of a stress intensity factor K_I, the fracture toughness K_{IC} of an interlayer intersected by the tip of the fracture. $K_{I\text{-}top}$ is a stress intensity factor for the upper tip of the fracture, $K_{I\text{-}bot}$ is a stress intensity factor for the lower tip of the fracture. The formula for the stress intensity factor in the two-dimensional case produces the following:

$$K_{I\_bot} = K_{IC\_0}, K_{I\_top} = K_{IC\_n-1}, \tag{1}$$

$$K_{I\_top} = \sqrt{\frac{\pi h}{2}}\left[p_{cp} - \sigma_n + \rho_f g\left(h_{cp} - \frac{3}{4}h\right)\right] + \sqrt{\frac{2}{\pi h}}\sum_{i=0}^{n-2}(\sigma_{i+1} - \sigma_i)\left[\frac{h}{2}\cos^{-1}\left(\frac{h-2h_i}{h}\right) - \sqrt{h_i(h-h_i)}\right], \tag{2}$$

$$K_{I\_top} = \sqrt{\frac{\pi h}{2}}\left[p_{cp} - \sigma_n + \rho_f g\left(h_{cp} - \frac{1}{4}h\right)\right] + \sqrt{\frac{2}{\pi h}}\sum_{i=0}^{n-2}(\sigma_{i+1} - \sigma_i)\left[\frac{h}{2}\cos^{-1}\left(\frac{h-2h_i}{h}\right) + \sqrt{h_i(h-h_i)}\right], \tag{3}$$

where p_{cp} is the fluid pressure inside the fracture at the middle of the initiation interval, ρ_f is the fracturing fluid density, and g is the gravitational acceleration.

2.2. *Rock Deformation*

The deformation is governed by a linear elasticity theory described by Hooke's law. The 2D theory of elasticity demonstrates the following solution [11-12] for the fracture opening $w(z)$:

$$w(z) = \frac{4}{E'}\left[p_{cp} - \sigma_n + \rho_f g\left(h_{cp} - \frac{h}{4} - \frac{z}{2}\right)\right]\sqrt{z(h-z)} +$$

$$+ \frac{4}{\pi E'}\sum_{i=0}^{n-2}(\sigma_{i+1} - \sigma_i)\left[(h_i - z)\cosh^{-1}\frac{z\left(\frac{h-2h_i}{h}\right)+h_i}{|z-h_i|} + \sqrt{z(h-z)}\arccos\left(\frac{h-2h_i}{h}\right)\right], \tag{4}$$

where $E' = E/(1-\upsilon^2)$ is the plane strain modulus averaged by the fracture height, E is the Young modulus, and υ is the Poisson ratio.

2.3. *Flow in Hydraulic Fracture*

Since the fracture opening is much smaller than the fracture height and length, the flow in a fracture can be described by the lubrication theory equation. For the case of a Newtonian fluid with viscosity μ:

$$\vec{q} = -\vec{\nabla} p \, \frac{w^3}{12\mu}, \tag{5}$$

where w is fracture opening at a point, and $\vec{q}$ is a thickness-averaged fluid flow vector.

3. The traditional approach of solving the Pseudo3D model. The governing equation.

As with the PKN model, the equation of continuity (the law of local volume balance) is traditionally used in the Pseudo3D model as the basic equation.

$$\frac{1}{H}\frac{\partial Q}{\partial x} + \frac{\partial \bar{w}}{\partial t} + \frac{C_L}{\sqrt{t - t_0(x)}} = 0, \tag{6}$$

where the $Q = \int_h q(z)dz$ is the average volume flow, H is the height of the permeable (leak-off) interval, $\bar{w}$ is the average-height opening, $t_0(x)$ is the time at which the front edge of the fracture will reach the x coordinate, C_L is the Carter leak-off coefficient. Thus, the classical approach of Pseudo3D model solves a system of equations (1-6).

4. The Proposed Proxy Pseudo3D Model

4.1. *The idea of the proposed approach is to address the shortcomings of the traditional approach.*

The traditional Cell-based Pseudo3D approach focuses on the numerical model solution characterized by the following features.

4.1.1. Disadvantages of the numerical grid in the Cell-based Pseudo3D model. The fracture geometry is divided into cells of fixed size lengthwise (see Figure 1). The fracture horizontal propagation is modeled by adding new cells to the right.

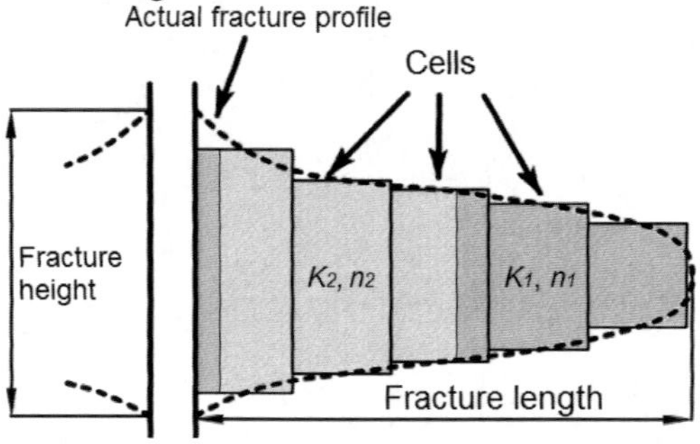

Figure 1. Cell-based Pseudo3D (several fluids).

The accuracy of tracking the front edge of the fracture represents the first drawback of the traditional approach. The fact is that the fracture length increases differently at each time step and can not be a multiple of a fixed cell size, which, with a sufficiently coarse grid, can lead to a significant accumulation of numerical error along the fracture length.

4.1.2. Tracking the front between fracturing fluids in Cell-based Pseudo3D model. If a number of fracturing fluids (see Figure 1) and a proppant are present, there is a challenge similar to that related the fracture front: the fluids interface does not coincide with computational nodes at the boundaries of cells. In that case, in order to keep the material balance between the injected volumes, very fine-pitch grids should be used when running the models. There is another approach which uses a saturation parameter, but this requires extra equations to be solved.

4.1.3. The sequence of the Cell-based Pseudo3D equation. Since the flow rate Q is a pressure derivative in the x coordinate, after substituting expression (5) into (6), the classical approach will produce the final second-order partial-derivative equation. It must be solved iteratively at each step in time.

Next, an approach to obtain a first-order equation which is solved by shock capturing method will be described.

4.2. The system of equations of the suggested model

The reason for solution complexity in the Cell-based Pseudo3D model is due to the coordinate system which is related to perforation interval at the bottom hole. Therefore, the authors of this paper suggest to address the problem in a fracture-related coordinate system, namely the tip-related system.

The fracturing fluid rheology can be of Newtonian type - in case of water-based frac (SlickWater technology). However, the fracturing fluid is typically a polymer system and is described by a power-law model

$$\tau = K\dot{\gamma}^n, \tag{7}$$

where K is the consistence index, n is a non-Newtonian behavior factor (the fluid is Newtonian at n = 1). If proppant is present, $K = K_f(1 - c/c^*)^{-2.5}$, where K_f is the consistence index of pure fluid without proppant, c is the volume concentration of proppant, c^* is the critical concentration (usually $c^* = 0.64$).

For a power-law fluid, the lubrication layer equation is as follows:

$$q|q|^{n-1} = -\frac{1}{\phi K} w^{2n+1}\frac{\partial p}{\partial x}, \tag{8}$$

$$\phi = 2^{2n+1}(2n+1)^n n^{-n}. \tag{9}$$

We assume that the abscissa axis points from the tip of the fracture to the well, the $x = 0$ coordinate determines the tip of the fracture. Then the total flow through a fracture cross-section Q from (8) will be:

$$q(z) = \left(\frac{1}{\phi K} w(z)^{2n+1}\frac{\partial p}{\partial x}\right)^{1/n}, \tag{10}$$

$$Q = \int_h q(z)dz = \left(\frac{1}{\phi K}\frac{\partial p}{\partial x}\right)^{1/n} \int_h w^{(2n+1)/n} q(z)dz, \tag{11}$$

the fracture shape is related to the friction pressure loss through the following expression:

$$\frac{\partial p}{\partial x} = \phi K Q^n \left(\int_h w^{\frac{2n+1}{n}} dz\right)^{-n}. \tag{12}$$

Here the height and width of the fracture are the pressure functions $h=h(p)$, $w=w(p,z)$, calculated from the system (1-5).

If we assume that there are no leak-offs, then the fracture volume V at time point t is equal to the volume of the injected fluid

$$V = Qt. \tag{13}$$

The fracture half-length $L(t)$ is determined by the fracture volume $V(t)$:

$$V(t) = \int_0^{L(t)} \int_{h(p)} w(p,z)dzdx. \tag{14}$$

Thus, equations (1-5), (9), and (11-14) form a closed system of equations for solving the Pseudo3D model for the case with no leak-off.

4.3. Numerical Algorithm

The numerical algorithm consists of a number of consecutive stages:

1. A preliminary computation of a geomechanical problem: the height $h = h(p)$ and the fracture opening profile $w(z, p)$ as a function of fluid pressure p are determined from equations (1-4). And the minimum pressure $Min\ p = p_0$ corresponds to the value at which the fracture height is equal to the height of the initiation interval $h(p_0) = h_{init}$.

2. A preliminary computation of the integrals (15-16) using the computed opening profiles determined at the first stage of the numerical algorithm:

$$J(p) = \int_{h(p)} w(p,z)dz, \tag{15}$$

$$F_n(p) = \int_{h(p)} \big(w(p,z)\big)^{\frac{2n+1}{n}} dz. \tag{16}$$

3. Then we used a Cauchy problem for the first-order ordinary differential equation

$$\frac{dp}{dx}(x) = \phi K Q^n \big(F_n(p)\big)^{-n}, p(0) = p_0 \tag{17}$$

to determine the $p(x)$.

4. In order to determine the complete fracture geometry, we only need to determine its boundary $L(t)$. For this, $V = Qt$ is used to reformulate (14) into a Cauchy problem for a first-order ODE:

$$\frac{dV}{dx}(x) = J\big(p(x)\big), V(0) = 0. \tag{18}$$

Taking into account that the inverse function $L^{-1}(x)=V(x)$ is found, the fracture front $x = L(V)$ can be determined.

Thus, the solution of the system of equations (1-5), (9), (11-14) was reduced to the pre-computation (15-16) and the solutions of two Cauchy problems for first-order ODEs (17-18) which are solved by any standard method through a number of operations $O(N)$, where N is the number of sampling points along the x coordinate

4.4. *Generalization for the case of several fracturing fluids*

Let us consider the case when several fluids are injected during fracturing operations. This does not affect the preliminary geomechanical computation. Let us consider how the form of the Cauchy problem (17), (18) will change.

We assume that M fracturing fluids with rheological parameters K_m, n_m, $m=1,..,M$, are consistently injected into a hydraulic fracture starting from the time $t = 0$; t_m is the end time of injection fluid m with a volume of a single wing $V_m=Qt_m$; then $p(V)$, $L(V)$ are piecewise smooth functions consisting of smooth functions $p_m(V)$, $L_m(V)$ on the intervals $V_{(m-1)}<V<V_m$. Thus, the Cauchy problem domain to determine the pressure (17) is divided into M intervals:

$$\frac{dp_m}{dx}(x) = \phi_m K_m Q^{n_m} \Big(F_{n_m}\big(p_m(x)\big)\Big)^{-n_m}, L_{m-1}(V_{m-1}) < x < L_m(V_m), \tag{19}$$

$$p_m(L_{m-1}(V_{m-1})) = p_{m-1}(L_{m-1}(V_{m-1})),\ p(L_0) = p(0) = p_0,\ m = 1, \dots, M, \tag{20}$$

$$\phi_m = 2^{2n_m+1}(2n_m+1)^{n_m} {n_m}^{-n_m}. \tag{21}$$

Moreover, the boundaries of the intervals $L_m(V_m)$ are determined from the solution of the Cauchy problem (18) which in the case of several fracturing fluids has the form:

$$\frac{dV_m}{dx}(x) = J\big(p_m(x)\big),\ L_{m-1} < x < L_m, \tag{22}$$

$$V_m(L_{m-1}) = V_{m-1}(L_{m-1}),\ V(0) = 0,\ m = 1, \dots, M. \tag{23}$$

Thus, in the case of M fracturing fluids, we have two Cauchy problems (19-21) and (22), which must be solved in sequence for each of the intervals $V_{(m-1)}<V<V_m$, $m=1,..,M$. Moreover, the overall complexity of the problem solution remains the same $O(N)$, where N is the number of sampling points at the x coordinate for solving the Cauchy problems (19-21) and (22-23). In this case, the boundaries

of various fluids $L_m(V_m)$ are determined quite accurately. Figure 2 below shows a computation example for two fracturing fluids.

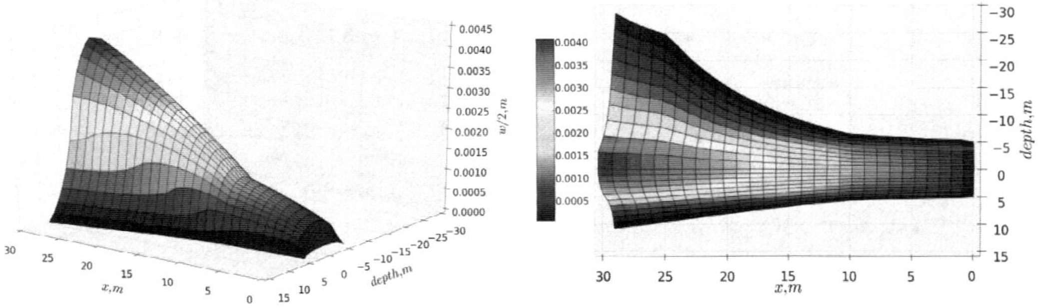

Figure 2. The example of Proxy Pseudo 3D calculations: 2 fluids, 4 layers.

5. Summary

In this paper, the authors described an efficient approach to calculate the geometry of a hydraulic fracture in a Proxy Pseudo3D setting. By reformulating the problem, the authors succeeded in reducing it to two ordinary first-order differential equations instead of a nonlinear second-order partial derivative equation, which fundamentally affects the computational complexity. Also, this method allows a more correct computation of the fracture front and fracturing fluids interface versus the traditional Cell-Based Pseudo3D model. Thus, this model meets the requirements of practically meaningful computations and significantly beats its counterparts in running speed, which is critically important for serial calculations. The developed proxy model can also be effectively used as an integral part of the fracture network model in fractured reservoirs, which will be a focus of further studies.

Acknowledgments
This work was supported by Ministry of Education and Science of the Russian Federation within the framework of the Federal Program Research and development in priority areas for the development of the scientific and technological complex of Russia for 2014 2020 (activity 1.4), grant No. 14.581.21.0027 of 03.10.2017, unique identifier: RFMEFI58117X0027.
The industrial partner of the grant is Gazpromneft Science & Technology Centre LLC.

References

[1] Khasanov M M, Paderin G V, Shel E V, Yakovlev A A and Pustovskikh A A 2017 Approaches to modeling hydraulic fracturing and their development *Oil Industry Journal* **2017(12)** 1–5 (in Russian).
[2] Perkins T K and Kern L R 1961 Widths of hydraulic fractures *Journal of Petroleum Technology* **222** 937–49
[3] Nordgren R P 1972 Propagation of vertical hydraulic fractures *Journal of Petroleum Technology* **253** 306–14
[4] Cleary M P 1980 Comprehensive design formulae for hydraulic fracturing *SPE Annual Technical Conference and Exhibition* SPE-9259-MS
[5] Palmer I D and Carroll H B 1982 Three-dimensional hydraulic fracture propagation in the presence of stress variation *Proceedings of the SPE/DOE/GRI Unconventional Gas Recovery Symposium* 870–8
[6] Palmer I D and Craig H R 1984 Modeling of asymmetric vertical growth in elongated hydraulic fractures and application to first MWX stimulation *Proceedings of the SPE/DOE/GRI Unconventional Gas Recovery Symposium* 453–62
[7] Settari A and Cleary M P 1986 Development and testing of a pseudo-three-dimensional model of hydraulic fracture geometry *SPE Production Engineering* 449–66
[8] Meyer B R 1986 Design formulae for 2d and 3-d vertical hydraulic fractures: model comparison

and parametric studies *Proceedings of the Unconventional Gas Technology Symposium* 391–408

[9] Meyer B R, Cooper G D and Nelson S G 1990 Real-time 3-D hydraulic fracturing simulation: theory and field case studies *SPE Annual Technical Conference and Exhibition* SPE-20658-MS

[10] Adachi J I, Detournay E and Peirce A P 2010 Analysis of the classical pseudo-3D model for hydraulic fracture with equilibrium height growth across stress barriers *International Journal of Rock Mechanics and Mining Sciences* **47(4)** 625–39

[11] Weng X, Kresse O, Cohen C-E, Wu R and Gu H 2011 Modeling of hydraulic-fracture-network propagation in a naturally fractured formation *SPE Production & Operations* **26(4)** 368–80

[12] Cohen C-E, Kresse O and Weng X 2015 A new stacked height growth model for hydraulic fracturing Simulation *49th U.S. Rock Mechanics/Geomechanics Symposium* ARMA-2015-073

[13] Dontsov E V and Peirce A 2015 An enhanced pseudo-3D model for hydraulic fracturing accounting for viscous height growth, non-local elasticity, and lateral toughness *Engineering Fracture Mechanics* **142** 116–39

[14] Paderin G V 2016 Modified approach to incorporating hydraulic fracture width profile in unified fracture design model *SPE Russian Petroleum Technology Conference* SPE 182034-MS

Problems of heat and mass transfer in the snow-ice cover

A A Papin and M A Tokareva

Department of Mathematics and Information Technology, Altai State University, Barnaul, Russia

E-mail: papin@math.asu.ru, tma25@mail.ru

Abstract. The snow-ice cover is considered as a three-phase continuous medium consisting of water, air, and ice. The mathematical model is based on the equations of mass conservation for each phase, the two-phase filtration equation for water and air in a movable porous ice skeleton, the rheological equation for porosity, the balance of forces and the equation of energy conservation for the ice-water-air system. The paper presents the results on analytic and numerical solutions of the problems under consideration.

1. Introduction

The urgency of the study of heat and mass transfer in multiphase media is due to the fact that numerous problems that arise, along with technological processes also in ecology and nature use lead to the need to model the processes of interpenetrating motion of continuous media. The constructed in this case mathematical models, as a rule, are nonclassical, and require the development of new approaches to their justification and numerical modeling.

A large class of filtration problems in a deformable porous medium with phase transitions is the problem of heat and mass transfer in a melting snow-ice cover [1]. The predominant part of the runoff of the northern rivers is formed due to the melting of the seasonal snow cover. Snowmelt conditions have a decisive influence not only on the amount of thawed water entering the reservoirs, but also on their quality. In addition, the amount of snow cover (snowfall) affects the freezing of the surface layer of soils and, consequently, its absorbency and determines the relationship between slope and ground runoff. Therefore, modeling the state of the snow cover and salt transfer during the snow melting period is of great importance in the development of methods for calculating and forecasting the hydrographers of spring flood and water quality in reservoir-receivers. There is a large number of works on salt-mass transfer in melting snow, in which the data of observations and empirical dependences are used [2]. Most empirical models are one-dimensional and do not allow to calculate the fluid filtration rate, and models that calculate the fluid filtration rate usually do not take into account phase transitions or are suitable only for specific modes of water movement in the snow cover, they also do not provide the necessary information on the velocity field and the saturation of the fluid phase necessary for the assessment of the flow of pollutants.

Thus, for a reliable prediction of pollutant discharge, it is necessary to know the velocity field and the saturation of the water phase, i.e., it is proposed to use complex models describing the joint movement of pollutants and water in the snow cover, taking into account different boundary conditions, phase transitions and sublimation process. These models will allow calculating the non-stationary movement of pollutants within the snow cover and assess the surface and underground runoff of substances. They must take into account a number of important factors, including variable snow cover porosity, phase transitions, specificity of boundary conditions (in particular, the presence of frozen or not frozen ground). The fundamentals of the theory of the motion of water and air in melting snow are laid in the works of S.C. Colbeck [3] and his followers. However, snow in these works, although considered as a multiphase medium, deformation of ice and phase transitions were not taken into account.

The purpose of this work is: to build a model on the joint movement of water and air in melting snow, taking into account phase transitions and deformation of ice; the statement of the problem of ablation of deformed ice cover; statement of the problem of distribution of water flow of melting snow between ground and surface waters; the construction of a model of groundwater movement in contact with frozen ground.

2. Mathematical models of the snow cover

Due to the high demand for data on snow cover, for practical purposes, there is a significant number of works devoted to field observations (see, for example, [4], [5]). For estimating calculations, simple balance models (see [6], [7]) and more complex models are used [3].

Following [8, 9, 10] we will consider melting snow as a continuous medium consisting of water ($i = 1$), air ($i = 2$) and ice ($i = 3$) constituting a solid porous skeleton. The filtration of water and air in a porous ice skeleton is described by the equations of mass conservation for each of the phases, taking into account phase transitions, the equations of two-phase filtration and the heat balance equation for a three-phase medium

$$\frac{\partial \rho_i}{\partial t} + div(\rho_i \vec{u}_i) = \sum_{j=1}^{3} I_{ji}, \quad i = 1, 2, 3, \quad I_{ji} = -I_{ij}, \quad \sum_{i,j=1}^{3} I_{ij} = 0, \tag{1}$$

$$\phi(\vec{u}_i - \vec{u}_3) = -K_0 \frac{k_{0i}}{\mu_i}(\nabla p_i + \rho_i^0 \vec{g}), \quad i = 1, 2, \quad p_2 - p_1 = p_c(s_1, \theta), \quad \sum_{i=1}^{2} s_i = 1, \tag{2}$$

$$(\sum_{i=1}^{3} \rho_i^0 c_i \alpha_i) \frac{\partial \theta}{\partial t} + (\sum_{i=1}^{3} \rho_i^0 c_i \alpha_i \vec{u}_i) \nabla \theta = div(\lambda_c \nabla \theta) - L_s I_{23} + L_i I_{13}. \tag{3}$$

Here $\vec{u}_i$ is the speed of the i phase; ρ_i is the reduced density associated with the true density ρ_i^0 and the volume concentration α_i by the relation $\rho_i = \alpha_i \rho_i^0$ (the condition $\sum_{i=1}^{3} \alpha_i = 1$ is a consequence of the definition of ρ_i); I_{ji} is the intensity of the mass transition from j into the i component per unit volume per unit time; ϕ is the porosity of the snow; s_1, s_2 - saturation of water and air ($\alpha_1 = \phi s_1$, $\alpha_2 = \phi s_2$, $\alpha_3 = 1 - \phi$); $K_0(\phi)$ is the filtration tensor; k_{0i} - phase permeabilities ($k_{0i} = k_{0i}(s_i) \geq 0$, $k_{0i}|_{s_i=0} = 0$); μ_i - dynamic viscosity; p_i is the phase pressure; p_c is the capillary pressure, $\vec{g}$ - acceleration vector of gravity; θ is the temperature of the medium ($\theta_i = \theta$, $i = 1, 2, 3$), $c_i = const > 0$ - the heat capacity of the i phase at constant volume; $L_i = const > 0$ - specific heat of melting of ice; $L_s = const > 0$ - specific heat of ice sublimation; λ_c is the thermal conductivity of snow.

The system of equations (1) - (3)with respect to the characteristics $\vec{u}_i$, p_i, s_i and θ of immiscible liquids moving in a non-deformable porous medium is closed either by the assumption of incompressibility of liquids, i.e. $\rho_i^0 = const$, or by the condition $\rho_i^0 = \rho_i^0(p_i)$. The obtained mathematical model in the case of a stationary porous medium ($\vec{u}_3 = 0$) and for a given porosity ϕ is called the Musket-Leverett model.

The principal point is to take into account the compressibility of the porous medium. Following [11], we supplement the system (1) - (3)

$$div\vec{u}_3 = -\frac{1}{\xi(\phi)}p_e - \beta_t(\phi)(\frac{\partial p_e}{\partial t} + \vec{u}_3 \cdot \nabla p_e), \tag{4}$$

$$\rho_{tot}\vec{g} + div((1-\phi)\eta(\frac{\partial \vec{u}_3}{\partial \vec{x}} + (\frac{\partial \vec{u}_3}{\partial \vec{x}})^*)) - \nabla p_{tot} = 0, \tag{5}$$

where $p_e = p_{tot} - p_f$ is the effective pressure, $p_{tot} = \phi p_f + (1-\phi)p_3$ is the total pressure, $p_f = s_1p_1 + s_2p_2$, p_3 are, respectively, the pressures of the liquid and solid phases, $\rho_{tot} = (1-\phi)\rho_3^0 + \phi(s_1\rho_1^0 + s_2\rho_2^0)$ is the total density; η is the viscosity of a solid skeleton, $\xi(\phi)$ and $\beta_t(\phi)$ are the coefficients volumetric viscosity and volume compressibility of a porous medium are given functions.

After determining the saturation of water and air s_i, temperature θ and the filtration rates $\vec{v}_i = \phi s_i(\vec{u}_i - \vec{u}_3)$ we can consider the problem of the motion of a conservative impurity due to the transfer of the water phase and diffusion. This process describes the convective diffusion equation [12]:

$$R + \frac{\partial}{\partial t}(\phi s_1\sigma) + div(\sigma\vec{v}_1 - D\nabla\sigma) = 0. \tag{6}$$

Here σ is the impurity concentration, $\vec{v}_1$ is the water filtration rate, R is the source that takes into account the possible deposition of the impurity. The following dependencies are used: $D = \eta_1 + \lambda_0|\vec{v}_1|$, $\eta_1 = const > 0$ is the coefficient of molecular diffusion, $\lambda_0 = const > 0$ is the dispersion parameter; $R = -\Gamma s_1(\sigma_* - \sigma)$, $\Gamma = const > 0$, $\sigma_* = const \in [0,1]$. System (1) - (5) is very complex and has been studied only in individual cases [13, 14, 15].

2.1. *Self-similar solution*

On the basis of [13], a self-similar solution of the traveling wave type for the problem of salt transfer in melting snow is constructed under the following assumptions:

$$\vec{u}_3 = 0, \quad I_{13} = I_{13}(\theta), \quad I_{12} = 0, \quad I_{23} = 0, \quad L_s = 0, \quad \rho_i^0 = const\ . \tag{7}$$

By (7), the continuity equation for the solid phase implies $\frac{\partial\rho_3^0(1-\phi)}{\partial t} = I_{31}(\theta)$. In particular, we can assume that porosity is a function of temperature.

The following dependence is accepted: $\alpha_3(\theta) = 0, \theta \geq \theta^+$ (ice melting temperature); $\alpha_3(\theta) = 1 - \phi^- - \phi_1(\theta - \theta_1), \theta_1 \leq \theta \leq \theta^+; \alpha_3(\theta) = 1 - \phi^-, \theta \leq \theta_1$. Here $0 < \theta^- < \theta_1 < \theta^+, \phi^- = \phi(\theta^-), \phi_1 = (1-\phi^-)/(\theta^+ - \theta_1)$ – specified parameters. Also assumed $K_0 = const > 0, \lambda_c = a_c + b_c\rho_c^2, \rho_c = \sum_{i=1}^3 \rho_i^0\alpha_i, (a_c, b_c) = const > 0, \vec{g} = (0,0,-g)$; the unknown functions depend only on the variable $\xi = x_3 - ct \in (-\infty, 0), c = const < 0$.

The system of equations (1) - (5) reduces to a system of equations for $s(\xi) \equiv s_1(\xi)$ and $\theta(\xi)$ of the form

$$\lambda_c\frac{d\theta}{d\xi} = f_1(\theta), \quad a_0(s) = \frac{ds}{d\xi} = f_2(s,\theta),$$

and s, θ satisfy the following conditions $\theta(0) = \theta^+, s(0) = s^+, (\theta, s, \frac{d\theta}{d\xi})_{\xi\to-\infty} = 0$. Here functions $f_1(\theta), f_2(s,\theta), a_0(s)(a_0(0) = a_0(1) = 0)$ can easily be recalculated through the initial parameters of the source system [13]. The solution of the formulated problem exists and in particular has properties: $0 \leq s(\xi) \leq 1$, there exists a point $\xi_* \in (-\infty, 0)$ such that $s(\xi) = 0$ for all $\xi \leq \xi_*$.

IOP Conf. Series: Earth and Environmental Science **193** (2018) 012055 doi:10.1088/1755-1315/193/1/012055

In the self-similar case, the following problem arises for finding the impurity concentration σ:

$$\begin{gathered} -\Gamma s(\sigma_* - \sigma) + \tfrac{d}{d\xi}(|c|\tfrac{\rho_3^0}{\rho_1^0}(\phi - \phi^-)\sigma - D\tfrac{d\sigma}{d\xi}) = 0, \\ \sigma|_{\xi\to-\infty} = 0, \quad \tfrac{\partial\sigma}{\partial\xi}|_{\xi\to-\infty} = 0, \quad \sigma(0) = \sigma^+ \in (0, 1]. \end{gathered} \tag{8}$$

Since $s(\xi) = 0$ for $\xi \le \xi_*$, then $\sigma(\xi) = 0$, $\xi \le \xi_*$. Thus for $\sigma(\xi)$ we consider the problem

$$\begin{gathered} \tfrac{d}{d\xi}(D\tfrac{d\sigma}{d\xi} - |c|r\sigma) + \Gamma s(\sigma_* - \sigma) = 0, \quad \xi_* < \xi < 0, \\ \sigma(\xi_*) = 0, \quad \sigma(0) = \sigma^+ \in (0, 1], \quad \sigma_* \in [0, 1]. \end{gathered} \tag{9}$$

Here $r = 0$, $D = \eta_1 + \lambda_0|c|\phi^- s$ for $\xi \in [\xi_*, \xi_1]$; $r = \frac{\rho_3^0}{\rho_1^0}(\phi - \phi^-)$, $D = \eta_1 + \lambda_0|v_1|$, $v_1 = |c|(\phi s - r)$ for $\xi \in [\xi_1, 0]$.

The solution of problem (8), (9) exists and satisfies inequality $0 \le s \le 1$. The numerical investigations of problem (8), (9) were carried out in [16].

2.2. The numerical study of the two-dimensional problem

Based on the hypotheses (7) of paragraph 2.1, the two-dimensional problem of the motion of water and air in melting snow is considered. The system (1) - (5) reduces to a system of three equations with respect to the saturation of the water phase s, the temperature θ and the reduced pressure p [1]:

$$\frac{\partial}{\partial t}(\rho_1^0\phi s) = div(\rho_1^0(K_0 a\nabla s + K_1\nabla p + \vec{f_0})) + \frac{\partial}{\partial t}(\rho_3^0\phi(1 - s)),$$

$$(\sum_{i=1}^{3}\rho_i^0 c_i\alpha_i)\frac{\partial\theta}{\partial t} + (\sum_{i=1}^{2}\rho_i^0 c_i\vec{v_i})\nabla\theta = div(\lambda_c\nabla\theta) + \nu\frac{\partial\rho_3}{\partial t},$$

$$div(K\nabla p + \vec{f}) = -\frac{\partial}{\partial t}((1 - \frac{\rho_3^0}{\rho_1^0})\phi),$$

where ν – specific heat of melting ice,

$$p = p_1 - \int_s^1 \frac{\partial p_c}{\partial\xi}\frac{\bar{k}_{02}}{k}d\xi, \quad k = \bar{k}_{01} + \bar{k}_{02}, \bar{k}_{0i} = \frac{k_{0i}}{\mu_i},$$

$$K = kk_0, K_i = K_0\bar{k}_{0i}, \quad a = -\frac{\partial p_c}{\partial s}\frac{\bar{k}_{01}\bar{k}_{02}}{k} \ge 0,$$

$$\vec{f} = \sum_{i=1}^{2}(K_i\rho_i^0\vec{g}), \quad \vec{f_0} = K_1\rho_1^0\vec{g}.$$

The solution of this system is sought in the region $\Omega = 0 \le x_1 \le L, 0 \le x_2 \le H$ under the following conditions

$$s(x_1, x_2, 0) = s_0(x_1, x_2), \quad \theta(x_1, x_2, 0) = \theta_0(x_1, x_2),$$

$$s(x_1, 0, t) = 0, \quad \theta(x_1, 0, t) = \theta^-, \quad s(x_1, H, t) = 0, \quad \theta(x_1, H, t) = \theta^+,$$

$$p(x_1, H, t) = p_H(x_1, t), \quad p(x_1, 0, t) = p_0(x_1, t),$$

$$\frac{\partial s}{\partial x} = \frac{\partial\theta}{\partial x} = \frac{\partial p}{\partial x} = 0, \quad x_1 = 0, x_1 = L.$$

The dependence of porosity on temperature is described in paragraph 2.1. For the numerical solution of this initial-boundary value problem, the variable direction method is used. After finding the temperature θ, the filtration rate $\vec{v_1}$ and the saturation s on the basis of equation (6), we can consider the problem of the motion of a conservative impurity in melting snow.

2.3. The problem of sublimation of the snow-ice cover

In addition to thawing, a high role in the balance of the snow cover is played by the sublimation process. Under certain conditions, a significant mass of snow passes into the gas phase, bypassing the liquid phase. The sublimation process is significantly accelerated by the effect of the wind impaction the snow. A mathematical study of the windless sublimation of snow was carried out in work [17] in which, in analogy with [11], a self-similar solution is constructed with allowance for the phase transition of ice into moist air. The formula for the intensity of the phase transition I_{32} (ice-moist air) is obtained in work [18] under wind-induced conditions:

$$I_{32}(x,t) = \frac{dm}{dt}\frac{c(x,t)}{\frac{4}{3}\pi r^3}, \quad \frac{dm}{dt} = \frac{2\pi r(\frac{\rho}{\rho_n(\theta)} - 1)}{\frac{L_s}{K\theta Nu}(\frac{L_s M}{R\theta} - 1) + \frac{1}{D\rho_n(\theta)Sh}}. \tag{10}$$

Here $c(x,t)$ is the concentration of ice, dm/dt is the rate of change in mass, $r(m)$ is the radius of the particle, ρ is the air density, θ is the absolute temperature, $\rho_n(\theta)$ is the saturated water vapor density, L_s is the heat of sublimation of ice, M is the molecular mass of water, D is the diffusion coefficient, R is the universal gas constant, K is the molecular thermal conductivity in the atmosphere, Nu is the Nusselt number, Sh is the Sherwood number.

Formula (10) is actively used to describe the processes of sublimation in the snow cover [19], [20]. As an example, let us consider the problem of the motion of moist air in a mobile porous ice skeleton. The system (1) - (5) in the case under consideration takes the form

$$\frac{\partial \phi\rho_2^0}{\partial t} + div(\phi\rho_2^0\vec{u}_2) = -I_{32}, \tag{11}$$

$$\frac{\partial(1-\phi)\rho_3^0}{\partial t} + div((1-\phi)\rho_3^0\vec{u}_3) = I_{32}, \tag{12}$$

$$\phi(\vec{u}_2 - \vec{u}_3) = -K_0\frac{k}{\mu}(\nabla p_2 + \rho_2^0\vec{g}), \quad p_2 = R\theta\rho_2^0, \tag{13}$$

$$(\sum_{i=2}^{3}\rho_i^0 c_i\alpha_i)\frac{\partial\theta}{\partial t} + (\sum_{i=2}^{3} c_i\rho_i^0\alpha_i\vec{u}_i)\nabla\theta = div(\lambda_c\nabla\theta) - L_s I_{32}. \tag{14}$$

Here we use the notations adopted for the system (1) - (5) (taking into account the absence of the liquid phase: $\alpha_2 = \phi, \alpha_3 = 1-\phi, p_{tot} = \phi p_2 + (1-\phi)p_3, p_e = p_{tot} - p_2$). In case $I_{32} = 0$ and $\theta = const$ the system (11) - (14) was used in [15]. In case $I_{32} \neq 0$, a self-similar solution of the isothermal motion of a compressible gas in a deformable porous medium is constructed, and it is established that $0 \leq \rho_2 \leq \rho_n, 0 \leq \phi \leq 1$ [21].

2.4. Distribution of water flow of melting snow between groundwater and surface water

The slope current of surface and ground waters, caused by intensive snowmelt, is considered. Such flows are characteristic of hilly regions. Consider a two-dimensional (in the vertical plane x, z) flow of groundwater to the river bank in a layer of soil between a horizontal water column and a land surface inclined to the river. The current is caused by the melting of snow, the intensity of which is greater than or equal to the filtration coefficient of the upper layer of the soil, as a result of which a slope runoff over the surface of the earth can also be formed. The complex aspects of freezing of such water are not yet considered, i.e. the case of positive surface temperature is modeled. The underground part of the current is modeled by the equation of water content:

$$div(K\nabla\Phi) = \frac{\partial W}{\partial t},$$

where the pressure $\Phi = p - z$ (the z axis is directed downwards), p is the capillary pressure. At full saturation of the pores $p \geq 0$ (it is assumed that the atmospheric pressure is 0), coefficient of moisture conductivity $K = K_s$, volume moisture $W = W_s$. For incomplete saturation, when $p < 0$, it is assumed that the dependences of K and W on p can be described by the expressions [22]

$$K = K_s \left(\frac{W - W_r}{W_s - W_r} \right)^n, \quad W = W_r + (W_s + W_r) \exp(p/p_0).$$

Here K_s is the filtration coefficient (at full saturation); W_s is the total saturation humidity, W_r is the residual water content; p_0 is the conditional height of capillary uplift; $n \approx 3.5$. The slope runoff on the surface of the earth Γ in the presence of a water layer ($p \geq 0$) is modeled by the diffusion wave equation [23] in the following form:

$$\frac{\beta}{\alpha} \frac{\partial}{\partial l} \left(p^{5/3} \frac{\partial \Phi}{\partial l} \right) + R - Q = \frac{\partial p}{\partial t},$$

where p, Φ are the sought in points of Γ; R is the specified snowmelt intensity; Q is the flow inside the domain; γ is the inverse coefficient of the roughness coefficient, $\beta = \cos^2 \alpha t g^{1/2} \alpha$, where α is the slope angle of the earth's surface; $l = x / \cos \alpha$ is a coordinate along Γ. The boundary conditions for W and ϕ are described in [22].

2.5. Movement of groundwaters in contact with frozen ground

If the snowmelt passes in the conditions of the frozen ground, the problem of distributing the water flow of melting snow between ground and surface waters becomes much more complicated. In particular, there is a need to describe the movement of groundwater in the aquifer, which is in contact with the frozen ground.

Following [24], frozen ground will be considered as a thermoelastic porous medium, the porous space of which fills the ice. For the stress tensor of the frozen soil skeleton P^{si}, the Dumel-Neumann law is adopted. For the stress tensor P of a frozen ground, it is natural to take $P = (1 - \phi) P^{si} + \phi P^i$, where ϕ is the porosity, P^i is the stress tensor ice.

In the region of groundwater movement, which have a positive temperature, the system of equations (1) - (5) is considered. The boundary between the frozen ground and the groundwater area is determined by the phase diagram of the frozen soil and by the strong discontinuity equation.

3. Conclusion

An analysis of mathematical models of the melting snow-ice cover, taking into account phase transitions and deformation of the ice skeleton under different rheological ratios, is carried out.

Acknowledgments

This work was partially supported by the grants RFBR 16-08-00291 "Hydroelastic and thermodynamic effects with inter-action of poroelastic ice".

References

[1] Papin A A, Sibin A N and Shishmarev K A 2016 *Mathematical models of melting snow-ice cover and thawing soils* (Altai State University Publishing, 2016) p 96 (in Russian)
[2] Lehning M et al 2002 *Cold Regions Science and Technology* **35(3)** 14–67
[3] Colbeck S C 1972 *J.Glaciol.* **11(63)** 369 - 85
[4] Mundy C J, Barber D G and Michel C 2005 *Journal of Marine Systems* **58** 107 20
[5] Thériault J M, Stewart R E, Milbrandt J A and Yau M K 2006 *Journal of Geophysical Research* **111**
[6] Gelfan A N, Pomeroy J W and Kuchment L S 2004 *Journal of Hydrometeorology* **5** 785–803
[7] Hock R and Holmgren B 2005 *Journal of Glaciology* **51(172)** 25–36

[8] Kuchment L S, Demidov V N and Motovilov Yu G 1983 *Formation of River Flow. Physical and Mathematical Models.* (Moscow: Nauka) (in Russian)
[9] Gray J M N T 1996 *Philosophical Transactions: Mathematical, Physical and Engineering Sciences* **354(1707)** 465–500
[10] Sellers S 2000 *Cold Regions Science and Technology* **31** 47-57
[11] Connolly J A D and Podladchikov Y Y 1998 *Geodinamica Acta* **11** 55–84
[12] Bear J and Cheng A H-D 2010 *Modeling Groundwater Flow and Contaminant Transport* (Springer) p 834
[13] Papin A A 2008 *Journal of Applied Mechanics and Technical Physics* **49(4)** 527–536
[14] Papin A A and Tokareva M A 2012 *Vestnik of Novosibirsk State University* **12(4)** 107–13
[15] Papin A A and Tokareva M A 2017 *Journal of Physics: Conference Series* **894(1)** 012070
[16] Korobkin A A, Papin A A and Khabakhpasheva T I 2013 *Mathematical Models of Snow and Ice Cover* (Barnaul: Altai State University Publishing) (in Russian)
[17] Papin A A and Yust E S 2017 *Izvestiya of Altai State University* **1(93)** 119–22 (in Russian)
[18] Thorpe A D and Mason B J 1966 *British Journal of Applied Physics* **17** 541–8
[19] Groot Zwaaftink C D, Löwe H, Mott R, Bavay M and Lehnin M 2011 *Journal of Geophysical Research* **116**
[20] Xiaoqing Dai and Ning Huang 2014 *Scientific Reports* **4**
[21] Glebova N Y and Papin A A 2017 *Materials of the international school-seminar Lomonosov Readings in the Altai* (Barnaul: Altai State University Publishing) pp 404–13
[22] Sabinin V I 1999 *Mathematical models of filtration and their applications* (Novosibirsk: Lavrentiev Institute of Hydrodynamics) p 204
[23] Antontsev S N, Epikhov G P and Kashevarov A A 1986 *System Modeling of Water Exchange Processes* (Novosibirsk: Science) p 216
[24] Khabirov V V and Khabirov S V 2010 *Proceedings of the Institute of Mechanics of the UNSC RAS* 202-10

On the issue of climate change in the Arctic zone of the Russian Federation

S Petrov[1], N Mamaeva[1,2] and M Narushko[1]

[1]Tyumen Scientific Center, SB RAS, Tyumen, Russia
[2]Industrial University of Tyumen, Tyumen, Russia

Abstract. The Arctic is one of the four regions of the world, classified by the Intergovernmental Panel on Climate Change as the most vulnerable region to climate change. A comparative analysis of the climate in almost half a century of the Purovsky District of the Yamalo-Nenets Autonomous District was carried out. We have established that the average annual air temperature and the amplitude of temperature fluctuations remained practically unchanged. In recent years, there has been an increase in precipitation in the cold season. There is a decrease in wind speed in the cold (except February) and in warm seasons. The number of cases without wind has decreased 2.8 times.

1. Introduction

One of the global problems facing humanity today is the changing planetary climate. Despite the fact that the views of scientists on the root causes of the observed climate changes are very different, the fact of the existence of the changes is beyond doubt [1]. Thus, at present it is possible to consider the established fact of the temporal dynamics of the Earth's climate occurring in separate regions in different scales and intensity [2]. At the same time, there is no single point of view on the main causes and trends in the dynamics of the modern climate, as well as its consequences for individual components of the natural environment. Therefore, recently in scientific circles [3-7] the dynamics of the climate, especially of the polar zones of the Earth, is actively studied. In modern studies, climate change, including in the Arctic, is taking an increasingly prominent place [6]. The Arctic is of great scientific and practical interest for both Russian and foreign scientists. First, as a powerful environmental regulator that can have a significant impact on the global climate system, in particular on ecosystems, people's way of life, their economies, etc. [8], since the Arctic is a "weather kitchen". Secondly, the Arctic is one of the four regions of the world assigned by the Intergovernmental Panel on Climate Change (IPCC) to the regions most vulnerable to climate change [6,9].

The urgency of the problem is exacerbated by the fact that the climatic significant processes and feedbacks operating in the Arctic climate system, the factors determining the regional climate changes in the region, the functioning of the climate system as a whole [9].

The aim of our work was to conduct a comparative analysis of the climate in almost half a century (1938 - 1950 and 2006-2010) using the example of the Purovsky District of the Yamalo-Nenets Autonomous District (YNAO).

2. Materials and methods

Climatic data were obtained according to the Climatological directories of the USSR for 1938 - 1950 [10,11] and on the site materials https://rp5.ru/Погода for 2006-2010. [12] from the meteorological station of Tarko-Sale. To study the climatic conditions of the Purovsky region of the Yamalo-Nenets Autonomous District, the temperature features were analyzed: the average monthly and mean annual air temperature, the maximum and minimum air temperature, and the amplitude of the temperature variations in the atmospheric air. In addition to the temperature characteristics of this region, the amount of precipitation, wind speed and direction, and the number of cases of calm are also considered.

Qualitative and quantitative characteristics were subjected to statistical processing using the integrated software package "IBM SPSS Statistics 21". In the normal distribution, the following statistical parameters were used: mean value (mean arithmetic mean, median, mode), variance and its derivative (mean square deviation). Comparison of the reliability of differences or similarity between the statistical characteristics obtained in the study of compared samples (by the Student's criterion) was made. To calculate the reliability of the differences between the mean values, the standard error of the arithmetic mean was calculated. In the absence of a normal distribution, nonparametric methods were used to compare two samples with the calculation of the Wilcoxon pair test.

3. Results of the study

The average monthly and average annual air temperature is given in table 1.

Table 1. Average monthly, average annual air temperature and amplitude of temperature fluctuations for the considered periods of time, °C.

Name	Average monthly and mean annual air temperature		Amplitude of temperature fluctuations	
	1938 – 1950	2006 – 2010	1938 – 1950	2006 – 2010
January	-24.96±1.57	-24.18±0.35	34.25±2.71	39.42±1.89
February	-22.2±1.22	-25.72±0.31*	37.38±1.61	32.84±2.92
March	-17.77±0.82	-13.5±0.25***	41.18±1.22	41.5±2.6
April	-5.92±0.79	-7.11±0.27	32.81±1.4	35.2±1.03
May	0.39±0.59	-0.51±0.16	31.15±1.59	26.24±2.36
June	10.25±0.71	11.24±0.18	26.6±1.02	27.96±0.93
July	14.92±0.49	16.90±0.16**	24.39±0.87	25.9±1.44
August	12.95±0.39	12.19±0.13	27±0.88	22.44±1.25#
September	6.48±0.49	6.53±0.14	19.82±1.75	25.64±1.82#
October	-3.42±0.83	-2.44±0.2	28.11±2.61	28.72±4.87
November	-15.59±1.15	-17.61±0.29	36.46±1.46	37.94±2.16
December	-22.52±1.32	-22.54±0.31	42.38±1.02	36.34±1.28##
Year	-5.51±0.4	-5.31±0.14	76.85±0.83	76.54±3.43

* - the reliability of differences with the average monthly and mean annual air temperature for 1938 - 1950. (* - $p < 0.05$; ** - $p < 0.01$; *** - $p<0.001$);

- the reliability of differences with the amplitude of temperature fluctuations for 1938 - 1950. (# - $p < 0.05$; ## - $p < 0.01$).

It has been established that in recent years there has been a decrease in the mean monthly temperature in the cold period of the year (in February, at $p < 0.05$) and an increase in the warm period (in July at $p < 0.01$). As for the average annual air temperature for 1938 - 1950 and 2006 - 2010, it practically did not change (-5.51 ± 0.4 °C and -5.31 ± 0.14 °C, respectively).

We calculated an important characteristic of weather and climate - the amplitude of temperature fluctuations (table 1), which is defined as the difference between the maximum and minimum air temperature (table 2).

Thus, the annual amplitude of temperature fluctuations did not practically change during the considered periods of time, with the exception of August ($p < 0.05$) and December ($p < 0.01$), where the temperature difference was higher for 1938 - 1950 and September, where the temperature difference is lower ($p < 0.05$).

As can be seen from table 2, there were no significant changes in the minimum values of air temperature during the analyzed periods. However, the maximum values of air temperature in recent years have become higher, especially in January, March, July, November. This confirms the presence of a change in the Arctic climate towards its warming.

Table 2. Minimum and maximum air temperature for the considered periods of time, °C.

Name	Minimum air temperature		Maximum air temperature	
	1938 – 1950	2006 – 2010	1938 – 1950	2006 – 2010
January	-41.72±2.77	-42.92±3.05	-7.46±1.24	-3.5±1.64#
February	-42.84±1.36	-42.38±1.35	-5.46±1.19	-9.54±3.15
March	-41.25±0.93	-38.48±2.3	-0.06±0.67	3.02±1.32#
April	-27.21±1.15	-27.24±1.24	5.6±0.67	7.96±1.72
May	-16.13±1.24	-13.28±1.91	15.02±1.48	12.96±1.99
June	-0.55±0.42	-1.08±0.96	26.05±1.18	26.88±0.86
July	2.78±0.39	4.66±1.21	27.17±0.83	30.56±1.24#
August	0.25±0.49	2.36±1.05	27.25±0.66	24.8±1.17
September	-3.02±0.47	-4.06±0.66	18.45±0.75	21.58±1.82
October	-22.95±1.31	-19.1±3.86	7.5±1.25	9.62±2.49
November	-37.18±1.61	-35.72±2.2	-0.72±0.85	2.2±0.67#
December	-45.66±1.06	-39.92±1.45*	-3.28±0.59	-3.58±1.65
Year	-47.72±1.01	-45.98±2.44	29.12±0.53	30.56±1.24

* - reliability of differences with a minimum of air temperature for 1938 - 1950 ($p < 0.01$);

\# - the reliability of the differences with the maximum air temperature for 1938 - 1950 ($p < 0.05$).

Also one of the characteristics of the climate are atmospheric precipitation - one of the links of moisture exchange on Earth. The average daily rainfall for the years 1938 - 1950 and 2006-2010 is presented in table 3.

Table 3. Average daily rainfall, mm.

Name	Average daily rainfall, mm	
	1938 – 1950	2006 – 2010
January	0.43±0.05	0.76±0.06***
February	0.42±0.05	0.68±0.05**
March	0.43±0.03	0.84±0.06***
April	0.76±0.11	1.61±0.17***
May	1.26±0.17	0.87±0.1*
June	2.15±0.21	1.87±0.19
July	1.97±0.33	2.28±0.24
August	2.13±0.32	2.65±0.27
September	2.04±0.27	1.8±0.17
October	1.25±0.14	1.36±0.09
November	0.71±0.07	1.22±0.08***
December	0.55±0.08	0.84±0.06*
Year	1.2±0.05	1.34±0.04*

* $p<0.05$; ** $p<0.01$; *** $p<0.001$.

When analyzing the average daily rainfall for 2006-2010 in comparison with 1938 - 1950 their increase was observed in January, March, April, November ($p < 0.001$), in February ($p < 0.01$), in December, and also for the year ($p < 0.05$). And for the cold period of 2006 - 2010 there is an increase of about 1.4 times in comparison with 1938 - 1950.

Another climatic characteristic is the wind regime, which has a significant influence on the dispersion of pollutants in the atmosphere in determining the maximum single concentrations, the calculation of which takes into account the corresponding adverse meteorological conditions in the region, including the dangerous wind speed [13]. The climatic parameters necessary for the realization of these calculations are determined by climatic data, the quality, volume and methods of processing the meteorological observations used while doing so must comply with the mandatory metrological requirements for measurements in accordance with Order No. 436 of the Ministry of Natural Resources of Russia of 19.10.2015 "On Approving the List of Measurements Relating to to the sphere of state regulation of ensuring the uniformity of measurements and performed in the implementation

of activities in the field of hydrometeorology and related her areas, and obligatory metrological requirements to them, including indicators measuring accuracy " [14].
The average annual wind speed is shown in table 4.

Table 4. Average annual wind speed, m/s.

Name	Average annual wind speed, m/s	
	1938 – 1950	2006 – 2010
January	3.22±0.24	2.48±0.05**
February	3.52±0.25	3.48±0.09
March	3.94±0.29	2.47±0.05***
April	4.29±0.19	2.69±0.05***
May	4.25±0.13	2.93±0.04***
June	4.25±0.14	2.58±0.05***
July	3.58±0.11	2.38±0.05***
August	3.25±0.12	2.04±0.04***
September	3.23±0.16	2.26±0.04***
October	3.89±0.16	2.53±0.04***
November	3.52±0.24	1.95±0.03***
December	2.78±0.28	2.05±0.04*
Year	3.64±0.09	2.38±0.01***

* p<0.05; ** p<0.01; *** p<0.001.

We found that in recent years of our studies, the wind speed both for cold (except February) and for warm periods has significantly decreased. The number of cases without wind (calm) became 2.8 times less.

In order to more fully characterize the wind regime of this region, we studied the direction of the wind. So for the years 1938 - 1950 the southern, north-western winds prevailed, and for 2006-2010 - south-western, north-western. Apparently, this is due to temperature differences between not very distant areas, which can be caused by different absorption coefficients of sunlight or different heat capacities of the surface.

4. Conclusions and suggestions

In recent years, due to technogenesis and general circulation of the atmosphere in the Arctic zone of Russia, there have been significant changes in the mean monthly temperature and amplitude of its fluctuations in some period of the year without changing the average annual indicators. Polar day and night cause an extremely uneven supply of solar radiation during the year. This loss was previously compensated by the inflow of warm air and water masses (southern and north-western directions) in the Arctic. In recent years, there has been a trend towards a decrease in the influx of warm air masses due to the reduction of southern winds. In recent years, winds in the region have become predominantly weak in the south-west and north-west directions, and their frequency has increased 2.8 times. Apparently, this is due to temperature differences between not very distant areas, which can be caused by different absorption coefficients of sunlight or different heat capacities of the surface. It is known that the anticyclonic circulation in the Siberian region of the Arctic develops in the winter. The average air temperature in February (the coldest winter month) has become significantly lower and ranges from -26.03 to -25.41 °C. The minimum temperatures did not change significantly. The average air temperature in July (the warmest summer month) became significantly higher and ranged from 16.74 to 17.06 °C. The maximum temperature was also higher and averaged 30.56 ± 1.24 °C. The increase in the maximum air temperature in 2006-2010 points to climate change in the Arctic zone of Russia towards its warming. Obviously, smoothing out the inversion of the air temperature can reduce the cooling of the surface in these areas. Here, in recent years, there has been an increase in the amount of precipitation to a greater extent due to snowfalls during the cold period of the year.

Thus, Purovsky region of Yamalo-Nenets Autonomous District for the analyzed period (≈ 50 years) there is no evidence of global warming. Nevertheless, the climate has become milder with increasing both the amount of precipitation and the redistribution of them due to the cold season.
In this direction the work will continue.

Acknowledgments
The work was carried out according to the state project, according to the Research Plan of the Tyumen Scientific Center of the SB RAS for 2018-2020, Protocol No. 2 of December 8, 2017 (Priority IX.133 Program IX.133.1 Project: IX.133.1.4 Cryobiological processes on land and in the coastal part of the Kara Sea in conditions of increasing average annual temperatures).

References

[1] Ivanov V V, Alekseev V A, Alekseeva T A, Koldunov N V, Repina I A and Smirnov A V 2013 Arctic ice cover becomes seasonal *Earth exploration from space* **4** 50-65
[2] Shatz M M and Skachkov Yu B 2016 The climate of the North: warming or cold snap *Climate and nature* **2**(19) 27-37
[3] Latysheva I V, Loschenko K A and Shakhaeva E V 2013 Circulation conditions of sudden stratospheric warming in the Northern Hemisphere in the 21st century *News of Irkutsk State University* **6**(1) 106-21
[4] Klimenko V V and Astrina N A 2006 Documentary evidence of strong climate fluctuations in the Russian Arctic in the 15th-20th centuries *History and Modernity* **1** 179-217
[5] Kislov A V, Grebenets V I, Evstigneev V M, Konishchev V N, Sidorova M V, Surkova G V and Tumel N V 2011 Consequences of possible climate warming in the 21st century in the north Eurasia *Bulletin of Moscow University. Series 5. Geography* **3** 3-8
[6] Katsov V M and Porfiryev B N 2012 Climatic changes in the Arctic: consequences for the environment and the economy *Arctic: ecology and economics* **2**(6) 66-79
[7] Klimenko V V, Matskovsky V V and Dalmann D 2013 Complex reconstruction of the temperature of the Russian Arctic over the past two millennia *Arctic: ecology and economics* **4**(12) 84-95.
[8] Davydov A N and Mikhailova G V 2013 Climate change and living conditions in the Arctic in the perception of the Nenets of the Vaigach Island *Ecology of man* **2** 29-34
[9] Panin G N, Diansky N A, Solomonova I V, Gusev A V and Vyruchalkina T Yu 2017 Assessment of climatic changes in the Arctic in the 21st century on the basis of a combined prognostic scenario *Arctic: ecology and economics* **2**(26) 35-52
[10] 1956 Climatological reference book of the USSR. Issue 17. Meteorological data for individual years. Part I: Air temperature. Part II: Precipitation. Part III: Snow Cover ed A A Shumanova (Leningrad: Hydrometeorological Publishing House) p 154
[11] 1961 Climatological reference book of the USSR. Issue 17. Meteorological data for individual years. Part IV: Wind ed A A Shumanova (Leningrad: Hydrometeorological Publishing House) p 581
[12] Reliable Prognosis. Access mode: https://rp5.ru/Погода
[13] Order of the Ministry of Natural Resources of Russia of 06.06.2017 No. 273 "On the approval of methods for calculating the dispersion of harmful (polluting) substances in the atmosphere" (Registered with the Ministry of Justice of Russia on August 10, 2017 No. 47734).
[14] Order of the Ministry of Natural Resources of Russia of 19.10.2015 No. 436 "On approval of the list of measurements related to the sphere of state regulation of ensuring the uniformity of measurements and performed in the implementation of activities in the field of hydrometeorology and related areas, and mandatory metrological requirements for them, including indicators of measurement accuracy "(Registered in the Ministry of Justice of Russia on 03.11.2015 No. 39593).

External ice reinforcements for ships in operation

Nikolay Pets

Faculty of Shipbuilding and Ocean engineering, Far eastern federal university, corp. 6.2-457, prosp. FEFU, Vladivostok, Russia

E-mail: pecng85@gmail.com

Abstract. The subject of the proposed article is the external ice reinforcements of the ship hull. Ice reinforcements for the Russian commercial fleet are an urgent topic, but the traditional measures of such matter (replacement of the shell plating, installation of an intermediate frame) are expensive and require a large amount of preparatory work. External reinforcement are an alternative, which is much cheaper than their installation. As a rule, it is performed in the form of a rolling profile of various shapes (most often a semicircle) welded to the outer skin. In addition to the lower cost, they have a number of other advantages.

1. Introduction

The relevance of ice hull reinforcement for already operating vessels is due to several reasons. First, the renovation of the domestic commercial fleet is usually through the purchase of foreign vessels that already have significant service lives. These ships are generally not designed for navigation in ice conditions, and therefore for full navigation in the Russian seas (and they all belong to the freezing), their hulls needs reinforcements. For example, in the FESCO group, the share of foreign-built vessels is 75%, mainly ships without ice categories, i.e. their navigation will be difficult in the winter-spring period, even in the water area of the Amur Bay. Other paragraphs are indented (BodytextIndented style).

The second reason is connected with the expansion of Arctic navigation due to the activation of domestic cargo transportation along the Northern Sea Route, the development of mineral deposits and hydrocarbons in high latitudes. The increase in cargo traffic according to the FGUU "Administration of the Northern Sea Route" and FSUE "Atomflot" from 2011 to 2017 was from 3111 thousand tons to 10,691 tons, i.e. for 7 years the freight traffic increased by 343.7%, on average by 19.3% per year. The increased number of transportations on Arctic routes requires a greater number of vessels of high ice categories. For this, it is possible to build new vessels or (which is much cheaper) to modernize existing ones.

Traditionally, the following measures are used to increase ice hull strength:

- installation of intermediate set - frames, stringer;
- replacement of the outer shell plating with thicker ones;
- complete modernization of the sections of the ship (as a rule, the fore end);
- a combination of several methods.

The installation of the intermediate set represents a reliable and the most frequently used method of improving the ice qualities of the vessel. In this case, the strength of not only the set but also of the outer shell increases, due to the reduction in the dimensions of the plate. However, most often the installation of additional frames is required in the area of bulbous fore end contours. Here the

IOP Conf. Series: Earth and Environmental Science **193** (2018) 012057 doi:10.1088/1755-1315/193/1/012057

execution of such work has technological difficulties, since The fore area of the hull is a cramped space, installation works in which it is difficult and requires a large amount of preparatory work.

Increasing the thickness of the outer shell plates is advisable to be used in combination with other measures of reinforcement of the shell or to increase the strength of the shell due to considerable wear. This is just a laborious and expensive method.

Modernization of the bow (cutting of the bulb) consists in the complete replacement of part of the underwater hull of the ship in the fore area. This is the most expensive measure, the use of which is justified only in rare cases. With an increase in ice hull strength and patency, this worsens the running and seaworthiness of the vessel on free water.

As a result of a brief review of traditional ice reinforcements, the following shortcomings should be noted:

- high cost of installation work because of the need to conduct them in cramped bow rooms;

- to achieve high levels of ice hull strength it is necessary to combine activities, as a rule, it is not enough just to install an intermediate set or change the shell plating.

Thus, there is a need for an alternative to traditional measures.

One option for this alternative is external ice hull reinforcement. They are rolled segments welded from the outside to the lining in parallel or at an angle to the waterline (figure 1 and 2).

Figure 1. External ice reinforcements in the bow of the ship.

Figure 2. External ice reinforcements.

In Russia, the concept of external reinforcements was developed at the Faculty of ship design of FENU (since 2012 it is part of FEFU) in 2008. Employees of the department developed projects of external amplification for the tanker-bunker "Flagman" and the refrigerator "Gulf of Raduga". Later, such gains were applied to other vessels.

In [1] the merits of this reinforcement method were called:

- in contrast to the internal set, the distance between the outer weld segments can vary over a wider range, which allows achieving the required strength level of the structure;
- such reinforcements are less labor intensive and the cost of installation is lower than that of traditional options, in addition, it requires considerably less preparatory work;
- welded segments serve as the protection of the outer shell plating, selecting some of the contact interactions with the medium.

Some aspects of this technology will be discussed below.

2. Effect of segments on the strength of the shell

The effect of segments on the strength of the hull was analyzed in [3]. As the object of analysis, a semicircular segment is taken (figure 3). The effect of segments on the strength of the skin was evaluated by two criteria: fiber yield and ultimate strength.

To assess the increase in the strength of the shell plating, the following indicators are proposed:

k_{min} - coefficient of reinforcement by the criterion of fiber yield,

k_0 - is the reinforcement coefficient by the criterion of ultimate strength.

Both coefficients represent the ratio of the moment of resistance of the shell plating to the reinforcement at the moment of the resistance of the shell plating without reinforcements, calculated according to the corresponding strength criteria. Below are the formulas for determining them.

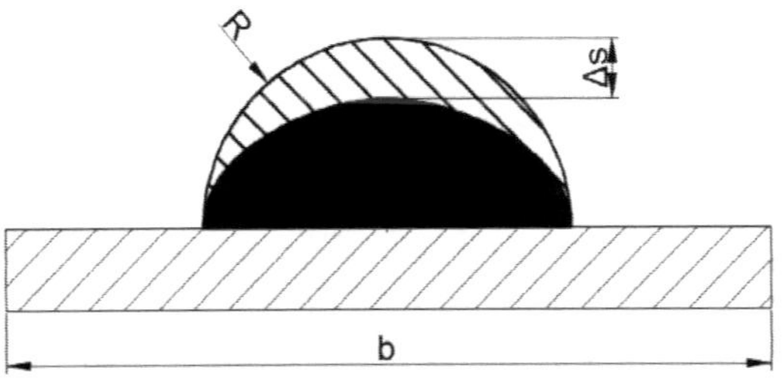

Figure 3. Symmetrical wear pattern of the segment.

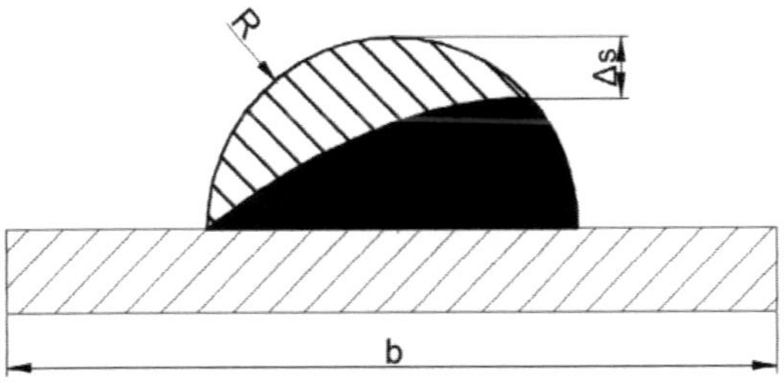

Figure 4. The asymmetric diagram of segment wear.

$$k_{\min} = W_{RS} / W_S = 6\frac{\overline{C} - \overline{B}^2 / \overline{A}}{R_s - \overline{B} / \overline{A}}, \tag{1}$$

where W_{RS}- the minimum moment of resistance of the shell plating with the reinforcements installed; W_S - the minimum moment of resistance of the shell plating without reinforcements;

$\overline{A} = A / sb_c = 0{,}5\pi \overline{R}_s \overline{R}_b + 1$; $\overline{B} = 2R_s^2 R_b / 3 - 0{,}5$; $\overline{C} = C / s^3 b_c = R_s^3 R_b / 8\pi + 1/3$; $R_b = R / b$; $R_s = R / s$;

R – radius of the reinforcement segment;

b – segmentation interval;

s – shell plating thickness.

$$k_0 = \frac{W_{ORS}}{W_{OS}} = 2\pi \overline{R}_s \overline{R}_b (4\overline{R}_s / 3\pi - \overline{e}) + 2\overline{e}^2 + 2(1 + \overline{e})^2, \tag{2}$$

where W_{ORS} - the limiting moment of the resistance of the shell plating with the reinforcements installed; W_{OS} is the limiting resistance moment of the shell plating without reinforcements, $\overline{e} = 0{,}25\overline{R}_s \overline{R}_b - 0{,}5$.

In [3], it was concluded that the segments had an ambiguous effect on the strength of the skin: profiles with a sufficiently low value of the minimum moment of resistance could reduce the strength of the shell plating. Those to have a positive effect on the strength, the moment of resistance of the segment to be installed must be at least a certain value.

3. Influence of the outer segments on the wear of the outer shell plating

In this area, certain results have also been achieved - based on theoretical consideration of the problem, the calculated dependences for determining the wear of the reinforced shell plating are proposed.

It should be said that there is no operational data on the wear of the outer segments. Therefore, all the results obtained in this area are purely theoretical and require verification by practice.

Two theoretical approaches can be distinguished in considering a plate with reinforcements. The first relates to the case of installing reinforcements in parallel to the waterlines and assumes uniform wear of the segment (see Figure 3). According to him the following assumptions are proposed:

- the total weight wear of the shell with reinforcements does not exceed the wear of the shell without reinforcements; their equality is the most dangerous case, from which the calculation should be based;
- the reduction in the cross-sectional area is calculated from the allowance for wear set in the classification survey rules.

To show the effect of segments on the wear of the shell plating, the reinforcement factor is used, representing the ratio of the wear of the shell with the established segments to the wear of the shell without them. Taking into account the accepted assumptions, the following formulas were obtained:

$$k_s = \frac{\Delta S_s}{\Delta S}, \tag{3}$$

where

k_s - reinforcement factor,
ΔS_s – surcharge for reinforced plating,
ΔS –surcharge for the shell plating without reinforcement;

$$k_s = 1 - \frac{\pi(1-m_s)}{2(1-m_0)} \cdot \frac{\bar{R}^2}{\bar{b}} \geq 0, \tag{4}$$

where m_s is the coefficient of the allowed residual sectional area of the segment; m_0 - coefficient of allowable residual thickness of the outer skin according to the Rules

$\bar{R} = R / S_0$, $\bar{b} = b / S_0$,

s_0 - the initial or required thickness of the outer shell plating, without taking into account the reinforcements.

The disadvantage of this approach is that wear depends on the thickness of the shell. For general reasons, it must depend on the intensity of the external loads.

Another approach is built for the case of the segment at an angle to the waterlines and assumes an asymmetrical wear (see Fig. 4). He proposes the following assumptions:

- reducing the thickness of the reinforcing segment does not exceed the reduction of the thickness of the shell without reinforcement as a result of wear;
- segment wear is uneven, resulting in a parabolic segment.

These assumptions made it possible to obtain the following formulas:

$$k_s \approx \frac{1}{1-2\bar{R}}\left(1 - \frac{4}{3}\bar{R} - 0{,}238\bar{R^2}\frac{b}{\Delta s}\right) \geq 0, \tag{5}$$

where $\bar{R} = \frac{R}{b}$. Table 1 shows the coefficient values for some combinations of R and b.

Table 1. The values of the skin gain factor for Δs = 5 mm and for various R and b.

R, mm	b, mm				
	250	300	350	400	450
30	0,88	0,90	0,92	0,93	0,94
35	0,81	0,85	0,88	0,89	0,91
40	0,71	0,78	0,82	0,85	0,87
45	0,59	0,68	0,74	0,79	0,82
50	0,43	0,57	0,66	0,71	0,76
55	0,23	0,44	0,55	0,63	0,68
60	0,00	0,27	0,43	0,53	0,60

As a rule, when installing reinforcements, b is selected in the range 250 ... 350 mm, R - 40-60 mm. The average value of the coefficient in this case is 0.67. Those. the amount of the wear allowance for the reinforced plating can be reduced by 30%.

4. Conclusion

The features of external ice reinforcement described above make it possible to consider them as a very effective and relatively inexpensive tool for reinforcing the hull. The advantages of this method were listed above. However, it is necessary to note some points that hinder the wide spread of this technology:

- there is no regulation on the part of the Register, and therefore the development of a reinforcement scheme and the rationale for increasing the ice class is a non-trivial task requiring high qualification of the performer;
- there is no systematic study of the impact of such reinforcements on the strength of the hull;
- there is insufficient operational information on the operation of the skin with such enhancements in ice conditions

In general, the wide spread of external amplifications is hindered by the poor knowledge of this method. The accumulation of experience in the realization of such amplifications, given thickness gauges on vessels with realized reinforcements and theoretical studies should accelerate its popularization.

References

[1] Kazakov E V, Kalenchuk S V, Kotlyarova I A and Sidorenko I S 2008 Increased ice qualities of reclassified vessels *Vologdin read.* 113-18
[2] Kulesh V A, Pets N G and Ohio S A 2017 The project of ice reinforcements of a dry cargo vessel *Marine intellectual technology* **37** p 64-70
[3] Kulesh V A 2013 Improving operation realiability of ice reinforcement of vessels hulls *Asia-Pacific Journal of Marine Science&Education* **3** pp 9-16

Analysis of elastic waves generated in frozen grounds by means of the electromagnetic pulse source "Yenisei"

V M Sadovskii[1,2], O V Sadovskaya[1,2] and V A Detkov[1,2]

[1] Institute of Computational Modeling SB RAS, Akademgorodok 50/44, Krasnoyarsk, Russia
[2] LLC "Evenkiyageofizika", Leningradskaya 66, Krasnoyarsk, Russia

E-mail: sadov@icm.krasn.ru, o_sadov@icm.krasn.ru, v-detkov@yandex.ru

Abstract. Computational technology for multiprocessor computing systems of cluster architecture is developed, the ultimate goal of which is numerical modeling of wave fields generated by the electromagnetic pulse source "Yenisei" in blocky-layered geomedia. To describe the wave processes, mathematical models of the dynamics of elastic and elastic-plastic media, of porous and granular materials are applied. The algorithms of numerical realization are constructed based on the method of two-cyclic splitting with respect to spatial variables. Computational experiments showed that the proposed technology allows reproducing the system of waves near the region of excitation of seismic oscillations in 3D setting with a high degree of details and accuracy. The results of computations can be used in working out the optimal modes of functioning the source "Yenisei", when mechanical characteristics of the layers vary in a wide range from solid and frozen grounds with inclusions of rock till granular and clayey water-saturated grounds. Numerical analysis makes also possible to obtain the averaged data, necessary for the adequate simulation of the localized pulse action from the source, using simplified mathematical models for calculating the synthetic seismograms of reflected waves over large scale and at great depth of layers bedding in complex geomedia.

1. Introduction

Electromagnetic source of seismic oscillations "Yenisei" is a non-explosive surface pulse seismic source with an electromagnetic actuator that contains one, two or four short-stroke electromagnets, working synchronously, with an autonomous power supply system from a capacitive storage of electric energy and a device for charging and discharging [1, 2, 3]. The source exists in wheeled, sledge, mobile and water variants (see Figure 1). "Yenisei" is quite competitive in comparison with sources of explosive and vibratory types by efficiency and quality of exploration works, and it has undeniable advantages in the economical and environmental aspects. The use of this source is incomparably cheaper, and it is almost the only possible means when working near buildings and constructions, in water protection zones and in areas where there are a lot of rivers and lakes. It is also applied on ice coverings of reservoirs, in shallow waters and on offshore.

In the process of creation, modification and improvement of operational and technical characteristics, the electromagnetic pulse source "Yenisei" was subjected to careful experimental analysis and testing [4, 5]. In the present work, computational technology is presented, the goal

Figure 1. Seismic sources of the "Yenisei" series: http://gseis.ru/our-business/field-seismic-works/impulse-technique/.

of which is detailed mathematical modeling of wave fields excited by the "Yenisei" source in blocky-layered geomaterials with different mechanical characteristics of blocks (ice cover, water, clay soils, granular and porous media). The results of modeling will be used to optimize regimes of functioning the source during seismic surveys.

2. Mathematical model

For numerical modeling of the processes of propagation of strain waves in rheologically complex media, we developed computational algorithms and software complexes, oriented to multiprocessor computing systems of cluster architecture. Mathematical models of elastic-plastic, granular and porous media, taking into account different resistance of materials to tension and compression, were worked out. Equations of the Cosserat continuum describing wave motion of structurally inhomogeneous media, in which along with translational degrees of freedom, independent rotations of the material particles are taking into account, were also realized numerically.

It is assumed that the structure of a medium is known and represented by a set of heterogeneous blocks with curvilinear boundaries. Each block is characterized by its homogeneous material with corresponding governing equations. In the simplest case of an elastic block, a system of equations of the linear dynamic elasticity, written in terms of velocities v and stresses σ, is fulfilled:

$$\rho\,\frac{\partial v}{\partial t} = \nabla\cdot\sigma, \quad \frac{\partial \sigma}{\partial t} = \rho(c_1^2 - 2\,c_2^2)(\nabla\cdot v)\,I + \rho\,c_2^2\,(\nabla v + \nabla v^*). \qquad (1)$$

Here ρ is the density, c_1 and c_2 are the velocities of longitudinal and transverse elastic waves, respectively, ∇ is the gradient over spatial coordinates, I is the unit tensor. Asterisk denotes conjugate tensor, conventional notations of tensor analysis are used.

The initial velocities and stresses are assumed to be zero. On a part of the boundary, external stresses caused by the action of concentrated load are preset. On the planes of symmetry, if any, the symmetry conditions are formulated. A part of the boundary can be a non-reflecting surface, on which the conditions for passage of waves without appreciable reflection are simulated. At the internal interfaces, the continuity conditions are set for vectors of velocities and stresses at the contact areas of the blocks.

Under numerical implementation, independent computational grids are constructed in the blocks. An algebraic method, consisting in the finding a one-to-one mapping of computational domain in the form of a unit cube with uniform grid onto the physical domain, is used.

The system of equations (1) during the construction of finite-difference scheme is represented in the matrix form:

$$\frac{\partial U}{\partial t} = \sum_{i=1}^{3} A_i \frac{\partial U}{\partial x_i}, \tag{2}$$

where U is the vector–function composed of the components of velocity vector and stress tensor, A_i are the matrices–coefficients of the equations. To solve this system, the scheme of integration based on the two-cyclic splitting method is applied [6]. The splitting is performed not in physical space, but in parametric one. The system is transformed to the following form:

$$\frac{\partial U}{\partial t} = \sum_{i=1}^{3} \widetilde{A}_i \frac{\partial U}{\partial \xi_i}, \quad \widetilde{A}_i = \sum_{j=1}^{3} \frac{\partial \xi_i}{\partial x_j} A_j. \tag{3}$$

The splitting method leads to a series of 1D problems, which are solved based on the explicit monotone ENO–scheme of the "predictor–corrector" type with the limit reconstruction of solution. To preserve the conservativity of scheme on curvilinear grids, when the matrices–coefficients $\widetilde{A}_i$ depend on coordinates, the approximation of 1D systems of equations on the "corrector" step is performed by means of the integro-interpolation method. As a result of integrating the system (2) over a curvilinear cell in the physical area with the subsequent application of the Green formula, we obtain the following equality:

$$\frac{\partial \overline{U}}{\partial t} = \frac{1}{\omega} \sum_{k=1}^{6} \gamma^k \left(n_1^k A_1 + n_2^k A_2 + n_3^k A_3\right) U^k, \tag{4}$$

where n_i are the directing cosines of the external normal, ω is the cell volume, $\overline{U}$ is the average integral value of the solution vector, and the index k is related to the faces of the cell, in particular γ^k is the area of corresponding face. The sum in the right-hand side of (4) is divided into three pairs of terms on opposite faces, each of which corresponds to the approximation of spatial derivatives in 1D systems.

Described algorithm was generalized for the cases of a granular medium [7], a porous material [8], a Cosserat continuum and a multiblocky medium [7, 9]. The developed method of solution was implemented as a software package in the Fortran language by means of the MPI (Message Passing Interface) library. Technology of parallelization is based on the uniform distribution of computational domain between the nodes of a cluster. Verification of the software was performed on exact solutions – formulas of geometric seismics for the hodographs of reflected and refracted waves. Parallel program system was registered in Rospatent [10].

3. Results of computations

Preliminary series of computations was fulfilled with the goal of validating the software package by basic parameters of the source – the frequencies and amplitudes of oscillations. Comparison of the numerical results with the available experimental data showed a satisfactory quantitative correspondence.

The problem was solved for a two-layered massif of an elastic medium of 60 m × 40 m × 40 m, in two variants, when the upper 10-meter layer is more compliant and, conversely, more rigid as compared with the lower 50-meter basic layer. Similar problem was considered for the case, when the upper layer is water, and for a three-layered medium consisting of the upper 3.5-meter layer from ice, the middle 10-meter layer of water and the lower thick layer of ground. To demonstrate the capabilities of the program, the interface between layers of ground with different properties (or between water and ground) was curved.

Figure 2 demonstrates the uniform distribution of two-layered computational domain between 96 computational nodes: 16 nodes in the upper layer and 80 nodes in the lower one. Each node of a cluster performs computations in parallel mode. The difference grid in the upper layer of a massif is $50 \times 200 \times 200$ cells, and in the lower layer – $250 \times 200 \times 200$ cells, that is, each cluster node performs computations on a grid of $50 \times 50 \times 50$ cells. For visibility, the difference grid in Figure 2 is thinned 5 times in each direction. In the case of a three-layered medium, computational domain is distributed uniformly between 16 nodes in the upper layer (ice), 16 nodes in the middle layer (water) and 64 nodes in the lower layer (ground).

At the upper boundary of computational domain a localized action from the source with four electromagnets is set. Taking into account the symmetry, computations are made for a quarter of the whole massif bounded by vertical coordinate planes. On the left and right boundaries of computational domain (see Figure 2) the symmetry conditions are given. The back boundaries and the lower base are considered as non-reflecting surfaces. At the stages of solving 1D systems of the splitting method, the boundary values of Riemann invariants corresponding to outgoing characteristics are assumed to be zero on these surfaces, that is equivalent to the absence of reflected waves in 1D problems.

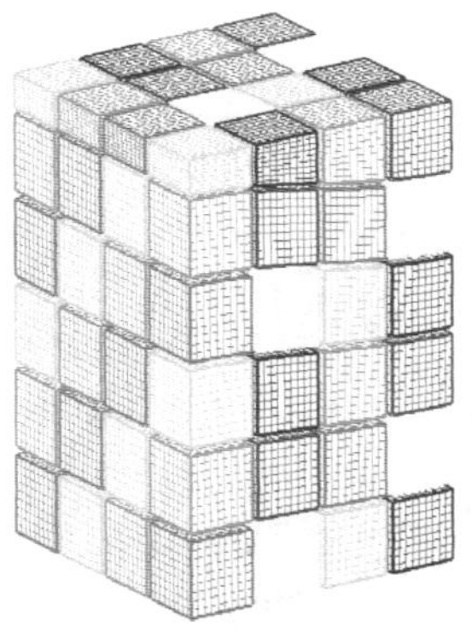

Figure 2. Uniform distribution of computational load between 96 cluster nodes.

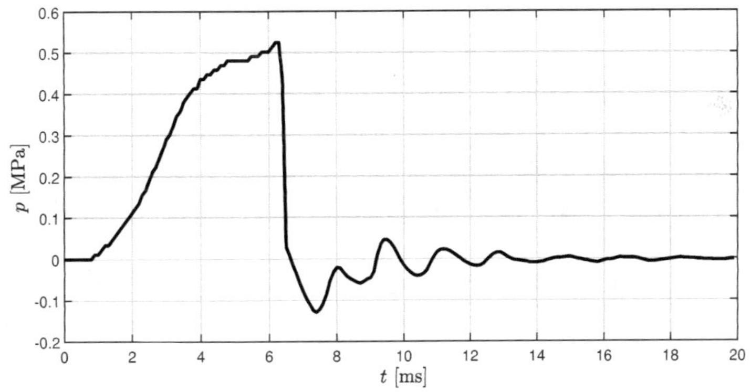

Figure 3. Dependence on time of pressure from the source in the localization zone.

In Figure 3 one can see the dependence on time of pressure from the source in the localization zone, a circle of area $1\,\mathrm{m}^2$, which is located at the distance of 2.5 and 1.25 m from the left and right boundaries of symmetry, respectively. The pressure in this form was determined on the basis of experimental measurements of the acceleration of reactive mass of the electromagnet. Densities and velocities of elastic waves in the layers from different materials are given in Table 1.

Table 1. Mechanical parameters of materials.

	$\rho\,[\mathrm{kg/m^3}]$	$c_1\,[\mathrm{m/s}]$	$c_2\,[\mathrm{m/s}]$
ice	900	3000	1800
water	1000	1450	0
clay	2100	1800	1100
ground	2400	4500	2700
rigid ground	2600	6000	3500

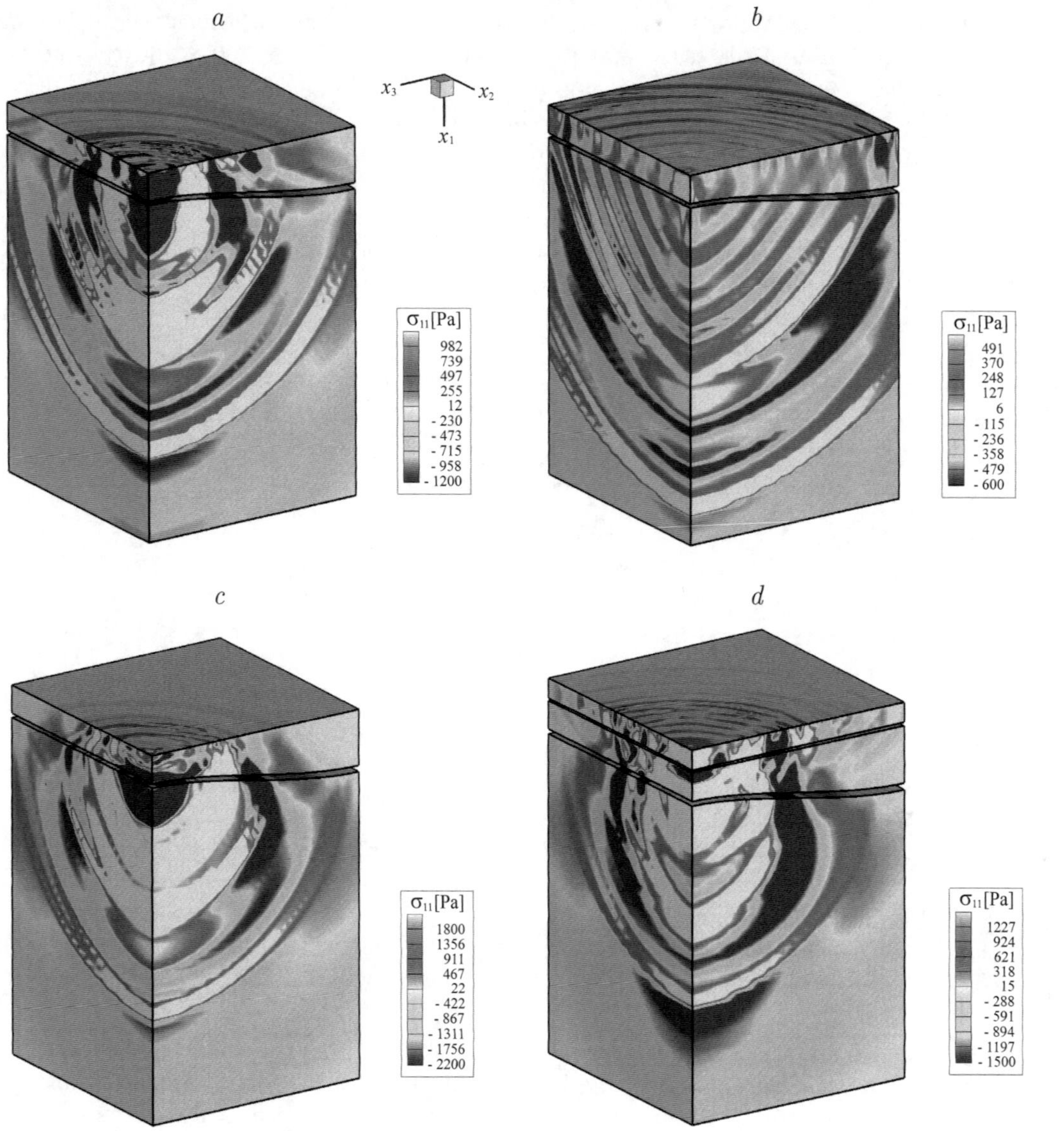

Figure 4. Level surfaces of the normal stress in the vertical direction. Lower layer is ground, upper layer is: clay (*a*), rigid ground (*b*), water (*c*), upper layer is ice, middle layer is water (*d*).

Figure 4 shows the characteristic level surfaces of the normal stress σ_{11} in the cases of: compliant upper layer of clay (Figure 4*a*), rigid upper layer of rigid ground (Figure 4*b*), upper layer of water (Figure 4*c*), upper layer of ice and middle layer of water (Figure 4*d*) at the time moment $t = 18\,\text{ms}$. The rest part of massif has the parameters of ground from the Table 1.

It can be seen that perceptible reflection of waves from the back surfaces does not occur. On seismograms of reflected waves the errors in modeling the boundary conditions also appear weakly. In computations for a three-layered massif with a layer of water covered by ice, the thickness of the ice cover was taken as 3.5 m, which is typical for the Arctic shelf.

IOP Conf. Series: Earth and Environmental Science **193** (2018) 012058 doi:10.1088/1755-1315/193/1/012058

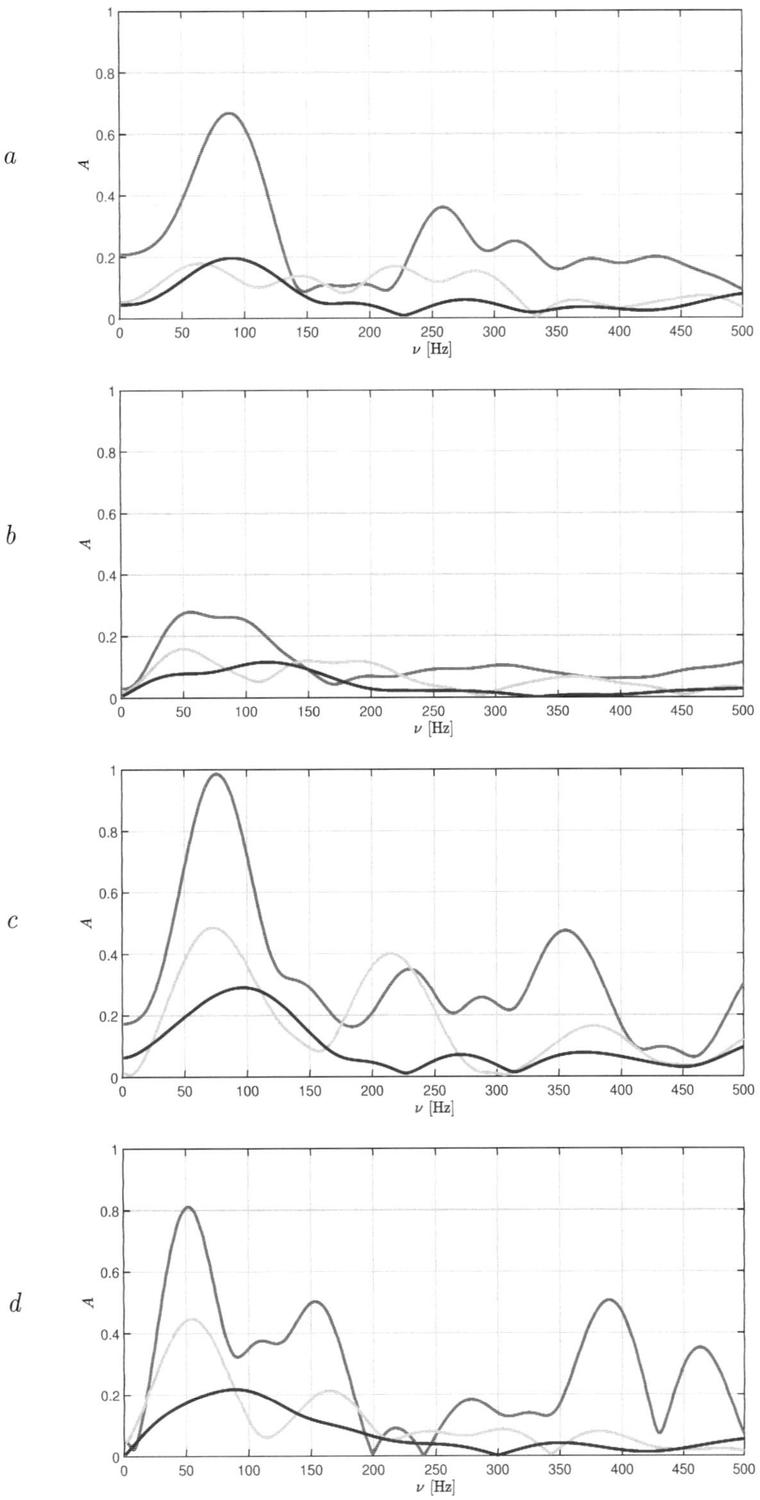

Figure 5. Amplitude-frequency dependencies for vertical velocity at the depth of 30 m. Lower layer is ground, upper layer is clay (*a*), upper layer is rigid ground (*b*), upper layer is water (*c*), middle and upper layers – water covered by ice (*d*).

The amplitude–frequency dependences for the velocity v_1 of the particles of a medium at the depth of 30 m in a system of points under the trace passing near the symmetry plane in the direction of x_2 axis are represented in Figure 5. Red line refers to the point under the source, green line – to the middle point to the left, and blue one – to the edge left point. Fourier analysis of the seismograms of velocities shows that the main frequency, which depends weakly on the point of registration, is about 50 – 100 Hz. At the same time, the received signal contains high-frequency oscillations, the degree of damping of which with the passage of waves due to the dissipation of mechanical energy must be analyzed using more general equations of viscoelastic media.

Due to the properties of the Fourier transform, the amplitude–frequency dependences for acceleration a_1 can be obtained from Figure 5 by means of multiplying the amplitude of the velocity by the frequency, and for displacement u_1 – by means of dividing by the frequency. The main frequencies of displacement turn out to be close to zero, and the main acceleration frequencies exceed hundreds of Hertz. This circumstance can serve as a basic information for choosing the new types of geophones (see, e.g., [11]) for exploration by means of the electromagnetic pulse sources "Yenisei".

4. Conclusions

The Arctic territory of Russian Federation is characterized by a permafrost structure of the surface layer of soil and the ice covered shelf structure, which reduces the efficiency of geological exploration using traditional seismic sources of explosive and vibratory types. Therefore, Geotech Holding Company developed a special eco-friendly electromagnetic pulse source "Yenisei", which acoustic waves will be the subject of analysis using high-performance computing in our further research. Now we worked out the computational technology and software complex for a detailed study of wave fields excited by the source in blocky-layered massifs of geomedium with different properties of layers. The conducted computational experiments showed that the main frequency of vertical velocity, regardless of the material of the surface layer (clay, rigid frozen ground, water, ice-covered water), lies in the characteristic range for seismic exploration from 30 to 100 Hz.

Acknowledgments

The reported study was funded by RFBR and the Government of Krasnoyarsk Region according to the research project No. 18-41-242001.

References

[1] Smirnov V P 2003 *J. Devices Systems Explor. Geophys.* **3**(1) 21 (in Russian)
[2] Detkov V A, Schadin P Y and Ivashin V V 2007 *J. Devices Systems Explor. Geophys.* **22**(4) 30 (in Russian)
[3] Detkov V A 2009 *J. Sib. Fed. Univ.: Math. Phys.* **2**(3) 298 (in Russian)
[4] Smirnov V P 2005 *J. Devices Systems Explor. Geophys.* **12**(2) 45 (in Russian)
[5] Schadin P Y 2013 *Seismic Technol.* **10**(4) 74 (in Russian)
[6] Sadovskaya O V and Sadovskii V M 2005 *Numer. Methods Program.* **6**(2) 86 (in Russian)
[7] Sadovskaya O and Sadovskii V 2012 *Mathematical Modeling in Mechanics of Granular Materials* (*Advanced Structured Materials* vol 21) (Heidelberg – New York – Dordrecht – London: Springer)
[8] Sadovskii V M and Sadovskaya O V 2016 *J. Appl. Mech. Tech. Phys.* **57**(5) 808
[9] Sadovskii V M and Sadovskaya O V 2015 *Wave Motion* **52** 138
[10] Sadovskii V M and Sadovskaya O V 2012 *Parallel program system for the solution of three-dimensional elastic-plastic problems of the dynamics of granular media (3Dyn_Granular), Certificate of state registration of the computer program no. 2012613990 from 28.04.2012 (Rospatent), RU OBPBT* vol 3 (Moscow: FIPS) p 209 (in Russian)
[11] Agafonov V M, Egorov I V and Shabalina A S 2014 *Seismic Instruments* **50**(1) 1

IOP Conf. Series: Earth and Environmental Science **193** (2018) 012059 doi:10.1088/1755-1315/193/1/012059

Singenetic waters composition reconstruction methods of mesozoic sedimentary basins

Ya V Sadykova

Trofimuk Institute of Petroleum Geology and Geophysics SB RAS, Novosibirsk, Russia

E-mail: SadykovaYV@ipgg.sbras.ru

Abstract. In the article the methods of singenetic waters composition reconstruction of Mesozoic and Cainozoic sedimentary basins based on naturalhistorical approach is presented. The average concentration of the main macrocomponents of waters buried both sea, and continental conditions are given. Example of application of the methods in Arctic basin is done.

1. Introduction

The analysis of macro- and microcomponent composition of groundwater makes it possible to reveal the hydrogeochemical zonality, to determine the paths of vertical migration of fluids, and to tentatively determine their genesis. However, present-day groundwater chemistry radically differs from that of syngenetic waters, buried simultaneously with sediments at the elision and infiltration stages of basin development. To understand the nature of groundwaters chemical composition and hydrogeochemical zonality requires paleohydrogeochemical reconstructions, making it possible to assess the effect of diagenetic and catagenetic changes and the processes of interaction in the water–rock–gas–organic matter system on transformations in groundwater chemistry during the evolution of the sedimentary basin.

2. Materials and methods

The procedure for the reconstruction of water salt composition in ancient marine and lake–alluvial basins is based on paleogeographic reconstructions and comparative–lithological analysis with naturalhistorical approach, the essence of which is the assumption that the formation of ion–salt composition of water in continental and marine environments in the past geological epochs was governed by the same factors and determined by the same processes as in the present time. Two groups of factors determine the direction of changes in water chemistry in the course of geological evolution of a sedimentary basin: (1) the factors that determine the genetic type and composition of original water, i.e., the lithological–facial, paleographic, and paleoclimatic, and (2) the factors that govern the changes in groundwater chemistry after its burial, i.e.,the temperature, pressure, dynamics, and geochemistry of the environment, as well as the effect of infiltration recharge. To analyze the development of a confined groundwater system throughout the period of its existence requires a periodization of the hydrogeological history [14]. It is based on the assumption that, in any region, one or several hydrogeological cycles can be identified, consisting of elision and infiltration stages. At the initial (elision) stage, transgression of the sea basin is taking place. In this time, sedimentogenic water

IOP Conf. Series: Earth and Environmental Science **193** (2018) 012059 doi:10.1088/1755-1315/193/1/012059

is buried along with thalassogenous sediments and the processes of elision water exchange are predominant. At the final (infiltration) stage, which corresponds to regression, meteogenic (infiltrogenic) water enters the lithosphere and partially replaces and displaces the sedimentogenic water that has accumulated before. Starting from the stage of the burial of syngenetic water, the process of transformation of its chemistry begins, governed, first, by sedimentation environments and next, at the stages of diagenesis and katagenesis, the processes of vertical and lateral migration and interaction in the water–rock-gas–OM system. This means that the initial composition of sedimentation waters can be derived from the sediment formation conditions. The increase in the total salts in continental environments is taking place mostly due to hydrocarbonate ion and calcium; while that in marine environments, due to ions of chlorine, sodium, and magnesium. Magnesium is an indication to the deepwater environments of silt-water accumulation.

To reconstruct the composition of waters of ancient marine basins (elision stages), data on the modern chemistry of the ocean [1, 18], silt water [22], and absorbed clay complex [8, 9, 21] were generalized. The initial composition of infiltration water (continental sedimentation environments) was reconstructed by the average composition of river, bog, and lake water and the geochemistry of the hypergenesis zone [7, 23]. In addition, the results of paleontological and paleoclimatic studies were used to determine the mean annual temperature and salinity in the Boreal Basin [3–6, 10–13, 17, 19, 20, 24]. Paleoclimatic reconstructions are based, primarily, on studying the spore–pollen complexes, the occurrence of benthic forms, isotopic thermometry (δ18O and δ13C), and the Ca/Mg ratio in the shells of marine organisms. The present-day studies show that the Mesozoic era was a period of greenhouse climate, though, with some zonality. The zone under consideration lied in the moderately warm boreal climate zone with a passage into subtropic zone. The seas of high paleolatitudes, in addition to boreal species, were periodically inhabited by subtropic and tropic mollusks; and the land, by tropic plants [10]. Against the background of relatively high mean annual temperatures, typical of the Mesozoic, short-time but, sometimes, wide temperature variations were taking place (5–6°C). The temperature of surface layers and shallow water varied from +11 to +24°C, while the bottom waters of the deep parts of the shelf were cooler [24].

Paleontological data confirm that the salinity in the epicontinental shallow seas of the Arctic was close to that of the ocean (35 g/dm^3); however, water in coastal zones was freshened (<18 g/dm^3) by continental runoff, especially, in the periods of climate humidization.

Studies of the composition of the modern sea (the seas of Okhotsk, Japan, Baltic, White, and Caspian) and oceanic waters suggest that the maximal salinity of the buried waters of Mesozoic basin could not be higher than 38 g/dm^3 in the deepest marine environ ments, while in shallow and coastal zones, it could vary from 10 (freshwater inflow from rivers) to 25 g/dm^3.

In terms of composition, chloride sodium water dominated, with an increase in magnesium with deepening of the sea basin.

Detailed analysis of the results of studying continental and coastal-marine sediments in the environments of semiarid climate has shown that infiltration water exchange caused the formation of fresh, mostly hydrocarbonate calcium waters with a total mineralization of up to 0.5 g/dm^3 and the formation of brackish waters (up to 10 g/dm^3) in the coastal-continental environments. The characteristics of the mean composition of syngenetic waters of different facial zones of the West Siberian paleobasin are giver in Table 1.

3. Results and discussion

The methods has been successfully approved in the Arctic region (Gydan, Pur-Taz, Yenisei-Khatanga, Anabaro-Hatanga, Lena-Anabar and Laptev see oil-and-gas- bearing regions and in the southern region of the West Siberian basin (Kaymysovsky and Vasyugan regions).

Let's consider history of formation of sedimentary deposits during Jurassic and Cretaceous period of eastern part of Bol'shekhetskaya Megasyneclise (west part of Yenisei-Khatanga megatrough) by way of example.

The vast transgression, which started at the Carnian stage of Triassic, caused the formation of a shallow marine basin in the northern West Siberia. The examined region contained a vast transition zone—a coastal plain, inundated by the sea from time to time [2, 16].

Table 1. Average composition of syngenetic waters of marine and continental sedimentation environments in northern regions of the West Siberian basin in the Mesozoic and Cenozoic.

Facial environments	M_{av}*	Elements						
		SO_4^{2-}	Cl^-	HCO_3^-	Na^+	K^+	Ca^{2+}	Mg^{2+}
Continental		mg/dm^3						
Mountains, elevated plain	65.0	2.4	5.7	35.0	11.0	1.4	3.2	1.7
Floodplain–bog	44.0	3.0	3.5	25.0	3.0	0.8	6.5	1.5
Lacustrine–alluvial	150.0	10.0	7.0	55.0	5.1	1.3	14.0	4.0
Marine		g/dm^3						
Coastal-marine	17.5	1.2	8.2	0.3	6.2	0.1	0.4	1.1
Deep-sea–marine	34.5	2.7	18.9	0.14	10.5	0.3	0.4	1.3

*M_{av} is mean mineralization.

This zone with moving shoreline showed the formation of coastal-marine and continental sediments with remains of deciduous flora and plant detritus. In addition, because of the unstable marine regime, water salinity in the paleobasin varied from 2 to 20 g/dm^3.

The composition of sedimentary formations that formed in the Late Triassic in Siberia territory suggests that, in that period, the study area lied within the moderate belt (boreal) with relatively warm and wet climate and mean annual water temperature reaching +20°C, which persisted until the Early Jurassic [21].

Since the early Jurassic (Hettangian), the marginal Siberian basins were at the transgression stage of their development caused by level rise in the seas of the Boreal Basin [12]. The deepest northwestern part of the examined region was inundated by a shallow (up to 25 m) sea. In this area, marine sediments have deposited with predominantly chloride sodium water with a high concentration of hydrocarbonate ion and calcium and with mineralization of 10–20 g/dm^3. In the eastern direction, the basin became shallower, and the sedimentation environments changed from coastalmarine to continental, where the processes of fresh atmospheric water infiltration predominated (Fig. 1a).

Such environments persisted up to the beginning of Toarcian, when, after an insignificant Late Pliensbachian regression, a transgression of sea basin, largest in the Middle Jurassic, took place. In this period, fine washed clays of kiterbyut formation formed [26]. The major portion of the examined region was a shallowsea basin with depths to 100 m; water salinity could reach 25 g/dm^3. The Early Toarcian showed warm climate with mean annual temperature estimated by oxygen isotope composition in belemnite rostra at +16.9– +24.5°C; this temperature caused a bloom of various marine fauna, especially, belemnites and ammonites, in the Arctic seas [19].

The second part of the Early Toarcian and the beginning of Aalenian featured a general short but vast regression of Siberian seas, resulting in a shallowing and freshening of the marine basin. Nevertheless, the contours of the main paleogeographic zones kept unchanged up to Callovian. The gradual cooling (down to +15°C) and moistening of the climate in the second part of the Aalenian caused the death of vegetation and, consequently, a wide occurrence of carbonaceous strata and the appearance of arctic genera of ammonites and belemnites in the basin [3, 12, 19].

In the Bajocian, short-time rises (in the beginning) and lowering (in the end of the century) of the sea level took place, which have no considerable effect on the position of paleogeographic zones [16]. The climate in the Bajocian was also cool and humid, the mean annual temperatures were +14–+16°C [3].

IOP Conf. Series: Earth and Environmental Science **193** (2018) 012059 doi:10.1088/1755-1315/193/1/012059

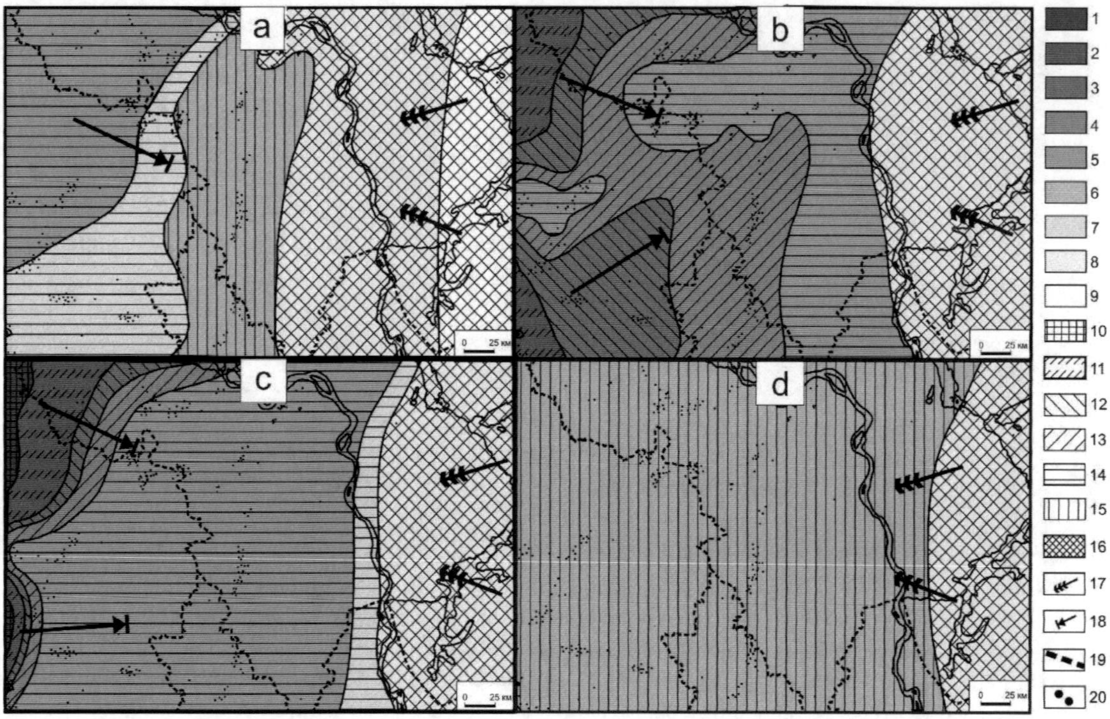

Figure 1. Paleohydrogeochemical scheme with elements of facial zonality and the modern hydrogeochemistry of the northeastern part of Bol'shekhetskaya Megasyneclise at the beginning of (a) Pliensbachian, (b) Volgian, (c) Valangimian, and (d) Aptian ages (paleogeographic data are based on materials of [2, 12, 15, 16]).

Sedimentation environments: (*1*) deep sea, continental slope (>400 m); (*2*) sea, deep part of the shelf (200–400 m); (*3*) sea, outer part of the shelf (100–200); (*4*) sea, internal part of shelf (25–100); (*5*) coastal zone, insular and beach parts of the shelf (0–25 m); (*6*) lowland accumulative plain; (*7*) coastal plain, inundated by the sea from time to time; (*8*) denudation elevated plain; (*9*) low maintains, plateau; paleohydrogeochemical burial zones of (*10*) salt (talassogenic) waters with mineralization of 35.0–38.0 g/dm^3 with the predominance in the composition of Cl^- and Na^+ at higher Mg^{2+} concentration (deep-water marine environments); (*11*) salt (talassogenic) water with mineralization of 25.0–35.0 g/dm^3 with the predominance in the composition of Cl^- and Na^+ ions at higher Mg^{2+} concentrations (marine environments); (*12*) salt (talassogenic) water with mineralization of 20.0–25.0 g/dm^3 with the predominance in the composition of Cl^- and Na^+(shallow-water marine environments); (*13*) salt water with mineralization of 10.0–20.0 g/dm^3 with the predominance in the composition of Na^+ and Cl^- at higher concentration of Ca^{2+} ions (coastal-marine environments); (*14*) salt waters with mineralization of 2.0–10.0 g/dm^3 with the predominance in the composition of Cl^- and Na^+ at higher concentration of and Ca^{2+} (coastal-continental environments); (*15*) fresh and brackish infiltrogenic water with a mineralization of 0.5–2.0 g/dm^3 and with the predominance in the composition of and Ca^{2+} ions at higher concentration of Cl^- and Na^+ ions (continental environments); (*16*) fresh infiltrogenic water (with the predominance of meteogenic) with mineralization from 0.5 g/dm^3 with the predominance in the composition of and Ca^{2+} ions (continental environments); the main directions of motion of (*17*) infiltration water and (*18*) elision water; other: (*19*) administrative boundaries; (*20*) wells.

In the Bathonian age, the positions of the zone of marine sedimentation did not change, but the periods of sea ingressions increased. In addition, the paleontological data suggest a new wave of warming, which led to the formation of subtropic humid climate on the northern West-Siberian

Lowland. This can be seen from the development of glauconitic sediments and coccolithophorids among algae [19].

The Callovian stage shows the beginning of a vast sea transgression, resulting in the deepening of the sea basin down to 200 m in the west of the region, where fine clay sediments with abundant fauna were accumulating. In this part of the sea basin, water salinity reached 30 g/dm^3. The fauna that inhabits the arctic basin becomes more diverse because of the continuing warming (water temperature increased by 2–3°C), and thermophilic species appear. At the passage from the Callovian to Oxfordian ages, the transgression gave place to regression, which continued until the mid-Oxfordian stage. The deeper zone shifted into the extreme northeastern part of the region, and the shallow zone (25–100 m) extended considerably with clay and sand–aleurite–clay sediments accumulating within it. The morphology and boundaries of the coastal–marine zone (with the sea depth of < 25 m) has changed insignificantly since the

Callovian age. The coastal plain, located in the southern part, was inundated by the sea [2, 16].

The transgression continued over the Volgian age and reached its maximum in the middle of the century (Fig. 1b). The climate in the Volgian age was nearly semiarid, resulting in the predominance of chemical weathering processes on the continent and the accumulation of mostly clay and organogenic rocks on the shelf [9, 12, 16, 21]. By the end of the Volgian age, a deep paleobasin of underbalance type with signs of hydrogen-sulfide pollution has formed in West Siberia. The study area was inundated by a sea of Late Jurassic transgression with depths of 400 m in the extreme northwestern and southwestern parts, where water salinity reached 35–38 g/dm^3. A zone with depths of 100–200 m formed near the Khal'merpayutinskaya area, where salt thalassic waters with mineralization of 20–25 g/dm^3 and the predominance in the composition of chlorine and sodium ions and higher magnesium concentration were buried along with marine sediments. The major area of the Khatanga marine basin with depths from 25 to 100 m has decreased from the Bathonian to the Oxfordian and, in the Volgian time, it lied as narrow belts along the contour of the deeper domain. The contours of the coastal zones have not changed. In the end of the Volga and the beginning of the Berriasian, a temperature drop by 4–5°C took place, maybe locally, in Northern Siberia. In the Late Berriasian, Valanginian, and early Hauterivian, the seas were still warm, somewhat cooler than they were in the Kimmeridge–Volgian time. This is confirmed by the diversity of the fauna (large bivalves and oysters and the spore–pollen complex (tropical ferns), suggesting subtropic humid climate [19]. Paleobiogeochemical analysis of oyster shells shows that water paleosalinity of the Khatanga Sea in the Early Berriasian age was close to that of the ocean (35 g/dm^3), reaching 32 in the Valanginian period and 31 g/dm^3 in the Hauterivian [18].

In the early Berriasian age, a depression with depths not greater than 400 m formed in the southwestern part of the region because of an increase in the rate of basin warping and not enough precipitation for its compensation [15]. The contours of other zones remained almost unchanged. In this time, the sea basin was connected with boreal seas and its water salinity in the deepest parts was close to that of the ocean.

In the Early Valanginian, the regression that began in the late Jurassic still kept developing (Fig. 1c). This has led to the formation of a vast shallow zone with depths <25 m. In the coastal zone, along with beach sediments, brackish water with mineralization of 10–20 g/dm^3, with the predominance in its composition of ions of chlorine and sodium and with higher calcium concentration because of the inflow of fresh meteogenic waters from the continent, was buried. A deep-water depression with a small area persisted in the northwestern part of the region. A coastal plain, inundated by the sea from time to time, formed again in the eastern part. The climate in the Hauterivian was humid and moderate, with mean annual water temperatures reaching +17°C. The West-Siberian Basin lost its connection with the boreal seas because of its further shallowing; its water also became much fresher [15]. In the examined area, the shallow zone became a vast coastal plain where floodplain, delta, and beach environments alternated, depending on the sea level, and brackish water with mineralization of 2–10 g/dm^3, chloride sodium composition, and higher concentration of hydrocarbonate ion and calcium was buried. Warm subtropical climate also existed in the West Siberian Plain in the second half of the Early Cretaceous, though with much higher humidity, facilitating the deeper decay of sedimentary material [19]. Under such conditions, bauxite deposits and thick coal-bearing strata were forming. Starting from Barreme and up to the Late Cenomanian, environments of a lowland accumulative plain and a coastal plain were alternating [15]. Under such conditions, infiltration meteogenic hydrocarbonate-calcium water with mineralization of up to 2 g/dm^3 was mixing in the zone of active water exchange with sedimentogenic waters, partially diluting and freshening them (Fig. 1d).

The further transgression, which started in the Turonian age, caused the formation of a vast shallow zone, covering the entire examined area up to the Late Campanian. Seawater salinity could reach 20 g/m^3 in deepest parts in the southwestern part of the region. In the Late Cretaceous epoch, the climate conditions generally somewhat deteriorated, as can be seen from the composition of marine fauna inhabiting the Arctic Basin, a single species of bivalve dominated, and rare belemnites and ammonites could be met [19]. The further temperature drop contributed to the formation of siliceous rocks (gaizes) in the Coniacian– Campanian. The Late Campanian showed a regression of the sea basin and warming of the West Siberian Sea, resulting in the deposition of calcareous sediments and an increase in the coastal plain area. In this period, dominating in the syngenetic waters in the coastal–continental environments were infiltrogenic hydrocarbonate calcium waters with higher concentration ofsodium and chlorine ions and mineralization reaching 10 g/dm^3.

4. Conclusion

The present-day formation water is a product of the geological evolution of the sedimentary basin. The sedimentogenous and ancient infiltrogenic water serve as a source of two major geochemical branches of waters widespread in the Arctic region. Further, their composition experiences radical changes; once in the sedimentary basin, they become involved in many processes of interaction in the water–rock–gas–organic matter system, experience diagenetic and katagenetic transformations, and mix with one another in different proportions. The genetic coefficients and paleohydrogeochemical reconstructions can be used to determine the nature of groundwater and to assess the changes that have taken place in its chemistry in the geologic time.

Acknowledgments

This study was supported by the Russian Foundation for Basic Research, project no. 17-45-543249 r_mol_a and 18-05-70074 «Arctic resources».

References

[1] Alekin O and Lyakhin Yu 1984 *Chemistry of the Ocean* (Leningrad: Gidrometeoizdat) p 344

[2] Vinogradov A 1986 *Atlas of Lithological–Paleogeographic Maps of the USSR* (Moscow: Vsesoyuznyi aerogeologicheskii trest ministerstva geologii SSSR) p 79

[3] Berlin T, Kiprikova E and Naidin D 1970 Some problems of paleotemperature analysis (by belemnite rostra) *Geol. Geofiz.* **4** 36–43

[4] Bulynnikova S, Gol'bert A, Klimova I et al 1978 *Paleobiofacies of Oil-and-Gas-Bearing Volga and Neocomian Deposits of the West-Siberian Plate* (Moscow: Nedra) p 87

[5] Volkova V 2014 Geological development stages of Priobskii Arctic shelf in West Siberia in Paleogene and Neogene *Geol. Geofiz.* **4** 619–633

[6] Volkova V 2011 Stratigraphy and trend of paleotemperatures in Paleogene and Neogene in West Siberia (by palinological data) *Geol. Geofiz.* **7** 906–915

[7] Gordeev V and Lisitsyn A 2014 Geochemical interaction between freshwater and marine hydrosphere *Geol. Geofiz.* **56** 721–744

[8] Gramberg I 1973 *Paleohydrogeochemistry of Terrigenous Strata* (Leningrad: Nedra) p 172

[9] Gramberg I and Spiro N 1965 *Paleohydrochemistry of the Northern Middle Siberia in the Late Paleozoic and Mesozoic Tr. NIIGA* (Moscow: Nedra) **142** p 120

[10] Zakharov V 2010 Boreal climate in the Mesozoic *Priroda* **4** 37–42

[11] Zakharov V, Marinov V and Agalakov S 2000 Albian Stage of the West Siberia *Geol. Geofiz.* **6** 769–791
[12] Zakharov V, Mesezhnikov M, Ronkina Z et al 1983 *Paleogeography of the Northern USSR in the Jurassic* (Novosibirsk: Nauka) p 192
[13] Zakharov V and Radostev I 1975 Salinity of the Early Cretaceous Sea in Northern Siberia by Paleogeochemical Data *Geol. Geofiz.* **2** 37–43
[14] Kartsev A , Vagin S and Baskov E 1969 *Paleohydrogeology* (Moscow: Nedra) p152
[15] Kontorovich A, Ershov S, Kazanenkov V et al 2014 Paleogeography of the West Siberian Sedimentary Basin in the Cretaceous Period *Geol. Geofiz.* **56** 745–776
[16] Kontorovich A, Kontorovich V, Ryzhkova S et al 2013 Paleogeography of the West Siberian Sedimentary Basin in the Jurassic Period *Geol. Geofiz.* **8** 972–1012
[17] Nikitenko B 2009 *Stratigraphy, Paleobiogeology, and Jurassic Biofacies of Siberia by Microfauna: Foraminifers and Seed Shrimps* (Novosibirsk: Parallel') p 680
[18] Popov N, Fedorov K and Orlov V 1979 *Seawater* (Moscow: Nauka) p 328
[19] Saks V 2007 *Selected Works. Biostratigraphy and Paleobiogeography of Mesozoic in Siberia* (Novosibirsk: GEO) **2** p 643
[20] Saks V and Ronkina Z 1960 Development of Siberian Relief in the Mesozoic Age *Geol. Geofiz.* **1** 58–73
[21] Ushatinskii I and Zaripov O 1978 *Mineralogical and Geochemical Characteristics of Oil-and-Gas Productivity of Mesozoic Deposits in the West Siberian Plate* (Sverdlovsk: Sredne-Ural'skoe kn. izd.) p 206
[22] Shishkina O 1972 *Geochemistry of Marine and Oceanic Silt Waters* (Moscow: Nauka) p 228
[23] Shvartsev S 1998 *Hydrogeochemistry of the Hypergene Zone* (Moscow: Nedra) p 366
[24] Shurygin B 2005 *Biogeography: Facies and Stratigraphy of the Lower and Middle Jurassic in Siberia by Bivalved* (Novosibirsk: GEO) p 156
[25] Shurygin B, Nikiteno B, Devyatov V et al *2000 Stratigraphy of Oil-and-Gas-Bearing Basins in Siberia. Jurassic System* (Novosibirsk: Izd. SO RAN, fil. GEO) p 480

IOP Conf. Series: Earth and Environmental Science **193** (2018) 012060 doi:10.1088/1755-1315/193/1/012060

The history of hydrocarbon generation in the Mesozoic-Cenozoic sedimentary cover in the northeast of the Bolshaya Kheta megasyneclise

P I Safronov, A A Deshin and L M Burshtein

Trofimuk Institute of Petoleum Geology and Geophysics SB RAS, Novosibirsk, Russian

Novosibirsk State University, Novosibirsk, Russian

E-mail: SafronovPI@ipgg.sbras.ru, DeshinAA@ipgg.sbras.ru, BurshteinLM@ipgg.sbras.ru

Abstract. The dynamics of hydrocarbon generation by the Jurassic and Cretaceous source rocks in the northeastern part of the Bolshaya Kheta megasyneclise was reconstructed by historical and geological modelling. The amounts of liquid and gaseous hydrocarbons generated by these sources rocks were estimated for each time period and catagenetic substage.

1. Introduction

The geology of the northern West Siberian petroleum province has been the focus of extensive research efforts since the 1950s and remains one of the highest priority exploration targets in the Russian Arctic zone. Most of proved reserves in the study area came from Cretaceous reservoirs. Additional exploration efforts are needed to ensure the development of Jurassic reservoirs. To further improve exploration efficiency and predict hydrocarbon potential, a historical-geological modeling has been used to delineate the spatial and temporal distribution of zones of different thermal maturity grades within the main source rock sequences (SRS).

2. The present state of the problem

The genetic methods for evaluating the sedimentary basins for hydrocarbon potential have been developed independently in numerous research papers of Russian [1–10] and Western authors [11–18]. The differences in approaches lie primarily in the methods used to describe the processes of hydrocarbon generation and expulsion. The genetic approach and its modern modifications is based on quantitative modeling of the formation of a sedimentary cover, heat and mass transfer in the rocks, as well as on chemical and kinetic modeling of organic matter transformation. The above became the basis for the development of integrated software packages intended for use with numerical three-dimensional modeling of the evolution of the sedimentary cover and the processes of petroleum generation. This approach has become known as basin modeling in the Western and, partly, Russian scientific literature. Along with this, the terms historic-geological or historical-genetic method coined by Russian authors [4, 19] are widely used.

Basin modeling has been recently widely applied in assessing hydrocarbon potential and restoring the history of hydrocarbon generation in the northern regions of Western Siberia [20–23], South Kara megasyneclise [24], Yenisei-Khatanga regional trough [25], Caspian basin [26], northeastern Sakhalin Island shelf [27], etc.

At the same time, basin modeling has been used to solve different problems with respect to the Bolshaya Kheta megasyneclise [28].

As regards hydrocarbon potential predictions, the reconstruction of the extent and dynamics of hydrocarbon generation can be of particular interest. In a generalized form, of great importance are the duration of burial, residence of the SRS within the oil window, estimated amounts of hydrocarbons that were generated in each SRS depending on a particular catagenetic grade of source rocks.

3. Problem statement

The section of the Mesozoic-Cenozoic sedimentary cover in the northeastern part of the Bolshaya Kheta megasyneclise represents an alternation of shale- and sand-dominated sedimentary sequences. The sedimentary cover reaches a maximum thickness of 9000 m and thins out eastward along the basin periphery.

The tectonic map of the Jurassic [29] shows that the North Taz megadepression is the deepest part acting as the main source kitchen (Figure. 1). It is bounded by several plays, such as the Messoyakha inclined ridge in the north, Tagul promontory and a series of large dome-shaped uplifts in the south. Of particular interest is the unique large-size fields of the Vankor group (Suzunskoe, Vankorskoe, Lodochnoe, Tagulskoe) in the east of the region, which were confined to a series of uplifts of the same name within the Togul promontory and Dolgan mesomonocline.

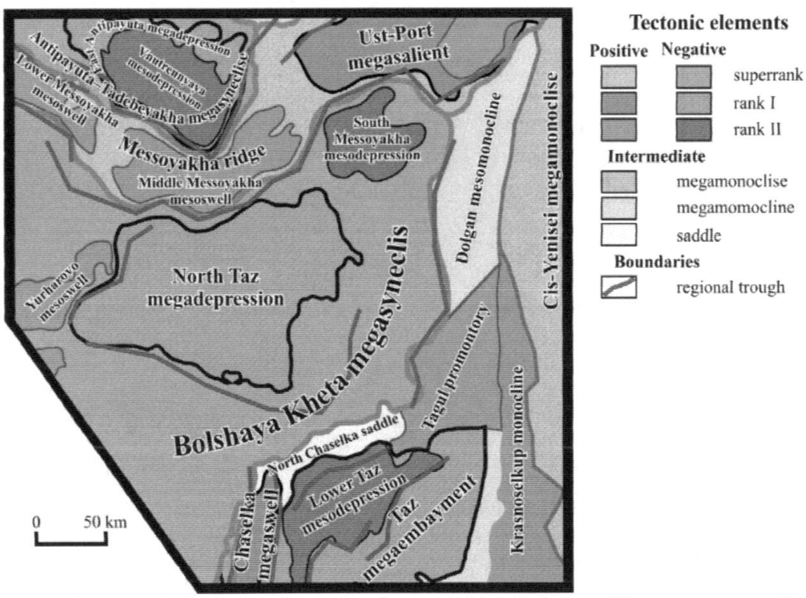

Figure 1. Tectonic map of the Jurassic structural stage of the Bolshaya Kheta megasyneclise [29].

Most of identified resources in the study area came from Cretaceous reservoirs, whereas Jurassic reservoirs have been penetrated only by a few wells. The hydrocarbon potential of Jurassic sediments remains poorly explored. In view of the above, the potential Jurassic sources rocks are of particular interest. These include primarily the Bazhenov horizon, the main source rocks of the central and southern West Siberian sedimentary basin. Thus, the following Jurassic SRS were selected as objects of research: Levin, Kiterbyut, Laida, Malyshev, Bazhenov (including the Bazhenov Formation and upper units of the Yanov Stan and Golchikha Formations). The combined Cretaceous Malaya Kheta-Yakovlev SRS was considered additionally.

IOP Conf. Series: Earth and Environmental Science **193** (2018) 012060 doi:10.1088/1755-1315/193/1/012060

Modeling of petroleum generation processes in the main Jurassic and Cretaceous source rock sequences of the northeastern part of the Bolshaya Kheta megasyneclise was performed using the most recent geological, geophysical, geothermal and geochemical data, as well as the experience in regional basin modeling gained by the Institute of Petroleum Geology and Geophysics, Siberian Branch, Russian Academy of Science.

4. 3D model of the sedimentary cover in the northeastern part of the Bolshaya Kheta mega syneclise

A three-dimensional model of the Mesozoic-Cenozoic sedimentary cover was generated using the Temis software package (Beicip-Franlab). The study area some 340·360 km in size was structurally and stratigraphically subdivided into 20 layers. The model uses 1 km grid. The model is limited to the structural map for the base of the Triassic sedimentary section. The lithological model of the study area comprises a set of lithofacies maps.

Calibration of the thermal history of sediments is based on vitrinite reflectance measurements on samples from wells and maps of organic matter thermal maturity (catagenesis) for the top of the Jurassic [30–32]. The heat flow through the base of the sedimentary cover was assumed to be constant.

The selection of SRS and justification of the types of organic matter were performed based on geochemical data (Institute of Petroleum Geology and Geophysics, Siberian Branch, Russian Academy of Science): present-day C_{org} content of rocks, hydrogen index, T_{max}, etc.

The following source sequences were selected: Levin (type III kerogen), Kiterbyut (type III kerogen), Laida (type III kerogen), Malyshev (mixed type II/III kerogen), Bazhenov horizon (mixed type II/III kerogen), Malaya Kheta-Yakovlev complex (type IV kerogen).

All the types of kerogens were modified in accordance with the modern understanding of the ratios between liquid and gaseous generated hydrocarbons. The calculated volumes of gas generation from the Malaya Kheta-Yakovlev complex in the early stages of catagenesis were corrected for balance models of the change in kerogen compositions [33–35].

The original concentrations of organic matter in each SRS were restored by calculating the degree of kerogen transformation after calibration of the rock thermal history.

5. Results

The results of the historical-geological modeling were used to reconstruct the dynamics of generation of liquid and gaseous hydrocarbons in different time (Figure 2) and different stages of catagenesis [36] (Table 1). Of primary importance are maps showing the time at which each SRS entered the oil/gas window.

5.1. Levin SRS

In the southern part of Messoyakha inclined ridge, the Levin SRS reached the mesocatagenetic substage (MC_1^1 grade) at 169 Ma in the mid-Bajocian, MC_1^2 at 160 Ma in the mid- Oxfordian, MC_2 at 139 Ma in the Late Berriasian, MC_3^1 at 126 Ma in the Late Barremian, MC_3^2 at 110 Ma in the Early Albian, AC_1 at 89 Ma in the Late Turonian, AC_2 at 65 Ma in the early Paleocene.The total hydrocarbon generation from Levin source rocks was 70,889 billion m^3 of gas and 16,589 million t of oil.

5.2. Kiterbyut SRS

In the eastern parts of the Vnutrennyaya and South Messoyakha mesodepressions, the Kiterbyut SRS reached the mesocatagenetic substage (MC11 grade) at 165 Ma in the early Callovian, MC12 at 150 Ma in the Tithonian, MC2 at 134 Ma in the Valanginian, MC31 at 119 Ma in the mid-Aptian, MC32 at 101 Ma in the late Albian, AC1 at 80 Ma in the mid-Campanian, AC2 at 52 Ma in the early Eocene. The total hydrocarbon generation from this SRS was 188,891 billion m3 of gas and 166,725 million t of oil. The most intensive hydrocarbon generation took place during MC2 and MC31 substages.

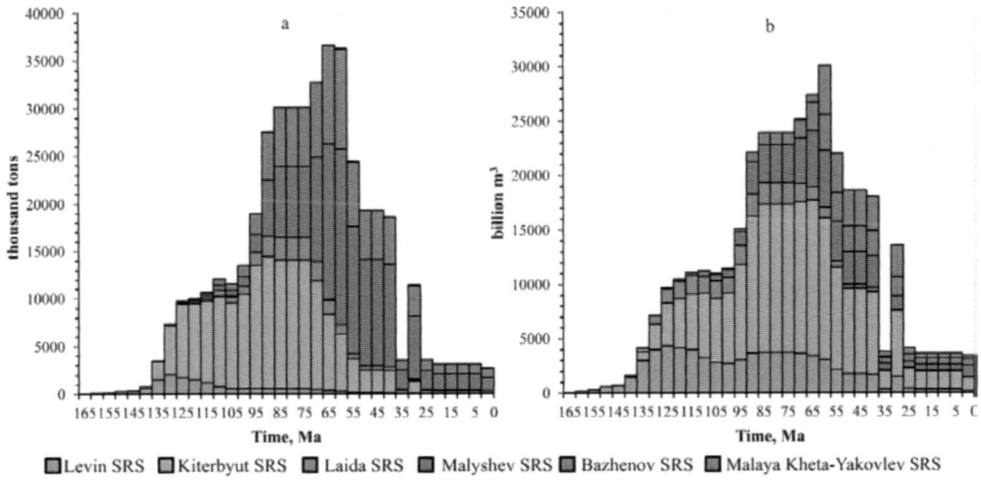

Figure 2. Mean generation rates of liquid (a) and gaseous (b) hydrocarbons in the northeast of the Bolshaya Kheta megasyneclise.

Table 1. Generation potential of different source rock sequences of the northeastern part of the Bolshaya Kheta megasyneclise divided into stages of catagenesis.

SRS	PC_1	PC_2	PC_3	MC_1^1	MC_1^2	MC_2	MC_3^1	MC_3^2	AC_1	AC_1	AC_2	Total
Oil ($\times 10^6$ t)												
Malaya Kheta-Yakovlev	64	151	284	136								695
Bazhenov			57	11,109	53,352	22,390	1500	27	1			89,323
Malyshev				331	11,657	64,941	58,936	8822	129			144,818
Laida					1783	15,759	2405	42				19,990
Kiterbyut					2472	104,465	51,249	8128	408			166,725
Levin					1206	13,814	1517	52				16,589
Gas ($\times 10^9$ m^3)												
Malaya Kheta-Yakovlev	3107	6707	9043	2597								24,532
Bazhenov			8	1395	11,204	13,943	1316	61	9			28,122
Malyshev				215	10,266	25,191	11,319	3065	112			50,169
Laida	4	15	57	1097	12,106	11,035	1388					25,705
Kiterbyut					831	48,432	55,968	42,076	34,848	6657	79	188,891
Levin					894	24,856	18,821	12,427	9022	4412	456	70,889

5.3. Laida SRS

In the northeastern part of the Bolshaya Kheta megasyneclise and eastern part of the East Antipayuta megadepression, the Laida SRS entered the mesocatagenetic substage (MC11 grade) at 149 Ma in the mid-Tithonian, MC12 at 135 Ma in the mid-Valanginian, MC2 at 116 Ma in the mid-Aptian, MC31 at 91 Ma in the late Turonian, MC32 at 73 Ma in the late Campanian, AC1 at 51 Ma in the early Eocene, AC2 at 3 Ma in the late Neogene. The total hydrocarbon generation from this SRS was 19,990 million t of oil and 25,705 billion m3 of gas. The most intensive hydrocarbon generation took place during MC12 and MC2 substages, at 90-50 Ma, from the late Turonian to the early Eocene.

5.4. Malyshev SRS

In the Antipayuta-Tadebeyakha and Bolshaya Kheta megasyneclises, the Malyshev SRS reached the mesocatagenetic substage (MC_1^1 grade) at 138 Ma in the early Valanginian, MC_1^2 at 127 Ma in the mid-Barremian, MC_2 at 99 Ma in the early Cenomanian, MC_3^1 at 74 Ma in the mid-Campanian, MC_3^2 at 55 Ma in the early Eocene, AC_1 at 18 Ma in the first half of the Miocene. At present, these rocks have reached the apocatagenetic substage (AC_2 grade) in the deepest parts of the North Taz megadepression.

The total hydrocarbon generation from this SRS was 50,169 billion m^3 of gas and 144,818 million t of oil. The most intensive hydrocarbon generation took place during MC_2 and MC_3^1 substages.

5.5. Bazhenov SRS

The Bazhenov SRS reached the MK_1^1 grade at 126 Ma in the late Barremian, MC_1^2 at 100 Ma in the late Albian, MC_2 at 71 Ma in the early Maastrichtian, MC_3^1 at 36 Ma in the late Eocene. Only the deepest parts of this SRS, within the North Taz megadepression are currently within the MC_3^2 and AC_1 grades. The total hydrocarbon generation from this SRS was 28,122 billion m^3 of gas and 89,323 million t of oil. The most intensive hydrocarbon generation took place during MC_1^2 and MC_2 substages.

5.6. Malaya Kheta-Yakovlev SRS

The Malaya Kheta-Yakovlev SRS reached the PC_1 grade at 67 Ma in the late Late Cretaceous, PC_2 at 64 Ma in the early Paleocene, PC_3 and MC_1^1 at 60 Ma in the mid-Paleocene, within the Ust-Porttovaya megasalient, Middle Messoyakha mesoswell, Dolgan mesomonocline, Chaselka megaswell, North Chaselka saddle, Taz and the Zapolyarnoe dome-shaped uplifts these rocks have not yet entered PC_3 and MC_1^1 grades. The total hydrocarbon generation from this SRS was 695 million t of oil and 24,532 billion m^3 of gas. The most intensive hydrocarbon generation took place during PC_3 substage.

6. Petroleum systems evolution

An analysis of the evolution of all petroleum systems shows that the process of intense hydrocarbon generation began at 165 Ma in the mid-Callovian. The oil/gas windows were reached at 65–60 Ma in the late Cretaceous—early Paleogene.The total generation of gaseous hydrocarbons from all SRS was 388,308 billion m^3. The Kiterbyut SRS account for 49%, Levin and Malyshev SRS for 18% and 13%, respectively. The total generation of liquid hydrocarbons from all SRS was 438,140 million t. The Kiterbyut, Malyshev, Bazhenov, Laida, and Levin SRS account for 38%, 33%, 20%, 5%, and 4%, respectively. The Malaya Kheta-Yakovlev SRS accounts for only 0.2% of total amount of liquid hydrocarbons generated by all SRS.

During the evolution of the sedimentary cover, the Jurassic source rocks have entered the oil window and subsequently gas window. As a result, gaseous hydrocarbons were charged into traps already filled with liquid hydrocarbons, which led to changes in the phase composition of hydrocarbon accumulations. Secondary cracking of hydrocarbons in the deeply buried Lower-Middle Jurassic source rocks also played a role. Under high temperature and pressure conditions, liquid hydrocarbons were decomposed into gaseous components. This process led to inflows of gaseous hydrocarbons into traps, and subsequent expulsion and dissipation of liquid hydrocarbons. The discovery of new large oil fields in the Jurassic part of the section can be regarded as highly unlikely.

7. Conclusions

The historical and geological modeling allowed us to reconstruct the history of hydrocarbon generation from the main Jurassic and Cretaceous source rock sequences and estimate the amounts of gaseous and liquid hydrocarbons generated from these source rocks in a particular period of time and catagenetic substage.

The intense hydrocarbon generation took place during the entire Mesozoic-Cenozoic in the northeastern parts of the Bolshaya Kheta megasyneclise in the Jurassic and later in the Cretaceous source rocks.

The Jurassic source rocks reached the oil window and subsequently gas window. The Malaya Kheta-Yakovlev sources rocks reached the early catagenetic stage of gas generation.

In the study area, the main potential for discovering new gas and gas condensate fields is associated with the Jurassic.

Acknowledgments

This study was performed as part of the basic project of the Russian Academy of Sciences (no. IX.131.4.1) and scientific project no. 18-35-00436 mol_a of the Russian Foundation for Basic Research with the technical support of the Russian Representative Office of BeicipFranlab.

References

[1] Sokolov V A 1948 *Outlines of the origin of petroleum* (Costoptechizdat. Moskow) (in Russian).
[2] Uspensky V A 1970 *An introduction in petroleum geochemistry* (in Russian). Nedra. Leningrad.
[3] Lopatin N V 1971 *Temperature and geologic time as factors in coalification* Akad. Nauk SSSR Izvestiya. Ser.Geol. **3** 95–196(in Russian)
[4] Vassoevich N B, Vysotskiy I V, Korchagina Yu I, Sokolov B A 1971 *The historic-geological-geochemical method for reservoir appraisal (example from the Central Russian petroleum basin)* Izv. AN SSSR Ser.Geol **11** 56–60 (in Russian)
[5] Vassoevich N B, Nazarevich B P, Nazarevich I A 1980. *An experience of applying the historic-genetic method for timing the formation of oil and gas fields in the East Manych basin*. Proc. IGC Session 26 Presentations of Soviet Geologists. Resources of Raw Materials for Energy. Fossil Fuels. (Moscow: Nauka) 149–61 (in Russian)
[6] Vassoevich N B, Sokolov B A 1976 *Modern achievements in the development of the sedimentary-migration theory of formation and accumulation of oil and gas* Geologiya i razvedka **5** 62–7 (in Russian)
[7] Kontorovich A E 1976 *Geochemical methods for quantitative petroleumprediction* (Moscow: Nedra) 229 (in Russian)
[8] Kontorovich A E, Melenevskiy V N 1988 *The theory of peak oil generation and its place in the sedimentary-migration oil generation theory* Izv. AN SSSR. Ser. Geol 1 3–13(in Russian)
[9] Galushkin Yu I 2007 *Sedimentary basins modeling and asessment their oil-gas generation.* (Moskow: Scientific world) (in Russian)
[10] Astahov S M 2015 *Georeactor. Oil and gas formation algorithms.* (Rostov on Don: Kontiki) (in Russian)
[11] Jungten H 1964 *Reactionskinetische uberlengungen zur deutung von pyrolise reactionen* Erdol Kohle-Erdgas Petrochem.Bd. 17 180–6
[12] Hanbaba P, Jungten H, Peters W 1968 *Nonisothermal reaction kinetics of coal pyrolysis. Part II: Extension of the theory of gas cracking and experimental confirmation on (bituminous) coals* Brennstoff-Chemie. **49**(12) 368–79
[13] Tissot B 1969 *Premières données sur le mécanismes et la cinétique de la formation du pétrole dans les sédiments; simulation d'un schéma réactionnel sur ordinateur* Rev. Inst. Franc. Pétrole 24 470–501
[14] Tissot B P, Espitalie J 1975 *L'evolution thermique de la matiere organique des sediments: Applications d'une simulation mathematique* Rev. Inst. Franc. Pétrole **24** 743–77
[15] Burnham A K, Sweeney J J 1989 *A chemical kinetic model of vitrinite maturation and reflectance* Geochimica et Cosmochimica Acta **53/10** 2649–57
[16] Behar F, Beaumont V, De Barros-Penteado H L 2001 *Rock-Eval 6 technology: performances and developments Oil & Gas Science and Technology* Rev. Inst. Franc. Pétrole 56 111–34
[17] Hantschel T, Kauerauf I A 2009. *Fundamentals of basin and petroleum systems modeling* Springer Verlag. Berlin–Heidelberg.
[18] Burnham A K 2017 Global Chemical kinetics of fossil fuels. How to model maturation and pyrolysis Springer International Publishing AG
[19] Kontorovich A E, Trofimuk A A 1973 *Methods for historic investigation of oil and gas fields* Geologiya Nefti i Gaza 7 18–24 (in Russian)
[20] Littke R, Cramer B, Gerling P, Lopatin N V, Poelchau H S, Schaefer R G, Welte D H 1999 *Gas Generation and Accumulation in the West Siberian Basin* AAPG Bulletin **83**(10) 1642–65
[21] Schaefer R G, Galushkin Y I, Kolloff A, Littke R 1999 *Reaction kinetics of gas generation in*

selected source rocks of the West Siberian Basin: implications for the mass balance of early-thermogenic methane Chemical Geology **156** chapter 1-4 41–65
[22] Malysheva S V, Vasilyev V E, Komissarov D K, San-Germes M L, Laigle Zh M, Pereslegin M V, Khafizov S F 2012 *Modeling of Bazhenov formation of Western Siberia as an unconventional hydrocarbon source* Oil Industry 12 14–47
[23] Deshin A A, Safronov P I, Burshtein L M 2017 *Evaluation of the time of the main phase of oil formation in the Middle Upper Jurassic deposits of the north of Western Siberia* Oil and Gas Geology 4 33–44. (in Russian)
[24] Malysheva S V, Vasiliev V E, Kosenkova N N, Ananiev V V, Komisarov D K, Murzin R R, Roslov Yu V 2011 *Modeling of petroleum systems of South Kara basin: principles and* results Oil Industry 3 13–7. (in Russian)
[25] Safronov P I, Ershov S V, Kim N S, Fomin A N 2011 *Modeling of processes of generation, Migration and accumulation of hydrocarbons in Jurassic and Cretaceous complexes of Yenisei-Khatanga basin* (in Russian) Geology of oil and gas 5 48–55.
[26] Kerimov V Yu, Guliev I S, Gusejnov D A, Lavrenova E A, Mustaev R.N., Osipov A V, Serikova U S 2015 *Forecasting of oil and gas potential in regions with a complex geological structure* (Moskow: Nedra) (in Russian).
[27] Galushkin Yu I, Sitar K A, Kunitsyna A V 2015 *Maturation level of organic matter and realization of hydrocarbon potential during burial history of deep formations of the basin in the north-eastern shelf of Sakhalin* Oil and Gas Geology 2 60–6 (in Russian)
[28] Bogdanov O A, Koloskov V N, Shuvayev A O, Musikhin K V, Istomin S B, Bobrova M A, Maglevannaya P S 2017 *Analysis of evolution of the Bolshehetsk depression hydrocarbon systems based on the use of basin modeling method* Oil and Gas Geology 2 55–64 (in Russian)
[29] Kontorovich V A, Belyaev S Yu, Kontorovich A E, Krasavchikov V O, Kontorovich A A, Suprunenko O I 2001 *Tectonic structure and evolution of the West Siberian geosyneclise in the Mesozoic and Cenozoic* Geologiya i Geofizika (Russian Geology and Geophysics) 42(11–12) 1832–45 (1740–53)
[30] Fomin A N, Kontorovich A E, Krasavchikov V O 2001 *Catagenesis of organic matter and petroleum potential of the Jurassic, Triassic, and Paleozoic deposits in the northern areas of the West Siberian megabasin Geologiya i Geofizika* (Russian Geology and Geophysics) 42 (11) 1875–87 (1783–96).
[31] Kontorovich A E, Fomin A N, Krasavchikov V O, Istomin A V 2009 *Catagenesis of organic matter at the top and base of the Jurassic complex in the West Siberian megabasin Geologiya* i Geofizika (Russian Geology and Geophysics) 50 917–29.
[32] Fomin A N 2012 *Catagenesis of Organic Matter. Diagnostic Methods* LAP Lambert Academic Publishing (in Russian)
[33] Uspenskiy V A 1954 *Mass balance processes associated with coalbed metamorphism: An experience of study* Akad. Nauk SSSR Izvestiya, Ser. Geol. 6 94–101(in Russian)
[34] Kontorovich A E, Rogozina E A 1967 *The extent of HC gas generation in Mesozoic Sediments of the West Siberian Plain* Geology and Petroleum Potential of the Southeastern West Siberian Plate. SNIIGGiMS. Novosibirsk 13–25 (in Russian)
[35] Kontorovich A E, Burstein L M, Malyshev N A, Safronov P I, Guskov S A, Ershov S V, Kazanenkov V A, Kim N S, Kontorovich V A, Kostyreva E A, Melenevsky V N, Livshits V R, Polyakov A A, Skvortsov M B 2013 *Historical-geological modeling of processes of naphthyogenesis in the Mesozoic-Cenozoic sedimentary basin of the Kara Sea* Geology and Geophysics **54**(8) 1179–26.
[36] Kontorovich A E, Trofimuk A A 1976 *Lithogenesis and petroleum formation* (in Russian) Fossil Fuels. Problems of Geology and Geochemistry of Naphthides and Bituminous Rocks. (Moscow: Nauka) 19–36

Peculiarities of natural gas hydrate formation from ice in reactors under high pressure

M Ye Semenov[1], I K Ivanova[1,2] and V V Koryakina[1]
[1] Institute of Oil and Gas Problems, Siberian Branch, Russian Academy of Sciences, Yakutsk, Russia
[2] M. Ammosov North-Eastern Federal University, Yakutsk

E-mail: xotoy_82@mail.ru

Abstract. Experimental results of natural gas hydrates formation from ice obtained under static conditions in the batch reactor applying the method of thermal cycling have been given. Hydrate forms subdivided according to their morphological features have been shown: powder and layered growth, retained ice shapes, and whisker crystals. The conducted experiments have defined effects of the thermal cycling regime on forms of obtained hydrates: decreased rates of heating/cooling results in an increase of a share of whisker crystals. The rate of decomposition of loose retained ice shapes in the form of hydrate has been defined to be 2.3 times faster than that of powder and dense layered hydrates.

1. Introduction

Gas hydrates are known to be solid crystalline compounds formed under certain thermobaric conditions from water and gas with low molecular weight and characterized by the ability to contain gas in a concentrated state [1, 2]. The rationale and comparative calculations of the prospects for natural gas use in a hydrate (solid) state for transportation and storage in oil and gas industry are given in [3].

From the technological point of view, an important parameter of gas hydrates is the specific gas content that depends on a molecular weight of the gas forming agent, type of structure and content, degree of hydrate cavity filling, as well as the presence of foreign inclusions and crystal lattice defects [4, 5]. At a 100% of the hydrate structure filling with methane molecules (CH_4*5,75H_20), the specific gas content of hydrate makes 164 m^3/m^3. [2].

Transportation and storage of natural/associated gas in the northern regions of Russia are promising technologies due to the self-preservation effect of gas hydrates, i.e. slow decomposition of hydrates at negative temperatures [6]. The most stable, under the same storage conditions, are massive forms of gas hydrates [7]. The following regularity has emerged: at a specific gas content of more than 130 cm^3/g, the hydrate maintains a high stability under non-equilibrium conditions at negative temperatures. The temperature boundaries of "the anomalous conservation area" of methane hydrate are determined in [8].

Despite existing technologies for industrial production of gas hydrates [9, 10], the issue of production of the gas hydrate ready for storage and transportation directly in the reactor itself is still unsolved [11]. Therefore, study of mechanisms of growth and forms of hydrate formation is a significant aspect of development of the gas hydrate production technology, and it is of fundamental importance [12]. It is assumed that [13] compensation is possible for the heat of phase transitions during hydrate formation and with maintenance of low temperatures during their accumulation, making use of naturally low temperatures of the North of Russia. Hence, there are prerequisites for simplification of the means of formation of synthetic gas hydrates and reduction of energy costs in the technological cycle of their production [14].

It is known [15-20] that at elevated pressures of a gas-hydrate forming agent, it is possible to transfer practically the entire solid-ice phase (0.18÷0.5 mm) into hydrates via thermal cycling in the batch reactor of high pressure. However, obtaining and preparation of finely ground ice particles is a laborious process, and the obtained particles of synthetic hydrates are unstable and rapidly decomposing products [11]. Currently, the review of literature has showed that, an absence of experimental data and quantitative parameters of natural gas hydrate (NGH) synthesis from larger ice fractions (> 5 mm).

Thus, the objective of this paper is an experimental study of formation of natural gas hydrates (NGH) from formed ice at various temperature regimes of synthesis and estimation of rates of the obtained hydrate decomposition.

2. Experimental part

The natural gas from the Srednevilyuskoe field with the following composition CH_4 – 92,89, C_2H_6 – 5,25, C_3H_8 – 1,21, i-C_4H_{10} – 0,102, n-C_4H_{10} – 0,12, N_2 – 0,38, CO_2 – 0,048 (total %). has been used in this research. The equilibrium condition of hydrate formation has been calculated on the basis of the natural gas composition applying the Sloan method [4]. Ice samples, 30 mm in diameter, were obtained by freezing distilled water in 10 ml forms (Fig. 1a). Four identical cylindrical reactor chambers with a volume of 1113 cm^3 were used. Cooling of reagent-loaded reactors was carried out in a SANYO MIR-254 refrigerated incubator (Fig. 1b), where the thermal cycling mode could be set up.

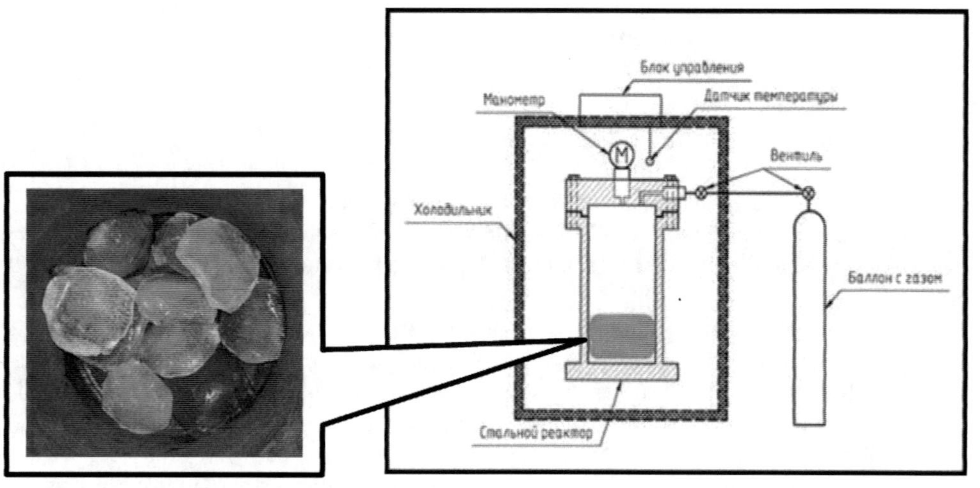

a b

Figure 1. Formed ice samples (a) and the scheme of experimental installation for gas hydrate production (b).

Since it is during the ice melting process [17-19], when the intensive hydrate formation occurs, the temperature range of thermal cycling is selected with transition through the melting point of ice from -5 to + 5°C. To select the optimal temperature regime for production of NGH from formed ice, five modes of 96 hour temperature effects on the system were used (from 4 to ½ cycles of heating/cooling for a day), according to the procedure described in [21, 22].

After the hydrate formation process was over, the reactor was opened in a freezer at a temperature of -10°C to fix the native structure and select hydrate samples for subsequent weighing and decomposition. The morphology of gas hydrates was recorded with a Pentax K200D digital camera. Kinetics of hydrate decomposition was studied applying a volumetric method at + 25°C , for this purpose, a sample of the hydrate with a given weight was placed under a glass dome located in the refrigerated heated bath circulator Huber CC 410, the dome, in its turn, was connected to a gas meter.

The specific gas content of the obtained hydrate (v, sm3/g) was calculated as a ratio of the measured volume of the gas released from the hydrate to the amount of the sample from the obtained gas [7]:

$$v = \frac{V_{gas}}{m_{hydrate}}; \qquad (1)$$

v is the specific gas content of the sample, V_{gas} is a volume of the gas released from the hydrate, $m_{hydrate}$ is the weight of the hydrate sample.

Kinetic characteristics of hydrate decomposition were calculated applying the Avrami-Erofeev-Kolmogorov equation [23], most often used for analysis of isothermal solid phase reactions [24:]

$$\alpha = 1 - \exp(-k\tau^n), \qquad (2)$$

where α is a degree of conversion, n is the kinetic parameter characterizing the mechanism of the process, k is the generalized rate constant, and τ is time. Kinetic parameters n and k can be determined by double-logarithm of the equation and presented as follows:

$$\lg[-\lg(1-\alpha)] = n\lg t + \lg k + \lg(\lg e). \qquad (3)$$

The equation (2) in this form (3) makes a direct line, thus allowing us to calculate values of the parameters n and k constructing logarithmic anamorphoses. And, the parameter n characterizes the mechanism of the reaction under study, thus, with n>1, the process predominantly occurs in the kinetic area, and with n<1 – in the diffusional one.

Substituting the determined kinetic parameters (n and k) into the Sakovich formula (4), the rate constants of the decomposition reaction of natural gas hydrates (K) have been calculated as:

$$K = nk^{1/n}. \qquad (4)$$

3. Results and discussions

Under the temperature regime with 4 heating/cooling cycles per day, only hydrate inclusions are distinct on the well preserved ice surface, which evidences that the ice does have time to melt under the given regime of thermal cycling. Under the temperature regime with 3 cycles per day, the major mass of ice retains its shape, and thin branched growths occur on the ice surface. Hence, under the above-mentioned temperature regimes, following a rapid temperature change in the refrigerated incubator, the temperature inside the reactor does not have time to rise to positive degrees, so ice samples in the reactor do not melt.

Under the temperature regime with 2 heating/cooling cycles per day, destruction of the original ice forms occurs, and white porous snow-like structures of NGH with a small amount of thin branched growths (Fig. 2) are observed.

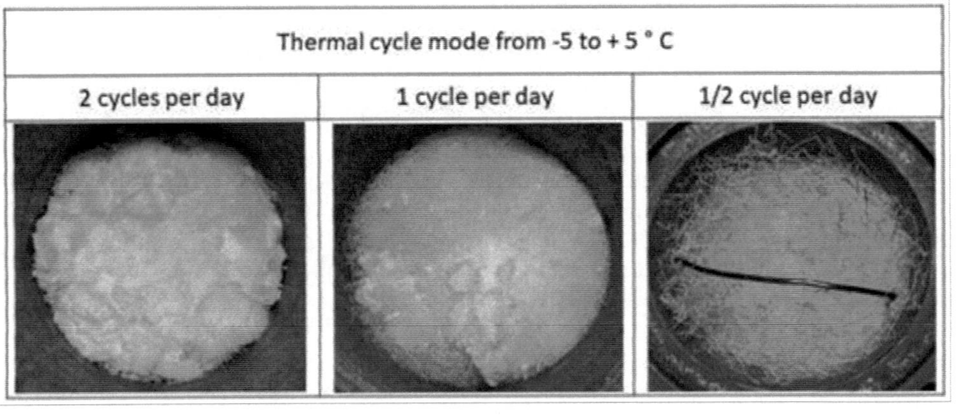

Figure 2. Images of the change in shapes of NGH depending on the thermal cycle of synthesis.

It has been identified that upon NGH synthesis from formed ice, elongation and enlargement of branched growths with the direction of crystal growth upwards from the ice surface (Fig. 2) occur, as the number of heating/cooling cycles decreases (from 2 cycles to ½ cycle).

Upon obtaining NGH from ice under the temperature regimes with 1 and ½ heating/cooling cycles, three major shapes of ice-hydrate associates are found to be formed: fine powder and layered growths at the chamber bottom, a retained ice form, and whisker crystals of hydrates (Fig. 3).

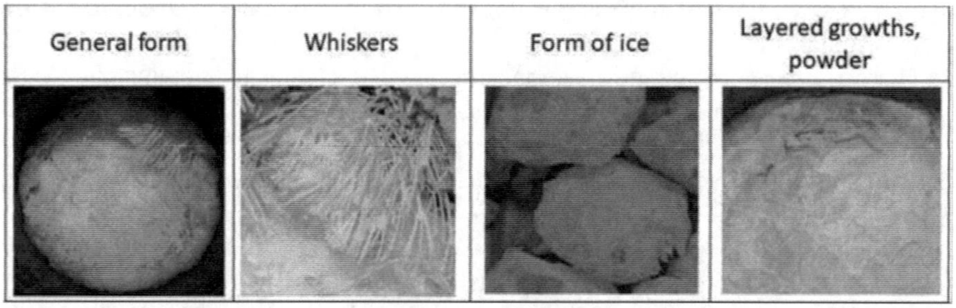

Figure 3. Images of various forms of NGH obtained from ice: whiskers, form of ice, powder and layered growths.

The specific gas content of synthesized NGH (Table1) has been calculated on the basis of measurements of the volume of the emitted gas from the hydrate samples obtained under the temperature regime with one heating/cooling cycle per day during 96 hours.

Table 1. Specific gas content of synthesized natural gas hydrates in accordance with their forms

Forms of hydrate	m, g	Vgas, cm^3	v, cm^3/g	χ, %
Whiskers	4,8	-*		
Forms of ice	24	2220	92.5	60
Hydrate from reactor bottom (layered growths, powder)	80	9000	112.5	73.5

It is worthwhile to note here that, regretfully, the precision of the measurement equipment did not allow us to measure a small amount of gas emitted from whiskers. The value of the specific gas content of the retained ice forms is 92.5 cm^3/g, and that of hydrates in the form of powder and dense layered growth at the reactor bottom is 112.5 cm^3/g. It has been found out that 75 wt% of hydrates have been accumulated at the bottom of the reactor as a powder and dense layered growth, with 22.2 wt % being in the well-preserved original ice form, and 2.8 wt% of the hydrates are growing as whiskers (Table 1).

Based on the measurement of a gas volume evolved during decomposition of various forms of hydrates, kinetic parameters of disassociation of natural gas hydrates have been calculated (Table 2), and logarithmic anamorphoses have been constructed according to the equation (3) (Fig. 4). High values of reliability approximation (0.99) demonstrate the validity of this equation for description of kinetics of decomposition of natural gas hydrates.

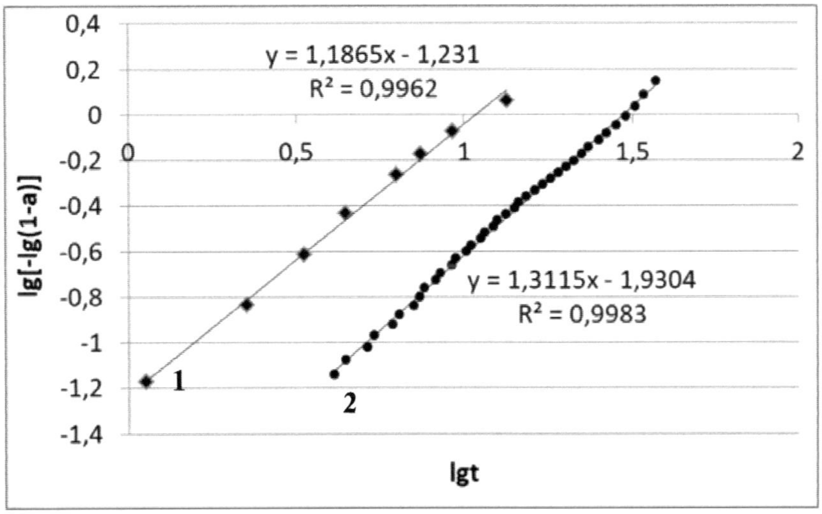

Figure 4. Kinetics of decomposition of various forms of hydrates at T=25°C: 1 – hydrates in the form of ice; 2- hydrates in the form of powder and layered growths.

At the atmospheric pressure and room temperature, hydrates in the form of powder and layered growth from the bottom of the reactor are identified to be completely decomposed within 50 min, and hydrates as preserved forms of ice completely decompose within 25 minutes.

Decomposition of synthesized forms of hydrates has been determined to occur in the kinetic area ($n>1$). Wherein, the rate of decomposition of hydrates in the form of ice is 2.3 times higher than that of hydrates from the bottom of the reactor in the form of layered growth and powder. The half-life of hydrates in the form of ice is 3.5 minutes, and that of hydrates from the bottom of the reactor makes 13 minutes (Table 2).

Table 2. Kinetic parameters of decomposition of various forms of natural gas hydrates, obtained from formed ice

Sample	Melting			
	n_{squ}	K_{squ}, min^{-1}	$\tau^{1/2}$	R^2
Hydrate in the form of ice	1.004	0.207	3.5	0.99
Hydrate from the reactor bottom (lawyered growth, powder)	1.414	0.089	13	0.99

Thus, it has been identified that the most thermodynamically stable forms of hydrates at the air pressure and at 25°C are hydrates from the bottom of the reactor in the form of powder and a dense layered growth.

4. Conclusions

Thus, it has been defined that natural gas hydrates with gas saturation over 70% can be obtained by thermal cycling of the "formed ice + natural gas" system during 96 hours. While producing NGH from formed ice at the temperature regime with 1 heating/cooling cycle from -5 to +5°C, hydrates are shown to be formed in the form of powder and a layered growth, in the form of ice, as well as whiskered crystal hydrates. Their mass redistribution in the reactor has been determined. It has been concluded that hydrates in the form of powder and dense layered growth at the bottom of the reactor possess the greatest thermodynamic stability.

IOP Conf. Series: Earth and Environmental Science **193** (2018) 012061 doi:10.1088/1755-1315/193/1/012061

Acknowledgments
The study has been conducted within the framework of the State Task №0377-2016-003. The research has been performed with the financial support provided by Russian Foundation for Basic Research (RFBR) and the Government of the Sakha Republic (Yakutia) within the framework of the Scientific Project № 18-45-140035.

References

1. Byk S SH 1980 *Gazovye gidraty* (Moscow: Himiya) p 296
2. Dyadin Yu A and Bondaryuk I V 1995 Stoichiometry of clathrates *Journal of Structural Chemistry* 1088-1141
3. Gudmundsson J 2002 *Proc. 4 Int. Conf. on Gas Hydrates* (Japan) pp 997-1002
4. Havkin A YA 2010 Perspektivy sozdaniya gazogidratnoj promyshlennosti *Vestnik RAEN* 43-45
5. Sloan E D and Koh C A 2008 *Clathrate hydrates of natural gases* (CRC Press) p 720
6. Makogon Yu F 1974 *Hydrates of natural gases* (Moscow: Subsoil) p 204
7. Istomin V A and Yakushev V S 2006 *Effekt samokonservacii gazovyh gidratov* (Moscow: Gazovaya promyshlennost) pp 36-46
8. Yakushev V S 2009 *Prirodnyj gaz i gazovye gidraty v kriolitozone* (Moscow: Gazprom VNIIGAZ) p 192
9. Stern L A, Circone S, Kirby S H and Durham W B 2001 Anomalous preservation of pure methane hydrate at 1 atm *J. Phys. Chem. B.* **105** 1756-1762
10. Horiguchi K and Watanabe S 2011 *Proc. Int. Conf. on Gas Hydrates (Edinburgh United Kingdom)* vol 1 (Electronic Materials) p 75
11. Murayama T and Iwabuchi W 2011 *Proc. Int. Conf. on Gas Hydrates (Edinburgh United Kingdom)* vol 1 (Electronic Materials) p 187
12. Yakushev V S, Kvon V G, Gerasimov Yu A and Istomin V A 2008 *Sovremennoe sostoyanie gazogidratnyh tekhnologij* (Moscow: IRC Gazprom) p 88
13. Yakushev B C, Kvon V G, Dolgaev S I, Podenok S E and Istomin V A 2009 *Gidratnye tekhnologii dlya gazifikacii regionov Rossii* (Moscow: Gazovaya promyshlennost) pp 75-79
14. Istomin V A 2010 *Fiziko-himicheskie osnovy i problemy razrabotki gazogidratnyh tekhnologij* (Moscow: OOO Gazprom ehkspo) p 48
15. Semenov M E, Shitz E Yu and Portnyagin A S 2016 *Sposob polucheniya gidratov prirodnogo gaza i razrabotka na ego osnove konceptual'noj tekhnologicheskoj skhemy processa ih proizvodstva* (Transport i hranenie nefteproduktov i uglevodorodnogo syr'ya) vol 3 pp 53-58
16. Grojsman A G 1985 *Teplofizicheskie svojstva gazovyh gidratov* (Novosibirsk: Nauka) p 94
17. Stupin D Yu 1979 Diehlektricheskie svojstva CCl_2F_2*$17H_2O$ i CCl_2F_2*$17D_2O$ *Izv. vuzov. Himiya i him Tekhnologiya* **2** 170-173
18. Stern L A 1996 Peculiarities of methane clathrate hydrate formation and solid-state deformation, including possible superheating of water ice *Science* **273** (5283) 1843-1848
19. Stern L A, Kirby S H, Circone S and Durham W B 2004 Scanning Electron Microscopy investigations of laboratory-grown gas clathrate hydrates formed from melting ice, and comparison to natural hydrates *American Mineralogist* **89** 1162–1175
20. Liu W and Wang L 2014 *Proc. Int. Conf. on Applied Energy (ICAE2014)* vol 61 pp 619–623
21. Nesterov A N 2006 *Dis. d-ra him. nauk: 02.00.04 Nesterov Anatolij Nikolaevich* (Tyumen) p 280
22. Semenov M E, Portnyagin A S and Shitz E Yu 2017 Poluchenie sinteticheskih gidratov prirodnogo gaza iz l'da v zakrytyh reaktorah pri termociklirovanii *Nauka i obrazovanie* **3** 76-81
23. Semenov M E and Shitz E Yu 2015 Vliyanie fazovogo sostoyaniya i sostava odnogo iz komponentov na morfologiyu sinteticheskih gidratov prirodnogo gaza *Mezhd. nauchno-issl. Zhurnal* **38** (7) 96-101
24. Rozovskij A YA 1974 *Kinetika topohimicheskih reakcij* (Moskva: Himiya) p 224
25. Talatory Sh and Barth T 2012 Rate of hydrate formation in crude oil/gas/water emulsions with different water cuts *Journal of Petroleum Science and Engineering* 32-40

Methane hydrate self-preservation features in oil suspensions

A A Sizikov[1], A S Stoporev[1,2,3], L K Altunina[4] and A Yu Manakov[1,3]

[1]Nikolaev Institute of Inorganic Chemistry SB RAS, Novosibirsk
[2]Gubkin University, Department of Physical and Colloidal Chemistry, Moscow
[3]Novosibirsk State University, Novosibirsk
[4]Institute of Petroleum Chemistry SB RAS, Tomsk

E-mail: manakov@niic.nsc.ru

Abstract. Features of methane hydrate self-preservation in the dispersions prepared from water in oil emulsions from two deposits in East Siberia were studied. Under experimental conditions, substantial self-preservation of the hydrate is observed in the dispersion based on oil 1, while the effect is weakened in the dispersions based on oil 2. It is demonstrated that the observed difference is connected with the presence of Span 80 surfactant in water in oil 2 emulsion.

1. Introduction

The structure of gas hydrates is a framework composed of hydrogen-bonded water molecules with the polyhedral cavities of molecular size, in which the guest molecules are included [1]. Clathrate hydrates are interesting from the scientific and practical viewpoints. Due to the ability to accumulate large amounts of gas, they are considered as promising materials for the storage and transportation of such gases as hydrogen and methane (for example, see [2, 3]). The ability of gas hydrates to include guest molecules only of definite size into cavities of the framework found application for the separation of gas mixtures [4], utilization of carbon dioxide [5, 6] and its extraction from exhaust gases [7, 8]. Gas hydrate technologies for concentrating the diluted aqueous solutions are extensively developing (in fact, these are the technologies of water freshening/purification) [9], air conditioning systems involving hydrate suspension as the coolant are under development [10, 11]. It has been known for a long time that gas hydrates are the reason of plug formation in producing wells and field pipelines [12], so much effort is made in gas and oil industry to take measures against hydrate formation [13]. During the recent decade, special attention is given to the problems connected with the prevention of hydrate formation in oil production. The properties of hydrate in oil dispersions are under intense investigation [14-17]. At the same time, problems connected with self-preservation of methane hydrate obtained from oil emulsions remain only poorly studied yet. It was demonstrated previously [18, 19] that the joint deposits of hydrates and oil components may exist for a long time under the conditions of negative Celsius temperatures due to self-preservation. The phenomena may manifest themselves in the production and pipeline transportation of oil in the regions with cold climate, which makes the investigation of the phenomenon especially urgent for the conditions existing in Russia. In the present work we studied the new features of methane hydrate formation and decomposition that has undergone self-preservation in water/oil emulsions taking into account layer thickness and the presence or absence of surfactants during the preparation of the emulsion.

IOP Conf. Series: Earth and Environmental Science **193** (2018) 012062 doi:10.1088/1755-1315/193/1/012062

2. Experimental section

Table 1. Composition and properties of the crude oils used in this work.

	Oil	
	O1=Mamontovskaya	**O2**=Snezhnaya
content of asphaltenes, wt%	0.10	0.67
content of paraffines, wt%	2.30	4.39
content of resins, wt%	19.70	5.07
content of saturates + aromatics, wt%	-	94.26
pour point, °C	-43.0	-10.4
density, kg/m^3	0.858	0.835
viscosity, mPa∗s (20°C)	19.3	7.29

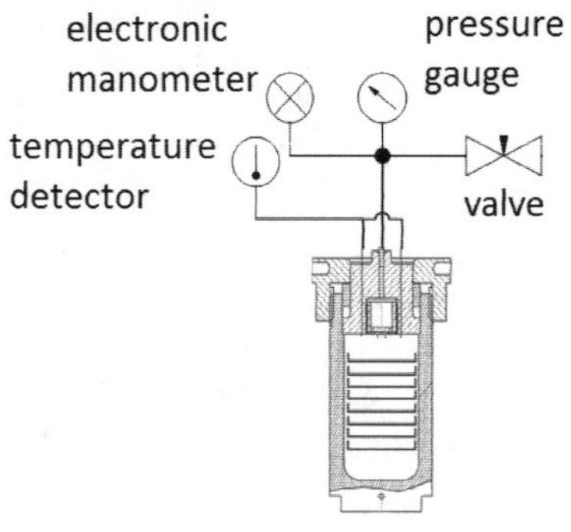

Figure 1. Scheme of an autoclave with 8 sample holders: a steel cup with a cover-obturator with two sealed thermocouples. 8 sample holders are attached to the obturator through the guide bolt. The pressure monitoring and recording system is also attached to the obturator.

Reagents used: methane (purity 99.98%), distilled water, two crude oils (table 1), surfactant Span 80 (Sigma-Aldrich). Emulsions of water in oil (50 mass %) were used in the work. The emulsions were prepared from weighted amounts of water and oil by stirring with a mixer (800 r.p.m.) for 20 min. The emulsion of water in **O1** oil was prepared without using any additives. The emulsion of water in oil **O2** decomposed rapidly, so it was stabilized by adding Span-80 surfactant (1 mass % of the amount of oil). Thus prepared emulsions were stable for a long time (several months). The kinetics of decomposition of methane hydrate that had undergone self-preservation was studied in autoclaves with 1 or 8 plate holders for emulsion samples; the amounts of emulsion were the same in both cases. A scheme of the autoclave with eight holders is shown in figure 1. The experiment proceeded as follows. Sample holders with the weighted portions of the emulsion under study (1 or 8 holders) were placed in a high-pressure cell (figure 1). The autoclave was washed with methane, then gas pressure was set at 125±5 bar at room temperature. After that, the autoclave was placed in a thermostat maintaining the temperature of 1°C, and kept until pressure stabilization, which is the evidence of the completion of hydrate formation (4-20 days, depending on the number of holders and the amount of emulsion under study). Decomposition was carried out in the same cell. After the completion of the synthesis, the autoclave, with methane hydrate formed in it, was cooled to the chosen experimental temperature (-5 – -20°C). After cell cooling, the pressure was released to the level close to the equilibrium for methane hydrate (about 20-25 bar, depending on chosen temperature) for maximal oil degassing in order to prevent sample from escaping out of the holder because of airing (foaming). Then pressure was sharply unloaded to 0.2-0.4 MPa, and then valve was closed immediately. Then, at the same temperature (-5 – -20°C), methane hydrate decomposition was followed by recording the dependence of pressure on time (for at least 300-400 minutes since the complete pressure release). Then the autoclave was heated at room temperature at a rate of (~3,3°C/hour), and the dependence of pressure on time was recorded.

To prepare the samples of hydrate suspension in **O2** oil without surfactants, the powders of frozen hydrate and ground oil were mixed at the liquid nitrogen temperature, loaded into the autoclave at this temperature, then methane pressure was set in the autoclave, and the autoclave was heated to slightly positive temperature (the conditions of hydrate stability were maintained). Then the autoclave was again cooled to the temperature of liquid nitrogen, and the sample of suspension was taken out from the autoclave. Thus homogeneous samples of the suspensions of methane hydrate in oil were prepared successfully, with oil kinds in which stable water emulsions are not formed. X-ray studies of the samples of quenched methane hydrate obtained from water/oil emulsions were carried out using a Bruker D8 Advance powder diffractometer equipped with a low-temperature attachment TTK 450 AntonPaar. Thermovolumetric studies of the hydrates were carried out with a set-up assembled at the laboratory of clathrate compounds of the Institute of Inorganic Chemistry SB RAS. The principle circuit diagram and a detailed description were presented in [20].

3. Results and discussion

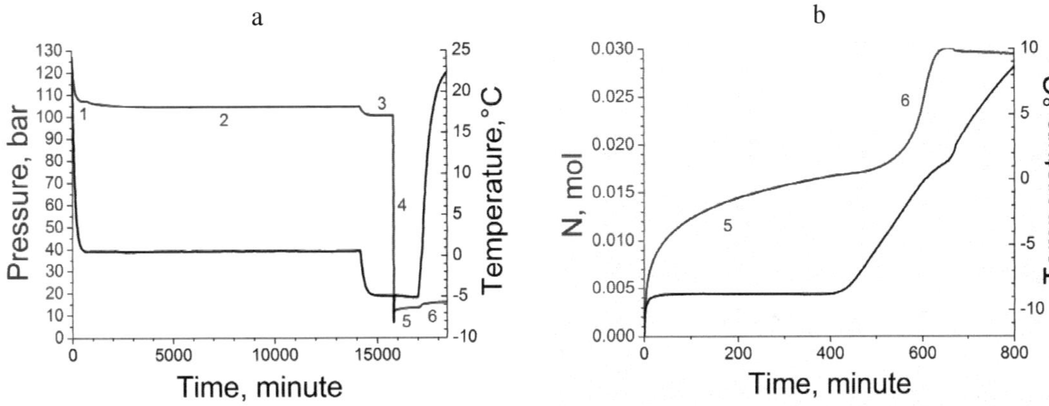

Figure 2. (**a**) Typical for the process of formation and decomposition of methane hydrate from oil emulsion temperature and pressure dependence on time; (**b**) Steps 5 and 6, reflecting the amount of methane hydrate decomposed during experiment.

Now the behavior of pressure and temperature during a typical experiment is to be considered (figure 2a). At stage (1), the autoclave is cooled to 1°C (temperature at which the hydrate is obtained), while pressure decreases due to methane dissolution in the emulsion and due to thermal compression of the gas phase. Stage (2) is characterized by a slow pressure decrease at a constant temperature, which is connected with hydrate formation. Pressure in the autoclave is stabilized after the completion of hydrate formation. At stage (3), the autoclave is cooled to the temperature at which it is planned to study self-preservation of the formed hydrates. Stage (4) involves a stepwise gas release from the autoclave to reach the pressure P_0 = 0.2-0.4 MPa. Then, at stage (5), the hydrate was decomposed at a constant temperature and variable pressure; the kinetic curve of hydrate decomposition was recorded. Finally, at stage (6) the sample was heated to room temperature, and the decomposition of self-preserved hydrate occurred. Stage (5) is to be considered in more detail. At first, as a result of the partial decomposition of methane hydrate, a sharp increase in pressure occurs (figure 2b). This region also includes self-preservation of the hydrate (if it is already present). Estimations show that pressure change due to solubility in oil can be neglected for this region. Further on, decomposition rate decreases. In all experiments, here we observed an extended region of the slow hydrate decomposition; this region was used to determine the rate of decomposition of methane hydrate that had undergone self-preservation. In the case if a noticeable part of hydrate underwent self-preservation, pressure jump was observed at stage (6) while crossing the zero (Celsius) temperature. This pressure jump was connected with the decomposition of self-preserved hydrate. The value of this pressure jump was used to calculate the total amount of hydrate in the sample and to estimate quantitatively the fraction of the hydrate that underwent self-preservation.

The recorded thermobaric curves were used to calculate the number of moles of the gas evolved/absorbed at different stages of the experiment. It was assumed that the degree of water transformation into hydrate was equal to 100%, and in all cases methane hydrate having the composition $CH_4 \cdot 6H_2O$ is formed from the emulsions of water in oil [20,21]. Methane ($N_{methane}$, mol) under experimental conditions could be present in 4 phases: (1): methane in the gas phase (N_g, mol); methane dissolved in oil (N_o, mol); methane dissolved in water (N_w, mol); methane in the form of hydrate (N_h, mol).

$$N_{methane} = N_g + N_h + N_o + N_w. \quad (1)$$

The amount of methane in the gas phase is calculated using equation 2:

$$N_g = \frac{PV}{ZRT}. \quad (2)$$

Here $Z = \frac{PV_m}{RT}$ – is compressibility factor which is determined from Peng-Robinson gas equation of state. To determine the amount of gas evolved during the decomposition of methane hydrate, we used equation (3) in which indexes i and f stand for the initial and current state of the system.

$$N_h = \Delta N_g = \frac{V}{R}\left(\frac{P_i}{z_i T_i} - \frac{P_f}{z_f T_f}\right). \quad (3)$$

Self-preservation of hydrate was studies directly at stage (5) (figure 2b); the moment of pressure release to the minimum is designated as t_0. The rates of decomposition of self-preserved methane hydrate were calculated at the initial stage (5 points starting from moment t_0), an hour later and 5 hours later, after pressure release (table 2). Calculating the amount of methane at stage (5) we neglected the changes in methane solubility in oil and in water because estimations showed that these values are small in comparison with the gas evolved from the hydrate (less than 1%). So, pressure increase at stage (5) was due only to methane hydrate decomposition.

Table 2. Decomposition rates.

Experiment	T_{decomp}, °C	v^0_{decomp} [a], mol/min	v^1_{decomp} [b], mol/min	$\alpha^1_{selfpres}$ [c], %	v^5_{decomp} [b], mol/min	$\alpha^5_{selfpres}$ [c], %
O1/8p2gmin10	-10.1	0.00525	$16.2 \cdot 10^{-5}$	51.27	$3.9 \cdot 10^{-5}$	26.32
O2/8p1gmin5	-4.8	0.01090	$1.9 \cdot 10^{-5}$	2.95	-	-
O2/8p1gmin10	-10.2	0.00428	$8.3 \cdot 10^{-5}$	20.26	$0.94 \cdot 10^{-5}$	1.69
O2/8p1gmin19	-18.8	0.01873	$7.3 \cdot 10^{-5}$	13.53	$0.14 \cdot 10^{-5}$	2.87
O2/1p8gmin11	-10.8	0.00304	$3.0 \cdot 10^{-5}$	3.56	-	-

[a] Initial decomposition rate (0-5 points from t_0).
[b] Decomposition rate at n^{th} hour.
[c] Relative amount of remaining hydrate at n^{th} hour.

The data shown in table 2 are to be considered in more detail. Each sample is encoded as **L/xpy**gmin**z**. Here **L** means the kind of oil used (table 1), **x** is the number of holders; **y** is the mass of the portion of emulsion per one holder; **z** is round temperature at which the self-preservation of methane hydrate was studied. One can see in the data presented in table 2 that hydrate sample **O1**/8p2g min10 has undergone self-preservation; the rate of its decomposition was rapidly decreasing with time. At a moment 5 hours later than the start of decomposition, more than 26% of initial hydrate was still present in the sample. In general, this picture corresponds completely to the situation expected for the sample that has undergone self-preservation. At the same time, all samples of the **O2**/8p1min**z** series have been decomposed almost completely by this moment of time. The observed decrease in decomposition rate in this case is connected with the exhaustion of hydrate in the sample. This result confirms that decelerated hydrate decomposition in experiment **O1**/8p2gmin10 is connected exactly with self-preservation rather than hindered heat and mass transfer because the conditions of heat and mass transfer were close in all experiments. It should be stressed that in different experiments hydrate decomposition in this series proceeded in different manners, and we did not observe any close correlation between experimental conditions and decomposition rates. The maximal initial decomposition rate was observed in experiment **O2/**8p1gmin19 which was carried out at minimal temperature. In experiment **O2/**1p8gmin11, the thickest suspension layer was used, so one might expect the slowest hydrate decomposition. Indeed, the initial decomposition rate is minimal in this experiment (table 2). However, hydrate had decomposed almost completely within one hour after the start of the experiment. Parameters v^1 and α^1 allow us to state that a thicker emulsion layer and lower temperatures do not ensure more efficient self-preservation and even its occurrence (in experiments **O2/**1p8gmin11 and **O2/**8p1min5 self-preservation was almost absent). In general, self-preservation is only weakly pronounced in hydrate dispersions based on **O2** oil; most likely, self-preservation here depends on small variations and pre-history of samples. Analysis of the results showed that the most probable reason of so different efficiencies of self-preservation of hydrate suspensions in oil **O1** and **O2** may be the presence of surfactants in suspensions based on **O2** oil. It should be stressed that hydrate self-preservation in oil dispersions described in [18,19] proceeded after preliminary deep (to liquid nitrogen temperature) freezing of the dispersion.

To test this assumption, we carried out experiments in which the efficiencies of methane hydrate self-preservation were compared by means of thermovolumetry for hydrate dispersions in **O2** oil in the presence and in the absence of Span 80 surfactant. First of all, dispersions prepared according to the above-described procedure in sample holders with thin and thick emulsion layers were studied. For this purpose, after stage (2) (figure 2a), the autoclaves under pressure were rapidly frozen to the temperature of liquid nitrogen boiling. The resulting hydrate samples were taken out, averaged and studied by means of X-ray powder diffraction (figure 3) and thermovolumetry (figure 4).

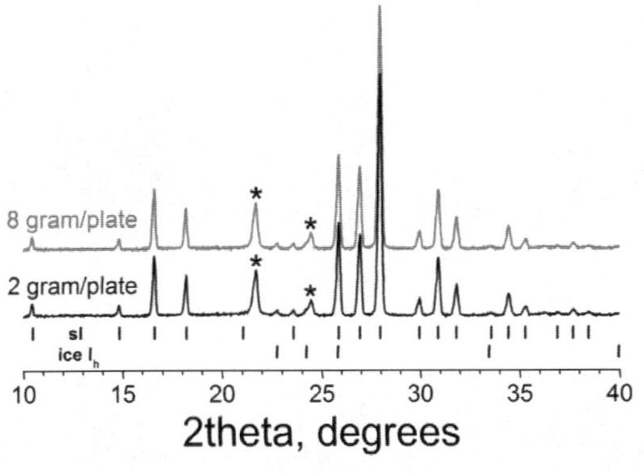

Figure 3. Methane hydrate X-ray diffraction patterns obtained from water/**O2** emulsion. Positions of the corresponding phases reflexes are marked by vertical risks; * – paraffins crystalline phase reflexes positions [22]. Both diffractograms are recorded at -100 ° C.

One can see in figure 3 that for hydrate samples obtained from water/**O2** emulsion in both cases we observe reflections corresponding to methane hydrate sI (the major phase) and the crystalline phase of paraffins [22]. Reflections corresponding to the phase of hexagonal ice I_h can hardly be distinguished. This fact confirms that the degree of water transformation into hydrate is close to the maximal possible one in all our experiments.

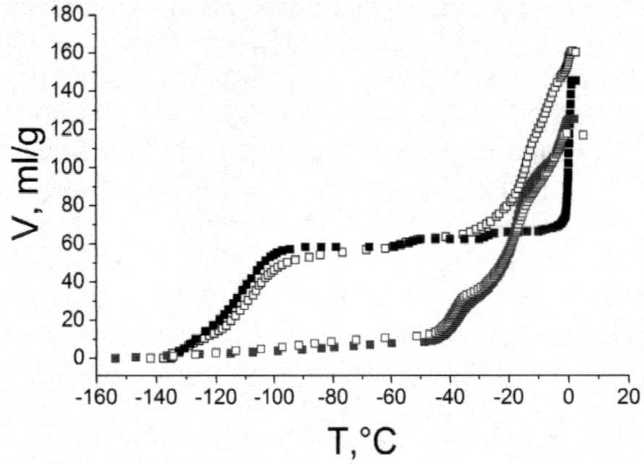

Figure 4. Methane hydrate gas emission curves obtained from water/**O2** emulsion. ■ – 2 grams/holder; □ – 8 grams/holder; ■ – pure methane hydrate mixed with **O2** (1: 1 by weight); □ – pure methane hydrate mixed with **O2** (1: 1 by weight, 1% Span 80 per weight of oil).

Thermovolumetric data show that the samples are gradually decomposed within temperature range -50 to 0°C (figure 4). The equilibrium temperature of methane hydrate decomposition is -75°C; in thermovolumetric experiments in the absence of self-preservation it always decomposes within temperature range -75 - -60°C [18]. So, thermovolumetric data are in good agreement with the above-driven conclusion concerning weak manifestation of self-preservation in dispersions based on **O2** oil, One can see in figure 4 that due to self-preservation curves ■ and □ do not have a step of methane hydrate decomposition under equilibrium conditions. At the same time, for samples ■ and □, intense decomposition starts at -45 – 40°C, which we suppose to be connected with the presence of Span 80 surfactant in the initial emulsion (1 mass % with respect to oil mass). The curves of gas evolution ■ and □ have a bend within temperature range -20 to -10°C (within this temperature range, decomposition of self-preserved methane kind of decelerates, which agrees with the data shown in table 2). So, self-preservation of methane hydrate occurs indeed, but one can hardly call it efficient. For comparison, experiments were carried out in which the dispersions of methane hydrate in **O2** oil were prepared by mixing hydrate powder with the powder of frozen oil (see Experimental section). One of the dispersions was prepared using pure **O2** oil, while another one was prepared by adding 1 mass % solution of Span-80 surfactant to the sample kind of oil. The results of thermovolumetric experiment with methane hydrate mixed with **O2** oil (1:1 by mass, points ■) are shown in figure 4. At -70°C, partial decomposition of methane hydrate occurs; the formed ice preserves the major part of hydrate efficiently (the major amount of gas evolves at the point of ice melting). In the experiment with methane hydrate mixed with **O2** oil + Span 80 surfactant, the results different substantially (figure 4, □). One can see that here the behavior of the curve coincides with that for the samples of suspensions obtained from water/**O2** oil emulsion (■ and □). So, the decisive contribution into the efficiency of self-preservation is made by the presence or absence of a surfactant in the initial oil emulsion. In addition, the data obtained demonstrate that self-preservation may manifest itself for the samples of hydrate suspensions in oil that had not been preliminarily exposed to deep freezing.

Acknowledgments
The investigation was supported by the Russian Scientific Foundation under Project 17-17-01085.

References

[1] Sloan E D and Koh C A 2008 *Clathrate Hydrates of Natural Gases 3rd Edition* (Boca Raton, London, New-York: CRC Press) p 752

[2] Chapoy A, Anderson R and Tohidi B 2007 Low-Pressure Hydrogen Storage in Semi-clathrate Hydrates of Quaternary Ammonium Compounds *JACS* **129** 746-7

[3] Wang W, Carter B, Bray C, Steiner A, Bacsa J, Jones J, Cropper C, Khimyak Ya, Adams D and Cooper A 2009 Reversible Methane Storage in a Polymer-Supported Semi-Clathrate Hydrate at Ambient Temperature and Pressure *Chem. Mater.* **21** 3810-5

[4] Eslamimanesh A, Mohammadi A H, Richon D, Naidoo P and Ramjugernath D 2012 Application of gas hydrate formation in separation processes: A review of experimental studies *J. Chem. Thermodyn.* **46** 62-71
[5] Lee Y, Lee D, Lee J W and Seo Y 2016 Enclathration of CO_2 as a co-guest of structure H hydrates and its implications for CO_2 capture and sequestration *Applied Energy* **163** 51-9
[6] Kim E, Lee S, Lee J D and Seo Y 2016 Enclathration of tert-butyl alcohol in sII hydrates and its implications in gas storage and CO_2 sequestration *Fuel* **164** 237-44
[7] Li X S, Xia Z M, Chen Z Y, Yan K F, Li G and Wu H J 2010 Gas hydrate formation process for capture of carbon dioxide from fuel gas mixture *Ind. Eng. Chem. Res.* **49** 11614-9
[8] Park S, Lee S, Lee Y and Seo Y 2013 CO_2 capture from simulated fuel gas mixtures using semiclathrate hydrates formed by quaternary ammonium salts *Environ. Sci. Technol.* **47** 7571-7
[9] Park K N, Hong S Y, Lee J W, Kang K C, Lee Y C, Ha M G and Lee J D 2011 A new apparatus for seawater desalination by gas hydrate process and removal characteristics of dissolved minerals (Na^+, Mg^{2+}, Ca^{2+}, K^+, B^{3+}) *Desalination* **274** 91-6
[10] Mahmoudi B, Naeiji P and Varaminian F 2016 Study of tetra-n-butylammonium bromide and tetrahydrofuran hydrate formation kinetics as a cold storage material for air conditioning system *J. Mol. Liq.* **214** 96-100
[11] Daitoku T and Utaka Y 2010 Separation characteristics of clathrate hydrates from a cooling plate for efficient cold energy storage *Applied Energy* **87** 2682-9
[12] Hammerschmidt E G 1934 Formation of Gas Hydrates in Natural Gas Transmission Lines *Ind. Eng. Chem.* **26** 851-5
[13] Sloan E D 2000 *Hydrate Engineering (Digital Edition)* (Society of Petroleum Engineers: Richardson, TX, USA)
[14] Choudhary N, Das S, Roy S and Kumar R 2016 Effect of polyvinylpyrrolidone at methane hydrate-liquid water interface. Application in flow assurance and natural gas hydrate exploitation *Fuel* **186** 613-22
[15] Park J, Kim H, Sheng Q, Wood C D and Seo Y 2017 Kinetic Hydrate Inhibition Performance of Poly (vinyl caprolactam) Modified with Corrosion Inhibitor Groups *Energy Fuels* **31** 9363-73
[16] Ding L, Shi B, Wang J, Liu Y, Lv X, Wu H, Wang W, Lou X and Gong J 2017 Hydrate Deposition on Cold Pipe Walls in Water-in-Oil (W/O) Emulsion Systems *Energy Fuels* **31** 8865-76
[17] Aspenes G, Høiland S, Borgund A E and Barth T 2009 Wettability of petroleum pipelines: Influence of crude oil and pipeline material in relation to hydrate deposition *Energy Fuels* **24** 483-91
[18] Stoporev A S, Manakov A Yu, Altunina L K, Bogoslovsky A V, Strelets L A and Aladko E Y 2014 Unusual Self-preservation of Methane Hydrate Particles in Oil Suspensions *Energy Fuels* **28** 794-802
[19] Stoporev A S, Manakov A Yu, Altunina L K and Strelets L A 2016 Self-Preservation of Gas Hydrate Particles Suspended in Crude Oils and Liquid Hydrocarbons: Role of Preparation Method, Dispersion Media, and Hydrate Former *Energy Fuels* **30** 9014-21
[20] Ogienko A G, Kurnosov A V, Manakov A Y, Larionov E G, Ancharov A I, Sheromov M A and Nesterov A N 2006 Gas hydrates of argon and methane synthesized at high pressures: composition, thermal expansion, and self-preservation *J. Phys. Chem.* B **110** 2840-6
[21] Collins M J, Ratcliffe C I, Ripmeester J A 1990 Nuclear magnetic resonance studies of guest species in clathrate hydrates: line-shape anisotropies, chemical shifts, and the determination of cage occupancy ratios and hydration numbers *J. Phys. Chem.* **94** 157-62
[22] Heyding R, Russell K, Varty T and St-Cyr D 1990 The Normal Paraffins Revisited *Powder Diffr.* **5** 93-100

Freezing of mineralized water droplets in winter sprinkling

A V Sosnovsky and A F Glazovsky

Institute of Geography of the Russian Academy of Sciences,
Staromonetny pereulok, 29, Moscow 119017 Russia

E-mail: sosnovsky@igras.ru

Abstract. Drip congelation is an effective method of desalination of salt water. The results of mathematical modelling of freezing of fresh and salt water droplets in conditions of free fall at negative air temperatures are presented. Different options of salt rejection from the frost boundary and mineralization growth in the unfrozen part of the drop are considered.

1. Introduction

Intensive economic development of Arctics has resulted in pollution of sewage waters of water bodies. For purification and desalination of large volumes of natural and technogenic waters, including for the purposes of water supply of the population, it is necessary to use lean processing based on the use of renewable natural energy. Our research has shown [1] that one of the most efficient method of desalination and purification of large volumes of mineralized water is the method of drip freezing. The method is based on the use of long-range sprinklers plants to create a drip torch and the formation of porous ice bodies. Water consumption of the most common sprinklers, based on DTT-70 system with nozzle diameter 55 mm is 240 m^3/hour, with a jet range 70 m. In the Arctic zone the performance of freezing with winter sprinkling can reach 10^5 tons of porous ice per month. Currently, winter sprinkling is widely used for the construction of ice crossings and ice roads. Experiments with salt water and estimations show that this method allows, even at the freezing stage, to reduce the mineralization of porous ice body by 10–12 times. When the water droplets are freezing, the salt pushed away off into the central unfrozen drop core. When the water drops fall on the ground and their ice shells destruct, the unfrozen brine is filtered away from the porous ice body that considerably reduces its mineralization. With increase of mineralization of initial water, the mineralization of the unfrozen water drop core grows, and, as a result, the desalination productivity decreases [2].

The aim of the paper is to assess the influence of water mineralization on freezing of droplets in winter sprinkling.

2. Problem statement

At winter sprinkling the height of the drip torch reaches 20 meters. The average diameter of drops of modern sprinklers units of the "Grad" series applied for winter sprinkling is 1.5 mm with diameter range 1-2 mm. The fall time of water droplets 1 to 2 mm in diameter from a height of 20 m is 3 – 5 seconds. Application of winter sprinkling allows to create in a drip torch a water/ice mixture which forms the monolithic ice used for construction of ice crossings and dams. The water drops in the mixture should not freeze completely, and at falling the ice shell of a drops should be destroyed and release unfrozen water. Otherwise, dry granulated ice will be formed, which is not suitable for the construction of ice crossings. Experimental studies have shown that for water droplets with a radius of

1.5 mm, the proportion of ice in a drop should not exceed 50 – 60 %. The same conditions are necessary at application of winter sprinkling for desalination of mineralized waters. This imposes restrictions on the parameters of the drip torch depending on the air temperature. Restrictions on the share of ice in freezing drops and the fall time will be used in calculating of water mineralization effect on freezing of drops at winter sprinkling.

3. Mathematical modelling

At freezing of drops of fresh water or water with low mineralization (up to 5-10 g/L) the ice shell is formed on a surface of a drop and it thickens, reducing the radius of the liquid part. For water with high mineralization, the growth of branchy crystals into the depth of the liquid part of the drop is possible. This increases the mineralization of the liquid part of the drop. However, the growth of branchy crystals is difficult to predict. Therefore, as well as in case of a drop of fresh water, the scheme of freezing is accepted with the formation of the ice shell and its symmetrical development to the center of the drop. It is assumed that the initial drop temperature is equal to the freezing temperature of water at initial salinity.

The transfer of heat and freezing water is described by a boundary problem with a moving phase boundary

$$\frac{\partial T_i}{\partial \tau} = a_i \left(\frac{\partial^2 T_i}{\partial r^2} + \frac{2}{r} \frac{\partial T_i}{\partial r} \right), \ \xi(\tau) < r < R,$$

$$\frac{\partial T_w}{\partial \tau} = a_w \left(\frac{\partial^2 T_w}{\partial r^2} + \frac{2}{r} \frac{\partial T_w}{\partial r} \right), \ 0 < r < \xi(\tau).$$

The condition of heat exchange at a drop boundary is

$$-\lambda_i \frac{\partial T_i}{\partial r}\Big|_{r=R} = \alpha_{ai} (T_i(R) - T_{ae}).$$

Stefan's condition at the freezing boundary at $r = \xi(\tau)$ is

$$\lambda_i \frac{\partial T_i}{\partial r} - \lambda_w \frac{\partial T_w}{\partial r} = -L \frac{d\xi}{d\tau}.$$

At freezing boundary when $r = \xi$ the temperature in solid and liquid parts suggested to be equal to the freezing temperature of saline water $T_0(s)$ with different salinity $T_i = T_w = T_0(s)$.
In the central part of the drop when $r = 0$ the following condition is accepted

$$\frac{\partial T_w}{\partial r} = 0.$$

In the initial period when $\tau = 0$ the water temperature is assumed to be equal to the freezing temperature $T_w = T_0$.

Accepted designations: T is temperature, τ is time, a and λ are coefficients of thermal diffusivity and thermal conductivity; L is the volumetric heat of the water-ice phase transition, R is the radius of the drop; α_{Ai} is a heat-exchange coefficient of falling drops; ξ is the position of the crystallization front (radius of the liquid core). Indexes w and i correspond to water and ice; T_{Ae} is an effective air temperature.

Thermal diffusivity is $a = \lambda / (c\rho)$, where c and ρ are heat capacity and density, respectively. Factor heat-exchange coefficient on the drop surface and effective air temperature are taken from [3] They were obtained using criterial dependencies of heat and mass exchange in relation with water vapor pressure and temperature.

The phase composition of the salt ice depends on the temperature and mineralization. At the temperature of salt ice above –8... –10 °C, as a first approximation one can take a linear dependence between the concentration of brine S_b (kg/m^3) in ice and ice temperature t_i [4]

$$S_b = \mathrm{S}\, t_i , \qquad (1)$$

where is $t_i = T_i - 273$; values of factor σ depend on the freezing temperature of different salts, and for sea water σ = –18.2 kg/(m^3 grad).

The efficient heat capacity of salt ice c_{Ie}, is taken into account for heat-physical calculations and is equal to the average value of the sum of the heat capacities of the crystal ice c_i and brine c_b,, and the heat of phase transitions L. The value of c_{Ie} is determined by the equation [4]

$$c_{ie}(T_i) = c_i + (c_b - c_i)\frac{S}{\sigma(T_i - 273)} - L\frac{S}{\sigma(T_i - 273)^2}. \qquad (2)$$

4. Calculation results

Calculations were carried out for fresh water and water with mineralization 35 g/L with freezing temperature –1.8 °C. For calculations, the drop diameters accepted as 1, 1.5, and 2 mm. On the freezing boundary of a saline water drop a layer of water with increased mineralization can develop. This process is difficult to predict. Therefore, as a first approximation, extreme cases of salt rejection from the freezing boundary into the liquid part of the drop were considered: 1) salt at drops freezing is not rejected in the liquid part of the drop, the salinity of the ice shell and the liquid core are equal to the freezing temperature; and at the same time the effective heat capacity of ice is determined by the equation (2); 2) salt during freezing is completely rejected into the liquid part of the drop, and the heat capacity of the ice shell is equal to the heat capacity of crystal Ice c_i, and the temperature of the phase transition is calculated according to the equation (1) taking into account the growing salinity of the liquid core of the drop; 3) the ice shell contains 1/3 part of the salt what often corresponds to the salinity of the young sea ice, and equations (1) and (2) are used to calculate the effective heat capacity of the ice and the freezing temperature of the liquid core.

The intensity of drops freezing depends on temperature difference between air and ice shell surface. Calculations showed, that in the first case the surface temperature of drops with radius of 0.5 and 1 mm when half of its volume frozen are –2.0 and –2.2 °C; and in second case –3.6 and –3.9 °C, respectively (Figure 1).

When half of a drop volume is frozen, radius of a liquid part is about 0.8 from a radius of a drop. For a drop of 1.5 mm in diameter the thickness of the ice shell is 0.15 mm. For this ice thickness the temperature difference between the freezing boundary and the drop surface (ice shell) is usually not more than 0.3 °C.

For heat exchange between the air and the falling droplet, an important parameter is the time-weighted value of drop surface temperature. For instance, at the time of half-volume freezing of the droplet 1.5 mm in radius, its surface temperature is about –3.8 °C (second case of calculations), whereas the average time-weighted temperature of its surface is C with air temperature –20 °C. Meanwhile the effective air temperature reaches –29 °C because of heat exchange by evaporation. The average time-weighted temperature of droplet surface (in first case) is –2.0 °C and (in third case) –2.6 °C. The temperature of the surface depends mainly on the ice share freezing drop and, in a lesser degree, on air temperature. For example, for a drop of 1.5 mm in diameter, its surface temperature at the moment of half-volume freezing is –3.7 °C, –3.9 °C, and –4.1 °C at air temperatures –10 °C, –20 °C, and –40 °C accordingly (second case of calculations).

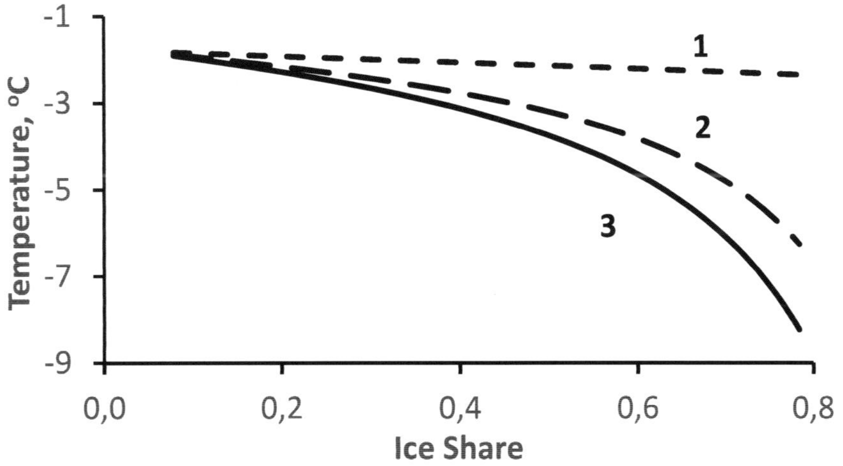

Figure 1. Surface temperature of a droplet depending on the share of ice in its volume. Curves 1, 2 and 3 correspond to three cases of salt rejection from the freezing boundary into the liquid part (details in text).

The results of calculations (for second case) of freezing time of a drop of 1.5 mm diameter depending on ice fraction are given in Figure 2. The fall time of the drop from a height of 18 m is 3.3 sec. During this time 0.14, 0.20, 0.24, or 0.39 of its volume freezes at air temperatures –10, –15, –20 , or – 40 °C, respectively (Figure 2).

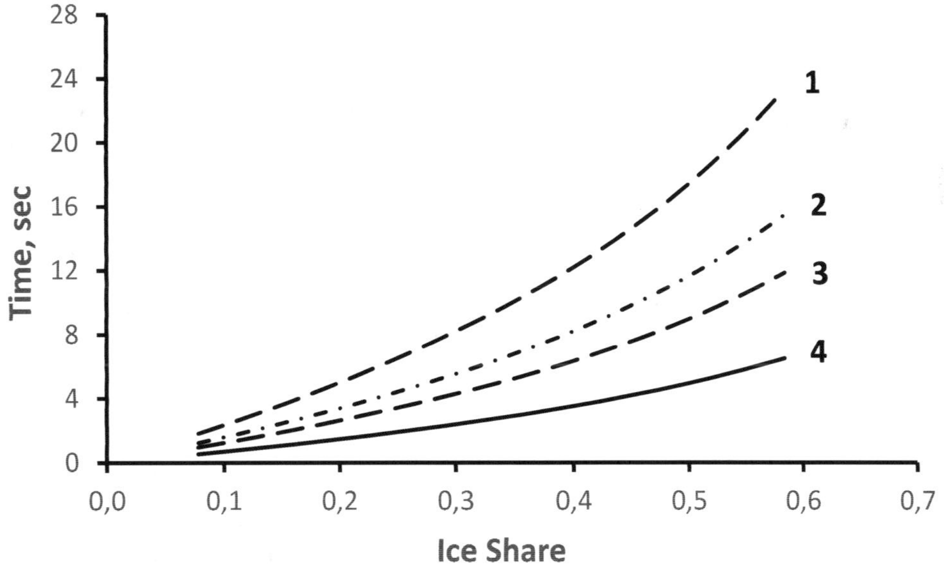

Figure 2. Freezing time of a droplet of 1.5 mm in diameter depending on ice volume fraction in it. The time is estimated for second case (see text) at different air temperatures: curve 1 is for –10 °C, curve 2 is for –15 °C, curve 3 is for –20 °C, and curve 4 is for –40 °C.

Effect of droplet size on freezing is shown in Figure 3. It can be seen that at air temperature –20 °C the half-volume freezing time (second case of calculations) is 4.9, 8.4, and 13.4 sec for drops of 1, 1.5, and 2 mm in diameter, respectively.

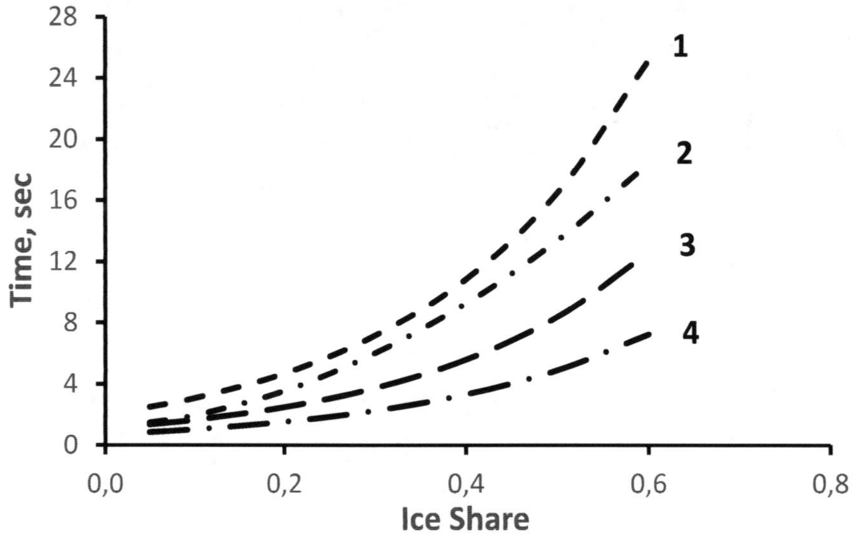

Figure 3. Freezing time of a droplet with a diameter of 1 mm (4), 1.5 mm (1, 3), and 2 mm (2) depending on ice volume fraction. The time is estimated for second case (see text) at different air temperatures: curve 1 is for –10 °C, curves 2, 3 and 4 are for –20 °C.

Freezing of half volume of a drop of 1.5 mm in diameter at air temperature –20 (–10) °C requires 8.5 (15.2) seconds for fresh water and 9.9 (19.0) seconds for water with mineralization of 35 g/L (third case of calculations). On the first case of calculations it takes 9.7 (17.8) seconds and on the second case 9.4 (18.2) seconds. This result is explained by the fact that the average time-weighted surface temperature differs slightly (relatively to effective air temperature) at freezing to the half of the drop volume. While retaining some salt in the ice in the form of brine cells significantly increases the effective heat capacity of ice, according to the equation (2). Thus, for various variants of salt rejection, the results of calculating the freezing time of the droplet to half its volume are determined both by the effective heat capacity of the ice and by the freezing temperature of the liquid part of the drop. The latter depends on the mineralization of the liquid core, which increases during freezing and is determined by the mechanism of rejection of salt ions from the freezing boundary.

5. Conclusions

Results of mathematical modeling of freezing of drops of fresh and salt water in conditions of free fall at negative temperatures of atmospheric air were presented. The Stefan condition was adopted at the freezing front. Various variants of the growth of mineralization of the unfrozen part of the drop during its freezing were considered. For drops of water with mineralization of 35 g/L and diameter of 1.5 mm, the freezing time of half the volume of the droplet at an air temperature of –10 °C is 25 % longer than for fresh water, and 17 % at an air temperature of –20 °C, when ice retains 1/3 of the salt.

Acknowledgments

The study was supported by the Presidium RAS program No 56 "Fundamental Basis of Breakthrough Technologies in the Interests of National Security".

References

[1] Sosnovskii A V and Khodakov V G 1995 Artificial ice formation in natural conditions for handling environmental problems *Data of Glaciological Studies* **79** 3–6 (in Russian)

[2] Sosnovskiy A V 1988 Estimation of the rate of spray-cone freezing from saline water *Data of Glaciological Studies* **61** 49–154 (in Russian)

[3] Sosnovsky A V 1980 Freezing of drops of the artificial rain *Data of Glaciological Studies* **38**

54–9 (in Russian)
[4] Doronin Yu P 1969 *Thermal Interaction of the Atmosphere and Hydrosphere in the Arctic* (Leningrad: Gidrometeoizdat) p 300 (in Russian)

Study of oil sludge, waste oil and other auxiliary substances influence on the methane hydrate dissociation

A S Stoporev [1,2,3], A P Semenov [1], A Y Manakov [2,3] and V A Vinokurov [1]

[1] Gubkin University, Department of Physical and Colloid Chemistry, Leninsky prosp. 65, Building 1, Moscow, 119991, Russian Federation
[2] Nikolaev Institute of Inorganic Chemistry SB RAS, Ac. Lavrentiev ave. 3, Novosibirsk, 630090, Russian Federation
[3] Novosibirsk State University, Pirogova Str. 2, Novosibirsk, 630090, Russian Federation

E-mail: stopor89@bk.ru

Abstract. In this paper, classes of oil medium that ensure the methane hydrate self-preservation were identified. The following compounds and substances were studied: asphaltenes, resins, asphaltene-resin-paraffin deposits (ARPD), multiwall carbon nanotubes, meloxicam ($C_{14}H_{13}N_3O_4S_2$), kinetic hydrate inhibitors (KHI) Luvicap 55W (copolymer of N-vinylcaprolactam and N-vinylpyrrolidone) and Luvicap EG (poly(N-vinylcaprolactam)), refined petroleum products (initial and waste) and 2,6-Di-*tert*-butyl-4-methylphenol (antioxidant fuel additive AO-29). Isolated oil components had practically no effect on the decomposition process of methane hydrate. It was observed, that the presence of surface-active additives in refined petroleum products leads to suppression of the investigated effect (self-preservation). It seems to be that in the case of used engine oils, the degree of manifestation of methane hydrate self-preservation effect depends on the duration of its operation (mileage). The obtained results can be the basis of recycling of ARPD and hazardous waste petroleum products in technologies for storage/transportation of natural gas.

1. Introduction

Gas hydrates are inclusion compounds in which the host framework is formed by water molecules bound to each other with hydrogen bonds, while the guest components fill the cavities in this framework [1]. When it comes to the storage and transportation of gas in hydrate form, it may be based on the self-preservation effect [2]. This phenomenon consists in a sharp decrease in the hydrate decomposition rate at temperature below 0 °C [3]. The following models are currently being considered. (1) Hydrate powder-in-liquid hydrocarbon suspensions transported through pipelines at the temperature and pressure within hydrate stability zone. (e.g., [4,5]), and (2) gas hydrate-ice pellets with size range from a few millimeters to several centimeters transported in metastable state at atmospheric pressure and temperature below 0 °C. (e.g., [6,7]). The pellets application for gas storage and transportation by the use of self-preservation effect was patented (see [8], for example). A hydrate granular form creates favorable conditions for its long-term existence outside thermodynamic stability zone in comparison with a hydrate powder. The overland transportation of natural gas hydrate to small consumers located within several hundred kilometers from the gas terminal was approved in the demo project launched by Mitsui Engineering & Shipbuilding CO [9]. The gas hydrates were transported in the form of pellets sized from 0.50 mm to 30 mm. Their storage at -20 °C led to hydrate preservation for two weeks.

Study of gas hydrates in an oil at temperature below 0°C showed that the self-preservation effect manifestation occurs due to some oil components forming "reinforced shell" on hydrate particles (see [10], for example). A variation in the oil composition allows the hydrate formation and decomposition processes to be controlled [10]. Thus, the hydrates can exist in some oil content systems for a long time in the *p-T* region, where they are thermodynamically unstable. It is known that some oil components can form adsorption layers ("reinforced shell") on the surface of water droplets dispersed in the oil [11]. The layers play an important role in the colloidal chemistry and rheology of oils (in particular they determine stability of oil emulsions). An influence of the adsorption layers on hydrate-in-oil suspensions behavior has almost not been investigated to date. At the same time, temperature interval (below 0 °C) favor the manifestation of self-preservation effect, which is important from both practical and theoretical point of view. This allows one to say that the study performed in this work are of a fundamental nature and offers the challenge to manage the effect of self-preservation in applications to technological processes with the use of waste oils and deposits formed on the oil pipelines walls.

2. Experimental section

TO, EO1, EO2, SOF and T_ARPD (Table 1) were selected as objects of study. TO is a high-quality product that is produced from a mineral base oil with the use of hydrocatalytic processes and contains an antioxidant additive (AO-29); engine oils have high antioxidant and dispersant properties (prevents formation of high- and low-temperature deposits on engine parts). Thus, the oils used contain various additives (antioxidant and dispersant). TO, EO1, EO2, SOF and T_ARPD were taken in pure form, while asphaltenes and resins isolated from the oil of the South-Philippine deposit, multiwall carbon nanotubes, meloxicam (4-hydroxy-2-methyl-N-(5-methyl-2-thiazolyl)-2H-1,2-benzothiazine-3-carboxamide-1,1-dioxide), Luvicap 55W (dry polymer), Luvicap EG (dry polymer) and antioxidant fuel additive AO-29 (2,6-di-*tert*-butyl-4-methylphenol) were dissolved in toluene. Content of KHI Luvicap 55W and Luvicap EG in the solutions obtained was 0.5 mass % (standard concentration of KHI in aqueous phase for hydrate inhibition); content of other auxiliary substances was 1 mass %.

Table 1. Characteristics of the initial liquids.

Property	Sample				
	TO[a]	EO1[a]	EO2[a]	SOF[a]	T_ARPD[a]
Content of asphaltenes (mass %)	< 0.1	–	–	1.5	1.8
Content of resins (mass %)	0.69	–	–	10.7	5.5
Pour point (°C)	–	-36	-39	–	–
Density at 20°C (kg/m^3)	837	878	856	865	859
Viscosity at 20°C (mPa·s)	4.9	127.5	90.9	17.3	–

[a] TO, EO1, EO2, SOF, T_ARPD – are acronyms of transformer oil (LUKOIL VG), engine oil 1 (LUKOIL SUPER 10W-40, semi-synthetic), engine oil 2 (IDEMITSU 5W-40, fully-synthetic), petroleum from Sovetskoe oil field, asphaltene-resin-paraffin deposits from Irelyakhskoe oil field diluted with toluene respectively.

The approach to control the hydrate decomposition process proposed in this work was not previously used in practice. The pure methane hydrate with stoichiometry of $CH_4{\cdot}6.1H_2O$ synthesized from distilled water and methane (Moscow Gas Refinery Plant, >99.99 % pure) was mixed with additives either at 1 °C under methane pressure (10 MPa) or at liquid nitrogen temperature (atmospheric pressure). The mixing process under methane pressure was described in [10]. As for mixing at the boiling point of liquid nitrogen, the procedure was performed by careful grinding of

IOP Conf. Series: Earth and Environmental Science **193** (2018) 012064 doi:10.1088/1755-1315/193/1/012064

hydrate and a frozen organic liquid. The obtained mixture was put into an autoclave and heated to 10 °C at 10 MPa. The samples were held at these *p-T* conditions for day to provide redistribution of components in the system. Then the obtained sample of methane hydrate-in-organic liquid suspension was pulled out at liquid nitrogen temperature. The synthesized samples were carefully powdered (the particle size was less than 250 μm [10]) and studied by thermovolumetry at atmospheric pressure and temperature below 0 °C [10] (two measurements were carried out for each sample). The average hydrate content was 40.1 ± 0.1 %. The procedure of pure methane hydrate synthesis was described in [10].

3. Results and discussion

Comparison of the curves of gas evolution from the samples obtained by mixing pure methane hydrate and organic liquid under different conditions (at 1 °C under methane pressure (10 MPa) or at liquid nitrogen temperature, followed by heating the sample) showed that the decomposition of these samples proceeds in the same way (Figure 1). The part of hydrate self-preserved in T_ARPD was 21.0 ± 1.0 and 19.0 ± 2.8%, for different preparation modes respectively. However, the preparation of a hydrate suspension by grinding pure hydrate with an organic liquid at nitrogen boiling point can significantly expand the range of test substances as mixing liquids under pressure is limited to the ones with low viscosities.

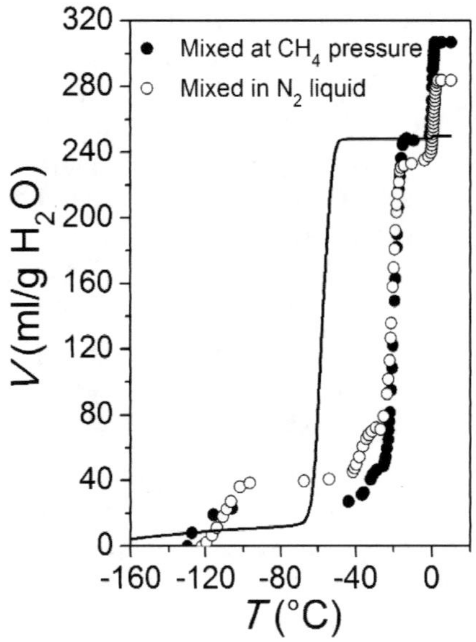

Figure 1. Dissociation of methane hydrate in T_ARPD; method of suspension preparation is indicated on the graph; the black line corresponds to the pure hydrate dissociation.

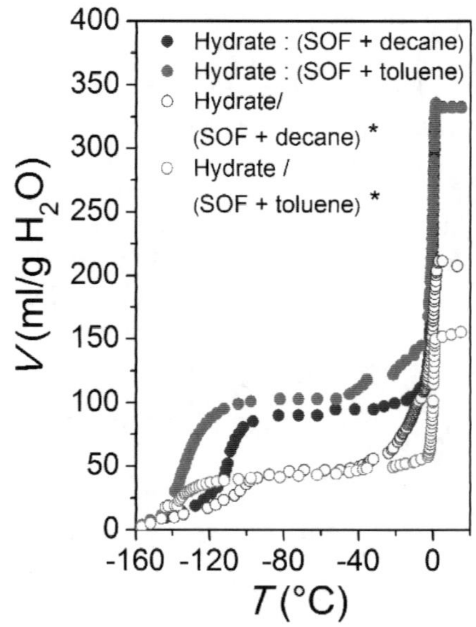

Figure 2. Dissociation of methane hydrate in SOF oil diluted with decane or toluene (1:1 by weight); filled circles indicate suspension prepared by mixing at liquid nitrogen temperature; open circles * correspond to data of [10] on dissociation of hydrate formed directly from water emulsified in the oil.

At the same time, Figure 2 clearly shows that the decomposition of methane hydrate in suspensions obtained by either mixing at liquid nitrogen temperature with subsequent heating or direct synthesis from the emulsion had some differences. This can be explained by the different morphology of hydrate particles. It was shown that the morphology of hydrate growing at a water–organic interface was varied considerably depending on the composition of organic phase [12]. Thus, the method chosen for preparing of hydrate suspensions allowed one to study the effect of an organic liquid composition on the hydrate dissociation at a constant composition, size, and morphology of hydrate particles. The first experiments showed that TO promoted the effective self-preservation of methane hydrate (96.3 ± 1.9%), whereas EO had no effect and the decomposition began at the equilibrium temperature (-80°C) as well as in the case of pure methane hydrate (Figure 3; the dissociation of the hydrate in EO1 and EO2 proceeded in the same way). It is known that various types of oils are prepared from the base oil by the addition of different additives. There is a wide range of additives in engine oils (anticorrosion, antioxidant, antifriction, detergent and dispersant, etc.), whereas in the transformer oil only an antioxidant additive is present. Thereby, there are types of additives capable of suppressing the self-preservation effect (most likely, surfactants providing a detergent and dispersant effect).

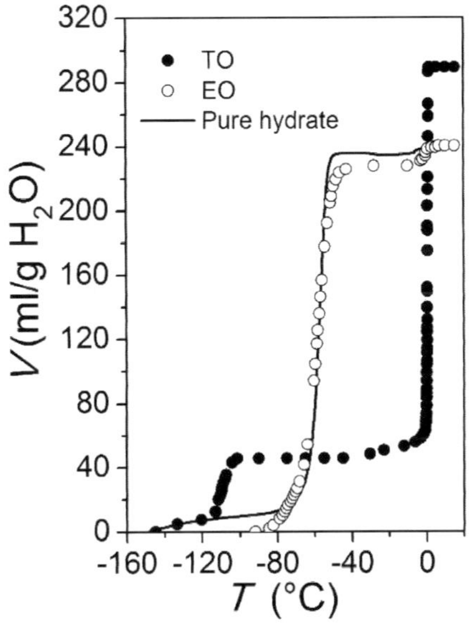

Figure 3. Dissociation of hydrate in the initial oils; dispersion media are indicated on the graph.

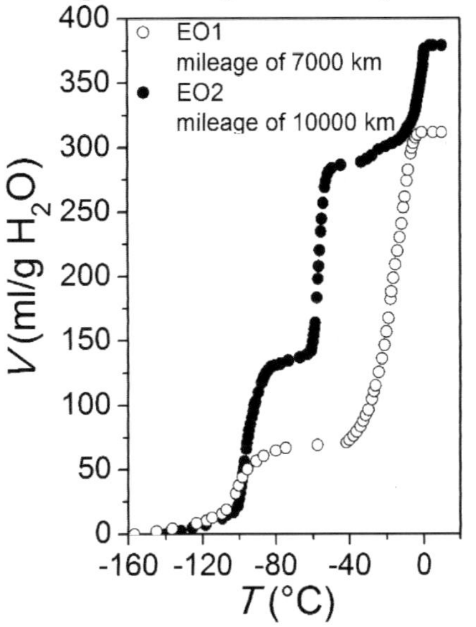

Figure 4. Dissociation of hydrate in the used oils; dispersion media are indicated on the graph.

Similar experiments with these engine oils differed in mileage (7000 and 10000 km in the case of EO1 and EO2, respectively) showed that the oils modified in this way had some influence on the process of hydrate decomposition (Figure 4). It seems to be that the part of methane hydrate, which decomposes in used engine oils at a temperature close to 0°C, depends on the duration of its operation (mileage). Dispersant additives are most likely destroyed or bound with solid deposits formed during cocking. These observations can contribute to the development of a new trend in the use of waste oils to stabilize the hydrate phase for storage and transportation of gas in hydrate form. Discussion of this possibility requires to provide additional experiments and goes beyond the scope of this paper.

In addition, the effect of the following compounds on the decomposition process of methane hydrate was studied: KHI Luvicap 55W and Luvicap EG, polyfunctional compound meloxicam, AO-29 (antioxidant additive in TO), asphaltenes and resins isolated from the oil of South-Philippine field and multiwall carbon nanotubes (MWCN) (Figure 5, on the examples of asphaltenes, MWCN and Luvicap 55W). The listed compounds had practically no effect on the decomposition process of methane hydrate. Samples with the addition of these compounds behaved themselves in a similar manner and decomposed without self-preservation. Since the substances were dissolved in toluene to mix it with hydrate, the similarity of decomposition curves for methane hydrate dispersed in pure toluene and with the addition of any substances showed that only toluene influenced in these cases (Figure 5).

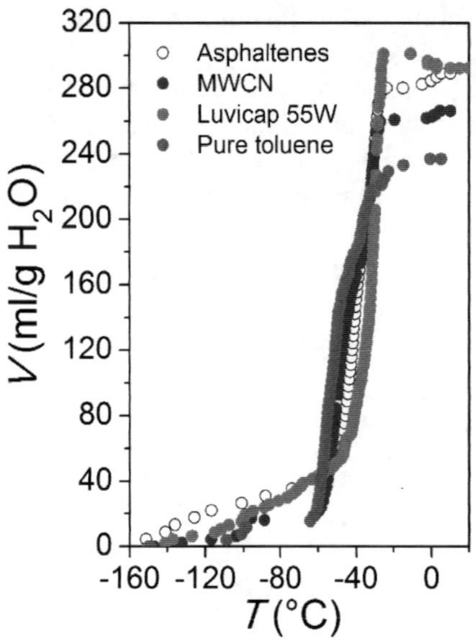

Figure 5. Dissociation of methane hydrate in the presence of different substances indicated on the graph.

Thus, it was demonstrated experimentally that self-preservation effect in TO and ARPD is possible. The increase in the duration of EO operation may also lead to the observation of this effect in the waste oils. The obtained data can be used to develop gas hydrate technologies for storage and transportation of hydrate-forming agents by reducing gas losses when placing hydrates particles in a medium that ensures their slow decomposition. It has been established that artificial additives (kinetic hydrate inhibitors Luvicap 55W and Luvicap EG, polyfunctional compound meloxicam and antioxidant additive) had practically no effect on the kinetic characteristics of gas hydrate decomposition process. The use of ARPD, residual heavy fractions after refining or waste petroleum products for hydrates treatment will not only increase the stability of the hydrate phase during storage, but will also open up a new way to use the listed hazardous wastes accumulated during the oil transportation and refining.

Acknowledgements
The work was supported by the Russian Scientific Foundation Project № 17-77-10051.

References

[1] Sloan E D and Koh C A 2008 *Clathrate Hydrates of Natural Gases* 3rd ed (Boca Raton, London, New-York: CRC Press) p 752
[2] Ershov E D, Lebedenko Y P, Chuvilin E M, Istomin V A and Yakushev V S 1991 *Dokl. Akad. Nauk SSSR* **321** 788–91
[3] Stern L A, Circone S, Kirby S H and Durham W B 2001 *J. Phys. Chem. B* **105** 1756–62
[4] Fitzgerald A and Taylor M 2001 Offshore gas-to-solids technology *Offshore Europe. Soc. Petrol. Eng.* SPE 71805
[5] Lv Y N, Jia M L, Chen J, Sun C Y, Gong J, Chen G J, Liu B, Ren N, Guo S D and Li Q P 2015 *Energy Fuels* **29** 5563–72
[6] Rehder G, Eckl R, Elfgen M, Falenty A, Hamann R, Kähler N, Kuhs W F, Osterkamp H and Windmeier C 2012 *Energies* **5** 2499–523
[7] Mimachi H, Takahashi M, Takeya S, Gotoh Y, Yoneyama A, Hyodo K, Takeda T and Murayama T 2015 *Energy Fuels* **29** 4827–34
[8] Yanagisawa N, Takanashi S, Yamaguchi T and Yamazaki T 2009 *Method of Production, Storage and Transportation for Gas Hydrate* EP2128228 A1 (Mitsui Engineering & Shipbuilding CO., LTD)

[9] Mimachi H, Takeya S, Yoneyama A, Hyodo K, Takeda T, Gotoh Y and Murayama T 2014 *Chem. Eng. Sci.* **118** 208–13
[10] Stoporev A S, Manakov A Yu, Altunina L K and Strelets L A 2016 *Energy Fuels* **30** 9014–21
[11] Mullins O C, Sheu E Y, Hammami A and Marshall A G 2007 *Asphaltenes, Heavy Oils, and Petroleomics* (New York: Springer Science & Business Media) p 669
[12] Stoporev A S, Semenov A P, Medvedev V I, Sizikov A A, Gushchin P A, Vinokurov V A and Manakov A Yu 2018 *J. Cryst. Growth* **485** 54–68

IOP Conf. Series: Earth and Environmental Science **193** (2018) 012065 doi:10.1088/1755-1315/193/1/012065

The motion of pressure distribution over a free surface near the edge of ice sheet

I V Sturova and L A Tkacheva

Lavrentyev Institute of Hydrodynamics SB RAS, Novosibirsk, Russia

E-mail: sturova@hydro.nsc.ru, tkacheva@hydro.nsc.ru

Abstract. The solution of a steady three-dimensional problem for the wave disturbances induced by a pressure distribution moving with uniform speed along the rectilinear edge of semi-infinite ice sheet is presented. This external load simulates the air-cushion vehicle. The problem is formulated within linear hydroelastic theory. The fluid is assumed to be inviscid and incompressible and its motion is potential. The ice sheet is treated as an elastic thin plate using the Kirchhoff-Love model. The solution of this problem is constructed using the Fourier transform and the Wiener-Hopf technique. The displacements of free surface and ice cover are determined, as well as power characteristics (wave resistance and side force) acting on the vehicle at various speeds of its movement: subcritical and supercritical relative to the minimum phase velocity of flexural-gravity waves in the ice cover. It is found that at speeds close to the critical velocity of flexural-gravity waves, the wave forces undergo sharp changes. It is shown that for some values of load speed, ice thickness and external pressure, the ice fracture near the edge is possible.

1. Introduction

The hydrodynamic aspects of an air cushion vehicle (ACV) can be studied by assuming its action to be equivalent to that of a pressure distribution acting on the free surface of water [1]. The disturbance induced by a pressure distribution moving over infinitely extended free surface has been thoroughly studied (see, e.g. [2]). However, in polar regions of the World Ocean, there are situations when the vehicle moves along the edge of ice sheet. Previously we obtained the solution of a steady three-dimensional problem for flexural-gravity waves generated by a local pressure distribution moving with uniform speed over semi-infinite ice sheet along its rectilinear edge [3] - [6]. Three configurations were considered: (i) the surface of fluid is free outside of ice sheet; (ii) two semi-infinite ice sheets (may be of different thickness) divided by a crack with free edges; (iii) the fluid is bounded by a rigid vertical wall and the edge of an ice sheet can be both free or clamped.

In this paper, the steady problem for the fluid and semi-infinite ice sheet under the action of an external load moving with uniform speed over a free surface along the edge of ice sheet is considered.

2. Mathematical formulation

The water is taken to be of constant density ρ_0 and uniform depth H. The pressure distribution $P(x, y)$ moves with constant speed U over a free surface along the rectilinear edge of the ice sheet. We consider the moving together with the load Cartesian coordinate system (x, y, z) with

x -axis directed perpendicular to the edge of the plate, the y -axis along the edge, and the z -axis vertically upwards. The upper boundary of fluid is covered by an ice sheet in the region $x < 0$ and the surface of fluid is free outside of ice sheet in the region $x > 0$. The fluid motion is irrotational and can be described by a velocity potential $\varphi(x, y, z)$. The ice sheet is assumed isotropic and homogeneous and is treated as an elastic thin plate. It is assumed that the plate is in contact with the water at all points and the plate draft is ignored.

The boundary-value problem for the velocity potential and the free surface elevation or plate deflection $w(x, y)$ can be written as

$$\Delta_3\varphi = 0 \quad (|x|, |y| < \infty, \ -H \le z \le 0), \quad \Delta_3 \equiv \Delta_2 + \partial^2/\partial z^2, \quad \Delta_2 \equiv \partial^2/\partial x^2 + \partial^2/\partial y^2 \ , \tag{1}$$

$$D\Delta_2^2 w + \rho h U^2 \frac{\partial^2 w}{\partial y^2} + g\rho_0 w - \rho_0 U \frac{\partial \varphi}{\partial y}\Big|_{z=0} = 0 \ (x < 0), \quad gw - U\frac{\partial \varphi}{\partial y}\Big|_{z=0} = -\frac{P(x,y)}{\rho_0} \ (x > 0), \tag{2}$$

$$\partial\varphi/\partial z\big|_{z=0} = -U\partial w/\partial y, \qquad \partial\varphi/\partial z\big|_{z=-H} = 0, \tag{3}$$

$$\Big(\frac{\partial^2}{\partial x^2} + \nu\frac{\partial^2}{\partial y^2}\Big)w = 0, \quad \frac{\partial}{\partial x}\Big[\frac{\partial^2}{\partial x^2} + (2-\nu)\frac{\partial^2}{\partial y^2}\Big]w = 0 \ \ (x = 0-, \ |y| < \infty). \tag{4}$$

Here $D = Eh^3/[12(1-\nu^2)]$; E, ν, h, ρ are Young's modulus, Poisson's ratio, the thickness and density of ice sheet, respectively; g is the acceleration due to gravity. For wave motion the decaying conditions should be satisfied far from the pressure region.

We restrict our consideration to the pressure distribution in the form

$$P(x, y) = \frac{P_0}{2}\{\tanh[\varkappa(y + b)] - \tanh[\varkappa(y - b)]\}[\mathcal{H}(x - x_0 + a) - \mathcal{H}(x - x_0 - a)], \tag{5}$$

where P_0 is the nominal pressure, and a and b are respectively the half-beam and half-length of pressure region whose center is located at the point $(x = x_0 > a, \ y = 0)$, and $\mathcal{H}(\cdot)$ is the Heaviside function. The rate of pressure fall-off at the edges is controlled by the parameter $\varkappa$. As a special case, $\varkappa \to \infty$ is equivalent to a uniform pressure acting on a rectangular area: $|x - x_0| \le a$, $|y| \le b$. This representation was used in [3] - [6]. However, when the load moves over a free surface, the constant pressure distribution in the rectangular planform leads to unrealistic oscillations in the wave resistance curve at the low Froude numbers [1].

We are interested in bending stresses in the ice cover. In particular, it is of practical interest to know whether the moving load can lead to stresses large enough to break the ice near the edge. The strain tensor $\varepsilon(x, y)$ is given by the matrix

$$\varepsilon(x, y) = -\frac{h}{2}\begin{Vmatrix} \partial^2 w/\partial x^2 & \partial^2 w/\partial x\partial y \\ \partial^2 w/\partial x\partial y & \partial^2 w/\partial y^2 \end{Vmatrix}. \tag{6}$$

This tensor describes the strain field in the ice sheet.

The forces F_x (side force) and F_y (wave resistance) acting on ACV and its non-dimensional values A_x, A_y are determined by formulas

$$(R_x, R_y) = -\int_{x_0-a}^{x_0+a}\int_{-\infty}^{\infty} P(x, y)\Big(\frac{\partial w}{\partial x}, \frac{\partial w}{\partial y}\Big)dydx, \quad (A_x, A_y) = -\frac{g\rho_0}{2aP_0^2}(R_x, R_y). \tag{7}$$

3. Method of solution

We describe briefly the solution of problem (1)~(4) by the Wiener-Hopf technique. The dimensionless variables and parameters are introduced

$$(x', y', z', a', b', x_0') = (x, y, z, a, b, x_0)/H, \ \varkappa' = H\varkappa,$$

$$\beta = D/(\rho_0 g H^4),\ F = U/\sqrt{gH},\ \sigma = \rho h/(\rho_0 H),\ P_0' = P_0/(\rho_0 g H).$$

Below, the primes are omitted. We will seek the velocity potential and the displacement in the form $\varphi = UH\phi(x,y,z),\ \ w = HW(x,y)$.

We use the Fourier transforms to the variables x and y in the form

$$\Phi_-(\alpha,s,z) = \int_{-\infty}^{\infty} \mathrm{e}^{-isy} \int_{-\infty}^{0} \phi(x,y,z)\mathrm{e}^{i\alpha x} dxdy,\ \Phi_+(\alpha,s,z) = \int_{-\infty}^{\infty} \mathrm{e}^{-isy} \int_{0}^{\infty} \phi(x,y,z)\mathrm{e}^{i\alpha x} dxdy.$$

From the Laplace equation (1) and no-flux bottom condition (3), we have

$$\Phi(\alpha,s,z) = \Phi_- + \Phi_+ = C(\alpha,s)Z(\alpha,s,z),\quad Z = \cosh[(z+1)\sqrt{\alpha^2+s^2}]/\cosh\sqrt{\alpha^2+s^2}, \tag{8}$$

where $C(\alpha,s)$ is unknown function. We introduce the functions $F_\pm(\alpha,s),\ G_\pm(\alpha,s)$ in the following manner:

$$F_- = \int\limits_{-\infty}^{\infty} \mathrm{e}^{-isy} \int\limits_{-\infty}^{0} \left(\frac{\partial\phi}{\partial z} + F^2\frac{\partial^2\phi}{\partial y^2}\right)\Big|_{z=0} \mathrm{e}^{i\alpha x} dxdy,\ F_+ = \int\limits_{-\infty}^{\infty} \mathrm{e}^{-isy} \int\limits_{0}^{\infty} \left(\frac{\partial\phi}{\partial z} + F^2\frac{\partial^2\phi}{\partial y^2}\right)\Big|_{z=0} \mathrm{e}^{i\alpha x} dxdy, \tag{9}$$

$$G_- = \int_{-\infty}^{\infty} \mathrm{e}^{-isy} \int_{-\infty}^{0} \left[\left(\beta\Delta_2^2 + 1 + \sigma F^2\frac{\partial^2}{\partial y^2}\right)\frac{\partial\phi}{\partial z} + F^2\frac{\partial^2\phi}{\partial y^2}\right]_{z=0} \mathrm{e}^{i\alpha x} dxdy, \tag{10}$$

$$G_+ = \int_{-\infty}^{\infty} \mathrm{e}^{-isy} \int_{0}^{\infty} \left[\left(\beta\Delta_2^2 + 1 + \sigma F^2\frac{\partial^2}{\partial y^2}\right)\frac{\partial\phi}{\partial z} + F^2\frac{\partial^2\phi}{\partial y^2}\right]_{z=0} \mathrm{e}^{i\alpha x} dxdy. \tag{11}$$

The functions with indexes $+/-$ are analytical on α in the upper/lower half-plane, respectively. From boundary conditions (2), we have

$$G_-(\alpha,s) = 0,\quad F_+(\alpha,s) = isQ(\alpha,s), \tag{12}$$

where

$$Q(\alpha,s) = \int_{-\infty}^{\infty} \int_{x_0-a}^{x_0+a} P(x,y)\mathrm{e}^{i(\alpha x - sy)} dxdy = \frac{2\pi P_0 \sin(\alpha a)\sin(sb)}{\alpha\varkappa \sinh[\pi s/(2\varkappa)]}\mathrm{e}^{i\alpha x_0}.$$

Using (8)$\sim$(12), we can write

$$F_-(\alpha,s) + isQ(\alpha,s) = C(\alpha,s)K_1(\alpha,s),\quad G_+(\alpha,s) = C(\alpha,s)K_2(\alpha,s), \tag{13}$$

where $K_1(\alpha,s)$ and $K_2(\alpha,s)$ are the dispersion functions for the free surface waves and the flexural-gravity ones in a moving coordinate system, respectively:

$$K_1(\alpha,s) = \sqrt{\alpha^2+s^2}\tanh\sqrt{\alpha^2+s^2} - F^2s^2,$$

$$K_2(\alpha,s) = [\beta(\alpha^2+s^2)^2 + 1 - \sigma F^2s^2]\sqrt{\alpha^2+s^2}\tanh\sqrt{\alpha^2+s^2} - F^2s^2.$$

The dispersion relation for free surface waves $\mathcal{K}_1(\gamma) \equiv \gamma\tanh\gamma - F^2s^2 = 0$ has two real roots $\pm\gamma_0$ and the countable set of imaginary roots $\pm\gamma_m$ $(m = 1,2,...)$. The dispersion relation for flexural-gravity waves $\mathcal{K}_2(\mu) \equiv (\beta\mu^4 + 1 - \sigma F^2s^2)\mu\mathrm{th}\mu - F^2s^2 = 0$ has two real roots $\pm\mu_0$, four complex roots $\pm\mu_{-1}$, $\pm\mu_{-2}$, $\mu_{-2} = -\bar{\mu}_{-1}$ (the bar denotes complex conjugation), and the countable set of imaginary roots $\pm\mu_m$ $(m = 1,2,...)$. Then the roots of dispersion functions $K_n(\alpha,s) = 0$ $(n = 1,2)$ are $\chi_m(s) = \sqrt{\gamma_m^2(s) - s^2}$ $(n = 1)$ and $\alpha_m(s) = \sqrt{\mu_m^2(s) - s^2}$ $(n = 2)$. We take the values of these roots from the upper half-plane.

Excluding the function $C(\alpha, s)$ from equations (13), we obtain

$$F_-(\alpha, s) + isQ(\alpha, s) = G_+(\alpha, s)K(\alpha, s), \quad K(\alpha, s) = K_1(\alpha, s)/K_2(\alpha, s). \tag{14}$$

In accordance with the Wiener-Hopf technique, we factorize the function $K(\alpha, s)$:

$$K(\alpha, s) = K_-(\alpha, s)K_+(\alpha, s), \; K_\pm(\alpha, s) = \frac{\mu_{-1}\mu_{-2}}{(\alpha \pm \alpha_{-1})(\alpha \pm \alpha_{-2})}N_\pm(\alpha, s), \; N_\pm = \prod_{j=0}^{\infty} \frac{(\alpha \pm \chi_j)\mu_j}{(\alpha \pm \alpha_j)\gamma_j},$$

where the functions $K_\pm$ are analytical in the upper/lower parts of complex plane α, respectively. Dividing (14) by $K_-(\alpha, s)$, we have

$$\frac{F_-(\alpha, s)}{K_-(\alpha, s)} + \frac{\Omega(s)\psi(\alpha)}{K_-(\alpha, s)} = G_+(\alpha, s)K_+(\alpha, s),$$

where

$$\Omega(s) = \frac{\pi s P_0 \sin(sb)}{\varkappa \sinh[\pi s/(2\varkappa)]}, \; \psi(\alpha) = \frac{\mathrm{e}^{i\alpha(x_0+a)} - \mathrm{e}^{i\alpha(x_0-a)}}{\alpha}.$$

Then we use the representation

$$\frac{\psi(\alpha)}{N_-(\alpha, s)} = L_+(\alpha, s) + L_-(\alpha, s), \quad L_\pm(\alpha, s) = \pm\frac{1}{2\pi i}\int_{-\infty\mp i\lambda}^{\infty\mp i\lambda} \frac{\psi(\zeta)d\zeta}{(\zeta-\alpha)N_-(\zeta, s)},$$

and as a result we obtain the equation

$$\frac{F_-(\alpha, s)}{K_-(\alpha, s)} + \Omega(s)\frac{(\alpha-\alpha_{-1})(\alpha-\alpha_{-2})}{\mu_{-1}\mu_{-2}}L_-(\alpha, s) =$$

$$G_+(\alpha, s)K_+(\alpha, s) - \Omega(s)\frac{(\alpha-\alpha_{-1})(\alpha-\alpha_{-2})}{\mu_{-1}\mu_{-2}}L_+(\alpha, s).$$

The functions on the left-hand and right-hand sides of this equation are analytical in the lower and upper parts of complex plane α, respectively. Then we have analytical function over the entire complex plane α. By Liouville's theorem, this function is a polynomial. The degree of the polynomial is determined by the behavior of this function as $|\alpha| \to \infty$ and is equal to one. Consequently, we can write

$$G_+(\alpha, s)K_+(\alpha, s) - \Omega(s)\frac{(\alpha-\alpha_{-1})(\alpha-\alpha_{-2})}{\mu_{-1}\mu_{-2}}L_+(\alpha, s) = \Omega(s)[a_0(s) + a_1(s)\alpha],$$

where $a_0(s)$ and $a_1(s)$ are unknown functions which are defined from edge conditions (4).

The deflection of the plate and the free surface elevation are determined by performing the inverse Fourier transform:
at $x < 0$

$$W(x, y) = -\frac{1}{2}\int_{-\infty}^{\infty} \mathrm{e}^{isy}\Omega_1(s)\sum_{j=-2}^{\infty} \frac{\mu_j \mathrm{th}\mu_j \mathrm{e}^{-i\alpha_j x}}{K_+(\alpha_j, s)K_2'(\alpha_j, s)}\Bigg[a_0(s) + a_1(s)\alpha_j +$$

$$\frac{(\alpha_j-\alpha_{-1})(\alpha_j-\alpha_{-2})}{\mu_{-1}\mu_{-2}}\sum_{m=0}^{\infty}\frac{\psi(\chi_m)}{(\chi_m-\alpha_j)N_-'(\chi_m, s)}\Bigg]ds, \quad \Omega_1(s) = \frac{\Omega(s)}{\pi s},$$

IOP Conf. Series: Earth and Environmental Science **193** (2018) 012065 doi:10.1088/1755-1315/193/1/012065

at $x > 0$

$$W(x,y) = -\frac{F^2}{2}\int_{-\infty}^{\infty} \mathrm{e}^{isy} s^2 \Omega_1(s) \sum_{j=0}^{\infty} \frac{\mathrm{e}^{i\chi_j x} K_+(\chi_j,s)}{K_1'(\chi_j,s)} \Bigg[a_0(s) - a_1(s)\chi_j +$$

$$\frac{(\chi_j+\alpha_{-1})(\chi_j+\alpha_{-2})}{\mu_{-1}\mu_{-2}} \sum_{m=0}^{\infty} \frac{\psi(\chi_m)}{(\chi_m+\chi_j)N_-'(\chi_m,s)} \Bigg] ds - \frac{F^2}{2}\int_{-\infty}^{\infty} \mathrm{e}^{isy} s^2 \Omega_1(s)\Lambda(x,s) ds,$$

where

$$\Lambda(x,s) = \sum_{j=0}^{\infty} \frac{1}{\chi_j K_1'(\chi_j,s)} \begin{cases} \exp[i\chi_j(x_0-x+a)] - \exp[i\chi_j(x_0-x-a)] & (0<x<x_0-a), \\ \exp[i\chi_j(x_0-x+a)] + \exp[i\chi_j(x-x_0+a)] - 2 & (|x-x_0|<a), \\ \exp[i\chi_j(x-x_0+a)] - \exp[i\chi_j(x-x_0-a)] & (x>x_0+a), \end{cases}$$

where the prime denotes the partial derivative of a function with respect to its first variable.

Using free-edge conditions (4), we obtain the system of two linear algebraic equations to define the coefficients $a_0(s)$ and $a_1(s)$. All integrals on α are evaluated by the residue method. The analysis of dispersion relation for free surface waves shows that at any speed of the load there is a value s_0, that $\gamma_0(s) < |s|$ at $|s| < s_0$ and $\gamma_0(s) > |s|$ at $|s| > s_0$, in this case χ_0 is real. Gravity waves spread at $|s| > s_0$. At $U \geq U_c \equiv \sqrt{gH}$, we have $s_0 = 0$. The velocity U_c is called the limit of long waves.

For the infinitely extended elastic plate, there is a minimum phase velocity of flexural-gravity waves c_m and always $c_m < U_c$ (see, e.g. [7]). If $U < c_m$, then $\mu_0(s) < |s|$ for any values of the parameter s, and the real roots of dispersion relation $K_2(\alpha,s) = 0$ are absent. If $c_m < U < U_c$, then there are two values s_1 and s_2 such that $\mu_0(s) > |s|$ at $s_1 < s < s_2$, in this case α_0 is real. Flexural-gravity waves extend back and forward in the plate for $s_1 < |s| < s_2$. With increasing speed of the load, the value s_1 decreases, and the value s_2 increases. At a speed greater than the long-wave limit U_c, we have $s_1 = 0$.

4. Numerical results

The following input data are used for water, ice sheet and external load: $E = 5GPa$, $\nu = 1/3$, $\rho_0 = 10^3$ kg/m^3, $\rho = 900$ kg/m^3, $P_0 = 10^3 Pa$, $a = 10$ m, $b = 20$ m, $x_0 = 50$ m, $\varkappa = 5/b$, $H = 100$ m. The speed of load U and the thickness of ice sheet h vary in the calculations. The minimum phase velocity of the flexural-gravity waves in infinitely extended ice cover increases with increasing ice thickness and for $h = 0.5$, 1, 2 m is equal to $c_m \approx 12.06$, 15.59, 20.09 m/s, respectively. The long-wave limit U_c is approximately equal to 31.32 m/s.

Figure 1 shows the three-dimensional plots for the vertical displacements of free surface at $x > 0$ and ice sheet at $x < 0$ for the motion of load at speed $U = 15$ m/s and different ice thicknesses: $h = 0.5$ m (Fig. 1(a)) and $h = 1$ m (Fig. 1(b)). The load moves from left to right. The load speed is supercritical for an ice cover 0.5 m thick and wave motions are excited in the ice cover, extending to a sufficiently large distance from the edge of ice cover. In the case of thicker ice, $h = 1$ m, the load speed is subcritical and in the ice cover only weak wave motions appear decaying rapidly away from the edge. The waves in fluid near the ice cover are weaker for the thinner ice because waves in the plate are greater and absorb stronger energy.

The free surface elevation and ice deflection at $x = 0$ are shown in Fig. 2(a) for $U = 15$ m/s and $h = 1$ m. We can see that the lengths of surface waves and flexural-gravity ones differ substantially near the edge of ice cover. On the free surface, the superposition of waves reflected from the edge of ice cover and generated by the pressure region occurs.

The strains along the edge of ice sheet at $x = 0$ for $h = 1$ m and two different speeds of load $U = 10$, 15 m/s are shown in Fig. 2 (b). To find the maximum strain in the ice sheet, we need to find the largest eigenvalue of the strain tensor (6) at each location. The strains

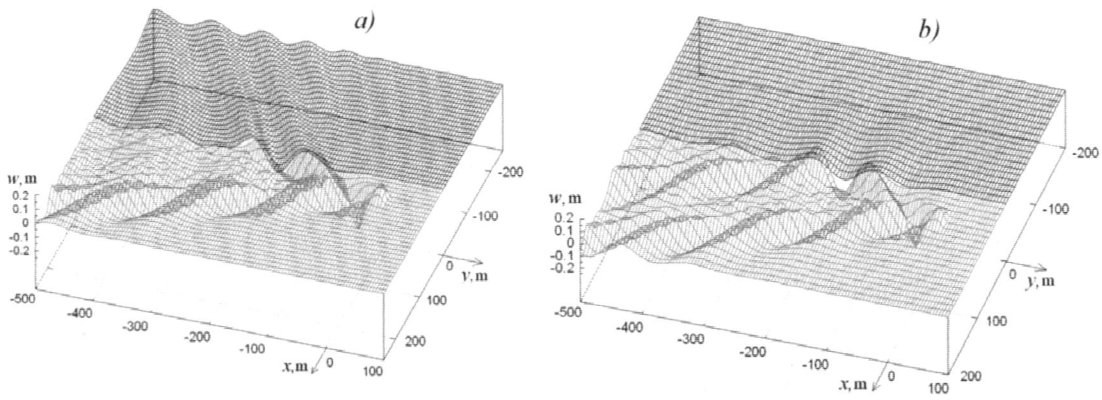

Figure 1. Wave patterns of ice cover ($x < 0$) (gray) and free surface ($x > 0$) (blue) for the load moving with speed $U = 15$ m/s and ice thickness $h = 0.5$ m (a) and $h = 1$ m (b).

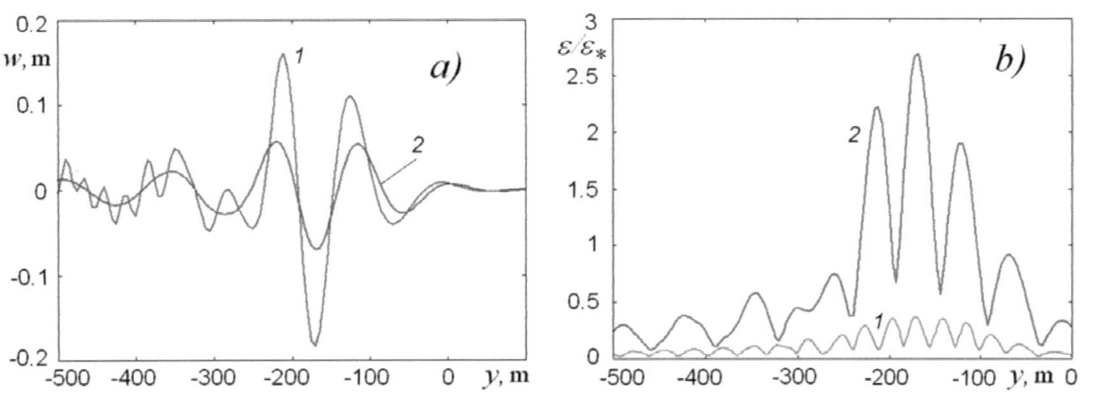

Figure 2. (a) Free surface elevation (*1*) and the deflection of ice sheet (*2*) along the ice edge at $x = 0$, $U = 15$ m/s, $h = 1$ m. (b) Strains $\varepsilon/\varepsilon_*$ along the ice edge at $x = 0$ for $h = 1$ m: $U = 10$ m/s (*1*) and $U = 15$ m/s (*2*).

are proportional to the magnitude P_0 of external load within the linear theory. The strains are scaled with $\varepsilon_* = 8 \cdot 10^{-5}$, where ε_* is the yield strain at which a material begins to deform plastically. According to [8], any strains greater than the yield strain ε_* are assumed to lead to ice fracture. We can see that at $U = 10$ m/s, $h = 1$ m and at given value of the external load P_0, ice does not break whereas at speed $U = 15$ m/s and the same values h and P_0 the arising strains exceed the value ε_* and the destruction of ice can be possible.

Non-dimensional values of wave forces A_x and A_y in (7) acting on moving vehicle are presented in Fig. 3 as functions of the load speed. It is known (see e.g. [1]) that when the pressure region (5) moves along an unbounded free surface, side force R_x in (7) is identically zero and wave resistance R_y is

$$R_y = \frac{2\pi P_0^2 F}{\rho_0 g H \varkappa^2} \int_{K_0}^{\infty} \frac{\sin^2(\frac{b}{FH}\sqrt{k \tanh k})\sqrt{k}\tanh k}{(F^2 k - \tanh k)^{3/2} \sinh^2(\frac{\pi}{2FH\varkappa}\sqrt{k \tanh k})} \sin^2\Big(\frac{a}{FH}\sqrt{k(F^2 k - \tanh k)}\Big) dk, \tag{15}$$

where K_0 is the real positive root of equation $K_0 F^2 = \tanh K_0$ if $F < 1$ while $K_0 = 0$ if $F > 1$.

When the pressure region moves along the edge of ice cover, the side force arises at speeds of load movement close to critical velocity c_m for given ice thickness. As the speed of load increases,

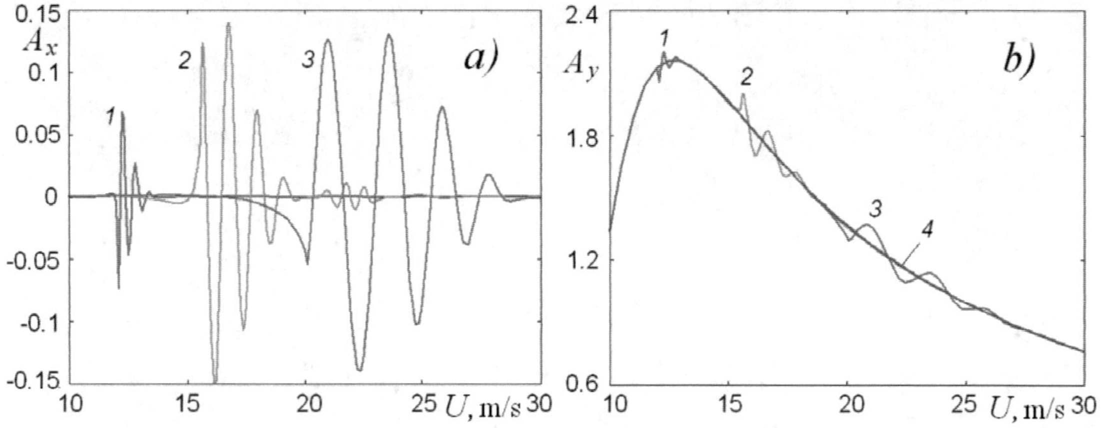

Figure 3. The non-dimensional values of side force A_x (a) and wave resistance A_y (b) for semi-infinite ice sheet with different thickness: curves *(1-3)* correspond to $h = 0.5,\ 1,\ 2$ m. Curve *4* shows the wave resistance for infinite free surface (15).

the side force decreases and tends to zero in an oscillatory manner. The greatest difference of the wave resistance from its value for infinite free surface is observed also at near-critical values of load speed. For subcritical speeds, the disturbances spread only behind the load and do not influence on the vehicle. If $U > c_m$, flexural-gravity waves extend forward in the plate and disturb the fluid near the vehicle. With increasing of speed, the waves in the plate spreading forward become shorter and of smaller amplitude and the influence of ice cover on wave forces decreases.

5. Conclusion

The three-dimensional linear hydroelastic problem on the waves induced by a uniformly moving load over a free surface along the rectilinear edge of a semi-infinite ice sheet is solved. This problem arises, for example, for the air cushion vehicle. The solution is obtained by using the Fourier transform and the Wiener-Hopf technique. The vertical displacements of ice cover and fluid, and wave forces acting on a moving vehicle are defined for different values of the load speed. The speed of the load is shown to affect strongly the response of ice cover. The response is maximum for the load speeds close to the critical speed of hydroelastic waves in infinitely extended ice cover. Wave forces sharply oscillate when the speed of motion is near-critical. It is shown that under certain conditions, the ice at the edge of ice sheet can fracture.

References

[1] Doctors L J and Sharma S D 1972 *J. Ship Research* **16** 248
[2] Wehausen J V and Laitone E V 1960 *Surface Waves* (Handbuch der Physik) **9** 446
[3] Sturova I V and Tkacheva L A 2017 *J. Phys.: Conf. Series* **894**
[4] Sturova I V 2018 *Fluid Dyn.* **53** 49
[5] Tkacheva L A 2018 *J. Appl. Mech. Tech. Phys* **59** 258
[6] Tkacheva L A 2018 *Fluid Dyn.* (accepted)
[7] Kheysin D Ye 1967 *Dynamics of Floating Ice Covers* (Leningrad, Gidrometeoizdat) (in Russian)
[8] Brocklehurst P, Korobkin A A and Parau E I 2010 *J. Eng. Math* **68** 215

Ductile-Brittle Transition in Geomaterials, Caused by Tensile Stress Concentration

S V Suknev

Chersky Institute of Mining of the North, Siberian Branch, Russian Academy of Sciences, Lenin Ave 43, 677980 Yakutsk, Russia

E-mail: suknyov@igds.ysn.ru

Abstract. The possibility of using known nonlocal and stress gradient fracture criteria to describe brittle, quasi-brittle, and ductile fracture of materials with openings is analyzed. A common property of these criteria is the introduction of intrinsic material length characterizing its structure, which allows one to describe the size effect in conditions of stress concentration. At the same time, the scope of application of nonlocal criteria is limited to cases of brittle or quasi-brittle fracture with a small fracture process zone. To expand the scope of application, new fracture criteria are proposed, which are the development of the average stress criterion, and the stress gradient criterion, and which contain a complex parameter that characterizes the size of the fracture process zone and accounts not only for the material structure, but also plastic properties of the material, geometry of the sample, and its loading conditions. Expressions are obtained for the critical pressure in the problem of the formation of tensile cracks under compression in the samples of geomaterials with a circular hole. The calculation results are in good agreement with the experimental data on the fracture of drilled gypsum plates.

1. Introduction

Ductile-brittle transition is a key problem in arctic materials science. It is directly related to changes in the strength and plastic properties of the material in the case of a low temperature. However, the transition from ductile to brittle fracture can be associated not only with temperature effects, but also with the effect of such structural factors as stress concentration, geometry of structural members, and their loading conditions. Nonlocal fracture criteria are based on the concept of the formation of a fracture process zone in the material, in which there is local redistribution of stresses, whereas the main material is deformed elastically until fracture [1–11]. Typical representatives of such quasi-brittle materials are geomaterials (concrete, gypsum, and rock), composites, high-strength metal alloys, cast iron, graphite, etc. A general property of these criteria is the introduction of intrinsic material length that characterizes its structure, which allows describing the size effect in conditions of stress concentration, thereby expanding the scope of application in comparison with conventional criteria. Pointing to the general property of the nonlocal criteria, Taylor [12] proposes to consider them as partial expressions of some general "theory of critical distances". Recently, there have been a large number of papers that, in order to describe the fracture of materials with openings, appeal to the theory of critical distances. In 2008, a special issue of the Engineering Fracture Mechanics was devoted to this theory [13]. So-called stress gradient criteria can be classified as a separate group of nonlocal fracture criteria. The most interesting are the criteria developed on the basis of the hypothesis of the effect of the stress gradient on the local failure stress. Thus, a stress gradient criterion for the tensile cracks initiation under compression is described in [14].

The scope of application of nonlocal criteria is, first of all, the brittle fracture of materials with openings. To describe ductile fracture, additional information is needed on the inelastic behavior of the material in the fracture process zone or on the material constants characterizing its plastic properties. An intermediate position between brittle and ductile fracture is occupied by quasi-brittle fracture region. Paper considers the possibility of expanding the scope of application of nonlocal criteria for cases of fracture with a developed process zone with the aim of predicting the ductile-brittle transition in the fracture, caused by stress concentration.

2. Theoretical approaches

Below, quasi-brittle fracture is understood as a sudden propagation of an unstable crack (also characteristic of brittle fracture), accompanied by the formation of a large fracture process zone. The size d of the fracture process zone is not associated with the size of the crack, as is customary in fracture mechanics, but with the characteristic size d_0 of the material structure. We talk about brittle fracture if $d = d_0$ and ductile fracture if $d >> d_0$. The redistribution of stresses within the limits of d_0 is not related with the plastic (in the macroscopic sense) deformation of the material. The plastic properties of the material begin to appear when $d > d_0$, and the larger d with respect to d_0, the more they manifest themselves. Taking this into account, we represent d in the following form:

$$d = d_0 + \beta L_e . \qquad (1)$$

Here L_e is the size of the stress concentration zone, β is the dimensionless parameter that characterizes the plasticity of the material. For brittle materials, $\beta = 0$; for plastic materials, $\beta >> 1$. In the case where $\beta \sim 1$, the material is characterized by moderate plastic properties.

For ductile fracture, the critical stress does not depend on the size of the opening, so the size of the damage zone is proportional to the size of the opening and, accordingly, to the size L_e (under unchanged boundary conditions). In brittle fracture, on the contrary, the size of the damage zone does not depend on the size of the opening and is determined by the structure of the material.

During compression, the behavior of the failure stress characterizing the appearance of tensile cracks at the opening has the form shown in Figure 1. For small values of L_e, the material does not feel the stress concentration and collapses as a smooth sample when the applied pressure reaches the compressive strength C_0. After reaching the critical size of the opening, the failure pressure p_c decreases, asymptotically approaching the tensile strength of the material T_0 in the case of brittle failure and the stress T_s ($C_0 > T_s > T_0$) in the case of ductile fracture.

This behavior of the failure stress agrees with the current understanding of the real solid as containing primary, inherent defects. Accordingly, small artificial defects, whose sizes are comparable with the sizes of the microstructural components of the material, do not affect its strength until their sizes reach a certain (critical) value.

IOP Conf. Series: Earth and Environmental Science **193** (2018) 012066 doi:10.1088/1755-1315/193/1/012066

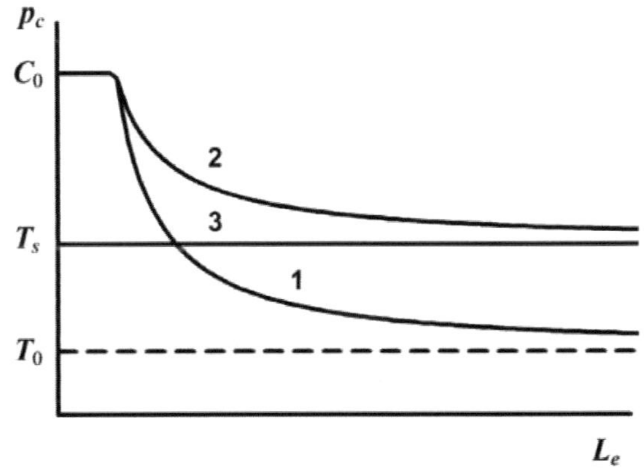

Figure 1. Fracture diagrams: 1 – brittle fracture; 2 – quasi-brittle fracture; 3 – ductile fracture.

2.1. Average stress criterion

The most well-known nonlocal criterion is the average stress criterion (ASC), or integral criterion, which has the form $\langle\sigma_e\rangle_d < \sigma_0$, where $\langle\sigma_e\rangle_d$ is the equivalent stress averaged at the distance d along the dangerous cross section and characterizing the internal stressed state of the deformed body. For brittle materials, the size of averaging of d is assumed as a material constant, which characterizes the structure of the material: $d = d_0 = \text{const}$, the material strength σ_0 is also assumed to be a constant.

Critical pressure for the sample with a circular hole [15]:

$$p_c = \chi C_0 \frac{(1+2d/l)^3}{1+d/l}, \quad l > l_c. \tag{2}$$

Here $\chi = T_0 / C_0$, l is the hole diameter, and l_c is the critical diameter of the hole.

To describe the quasi-brittle fracture, we determine the size of averaging according to equation (1), in which the size of the stress concentration zone has the form $L_e = \dfrac{\sigma_e}{|\text{grad}\,\sigma_e|}$. For the problem under consideration, $L_e = l/10$. The critical pressure is determined by substituting equation (1) into equation (2) with account for the obtained estimate for L_e:

$$p_c = \chi C_0 \frac{(1+2d_0/l+0.2\beta)^3}{1+d_0/l+0.1\beta}. \tag{3}$$

2.2. Stress gradient criterion

An alternative point of view on the size effect is that, in brittle materials, it is not associated with redistribution of stresses, but with a real increase in the local failure stress. This hypothesis, in particular, underlies the statistical theories of strength.

On the basis of this hypothesis, an approach was proposed to describe brittle fracture [16], according to which the local strength of the material is assumed to be dependent on the size of the stress concentration zone L_e. If the size L_e is sufficiently large in comparison with the size of the structural features of the material, the value of the local strength differs slightly from the value of the ultimate stress σ_0 determined under the conditions of uniform stress. Conversely, if L_e is comparable with the size of structural features, their effect on the local strength becomes noticeable. Moreover, the smaller the size L_e with respect to the characteristic size of the material structure L_0, the larger this effect. Thus, the local strength of the material depends not only on the size of the stress concentration zone L_e, but also on the ratio L_0 / L_e that characterizes the scale in the problem under consideration. In view of this, the local strength condition is represented in the form $\sigma_e < f(\sigma_0, L_0 / L_e)$. The local strength function $f(\sigma_0, L_0 / L_e)$ is determined with account for additional conditions that reflect the specific features of the problem under consideration. For the problem of the tensile cracks initiation under compression, the critical pressure at which tensile cracks start from the hole are determined by the expression [14]:

$$p_c = \chi C_0 \left(1 + \left(\frac{L_0}{L_e} \right)^n \right). \tag{4}$$

For the structural parameter L_0, we write the expression similar to equation (1):

$$L_0 = d_0 + \beta L_e. \tag{5}$$

In the initial formulation of the criterion, the parameter L_0 has a meaning of the characteristic size of the material structure d_0 and is not associated with the fracture process zone. Taking into account equation (5), L_0 acquires the meaning of the parameter d in the nonlocal criteria and the increase in local strength is related not only with the real size effect characteristic of brittle materials, but also to the apparent increase due to the redistribution of elastic stresses in the fracture process zone.

Substitution of equation (5) into equation (4) with account for the fact that, $n = 1$ for brittle materials, gives the following result:

$$p_c = \chi C_0 \left(1 + \beta + 10 \frac{d_0}{l} \right). \tag{6}$$

The third term in the right side of equation (6) determines the contribution of a real increase in the local strength of the material to the size effect, and the second term in the right side of expression (6) determines the contribution of an apparent increase in the local strength of the material to the size effect.

3. Experimental verification

The applicability of the developed quasi-brittle fracture criteria can be verified by the experimental data on the tensile cracks initiation under compression in the samples of geomaterials with circular holes.

We test samples made of gypsum cement of various compositions. One series of samples is made from a gypsum cement with high (higher than 90 %) content of hemihydrate gypsum (Gypsum 1), and the second one from the gypsum cement having a low (within 60–70 %) content of hemihydrate gypsum in the initial composition (Gypsum 2). The peculiarities of sample preparation and the experimental procedure are given in [17, 18].

The samples from Gypsum 1 demonstrate brittle fracture. The formation of tensile cracks at the contour of the circular hole is sudden for all the investigated diameters of holes, and the crack length at the time of initiation is 5 to 6 cm. With an increase in the hole diameter, there is a decrease in the critical pressure at the time of crack initiation, and further development of cracks leads to failure of the sample in the form of cleaving into two pieces.

The formation of tensile cracks at the contour of the circular hole in the samples of Gypsum 2 have a different character for small and large holes. Their formation on the contour of a small diameter (up to 5 mm inclusive) is sudden, and the samples demonstrate brittle fracture as in the samples of Gypsum 1. The formation and propagation of cracks on the contour of a hole of large diameter (10 mm and larger) occurs gradually, which is typical for ductile fracture. After the formation of the remote cracks located at a distance from the hole, the opening of primary tensile cracks decreases, their growth stops, and they no longer affect the further fracture of the sample.

3.1. Average stress criterion

Figure 2 shows the experimental data (solid circles) on the load at the moment of tensile cracks initiation from the hole as a function of its diameter, obtained from the samples of Gypsum 1, and the results of the calculations of the critical pressure (curve 1), carried out according to equation (3) for $\beta = 0$. The size d_0 is 1.1 mm and comparable to the size of the largest pores. The dashed line is calculated according to the conventional approach. The experimental data (open circles) and the calculation results for Gypsum 2 for $\beta = 2.5$ (curve 2) are also given in Figure 2. The size d_0 is 2 mm. The stress T_s is equal to T_0 in the first case and $T_s = 2.7T_0$ (solid straight line) in the second case.

Figure 2 illustrates a significant size effect, i.e. the effect of the diameter of the hole on the local strength of the material. When it decreases, the critical pressure increases, reaching the compressive strength limit. When it increases, the critical pressure asymptotically approaches the tensile strength T_0 for Gypsum 1 and the stress T_s for Gypsum 2. This behavior is well described by the modified ASC, in which the averaging size *d* is determined by equation (1).

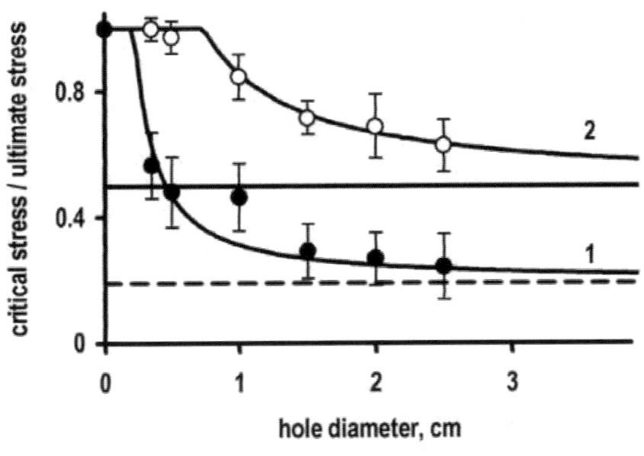

Figure 2. Relationship of critical pressure with hole diameter. Calculated by the modified ASC.

3.2. Stress gradient criterion

Figure 3 illustrates the experimental data (solid circles) on the value of the critical load, obtained from the samples of Gypsum 1, and the results of the calculations (curve 1), carried out according to equation (6) for $\beta = 0$. The size d_0 is 0.7 mm. The dashed line is calculated according to the conventional approach. The experimental data (open circles) and the results of the calculations for Gypsum 2, carried out according to equation (6) for $\beta = 1.45$ (curve 2), are also given in Figure 3. The size d_0 is 1.8 mm. The stress T_s is equal to T_0 in the first case and $T_s = 2.7T_0$ (solid straight line) in the second case.

The experimental results are well described by the stress gradient criterion (SGC) in which the size L_0 is determined by equation (5).

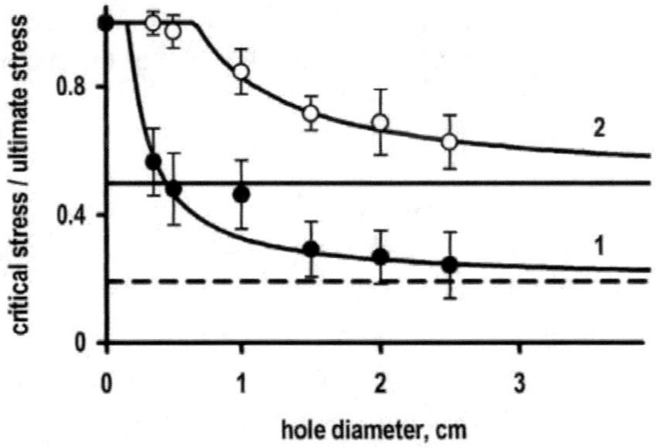

Figure 3. Relationship of critical pressure with hole diameter. Calculated by the modified SGC.

4. Discussion

As can be seen from Figures 2, 3, the experimental data obtained for brittle (Gypsum 1) and ductile (Gypsum 2) crack propagation are equally well described by the modified nonlocal criteria that take into account the change in the size of the fracture process zone in accordance with equations (1) or (5). The plasticity parameter β is zero in the first case, and it is determined by the form of the nonlocal criterion and depends on the ratio T_s / T_0 in the second case. The stress T_s is determined by the asymptotic form of the dependence $p_c(L_e)$ and equals $2.7T_0$ for both the criteria considered.

The representation of the structural parameter *d* in the ASC (or L_0 in the SGC) as a sum of two terms, one of which characterizes the material structure proper and the other one characterizes the zone of inelastic deformations, allows one to describe the process of material fracture more profoundly (in particular, the transition from brittle to ductile fracture).

For Gypsum 1, the plasticity parameter β turns out to be equal to zero, which agrees with the sudden formation of tensile cracks observed in the experiment, which is characteristic of brittle fracture. The redistribution of stresses within the limits of the size d_0 is associated with the discreteness of the structure of the material rather than its plastic properties. Gypsum 2 has plastic properties, and β is not zero. But these properties begin manifesting only when βL_e is larger than d_0. In the samples with small-diameter holes, the plastic zone is also small and does not affect the fracture pattern. With increasing diameter, the size of the plastic zone increases, and it determines the ductile character of the crack propagation, which is observed experimentally. Thus, the condition of brittle fracture of the samples with stress concentrations can be represented in the form $\beta L_e < d_0$. For $\beta = 0$, the material is brittle by default; for $\beta > 0$, the failure mode (brittle or ductile) is determined by the size and form of the opening, as well as loading conditions (boundary conditions). In the case of a circular hole, one can make the following estimate for the diameter of the hole l^*, starting with which the crack propagation becomes ductile:

$$l^* = 10 d_0 / \beta . \tag{7}$$

Diameter l^* calculated for Gypsum 2 according to equation (7) is 8.0 mm for the modified ASC, and 10.6 mm for the modified SGC. Although equation (7) is an estimation expression, the calculated diameters of the hole l^* are in good agreement with what is observed in the experiment.

5. Conclusions

The scope of application of known nonlocal fracture criteria is limited to cases of brittle or quasi-brittle fracture with a small fracture process zone. To expand the scope of application for cases of fracture with a developed fracture process zone, it is proposed to abandon the hypothesis of the size of the process zone as a material constant associated only with its structure. The structural parameter underlying the nonlocal criteria combined into the "theory of critical distances" should be considered as a material constant only in one particular case, which is brittle fracture. For quasi-brittle materials, the size of the fracture process zone is represented as the sum of two terms, one of which characterizes the material structure itself, and the other one refers to the zone of inelastic deformations, whose size is determined by the structural features of the member (notch shape, notch size, loading conditions) and plastic properties of the material. The ratio of these sizes determines which fracture mechanism (brittle or ductile) is to be implemented. The plastic properties that determine the ductile character of fracture manifest only when the size of the zone of inelastic deformations exceeds the characteristic size of the structure of the material. Otherwise, the plastic zone is small and the fracture is brittle.

The proposed approach is used in the development of new (modified) average stress criterion, and stress gradient criterion. The applicability of the developed fracture criteria is verified on the problem of the tensile cracks initiation under compression in the samples of geomaterials with circular holes. It is shown that the modified criteria describe well the experimental data on the fracture of quasi-brittle materials. In addition, the application of the developed criteria makes it possible to explain the change in the fracture mechanism from brittle to ductile with an increase in the size of the hole, observed in the experiment.

Acknowledgments
The research was supported by the Russian Foundation for Basic Research under grant number 18-05-00323.

References

[1] Neuber H 1937 *Kerbspannungslehre, Grundlagen für genaue Spannungsrechnung* (Berlin: Springer)
[2] Novozhilov V V 1969 *J. Appl. Math. Mech.* **33** 201–10
[3] Whitney J M and Nuismer R J 1974 *J. Compos. Mater.* **8** 253–65
[4] Carter B J, Lajtai E Z and Yuan Y 1992 *Int. J. Fract.* **57** 221–36
[5] Radaj D and Zhang S 1995 *Eng. Fract. Mech.* **50** 111–20
[6] Seweryn A and Mroz Z 1995 *Eng. Fract. Mech.* **51** 955–73
[7] Mikhailov S E 1995 *Eng. Fract. Mech.* **52** 731–54
[8] Toribio J 1997 *Eng. Fract. Mech.* **57** 391–404
[9] Pugno N M and Ruoff R S 2004 *Philos. Mag.* **84** 2829–45
[10] Yosibash Z, Bussiba A and Gilad I 2004 *Int. J. Fract.* **125** 307–33
[11] Taylor D, Cornetti P and Pugno N 2005 *Eng. Fract. Mech.* **72** 1021–38
[12] Taylor D 2007 *The Theory of Critical Distances: a New Perspective in Fracture Mechanics* (Oxford: Elsevier)
[13] Taylor D 2008 *Eng. Fract. Mech.* **75** 1695–890
[14] Suknev S V and Novopashin M D 2003 *J. Min. Sci.* **39** 132–8
[15] Suknev S V 2015 *J. Min. Sci.* **51** 435–41
[16] Suknev S V and Novopashin M D 2000 *Doklady Phys.* **45** 339–41
[17] Suknev S V, Elshin V K and Novopashin M D 2003 *J. Min. Sci.* **39** 460–6
[18] Suknev S V 2008 *J. Min. Sci.* **44** 43–50

IOP Conf. Series: Earth and Environmental Science **193** (2018) 012067 doi:10.1088/1755-1315/193/1/012067

Regional model of the geological structure of the Yamal and Gydan oil-and-gas areas

E S Surikova[1,2], A E Solmin[2] and S M Guseva[1]

[1]A.A. Trofimuk Institute of Petroleum Geology and Geophysics, Siberian Branch of the Russian Academy of Sciences, pr. Akademika Koptyuga 3, Novosibirsk, 630090, Russia
[2]Novosibirsk State University, Pirogova st. 2, Novosibirsk, 630090, Russia

Abstract. Despite the fact that a number of hydrocarbon fields have been discovered on the Yamal and Gydan Peninsula, including large gas fields, there is still no generalized regional model for geologists. In connection with the appearance in recent years of new geological information (2D, 3D seismic, logging, drilling of well Gydanskaya № 130), the authors believe it makes sense to reinterpret seismic data using new information too.

1. Introduction

In recent years, special attention has been paid to both the state geological programs and oil and gas companies to the industrial and resource base of the Northern and Arctic territories, the gas production centers of West Siberia are shifting to the Far North. It is here, within the Yamal and Tazov districts of the Yamal-Nenets Autonomous District, that the research area is located. According to modern schemes of oil and gas zoning, the region covers the Yamal and Gydan oil-and-gas areas (OGA). 32 hydrocarbon fields (HCF) were discovered on the Yamal Peninsula, the largest of which are the South Tambey, Bovanenkovo, Kharasaveyskoe, Krusenstern, Arctic, etc., 17 HCF were discovered on the Gydan Peninsula, including Salmanov, Gydanskoe, Geophysical, West-Messoyakh (Figure 1).

In all fields of the region, the traps are structural anticline, the reservoirs are mainly stratified or massive, limited by faults or impermeable lithology.

The stratigraphic interval of the industrial oil-gas content of both OGA includes the section from the Cenomanian top to the Jurassic bottom deposits, while in Yamal Paleozoic rocks and weathered contact zone of the top of Paleosoic are productive too, which are not exposed by drilling in Gydan. The main hydrocarbon resources are concentrated in the permeable Aptian-Cenomanian complex. There is a change down the section from the pure gas uncondensate deposits to pure condensate, while the oil rims are found in the interval Barreme - Middle Jurassic [2, 7, 10, 11].

In the tectonic map of the top Jurassic rocks, the territory of the Yamal peninsula is megamonoclysis (Paikhoi-Novaya Zemlya), complicated by positive closed high-amplitude structures of 2-3 ranks; the territory of the Gydan Peninsula is according to the Antipayutin-Tadebeyakh megasineclise (Figure 2).

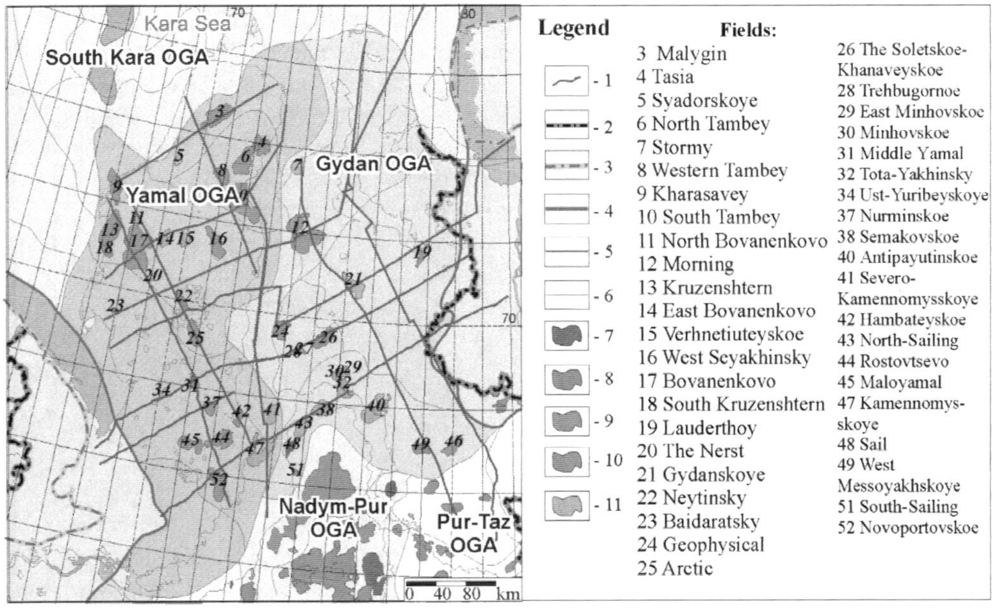

Figure 1. Yamal and Gydan oil-and-gas areas of the West Siberian oil-and-gas basin [13] Legend: 1 - regional seismic cross-sections; boundaries: 2 - administrative, 3 - West Siberian plane, 4 - West Siberian oil and gas basin, 5 - oil and gas areas, 6 - oil and gas subareas; fields: 7 - oil and gas, 8 - gas oil, 9 - oil and gas condensate, 10 - gas condensate, 11 - gas.

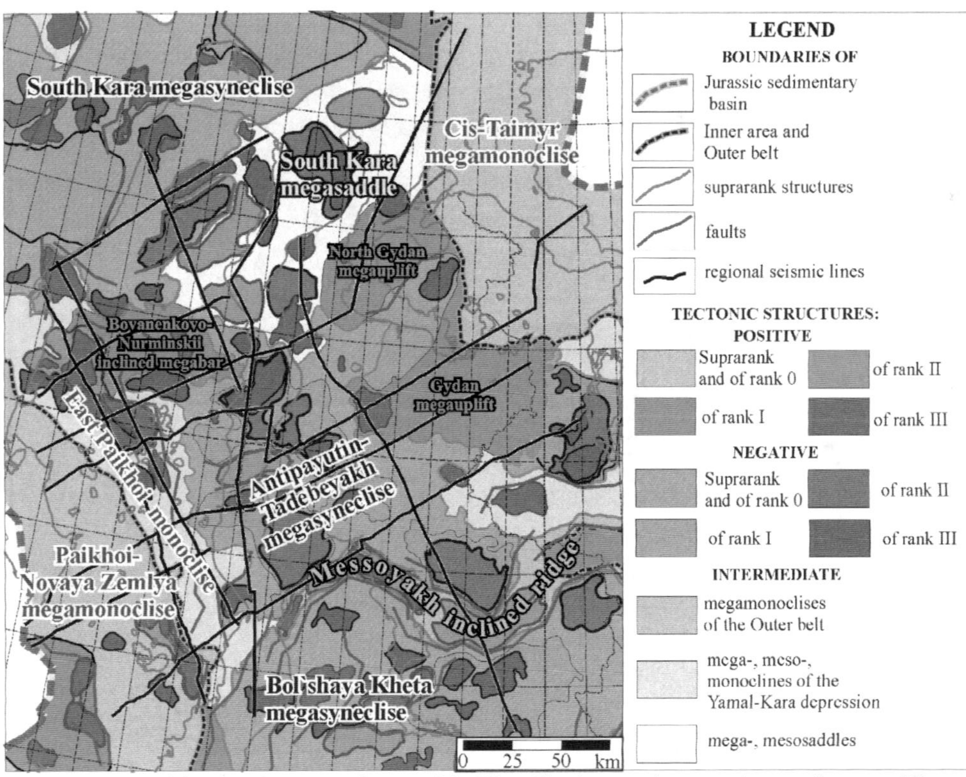

Figure 2. Tectonic map of the Jurassic structural stage of the West Siberian petroleum province northern part [6].

IOP Conf. Series: Earth and Environmental Science **193** (2018) 012067 doi:10.1088/1755-1315/193/1/012067

2. Methods of study

The paper deals with a comprehensive interpretation of seismic materials based on stratigraphic interpretation of seismic data [4, 5, 8, 12]. Analysis of regional time cross-sections and deep drilling data of the region made it possible to identify all the seismic geological complexes traditionally allocated in West Siberia: Paleozoic, Triassic, Jurassic, Berriasian-Lower-Aptian, Aptian-Cenomanian and Turonian-Cenozoic, which are limited in the top and bottom by seismic reflectors : A (the bottom of the sedimentary cover), Kt (the Kiterbut formation), B (the top of the Bazhenov formation), M (the Neitytin clay unit), and G (the Kuznetsov formation) (Figure 4). Structural maps for all reflectors, maps of thicknesses of seismic complexes were built.

Using paleosections, the reconstruction of the positive structures development during geological time was carried out using the thickness method [1, 3, 9]. Thus, for example, the section aligned along B reflector is the paleorelief of the lower horizons at the time of the formation of the reflector B.

3. Structural and tectonic characterization

Structural maps analysis on seismic reflectors showed that the sedimentary cover structure of all reflectors in the Yamal peninsula is a monoclysis complicated by high-amplitude positive structures of 2 and 3 ranks which amplitudes increase upward the section. The sedimentary cover structure of the Gydan Peninsula changes upward the section – landscape of the Jurassic and Lower Cretaceous subsurfaces is a megasineclise complicated by the Gydan, Geophysical and Messoyakh uplifts; landscape of the Upper Cretaceous and Cenozoic reflectors - monocline. The structures of the Yamal OGA are more contrasting in the Cenozoic subsurfaces, in comparison with the structures of the Gydan OGA. The positive tectonic structures of the Gydan Peninsula are flattened upward the section, since the Cenozoic structure formation processes within the Gydan OGA passed less intensively.

The main hydrocarbon reserves are concentrated in the Aptian-Cenomanian deposits, so we pay special attention to the structural plan for the Cenomanian top, where positive structures are marked (Figure 3). All the allocated structures are productive.

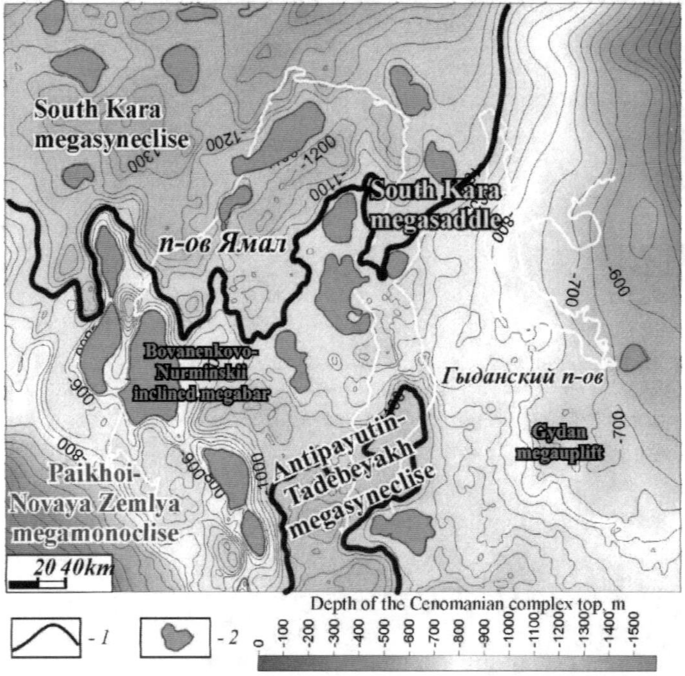

Figure 3. Structural map of senomanian top subsurface with allocated positive structures. Legend: 1 – boundaries of the megasyneclises, 2 – positive structures.

Figure 4 shows a modern time section and paleosections runs in the southern parts of the Yamal and Gydan OGA and crosses the Novii Port uplift, the Nizhnemessoyask and the Gydan uplifts. Within the both OGA, the same Jurassic, Cretaceous and Cenozoic reflectors are traced. Within the Gydan OGA, the total time thickness of the sedimentary cover is much larger than in the Yamal OGA (here we can distinguish extra 3-4 well-traced reflectors) due to a thick complex of sediments of pre-Toarcian age. It can be assumed that on the Yamal peninsula sedimentary deposits under the reflector Kt are represented by the Jurassic and Triassic, and on the Gydan Peninsula, possibly, in addition to the sedimentary deposits of the Triassic, Upper Paleozoic sediments are also widespread.

The paleosections analysis shows (Figure 4) that at the sedimentation time of the Bazhenov formation, the gentle relief of the Jurassic subsurfaces was complicated only by the contrasting Messoyakh uplift of the basement. During Berriasian-Aptian time the Novii Port uplift began to form and the Nizhnemessoyakh arch grew due to the basement block upward movement. During Aptian-Turonian, the South-Gydan uplift began to form. During post-Turonian time, the final formation of all structures, the maximum growth of the Novii Port uplift, the inherited development of the Nizhnemessoyakh arch take place.

It can be concluded that all 3 positive structures crossed by the line developed in different ways: Novii Port uplift formed due to the basement block upward movement mainly in Cenozoic time, Messoyakh inclined ridge formed over the basement block gradually throughout the whole Mesozoic-Cenozoic time, South-Gydan uplift is a rootless structure, which does not confined to the basement block, the formation of it took place mainly in Cenozoic time.

Figure 5 shows a section that also passes through the territories of the Yamal and Gydan Peninsulas, but considerably to the north. The line crosses the North-Arctic, East-Zelenomysovsky, Paesedskii uplifts.

The paleosections analysis shows the Paesedskii and the East Zelenomysovsky uplifts gradually grow throughout the whole Mesozoic-Cenozoic tectonic history, but the North Arctic uplift is formed over the basement block mainly during the Cenozoic period.

Each of the tectonic development stages is accompanied by faults formation. In the research, four types of faults were identified: separating basement blocks; penetrating into the Jurassic rocks, penetrating into the Cretaceous rocks, cutting the entire sedimentary cover.

The Cenozoic type of faults is important in terms of oil and gas potential, since they could serve as channels for migration of hydrocarbons from source rocks to the reservoirs upward the section.

Paleotectonic analysis shows before the early Cretaceous, the Mesozoic-Cenozoic sediments of the Gydan OGA tended to a relative subsidence, in the Cretaceous-Cenozoic time - to a relative uplift; the sediments of the Yamal OGA during the development history tend to a relative uplift, with the maximum growth and the formation of the modern uplifts (anticlinal structures) in Cenozoic.

During the research three formation models of positive structures of the 2-3 ranks, which the HCF are associated with, were identified. The first type is "root" (formed over the basement block) structures that have been growing gradually from early Cretaceous time. The second type is "root" structures, which grow maximum intensive in the post-Turonian period. The last type is "rootless" structures with predominantly Cretaceous-Cenozoic development time. The second and third types of uplifts, for which the most important post-Turonian stage of tectonic development, are common for structures located on the Yamal Peninsula, and the first type with a gradual growth of anticlines throughout the Mesozoic-Cenozoic history of tectonic development is common for the Gydan Peninsula.

It should be noted that the coincidence in the Cenozoic era of factors contributing to the formation of HCF, predetermined the extremely high gas potential of the studied region. Favorable factors for the formation of a large number of HCF in the study area are:

1) the presence of beds with good reservoir properties in the Jurassic-Cretaceous rocks;

2) the presence of reliable thick reservoir cap rocks;

3) gas generation by Jurassic oil-producing source formations and by gas-producing source coal layers of the Pokur formation;

4) the formation of high-amplitude structural traps in the Cenozoic;

5) the formation of faults cutting almost the whole sedimentary cover (from Triassic to Cenozoic deposits), contributing to the migration of hydrocarbons from source rocks to the Aptian-Cenomanian reservoirs.

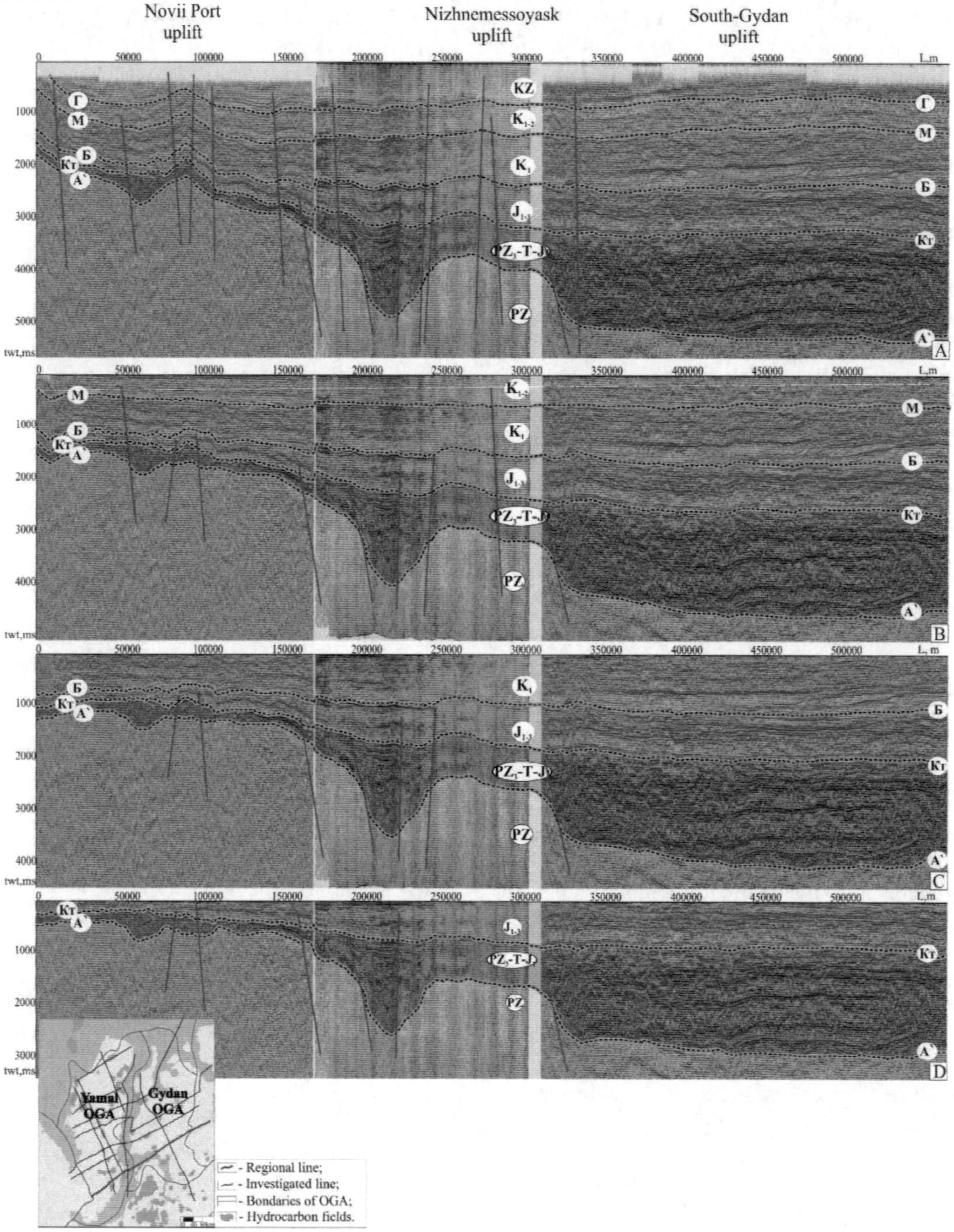

Figure 4. Present-day (A) and paleoseismic sections flattened on the G (B), M (C), B (D) reflectors along line Reg_5. Legend:1 – reflector, 2 - seismigeological complex, 3 – faults.

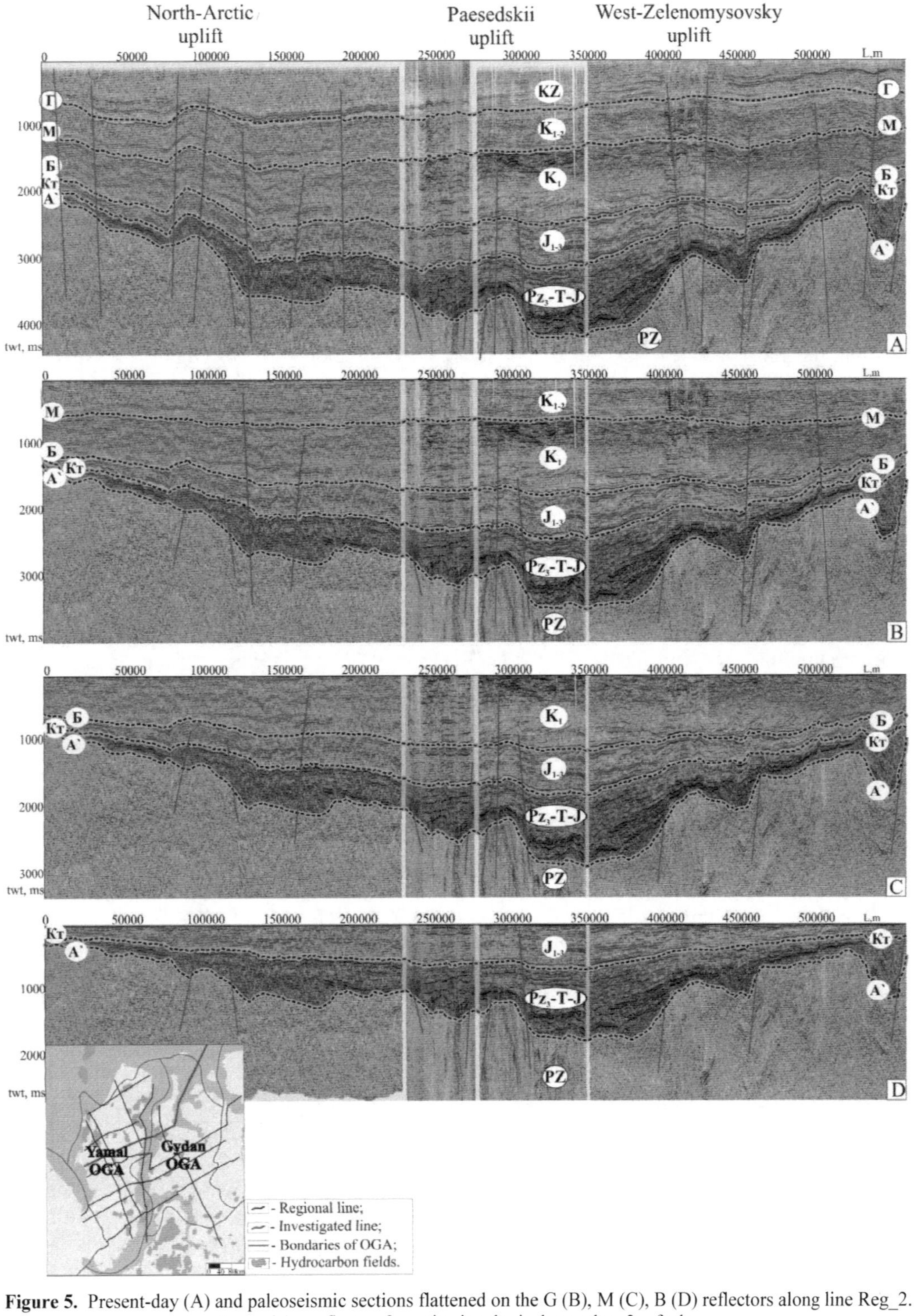

Figure 5. Present-day (A) and paleoseismic sections flattened on the G (B), M (C), B (D) reflectors along line Reg_2. Legend:1 – reflector, 2 - seismigeological complex, 3 – faults.

4. Conclusion

As a result of paleotectonic analysis, we can draw the following conclusions:

• Before the early Cretaceous, the Mesozoic-Cenozoic sediments of the Gydan OGA territory tended to a relative subsidence compared to the Yamal OGA territory. The sedimentary cover of the Gydan Peninsula tends to relative uplift, mainly in the Cretaceous-Cenozoic time, and the Yamal OGA territory tends to relative uplift throughout the history of development. Despite the difference in the history of tectonic development in the Yamal and Gydan OGA, the maximum growth and formation of the modern appearance of anticlinal structures in both regions occurred in the Cenozoic.

• Messoyakh inclined ridge tends to a relative uplift throughout the whole history of tectonic development.

• Most of the positive structures of the Yamal Peninsula are confined to the basement blocks and tended to grow relatively throughout the Mesozoic-Cenozoic history of tectonic development. At the same time, the most intensive growth of positive closed structures of this region took place in the Berrias-Aptian and Cenozoic.

• The formation of the depression zone separating the structures of the Yamal and Gydan peninsulas began in the middle Triassic and continues to this day.

• Analysis of Mesozoic-Cenozoic tectonic processes made it possible to distinguish four main stages of tectonic activation, accompanied by the faults formation: pre-Jurassic, early Jurassic, early Cretaceous and post-Turonian (Cenozoic), which allow us to distinguish five main classes of faults:

• Separating of basement blocks, penetrate the deposits of the sedimentary cover, but do not cut Jurassic deposits (the time of formation is pre-Jurassic, presumably Perm-Triassic);

• penetrating into the Jurassic (the time of formation of the Gettang-Aalen);

• penetrating into the Cretaceous deposits (Berrias-Valanginian);

• penetrating into the Cenozoic deposits (Late Paleocene) - cutting the entire sedimentary cover.

The last type of faults is important in terms of oil and gas potential, since they could serve as channels for migration of hydrocarbons from oil and gas source rocks upwards along the section to the reservoirs.

Acknowledgments

The reported study was funded by RFBR according to the research projects № 18-35-00492.

References

[1] Belousov V V 1940 The thickness of sediments as an expression of the regime of oscillatory movements of the earth's crust *Sovet geology* **2** 24-58

[2] Ermilov O M, Karogodin Yu N, Kontorovich A E et al 2004 *Features of the geological structure and development of unique gas deposits of the far North of West Siberia* ed I I Nesterov (Novosibirsk: SB RAS) p 140

[3] Garetsky R G and Yanshin A L 1960 Tectonic thickness analysis *Methods of studying tectonic structures* **1** 115-333

[4] Kontorovich V A, Ajunova D V, Gubin I A et al 2017 Tectonic evolution of the Arctic onshore and offshore regions of the West Siberian petroleum province *Russian Geology and Geophysics* **58** (3-4) 343-361

[5] Kontorovich V A, Ajunova D V, Gubin I A, Yershov S V, Kalinin A Yu, Kalinina L M, Kanak M S, Solov'ev M V, Surikova E S and Shestakova N I 2016 Seismic stratigraphy formation history and gas potential of the Nadym-Pur interfluve area *Geology and Geophysics* **57** (8) 1583-1595

[6] Kontorovich V A, Belyaev S Yu, Kontorovich A E, Krasavchikov V O, Kontorovich A A and Suprunenko A I 2001 Tectonic structure and history of evolution of the West Siberian geosyneclise in the Mesozoic and Cenozoic *Geology and Geophysics* **42** (11-12) 1832-1845

[7] Kontorovich A E, Nesterov I I, Salmanov F K, Surkov V S, Trofimuk A A and Erve Y G 1975 *Geology of oil and gas of Western Siberia* (Moscow: Nedra) p 680

[8] Kontorovich V A 2009 The Mesozoic-Cenozoic tectonics and petroleum potential of West

Siberia *Geology and Geophysics* **50** (4) 461-474
[9] Mashkovich K A 1976 *Methods of paleotectonic research in the practice of oil and gas exploration* (Moscow: Nedra) p 221
[10] Skorobogatov V A and Stroganov L V *2006 Gydan: geological structure, hydrocarbon resources and the future* (Moscow: Nedra) p 261
[11] Skorobogatov V A, Stroganov L V and Kopeev V D 2003 *The geological structure and gas-and-oil presence of Yamal* (Moscow: Nedra) p 352
[12] Vail P R, Gregory A P, Mitcham R M jr and Sheriff R 1982 *Seismic stratigraphy. Applications to Hydrocarbon Exploration* ed C E Payton (Moscow: MIR) p 847
[13] Grop of authors of INGG SB RAS 2006 *Scheme of oil and gas geological zoning oil-gas-bearing province of the West Siberian* ed A E Kontorovich

IOP Conf. Series: Earth and Environmental Science **193** (2018) 012068 doi:10.1088/1755-1315/193/1/012068

Simulation of influence of various technical systems on permafrost boundaries movement

M Yu Filimonov[1,2] **and N A Vaganova**[1,2]

[1] Ural Federal University, Yekaterinburg, Russia
[2] IMM UB RAS, Yekaterinburg, Russia

E-mail: fmy@imm.uran.ru

Abstract. Prediction of dynamics of cryogenic processes in soil and foundations is of great importance in the development and operation of oil and gas fields in the Arctic and Subarctic regions. These processes are influenced by the thermal effects of various technical systems available on the well pads. The thawing of permafrost due to such impacts will be accompanied by subsidence of the earth's surface around engineering facilities and development of dangerous geological process called thermokarst. This may lead to major technological accidents at well pads with great harm to the environment. To simulate cryogenic processes in a specific geographical area, algorithms and programs are developed, taking into account the climatic factors and lithology of the soil. An algorithm for determining the features of the area under consideration in the computer simulation of the propagation of thermal fields in frozen ground is described and the results of numerical calculations are presented.

1. Introduction

In Russia, more than 60% of its territory is occupied by permafrost. It is most widely distributed in Eastern Siberia and in Baikal regions. In these areas the main Russian oil and gas fields are located. The production in these fields leads to thawing of permafrost and to possible dangerous earth surface subsidence around engineering facilities at the well pads. Most of the industrial structures in the permafrost zones are constructed and operated on the base of principle of conserving the frozen state of the soils foundation. Studying the dynamics of permafrost boundaries and the movement is of great importance for design and construction of various structures in these territories and is also associated with climate change problems [1–3]. Therefore, the problem of reducing the intensity of the thermal interaction in the "heat source – permafrost" system is an urgent task and has a special significance for safety, saving costs and improving the operational reliability of engineering structures problems.

For long-term simulation of cryogenic processes, a mathematical models and approaches to solving such problems are described in Chapter 5 of [4], on the base of which are developed numerical algorithms [5–8], used in the arrangement of 12 Russian oil and gas fields, when various technical systems operate [9–11] taking into account specific climatic and geographical conditions [12]. Since detailed calculations of the long-term prediction of permafrost boundary changes require considerable computing power, a parallel approach to solving such problems is also developed [13].

This paper presents a mathematical model and numerical calculations for some technical systems used in northern oil and gas fields. A detailed description of the technique which allows to take into account the specific conditions in the model related with the well pad area, and take into account the most significant climatic and physical parameters that affect the propagation of thermal fields in the ground.

2. Object of simulation and mathematical model

Following [4, 12] we will state the problem. Let $T = T(t, x, y, z)$ be soil temperature at the point (x, y, z) at the time moment t. Simulation of unsteady three-dimensional thermal fields, such as oil and gas fields (the well pads) located in the area of permafrost, is required to take into account the different technological (figure 1a) and climatic (figure 1b) factors.

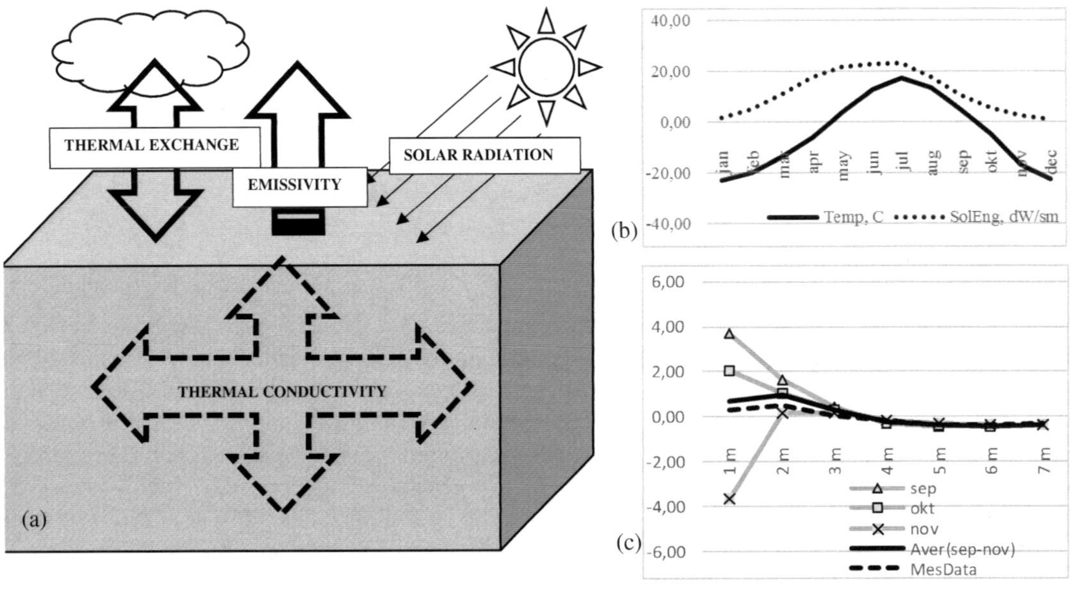

Figure 1. Thermal flows, which form temperature fields in a soil (a); an example of average annual temperatures (solid line) and solar radiation (dotted line) (b); temperatures in soil: measured (dashed line) and computed (solid lines) (c).

Let the modeling area in figure 1a is the box $\Omega = \{(x, y, z) : -L_x \leq x \leq L_x, -L_y \leq y \leq L_y, -L_z \leq z \leq 0\}$, which is defined by positive numbers L_x, L_y, L_z. Simulation of processes of heat distribution is reduced to solution of three-dimensional diffusivity equation with non-uniform coefficients including localized heat of phase transition — an approach to solve the problem of Stefan type, without the explicit separation of the phase transition in Ω [14]. The equation has the form

$$\rho(c_\nu(T) + k\delta(T - T^*))\frac{\partial T}{\partial t} = \nabla\left(\lambda(T)\Delta T\right), \tag{1}$$

with initial condition

$$T(0, x, y, z) = T_0(x, y, z). \tag{2}$$

Here ρ is density $[kg/m^3]$, T^* is temperature of phase transition $[K]$,

$$c_\nu(T) = \begin{cases} c_1(x, y, z), & T < T^*, \\ c_2(x, y, z), & T > T^*, \end{cases} \text{ is specific heat [J/kg K],}$$

$$\lambda(T) = \begin{cases} \lambda_1(x,y,z), & T < T^*, \\ \lambda_2(x,y,z), & T > T^*, \end{cases} \text{ is thermal conductivity coefficient [W/m K],}$$

$k = k(x, y, z)$ is specific heat of phase transition, δ is Dirac delta function.

The nonlinear condition on the surface of the ground has the form

$$\alpha q + b(T_{air} - T(x,y,0,t)) = \varepsilon\sigma(T^4(x,y,0,t) - T_{air}^4) + \lambda \frac{\partial T(x,y,0,t)}{\partial z}. \tag{3}$$

The possibility of adaptation of the numerical algorithm to a specific geographic area, taking into account various climatic and physical factors, and the importance of the correct setting of the boundary condition (3) are discussed in [12].

At the boundaries of the computational domain the boundary conditions are given

$$\left.\frac{\partial T}{\partial x}\right|_{x=\pm L_x} = \left.\frac{\partial T}{\partial y}\right|_{y=\pm L_y} = 0, \quad \left.\frac{\partial T}{\partial z}\right|_{z=-L_z} = \gamma. \tag{4}$$

In (4) γ is a positive number, corresponding to a geothermal flux value. As a rule γ is a small number and it is possible to be set zero in calculations.

3. Numerical results

With using ideas [14], to solve problem (1)–(4) in three-dimensional box a finite difference method is used with splitting by the spatial variables and taking into account the inner boundaries from different technical systems. Solvability of the same difference problems approximating (1)–(4) is proved in [15]. An experimental data and numerical estimations obtained during 3 years of monitoring the changes in the thawing zone boundary around an operating well and predicted on the basis of the developed software package for the problem (1)–(4), respectively, are in a good agreement (5% in position of thawing boundary).

To conserve the soil in the frozen state seasonal cooling devices (SCDs) are used [9], operating on the basis of physical laws, and creating zones of additional freezing on winter. Let consider a system of several SCDs located in frozen soil. The thermal fields around SCD's have been stabilized after 3–4 years of simulation. In figure 2 there are presented thermal fields on December (left figures) and September (right figures) from 9 SCDs after 3–4 years of simulation. In figure 3 there are presented thermal fields on December (left figures) and September (right figures) from 4 SCDs after 3–4 years of simulation.

December illustrates the soil freezing under SCDs operation and September shows the thermal trace of SCDs effect after the season the soil warmed up.

Let consider the operating mode of a horizontal flare system (HFS) when it works continuously, or when HFS runs periodically with the specified mode of switching. To simulate the thermal trace from the HFS on the surface $z = 0$, there was used the temperature or the heat flux. It is necessary to calculate the predicted change in the temperature regime of the foundation soils under HFS, taking into account the adopted design of the heat-insulating screen, which ensures the preservation of permafrost in the base for the period of operation of the HFS. The temperature fields are shown in figure 4 in the first and fifth years of operation of the HFS, respectively, in the xz plane.

4. Conclusion

Thus, the developed mathematical model and software product allow to carry out detailed numerical calculations on long-term forecasting of temperature field changes from different technical systems in the near-surface layer of soil. Numerical simulation of thermal fields around SCDs have shown that it is possible to choose an optimal number of SCDs for thermal

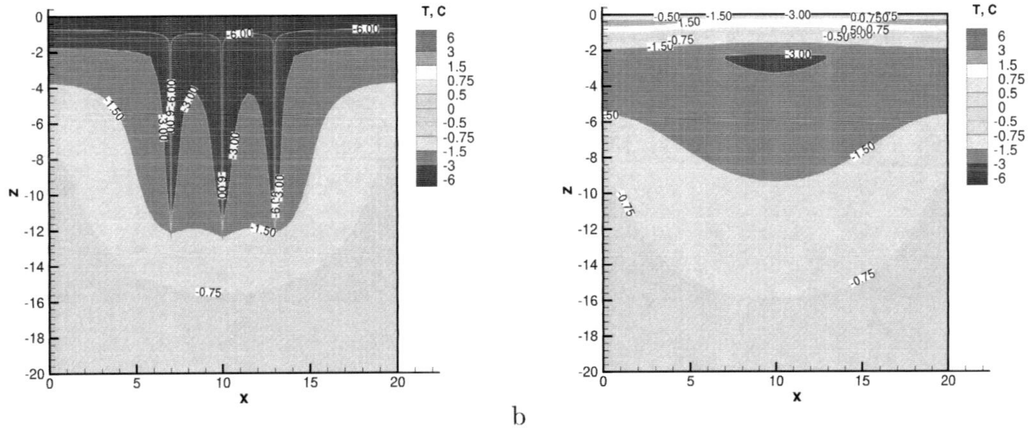

a b

Figure 2. Thermal flows, which form temperature fields in a soil from 9 SCDs with distance 3m between SCDs: a — in December, b — in September.

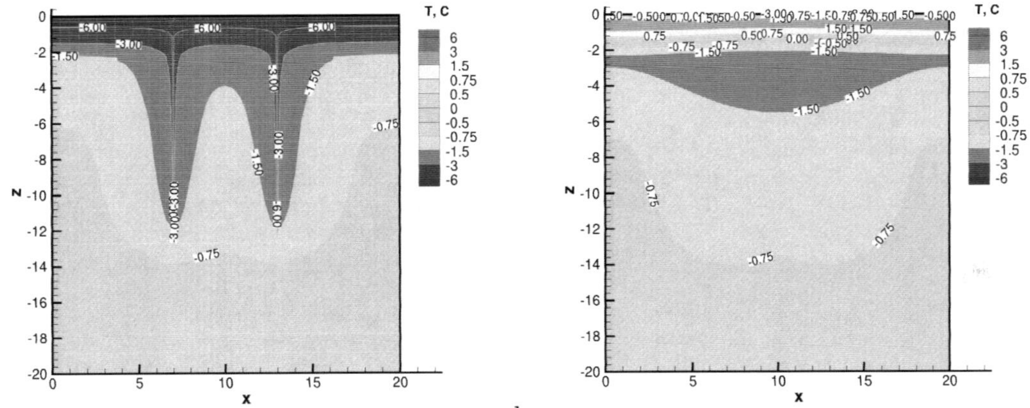

a b

Figure 3. Thermal flows, which form temperature fields in a soil from 4 SCDs with distance 6m between SCDs: a — in December, b — in September.

stabilization of soil depending on the technical conditions of operation of the technical systems or engineering structures. Since the stabilization of thermal fields around SCDs is achieved after 3 to 4 years of SCDs operating, it is advisable to provide thermal stabilization of the soil for three years, if necessary, or to use an additional thermal insulation before starting construction. The operating mode of HFS can be described by using a variable time step. Carried out computer simulation allows one to solve specific applications related to the design of the foundation construction under HFS, as well as the task of compiling an optimal schedule for switching on and off the HFS in order to reduce the thermal impact on the basis for HFS and permafrost.

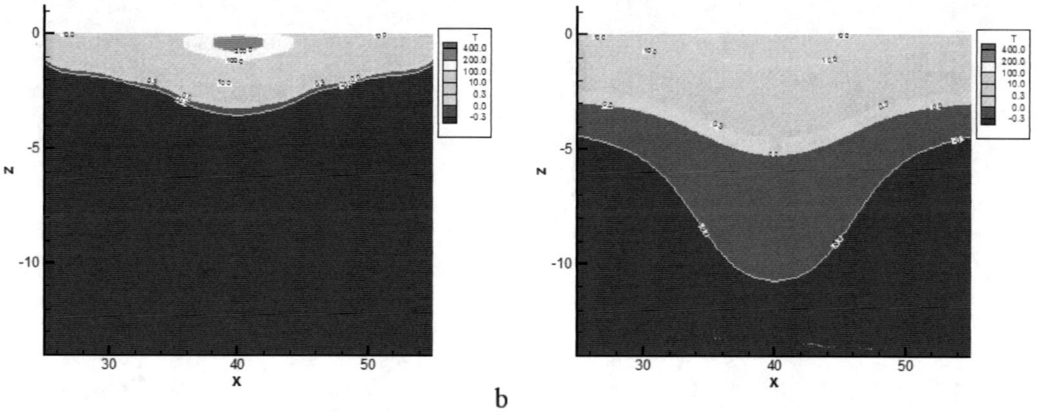

a b

Figure 4. Temperature fields from the HFS in the fifth year: a — in May, b — in October.

Acknowledgments
This work was supported by RFBR project 16–01–00401 and program of scientific research UrB RAS 18–1–1–8.

References

[1] Zhang T, Barry R G, Knowles K, Heginbottom J A and Brown J 2008 *Polar Geography* **31**(3) 47
[2] Vaganova N and Filimonov M 2017 *IOP Conference Series: Earth and Environmental Science* **72** 012005
[3] Filimonov M and Vaganova N 2017 *IOP Conference Series: Earth and Environmental Science* **72** 012006
[4] Filimonov M and Vaganova N 2017 *Advances in Environmental Research. Vol. 54. Chapter 5. Numerical Simulation of Technogenic and Climatic Influence on Permafrost* (New York: Nova Science Publishers)
[5] Vaganova N A 2017 *Applications of Mathematics in Engineering and Economics, (AMEE'2017)* (*AIP Conference Proceedings* vol 1910) ed Pasheva, Popivanov N and Venkov G (Melville, NY: American Institute of Physics) p 020006
[6] Filimonov M Y and Vaganova N A 2017 *Journal of Physics: Conference Series* **899** 092004
[7] Filimonov M Y and Vaganova N A 2016 *Journal of Physics: Conference Series* **754** 112004
[8] Vaganova N A and Filimonov M Y 2015 *6th International Conference on Finite Difference Methods - Theory and Applications* (*Lecture Notes in Computer Science* vol 9045) ed Dimov, Farago and Vulkov L (Germany: Springer Verlag) p 185
[9] Filimonov M Y and Vaganova N A 2013 *Applied Mathematical Sciences* **7** 7151
[10] Filimonov M and Vaganova N 2016 *Modern Problems in Mathematics and its Applications, (MPMA 2016)* (*CEUR Workshop Proceedings* vol 1662) ed Makhnev A and Pravdin S (CEUR-WS.org) p 253
[11] Vaganova N A and Filimonov M Y 2015 *Applications of Mathematics in Engineering and Economics, (AMEE'2015)* (*AIP Conference Proceedings* vol 1690) ed Pasheva, Popivanov N and Venkov G (Melville, NY: AIP Publishing) p 020016
[12] Filimonov M Y and Vaganova N A 2018 *Information Technologies and Mathematical Modeling for Efficient Development of Arctic Zone, (IT&MathAZ 2018)* (*CEUR Workshop Proceedings* vol 2109) ed Filimonov M, Blisorukova M and Kruglikov S (CEUR-WS.org) p 18
[13] Vaganova N and Filimonov M 2015 *Ural Workshop on Parallel, Distributed, and Cloud Computing for Young Scientists, (Ural-PDC 2015)* (*CEUR Workshop Proceedings* vol 1513) ed Sozykin A, Akimova E and Ustalov D (CEUR-WS.org) p 42
[14] Samarsky A A and Vabishchevich P N 1995 *Computational Heat Transfer, Volume 2, The Finite Difference Methodology* (Wiley)
[15] Vaganova N A 2008 *Proceedings of the Steklov Institute of Mathematics* **S1** S260

State instruments for the development stimulation of Arctic resources regions

I V Filimonova[1,2], A V Komarova[1,2], L V Eder[1,2] and I V Provornaya[1,2]

[1] Trofimuk Institute of Petroleum Geology and Geophysics, Novosibirsk, Russia
[2] Novosibirsk State University, Novosibirsk, Russia

E-mail: FilimonovaIV@list.ru

Abstract. Arctic regions contain significant amounts of natural resources; however, the level of hydrocarbon resources exploration in the region is quite low, which increases the prospect of geological exploration and discovery of new hydrocarbon deposits. The authors studied the features of socio-economic, infrastructural and resource-raw development of the Arctic regions using such indicators as population density, per capita incomes, gross regional product, investments in fixed assets, transport infrastructure, reserves and oil and gas production. It is shown that there is a high differentiation of the regions development in relation to the average Russian indicator. The article considers the state stimulation instruments of oil production in the Arctic regions, with special attention paying to tax benefits. With using cluster analysis, all regions of Russia are divided into groups on tax burden and benefits. The Arctic resource regions are shown to be distributed over two clusters: regions that have a high tax burden due to significant oil production (some of them are traditional and receive fewer benefits); regions that receive significant benefits from the government to intensify oil production. Thus, the analysis revealed that preferential rates for specific industry taxes can be an effective tool for government regulation of strategic industries.

1. Introduction

At present, the geography of oil production is shifting towards the Arctic regions [1,2]. This is due to the decline in production and the exhaustibility of oil reserves in traditional West Siberian and central regions [3,4]. The development of the Arctic regions natural resources is characterized by extremely difficult production conditions (natural and climatic, socio-economic, infrastructural, technological) [5], which causes a low investment attractiveness for subsoil users. In these conditions, instruments of state incentives aimed at creating a favorable investment climate are of crucial importance [6]. One of the most common instruments of state incentives is the provision of tax preferences. Therefore, the aim of the article is to study the features of taxation and the benefits provision to extractive industries (in a regional context) by the example of the oil and gas industry.

1.1. Socio-economic premises for the development of Arctic regions

In accordance with the Decree of the Russian Federation President No. 287 "On the Land Regions of the Arctic Zone of the Russian Federation", as amended on June 27, 2017, the Murmansk Region, Nenets Autonomous Okrug, Chukotka Autonomous Okrug, Yamalo-Nenets Autonomous District, as well as part of the municipal formations of the Komi Republic, the Sakha Republic (Yakutia), the

Table 1. Socio-economic and infrastructural indicators of the Arctic regions.

Subject	Density of population, number of people per km squared	Per capita incomes (per month), rubles	Gross regional product, thousand rubles per person.	Investments in fixed assets, %	Density of transport infrastructure, km per km squared
Arctic regions, total	**0,9**	**34667**	**803**	**15,6**	**8,8**
Yamalo-Nenets Autonomous Okrug	0,7	67521	3383	7,5	3,6
Nenets Autonomous Okrug	0,2	69956	4957	0,6	1,3
Komi Republic	2,0	31527	615	1,4	19,6
Krasnoyarsk Krai	1,2	28030	563	2,9	12,3
Sakha (Yakutia) Republic	0,3	38933	779	1,9	4,0
Chukotka Autonomous Okrug	0,1	63909	1283	0,1	1,0
Murmansk Oblast	5,2	36115	515	0,7	29,2
Arkhangelsk Oblast	2,7	31043	356	0,6	33,6
Russian Federation	**8,6**	**30744**	**443**	**100**	**72,9**

Krasnoyarsk Territory, and the Arkhangelsk Region are territories of the Arctic zone. Regions have a large area and a low population, which is typical for the Arctic regions around the world, primarily due to severe climatic conditions (table 1).

Investment and infrastructure indicators are significantly inferior to the average Russian values. So, investment in fixed assets is only about 16%, while, for example, the Yamalo-Nenets Okrug provides almost 80% of natural gas production in Russia. The density of roads and railways is almost 10 times lower than the national average, and in some regions, for example, the Nenets Autonomous District and the Chukotka Autonomous Area, there are no railways. High incomes of the population in the Yamal-Nenets Autonomous Okrug, the Nenets Autonomous District and the Chukotka Autonomous District are provided by the mining sector and almost double the average for Russia. In the remaining regions of the Arctic zone, this indicator is an average for the country level.

Insufficient volume of investments and weak infrastructural equipment of the Arctic zone regions necessitate a careful choice of instruments of state stimulation of development in order to avoid reduction of economic development indicators [7].

1.2. Resource prerequisites for the development of the Arctic regions

Arctic regions have a significant resource potential and are characterized by varying degrees of exploration. The traditional regions of oil and gas production can be attributed Yamalo-Nenets Autonomous Okrug, Nenets Autonomous District, Komi Republic (Table 2). Part of the Arctic regions, for example the Krasnoyarsk Territory and the Republic of Sakha (Yakutia), currently provides the main increase in oil production in Russia and is classified as promising. The development advantage of such regions as the Chukotka Autonomous District, the Murmansk and Arkhangelsk

Table 2. Resource and raw indicators of the Arctic regions.

Subject	Reserves of oil and gas,%	Accumulated production of oil and gas,%	Oil production in 2017, million tons	Gas production in 2017, cubic meters
Arctic regions, total	**53,8**	**44,7**	**136,3**	**576,2**
Yamalo-Nenets Autonomous Okrug	43,7	41,2	56,6	558,2
Nenets Autonomous Okrug	1,6	0,5	14,0	0,3
Komi Republic	1,1	2,4	32,0	2,2
Krasnoyarsk Krai	3,8	0,4	23,3	12,6
Sakha (Yakutia) Republic	3,7	0,3	10,4	2,9
Chukotka Autonomous Okrug	0,0	0,0	0,0	0,0
Murmansk Oblast	0,0	0,0	0,0	0,0
Arkhangelsk Oblast	0,0	0,0	0,0	0,0
Russian Federation	**100,0**	**100,0**	**546,8**	**691,1**

regions is not so much the energy potential as the proximity to the sea transport routes and the port infrastructure.

The development and production of minerals in the Arctic regions is associated with a number of difficulties, related not only to natural and climatic, but also to geological, infrastructural and other factors. In this connection, the mineral reserves in the Arctic regions can be classified as hard-to-recover. Since in recent years the deterioration in the quality and structure of the hydrocarbon resource base has been observed in Russia as a whole, the state is pursuing a large-scale policy of stimulating production, including in the regions and on the shelf of the Arctic zone [8,9]. So, in the Russian legislation in the RF Law "On Subsoil" and the Tax Code of the Russian Federation, the following benefits are provided [10,11]:

1. Benefits in Mineral extraction tax (MET) for the taxation of natural gas and gas condensate produced in the Yamalo-Nenets Autonomous District and used for the production of liquefied natural gas (LNG);
2. When calculating mineral extraction tax for oil and natural gas, a number of coefficients characterizing the geographic location of the subsoil plot are applied. If the subsoil area is located in the Arctic regions, the coefficient is calculated in a special way to reduce the tax burden.
3. Exemption or privileges for mineral extraction tax for oil with special physicochemical properties (viscosity, sulfur content).
4. When conducting geological study of subsoil within the subjects of the Russian Federation relating to the Arctic zone, the study period is 7 and 10 years respectively for the continental part and the shelf of the Arctic seas.

2. Method

To analyze the current and retrospective analysis of the tax burden of the main oil-producing regions, including the Arctic, the authors used cluster analysis. For the clustering of the oil-producing regions, two variables were chosen:

- Accumulated oil production;

- Average real tax burden for the period.

To delineate the boundaries of clusters, the median values of these indicators were calculated.

The clustering algorithm can be written in the following sequence:

1. Calculation of real income from mineral extraction tax for oil. For the analysis, the period from 2007 to 2017 was taken, therefore, tax revenues were cited by 2007. To bring tax revenues to 2007, the deflator, calculated on the basis of the base rate and the price coefficient, is used as equations (1), (2):

$$D_{t/t-1} = \frac{T_t}{T_{t-1}} \times \frac{K_{pt}}{K_{pt-1}}, \quad (1)$$

$$D_{t/2007} = D_{t/t-1} \times D_{t-1/t-2} \times \ldots \times D_{2008/2007}, \quad (2)$$

where $D_{t/t-1}$ – deflator, leading indicator of year t to year t-1, T_t – tax rate in year t, K_{pt} – price coefficient of year t.

Then real tax revenues in year t from oil production in region i, cited by 2007 (equation (3)):

$$N_{it}^{r} = \frac{N_{it}}{D_{t/2007}} \quad (3)$$

where N_{it}^{r} – real income tax in the year t from oil production in the region i, N_{it} – actual tax revenues in the year t from oil production in the region i, $D_{t/2007}$ – deflator, leading to the base year 2007.

2. Calculation of the real tax burden as in equation (4):

$$n_{it}^{r} = \frac{N_{it}^{r}}{B_{it}} \quad (4)$$

where n_{it}^{r} – real tax burden in the year t in the region i, N_{it}^{r} - real tax revenues in year t from oil production in the region i, B_{it} – tax base in the year t in the region i.

3. Calculation of the average deflated tax burden for the period from 2007 to 2017 as in equation (5):

$$n_{i}^{r} = \sum_{t=2007}^{2017} N_{it}^{r} \left(\sum_{t=2007}^{2017} B_{it} \right)^{-1} \quad (5)$$

where n_{i}^{r} – average deflated tax burden for the period, N_{it}^{r} - real tax revenues in year t from oil production in the region i, B_{it} – tax base in the year t in the region i.

4. Calculation of median levels of the average deflated tax burden and accumulated production by region.
5. Drawing up a table and a graph of regions division into 4 clusters.

The result of the algorithm is the clusterization of the oil-producing regions of Russia into groups with similar characteristics. This makes it possible to analyze groups of regions and develop common principles of state stimulation of oil production in regions from the relevant groups.

3. Result and discussion

The clustering method allowed the distribution of 32 oil-producing regions of Russia into 4 clusters, depending on the accumulated oil production for 2007-2017 and the average deflated tax burden on the severance tax for oil. The results of this distribution are presented in table 3.

Cluster I is represented by nine regions with low tax burden and small oil production, they can be called low-productivity, *cluster II* - seven regions with high tax burden and oil production below the median level. *Cluster III* included nine regions with high oil production and a significant tax burden, which can be described as highly productive.

Table 3. Clustering of regions on tax burden and accumulated production.

	The tax base is less than the median - 29.2 million tons			**The tax base is more than the median - 29.2 million tons**		
	Subject	Tax burden, rubles / t	Accumulated production (2007-2017), million tons	Subject	Tax burden, rubles / t	Accumulated production (2007-2017), million tons
	Cluster II (7 regions)			**Cluster III (9 regions)**		
The average deflated tax burden is above the median (more than 2043 rubles per ton)	Kirov Oblast	2483	0	Tyumen Oblast	2386	71
	Penza Oblast	2425	2	Udmurt Republic	2137	108
	Omsk Oblast	2446	7	Tomsk Oblast	2400	111
	Ulyanovsk Oblast	2142	7	Komi Republic	2067	138
	Stavropol Krai	2289	9	Perm Krai	2305	152
	Novosibirsk Oblast	2376	11	Samara Oblast	2046	163
	Kaliningrad Oblast	2276	12	Orenburg Oblast	2099	234
				Yamalo-Nenets Autonomous Okrug	2225	282
				Khanty–Mansi Autonomous Okrug – Yugra	2373	2 830
	Cluster I (9 regions)			**Claster IV (7 regions)**		
The average deflated tax burden is below the median (less than 2043 rubles per ton)	Kabardino-Balkar Republic	1874	0	Volgograd Oblast	1861	35
	Republic of Ingushetia	1066	1	Sakha (Yakutia) Republic	1261	67
	Republic of Kalmykia	1776	2	Irkutsk Oblast	1095	98
	Republic of Dagestan	2007	2	Nenets Autonomous Okrug	1631	143
	Astrakhan Oblast	552	7	Republic of Bashkortostan	1922	157
	Chechen Republic	2033	10	Krasnoyarsk Krai	1817	162
	Krasnodar Krai	1860	12	Republic of Tatarstan	2040	366
	Saratov Oblast	1924	15			
	Sakhalin Oblast	1495	24			

Seven regions located in the *cluster IV* are characterized by high oil production and low tax burden, that is, they are privileged.

The data obtained suggest that there is no direct correlation between the level of oil production and the tax burden. Approximately the same number of regions were included into all four clusters, that is, it can not be said that the level of the tax burden is growing (or decreasing) as production from the region to the region increases. There are regions with high tax burden and low production, such as the

Novosibirsk, Omsk, Penza regions, etc., and there are also subjects with low load and high production, such as the Irkutsk Region, the Krasnoyarsk Territory, and the Nenets Autonomous District.

Despite the fact that the Khanty-Mansiysk and Yamal-Nenets Autonomous Districts are one of the most favored regions (in 2017, the amount of tax lost from granting benefits in these regions amounted to about 40% of the total amount of benefits), the tax burden in them is anyway high.

At the next stage, clustering of 32 oil-producing regions was carried out according to the average for 2007-2017 real tax burden and accumulated production for the same period. Based on the data obtained, it can be assumed that the severance tax for oil is not characterized by a stable direct relationship between the tax burden and the tax base. In the cluster of low-performing regions, nine regions were included, among which the Astrakhan region with the lowest tax burden, the Republic of Ingushetia, Dagestan, Kalmykia and others. The cluster of high-performing regions includes the Khanty-Mansiysk Autonomous District, the Yamalo-Nenets Autonomous District, the Orenburg Region, the Tomsk Region with the highest tax burden, and others. Thus, despite the significant application of benefits for the sums and amount of preferential oil, Khanty-Mansiysk, Yamalo-Nenets Autonomous Districts, Orenburg region, etc. are regions with a relatively high tax burden.

The Arctic resource regions were divided into two clusters - III and IV: regions that have a high tax burden due to significant oil production, but some are traditional and receive fewer benefits; regions with significant benefits from the state to intensify oil production.

Thus, the analysis revealed that preferential rates for specific industry taxes can be an effective tool for government regulation of strategic industries.

Acknowledgments
The research was supported integration project № ИП-22ПН.

References

[1] Eder L V, Filimonova I V, Provornaya I V, Nemov V U and Nikitenko S M 2017 Regional smart specializations in fostering innovation development of resource regions of Russia *International Multidisciplinary Scientific GeoConference Surveying Geology and Mining Ecology Management, SGEM* **17** 727-34

[2] Eder L V, Filimonova I V, Provornaya I V and Nemov V Y 2017 The current state of the petroleum industry and the problems of the development of the Russian economy *IOP Conf. Ser.: Earth Environ. Sci.* **84** 012012

[3] Filimonova I V, Eder L V, Mishenin M V and Mamakhatov T M 2017 Current state and problems of integrated development of mineral resources base in Russia *IOP Conf. Ser.: Earth Environ. Sci.* **84** 012011

[4] Kontorovich A E, Eder L V and Filimonova I V 2017 Paradigm oil and gas complex of Russia at the present stage *IOP Conf. Ser.: Earth Environ. Sci.* **84** 012010

[5] Nikitenko S M and Goosen E V 2017 Socio-economic development of territories based on the principles of public-private partnership in the sphere of comprehensive mineral exploration *IOP Conf. Ser.: Earth Environ. Sci.* **84** 012013

[6] Nikitenko S M and Mesyats M A 2017 Objects of Industrial Property as an Instrument for Introducing Technological Innovations in Machine Building *IOP Conf. Ser.: Mater. Sci. Eng.* **253** 012024

[7] Peter C, Maxfield J, Miller A and Allen R M 2018 *Taxation of Mining Operations* p 856

[8] Sharf I, Tsibulnikova M and Dmitrieva N 2016 Economic evaluation of the approaches to associated petroleum gas utilization. International *Multidisciplinary Scientific GeoConference Surveying Geology and Mining Ecology Management, SGEM* **3** 153-60

[9] Sharf I V 2017 Financial and organizational aspects of the recovery of hydrocarbon resource base in the regional context *Economy of Region* **13** 628-40

[10] Kesler S E and Simon A C 2015 *Mineral Resources, Economics and the Environment* (Cambridge: Cambridge University Press) p 434

[11] Tagaeva T O, Baranov A O and Gilmundinov V M 2016 Assessment of the required changes of Russian ecological taxes *International Journal of Energy Economics and Policy* **6** 611-16

Structure of fracture of ice cover, finger rafting

N M Osipenko

Ishlinsky Institute for Problems in Mechanics, Moscow, Russia

E-mail: osipnm@mail.ru

Abstract. The mechanism of fracture in the vicinity of the region of mutual cutting of ice cover plates is considered on the basis of the analysis of sea ice cover compression structures development. A model of the cutting process based on the representation of this region as the longitudinal shear is proposed. The obtained structures of fracture at shear in the model materials containing capillary pores were analyzed. In particular, it was shown that during the formation of the cracks the distance between cracks along the shear front in the primary echelon structure of brittle fracture is linearly related to the incision tip radius. An estimate is obtained for the effective fracture toughness of ice at longitudinal shear.

1. Introduction

The mutual cutting of the contacting ice floes in the sea ice cover takes place at the front of the finger rafting. The mechanism of fracture is associated with a tendency to develop a shift along the grain axis and a system of capillary pores [1]. The presence of pores creates anisotropy of the shear strength [2, 3]. The least shear resistance is observed when the shear is oriented across the plane of the ice cover. At finger rafting main part of the plates on each side of the slot lie on each other. The concentration of longitudinal shear stresses is created on the interaction front in the mutual cutting region. The present study is devoted to simulation of the internal problem of longitudinal shear initiation in a capillary porous medium on model materials. As a result, a model is proposed for the process of finger rafting of ice, based on the concept of the shear crack front propagation.

2. Model experiments

To simulate this form of failure, a number of experiments were performed on samples with a system of holes simulating capillaries loaded with shear loads. Figure 1a shows the scheme of experiments. When loading by shear in the vicinity of the notch, whose apex has the shape of a capillary, a system of feathering cracks develops (figure 1b). It is analogous to fracture systems in the vicinity of the longitudinal shear crack [4]. On the surface of failure from two concentrators under shear, the structure of opposing fracture systems are observed (figure 1c). The following stages of the process can be identified (figure 2):

- initiation of the cracks on the contours of the nearest holes on the extension of the incision line (figure 2a) (stage I);
- fusion of feathering cracks between the holes on the extension of the incision (figure 2b) (stage II);
- the formation of a system of particles between the surfaces of the growing main shear (figure 2c) (stage III).

The particle system forms a moving layer (third body) between the contacting surfaces under shear. Figure 3 shows the test scheme and the shape of the sample when modeling the echelon of cracks. The location of the crack system (the primary echelon of cracks) formed in the shear crack tip region is given in figure 3, on the right.

The cracks formed on the continuation of the shear have the character of fractures of rupture. In particular, in the absence of normal stresses on the shear plane (stress intensity factor $K_I = 0$), the angle between the planes of cracks and the plane of shear is $\alpha \approx \pi / 4$. The interrelation of the distance between the cracks in the primary echelon (h) and the transverse dimension (radius) of the end region of the incision (R) for the materials used in experiments is shown in figure 4. We obtain the ratio: $h \approx 4R$. With a brittle mechanism of fracture, the structure in the fracture region does not depend on the deformation properties of the material.

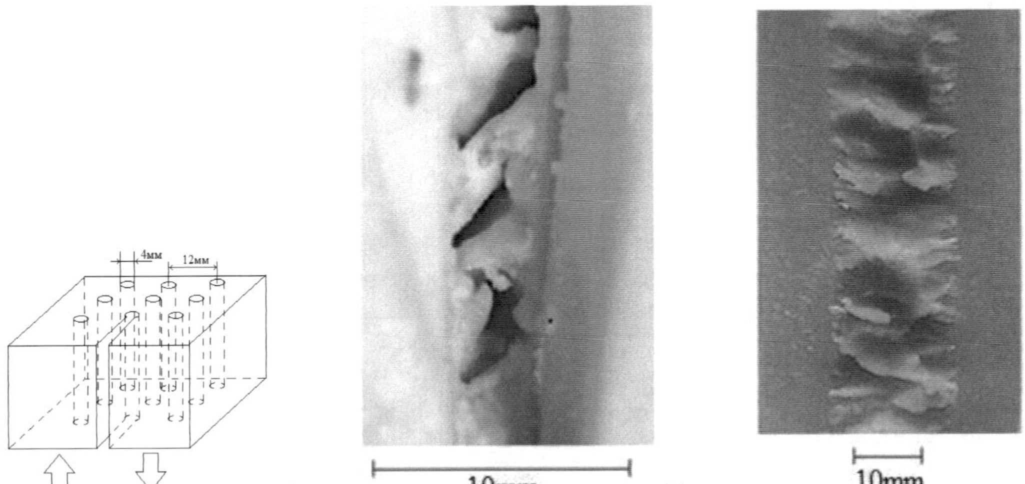

a) b) c)

Figure 1. Experiment on the fracture of model samples under shear.

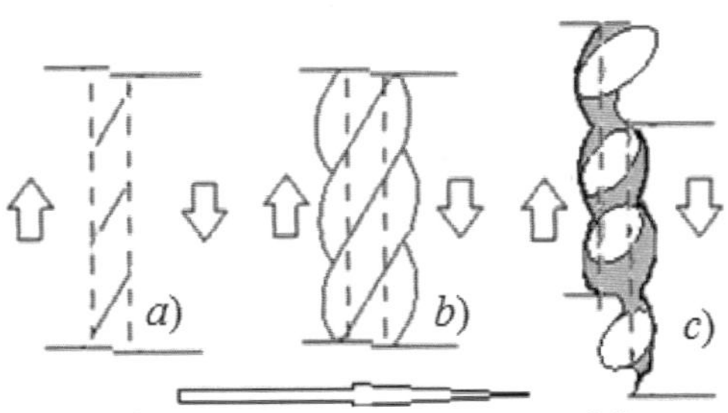

Figure 2. Development structure of fracture top of longitudinal shear.

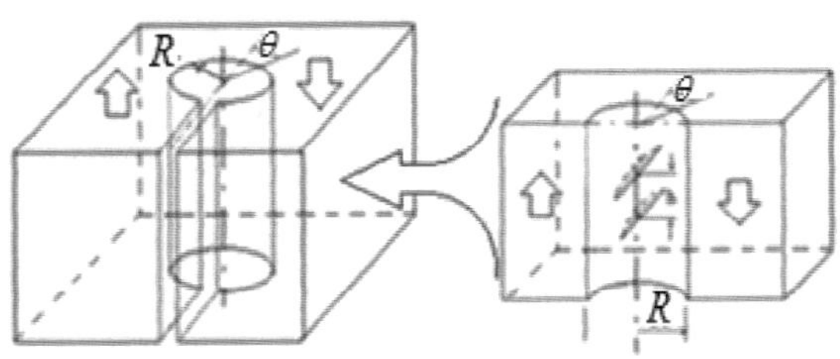

Figure 3. Initiation of fracture on the contour of the hole at the top of the shear.

3. Model representation of the end zone of the longitudinal shear

The initiation of the fracture occurs on the continuation of the initial incision on the internal surface of the hole which is the end region of the shear, where the stress concentration is created (stage I). Then the crack propagates in an unstable regime in a plane inclined to the shear axis at an angle of $\sim \pi / 4$ (stage II). Further crack growth together with adjacent cracks formation occurs in a stable regime (stage III). It should be noted that the main events in the initiation of the crack system are tied to certain planes of the vicinities of the end region of the shear. These are the plane of the crack, the plane of the macro shear and the plane touching the end region, transverse to it.

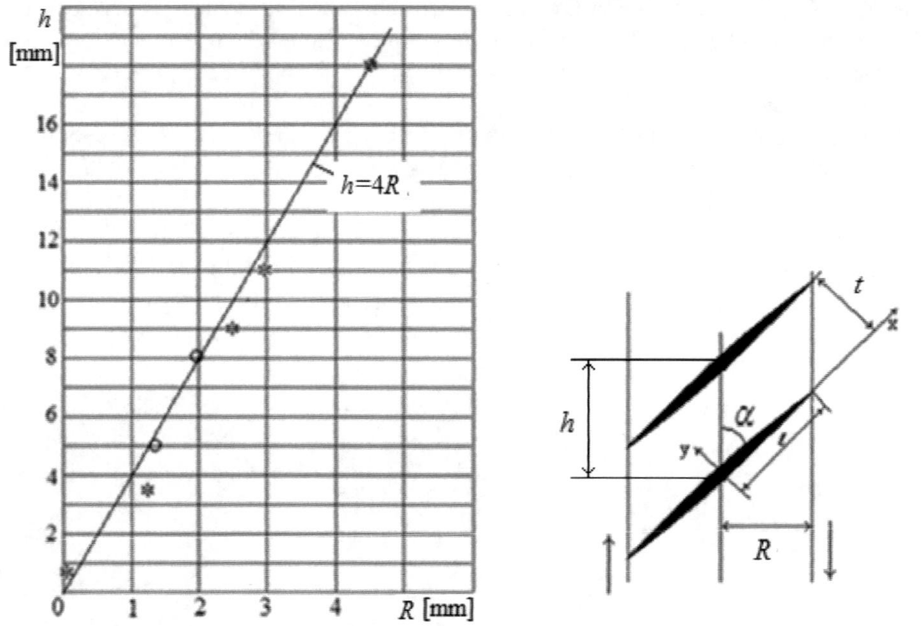

Figure 4. Interrelation of the distance between cracks in the primary echelon (h) and the radius of the end region of the incision (R) (* - cheese, o - gypsum).

We give some estimates. Consider a set of parallel cracks in the vicinity of the longitudinal shear that is cut with a finite radius R. We use the solution for the stress intensity factor in a system of closely spaced parallel rectilinear cracks, in a plane loaded uniformly by a uniform load at infinity [4]. For a plane stress state

$$K_I \approx \sigma_y \sqrt{t/2}\,, \tag{1}$$

where t - is the distance between the cracks, and σ_y - the stresses along the normal to the crack line.

The relation (1) coincides with the estimate from the numerical solution [5] for a plane with an infinite periodic system of parallel cracks of the same length, loaded with a uniform stress along the normal to the lines of cracks at $t \le \ell$

$$K_I \approx 0.71\sigma_y \sqrt{t}\,.$$

We also mention the coincidence with the solution given in [6] for a system of parallel cuts loaded with homogeneous internal pressure.

The stress $\sigma_y = \sigma_1$ in the vicinity of the end region of the shear is estimated from the asymptotic behavior of the stresses in the external problem of fracture mechanics for the end region crack of the longitudinal shear [6]. The principal stresses on the selected plane of the section at the end region of the shear are

$$\sigma_{1,2} = \pm \frac{K_{\mathrm{III}}}{\sqrt{2\pi r}} \cos\frac{\theta}{2} \sqrt{\cos\theta}\,, \tag{2}$$

where K_{III} is the stress intensity factor of longitudinal shear stresses for a cut as a crack of longitudinal shear. θ is the angle between the stress vector and the axis of the crack.

Assuming that the level of K_{III} and the level of stresses provided by it at the unstable initial phase of cracks formation change insignificantly at this phase, and also that the process is concentrated near the contour of the notch on its continuation (i.e. $r \to R, \theta \to 0$), we obtain from (1)

$$K_{III} \approx \sigma_1 \sqrt{2\pi R} \to \sigma_1 \approx K_{III} / \sqrt{2\pi R}.$$

We will assume that in the state of limiting equilibrium on the contour of a single crack of normal rupture in the scale of the external problem of cracks we have $K_I = K_{IC}$, where K_{IC} is the fracture toughness of the material. From here

$$t \approx \left(\frac{K_{1C}}{\sigma_1}\right)^2 = 4\pi R \left(\frac{K_{1C}}{K_{III}}\right)^2. \tag{3}$$

According to the accepted fracture scenario, the initial stage of cracks development (as isolated cracks) that takes place in the stress field (2) is unstable. Then the cracks are inhibited, as the stress intensity decreases when the front of the crack leaves the stress concentrator region [4]. The level K_I corresponding to the moment of initiation of the next crack is determined by the relation

$$K_I \approx \sigma_1 \sqrt{\pi \ell} \approx K_{IC},$$

where ℓ is the size of the primary equilibrium crack in the plane under consideration. The size of the primary equilibrium crack ℓ upon initiation of the next crack on the contour of the notch is controlled by the size (radius) of the tip of the cut, i.e. $\ell \sim fR, (f < 1)$. The condition relating the limiting parameters for the cut and cracks has the form

$$\frac{K_{IC}}{\sqrt{\pi f R}} \approx \frac{K_{III}}{\sqrt{2\pi R}}; K_{IC} \approx K_{III} \sqrt{\frac{f}{2}}. \tag{4}$$

From (3) and (4), we obtain a relation showing the linear relationship between the radius of the cut tip and the distance between the cracks in the primary echelon of cracks:

$$t \approx 2\pi f R.$$

The distance between the centers of the cracks h, taking into account the orientation of the cracks relative to the axis of the echelon in this approximation is determined by the relation

$$h = t / \cos\alpha \approx 2\pi f R / \cos\alpha,$$

where α is the angle between the shear axis and the plane of the crack. The result is correlated with the experimental data for $f = 0.45$, $\alpha = \pi / 4$.

The location of the foci of next crack initiation is determined by the maximum of the tensile stresses in the vicinity of the primary crack. The stress field in the vicinity of the initial crack is composed of a field created by the longitudinal shear and a local perturbation, the source of which is the initial crack. We give some asymptotic estimates in the framework of the plane problem, for a single initial crack. Let us write down the principal stresses at the site perpendicular to the front of the longitudinal shear along its continuation in the coordinates associated with the initial crack

$$\sigma_y = \frac{K_{III}}{2\sqrt{\pi r}} \sqrt{\left(1 + N^{-1/2}\right) N^{-1/2}}, \quad N = \left(\frac{x}{r}\right)^2 \sin^2\alpha + 1, \tag{5}$$

IOP Conf. Series: Earth and Environmental Science **193** (2018) 012070 doi:10.1088/1755-1315/193/1/012070

where K_{III} is the stress intensity factor at the longitudinal shear front, r - is the distance from the axis of the shear front to the site, and α - the orientation of the normal to the trajectory of the principal stresses (here, the orientation of the plane of the crack).

We assume that $r \sim R$, where R is the characteristic radius of curvature of the top of the shear. We also assume that, in the asymptotic approximation, the perturbation of the stress state in the vicinity of the macro-shear, associated with the presence of the initial crack, is provided by a singularity of the stress field in the region of the crack tip. In the end zones of the equilibrium initial crack of a normal discontinuity, the Irvine condition (K_I=K_{IC}) is fulfilled. For initiation of the next crack, the most interesting is the situation in which initial crack having a size close to the size of radius of curvature of the top of the sear. Stress σ_y, coinciding with the direction σ_1 of the external stress field at the site without cracks at the coordinates associated with the fracture plane is [6]

$$\sigma_y = \frac{K_1}{\sqrt{2\pi r}} \cos\left(\frac{\theta}{2}\right)\left(1+\sin\left(\frac{\theta}{2}\right)\sin\left(\frac{3\theta}{2}\right)\right). \quad (6)$$

Passing further to dimensionless coordinates $\bar{x} = x/\ell; \bar{y} = y/\ell$, we obtain from (5) and (6) the asymptotic formulas for estimating the normal stress σ_y at an arbitrary point of the site in the variant of K_I=K_{IC} and $\alpha = \pi/4$.

$$\sigma_y = \frac{K_I}{2\sqrt{\pi\ell}}\left(\frac{K_{III}}{K_{1C}}\sqrt{\left(1+N^{*-1/2}\right)N^{*-1/2}} + \frac{\sqrt{\sqrt{\bar{y}^2+(1-\bar{x})^2}-1+\bar{x}}}{\sqrt{\bar{y}^2+(1-\bar{x})^2}}\frac{\left(1.5-3\bar{x}+1.5\bar{x}^2+0.5\bar{y}^2\right)}{\left(\bar{y}^2+(1-\bar{x})^2\right)}\right)$$

$$N^* \approx 1+0.5(\bar{x}-\bar{y})^2$$

Figure 5 shows the isolines of the parameter characterizing the stresses $\beta = \sigma_y\sqrt{2\pi\ell}/K_{IC}$ in the vicinity of the crack for different values of the ratio $k = K_{III}/K_{IC}$. The figure gives an asymptotic estimate of the location of a maximum of tensile stresses in a distant neighborhood of the initial crack. It can be seen that if the shear is predominant (k = 10), the stress maximum σ_y (denoted by a cross) is observed on the shear axis outside the unloading zone at a distance of about 4R from the center of the primary crack along the shear axis. This estimate coincides with the data of the model experiment which coincides with the model experiment data

$$h \approx 4R.$$

This region can serve as a point for initiating the next crack in the developing echelon of cracks. Thus, the characteristic scale of the echelon of cracks in the vicinity of the top of the shear is given. As k decreases, the distance h decreases, and in the region $k \sim 2$ the local maximum of the parameter is practically absorbed by the stress concentration region near the crack tip. With a small fraction of the shear in the field of external stresses, a change in the fracture scenario takes place - instead of a crack echelon, a single zigzag crack is formed at the front of the cut.

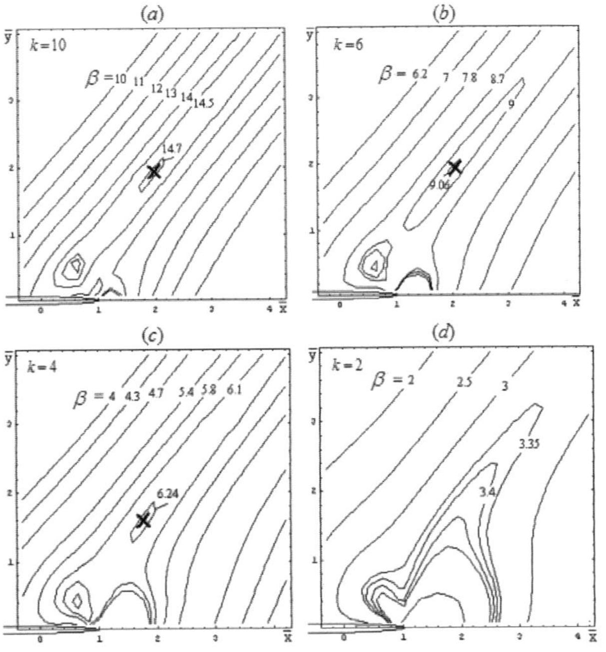

Figure 5. The isolines of the parameter $\beta = \sigma_y \sqrt{2\pi\ell} / K_{IC}$ in the neighborhood of the top of the shear.

4. Conclusion, effective crack resistance

As shown earlier, cracks surrounding the capillaries on the continuation of the main shear form an echelon of cracks between the capillaries. The contour of the individual crack at the top of the shear expands at an angle $\alpha = \pm\pi/2$ [4, 7]. With the growth and fusion of cracks between the two capillaries, a crack is formed in a gap inscribed in a square with a diagonal close to the distance between the centers of the capillaries (L) (figure 6). When the dimensions of the cracks in the echelon reach the boundaries of the square mentioned above, the limit equilibrium state of the system is attained. The subsequent integration of cracks and the formation of moving fragments (elements of the third body, figure 2) are not accompanied by the increase in the applied load, according to our experimental data. These estimates do not take into account the interaction with the third body in the implementation of the shear.

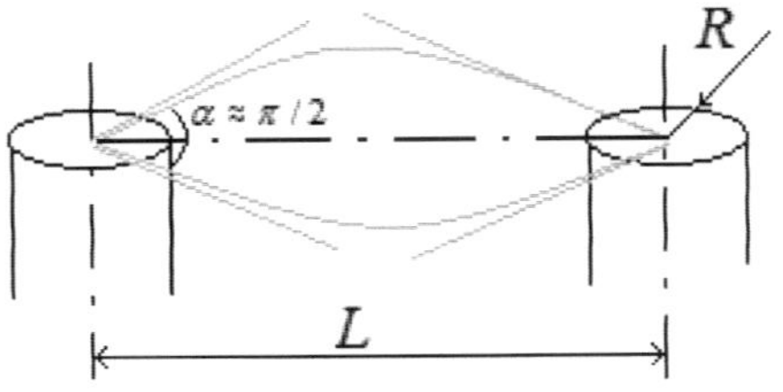

Figure 6. Scheme of development of a single crack between capillaries at shear along the axis of capillaries.

It is convenient to consider the model situation in a plane section normal to the plane of shear on its continuation. The principal stresses on the plane of the section in a small neighborhood of the top of the shear, as before, are estimated from (2). Suppose, as usual [8], that the fractures of the normal rupture formed on the continuation of the shear are oriented along the normal to the maximum principal stress. In the situation of shear dominance under the orientation of feathering cracks at an angle $(\alpha \approx \pi/4)$ to the axis of the echelon

$$\frac{K_{\mathrm{IC}}}{\sqrt{R\sqrt{2}}} \approx \frac{K_{\mathrm{III}}}{2\sqrt{(\pi L)}}.$$

From this we obtain the condition of limiting equilibrium in a system of cracks in a neighborhood of the shear along the capillaries

$$K_{\mathrm{IIIC}} \approx 2K_{\mathrm{IC}}\sqrt{\frac{\pi L}{\sqrt{2}R}}.$$

The capillary porosity (n) is related to the geometry of the porous space in the plane by the relation

$$R/L \approx \sqrt{n}.$$

From this we obtain an estimate of the level of effective crack resistance at longitudinal shear of the capillary porous medium

$$K_{\mathrm{IIIC}} \approx m\pi^{0.5} K_{\mathrm{IC}} n^{-0.25}; m \sim 1.7.$$

The described scenario of the finger rafting mainly refers to the case high porosity and to the situation of active ice movement, since when it stops the friction surfaces of ice floes may freeze and change the fracture mechanism.

Acknowledgments
This work was supported by the Fundamental Research Program of the Presidium of the Russian Academy of Sciences 1.3P "Fundamental Problems of Oceanology: Physics, Geology, Biology and Ecology".

References

[1] Osipenko N M and Chencow A B 2017 *Preprint IPMech RAS* **1151** p 28
[2] Tyshko K P, Cherepanov N V and Fedotov V I 2000 *Crystalline Structure of Sea Ice Cover* (St. Petersburg: Gidrometeoizdat) p 66
[3] Cherepanov N V 1976 *Proc. of AARI* **331** 77-9
[4] Goldstein R V and Osipenko N M 2012 *Mechanics of Solids* **47** 505-16
[5] Murakami Y 1987 *Stress Intensity Factors Handbook* (New York: Pergamon Press) p 1464
[6] Cherepanov G P 1978 *Mechanics of Brittle Fracture* (New York: McGrawHill) p 950
[7] Goldstein R V and Osipenko N M 2014 *Fatigue Fract. Eng. Mater. Struct.* **37** 1292–305
[8] Barenblatt G I 1961 *J. Appl. Mech. Tech. Phys.* **4** 3-56

IOP Conf. Series: Earth and Environmental Science **193** (2018) 012071 doi:10.1088/1755-1315/193/1/012071

Invariant submodels of model of nonlinear diffusion in inhomogeneous medium with nonstationary absorption or source

Yu A Chirkunov

Novosibirsk State University of Architecture and Civil Engineering (Sibstrin), 630008 Novosibirsk, Russia

E-mail: chr101@mail.ru

Abstract. We carried out a complete group classification of the nonlinear differential equation of the model, describing the nonlinear diffusion process (or the process of heat propagation) in an inhomogeneous medium with nonstationary absorption or source. We found all models whose equations admit a nontrivial continuous Lie group of the transformations. We researched invariant submodels of the obtained models. Some invariant solutions, describing these submodels were found either explicitly, or their search was reduced to solving of the nonlinear integral equations. In particular, we obtained an invariant solution that describes the nonlinear diffusion process (or heat distribution process) with two fixed "black holes", and invariant solution that describes the nonlinear diffusion process (or heat distribution process) with a fixed "black hole", and with a moving "black hole". The solutions with "black holes" can be used to describe the process of destruction of glaciers under the influence of the external heat source, for example a solar energy. Using other invariant solution, we studied a diffusion process (or heat distribution process) for which at the initial moment of the time at a fixed point a concentration (a temperature) and its gradient are specified. The solution of the boundary value problem describing this process reduces to the solution of nonlinear integral equations. We have established the existence and uniqueness of solutions of this boundary value problem under certain conditions. The mechanical significance of the obtained solutions is as follows: 1) they describe specific nonlinear diffusion processes (or heat distribution processes), 2) they can be used as tests in numerical calculations in studies of nonlinear diffusion (or heat distribution) in an inhomogeneous medium with nonstationary absorption or source, 3) they allow to assess the degree of adequacy of a given mathematical model to real physical processes, after carrying out experiments corresponding to these decisions, and estimating the arising deviations.

1. Introduction

Symmetry properties of the model of nonlinear diffusion (or heat distribution) in an inhomogeneous medium without absorption or a source and also with stationary absorption or a source were studied in many papers (see, for example, bibliographies given in [1]). Symmetry properties of the model of nonlinear diffusion (or heat distribution) in an inhomogeneous medium with nonstationary absorption or a source of a special type were studied in [2]. A model of nonlinear diffusion (or heat distribution) in an inhomogeneous medium with nonstationary absorption or a source of general form was investigated in [3]. In this report we will use the materials of this article.

IOP Conf. Series: Earth and Environmental Science **193** (2018) 012071 doi:10.1088/1755-1315/193/1/012071

A model describing the nonlinear process of diffusion (or heat distribution) in an inhomogeneous medium with nonstationary absorption or a source is determined by the equation:

$$\frac{\partial u}{\partial t} = \frac{\partial}{\partial x}\left(x^{\alpha} u^{\beta} \frac{\partial u}{\partial x}\right) - \varphi(t) u, \tag{1}$$

where $u = u(t, x)$ is the concentration of the substance (or temperature) at the point $x \in (-\infty, \infty)$ at time t; α and β are parameters of the diffusion coefficient $D = x^{\alpha} u^{\beta}$, which characterizes the diffusion rate (α and β are arbitrary real constants); $\varphi(t)$ is non-stationary absorption or source coefficient.

We assume further that $\varphi(t)$, α and β satisfy the condition

$$\alpha\beta\varphi'(t) \neq 0. \tag{2}$$

This condition means that the diffusion is nonlinear; the medium is inhomogeneous; for $\varphi(t) > 0$ there is an absorption, and for $\varphi(t) < 0$ there is an external source.

The model (1) under condition (2) will be studied by methods of group analysis [4, 5].

2. Group classification

We fulfilled group classification of equation (1). We solved the problem of the group classification of this equation using the algorithm proposed in [4–6]. This algorithm has been successfully used in [1–3, 7–10] for group classification of the various equations of mechanics and mathematical physics. The results of the group classification are as follows [3]:

The kernel of the main groups of the equation (4) consists only of the identity transformation of the space $R^3(t, x, u)$.

For each pair of non-zero model parameters (α, β) and

$$\varphi(t) = -\frac{1}{\beta}\left(\ln\left(\psi'(t)\right)\right)', \quad \psi'(t) > 0, \quad \left(\ln\left(\psi'(t)\right)\right)'' \neq 0 \tag{3}$$

the main group of the equation (1) is generated by the operators

$$X_1 = x\partial_x + \frac{2-\alpha}{\beta} u\partial_u, \quad X_2 = \beta\frac{\psi}{\psi'}\partial_t - \left(\frac{\psi}{\psi'}\right)' u\partial_u, \quad X_3 = \frac{\beta}{\psi'}\partial_t - \left(\frac{1}{\psi'}\right)' u\partial_u. \tag{4}$$

Note. *The function $\psi(t)$ is expressed via the function $\varphi(t)$, by the formula*

$$\psi(t) = k_1 \int \exp\left(\beta \int \varphi(t) dt\right) dt + k_2,$$

where k_1 and k_2 are real constants.

Let $\psi(t)$ be determined by the formula (3). Then:

for

$$\alpha = 1, \quad \beta = -1 \tag{5}$$

the main group of the equation (1) is generated by the operators (4) and operator

$$X_4 = x(\ln x - 2)\partial_x - u \ln x \partial_u$$

for

$$\alpha = \frac{3\beta+4}{2\beta+3}, \quad \beta(\beta+1)\left(\beta+\frac{3}{2}\right)\left(\beta+\frac{4}{3}\right) \neq 0 \tag{6}$$

the main group of the equation (1) is generated by the operators (4) and operator

$$X_5 = x^{-\frac{\beta+1}{2\beta+3}}\left((2\beta+3)x\partial_x - u\partial_u\right).$$

Thus, the set of the models (1), having essentially different group properties consists of models (3) - (6). We give some invariant solutions for these models.

3. Invariant solutions

Exact solutions for each mathematical model are important. They allow us to assess the degree of the adequacy of the mathematical model of real physical processes, after carrying out experiments appropriate to these solutions, and an evaluation of the arising deviations. Exact solutions can be used to describe of some physical processes. Exact solutions are good tests to check the approximate numerical solutions.

For $(\beta+1)(2-\alpha+\beta)\neq 0$ invariant $\left\langle \frac{\beta(\beta+1)}{2-\alpha+\beta}X_1 + X_2,\ X_3\right\rangle$ -solution is determined by the formula

$$u = c_0 x^{\frac{1-\alpha}{\beta+1}}\left(\psi'(t)\right)^{\frac{1}{\beta}} \quad \left(c_0 = \text{const}\right). \tag{7}$$

When $\alpha = 1$, this solution describes the nonlinear diffusion process (or the heat propagation process) with a concentration (or a temperature) which is independent of the inhomogeneity of the medium at any given time.

For $(\alpha-2)(2-\alpha+\beta)\neq 0$ invariant $\langle X_1,\ X_2\rangle$–solution is determined by the formula

$$u = \left(\frac{\beta}{(\alpha-2)(2-\alpha+\beta)}\right)^{\frac{1}{\beta}} x^{\frac{2-\alpha}{\beta}}\left(\left(\ln\psi(t)\right)'\right)^{\frac{1}{\beta}}. \tag{8}$$

Let $\left(\frac{\beta}{(\alpha-2)(2-\alpha+\beta)}\right)^{\frac{1}{\beta}} > 0$. When $\frac{2-\alpha}{\beta} > 0$, this solution describes the nonlinear diffusion process (or the heat propagation process) with a concentration (or a temperature) $u\to\infty$ when $x\to\infty$. When $\frac{2-\alpha}{\beta} < 0$, this solution describes the nonlinear diffusion process (the heat propagation process) with a concentration (or a temperature) $u\to 0$ when $x\to\infty$.

Model (5) has the following invariant $\langle X_2,\ X_4\rangle$–solution which is determined by the formula

$$u = \frac{2\psi(t)}{x(\ln x - 2)^2\psi'(t)}.$$

Let $\psi(t) > 0$. This solution describes the nonlinear diffusion process (or the heat distribution process) in which a concentration (or a temperature) infinitely increases when $x\to +0$ or $x\to \mathrm{e}^2$. The points $x = 0$ and $x = \mathrm{e}^2$ are similar to the fixed "black holes", well known in astrophysics.

Model (5) has the following invariant $\langle X_1 + X_3,\ X_4 - X_2\rangle$–solution which is determined by the formula

$$u = -\frac{1}{x\left(\ln x - 2 + \psi(t)\right)\psi'(t)}.$$

This solution describes the nonlinear diffusion process (or the heat distribution process) in which a concentration (or a temperature) infinitely increases when $x \to +0$. The point $x = 0$ is similar to the fixed "black hole", well known in an astrophysics. In the vicinity of the trajectory $x = \mathrm{e}^{2-\psi(t)}$ the concentration (or the temperature) also infinitely increases. This trajectory describes the evolution of the "black hole".

Model (6) has the following invariant $\langle X_3 + X_5 \rangle$–solution which is determined by the formula

$$u = x^{-\frac{1}{2\beta+3}} \left(-\frac{\beta^2 (2\beta+3)^2}{(\beta+1)^3} \left(x^{\frac{\beta+1}{2\beta+3}} - \frac{\beta+1}{\beta} \psi(t) \right) \psi'(t) \right)^{\frac{1}{\beta}}.$$

This solution describes the nonlinear diffusion process (or the heat propagation process) with a concentration (or a temperature) $u = 0$ along a trajectory $\psi(t) = \dfrac{\beta}{\beta+1} x^{\frac{\beta+1}{2\beta+3}}$.

Model (6) has the following invariant $\left\langle \dfrac{\beta(2\beta+3)}{\beta+1} X_1 + X_2,\ X_3 + X_5 \right\rangle$–solution which is determined by the formula

$$u = x^{-\frac{1}{2\beta+3}} \left(\psi'(t)\right)^{\frac{1}{\beta}} V(\xi),\quad \xi = x^{\frac{\beta+1}{2\beta+3}} - \frac{\beta+1}{\beta}\psi(t),$$

where the function $V(\xi)$ is implicitly determined by a quadrature

$$\beta \int \left(c_1 - (2\beta+3)^2 V^{\frac{1}{\beta+1}} \right)^{-\frac{1}{2}} \mathrm{d}V = \xi + c_2 \quad \left(c_1,\ c_2 = \mathrm{const}\right). \tag{9}$$

For some values of parameter β the integral in the left-hand side of (9) is expressed in terms of elementary functions. This solution describes the nonlinear diffusion process (or the heat propagation process), for which the value $x^{\frac{1}{2\beta+3}} \left(\psi'(t)\right)^{-\frac{1}{\beta}} u(t,\ x)$ is constant along each trajectory $x = \left(\dfrac{\beta+1}{\beta} \psi(t) + c_3 \right)^{\frac{2\beta+3}{\beta+1}} \quad \left(c_3 = \mathrm{const}\right)$.

Invariant $\langle \gamma X_1 + X_2 \ \ (\gamma = const) \rangle$–solution has the form

$$u = \left(\psi(t)\right)^{\frac{\gamma(2-\alpha)}{\beta^2}} \left(\left(\ln \psi(t) \right)' \right)^{\frac{1}{\beta}} U(\xi),\quad \xi = x \left(\psi(t)\right)^{-\frac{\gamma}{\beta}}. \tag{10}$$

Substitution (8) into (1) gives a factor-equation

$$\xi^{\alpha} U^{\beta} U'' + \beta \xi^{\alpha} U^{\beta-1} U'^2 + \left(\alpha \xi^{\alpha-1} U^{\beta} + \frac{\gamma}{\beta} \xi \right) U' + \frac{\gamma(\alpha-2)+\beta}{\beta^2} U = 0. \tag{11}$$

The equation (11) is reduced to the following integral equations:

if $(\alpha-1)(\beta+1) \neq 0,\ U = V^{\frac{1}{\beta+1}}(\xi)$, then

$$V(\xi)=c_4+c_5\xi^{1-\alpha}+ \\ +\frac{(\beta+1)}{\beta^2(1-\alpha)}\int_{\xi_0}^{\xi}\left(\left(\gamma(\alpha-2)(\beta+1)+\beta\right)\tau^{1-\alpha}-\left(\gamma(\alpha-2-\beta)+\beta\right)\xi^{1-\alpha}\right)V^{\frac{1}{\beta+1}}(\tau)\,\mathrm{d}\tau \tag{12}$$

if $\alpha=1,\ \beta\neq-1,\ U=V^{\frac{1}{\beta+1}}(\xi)$, then

$$V(\xi)=c_6+c_7\ln\xi+\frac{(\beta+1)}{\beta^2}\int_{\xi_0}^{\xi}\left(\left(\beta-\gamma(\beta+1)\right)\ln\frac{\tau}{\xi}-\beta\gamma\right)V^{\frac{1}{\beta+1}}(\tau)\,\mathrm{d}\tau \tag{13}$$

if $\alpha\neq1,\ \beta=-1,\ U=\mathrm{e}^{V(\xi)}$, then

$$V(\xi)=c_8+c_9\xi^{1-\alpha}+\frac{1}{(1-\alpha)}\int_{\xi_0}^{\xi}\left(\left(\gamma(1-\alpha)+1\right)\xi^{1-\alpha}-\tau^{1-\alpha}\right)\mathrm{e}^{V(\tau)}\,\mathrm{d}\tau \tag{14}$$

if $\alpha=1,\ \beta=-1,\ U=\mathrm{e}^{V(\xi)}$, then

$$V(\xi)=c_{10}+c_{11}\ln\xi+\int_{\xi_0}^{\xi}\left(\ln\frac{\xi}{\tau}+\gamma\right)\mathrm{e}^{V(\tau)}\,\mathrm{d}\tau, \tag{15}$$

where c_k $(k=4, 5, \ldots 11)$, $t_0>0$, $x_0>0$, $\psi_0>0$ are arbitrary real constants, $\xi_0=x_0\psi_0^{-\frac{\gamma}{\beta}}$, $\psi_0=\psi(t_0)$.

The function $V(\xi)=c_0^{\beta+1}\xi^{1-\alpha}$ is a particular solution of the nonlinear integral equation (12) for

$$(2-\alpha+\beta)(3+2\beta-2\alpha-\alpha\beta)\neq0,\ c_4=\frac{c_0(\beta+1)\left(\gamma(\alpha-2)(\beta+1)+\beta\right)}{\beta^2(1-\alpha)(2-\alpha+k)}\xi_0^{2-\alpha+k},$$

$$c_5=c_0^{\beta+1}-\frac{c_0(\beta+1)^2\left(\gamma(\alpha-2-\beta)+\beta\right)}{\beta^2(1-\alpha)(2-\alpha+\beta)},\ k=\frac{1-\alpha}{\beta+1},\ \gamma=\frac{\beta(\beta+1)}{2-\alpha+\beta}.$$

Substitution of this function into (10) gives a solution (7).

The function $V(\xi)=\left(\frac{-\beta\xi}{\beta+1}\right)^{\frac{\beta+1}{\beta}}$ is a particular solution of the nonlinear integral equation (13) for

$$c_6=\left(\frac{-\beta}{\beta+1}\right)^{\frac{1}{\beta}}\xi_0^{\frac{\beta+1}{\beta}}\left(\frac{\left(\beta-\gamma(\beta+1)\right)}{\beta}\left(\ln\xi_0-\frac{\beta+1}{\beta}\right)-\gamma\right),\ c_7=\left(\frac{-\beta}{\beta+1}\right)^{\frac{1}{\beta}}\frac{\left(\gamma(\beta+1)-\beta\right)}{\beta}\xi_0^{\frac{\beta+1}{\beta}}.$$

Substitution of this function into (10) gives a solution (8) for $\alpha=1$.

The function $V(\xi)=\left((\alpha-1)(\alpha-2)\xi^{\alpha-2}\right)^{\beta+1}$ is a particular solution of the nonlinear integral equation (14) for

$$c_8=\ln\left((\alpha-1)(\alpha-2)\right)+\frac{(\alpha-1)(\alpha-2)\left(\gamma(1-\alpha)+1\right)}{(1-\alpha)^2}-\frac{(\alpha-1)(\alpha-2)}{(1-\alpha)}\ln\xi_0,$$

$$c_9=-\frac{(\alpha-1)(\alpha-2)\left(\gamma(1-\alpha)+1\right)}{(1-\alpha)^2}\xi_0^{\alpha-1}.$$

Substitution of this function into (10) gives a solution (8) for $\beta=-1$.

The function $V(\xi)=c_{12}^{\beta+1}\xi^{-\frac{(\gamma+1)(\beta+1)}{\gamma}}$ $(\gamma\neq 0,\ c_{12}=\text{const}>0)$ is a particular solution of the nonlinear integral equation (15) for

$$c_{10}=\ln c_{12}-\gamma c_{12}\xi_0^{-\frac{1}{\gamma}}\left(\gamma-\ln\xi_0+1\right),\ c_{11}=-\gamma c_{12}\xi_0^{-\frac{1}{\gamma}}-\frac{\gamma+1}{\gamma}.$$

Substitution of this function into (10) gives a solution $u=\frac{c_{12}}{\psi'(t)}x^{-\frac{\gamma+1}{\gamma}}$ of the equation (1).

The presence of the arbitrary constants in the integral equations (12) - (15), allows one to use the obtained solutions for solving of boundary value problems. In particular, these solutions describe a nonlinear diffusion process (or a heat propagation process) for which a concentration (or a temperature) and its gradient are specified at an initial time $t=t_0\geq 0$ at a fixed point $x=x_0>0$:

$$u(t_0,x_0)=u_0>0,\ \frac{\partial u}{\partial x}(t_0,x_0)=u_1. \qquad (16)$$

In this case the constants c_k $(k=4,\ 5,...,11)$ are determined by the formulas $(\psi_1=\psi'(t_0))$:

$$c_4=\left(u_0\psi_0^{\frac{\gamma(\alpha-2)+\beta}{\beta^2}}\psi_1^{-\frac{1}{\beta}}\right)^{\beta+1}-c_5\left(x_0\psi_0^{-\frac{\gamma}{\beta}}\right)^{1-\alpha},$$

$$c_5=\frac{\beta+1}{\beta(1-\alpha)}\psi_0^{\frac{\beta-\gamma(\beta+2-\alpha)}{\beta^2}}\psi_1^{-\frac{\beta+1}{\beta}}\left(\gamma x_0u_0\psi_1+\beta x_0^{\alpha}u_0^{\beta}u_1\psi_0\right),$$

$$c_6=\left(u_0\psi_0^{\frac{\beta-\gamma}{\beta^2}}\psi_1^{-\frac{1}{\beta}}\right)^{\beta+1}-c_7\left(\ln x_0-\frac{\gamma}{\beta}\ln\psi_0\right),$$

$$c_7=\frac{\beta+1}{\beta}\psi_0^{\frac{\beta-\gamma(\beta+1)}{\beta^2}}\psi_1^{-\frac{\beta+1}{\beta}}x_0\left(\gamma u_0\psi_1+\beta u_0^{\beta}u_1\psi_0\right),$$

$$c_8=\ln u_0+\left(\gamma(\alpha-2)-1\right)\ln\psi_0+\ln\psi_1-c_9\left(x_0\psi_0^{\gamma}\right)^{1-\alpha},$$

$$c_9=\frac{1}{(1-\alpha)}\psi_0^{\gamma(\alpha-1)}\left(-\gamma x_0u_0\psi_0^{-1}\psi_1+x_0^{\alpha}u_0^{-1}u_1\right),$$

$$c_{10}=\ln u_0-(\gamma+1)\ln\psi_0+\ln\psi_1-c_{11}\left(\ln x_0+\gamma\ln\psi_0\right),$$

$$c_{11}=x_0\left(-\gamma u_0\psi_0^{-1}\psi_1+u_0^{-1}u_1\right). \qquad (17)$$

The unique solution of the equation (1), satisfying to the conditions (16), for which the value $(\psi(t))^{\frac{\gamma(\alpha-2)+\beta}{\beta^2}}\left((\ln\psi(t))'\right)^{-\frac{1}{\beta}}u(t,x)$ is constant along an each trajectory $x=c(\psi(t))^{\frac{\gamma}{\beta}}$ $(c=\text{const})$ exists at the neighborhood of the point $(t_0,\ x_0)$. This solution is given by the formulas (10), (12) - (15), (17).

4. Conclusion and discussion

We performed the full group classification of the nonlinear differential equation of the model, describing the nonlinear diffusion process (or the heat propagation process) in an inhomogeneous medium with non-stationary absorption or source. We found all values of the diffusion and absorption or source coefficients, for which the equation admits a continuous Lie group of the transformations, acting on the set of the solutions of this equation. We researched invariant submodels of the obtained models. Some invariant solutions, describing these submodels are found either explicitly, or their search is reduced to solving of the nonlinear integral equations. For example, we obtained the invariant solution describing the nonlinear diffusion process (or the heat propagation process) with two fixed "black holes", and the invariant solution describing the nonlinear diffusion process (or the heat distribution process) with the fixed "black hole" and the moving "black hole". The solutions with "black holes" can be used to describe the process of destruction of glaciers under the influence of the external heat source, for example a solar energy. Namely, along each ray in the glacier, under the action of the external heat source, the temperature at two points is much higher than the temperature at other points of the ray. Thus, within the glacier, the most strongly heated zones are formed in which melting of ice occurs. Over time, these zones are combined, which leads to the formation of cracks and the destruction of the glacier. In particular, this way it is possible to explain the occurrence of icebergs off the coast of Antarctic continent. With a help of the other invariant solution we have studied a diffusion process (or heat distribution process) for which at the initial moment of the time at a fixed point a concentration (a temperature) and its gradient are specified. Solving of boundary value problem describing this process is reduced to the solving of nonlinear integral equations. We are established the existence and uniqueness of solutions of these boundary value problem under some conditions.

Acknowledgments

The reported study was funded by RFBR according to the research project No. 16–01–00446 a.

References

[1] Chirkunov Yu A 2015 Submodels of the generalization of the Leith's model of the phenomenological theory of turbulence and of the model of nonlinear diffusion in the inhomogeneous media without absorption *J. Phys. A: Math. Theor.* **48(39)** 395501

[2] Chirkunov Yu A 2015 Submodels of model of nonlinear diffusion in the inhomogeneous medium involving absorption *J. Math. Phys.* **56(10)** 101502

[3] Chirkunov Yu A 2017 Submodels of model of nonlinear diffusion with non-stationary absorption *Int. J. Non–Linear Mech.* **91** 86–94

[4] Chirkunov Yu A 2007 *Group Analysis of Linear and Quasi–Linear Differential Equations* (Novosibirsk: NSUEM) (in Russian)

[5] Chirkunov Yu A and Khabirov S V 2012 *Elements of Symmetry Analysis of Differential Equations of Continuum Mechanics* (Novosibirsk: NSTU) (in Russian)

[6] Chirkunov Yu A 2012 Generalized equivalence transformations and group classification of systems of differential equations *J. Appl. Mech. and Techn. Phys.* **53(2)** 147–55

[7] Chirkunov Yu A 2015 Submodels of model of nonlinear diffusion in the inhomogeneous medium involving absorption *J. Math. Phys.* **56** 101502

[8] Chirkunov Yu A 2015 Nonlinear longitudinal oscillations of viscoelastic rod in Kelvin model *J. Appl. Math. and Mech.* **79** 506–13

[9] Chirkunov Yu A 2016 Invariant submodels of the Westervelt model with dissipation *Int. J. Non-Linear Mech.* **84** 139–44

[10] Chirkunov Yu A 2016 Invariant submodels of the Westervelt model of nonlinear hydroacoustics without dissipation *Int. J. Non-Linear Mech.* **85** 41–53

The ice response to an oscillating load moving along a frozen channel

K A Shishmarev[1], T I Khabakhpasheva[2] and A A Korobkin[3]
[1]Altai State University, Lenina avenue 61, Barnaul 656049, Russia
[2]Lavrentyev Institute of Hydrodynamics of SB RAS, Lavrentyeva avenue 15, Novosibirsk 630090, Russia
[3]University of East Anglia, Norwich Research Park, Norwich NR4 7TJ, UK
E-mail: shishmarev.k@mail.ru

Abstract. Unsteady response of an ice cover to an oscillating load moving along a frozen rectangular channel is studied for large times. The channel is filled with ideal incompressible fluid. The ice cover is modelled by a thin elastic plate. The flow caused by the deflection of the ice cover is potential. The problem is formulated within the linear theory of hydroelasticity. External load is modelled by a smooth localized pressure distribution. The load has periodic magnitude and moves along the channel with constant speed. Joint system of equations for the ice plate and the flow potential is closed by initial and boundary conditions: the ice plate is frozen to the walls of the channel, the flow velocity potential satisfies the impermeability condition at the rigid walls of the channel and linearized kinematic and dynamic conditions at the ice-liquid interface; at the initial time the load is stationary, the fluid in the channel is at rest and the stationary ice deflection is determined from the plate equation for the initial magnitude of the load. The problem is solved with the help of the Fourier transform along the channel. The ice deflection profile across the channel is sought in the form of the series of the eigenmodes of the ice cover oscillations in a channel. The solution of the problem is obtained in quadratures and consists of three parts: (1) symmetric with respect to the load deflection corresponding to the stationary load; (2) deflection corresponding to steady waves propagating at the load speed; (3) deflection corresponding to waves propagating from the load and caused by the oscillations of the load. The number of the last waves, depending on the parameters, can not exceed four for each eigenmode. In this article the results of the analytical and numerical analysis of the considered problem is presented.

1. Introduction

The problems of loads moving along an unbounded ice cover without vertical walls are well studied (see, for example, an excellent review in [1]). The problems are usually solved within the theory of linear hydroelasticity. It is known that the ice deflection is strongly dependent on the speed of the load. If the load speed is below a certain critical value, the ice deflection is localized near the load and quickly decays with the distance from the load. For higher speeds of the load, outgoing waves are formed in the far field if viscous damping in ice is not included in the mathematical model. At the critical speed, the linear theory without damping predicts unbounded ice response. To obtain the estimates of the ice response for the critical speed of the load either nonlinear effects [2] or viscous damping [3] or both are included in the ice model. The presence of the channel walls complicates the problem. There are an infinite number of critical speeds for a frozen channel, in contrast to the ice sheet of infinite extent, for which there are two simple critical values of the load speed. Each critical speed for an ice cover in a channel corresponds to a mode of propagating along the channel and sloshing across the channel waves studied by Korobkin et al. [4]. For a speed of the load different from the critical values, the linear theory of hydroelasticity can be used to estimate the strains in the ice cover. We restrict ourselves to a constant and different from the critical values speed of the load and the strain/deflection distributions which are stationary in the coordinate system moving together with the load. These stationary distributions can be obtained directly within ice models with viscous damping, where ice deflection decays exponentially with distance from the moving load, see [5]. Decreasing viscous damping, we obtain higher strains in the ice cover and longer region of significant deflections along the channel. Both deflections and strains are given by infinite series and integrals. Numerical integration becomes challenging for small damping. To estimate the strains, we need them for zero damping, where the approach from [5] does not work.

Another approach to stationary strain/deflection distributions caused by a load moving along the channel for zero damping in the ice cover [6] is employed in the present study. In this approach, the stationary distributions are approached in time t starting from the initial rest state. At $t = 0$ the load, which is modelled by an external pressure acting on the ice plate, is at rest. The initial ice deflection satisfies the stationary equation of thin elastic plate with proper boundary conditions on the walls of the channel. The pressure distribution generated by the load is symmetric with respect to the central line of the channel. Then the load starts to move at a constant speed U along the channel and the load magnitude starts to oscillate with the given constant frequency ω'. The ice deflection decays far ahead and far behind the moving load as $|x - Ut| \to \infty$ at any finite time t. The unsteady problem of hydroelasticity is solved with the help of the Fourier transform along the channel and the normal-mode method [7] for the ice deflection. A second-order differential equations in time for the principal coordinates of the normal modes are derived and solved analytically. As a result, the ice deflection is presented by an infinite series of regular Fourier integrals. The integrals are evaluated for large times by asymptotic methods, see [6]. The asymptotic behaviours of the integrals depend on the speed of the load with respect to the critical speeds of the propagating-sloshing wave modes. It is shown that for large times the ice deflection consists of symmetric deflection localized near the load and a system of waves in front and behind the load. The number of these waves is obtained and the wave amplitudes are evaluated numerically.

This study is motivated by experiments in ice tanks, operations on ice in rivers and channels such as cargo transportation or ice breaking to avoid flooding, and ice-structure interaction. The strains calculated with zero damping are higher than the real ones. For safe transportation on ice, one needs to compare the computed strains with a strain critical value and determine safe speed of transportation. Knowing amplitudes of the waves generated by the moving load, we can find places and estimate the values of the highest strains far ahead and behind the load. It is expected that there are such speeds of the load that the maximum strains behind and or in front of the load are achieved at the walls of the channel. On the other hand, oscillations of the load can create extra waves in the ice cover. These waves can increase maximum stresses in the ice cover, that leads to the destruction of the ice cover in the non-resonant cases for a load with a constant magnitude. Linear hydroelastic problem for steady forced vibrations of a semi-infinite ice cover under the effect of localized external load was studied in [8]. The problem was solved within the shallow water theory. It was shown that, in the case of a free edge for the ice plate, there are edge waves. The edge waves are localized near the edge of the plate and exponentially attenuate as they move away from the edge. In this article, in addition to the movement of the load, periodic oscillation of its magnitude and the effects of the oscillations on the formation of the ice deflections are studied.

IOP Conf. Series: Earth and Environmental Science **193** (2018) 012072 doi:10.1088/1755-1315/193/1/012072

2. Formulation of the problem

The unsteady hydroelastic waves caused by an oscillating load moving along a frozen channel are considered. The channel has a rectangular cross section with finite depth H, $-H < z < 0$, and width $2L$, $-L < y < L$. The channel is infinitely long, $-\infty < x < \infty$. The channel is filled with ideal incompressible fluid of density ρ_l. The flow caused by the ice deflection is potential. The thickness of the ice cover h_i, the rigidity of the ice D and the ice density ρ_i are constant. The vertical displacement of the ice sheet, $w(x, y, t)$, satisfies the equation of thin elastic plate with clamped boundary conditions at $y = \pm L$

$$Mw_{tt} + D\nabla_2^4 w = p(x, y, 0, t) - P(x - Ut, y), \qquad (-\infty < x < \infty,\ -L < y < L,\ z = 0), \quad (1)$$

$$w = 0, \quad w_y = 0 \qquad (-\infty < x < \infty, y = \pm L), \quad (2)$$

where $\nabla_2^4 = \nabla_2^2 \cdot \nabla_2^2 = \partial^4/\partial x^4 + 2\,\partial^4/(\partial x^2 \partial y^2) + \partial^4/\partial x^4$, $M = h_i \rho_i$ is the mass of the ice plate per unit area, $p(x, y, 0, t)$ is the hydrodynamic pressure acting on the lower surface of the ice plate and $P(x, y, t)$ is the external pressure distribution on the upper surface of the ice plate caused by the moving load. The flow velocity potential $\varphi(x, y, z, t)$ satisfies the Laplace equation, impermeability conditions at the rigid walls, and the kinematic and dynamic boundary conditions at the ice-liquid interface

$$\nabla^2 \varphi(z, y, z, t) = 0 \quad (-\infty < x < \infty,\ -L < y < L,\ -H < z < 0), \quad (3)$$

$$\varphi_y = 0 \quad (y = \pm L), \qquad \varphi_z = 0 \quad (z = -H), \qquad \varphi_z = w_t \quad (z = 0). \quad (4)$$

$$p(x, y, 0, t) = -\rho_l \varphi_t(x, y, 0, t) - \rho_l g\, w(x, y, t) \qquad (-\infty < x < \infty,\ -L < y < L,\ z = 0), \quad (5)$$

where g is the gravitational acceleration and $\varphi(x, y, z, t)$ is the velocity potential of the flow in the channel caused by the moving load.

The load is modelled by a localized smooth pressure distribution, $P(x, y, t)$, moving along a centre line of the channel at constant speed U with periodic magnitude, $P_0/2 \le P \le 3P_0/2$

$$P(x, y, t) = P(x - Ut, y, t) = P_0 P_1(X)\, P_2(y)\, (1 + \cos(\omega' t)/2) \quad (-\infty < x < \infty,\ -L < y < L), \quad (6)$$

$$P_1(X) = (\cos(\pi c_1 X) + 1)/2 \ \ (c_1|X| < L), \quad P_1(X) = 0 \ \ (c_1|X| \ge L),$$

$$P_2(y) = (\cos(\pi c_2 y) + 1)/2 \ \ (c_2|y| < L), \quad P_2(y) = 0 \ \ (c_2|y| \ge L),$$

where $X = x - Ut$, P_0 is the magnitude of the stationary load, c_1, c_2 – nondimensional parameters characterising area of the load and ω' is the oscillating frequency.

At the initial moment the flow is at rest and the ice deflection $w(x, y, t)$ satisfies initial conditions

$$w(x, y, 0) = w_0(x, y), \qquad w_t(x, y, 0) = 0. \quad (7)$$

The last condition is obtained from the kinematic condition (4) for $t = 0$. The RHS of the first condition in (7) is non zero. The equation of the plate (1) with $t = 0$ provides

$$D\nabla_2^4 w_0(x, y) = -\rho_l g\, w_0(x, y) - P(x, y, 0). \quad (8)$$

The solution of the problem (1) – (8) depends on the density of the liquid, ρ_l, parameters of the ice, ρ_i, h_i, D, parameters of the channel, H, L, parameters of the load, U, c_1, c_2, ω' and initial data. The problem and the solution are time-dependent. We shall find large time behaviour of the ice deflection, $w(x, y, t)$, and the effect of the load oscillations (term $\cos(\omega' t)$) on the ice deflections for some given values of the parameters of the problem. The solution will be given by the asymptotic formulas for amplitudes of the hydroelastic waves and consisted of three parts:

(1) symmetric ice response with respect to the load corresponding to the stationary load and (2) deflection corresponding to waves propagating at the load speed, both are established for $t \to \infty$; and (3) deflection corresponding to waves propagating from the load and caused by the oscillations of the load. The last waves propagate from the load with encounter frequency and their phase depend on time. It will be shown that number of these waves is limited. Obtained formulas provide largest possible strains in the ice cover for the given load. If one wants find the ice deflection for given value of time then other methods of the solution should be used. Asymptotic analysis is presented for the rectangular channel and the given form of the load, but can be extended to non-rectangular channels and different forms of the load.

Steady state solution of similar problem of the moving load on ice was studied in [5]. The problem is stationary and was formulated for a Kelvin-Voigt viscoelastic plate with non-zero retardation time, $\tau > 0$ s. The steady state solution of the problem, $w(x.y,t) = w(X,y) = w(x-Ut,y)$, was investigated in the coordinate system moving with the load. The solution did not depend on time and that approach did not require initial data. The problem was solved numerically. Due to the viscous effect of the plate all numerical results showed quick decay of the ice deflection away from the load. The effects of the channel walls, the plate thickness and speed of the load on the solution were studied. We should note that determination of the value of the retardation time is complicated and not established yet. In this article we shall find the asymptotic behaviour of the ice deflection for an elastic plate with no viscous effects in the nonstationary problem (1) – (8). Models of ice response without damping are less physical, however, they may provide helpful estimates of maximum strains and bearing capacity of the ice cover. Such estimates are known for loads moving on the ice cover of infinite extent.

3. Method of the solution

The considered initial-boundary value problem is solved with the help of the Fourier transform along the channel

$$w(X,y,t) = \frac{1}{\sqrt{2\pi}} \int_{-\infty}^{\infty} w^F(\xi,y,t)e^{-i\xi x}d\xi, \qquad w^F(\xi,y,t) = \frac{1}{\sqrt{2\pi}} \int_{-\infty}^{\infty} w(x,y,t)e^{-i\xi x}dx \tag{9}$$

where ξ is the parameter of the transform. The equation (1) provides

$$Mw^F_{tt} + D\left(w^F_{yyyy} - 2\xi^2 w^F_{yy} + \xi^4 w^F\right) = -\rho_l \varphi^F_t - \rho_l g w^F - P^F(\xi,y)e^{-i\xi Ut}(1+\cos(\omega' t)/2), \tag{10}$$

where $\varphi^F(\xi,y,z,t)$ and $P^F(\xi,y)e^{-i\xi Ut}$ are corresponding Fourier images of $\varphi(x,y,z,t)$ and $P_o P_1(X)P_2(y)$. Then the Fourier image of the ice deflection, $w^F(\xi,y,t)$, is sought in the form

$$w^F(\xi,y,t) = w^a + w^b = \sum_{n=1}^{\infty} \left[a_n(\xi,t)\psi_n(\xi,y) + b_n(\xi,t)\psi_n(\xi,y)\right]. \tag{11}$$

Here $\psi_n(\xi,y)$ are the eigenmodes of the ice cover oscillations in the channel, and $a_n(\xi,t)$ and $b_n(\xi,t)$ are the principle coordinates of the modes. The functions $\psi_n(k,y)$ were calculated in [4]. They describe the hydroelastic waves propagating along the channel

$$w(x,y,t) = \mathrm{Re}\left[A^n \psi_n(k,y)e^{i(kx-\omega_n t)}\right], \tag{12}$$

where $\omega_n(k)$ is the wave frequency, $k > 0$ is the wave number and A^n is the wave amplitude. The function $w^a = \sum_{n=1}^{\infty} a_n(\xi,t)\psi_n(\xi,y)$ in (11) describes established at large times hydroelastic waves caused by the load with constant magnitude

$$Mw^a_{tt} + D\left(w^a_{yyyy} - 2\xi^2 w^a_{yy} + \xi^4 w^a\right) + \rho_l g w^a = -\rho_l \varphi^a_t - P^F(\xi,y)e^{-i\xi Ut}, \tag{13}$$

and function w^b describes unsteady hydroelastic waves corresponding to the effect of the load oscillations

$$Mw_{tt}^b + D\left(w_{yyyy}^b - 2\xi^2 w_{yy}^b + \xi^4 w^b\right) + \rho_l g w^b = -\rho_l \varphi_t^b - \frac{P^F(\xi,y)e^{-i\xi Ut}\cos(\omega' t)}{2}, \quad (14)$$

where $\varphi = \varphi^a + \varphi^b$.

4. Deflection corresponding to the steady waves propagating at the load speed

By the method of separating variables in the equation (13) we arrive at the following differential equations for the principle coordinates $a_n(\xi,t)$,

$$\frac{d^2 a_n}{dt^2} + \omega_n^2(\xi)a_n = H_n(\xi)e^{-iU\xi t}, \quad H_n(\xi) = \frac{\int_{-L}^{L} P^F(\xi,y)\psi_n(\xi,y)dy}{\rho_l L \int_{-L}^{L} (\alpha\psi_n(\xi,y) + \varphi_n(\xi,y,0))\psi_n\, dy},$$

where $\alpha = (\rho_i h_i/\rho_l L)$ and $\varphi_n(\xi,y,z)$ is the flow potential corresponding to the mode $\psi_n(\xi,y)$. This equation is solved analytically subject to the initial conditions.

The ice deflection $w^a(x,y,t)$ is given by the inverse Fourier transform applied to (11), where

$$\int_{-\infty}^{\infty} a_n(\xi,t)\psi_n(\xi,y)e^{i\xi x}d\xi = \int_{-\infty}^{0} a_n(\xi,t)\psi_n(\xi,y)e^{i\xi x}d\xi + \int_{0}^{\infty} a_n(\xi,t)\psi_n(\xi,y)e^{i\xi x}d\xi =$$

$$= \int_0^{\infty} H_n\psi_n\left(\frac{\xi U}{2\omega_n^2}\left[\frac{e^{i(\xi x+\omega_n t)}}{\omega_n+\xi U} - \frac{e^{i(\xi x-\omega_n t)}}{\omega_n-\xi U} + \frac{e^{-i(\xi x+\omega_n t)}}{\omega_n+\xi U} - \frac{e^{-i(\xi x-\omega_n t)}}{\omega_n-\xi U}\right] + \frac{\cos(\xi(x-Ut))}{2(\omega_n^2-\xi^2U^2)}\right)d\xi. \quad (15)$$

In the numerical calculations, the series (11) for $w^a(\xi,y,t)$ is truncated to N_{mod} terms. Using moving coordinate system (X,y,z) we find asymptotic behaviour of each integral in (15) for each mode number n for large times, $t\to\infty$ and $X = O(1)$. Depending on the load speed U some integrals in (15) have poles at the points where $\omega_n(\xi) = \xi U$. Note that the integral (15) is regular and can be written as the sum of five integrals which are understood as Cauchy principal value integrals, in general. If an integral does not have a pole then it is regular and its contribution is of order $O(1/t)$ as $t\to\infty$. The large-time contributions of non-regular integrals consist of two parts: *(a)* a part which is even in X and decaying as $|X|\to\infty$, *(b)* waves (12) with amplitudes A^n propagating behind and in front of the load. In these waves, $k=\xi_m^n$, $m=1,2$, where ξ_m^n are solutions of the equation $\omega_n(\xi)=\xi U$, $\xi_1^n<\xi_2^n$, and the amplitudes A_m^n are given by

$$A_m^n = \frac{2\pi H_n(\xi_m^n)}{\xi_m U(c_g^n(\xi_m^n)-U)},$$

where c_g^n is the group velocity of the n-th mode of hydroelastic wave in the channel. The phase speeds of these waves is equal to the speed of the load, U. These waves do not propagate with respect to the load in the moving coordinate system. The long waves with ξ_1^n are placed behind the load, and the short waves with ξ_2^n are in front of the load.

5. Deflection corresponding to the unsteady waves propagating from the load

By the method of separating variables in the equation (14) we arrive at the following differential equations for the principle coordinates $b_n(\xi,t)$,

$$\frac{d^2 b_n}{dt^2} + \omega_n^2(\xi)b_n = H_n(\xi)\left(\frac{e^{i\omega' t-i\xi Ut}+e^{-i\omega' t-i\xi Ut}}{2}\right),$$

This equation is solved analytically subject to the initial conditions.

The ice deflection $w^b(x,y,t)$ is given by the inverse Fourier transform applied to (11), where

$$\int_{-\infty}^{\infty} b_n(\xi,t)\psi_n(\xi,y)e^{i\xi x}d\xi = \int_{-\infty}^{0} b_n(\xi,t)\psi_n(\xi,y)e^{i\xi x}d\xi + \int_{0}^{\infty} b_n(\xi,t)\psi_n(\xi,y)e^{i\xi x}d\xi =$$

$$= \int_0^{\infty} H_n\psi_n \left\{ \frac{1}{4\omega_n^2}\left(\left[\frac{\xi U+\omega'}{\omega_n+(\xi U+w')}+\frac{\xi U-\omega'}{\omega_n+(\xi U-\omega')}\right]e^{i(\xi x+\omega_n t)}- \right.\right.$$

$$-\left[\frac{\xi U-\omega'}{\omega_n-(\xi U-w')}+\frac{\xi U+\omega'}{\omega_n-(\xi U+\omega')}\right]e^{i(\xi x-\omega_n t)}+\left[\frac{\xi U+\omega'}{\omega_n+(\xi U+w')}+\frac{\xi U-\omega'}{\omega_n+(\xi U-\omega')}\right]e^{-i(\xi x+\omega_n t)}-$$

$$\left.-\left[\frac{\xi U-\omega'}{\omega_n-(\xi U-w')}+\frac{\xi U+\omega'}{\omega_n-(\xi U+\omega')}\right]e^{-i(\xi x-\omega_n t)}\right)+$$

$$\left.\frac{e^{i(\xi x-\xi Ut)+i\omega' t}}{2(\omega_n^2-(\xi U-\omega')^2)}+\frac{e^{i(\xi x-\xi Ut)-i\omega' t}}{2(\omega_n^2-(\xi U+\omega')^2)}+\frac{e^{-i(\xi x-\xi Ut)+i\omega' t}}{2(\omega_n^2-(\xi U+\omega')^2)}+\frac{e^{-i(\xi x-\xi Ut)-i\omega' t}}{2(\omega_n^2-(\xi U-\omega')^2)}\right\}d\xi. \tag{16}$$

The last four terms in the equation (16) describe unsteady hydroelastic waves propagating from the load with the wave number ξ and frequency ω'. The numerators of these terms can be written in the form $\cos(\xi X+\omega' t)$ and $\cos(\xi X-\omega' t)$.

Asymptotic analysis, done for the steady waves, can be repeated in this case. First, note that the integrals in the equation (16) can be singular in four cases: $\omega_n=-(\xi U+\omega')$, $\omega_n=-(\xi U-\omega')$, $\omega_n=(\xi U-\omega')$ and $\omega_n=(\xi U+\omega')$. We omit the first case since it is not physical. Other cases give

$$c_n+c'=U, \qquad c_n-c'=U, \qquad c'-c_n=U, \tag{17}$$

where c' is the phase speed of the wave propagating along the channel from the load, $c'=\omega'/\xi$. If there are singularities at the point ξ_* in some terms in the equation for $b_n(\xi,t)$, then these singularities will always be in two terms, one in term with $1/(4\omega_n^2)$ and one in the four last terms in the equation (16). These singularities counter each other and the total integral for $b_n(\xi,t)$ is regular. Following the same asymptotic analysis, equation (17) gives that, in contrast to the steady waves, waves caused by the load oscillations occur where the speed of the load is equal to the phase speed of the hydrodynamic wave shifted by the phase speed c', which is a hyperbola for $\omega'\neq 0$. In total there can be no more than four such waves for each mode ψ_n. Figure 1 shows the curves c_n+c', c_n-c' and $c'-c_n$ as functions of the wave number for the first mode in the considered channel. Calculations are performed for the freshwater ice with $h_i=10$cm, $L=10$m, $H=5$m and $\omega'=1$ s^{-1}. The dotted line shows the example of the speed of the load, for which there will be maximum number of the unsteady waves propagating from the load for the given mode. For an example, let us calculate the wave parameters in the case of a singular point ξ_*, $\omega_n(\xi_*)=-\xi_*U+\omega'$. In which direction this wave runs, forward or backward from the load, depends on the sign of the derivative $\omega_n+\xi U-\omega'$. The sign is greater than zero, hence, in this case the wave is formed before the load. This wave has the form

$$w(x,y,t)=\mathrm{Re}\left[A'\psi_n(\xi_*,y)e^{i(\xi_* X+\omega' t)}\right], \tag{18}$$

where the amplitude A is given by

$$A'=\frac{\pi H_n(\xi_*)}{(\omega'-\xi_* U)(c_g^n(\xi_*)+U)},$$

6. Conclusion

The obtained exact decomposition of the ice deflection to the symmetric, long and short wave parts for large times makes it possible to estimate the maximum strains under the load, far behind and in front of the load. The analytical formulas are obtained for these waves. It is shown that effect of the oscillation of the magnitude of the load leads to the unsteady waves propagating from the load with speeds different from the speed of the load. Number of these waves is limited by four for each mode.

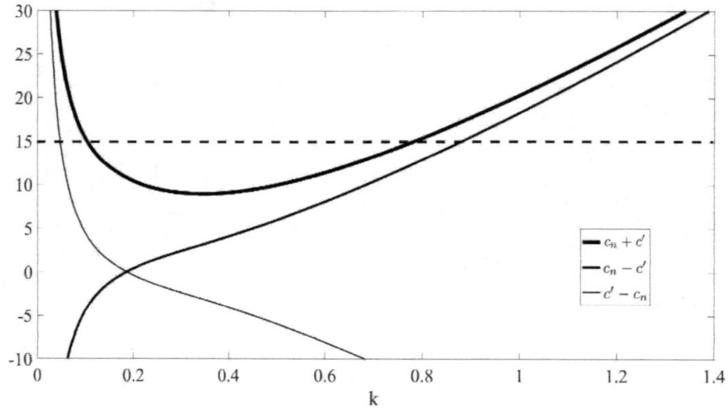

Figure 1. The curves $c_n + c' = U$, $c_n - c' = U$, and $c' - c_n = U$, where c_n is the phase speed of the first mode, $c' = \omega'/k$, $\omega' = 1\ \mathrm{s}^{-1}$. The dotted line shows the speed of the load.

Acknowledgments

This work was partially supported by the grant RFBR 16-08-00291 "Hydroelastic and thermodynamic effects with interaction of poroelastic ice and structures".

References

[1] Squire V, Hosking R, Kerr A and Langhorne P 1996 *Moving loads on ice* (Amsterdam: Kluwer Academic Publishers)
[2] Guyenne P and Parau E 2012 *Journal of Fluid Mechanics* **713** 307–29
[3] Hosking R J, Sneyd A D and Waugh D W 1988 *Journal of Fluid Mechanics* **196** 409–30
[4] Korobkin A, Khabakhpasheva T and Papin A 2014 *European Journal of Mechanics B/Fluids* **47** 166–75
[5] Shishmarev K, Khabakhpasheva T and Korobkin A 2016 *Applied Ocean Research* **59** 313–26
[6] Schulkes R and Sneyd A 1988) *Journal of Fluid Mechanics* **186** 25–46
[7] Khabakhpasheva T I 2006 *Fluid Dynamics* **41(3)** 424–33.
[8] Sturova I V 2017 *Journal of Applied Mechanics and Technical Physics* **58** 80–8

IOP Conf. Series: Earth and Environmental Science **193** (2018) 012073 doi:10.1088/1755-1315/193/1/012073

Exploration and production of oil and gas as a factor of economic development of the Arctic resource regions, resource curse

L V Eder[1,2], I V Provornaya[1,2], I V Filimonova[1,2] and A V Komarova[1,2]

[1] Center energy resources and economics, Trofimuk Institute of Petroleum Geology and Geophysics, Siberian Branch of Russian Academy of Sciences, 3 Prospekt Akademika Koptyuga, Novosibirsk 630090, Russia
[2] Novosibirsk State University, 1 Pirogova, Novosibirsk 630090, Russia

E-mail: Ederlv@yandex.ru

Abstract. The results of the cluster analysis showed that all the regions classified as the Arctic zone of Russia can be identified as resource regions. Identification was based on such factors as the role of the mining sectors in the region's economy, the share of mineral extraction tax and employment in the extractive industries, especially in oil and gas sector. Econometric analysis built on panel data showed that the production of energy resources in the Arctic regions directly affects the level of socio-economic development of the Arctic regions. At the same time, it is important to take into account that these regions are characterized by a single-industry type of economy, which does not allow for comprehensive innovative development of various sectors of the region's economy. As a result, the "resource curse" of the regions of the Arctic zone of Russia was revealed. This includes significant challenges and threats to the sustainable development of the Arctic regions.

1. Introduction

Traditionally, countries rich in natural resources, such as oil and gas, have been considered to be able to base their development on the use of these resources as a main path to sustainable economic growth. However, since the 1980s, a significant amount of literature has emerged that contradicts the traditional view that natural resources are a benefit to developing countries. Economists and sociologists have identified various cause-and-effect relationships through which the resource curse can function, and various performance variables that affect economic growth. Number of important studies has attempted to summarize and evaluate the literature on the "resource curse" including Frankel [1], Van der Ploeg [2], Ross [3], Deacon [4], Ramez [5].

The aim of the paper is to study the impact of the implementation of projects of exploration and production of hydrocarbons in the Arctic on the socio-economic development of resource regions, as well as their innovative potential.

In recent years, the Arctic regions have become increasingly important for the development of the oil and gas industry of Russia. There is an increase in investment at all stages of the exploration and production of hydrocarbon resources of this territory both onshore and offshore. The development of large-scale oil and gas production, transport and export projects has a significant impact on the socio-economic development of the Arctic regions of Russia. Ultimately, the extraction of hydrocarbon resources serves as a catalyst for the socio-economic development of the Arctic regions. However, the regions are significantly lagging behind in the innovation development from other regions of Russia, that, in general, adversely affects to the sustainable development of the resource regions of the Arctic.

2. Methods of research

In the previous works [6-9] in the enclave cluster are located in the Arctic: the Nenets Autonomous district, the Khanty-Mansi Autonomous district, the Yamal-Nenets Autonomous district, the Chukotka Autonomous district and the Magadan region, as well as the Sakhalin region of the Far East. In this regard, it is necessary to determine the mechanisms for sustainable development of resource regions of the Arctic zone.

To determine the impact of resource factors on the socio-economic indicators of the Arctic, a panel analysis was conducted using data from 2011-2016. Models based on panel data are among approaches for building models of the dependence of variables on various factors. The panel data combines both spatial type data and time series data.

In general, the regression model of panel data has the following form [10, 11] as in equation (1):

$$x_{it} = Z_{it} \propto_{it} + \varepsilon_{it} \quad (1)$$

where i denotes index of economic unit, t – time period, $\propto_{it}$ - the coefficients of the explanatory variables of vector Z_{it} in the time period t for the unit i, ε_{it} – vector of errors.

This model cannot be estimated, but it is possible to isolate specific factors that are related to a particular economic object at a certain point in time, consequently individual (specific) features can be evaluated as in equation (2):

$$x_{it} = Z_{it} \propto_{it} + \gamma_t + f_i + \varepsilon_{it} \quad (2)$$

with Z_{it} denoting n-dimensional vector of independent variables without constant, f_i – individual features that do not depend on time, γ_t– specific features that are not observed in all periods, but for all objects, ε_{it} – independent, identically distributed random variables (both for individuals and in time) with zero mathematical expectation and variance σ_ε^2 .

Depending on the assumptions about the nature of f_i, models with fixed (f_i are N constant unknown parameters) and random effects (f_i are random and not correlated with errors) are distinguished. To choose between models with fixed-effect, pooled regression, and model with random effects a Wald test, Breusch–Pagan test and Hausman test are performed. Once the relationship model of the dependent variable and factors has been obtained, it is necessary to estimate its accuracy with the help of the determination coefficient, showing how well the regression obtained describes the dependent variable and accounting to be at least 50%.

Further, it is necessary to check statistical significance of the obtained regression with the help of F-statistic on 5% significance level. To test the normal distribution of small samples with size from 3 to 50 elements, the Shapiro-Wilk test was developed. The presence of multicollinearity can be verified using the coefficients of pair correlation. If the factor correlation coefficient is large, then there is a connection between these factors. To determine the autocorrelation, the Durbin–Watson statistic is used.

3. Empirical results and discussion

The paper considered 6 factors in two groups: economic (investment per capita, GRP per capita, monetary income per capita) and resource factors (share of employment in the mining industry, the share of mineral extraction taxes in the structure of taxes, the share of mining in GRP). The impact of each resource factors on economic indicators was assessed (3 options). Further, according to the data for 2011-2016 three models were built on the basis of panel analysis for 27 resource regions: the model of pooled regression, the model with fixed effects and the model with random effects.

In general, all the models have approximately the same coefficients and show good quality.

The most adequate model was chosen on the basis of pairwise comparison of the estimated models (Tables 1-3):

- The fixed effects model was compared with the pooled regression (Wald test).
- The random effects model was compared with the pooled regression (Breusch–Pagan test).
- The random effects model was compared with fixed effects model (Hausman test).

Table 1. Tests of models (significance level 0.05) for the dependent variable of investment per capita.

Test	Chisq	p-value	Best model
Wald	15,10	p-value < 0,01	Fixed
Breusch-Pagan	116,52	p-value < 0,01	Random
Hausman	22,08	p-value < 0,01	Fixed

The results obtained allow for conclusion that in this case the model with fixed individual effects is suitable. This was expected, since specific regions were chosen for the study, their composition did not change in the analyzed period.

Table 2. Tests of models (significance level 0.05) for the dependent variable of GRP per capita.

Test	Chisq	p-value	Best model
Wald	28,03	p-value < 0,01	Fixed
Breusch-Pagan	158,01	p-value < 0,01	Random
Hausman	25,78	p-value < 0,01	Fixed

Table 3. Tests of models (significance level 0.05) for the dependent variable of per capita income.

Test	Chisq	p-value	Best model
Wald	13,89	p-value < 0,01	Fixed
Breusch-Pagan	111,91	p-value < 0,01	Random
Hausman	23,02	p-value < 0,01	Fixed

As a result, three equations (3), (4), (5) were obtained:

$$In = 0.033 * \ln(\mathrm{ER}) + 0.011 * \ln(MET) + 0.396 * \ln(ME) + 3.709 \quad (3)$$

$$GRP = 0.236 * \ln(\mathrm{ER}) + 0.016 * \ln(MET) + 0.134 * \ln(ME) + 6.224 \quad (4)$$

$$AI = 0.152 * \ln(\mathrm{ER}) + 0.034 * \ln(MET) + 0.091 * \ln(ME) + 3.469 \quad (5)$$

where, In - investments per capita, GRP - GRP per capita, AI - monetary income per capita, ER - share of employment in the mining industry, MET - share of mineral extraction tax in total taxes, ME - share of mining in GRP.

The share of extraction industries in GRP has a great impact on the volume of investments in the region. The coefficient for this factor is 0.396. It was shown that for resource regions the change in the share of mining significantly affects the volume of investment in the region. The level of GRP is significantly affected by the share of extractive industries in the economy and the share of

IOP Conf. Series: Earth and Environmental Science **193** (2018) 012073 doi:10.1088/1755-1315/193/1/012073

employment in the extractive industry, with a positive relationship. If the share of employment in the extractive industry changes, GRP will increase by 0,236. The share of employment in the extractive industries and share of extractive industries in the economy also has a significant impact on the average per capita income of the population. Shares of innovative activity in GRP of resource regions of Russia are presented in table 4.

Table 4. Share of innovative goods, works and services in GRP of resource regions of Russia in 2011-2015, %.

The name of the region	Ranking in the innovative development	Ranking in the resource specialization	2011	2012	2013	2014	2015
Republic of Tatarstan	2	8	15,4	19,0	20,8	20,2	20,4
Samara region	3	12	22,3	25,8	23,0	21,3	18,8
Sakhalin region	8	18	45,3	49,7	47,8	53,7	11,2
Republic of Bashkortostan	10	7	6,1	5,4	5,9	9,0	10,6
Perm region	13	10	9,6	9,3	20,9	11,3	9,2
Kemerovo region	34	3	0,6	0,3	0,5	2,9	3,8
Tomsk region	35	16	3,3	1,8	2,7	2,6	3,8
Krasnoyarsk region	39	14	1,0	3,0	4,3	3,5	3,6
Tyumen region	40	13	0,1	0,8	0,8	2,6	3,6
Republic of Komi	43	5	6,6	5,0	4,6	4,8	3,2
Irkutsk region	51	15	0,8	1,0	0,6	1,2	2,3
Amur region	54	24	1,7	2,2	2,5	2,2	2,0
Orenburg region	55	11	2,6	1,5	1,3	0,9	1,9
The Republic Of Sakha (Yakutia)	65	17	0,3	0,2	1,6	0,9	0,4
Murmansk region	59	23	0,1	0,1	0,6	2,6	1,1
Khanty-Mansiysk AD	67	1	2,4	0,4	0,3	0,3	0,3
Chukotka AD	70	20	0,0	1,0	1,3	0,1	0,2
Yamalo-Nenets AD	71	2	1,0	1,0	0,0	0,0	0,1
Republic of karelia	74	22	0,2	0,2	0,1	0,1	0,1
Nenets AD	84	4	0,0	0,0	0,0	0,0	0,0

As can be seen from table 4, the resource regions of the Arctic zone, as well as less developed regions with severe natural and climatic conditions, are the leaders in the production of oil and gas, as well as a number of other minerals in Russia. At the same time, the share of innovative production in the GRP structure of such regions is absent, or has a close value to zero. All resource regions of the Arctic are at the end of the rating of innovative regions of Russia.

4. Conclusion

In recent decades, economists note that the economy of resource-rich regions and countries (especially in Africa, Latin America, the Middle East, the former Soviet Union) tends to grow more slowly than countries In recent years, some resource-rich countries have seen the opposite situation - positive and rapid economic growth. As a result, it is not the availability of natural resources that affects the growth of production, but the efficiency of the use of revenues from the production and export of natural resources. In the end, at present, the researchers focus more and more on the impact of resource countries on such indicators as labor productivity, development of the financial sector, human capital, and growth of innovation in the economy.

Based on the use of panel data, the article showed that for the resource regions of Russia, first of all, with fewer natural resources. The special term "resource curse" was introduced. "Resource curse" refers to the inverse relationship between the dependence on natural resources and economic growth.

Arctic zone, there is a direct relationship between the resource factors and socio-economic indicators. Meanwhile, the resource regions, especially the Arctic zone, are significantly lagging behind in terms of innovative development, which determine the sustainability of the region's development.

Acknowledgements
The research was supported integration project № ИП-66ПН.

References

[1] Frankel J A 2010 The natural resource curse: a survey *National Bureau of Economic Research* w15836
[2] Van der Ploeg F 2011 Natural resources: curse or blessing? *J. EconLit* **49** pp 366-420.
[3] Ross M 2007 How Mineral-Rich States Can Reduce Inequality *Escaping the Resource Curse* ed M Humphreys, J Sachs and J Stiglitz (New York: Columbia University Press) chapter 9 pp 236–255
[4] Deacon R T 2011 The political economy of the natural resource curse: a survey of theory and evidence *Found. Trends Microecon.* **7** pp 111–208
[5] Badeeb R A, Lean H H and Clark J 2017 The evolution of the natural resource curse thesis: A critical literature survey *Resources Policy* **51** pp 123–134
[6] Eder L V, Filimonova I V, Provornaya I V, Nemov V U and Nikitenko S M 2017 Regional smart specializations in fostering innovation development of resource regions of Russia *International Multidisciplinary Scientific GeoConference Surveying Geology and Mining Ecology Management, SGEM* **17** pp 727-34
[7] Eder L V, Kontorovich A E, Filimonova I V, Provornaya I V and Goosen E V 2017 Regional innovation cluster: Environmental issues and efficient use of resources *International Multidisciplinary Scientific GeoConference Surveying Geology and Mining Ecology Management, SGEM* **17** pp 719-26
[8] Eder L V, Filimonova I V, Provornaya I V and Nemov V Y 2017 The current state of the petroleum industry and the problems of the development of the Russian economy *IOP Conf. Ser.: Earth Environ. Sci.* **84** 012012
[9] Kontorovich A E, Eder L V and Filimonova I V 2017 Paradigm oil and gas complex of Russia at the present stage *IOP Conf. Ser.: Earth Environ. Sci.* **84** 012010
[10] Kolomak E A 2011 *Panel data: methods of econometric analysis: textbook* (Novosibirsk: Editorial and publishing center of NSU)
[11] Ratnikova T A and Furmanov K K 2014 *Analysis of panel data and data on the duration of States: studies manual* (Moscow: Publishing house of the Higher School of Economics)

Hydrocarbon expulsion history in Kuonamka formation of North Tunguska Petroleum region

E S Yaroslavtseva[1] and L M Burshtein[2]

[1] Trofimuk Institute of Petroleum Geology and Geophysics SB RAS, Novosibirsk, Russia
[2] Novosibirsk State University, Novosibirsk, Russia

Abstract. The computer modelling results for Kuonamka source rock evolution in catagenesis are represented for the area of North Tunguska petroleum region for pre-igneous period. Petroleum generation hotbeds are determined. Petroleum resources indirect assessment is made for Kuonamka source rock for pre-igneous period.

1. Introduction

North Tunguska petroleum region occupies the north-west part of Lena-Tunguska petroleum province, which is known as one of the biggest oil and gas basins of the world. Investigations on the history and the amount of hydrocarbon generation of the area have been carried on since the 1960s of the XX century. However, the geological model of its structure, stated in the past is now sufficiently specified. Similarity of the main petroleum system of Cambrian section (Kuonamka source rock and the clinoform complex of Mayian stage) to that of the West Siberian Jurassic one (the Bazhenov formation and the Neokom clinoforms) significantly increases the interest in the territory under consideration and the prospects of its oil and gas potential [1]. The accumulated factual material makes it possible for the first time for this territory to reconstruct petroleum expulsion dynamics in Kuonamka complex of the North Tunguska petroleum region using methods of basin modeling.

Investigation on such regions as the North Tunguska region is accompanied by some difficulties. The first is that the area is not explored enough. The second is that western part of Siberian platform was completely involved into Permian – Triassic regional trap magmatizm. Igneous process is known to affect on petroleum potential of basins. This affection may be positive, consisting in source rock more rapidly getting into the main zone of oil generation (oil window). In contrast, negative influence of traps implies destruction of existing hydrocarbon accumulations [2]. According to existing criteria of oil and gas potential prognosis of petroleum basins one should actualize such a research in two steps:1) reconstruction of hydrocarbon expulsion dynamics at the time before trap magmatism, 2) trap affection assessment [3]. In the present study, the first group of problems was realized: for the first time, the dynamics of hydrocarbon generation was reconstructed for Kuonamka formation of the North Tunguska petroleum region at the time before Permian age, using the methods of basin modeling. The research is based on factual data and geological models, prepared in IPGG SBRAS.

2. Studing area

The North Tunguska region occupies the northern part of Kureyka syncline, the vast tectonic depression accomplished by Vendian-Lower Paleozoic sedimentary rocks. It is overlaid by Tunguska syncline composed by Higher Paleozoic and Triassic formations [1]. Kuonamka formation is expected to be the

IOP Conf. Series: Earth and Environmental Science **193** (2018) 012074 doi:10.1088/1755-1315/193/1/012074

main source rock of Cambrian section of the territory. The formation is spread commonly over the platform. It is characterized by high hydrogen index (varies between 352-557 mg HC/g TOC) and high total organic carbon (averages between 1.5–15% and sometimes reaches 30-35%) [4; 5]. Geochemical properties of Kuonamka formation are widely explored by the outcrop materials of the west and north-west parts of the platform. The results are published in many sources [4–9]. Hydrocarbons expelled by Kuonamka formation are suggested to accumulate in collectors of Cambrian reef barrier, developed in north-west of Siberian platform as well as clinoforms of Mayian age. [1; 10].

3. Methods and materials

Common principles of petroleum system modeling are represented in many studies [11–14]. In current investigation the methods of basin modelling are performed using TemisFlow software. 3D structural-lithological model is formed, and then, the sedimentary cover history is reconstructed by the method of backstripping. Hydrocarbon generation dynamics in Kuonamka formation of North Tunguska petroleum region is reconstructed on the basis of thermal data and kerogen kinetics.

3D structural-litological model is formed on the basis of structural maps of the main reflecting horizons (F, R_0, M_2, M_1, B, K_1, H_1, ev) and the well data for Kyramkynscaya-1, Kochechumsksaya-2, Khoshonskaya-256, Chyryndinskaya-271, Srednetaymurinskaya-272, Ledyanskaya-2, Ledyanskaya-3, prepared in IPGG SBRAS.

Temperature model typically includes surface temperature history, factual data on paleothermometers, present day temperature and sedimentary section heat flow distribution. Poor exploration of North Tunguska region, ancient age of its sedimentary boundary and the lack of reliable data on natural paleothermometers causes 1D modelling of heat field of the basin to be the only way to get information on sedimentary boundary paleothemperature. In order to carry on temperature field reconstruction structural-lithological models of Ledyanskaya-3 and Phokinskaya-225 cuts are applied. Heat flow value is presented as heatflow density through the lower margin of lithosphere and averages about 40 mVt/m^2 (to consider the influence of Siberian superplume heat flow density is suggested to be slightly increased to 70 mVt/m^2 at the end of Permian age) [15–17].

Geochemical model includes data on kerogen type and kinetics [18, 19]. According to pyrolysis parameters kerogen of Kuonamka source rock corresponds to II type of kerogen. Since the lack of data on geochemical properties of Kuonamka formation directly situated on studying area information was taken on the basis of adjacent territories. The width of Kuonamka formation, enriched by organic matter, was estimated as 15 m, initial total organic matter was taken as 15%, initial hydrogen index as 500 mg HC/g TOC.

4. Relsults

4.1 1D modeling

According to the results of 1D modelling of Ledyanskaya-3 cut, the Riphean, Vendian and Cambrian deposits reached more than 100°C. The Riphean deposits reached 100°C about 470-440 Ma, Vendian - about 440-420 MA, Cambrian - nearly about 420-450 Ma. Riphean and Vendian deposits reached 150°C. Before the trap magmatism period, the Riphean, Vendian and low-middle Cambrian deposits reached the main zone of oil generation (oil window) with vitrinite reflectance (Ro) about 0.7% [12] and all of them except for middle Cambrian left it (Figure 1).

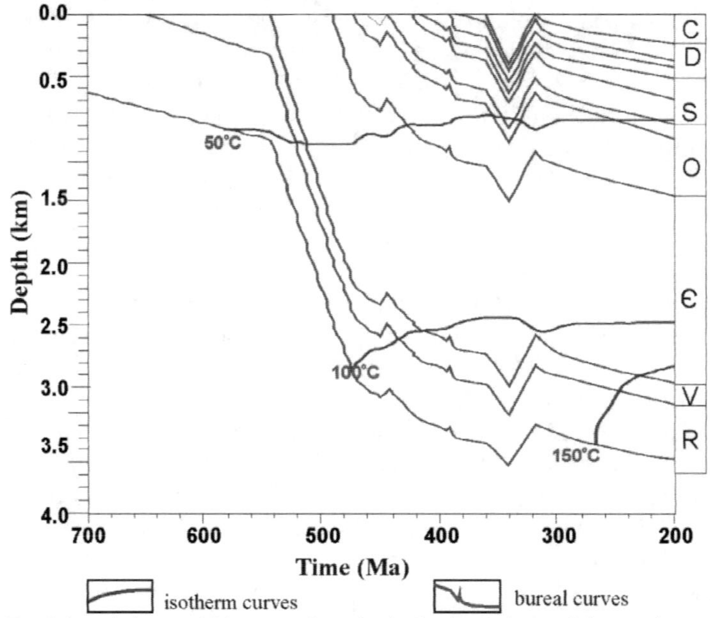

Figure 1. Burial and thermal history of rocks in Ledyanskaya-3 by preigneous period.

According to the results of 1D modeling of Phokinskaya-225 well, the Riphean, Vendian, Cambrian, Ordovician and Silurian deposits reached 100°C by the end of the Permian period. The Riphean deposits reached 100°C about 480-460 Ma, Vendian nearly about 460-450 Ma, Cambrian about 450-410 MA, Ordovician about 410-260 Ma, Silurian relatively about 260-240 Ma. The Riphean, Vendian, Cambrian and early Ordovician deposits reached 150°C. According to the results of vitrinite reflectance calculation, the Riphean, Vendian, Cambrian, Ordivician and Early Silurian deposits reached the higher margin of Middle Zone of Oil Generation (oil window) with Ro=0.7 % before trap period. The Riphean, Vendian, Cambrian deposits reached apocatagenesis (Ro=2.0 %), which means it totally exhausted it's hydrogen index.

4.2 3D modeling

Based on factual data on basin structure, Kuonamka source rock kerogen properties and by the results of 1D temperature history modeling regional-zonal schemes of kuonamka formation catagenesis are formed. According to the results of modeling, hydrocarbon generation in Kuonamka formation began in by Ordovician-Silurian age. This is how Tura oil and gas hotbed originated. In the Devonian period hydrocarbon generation reached its maximum and then Kuonamka formation of central part of the hotbed begun to emerge from the main oil and gas generation zones. Hydrocarbon generation of the hotbed periphery continued until the Higher Carbon and then oil and gas potential of the source rock was totally exhausted.

Since the lower Devonian age one more petroleum hotbed originated in the north-western part of North Tunguska region. This was Lama-Khantay hotbed. Kuonamka source rock slightly got into the main zone of oil generation and till the end of the Carbon reached the maximum of generation (Figure 2).

Cummulative mass of oil and gas generatied by the Ordovician-Carbonian age is shown on figure. Common mass of hydrocarbons accounts 570 billion tons. With accumulation coefficient taken into account resources of hydrocarbons, generated by Kuonamka formation might amount 5,7-57 billion tons.

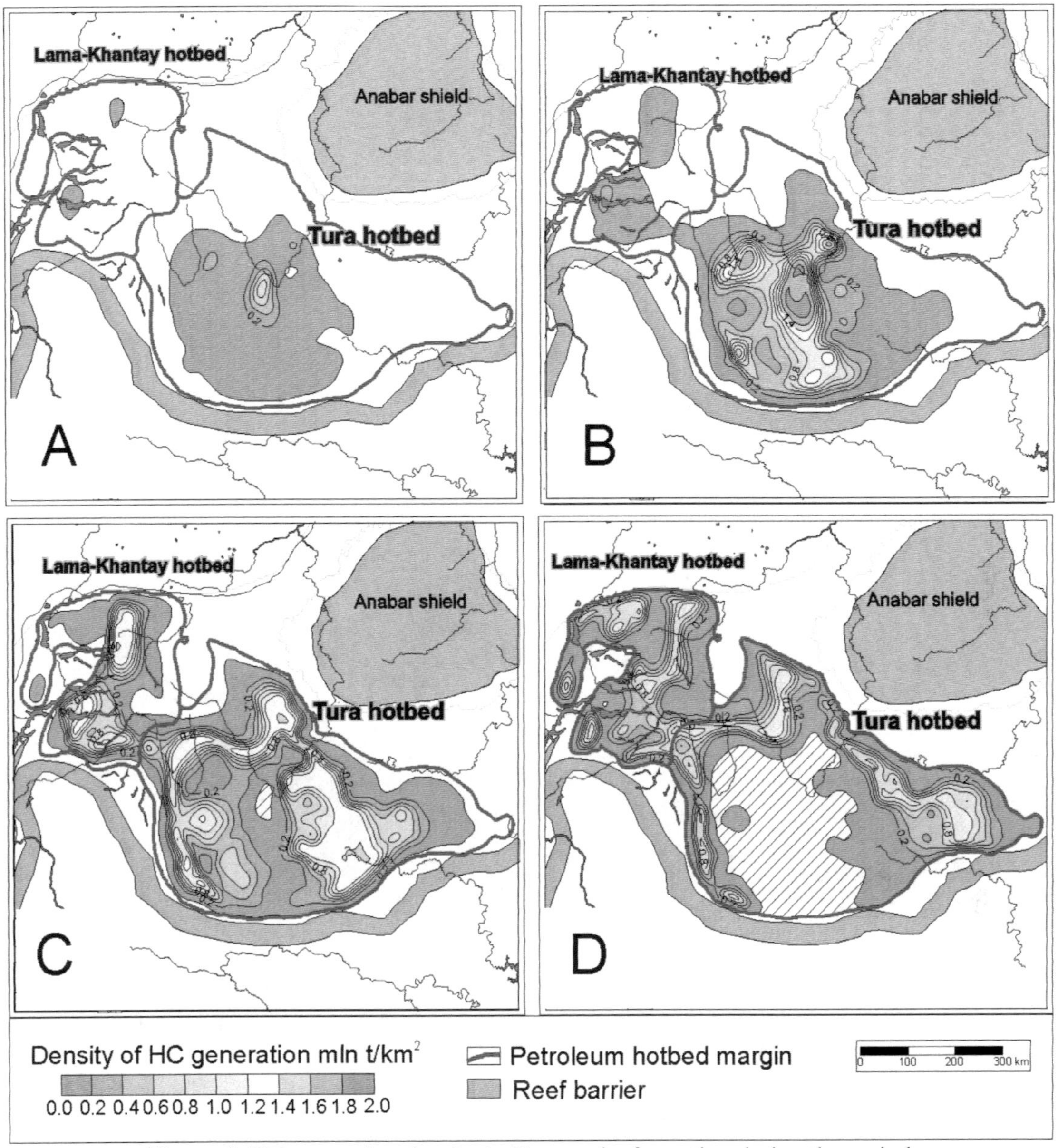

Figure 2. HC generation density in Kuonamka formation during the period:
A. 450-425 Ma; B. 425-400 Ma; C. 375-350 Ma; D. 325-299 Ma.

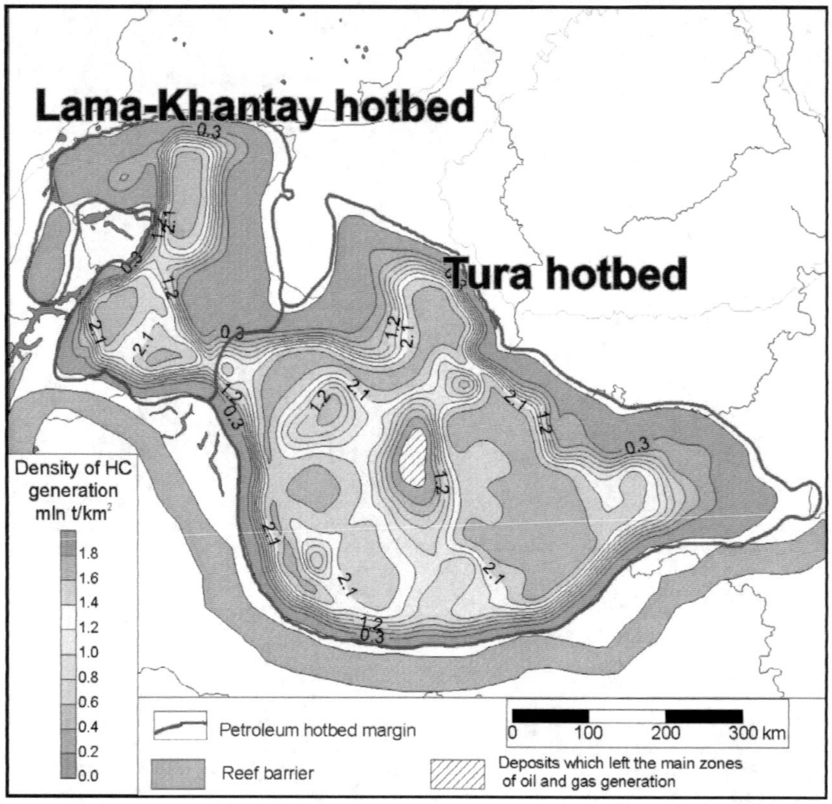

Figure 3. Cummulative mass of hydrocarbon generation in Kuonamka source rock during the period since Ordivician till Carbon.

5. Conclusion

Since the higher Ordovician till the beginning of the Permian period two petroleum hotbeds occupied the area of North Tunguska region. These are Tura and Lama-Khantay hotbeds. In Tura hotbed oil and gas potential of Kuonamka source rock was totally exhausted till the Permian period. Thus igneous intrusions could not control source rock potential in this area. In Lama-Khantay hotbed oil and gas generation has reached its maximum, which somehow increases the variety of magmatizm influence on hydrocarbon potential of source rock of the territory.

Acknowledgments

The results were obtained within the framework of basic project of RAS IX.131.4.1. and with technical support of Russian agency of BeicipFranlab Company.

References

[1] Antsiferov A S, Bakin V E, Varlamov I P 1981 *Geology of oil and gas of the Siberian platform.* ed. by Kontorovich A E, Surkov V S, Trofimuk A A (Moscow: Nedra) p 552

[2] Khomenko A V 1988 *Trap intrusions as the factor to define location of hydrocarbon fields (South Tunguska region)* in Prognozis of petroleum accumulation areas and local objects of Siberian platform (Leningrad: VNIGRI) 96–103

[3] Kontorivich A E, Melmikov N V, Staroseltsev V S, Khomenko A V 1987 *Intrusive traps influence on petroleum potential of Paleozoic deposits of Siberian platform* Geology and geophysics (5) 14

[4] Bazhenova T K, Dakhnova M V, Zheglova T P, Lebedev V S, Mozhegova S V, Lharkhin V N, Nazarova E S, Nechytaylo G S, Grazer A M, Kiselev S M, Kiseleva Y A, Goryunova E A,

Borisova L B 2014 *Source formations, oil and gas of Precambrian, lower and middle Cambrian of Siberian platform* (Moscow: VNIGRI) p 128

[5] Bakhturov S F, Yevtushenko V M, Pereladov V S 1988 *Kuonamka bituminous carbonate shale formation* (Novosibirsk: Nauka) p 160

[6] Kashirtsev V A 2003 *Organic geochemistry of Bitumens of the Eastern Siberian Platform* (Yakutsk: SB RAS) p 159

[7] Parfenova T M, Kontorovich A E, Borisova L S, Melenevsky V N 2010 *Kerogen of Kuonamka formation* Geology and geophysics **21**(3) 353–63

[8] Parfenova T M, Bahturov S F, Shabanov Y Y 2004 *Organic geochemistry of oil-producing rocks of the Cambrian Kuomanka formation (eastern Siberian platform)* Russian Geology and geophysics **45**(7) 862–74

[9] Savitskiy V E, Evtushenko V M, Egorova L I, Kontorovich A E, Shabanov Yu Ya 1972 The Cambrian of the Siberian platform (Yodima-Olenek section type) (Moscow: Nedra) SNIIGGiMS Transactions vol. 130 p 200

[10] Filiiptsov Yu A, Melnikov N V, Efimov A S, Valchak V I, Goryunov N A, Evgrafov A A, Simonov E V, Scherbakov V A, Kultyshev V Yu 2014 *Lower-Middle Cambrian reef barrier in the northern Siberian platform as a target of primary exploration for oil and gas* Geology and mineral resources of Siberia **2** (18) 25–35

[11] Galushkin Y I 2007 *Sedimentary basin modeling and estimation of its hydrocarbon potential* (Moscow: Nauchnyi mir) p 445

[12] Kontorovich A E 1976 *Geochemical methods of quantity prognosis of oil potential* (Moscow: Nedra) SNIIGGiMS Transactions (229) p 250

[13] Hantschel T, Kauerauf A I, Doedrecht, Heidelberg 2009 *Fundamentals of Basin and Petroleum Systems Modeling* (London, New York: Springer Science & Business Media) p 476

[14] Tissot B P Welte D H 1984 *Petroleum Formation and Occurrence* (Berlin-Heidelberg-New York : Springer-Verlag) p 538

[15] Dobretsov N L 2008 *Geologocal implicatinons of the thermochemical plume model* Russian Geology and geophysics **49**(7) 441–54

[16] Dobretsov N L 2010 *Global geodynamic evolution of the Earth and global geological models* Rissuan Geology and Geophysics **51**(6) 592–610

[17] Duchkov A D, Dobretsov N N, Ayunov D E, Sokolova L S 2011 *Permafrost-geothermic Atlas of Siberia and Far East* Dynamics of physical fields of the Earth (Moscow: Svetoch Plus) 207–21

[18] Bordenave M L 1993 *Applied petroleum geochemistry* (Paris) p 524

[19] Burnham A K, Sweeney J J 1989 *A chemical kinetic model of vitrinite maturation and reflectance* Geochimica et Cosmochimica Acta **53**(10) 2649–57

[20] Burnham A K 2015 *A Simple Kinetic Model of Oil Generation, Vaporization, Coking, and Cracking* Energy Fuels **29**(11) 7156–67

IOP Conf. Series: Earth and Environmental Science **193** (2018) 012075 doi:10.1088/1755-1315/193/1/012075

Modeling of multiple fractures growth during multistage hydraulic fracturing based on cell-based pseudo 3D model

S A Budennyy, R N Nikitin, A A Erofeev, R M Sitdikov[1] and G V Paderin[2]

[1] Center For Engineering and Technology of Moscow Institute of Physics and Technology, Dolgoprudny, Moscow Region, 141701, Russian Federation

[2] Gazpromneft Science and Technology Center, Saint-Petersburg, 190000, Russian Federation

E-mail: Budennyy.sa@cet-mipt.ru

Abstract. The paper presents a mathematical model describing the process of formation of multiple fractures in inhomogeneous rock formation. The features of its numerical implementation and the results of implementation are given. Modelling of fracture growth dynamics is carried out in the frame of cell-based pseudo 3D model, taking into account the associated solution of the problems of fracture mechanics and multiphase flow of a proppant mixture acting as a proppant dispersed phase and a power law rheology fluid. To calculate the flow parameters at the points of initiation of multiple fractures for given wellhead data, the model implements the calculation of hydraulics in the wellbore based on a homogeneous multiphase model. Pathway of fracture development is determined for arbitrary field of the strained deformed state of the formation and takes into account its change caused by stress shadow effect. The developed model allows to evaluate the efficiency of the application of the multi-stage fracturing technology with several perforation clusters or hydraulic fracturing ports in a single stage, as well as to optimize the completion of a horizontal well.

1. Introduction

Receiving information on fracture growth dynamics in the rock formation is limited by a few number of indirect methods, the most common of which is micro-seismic monitoring. As no direct methods for determining the fracture geometry during its growth in the formation exists we are to appeal to results of numerical modelling. Despite a relatively long period of commercial use of hydraulic fracturing (HF) technology, there is still no generally accepted physical and mathematical model capable to describe all key processes occurring during HF treatment: rock destruction with heterogeneous mechanical (elastic-strength) properties, mass transfer of a mixture of non-Newtonian fluid and solid phase within the fracture channel.

The complexity of joint modeling of all processes is due to their strong nonlinear relation on the one hand and the need for a sufficient set of data for numerical experiments — on the other. The problem is coming less trivial while examining the case of multiple fractures growth as one needs to take into account the interaction between fractures via induced pressure and the wellbore hydraulics as well [1].

IOP Conf. Series: Earth and Environmental Science **193** (2018) 012075 doi:10.1088/1755-1315/193/1/012075

To date, a great number of works are devoted to separated problems on the topic of HF, a much smaller number of works describe the joint solution of all processes. In this regard, there is a need for a comprehensive solution to the problem of determining the dynamics of fractures growth with reasonable simplifying assumptions and at the same time sufficient for efficient planning of fracture design for tackling real industrial problems.

2. Problem statement

To tackle the problem of multi-stage HF it is first worth making a problem statement for a single fracture growth within an inhomogeneous geological formation [2]. The problem of a single fracture growth is posed as follows. For given profile of the elastic-strength properties of formation (Young's modulus E, Poisson's ratio ν, the minimum horizontal stress σ_{min}, the coefficient of fracture toughness K_{IC}, the coefficient of leakage of C_l), the carrier phase properties (density ρ_f, the rate of non-Newtonian behavior of n, the index of consistency of K) and particles of the proppant (material density ρ_p, particle radius a), and mode download (volume flow rate of the mixture, q_m, the volume concentration of proppant β) we need to determine the dynamics of fracture growth, — time dependences of the fracture length L, the opening profile w, the fracture height h_f, and the fluid overpressure profile in the fracture channel $p_{net} = p_f - \sigma_{min}$.

While modeling numerically the HF processes, the mass transfer problem must be solved in a joint way for the "wellbore-fracture" system by linking solutions at the bottom of the well. This is especially essential when the problem of simultaneous growth of fractures on several cuts within a single cluster of perforations in the horizontal wellbore is considered. At the same time, during the transfer of the working fluid and proppant in the wellbore there are pressure losses due mainly to viscous friction. Thus, to determine the order of initiation of cracks and the distribution of costs for perforations, it is necessary to evaluate the pressure losses in the wellbore, solving the problem associated with fractures growth.

The problem of multistage HF (MHF) can be formulated as follows. For given properties of the rock formation, fluid and proppant particles, as well as the injection mode set at the wellhead, it is a goal to determine the fracture dynamics at each stage.

3. Multi-stage hydraulic fracturing model

3.1. Fracture model

The cell-based P3D model is chosen for linking the relation between the pressure inside the fracture channel and its width profile. In the frame of P3D model the fracture is represented with hydrodynamically associated volume cells (hereinafter cells) $A_f \Delta s$, where Δs — the length of the cell (s —curvilinear coordinate that determines the direction of fracture growth), A_f — cross-sectional area of the channel cracks that are defined for the specified profile disclosure $w(s,z)$ (z — vertical coordinate) and fracture height $h_f(s)$ expression:

$$A_f(s) = \int_{-h_f(s)/2}^{h_f(s)/2} w(s,z)dz. \tag{1}$$

The dynamics of the fracture channel cross-sectional area change A_f is determined by the fluid flow at the fracture inlet: continuity and momentum conservation equations are solved. The model implies the presence of a one-dimensional pressure profile inside the fracture $p(s)$, changing only along the direction of growth, i.e. along the curvilinear coordinate s. The fracture opening profile $W(z)$ for a given pressure in the p cell is determined in the frame of the geomechanics problem.

3.2. Mass transfer model

The multiphase flow problem in the fracture channel is based on approximation of the effective medium, where the suspension is considered as a viscous incompressible fluid with a density and viscosity depending explicitly on the volume fraction of the dispersed phase. From the theory of the lubricant layer we have the following expression for the local continuity of the liquid and dispersed (proppant particles) phases in the fracture channel of variable cross-section A_f:

$$\frac{\partial(1-\beta)A_f}{\partial t} + \frac{\partial(1-\beta)q}{\partial s} + (1-\beta)q_l = 0, \tag{2}$$

$$\frac{\partial \beta A_f}{\partial t} + \frac{\partial \beta q}{\partial s} = 0, \tag{3}$$

where β — volumetric proppant concentration , q — volumetric flow rate per unit of fracture, δ — the Kronecker's symbol , q_l — volume flow leakage from defined by Carter's model [4] as follows

$$q_l(s,t) = \frac{2C_l h_f}{\sqrt{t-\tau(s)}}, \tag{4}$$

where t — time elapsed since the beginning of the leakage process, $\tau(s)$ - time of arrival of the fracture tip boundary to the position s, C_l — empirical Carter's leakage coefficient.

The relation between the mixture flow q and the net pressure p_{net}, following from the solution of the problem of steady laminar fluid flow between two parallel planes of unit height spaced at a distance w, is given by the Poiseuille equation:

$$q = -\Lambda(w,p)\frac{\partial p_{net}}{\partial s}, \tag{5}$$

where Λ — fluid mobility operator.

As most of HF operations exploit gel-like working mixtures we consider a model of a non-Newtonian fluid whose shear stresses τ depend nonlinearly on the instantaneous shear rate $\dot{\gamma}$. The power law model is chosen as it describes the rheology of most standard HF fluids $\tau = K\dot{\gamma}^n$, where K — the flow consistency index, n — the flow behavior index.

For a model with power rheology, the mobility operator has the following relation:

$$\Lambda(w,p) = \frac{n}{(2n+1)2^{(n+1)/n}} K^{-1/n} w^{\frac{2n+1}{n}} \left| \frac{\partial p_{net}}{\partial s} \right|^{\frac{1-n}{n}}. \tag{6}$$

The influence of the solid phase on the viscosity behavior of the mixture was taken into account by using the empirical dependence of the consistency index on the volume concentration of proppant $K(\beta) = K_f(1-\beta/\beta^*)^{5n/2}$, where K_f — the consistency index of pure liquid, β^* — the critical concentration of the solid phase.

The boundary conditions are zero volume flow (degeneration of the Stefan's condition) and zero opening at the fracture tip ($s = L$):

$$q(t,L) = 0, w = 0, \ \forall t > 0, \tag{7}$$

as well as the conditions on the flow rate of the mixture and concentration of proppant at the fracture inlet:

$$q(0,t) = q_0(t) \ \beta(0,t) = \beta_0(t) \ \forall t > 0. \tag{8}$$

The initial condition is zero fracture opening and zero proppant concentration:

$$w(s,0) = 0, \beta(s,0) = 0. \tag{9}$$

3.3. Geomechanics problem

The geomechanics problem for each volume cell of the fracture is solved in two stages: first, the optimization problem of finding the equilibrium state (the positions of the lower and upper boundaries of the cell z_b, z_t) for a given pressure p is solved. After, the width profile $w(z)$ is calculated for given pressure.

The equilibrium height is achieved in the case when the stress intensity factor (SIF) at the tip of the fracture coincides with the coefficient of fracture toughness K_{IC}, i.e., $K_\pm = K_{IC}$. In a linearly elastic medium, the SIF at the upper and lower tips of the fracture (K_{I+} and K_{I-}, respectively) is defined by the following relation [3]:

$$K_\pm = \sqrt{\frac{2}{\pi h_f}} \int\limits_{-h_f/2}^{h_f/2} (p - \sigma(z)) \sqrt{\frac{h_f/2 \mp z}{h_f/2 \pm z}} dz, \tag{10}$$

where $\sigma(z)$ — the stress profile, h_f — the fracture height, p — pressure weighted in the vertical cross-section.

The problem of the relation between the pressure in the fracture cell p and its opening profile w is solved by finding the global minimum for the functional:

$$\min_{z_t, z_b} \{ |K_+(z_t, z_b) - K_{Ic}(z_t)| + |K_-(z_t, z_b) - K_{Ic}(z_b)| \}, \tag{11}$$

by limits arising from the failure criterion

$$K_+(z_t, z_b) \geq K_{Ic}(z_t), \;\; K_-(z_t, z_b) \geq K_{Ic}(z_b), \tag{12}$$

where $K_{Ic}(z)$ — a given fracture toughness profile.

The SIF is convenient to rewrite in the following form:

$$K_\pm(z_t, z_b) = \sqrt{\frac{2}{\pi(z_t - z_b)}} \int\limits_{z_b}^{z_t} (p - \sigma(z)) \left(\frac{z_t - z}{z - z_b} \right)^{\pm 1/2} dz. \tag{13}$$

The resulting functionality has a number of features. First, the functional includes the integral SIF representation for the stress profile in general case, while the required optimization parameters z_t, z_b are included both in the integral and in the integration boundaries. Second, as the stress profiles $\sigma(z)$ and the fracture toughness coefficient $K_{Ic}(z)$ are given by piecewise constant functions, the surface of the target functional suffers a first-order discontinuity.

Non-local relationship between the pressure p and profile disclosure $w(z)$ in a linearly elastic medium is given by relation:

$$w(z) = \frac{4}{\pi E'} \int\limits_0^{h_f} (p - \sigma(z')) K(z, z'; h_f) dz', \tag{14}$$

where $E' = E/(1 - \nu^2)$ — module of plane deformation, $K(z, z'; h_f)$ — singular kernel defined by expression:

$$K(z, z'; h_f) = \ln \left| \frac{\sqrt{h_f^2 - z^2} + \sqrt{h_f^2 - z'^2}}{\sqrt{h_f^2 - z^2} - \sqrt{h_f^2 - z'^2}} \right|. \tag{15}$$

3.4. Join solution of multi-stage hydraulics fracturing and wellbore hydraulics

The problem of transfer of a mixture of liquid and proppant in the wellbore of an arbitrary cross section $A_w(l)$ and an arbitrary geometric profile is considered in a non-stationary quasi-one-dimensional formulation by averaging the continuity equations and preserving the moment of pulses along the wellbore cross section: all values change only along the wellbore (along the curvilinear coordinate l, measured from the wellhead to the bottom of the well).

Given the flow rate of the working mixture $q_0(t)$ and the volume concentration of the solid phase $\beta_0(t)$ at the wellhead (where the wellbore intersects the earth's surface), we need to determine the same parameters at the fracture inlets, i.e. the boundary conditions for fracture inlets $q_i(t)$, $\beta_i(t)$, where the index i denotes i-th fracture (Figure 1).

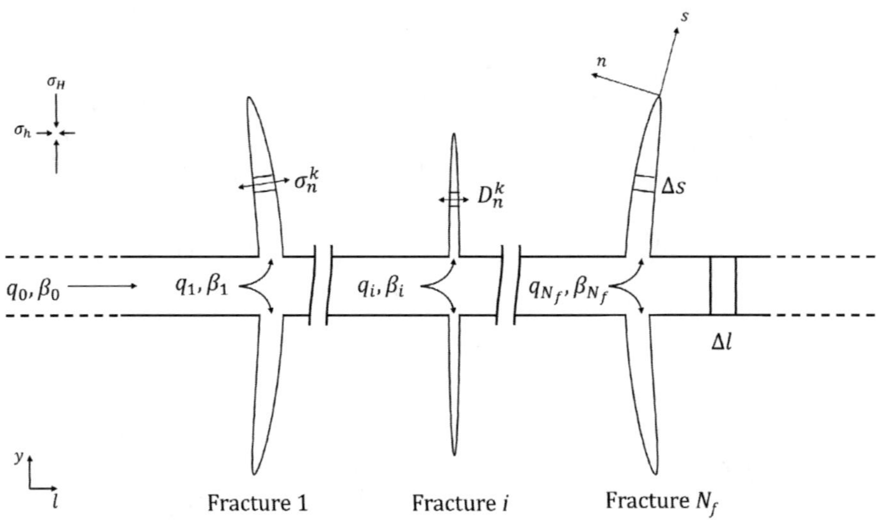

Figure 1. Multi-stage hydraulic fracturing scheme

The continuity equations for the mixture of liquid and proppant are obtained in the approximation of weakly compressible fluid. Taking into account the boundary conditions at the wellhead and effluents in the perforations (in the holes of the well column, where fracture growth is initiated), the equations for the pressure of p and the proppant concentration of β in the well bore are obtained:

$$A_w c\frac{\partial p}{\partial t} + \frac{\partial A_w v_w}{\partial l} + A_w v_w c\frac{\partial p}{\partial l} = \delta(l)q_{M0}(t) - \sum_{I=1}^{N_f}\delta(d-l_i)q_{mi}, \tag{16}$$

$$A_w\frac{\partial \beta}{\partial t} + \frac{\partial A_w \beta v_w}{\partial l} = \delta(l)\beta_0(t)q_{M0}(t) - \sum_{I=1}^{N_f}\delta(d-l_i)\beta_I q_{mi}, \tag{17}$$

where N_f — number of fractures, $A_w = \pi r_w^2$ — cross-sectional area of the bore of the table, q_{M0} and β_0 — volumetric flow rate of the mixture and concentration of proppant at the wellhead, respectively, q_{mi} and β_i — volumetric flow rate of the mixture and the volume concentration of proppant in the i-th perforation hole (for i-th fracture), respectively, v_w — the speed of the mixture, c — the liquid compressibility coefficient.

In the approximation of the rapidly setting pressure, the expression for pressure losses in the wellbore taking into account the variable cross-sectional area of the wellbore has the form:

$$\frac{\partial A_w p}{\partial l} = -2\pi r_w \tau + A_w \rho_m g \cos\phi(l), \quad \tau = \rho_m \frac{f(Re)}{4} \frac{v_w |v_w|}{2}, \tag{18}$$

where $\phi(l)$ is the inclinometry angle, $\rho_m = (1-\beta)\rho_f + \beta\rho_p$ is the phase weighted mixture density, τ is the tangential stresses, $f(Re)$ is the friction coefficient depending on the flow regime.

The Reynolds number Re for fluid flow with a power-law rheology:

$$Re = \frac{(2r_w)^n v_w^{2-n} \rho_f}{8^{n-1}(\frac{3n+1}{4n})^n K}. \tag{19}$$

The coefficient of friction f for laminar flow regime ($Re <= 3250 - 1150n$) is determined by the expression $f = 16/Re$, in turbulent flow regime we have a transcendental expression for the relation between the friction coefficient and Reynolds number for the fluid of power rheology:

$$1/\sqrt{f} = \frac{4}{n^{3/4}} \log\left(Re f^{1-n/2}\right) - 0.395 n^{-1.2}. \tag{20}$$

The system of equations is supplied with equations on mass transfer (2, 3) of geomechanics and each (11, 14) i-th fracture. The system of integral equations is closed the flow of mixture and the volume concentration:

$$q_{M0}(t) = \sum_{I=1}^{N_f} q_{mi}, \ \beta_0(t) = \sum_{I=1}^{N_f} \beta_I. \tag{21}$$

Boundary conditions at the wellhead are given :

$$q_m(0,t) = q_0(t), \ \beta(0,t) = \beta_0(t), \ \forall t > 0. \tag{22}$$

At the initial time, the well is filled with buffer fluid with zero proppant concentration, the pressure is determined by hydrostatics :

$$v_m(l, t=0) = 0, \ \beta(l, t=0) = 0, \ P(L, t=0) = \rho_f g z(l), \ \forall l \subset [0, L_w]. \tag{23}$$

3.5. Stress shadow

The problem of mutual influence of fractures on each other is solved by the method of discontinuous displacements proposed by S. L. Crouch, A. M. Starfield [5], [6]. The method is based on solving the problem of an infinite plane, the displacements of which suffer a constant gap within the elementary segment. The discontinuous displacement method makes it possible to replace discontinuities of displacements continuous along the fracture with a discrete approximation, namely, to divide the fracture into a discrete set of boundary elements, considering the fracture within each element to be rectilinear, and its opening to be constant.

We assume that the fracture opening width is equal to the normal displacement rupture, there is no shear stress on the fracture surface.

Induced stresses acting on the fracture element in the presence of discontinuous displacements in the computational domain are expressed explicitly:

$$\sigma_n^i = \sum_{j=1}^{N} A^{ij} C_{ns}^{ij} D_s^j + \sum_{j=1}^{N} A^{ij} C_{nn}^{ij} D_n^j, \ \sigma_s^i = \sum_{j=1}^{N} A^{ij} C_{ss}^{ij} D_s^j + \sum_{j=1}^{N} A^{ij} C_{sn}^{ij} D_n^j, \tag{24}$$

where D_s^j and D_n^j — tangential and normal components of displacement breaks the jth item and C^{ij} — boundary influence coefficient of the gap offsets the j-th element of tension in the ith element of A^{ij} — coefficients proposed by D. Olson [7] that account for the finite height of fractures.

4. Results

This section presents an example of a multi-stage fracturing simulation based on real data. In the example, a horizontal well with 3 stages of fracturing with a different number of ports for each stage is considered (1st stage - 1 port, 2 stage - 2 ports, 3 stage - 3 ports). Figure 2 shows the well trajectory (top view) and the port positions in the stages of the MHF.

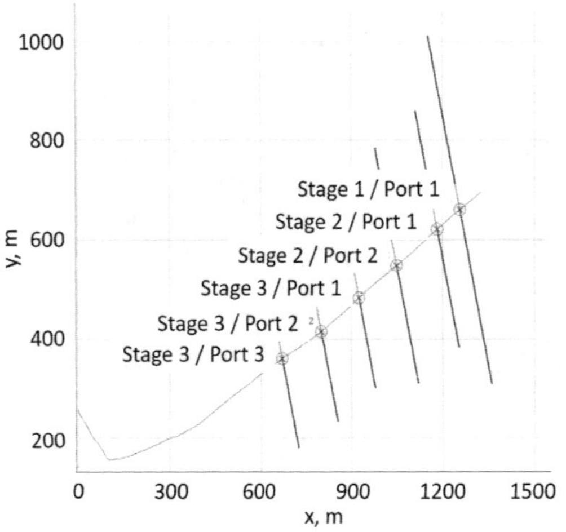

Figure 2. Results of MHF modeling based on real data. Fracture pathawys.

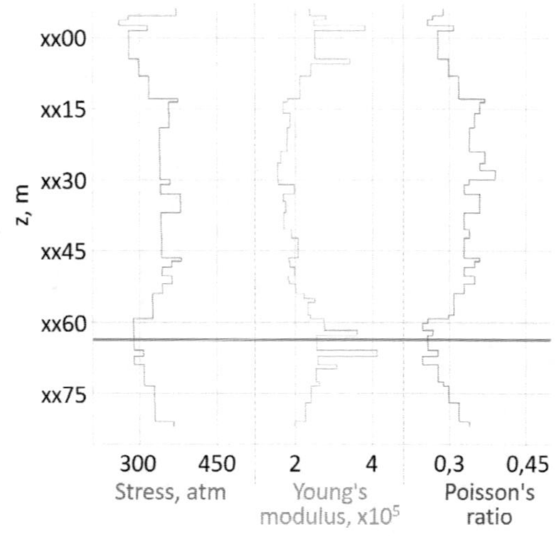

Figure 3. Mechanical properties profiles.

The values of the geomechanical properties for the depth of the target interval (minimum horizontal stresses, Young's modulus, Poisson's ratio) are shown in the Figure 3. The value of toughness is equal to 10 atm $m^{1/2}$. The leakage coefficient throughout the entire section is equal to 6.10^{-5} m $sec^{1/2}$.

In each stage of MHF a mixture of liquid with power-law rheology (K=0.1 Pa sec^n, n=0.7) and proppant 0.5 mm in diameter was pumped at a flow rate of 3.5 m^3/min. During the first 22 minutes of the stage pure fluid was pumped, then the proppant concentration increased from 125 kg/m^3 to 275 kg/m^3. For each stage: total injection time 38 min, total mass of proppant 9 800 kg, total fluid volume 133 m^3. The simulation results are shown in Figures 4, 5.

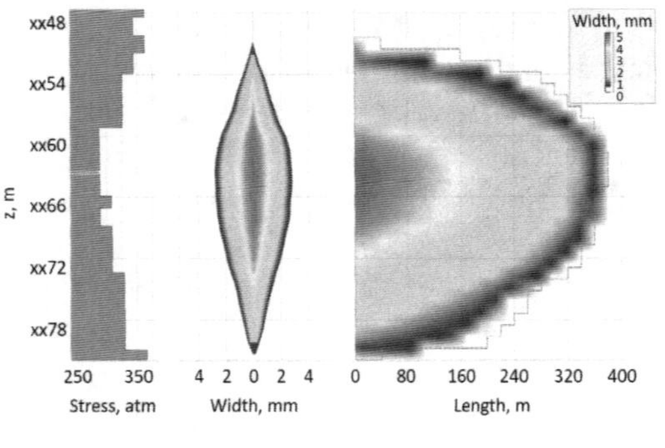

Figure 4. Width profile of hydraulic fracture at stage 1.

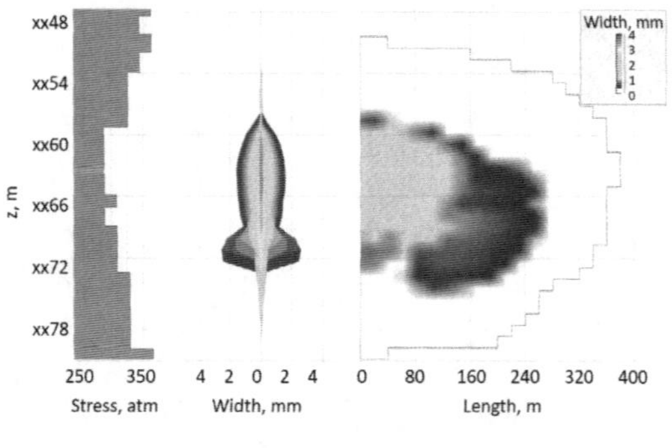

Figure 5. Width profile of closed fracture at stage 1.

The results show an expected difference in the total length of fractures which observed in the case of development of 1, 2 and 3 fractures simultaneously, caused by a decrease in the amount of fluid and proppant per stage. At a given distance between ports (more than 70 meters), an insignificant change in the path of the development of fractures is observed. Calculation of the proppant deposition after stopping the injection made it possible to estimate the conductivity of the fixed fractures. For example, for the first stage, the half-length of the fracture is reduced from 400 to 240 m, and the height of the conductive part of the fracture after depositing proppant was reduced from 30 to 10 m.

5. Conclusions

A mathematical model for multistage hydraulic fracturing in the inhomogeneous rock formation is given.

The constructed numerical solution allows separately and in aggregate way to carry out: analysis of the influence of the tip screen-out effect on the profile of the proppant-fixed fracture, analysis of the stress shadow effect on the fixed fracture profile and its propagation path in the rock, injection modes, properties of the carrier and dispersed phases on the dynamics of fracture growth.

On the basis of developed approach to MHF modeling it is possible to evaluate the efficiency of the application of MHF technology in the horizontal well, in particular to specify an injection rate at the wellhead, various types of proppant and liquid, a geomechanical section with an arbitrary number of layers, and a spatial profile of the well with an arbitrary number of stages of the MHRP and fracture initiation points in each of them.

Acknowledgments

This work is supported by the Ministry of Education and Science of the Russian Federation (Contract No. 14.581.21.0027 of 03.10.2017, Unique identifier RFMEFI58117X0027).

References

[1] Bochkarev A, Budennyy S, Nikitin R, Mitrushkin D, Erofeev A and Zhukov V 2017 Optimization of multi-stage hydraulic fracturing design in conditions of Bazhenov formation *Oil Industry* **3** 50–3

[2] Bochkarev A, Byddennyy S, Nikitin R and Mitrushkin D 2016 Pseudo-3D hydraulic fracture model with complex mechanism of proppant transport and tip screen out *ECMOR XV - 15th European Conference on the Mathematics of Oil Recovery* DOI: 10.3997/2214-4609.201601775

[3] Sneddon I N and Lowengrub M 1969 *Crack Problems in the Classical Theory of Elasticity* (New York: Wiley)

[4] Carter R D 1957 Derivation of the general equation for estimating of extent of the fracturing area, Appendix to optimum fluid characteristics for fracture extension *Drilling and Production Practice* (New York: American Petroleum Institute) pp 261–9

[5] Crouch S L, Starfield A M and Rizzo F J 1983 Boundary element methods in solid mechanics *J. Appl. Mech.* **50** 704-5

[6] Vostrikova V, Sitdikov R, Nikitin R, Erofeev A, Mitrushkin D and Demo V 2018 Modeling of stimulated reservoir volume by multistage hydraulic fracturing in formation with pre-existing natural fractures. *ECMOR XVI - 16th European Conference on the Mathematics of Oil Recovery* DOI: 10.3997/2214-4609.201802158

[7] Olson J E 2004 Predicting fracture swarms the influence of sub-critical crack growth and crack tip process zone on joints spacing in rock *Geol Soc London Spec Publ* **231** 73–87

Temperature of rock formation and fracturing fluid during the hydraulic fracturing process

M B Babenkov[1,2]

[1] Department of Theoretical Mechanics, Peter the Great St. Petersburg Polytechnic University (SPbPU), Politekhnicheskaja 29, 195251 Saint Petersburg, Russia
[2] Institute for Problems in Mechanical Engineering, Russian Academy of Sciences, Bolshoy pr. 61, V.O., 199178 Saint Petersburg, Russia

E-mail: babenkov_mb@spbstu.ru

Abstract. The difference in temperature at the surface and at the depth of hydraulic fracturing is of the order of 80 degrees, which imposes higher requirements on the rheological properties of hydraulic fracturing fluid. Under the influence of temperature, the viscosity changes that affects the final shape of the fracture and the dynamics of its propagation. It is necessary to build a model that makes it possible to calculate the temperature of the liquid and the rock layer in the vicinity of the crack, consuming minimum of computational resources and using available data on the thermal properties of the rock and hydraulic fracturing fluid.

1. Introduction
The temperature of the liquid $T_F(t)$ in the fracture and the temperature of the rock layer $T_R(t)$ are assumed to be homogeneous throughout the entire area under the consideration. The fluid exchanges thermal energy with the rock due to leakage into the rock and heat transfer according to the Newton-Richman law [1,2] through the growing crack surface area with a variable coefficient of heat transfer. The crack surface is assumed to be known and in the simplest case can be taken from the PKN model [3,4,5]. It is assumed that the rock is uniform.It is assumed that the rock is uniform. Thickness of the layer with which the liquid exchanges heat is found by using the approximate assumption proposed in [6] and it grows as a function of time as follows:

$$D(t) = \sqrt{\lambda_R \, t/(C_V^R \rho_R)},$$

where λ_R is the heat conductivity of rock formation, t is the time counted from the beginning of the pumping, C_V^R is the specific heat of rock, ρ_R is the rock density.

The crack into which the liquid is pumped is considered to be one cell and so the coordinate "x" is reduced, that is, the approximation used considers temperature to be equilibrium both in the direction of the "x" axis and in the direction of the "y" axis. The liquid enters the entire volume of the crack at the initial moment of time.

2. Thermomechanical data used in the work
The following are the thermomechanical parameters used in the calculation as an example. The average values are taken for sandstone and water at normal pressure and temperature [3,4].

Table 1. Example of possible thermomechanical data used in the present work for illustrative purposes.

Parameter	Value
Young's modulus, E	$40 \times 10^9\ Pa$
Plane strain Young's modulus, E_l	$E/(1-\nu^2)\ Pa$
Poisson's ratio, ν	0.25
Carter's leakoff coefficient, C_l	$3.54 \times 10^{-4}\ m/\sqrt{s}$
Viscosity of fracturing fluid, μ	$0.001\ Pa \cdot s$
Density of fracturing fluid, ρ_F	$1000\ kg/m^3$
Specific heat of fracturing fluid, C_v^F	$4200\ J/(kg \cdot K)$
Density of the rock formation, ρ_R	$2968\ kg/m^3$
Specific heat of rock formation, C_v^R	$850\ J/(kg \cdot K)$
Thermal conductivity of rock formation, λ_R	$3\ W/(m \cdot K)$
Newton-Richman heat exchange coefficient, L	$100\ W/(m^2 \cdot K)$
Fracturing fluid pumping time, t_P	$1000\ s$

The heat transfer coefficient L depends on the physical properties of the fluid and the physical conditions in which heat exchange takes place. Therefore, for each system, this coefficient must be obtained or found experimentally. For laminar flows, the heat transfer coefficient is rather small in comparison with one for turbulent flows [7].

The Perkins-Kern-Nordgren model (PKN) is another well-known two-dimensional classical hydraulic fracture model [5] with a long fracture length (hundreds of meters in length along the "x" axis), limited but constant height (tens of meters along the "y") and a small width (measured in millimeters along the "z" axis) propagating in an infinite homogeneous isotropic linear elastic reservoir that is characterized by Young's modulus E, Poisson's ratio ν and viscosity coefficient μ. Then the change of the crack length with time will have the following form [5]:

$$l = \left(\frac{625}{4096\pi^3}\right)^{\frac{1}{5}} \left(\frac{{Q_0}^3 E_l}{\mu H^4}\right)^{\frac{1}{5}} t^{\frac{4}{5}},$$

where the crack height H is chosen to be $40m$. Since the length of the crack is large in comparison with two other dimensions, the approximate condition of plane strain can be taken in all the planes orthogonal to the length (directed along the x axis).

Let us consider the rest of values used in calculations. The internal crack surface area is given by the expression:

$$A(t) = 4Hl.$$

The total volume of fracturing fluid pumped into the fracture increases linearly with time:

$$V_P(t) = Q_0 t.$$

The volume of fluid leakoff into the rock formation is determined by Carter's empirical law:

$$Q_L = \frac{2C_l}{\sqrt{t+t_0}} A(t).$$

The total volume of leakoff in time t is as follows:

$$V_L(t) = Q_L t.$$

An estimation of the volume of a rock layer exchanging heat with a fracturing fluid is given by:

$$V_R(t) = A(t)\sqrt{t}\sqrt{\frac{\lambda_R}{\rho_R C_v^R}}.$$

The heat transfer coefficient grows along with the surface area of the crack:

$$c(t) = LA(t)\,H(-(t-t_P)) + LA(t_P)\,H((t-t_P)), \tag{1}$$

where $H(t)$ is the Heaviside function, used to simulate the termination of the surface area grows. In all of the equations considered below, the heat transfer coefficient enters as a function of time.

3. Energy balance equations for the PKN crack

Equations of energy balance for a fluid in a fracture and its surrounding rock formation without taking into account the effect of leakoff will have the form:

$$\rho_F C_v^F T_F'(t) = -\rho_F C_v^F \frac{V_P'(t)}{V_P(t)}\left(T_F(t) - T_F^0\right) H(-(t-t_P)) + c\left(T_R(t) - T_F(t)\right), \tag{2}$$

$$\rho_R C_v^R T_R'(t) = -\rho_R C_v^R \frac{V_R'(t)}{V_R(t)}\left(T_R(t) - T_R^0\right) - c\left(T_R(t) - T_F(t)\right), \tag{3}$$

where C_V^F is the specific heat of fracturing fluid, ρ_V^F is the density of fracturing fluid, C_V^R is the specific heat of rock formation, ρ_V^R is the density of rock formation. Unknown functions that can be found from the solution of this system: $T_F(t)$, $T_R(t)$. Initial conditions are given:

$$T_R(0) = T_R^0, \quad T_F(0) = T_F^0 \tag{4}$$

4. The equilibrium temperature

The system of equations in a stationary form (the time derivatives of the temperature vanish) has the following form:

$$-\rho_F C_v^F \frac{V_P'(t)}{V_P(t)}\left(T_F(t) - T_F^0\right) + c\left(T_R(t) - T_F(t)\right) = 0,$$

$$-\rho_R C_v^R \frac{V_R'(t)}{V_R(t)}\left(T_R(t) - T_R^0\right) - c\left(T_R(t) - T_F(t)\right) = 0.$$

We write out the solution of the system under consideration that the both temperatures do not change, i.e. let us pass to the equilibrium state of temperatures provided that the working fluid is slowly pumped into the fracture:

$$T_F^S(t) = \frac{cC_v^F T_F^0 \rho_F V_R(t) V_P'(t) + C_v^R \rho_R\left(cT_R^0 V_P(t) + C_v^F T_F^0 \rho_F V_P'(t)\right) V_R'(t)}{cC_v^F \rho_F V_R(t) V_P'(t) + C_v^R \rho_R\left(cV_P(t) + C_v^F \rho_F V_P'(t)\right) V_R'(t)},$$

$$T_r^S(t) = T_R^0 + \frac{cC_v^F\left(T_F^0 - T_R^0\right)\rho_F V_R(t) V_P'(t)}{cC_v^F \rho_F V_R(t) V_P'(t) + C_v^R \rho_R\left(cV_P(t) + C_v^F \rho_F V_P'(t)\right) V_R'(t)}.$$

Let us find the limiting values for the solutions obtained, letting the time go to infinity. The equilibrium solution of the problem of heat exchange between the fluid and the rock layer (2)–(3) under initial conditions (4) is given by the following expression:

$$T_{eq} = \frac{10C_V^F \rho_F T_F^0 + 13C_V^R \rho_R T_R^0}{10C_V^F \rho_F + 13C_V^R \rho_R}. \tag{5}$$

If the fracturing fluid is water, and the rock formation in which the fracture propagates is the sandstone, the equilibrium temperature can be given by the following formula:

$$T_{eq} = (T_R^0 + T_F^0)/2.2, \quad (6)$$

which agrees well with practical engineering estimates.

The equations, taking into account Carter's leakoff coefficient, where the corresponding power of heat exchange of the fluid with the reservoir is calculated through the total leakage volume $V_L(t)$, will take the form:

$$\rho_F C_v^F T_F'(t) = -\rho_F C_v^F \frac{V_P'(t)}{V_P(t)} \left(T_F(t) - T_F^0\right) H(-(t - t_P)) + c(T_R(t) - T_F(t)), \quad (7)$$

$$\rho_R C_v^R T_R'(t) = -\rho_R C_v^R \frac{V_R'(t)}{V_R(t)} \left(T_R(t) - T_R^0\right) - c(T_R(t) - T_F(t)) + \rho_F C_v^F \frac{V_L'(t)}{V_R(t)} (T_F(t) - T_R(t)). \quad (8)$$

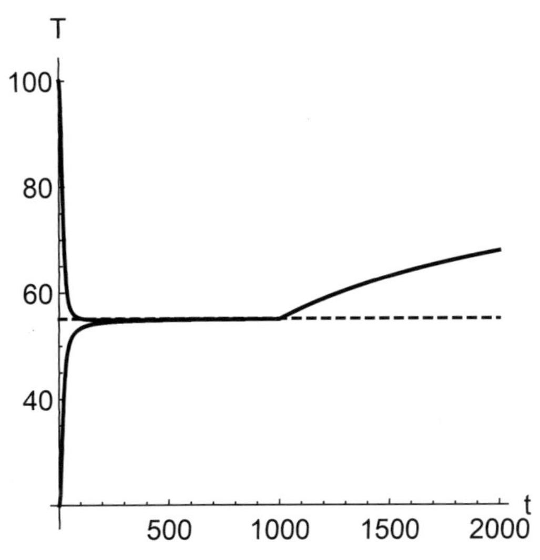

Figure 1. Temperature of the fracturing fluid (lower curve) and the rock formation (upper curve). Equillibrium temperature is shown by the dashed curve.

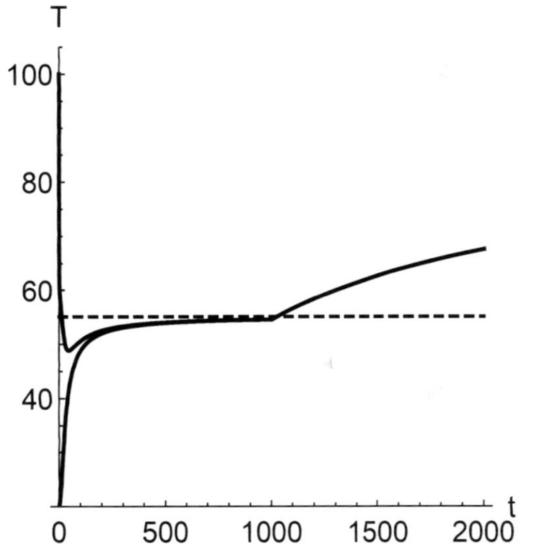

Figure 2. Temperature of the fracturing fluid (lower curve) and the rock formation (upper curve) calculated with regard to Carter's leakoff coefficient.

Plots in Figures 1 and 2 show the time dependence of the temperature, obtained from the solution of the equations (2)–(3) and (7)–(8) with the initial conditions (4). The upper curve shows the temperature of the formation, the lower curve shows the temperature of the working fluid. The dashed line shows the equilibrium temperature (6). The fluid pumping is stopped at moment of time $t_P = 1000$ s. For simplicity, it is assumed that the crack area ceases to increase immediately after the end of pumping, but the Heaviside function in the formula for the heat transfer coefficient (1) can be replaced by any monotonically decreasing function. If leakoff effects are taken into account (Figure 2), the temperatures of the rock formation and the fracturing fluid reach its equilibrium value (5) later than it is shown in Figure 1.

IOP Conf. Series: Earth and Environmental Science **193** (2018) 012076 doi:10.1088/1755-1315/193/1/012076

5. Three-dimensional models of crack propagation

The 3D models provide more realistic estimates of the geometry and size of the fracture, which can help to improve the design of hydraulic fracturing process and wellbores. The discussed model can be generalized to Pseudo 3D (P3D) and Planar 3D (PL3D) approaches. In the first case the energy balance equation needs to be formulated for an element of the allocated volume (for a cell).

Let us now calculate the non-stationary temperature of the rock formation and fracturing fluid using data obtained by the P3D algorithm: the power of fracturing fluid supply to each of the cells from its neighboring ones and the area of the internal surface of each cell into which the whole crack is subdivided. The given quantities has both time "t" and cell number "p" dependencies. Since time steps may not be sufficiently robust for the satisfactory convergence of the non stationary heat transfer equations, the table data can be interpolated and a smaller time step can be adopted.

The system of equations without taking into account the leakoff effect can be written as follows:

$$\rho_F C_v^F \partial_t T_F(p,t) = -\rho_F C_v^F \mu(t)\left(T_F(p,t) - T_F^0\right) + c(t)\left(T_R(p,t) - T_F(p,t)\right), \tag{9}$$

$$\rho_R C_v^R \partial_t T_R(p,t) = -\rho_R C_v^R \frac{V_R'(t)}{V_R(t)}\left(T_R(p,t) - T_R^0\right) - c(t)\left(T_R(p,t) - T_F(p,t)\right). \tag{10}$$

The solution of system (9)–(10) gives the temperature values in each cell. The initial conditions have the form:

$$T_R(p,0) = T_R^0, \quad T_F(p,0) = T_F^0. \tag{11}$$

Let us consider the leakoff effect on the non stationary temperature distribution. Then the modified equations taking into account Carter's leakoff coefficient will take the form:

$$\rho_F C_v^F \partial_t T_F(p,t) = -\rho_F C_v^F \mu(t+1)\left(T_F(p,t) - T_F^0\right) + c\left(T_R(p,t) - T_F(p,t)\right). \tag{12}$$

$$\rho_R C_v^R \partial_t T_R(p,t) = -\rho_R C_v^R \frac{V_R'(t)}{V_R(t)}\left(T_R(p,t) - T_R^0\right) - c\left(T_R(p,t) - T_F(p,t)\right) - \\ -\rho_F C_v^F \frac{V_L'(t)}{V_R(t)}\left(T_R(p,t) - T_F(p,t)\right) \tag{13}$$

The solutions of the equations (9)–(10) and (12)–(13) under the initial conditions (11) give the temperature of the fracturing fluid and rock formation as shown in Fig. 3 and Fig. 4. The family of curves from above shows the temperature of the rock formation, the family of curves from below shows the temperature of the fracturing fluid in the cells of the crack (the lower the curve in the plot, the closer the corresponding cell to the wellbore). The dashed line shows the equilibrium temperature (6).

Comparing Figures 3 and 4 one can conclude that under the given thermomechanical parameters, if the leakoff effect is neglected, the equilibrium temperature is reached in the fluid at the tip of the crack after 1000s of pumping. If the leakoff effect is taken into account, the equilibrium temperature is reached in the rock formation at first. This is due to the fact that the layer of rock is cooled more rapidly due to leakoff and the heat exchange power c (1), depending on the difference in the temperatures of the formation and fluid, decreases proportionally.

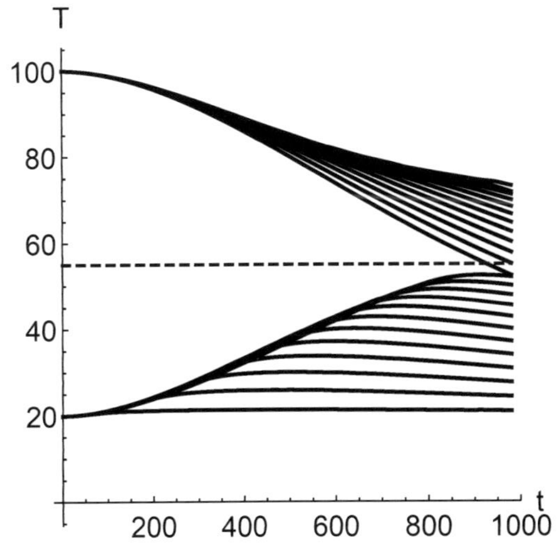

Figure 3. Temperature of the fracturing fluid (lower curves) and the rock formation (upper curves) in different fracture cells calculated using data of a P3D model. Equillibrium temperature is shown by the dashed curve.

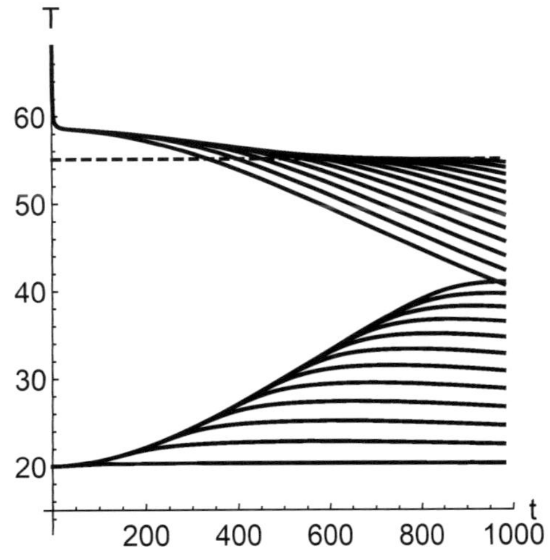

Figure 4. Temperature of the fracturing fluid (lower curves) and the rock formation (upper curves) in different fracture cells calculated with regard to Carter's leakoff coefficient.

Acknowledgments
This work was supported by Ministry of Education and Science of the Russian Federation within the framework of the Federal Program Research and development in priority areas for the development of the scientific and technological complex of Russia for 2014 2020 (activity 1.2), grant No. 14.575.21.0146 of September 26, 2017, unique identifier: RFMEFI57517X0146. The industrial partner of the grant is Gazpromneft Science & Technology Centre LLC.

References
[1] Lienhard J H 2013 *A Heat Transfer Textbook* (New York: Dover Publications)
[2] Zhao J and Tso C P 1993 Heat transfer by water flow in rock fractures and the application to hot dry rock geothermal systems *Int J Rock Mech Min Sci & Geomech Abstr* **30** 633–41
[3] Yew C H and Weng X 2014 *Mechanics of Hydraulic Fracturing* (New York: Gulf Professional Publishing)
[4] Valkó P and Economides M J 1995 *Hydraulic Fracture Mechanics* (New York: Wiley)
[5] Nordgren R P 1972 Propagation of a vertical hydraulic fracture *Soc Petrol Eng J* **12** 306–14
[6] Slepian L I 1972 *Niestacionarnyie Uprugie Volny* (Leningrad: Sudostroenie) (in Russian)
[7] Heinze T, Hamidi S and Galvan B 2017 A dynamic heat transfer coefficient between fractured rock and flowing fluid *Geothermics* **65** 10–6

Institute of Physics
Dirac House, Temple Back
Bristol BS1 6BE UK

ISSN: 1755-1307
ISBN 978-1-5108-7378-0

CURRAN ASSOCIATES INC.
proceedings.com

XVII International Workshop on Magneto-Plasma Aerodynamics 2018

Journal of Physics: Conference Series Volume 1112

Moscow, Russia
17 - 19 April 2018

IOP
Institute of Physics